Koritké/Sick

Atlas of Sectional Human Anatomy

Vol. 1: Head, Neck, Thorax

J. G. Koritké and H. Sick

Atlas of Sectional Human Anatomy

Frontal, Sagittal, and Horizontal Planes

Vol. 1: Head, Neck, Thorax

Drawings in collaboration with B. Lafleuriel
Photographs by R. Becker

152 Photographs and Drawings

Urban & Schwarzenberg · Baltimore – Munich 1983

Authors' addresses:

Jean Georges Koritké
Chairman
Department of Anatomy
Faculty of Medicine
Louis Pasteur University
4, rue Kirschleger
F-67085 Strasbourg CEDEX
France

Henri Sick
Professor of Anatomy
Department of Anatomy
Faculty of Medicine
Louis Pasteur University
4, rue Kirschleger
F-67085 Strasbourg CEDEX
France

This atlas consists of two seperate volumes:
Vol. 1: Head, Neck, Thorax (incl. Index)
Vol. 2: Abdomen, Pelvis (incl. Index)

Printed in Germany by Kösel, Kempten

Library of Congress Cataloging in Publication Data

Koritké, Jean Georges, 1928 –
Atlas of sectional human anatomy.
Bibliography: v. 1, p. 185; v. 2, p. 172
Includes indexes.
Contents: v. 1. Head, neck, and thorax –
v. 2. Abdomen and pelvis.
1. Anatomy, Human–Atlases. 2. Anatomy, Surgical and topographical–Atlases. I. Sick, Henri. II. Title.
[DNLM: 1. Anatomy, Regional–Atlases. QS 17 K84a]
QM25.K67 1983 611'.0022'2 82-13558
ISBN 0-8067-1030-6 (U.S. : set)
ISBN 0-8067-1031-4 (U.S. : v. 1)
ISBN 0-8067-1032-2 (U.S. : v. 2)

Vol. 1: ISBN 0-8067-1031-4 (Baltimore)
ISBN 3-541-71031-4 (Munich)
Complete set: ISBN 0-8067-1030-6 (Baltimore)
ISBN 3-541-71030-6 (Munich)

Preface

Illustrations of anatomical sections have existed for a very long time. Some of these sections which are considered characteristic have been presented in papers for essentially didactic purposes. Various parts of the body, in particular the head and the thorax, have been studied using series of sections which have usually been made in one plane in space. Only in exceptional cases have these been made in the three major planes.

New methods of radiologic and tomographic exploration have given a new thrust to the topographical anatomy of sections. It is no longer only the anatomists who deal with pictures of sections of the human body on a daily basis. Today, clinicians in ever-increasing numbers must do the same.

Topographical anatomy has been a part of the Institute of Anatomy at the Department of Medicine in Strasbourg for a long time. It was here that G. Joessel left the most remarkable reproductions of serial sections ever published in their entirety. It was also here that Ph. Bellocq was a lecturer. Here he published his doctoral thesis "Medical-Surgical Anatomy" which actually contained several sections identical to those prepared for this publication.

This atlas could not have been completed without the great help received from the staff at the Institute of Anatomy. We are also grateful to Mr. Robert Becker, the photographer at the Institute of Pathological Anatomy (Director: Prof. Y. Legal). Mr. Bernard Lafleuriel, the illustrator at the Department of Medicine in Strasbourg helped to prepare the diagrams. The control radiographs were made possible through the help of Prof. Francois Kempf.

The publishing house, Urban & Schwarzenberg, has done everything in its power to make the presentation clear and the quality of the print excellent. We are extremely grateful to them, and to Prof. Dr. H. J. Clemens especially, for the insight making it possible to publish an atlas of such high quality.

Strasbourg, April 1982

J. G. Koritké
H. Sick

Table of Contents

Text

Illustrations

Head and neck (cephalic and cervical regions)

Thorax

Table of Contents Vol. 2

(Condensation)

Introduction

(H. J. Clemens)

An atlas of anatomy normally presents illustrations of a specific anatomical region or organs that are superimposed within that region. Only rarely does it present horizontal sections of the trunk or the extremities, and other sections are even more rare. With the advent of transverse computerized tomography and ultrasonography (echography) and their increasingly common use in clinical examinations, it has become necessary to show serial sections of the various planes of the body. These two diagnostic imaging methods provide the clinician with representations of sections of the human body; interpreting them, however, raises a number of problems, due in part to the lack of literature on normal cross-sectional human anatomy.
Tomographic sections have been prepared in radiology since 1921 (Bocage). These present layers of the body shown by blurring the adjacent zones. The principle involved concerns moving either the X-ray tube and the sensitive plate with respect to the stationary patient or moving the patient and the sensitive plate with respect to the stationary source of radiation. Pictures of sections can be obtained from orientation in all three dimensions. Without minimizing the contribution of this diagnostic method, it should be mentioned that tomography cannot be used to present an exact reproduction of the anatomical substratum.
There has been considerable improvement in the imaging of sections of the human body resulting from computerized tomography (CT). Unlike the methods that were used before this development, the X-rays pass through only the layer to be studied, and yield an exact anatomical reproduction without any displacement. In conventional radiology, the absorption of the X-rays by the body is shown by the various degrees of blackening of the film or by the differences in the luminous intensity of a fluorescent screen. In CT, this absorption is recorded by highly sensitive detectors (scintillation crystals). The source of radiation as well as the system for detecting and measuring the absorption is located behind the patient, and moves horizontally around the body (360° in 2–5 seconds). The impulses detected by the system are amplified electronically and transformed into potential differences. These are sent to a computer where the difference in density is determined for each point of the picture. This information is then sent to a monitor showing the differences in light intensity. The resolving power for two adjacent structures is approximately 2 mm. The tomographic picture displayed on the monitor can be photographed and thus preserved. CT can show differences in absorption of approximately 0.5%, whereas X-ray film only shows differences of 4%. New computer programs will make it possible to produce frontal, sagittal, and oblique sections. In addition they will reduce exposure time and thus artefacts resulting from movement of the heart, the lungs, or other organs.
Two different devices are being used at present: the cranial computer tomograph and the whole body tomograph. The sections of the encephalon are generally made in the axial direction, but depending on the problems presented, the head is sometimes retroflexed (to examine the eye socket) or anteflexed (to examine the posterior fossa). The orbito-meatal line is used as the reference plane in standard examinations. Ten to twelve sections with a thickness of 1 cm are needed to examine the contents of the braincase.
Whole body CT presents transverse sections of the trunk viewed from the inferior side. The section is generally 1 cm thick and can be made at any height of the trunk with the patient supine.
Cranial CT has been of great benefit in medicine. CT of the trunk, however, is only used diagnostically in certain cases since its application overlaps with both ultrasonography and nuclear medicine. In addition to its great importance diagnostically, CT is also used to plan and monitor the progress of therapeutic measures.
An ultrasonic examination, like CT, also makes it possible to present slices of the body. This is based on a completely different physical principle: the reflection of an ultrasonic wave.
Ultrasonic waves emitted by a sender reach the surface of tissues that have varying degrees of permeability. The waves are thus reflected at different intensities. The reflected waves return to the sender, which serves as a receiver between emissions, and there are transformed electronically to luminous points of varying brightness

displayed on a screen. Unlike radiology or scintigraphy, this method makes it possible to obtain pictures of sections from any plane of the body. In general, the acoustic borders of the organs are identical to the anatomical borders. The quality of the ultrasonic picture as well as its usefulness depend on the experience and skill of the operator.
The penetration of an ultrasonic wave into organs depends on the degree to which these organs absorb certain frequencies. Generally, the higher the frequency emitted, the lower the penetration. Because bone almost completely reflects ultrasonic waves and air almost completely absorbs them, organs and tissues located behind bone or air cannot be examined using this method. Bone itself cannot be examined, air becomes an obstacle in the lungs, and intestinal gases block that region. Unlike CT, differences in organ density cannot readily be determined by ultrasonography. Liquids can be distinguished from solids with a high degree of certainty, but resolving power is far less that that of computed tomography.
The advantages of this method should be stressed in addition to the drawbacks:
It is a noninvasive method; neither the patient nor the operator is exposed to any dangerous radiation. The procedure is not expensive and can be repeated an unlimited number of times to follow the course of any process. In the same examination, several organs can be observed from various angles depending on the clinical problems involved and the anatomical factors in the region being examined. Less patient compliance is required than with CT (immobility, length of examination). Like classical radiology and CT, ultrasonic methods also make it possible to observe the dynamic phenomena in the human organism and to present views of topographical anatomy.
The "real time scan" is an orienting examination method that can produce series of sonograms at a rate of 16 to 30 per second. The examination can be performed ventrally, dorsally, or from the flanks, and the probe is moved continuously so that any dynamic changes appearing on the screen can be observed and interpreted without interruption. The equipment required is relatively small, easy to handle, and thus portable. This procedure can be compared with radioscopy.
The "storage scan" stores the information obtained by ultrasonography on transverse or longitudinal sections in a computer. This procedure has an advantage over the real time scan in that transverse, longitudinal, and oblique sections can be obtained from the entire epigastric region. These sections are similar to those obtained with CT and are easier to decipher than those obtained with the real time scan. The power of resolution is greater since the borders of the organs are analyzed from several sides.
The A scan presents a graphic reproduction and makes it possible, among other things, to recognize various parenchymatous organs and to make a distinction between tumors and cysts. Its main application is in echoencephalography – a one-dimensional ultrasonic method in which the amplitudes displayed on the screen do not depend on the intensity of the echos. New devices have been developed recently that make it possible to store a picture displayed during a real time scan without interrupting the examination. This picture can be photographed or entered into a computer memory and recalled whenever required. When the picture is being taken, the contrast, brightness, and positive or negative representation are adjusted automatically according to a preset program. Distances of 1 mm can be measured between two sliding marks on this picture. The result appears immediately when the desired function is called.
It is important that the dynamic phenomena can be recorded on video tape and that the static images from the real time scan can be reproduced on standard photographic equipment. Computerized tomography and ultrasonography can supply us with pictures of sections from nearly every part of the body. The sections can vary in thickness, but are generally 1 cm thick with CT, and a few millimeters thick with the ultrasonic method. There are great differences in the power of resolution for both methods, and improvement is a constant goal. Physical limitations are such that sonograms are more difficult to interpret anatomically than are CT scans.
Before the use of ultrasonics and CT became common, physicians had no noninvasive means of obtaining pictures of sections of the human body. For this reason, they did not have to study such sections in the dissecting room or the hospital. Problems involved in presenting these sections are becoming greater now that it is possible to show them in three dimensions.
It is important to note that the clinician should be wary of many CT atlases in which tomograms are placed next to sections of normal or pathological anatomy. The CT pictures are often not of high quality since they may have been taken with first or second generation equipment. They often give nothing more than a relatively rough orientation since details are lacking and the sections are spaced too far apart. As far as a comparison of sections of pathological anatomy and CT pictures is con-

cerned, knowledge concerning sections of normal anatomy is required before the pathological pictures encountered daily in clinical practice can be detected and then interpreted. As is the case in pathology, it is impossible to present one volume with all the forms, varieties, anomalies, and characteristics depending on age and physiological state. There are also varying degrees of modification due to the preparation methods.

This atlas, like other publications including anatomical illustrations, is a collection of "normal" pictures. Artefacts and pathological lesions have been omitted to the greatest possible extent.

Materials and Methods

We were able to progressively improve our techniques over the long period of time involved in the preparation of this atlas. If new series of sections had to be prepared, the report on the experimentation, which has also been adopted for one part of the sections shown in this atlas, could be defined as follows:

Selection and Preservation of the Material

The material must be normal, *i.e.,* free of any lesions, malformations, or deviations as compared to macroscopic standards. Our subjects were preferably mediolineal young adults whose history and cause of death allowed us to assume that the anatomical structures of the regions to be studied were complete. Approximately 30 subjects meeting these requirements were used in this study. The series of sections presented here were supplied by only 11 of them.

As soon as the cadaver is brought to the Institute of Anatomy it is embalmed by injecting a preserving fluid (consisting mainly of alcohol, formaldehyde solution, and glycerine) into one of the femoral arteries. The nerve centers are not sufficiently fixed by this method, so they are injected with 100 cc of a 20% formaldehyde solution into the leptomeningeal spaces. The cadaver is then immersed in a 30% solution of ordinary alcohol where it lies perfectly horizontally for three months. This length of fixation appears to be the most appropriate. If it is shorter, fixation is incomplete, resulting in soft tissues which are difficult to work with and preserve. If it is longer, the colors and contrasts between neighboring tissues are altered. The sections are also preserved in a 30% solution of alcohol.

Production and Preparation of Sections

After removing the cadaver from the fixing fluid, it is placed in the horizontal position in the freezing room. The body is placed and maintained in the anatomical position for the complete freezing period (one week at −20° C). After this period, the different hardened tissues present a uniform resistance to the progression of a saw blade. The low temperature of preparation and the great length of the saw blade reduce the heat generated, and thus prevent any alteration of the slices.

The body, frozen *in toto,* is cut into blocks along the transverse or horizontal planes perpendicular to the median sagittal plane or the frontal plane upon which the subject is lying. These cuts are made at a certain distance from the areas to be studied. Thus, the head and neck block includes the upper part of the thorax; the thorax block extends from the middle of the cervical column to the middle of the lumbar column and the abdomen block from the region above the diaphragm to the middle third of the thigh.

Each block is cut into sections along one of the following planes of reference.

The plane of reference for frontal sections is the plane upon which the subject is lying. The eye-ear plane (Frankfurt, horizontal, orbito-meatal line) is thus perpendicular to this frontal plane of reference.

The plane of reference for sagittal sections is the median plane. It is difficult to locate and follow since the spinal column presents lateral curves, pronounced to varying degrees, in nearly all cases.

The plane of reference for horizontal sections depends on the region to be studied. It is represented by the eye-ear plane (Frankfurt horizontal, orbito-meatal line) for the head and neck and by the plane passing through the two antero-superior iliac spines for the thorax, abdomen, and pelvis.

The sections made from the plane of reference are normally of uniform thickness for a given block. The thickness varies, however, for the different blocks. It ranges from 1.5 cm for horizontal sections from the cephalo-cervical region to 2.5 cm for frontal sections. The very large size of the frontal sections made it necessary to increase their thickness. The block is placed on an adjustable guiding plane parallel to the saw blade in order to ensure that the section is of uniform thickness.

Immediately after sectioning, each specimen is placed on a plate and covered with plastic material permitting easy handling. The accessible surface is placed on a similar plate by simply turning it over. The surfaces of the sections are washed with running water and a sponge. The contents of the

natural cavities are thus removed while the specimen is still in the frozen state. While these simple operations are being performed, the remarkable appearance of the shape and colors of the organs can be observed.
The washed sections are covered with plastic sheet material to prevent dehydration. They are then stored in the refrigerating room. Before photographing, they are studied separately to determine degree of normality, to touch up an organ of primary importance located at a very slight distance from the cutting plane, and to regularize certain contours.

Photography

The photographic equipment consists of a 9×12 Technika Linhof camera with a 4.5/150 mm Xenar F objective. Normal exposure is ½ sec with the aperture set at 16. Two 500 W lamps are placed 1.5 m away from the specimen.
Pictures are taken of both sides of each specimen next to a graduated rule. Enlargements are made in order to present horizontal sections of the trunk in their natural size, to present horizontal sections of the head and neck as 1.5 times their natural size, and to present all sagittal and frontal sections as 0.75 times their natural size. A ruler with the conversion data is contained in the back of this volume.
The photographs have been chosen to a large extent on the basis of their clinical importance. Thus, horizontal sections are viewed from the bottom side; when elements of primary importance only appear on the top side, however, this side has been added to complete the information. All frontal sections are viewed from the front. Some posterior views considered absolutely necessary complete each series. All sagittal sections are viewed from the exterior.
The frontal sections are numbered starting from the posterior plane of reference upon which the subject is lying. The letter "a" indicates that the specimen is viewed from the front and "p" that it is viewed from the back. The sagittal sections are numbered starting from the median plane and moving to the left and the right. The median plane divides the specimen into a right segment viewed from the left (MD) and a left segment viewed from the right (MSi). The horizontal sections are numbered from top to bottom within the same block and are marked with respect to a plane of reference.
Thus, each section is defined as follows: by a first letter indicating the general orientation (F=frontal; S=sagittal; H=horizontal), by a second letter qualifying the overall region to which it belongs (C=cephalo-cervical; T=thoracic; A=abdominal; P=pelvic), by a symbol, if required, indicating the sex (♀ or ♂), by a number marking it within the block, and perhaps by an indication as to the position of the observer with respect to the plane of the section (superior, inferior, right, left, anterior, posterior).

Analysis of Sections

During specimen preparation, the major objective is to keep the organs in their original position.
During the process of identification, however, the organs can be treated individually. A sufficient order must be retained, although, to allow for later verification.
First the sections are radiographed. Each radiologic negative is photographed. The photographic enlarger used to project the radiologic image of the skeleton makes it possible to set up a grid displaying the height of the center of the vertebral bodies. This grid is applied to each side of the photographs of frontal and sagittal sections.
The organ parts separated by the saw cut are sutured to each other and then to the specimen box wall. After all the elements of the section have been joined, it can be immersed in the fixing fluid and handled without being damaged.
Tracing paper is placed on each photograph, which is then compared with the original specimen. The radiologic negative makes it easier to recognize the skeletal elements, but quite frequently the entire anatomical region must be reconstructed. The muscle or aponeurotic layers are separated from each other. Each vascular lumen in all sections is injected systematically to determine the origin and area of the vessels. The full path of the nerves is followed from one section to the next. In some instances we had to revert to histological techniques to support, with certainty, some of our findings.

Drawings

Each drawing, traced from a photograph, is a true reflection of the actual anatomy. Only those elements that are considered absolutely necessary to interpret the photograph are added to the drawing using a conventional code. The photograph thus serves as the primary reference. The drawing directs the observer and stimulates his research, but should in no way limit his field of investigation. Because each photograph contains a vast amount of interesting elements, we were often faced with difficult decisions regarding the representation and references to the labels. Regrettably the multitude of direction lines made it impossible to place the nomenclature in a good readable way at the lines. To achieve a better clearness a code system has been employed whose abbreviations should be understandable internationally.

The references to labels are based on a letter and a number permitting rapid access to the terminology by referring to a system (A=angiology; M=myology; N=nervous system; O=osteology; OS=sense organs; S=splanchnology; Sy=syndesmology). Each reference is explained in the "Index of Labels" section.

The choice of nomenclature is always a difficult problem. To make the terms as accessible as possible, all labels have been reviewed for their appropriateness to the English-speaking clinician and strict adherence to Nomina Anatomica has been abandoned in favor of practical terminology.

Directions for Use

The most important document is the photograph, a true reflection of the actual anatomy. From the very beginning, the observer is able to record a large amount of information, with the progression of the analysis indicated by the diagram and then the labels.

Photographs

The photographs reproduced on the right-hand pages convey different types of information: general appearance of the section, shape of its surfaces, positions, proportions, impressions of the density of the organs. The entire section or each of its elements can be situated within the body by referring to the following: the letters characterizing each section with respect to the planes of reference; the height of the vertebra shown on the osseous section as soon as it appears; the height of the center of the vertebral body indicated on both sides of all photographs of frontal and sagittal sections; the level of the ribs; and the scale shown at the bottom of each photograph.

Drawings

Each diagram, an interpretation of the actual anatomy, brings out the essential characteristic elements of the section. The thickness of the features and the character of the grid representing the organs indicate to which system the organs shown belong. References are also made using keys. Symmetric organs are not necessarily presented in a bilateral manner. This makes the drawing clearer. It was not always possible to superimpose extremely thin planes. A single feature can give an outline of several structures. Cuts of vertebrae and ribs are designated in any case.

Labels

The labels supply further information and help to check the identification of the photograph and the interpretation of the drawing.
Abbreviations have been used for the identification and orientation of the section, the classification of organs referred to by large systems, and finally in the labels. The first examination of the photograph, the interpretation of the drawing, and then reading the labels should logically lead the observer to re-examine the photograph in order to record the numerous elements that could not be referred to in the space allowed. It should then be possible to imagine and compare the image that will be supplied by CT, ultrasound, or any other new technique technology brings to us.

Abbreviations

1. Identification and Orientation of the Sections

a: anterior – anterior
p: posterior – posterior
D: right – dexter
Si: left – sinister
F: frontal – frontalis
H: horizontal – horizontalis
S: sagittal – sagittalis
C: head, neck – caput, collum
T: thorax – thorax
A: abdomen – abdomen
P: pelvis – pelvis
♀: female – femininum
♂: male – masculinum

The vertebrae are indicated by the letters C (cervical – cervicalis), T (thoracic – thoracicus), L (lumbar – lumbalis), S (sacral – sacralis) and Co (coccygeal – coccygea). The number that corresponds to their level is added to the letter. The ribs are marked by a number indicating their level ("3" refers to the third rib).

2. Association of the Organ to a Large System

A: Angiology – Angiologia
M: Myology – Myologia
N: Nervous System – Systema Nervosum
O: Osteology – Osteologia
OS: Sense Organs – Organa Sensuum
S: Splanchnology – Splanchnologia
Sy: Syndesmology, Arthrology – Syndesmologia

3. Abbreviations used in the Labels

a.: artery – arteria
art.: articulation – articulatio
b.: bursa – bursa
C.: cervical (vertebra) – cervicalis (vertebra)
Co.: coccyx – os coccygis
gl.: gland – glandula
L.: lumbar (vertebra) – lumbalis (vertebra)
lig.: ligament – ligamentum
m.: muscle – musculus
n.: nerve – nervus
no.: node (lymphatic) – nodus (lyymphaticus)
r.: ramus – ramus
S.: sacral (vertebra) – sacralis (vertebra)
v.: vein – vena
( ): specifies the preceding term

References

Adachi, B.: Das Arteriensystem der Japaner. Verlag der Kaiserlich-Japanischen Universität zu Kyoto, Kyoto 1928.

Alfidi, R., J. Haaga, M. Weinstein, J. de Groot: Computed tomography of the human body. An atlas of normal anatomy. C. V. Mosby Co., St. Louis, 1977.

Aubaniac, R., J. Porot: Radio-anatomie générale de la tête. Masson, Paris, 1955.

Baert, A., L. Jeanmart, A. Wackenheim: Clinical computor tomography, Springer, Berlin, Heidelberg, New York, 1978.

Baert, A., L. Jeanmart, A. Wackenheim: Atlas of computor tomography, Springer, Berlin, Heidelberg, New York, 1980.

Bellocq, Ph.: Anatomie médico-chirurgicale. Anatomie des formes extérieures et Anatomie des régions. Masson, Paris 1925–1957, I–XI

Bo, W., I. Meschan, W. Krueger: Basic atlas of cross-sectional anatomy. W. B. Saunders Co., Philadelphia, London, Toronto, 1980.

Bocage, A. E. M.: Procédé et dispositifs de radiographie sur plaque en mouvement. Brevet n⁰ 536464 (1921/22).

Braune, W.: Topographisch–anatomischer Atlas nach Durchschnitten an gefrorenen Cadavern. Veit u. Comp., Leipzig, 1875.

Carter, B., J. Morehead, S. Wolpert, S. Hammerschlag, H. Griffiths, P. Kahn: Cross-Sectional anatomy. Computed tomography and ultrasound correlation. Appleton-Century-Crofts, New York 1977.

Delmas, A., B. Pertuiset: Topométrie crânio-encéphalique chez l'Homme. Masson, Paris/Thomas, Springfield, 1959.

Eycleshymer, A., D. Schoemaker: A cross-sectional anatomy. Appleton-Century-Crofts, New York 1970.

Feifenbaum, H.: Echography. Lea & Febiger, Philadelphia, 1972.

Friedemann, G., U. Mödder: Leistungsfähigkeit der Ganzkörpertomographie. Deutsches Ärzteblatt, 48, 2891–2899 (1978).

Gambarelli, J., G. Guerinel, L. Chevrot, M. Mattei: Coupes sériées du corps humain. Anatomie – Radiologie – Scanner. Springer, Berlin, Heidelberg, New York/Masson, Paris, New York, Barcelone, Milan, 1977

Gebauer, A., U. Scherer: Differentialindikation für die Computertomographie und die Ultraschalluntersuchung. Internist, 19, 568–578 (1978).

Gonzalesz, C., C. Grossman, E. Palacios: Computed brain and orbital tomography. Technique and interpretation. Wiley & Sons, New York, London, Sydney, Toronto, 1976

Hounsfield, G. H.: Computerized transverse axial scanning (tomography). I. Description of system. Br. J. Radiol., 46, 1016 (1973).

Hovelacque, A., O. Monod, H. Evrard: Treize coupes horizontales du thorax. Maloine, Paris, 1938.

Joessel, G.: Lehrbuch der topographisch-chirurgischen Anatomie. Max Cohen, Bonn, 1884.

Lanz, T. v., W. Wachsmuth: Praktische Anatomie. Hals. Springer, Berlin, Göttingen, Heidelberg, 1955.

Lanz, T. v., W. Wachsmuth: Praktische Anatomie. Kopf. Springer, Berlin, New York, Heidelberg, 1979.

Lissner, J., W. Schwarz: Technik und Apparaturen der Ganzkörper-Computertomographie. Internist, 19, 559–563 (1978).

Luschka, H. v.: Die Lage der Bauchorgane des Menschen. C. F. Müllersche Hofbuchhandlung, Carlsruhe, 1873.

Lyons, E. A.: A color atlas of sectional anatomy chest, abdomen and pelvis. C. V. Mosby Co., St. Louis, 1978.

Matsui, T., A. Hirano: An atlas of the human brain for computerized tomography. Fischer, Stuttgart, New York, 1978.

Nieuwenhuys, R., J. Voogd, C. V. Huijzen: The human central nervous system. Springer, Berlin, Heidelberg, New York, 1978.

Pernkopf, E.: Topographische Anatomie des Menschen. Urban & Schwarzenberg, Berlin, Wien, 1937–1960 I–IV.

Pernkopf, E.: Atlas der topographischen und angewandten Anatomie des Menschen. Urban & Schwarzenberg, München, Wien, Baltimore, 1980, I. u. II.

Rettenmeier, G.: Sonographischer Oberbauchstatus. Aussagefähigkeit und Indikationen der Ultraschall-Schnittbilduntersuchung des Oberbauches. Internist, 17, 549 (1976).

Roy-Camille, R.: Coupes horizontales du tronc. Masson, Paris, 1959.

Sarazin, R., J. Champetier et Ch. Contamin: Quatorze coupes sagittales du médiastin. Maloine, Paris, 1965.

Schaltenbrand, G.: Einführung in die stereotaktischen Operationen mit einem Atlas des menschlichen Gehirns. Thieme, Stuttgart, 1959.

Scherer, U.: Computertomographische Untersuchungen der Oberbauchorgane. Internist, 19, 579–585 (1978).

Sobotta, J.: Atlas d'anatomie humaine. Urban & Schwarzenberg, München, Wien, Baltimore/ Maloine, Paris, 1977.

Struve, C.: Ultraschalltomographie des Abdomens. Urban & Schwarzenberg, München, Wien, Baltimore, 2. Aufl., 1981.

Testut, L., O. Jacob: Traité d'anatomie topographique avec applications médico-chirurgicales. Doin, Paris, 1905.

Thompson, J. R. and A. N. Hasso: Correlative sectional anatomy of the head and neck. C. V. Mosby Co., St. Louis, 1980.

Töndury, G.: Angewandte und topographische Anatomie. Thieme, Stuttgart, 4. Aufl., 1970.

Wyman, A., T. Lawson, L. Goodman: Transverse anatomy of the human thorax, abdomen and pelvis. An atlas of anatomic, radiologic, computed tomographic and ultrasonic correlation. Little Brown, Boston, 1978.

FC

Si

D

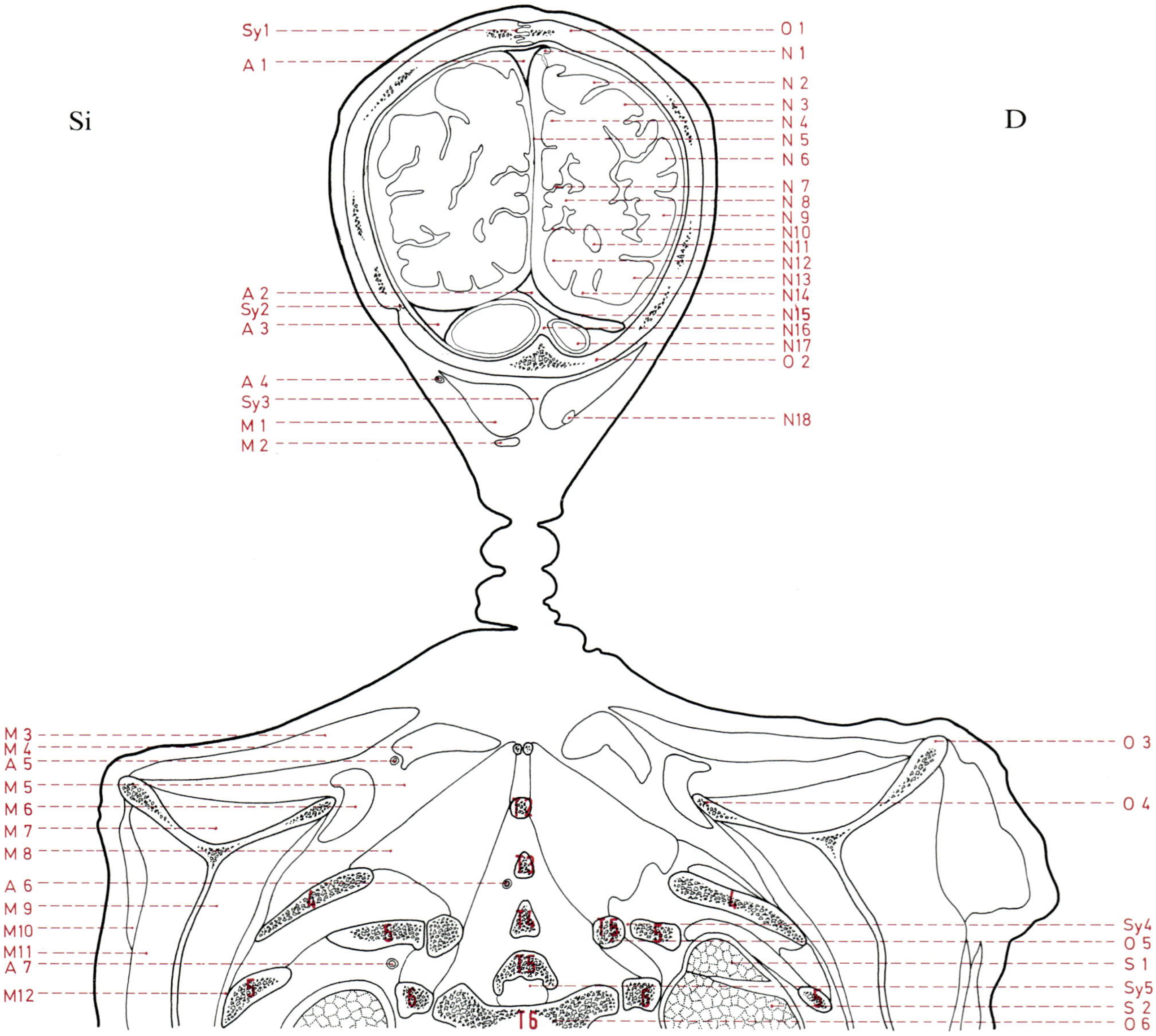

Sy1
A 1
A 2
Sy2
A 3
A 4
Sy3
M 1
M 2
O 1
N 1
N 2
N 3
N 4
N 5
N 6
N 7
N 8
N 9
N10
N11
N12
N13
N14
N15
N16
N17
O 2
N18
M 3
M 4
A 5
M 5
M 6
M 7
M 8
A 6
M 9
M10
M11
A 7
M12
O 3
O 4
Sy4
O 5
S 1
Sy5
S 2
O 6

Fig. FC 1

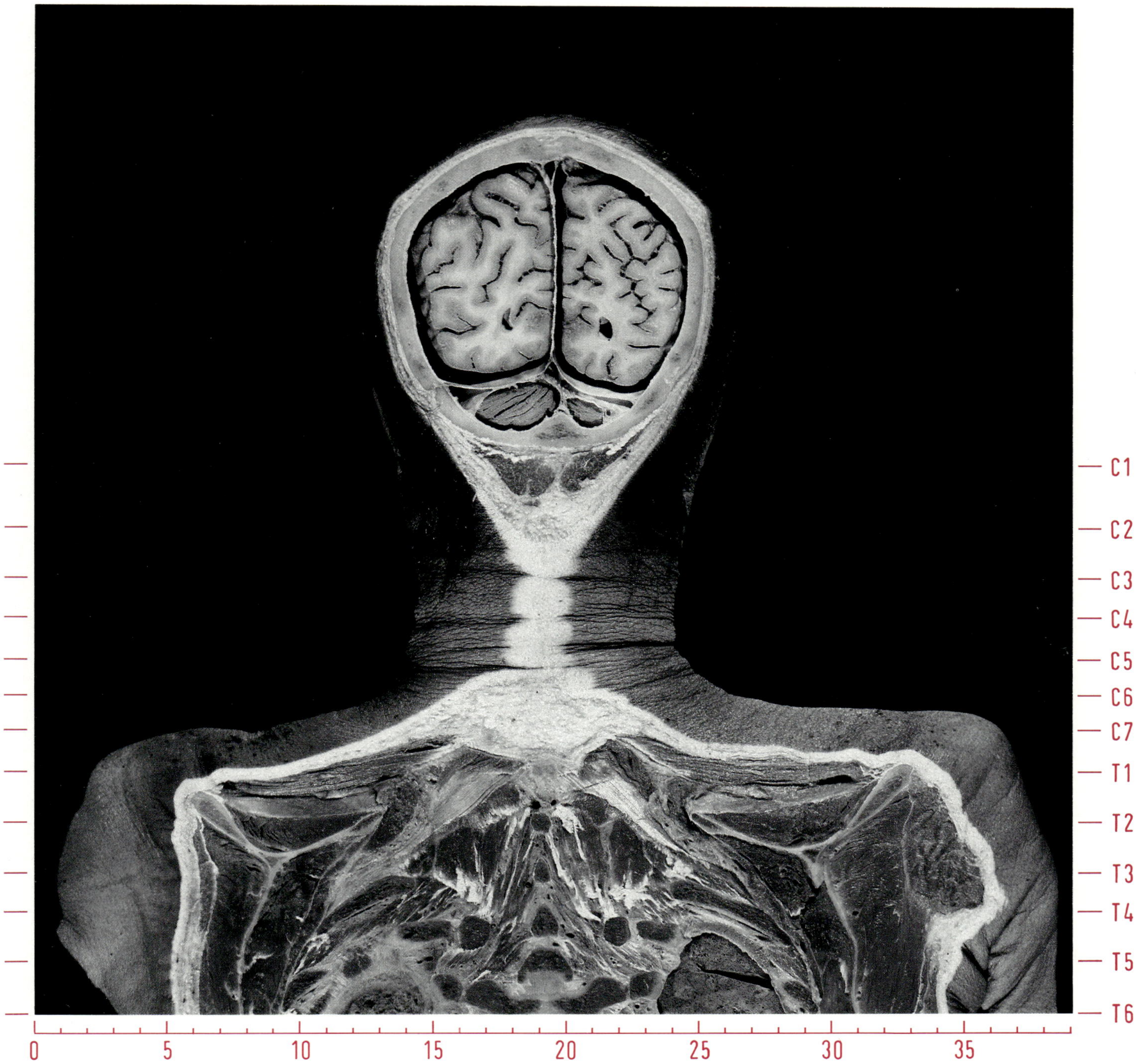

Fig. FC 1

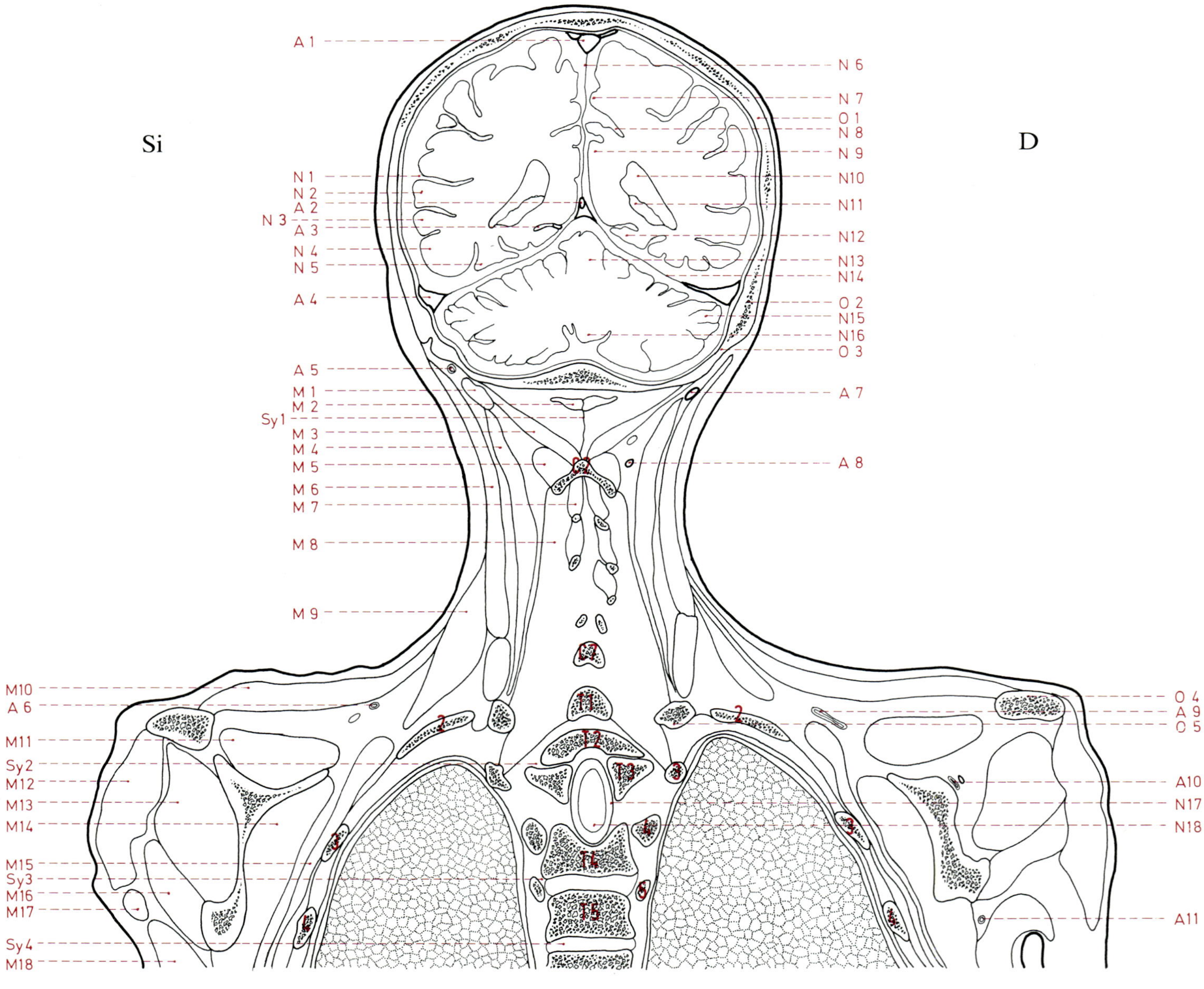
Si
D
A 1
N 6
N 7
O 1
N 8
N 9
N 1
N 2
A 2
N 3
A 3
N 4
N 5
N10
N11
N12
N13
N14
A 4
O 2
N15
N16
O 3
A 5
M 1
M 2
Sy1
M 3
M 4
M 5
M 6
M 7
M 8
M 9
A 7
A 8
M10
A 6
M11
Sy2
M12
M13
M14
M15
Sy3
M16
M17
Sy4
M18
O 4
A 9
O 5
A10
N17
N18
A11
T1
T2
T3
T4
T5

Fig. FC 2

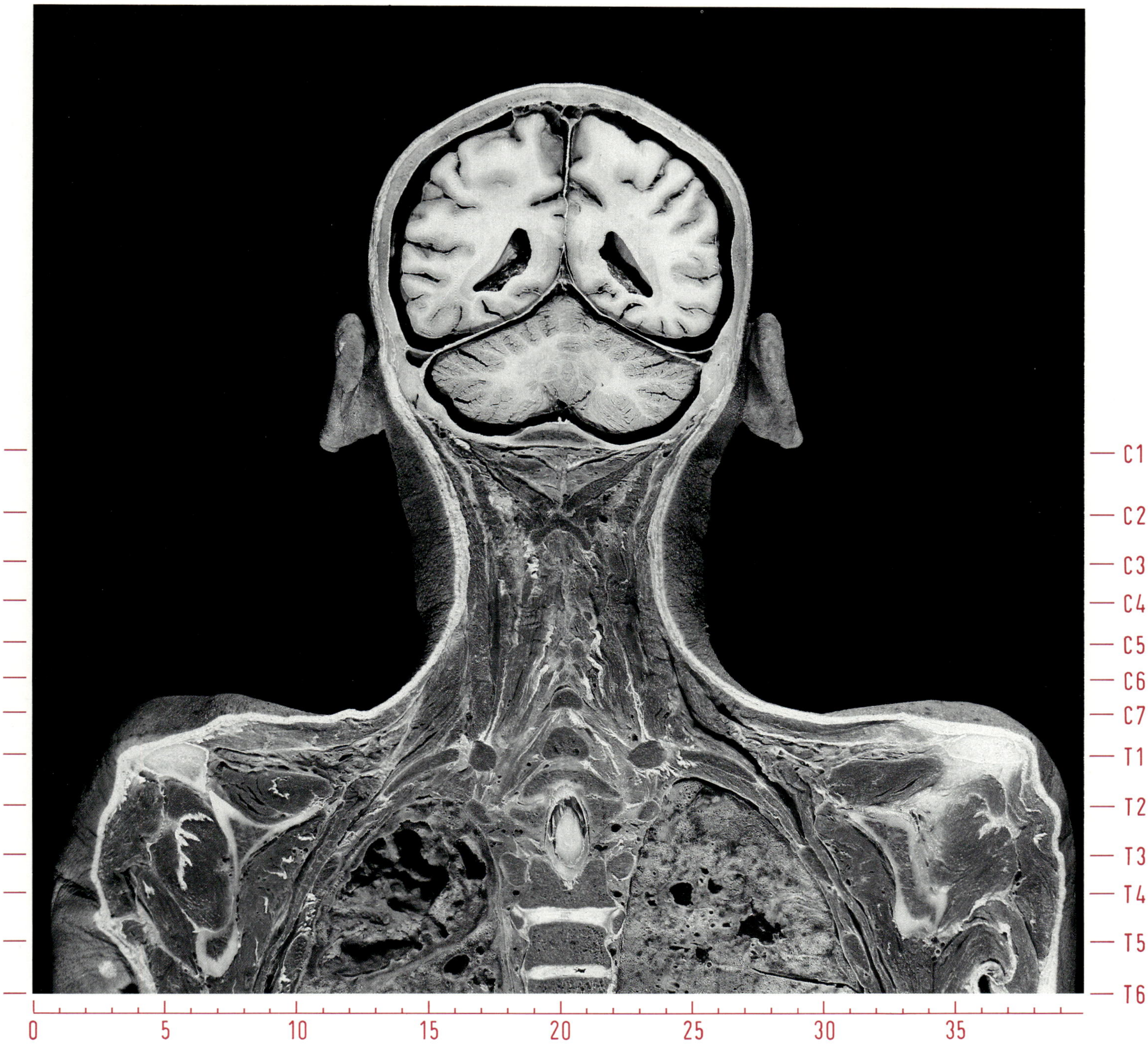

Fig. FC 2

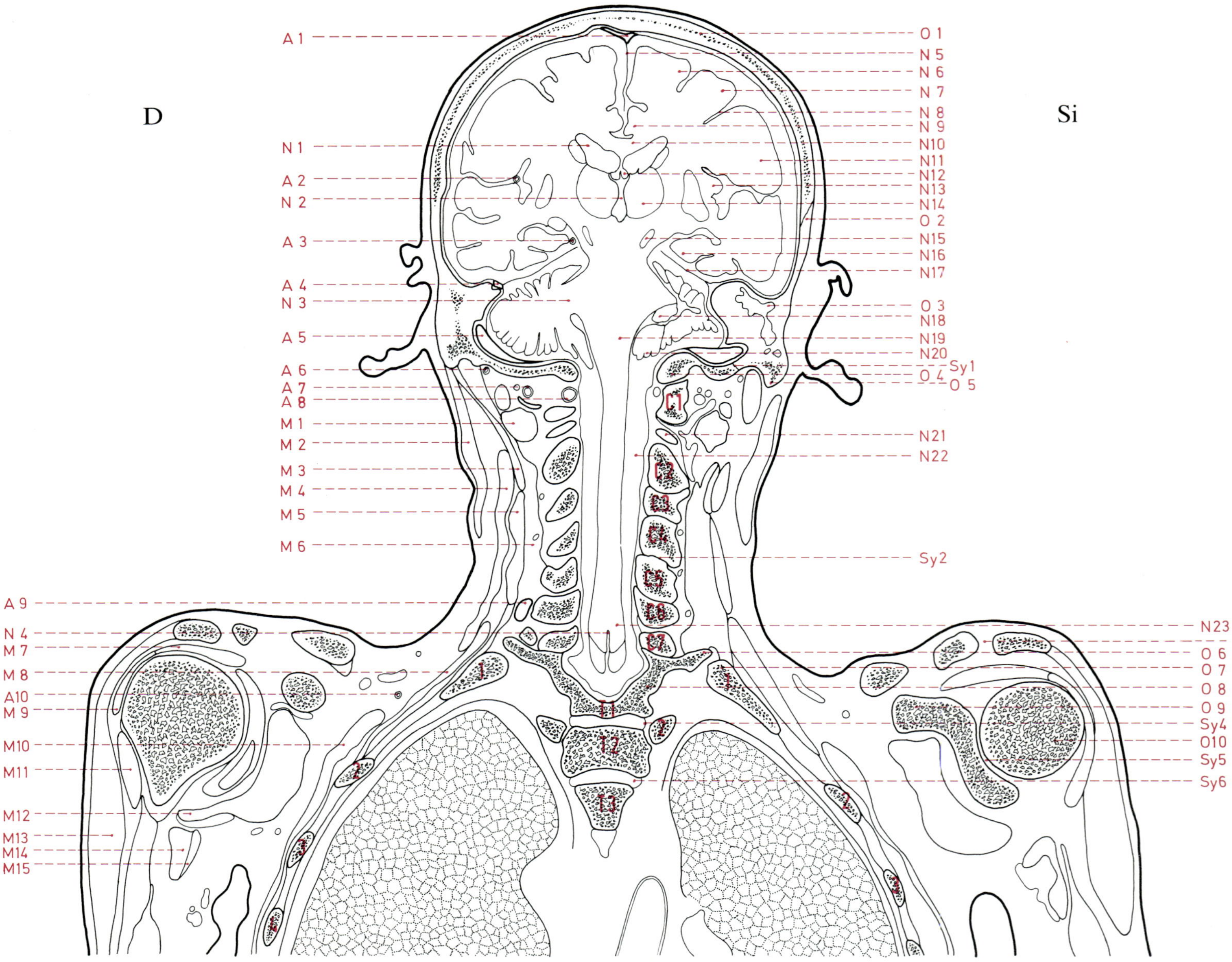

D
Si
A 1
N 1
A 2
N 2
A 3
A 4
N 3
A 5
A 6
A 7
A 8
M 1
M 2
M 3
M 4
M 5
M 6
A 9
N 4
M 7
M 8
A10
M 9
M10
M11
M12
M13
M14
M15
O 1
N 5
N 6
N 7
N 8
N 9
N10
N11
N12
N13
N14
O 2
N15
N16
N17
O 3
N18
N19
N20
Sy1
O 4
O 5
N21
N22
Sy2
N23
O 6
O 7
O 8
O 9
Sy4
O10
Sy5
Sy6
C1
C2
C3
C4
C5
C6
C7
T1
T2
T3

Fig. FC 3a

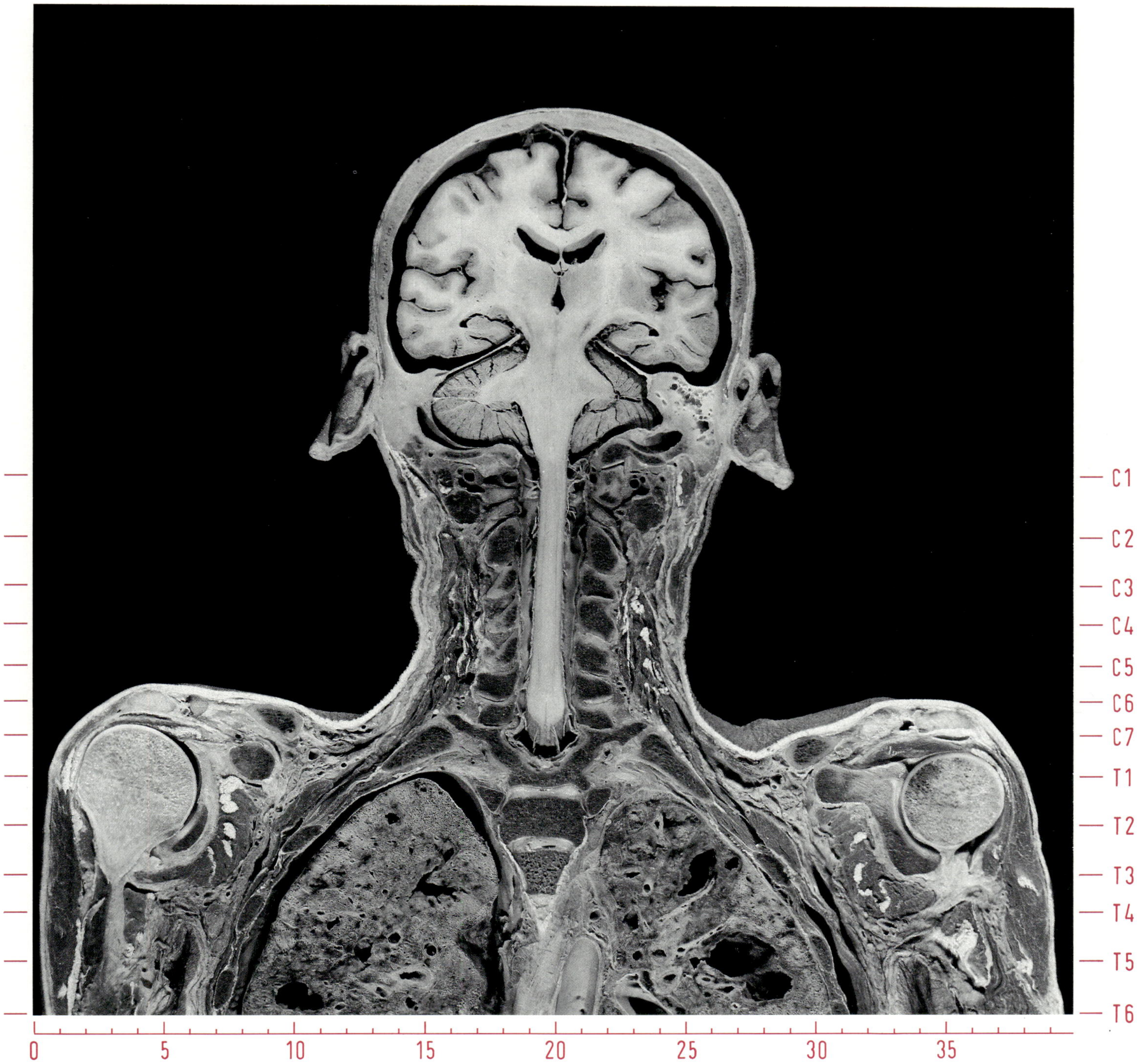

Fig. FC 3 a

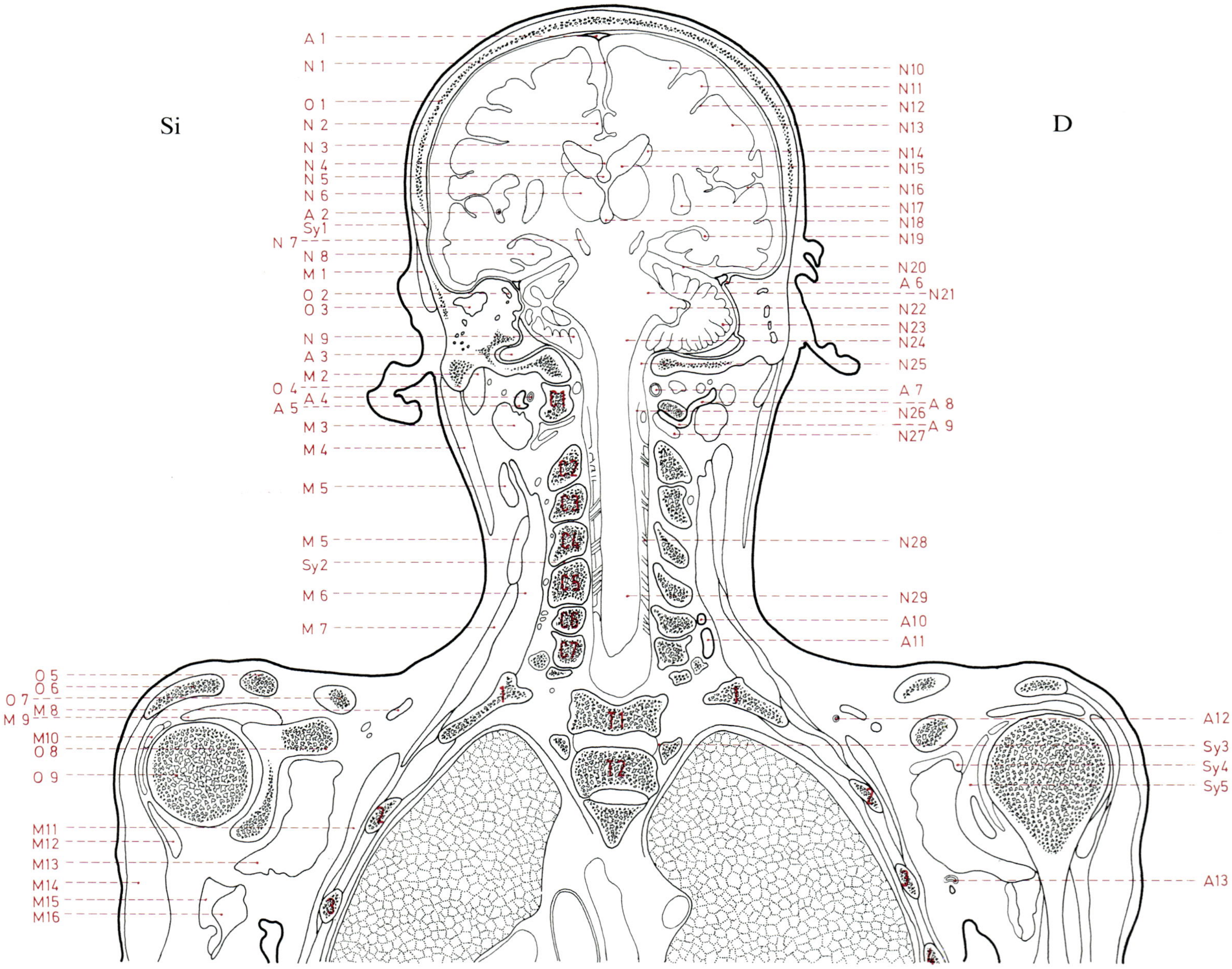
Si
D
A 1
N 1
O 1
N 2
N 3
N 4
N 5
N 6
A 2
Sy1
N 7
N 8
M 1
O 2
O 3
N 9
A 3
M 2
O 4
A 4
A 5
M 3
M 4
M 5
M 5
Sy2
M 6
M 7
N10
N11
N12
N13
N14
N15
N16
N17
N18
N19
N20
A 6
N21
N22
N23
N24
N25
A 7
N26
A 8
N27
A 9
N28
N29
A10
A11
O 5
O 6
O 7
M 8
M 9
M10
O 8
O 9
M11
M12
M13
M14
M15
M16
A12
Sy3
Sy4
Sy5
A13
C1
C2
C3
C4
C5
C6
C7
T1
T2

Fig. FC 3p

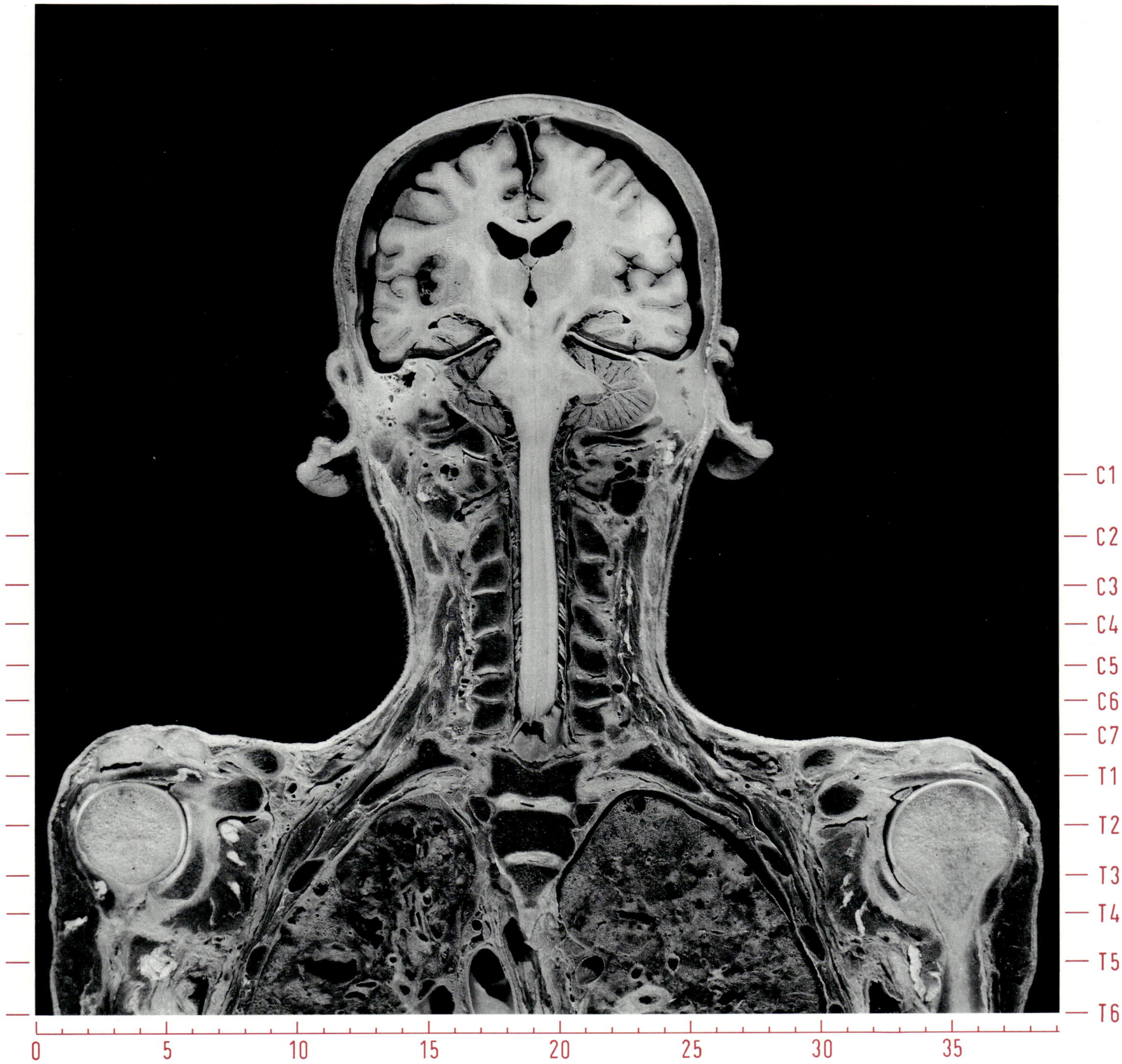

Fig. FC 3p

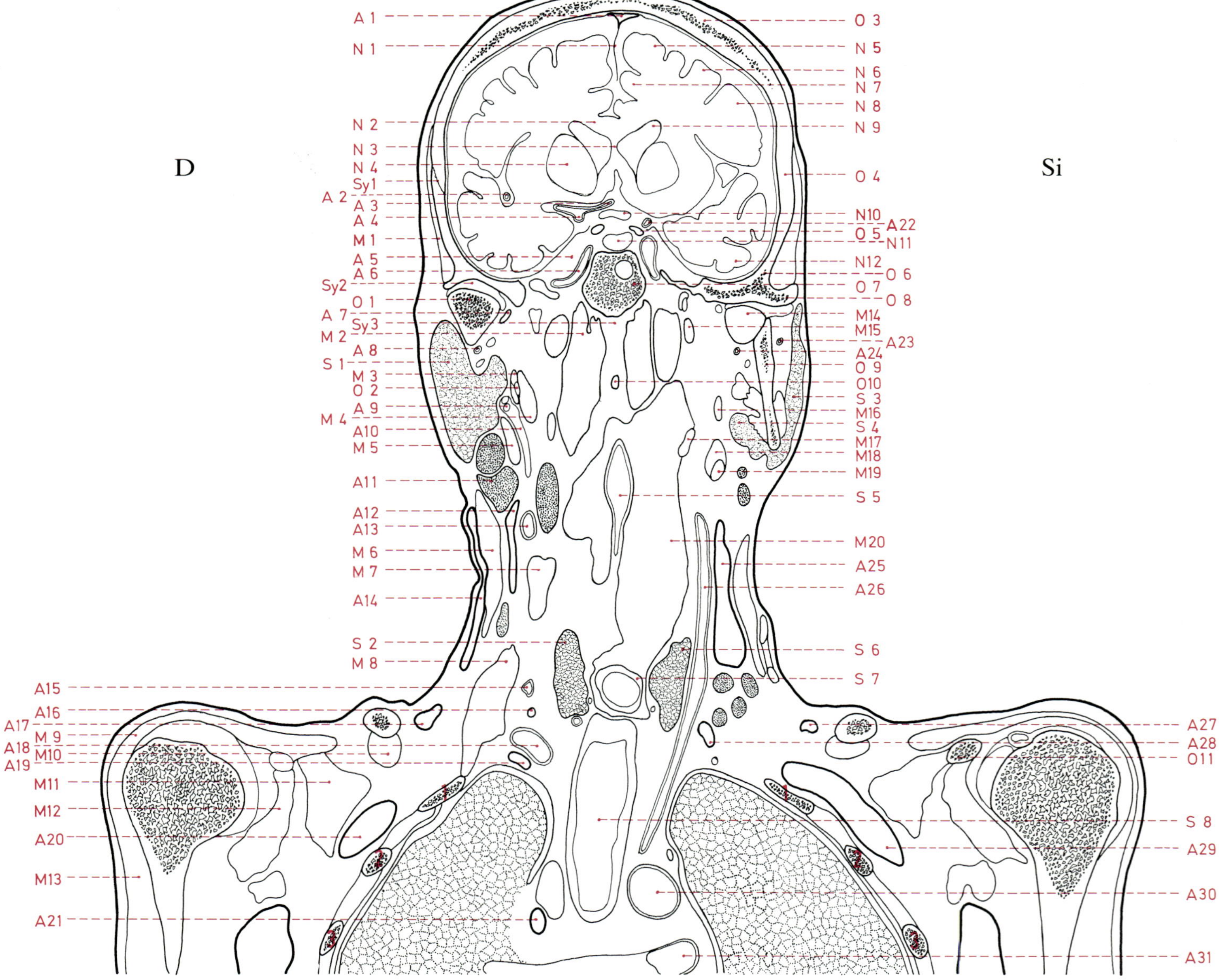
D
Si
A 1
N 1
N 2
N 3
N 4
Sy1
A 2
A 3
A 4
M 1
A 5
A 6
Sy2
O 1
A 7
Sy3
M 2
A 8
S 1
M 3
O 2
A 9
M 4
A10
M 5
A11
A12
A13
M 6
M 7
A14
S 2
M 8
A15
A16
A17
M 9
A18
M10
A19
M11
M12
A20
M13
A21
O 3
N 5
N 6
N 7
N 8
N 9
O 4
N10
A22
O 5
N11
N12
O 6
O 7
O 8
M14
M15
A23
A24
O 9
O10
S 3
M16
S 4
M17
M18
M19
S 5
M20
A25
A26
S 6
S 7
A27
A28
O11
S 8
A29
A30
A31

Fig. FC 4 a

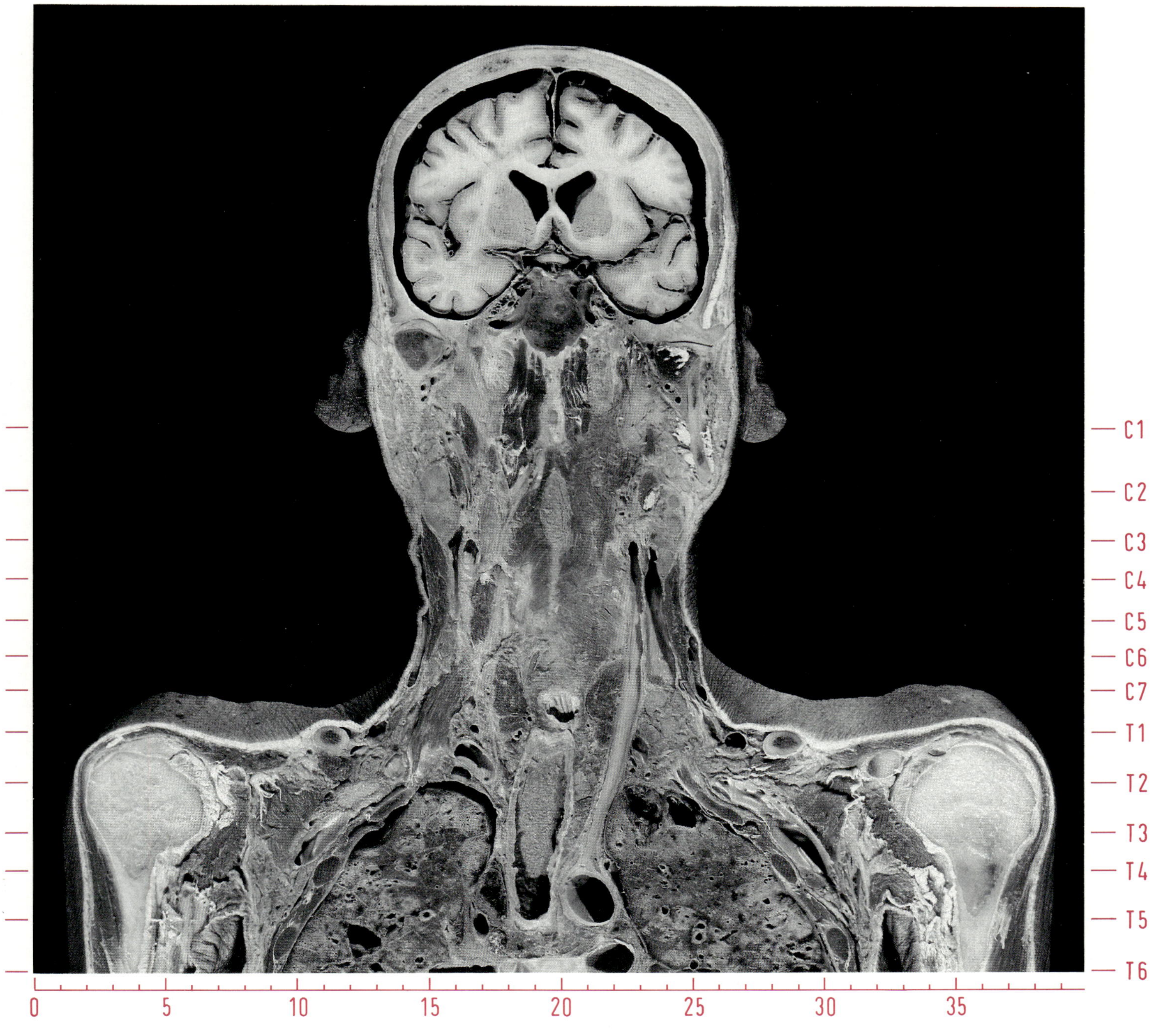

Fig. FC 4a

Si

D

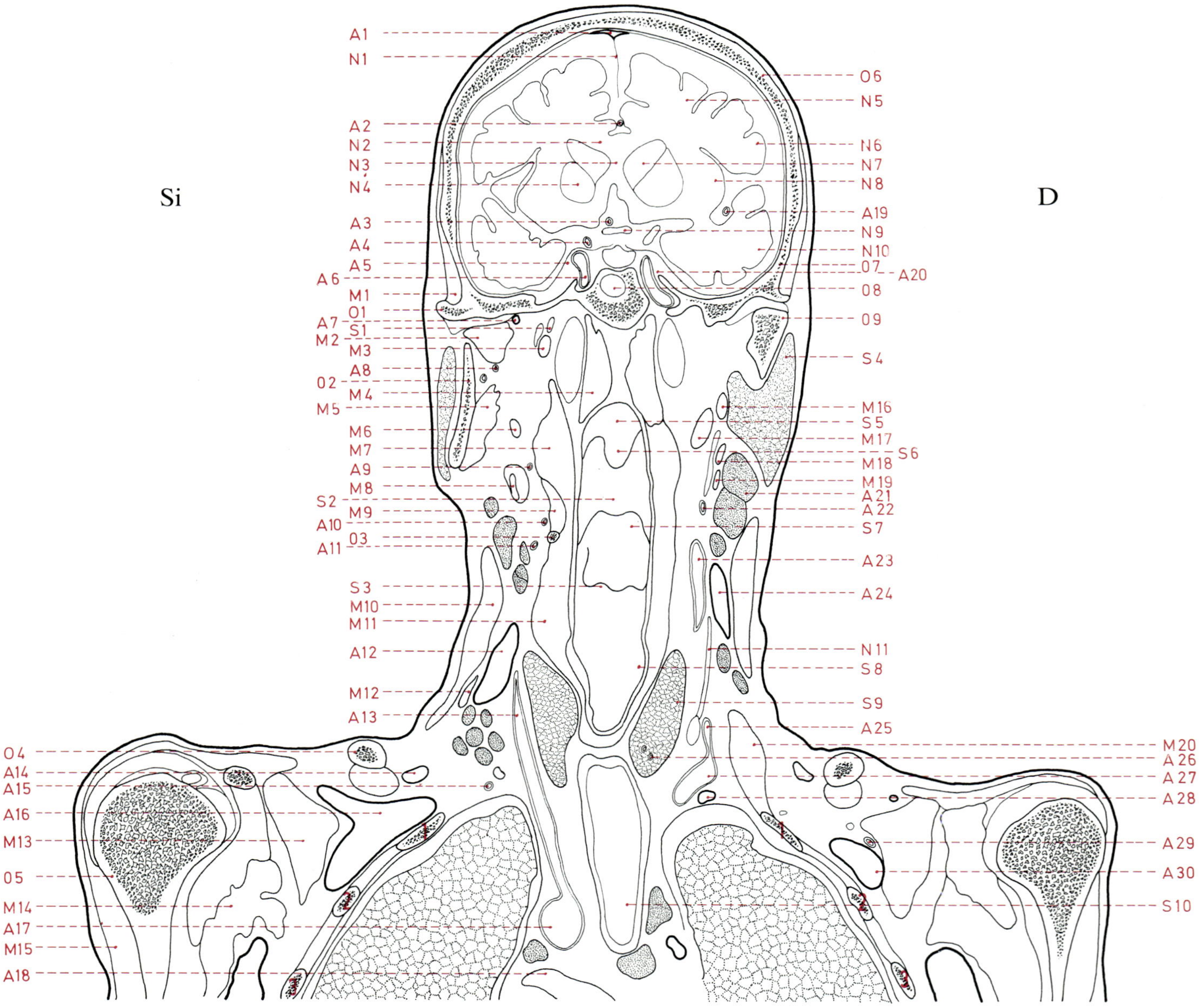
A1
N1
A2
N2
N3
N4
A3
A4
A5
A6
M1
O1
A7
S1
M2
M3
A8
O2
M4
M5
M6
M7
A9
M8
S2
M9
A10
O3
A11
S3
M10
M11
A12
M12
A13
O4
A14
A15
A16
M13
O5
M14
A17
M15
A18
O6
N5
N6
N7
N8
A19
N9
N10
O7
A20
O8
O9
S4
M16
S5
M17
S6
M18
M19
A21
A22
S7
A23
A24
N11
S8
S9
A25
M20
A26
A27
A28
A29
A30
S10

Fig. FC 4p

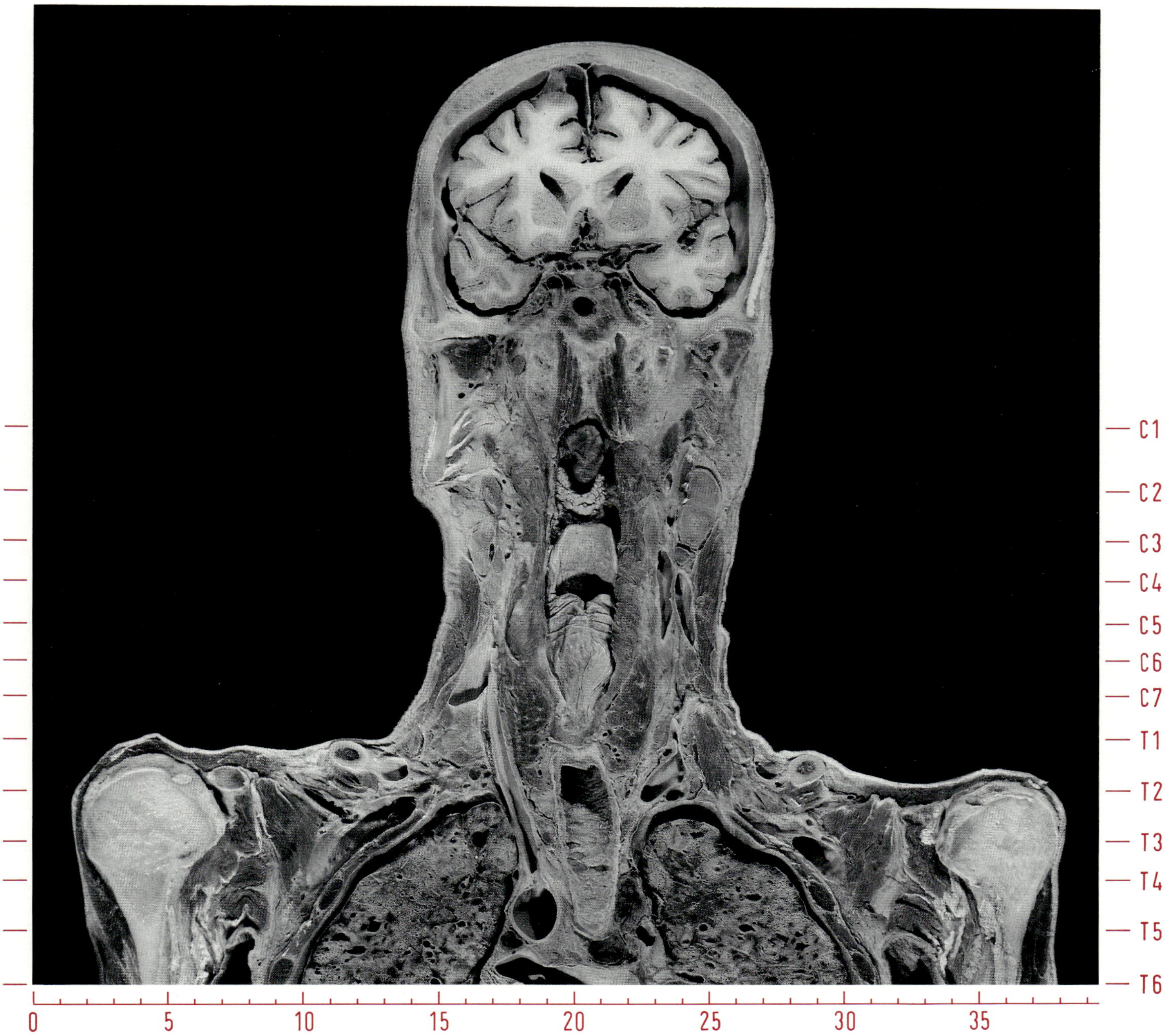

Fig. FC 4 p

D Si

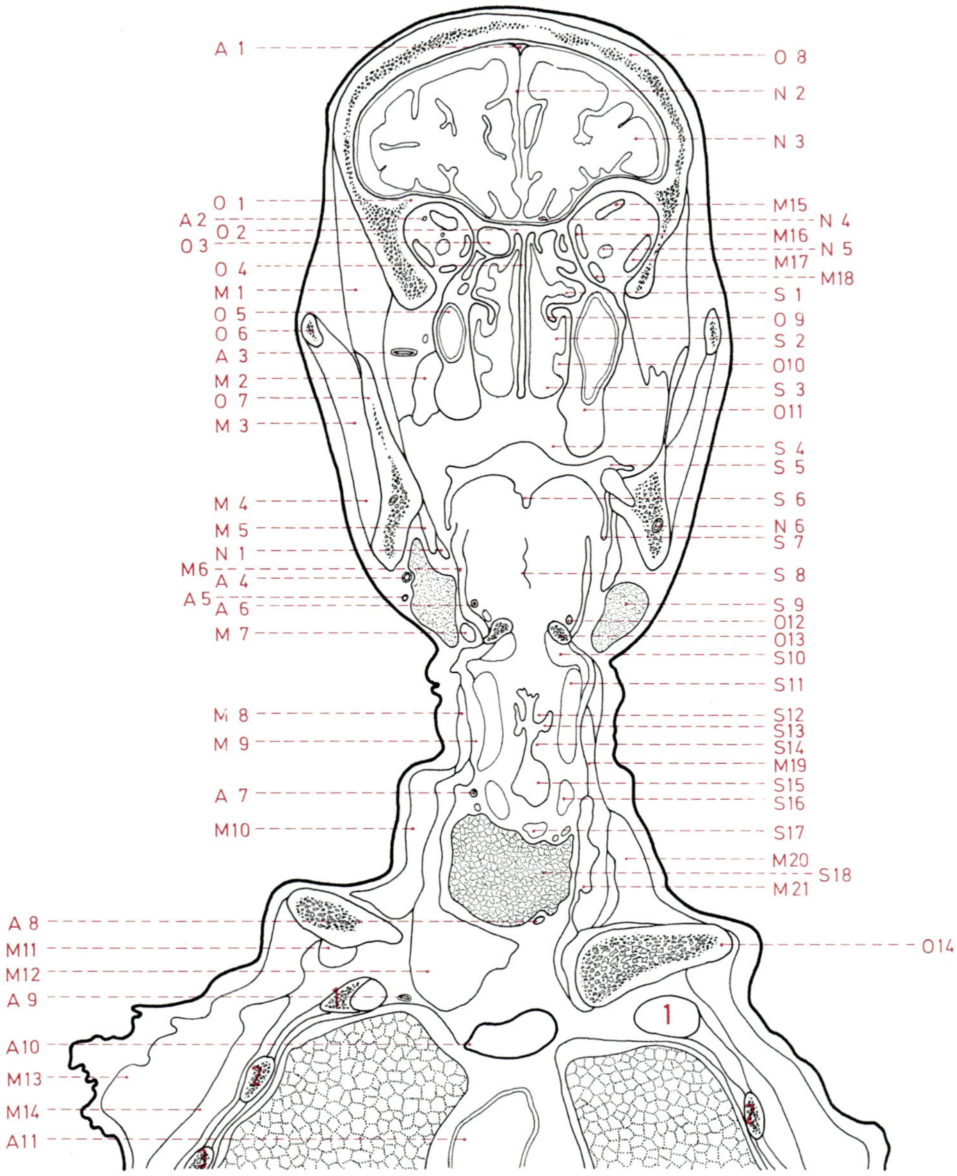

A 1
O 1
A 2
O 2
O 3
O 4
M 1
O 5
O 6
A 3
M 2
O 7
M 3
M 4
M 5
N 1
M6
A 4
A5
A 6
M 7
M 8
M 9
A 7
M10
A 8
M11
M12
A 9
A10
M13
M14
A11
O 8
N 2
N 3
M15
N 4
M16
N 5
M17
M18
S 1
O 9
S 2
O10
S 3
O11
S 4
S 5
S 6
N 6
S 7
S 8
S 9
O12
O13
S10
S11
S12
S13
S14
M19
S15
S16
S17
M20
S18
M21
O14
1

Fig. FC 5

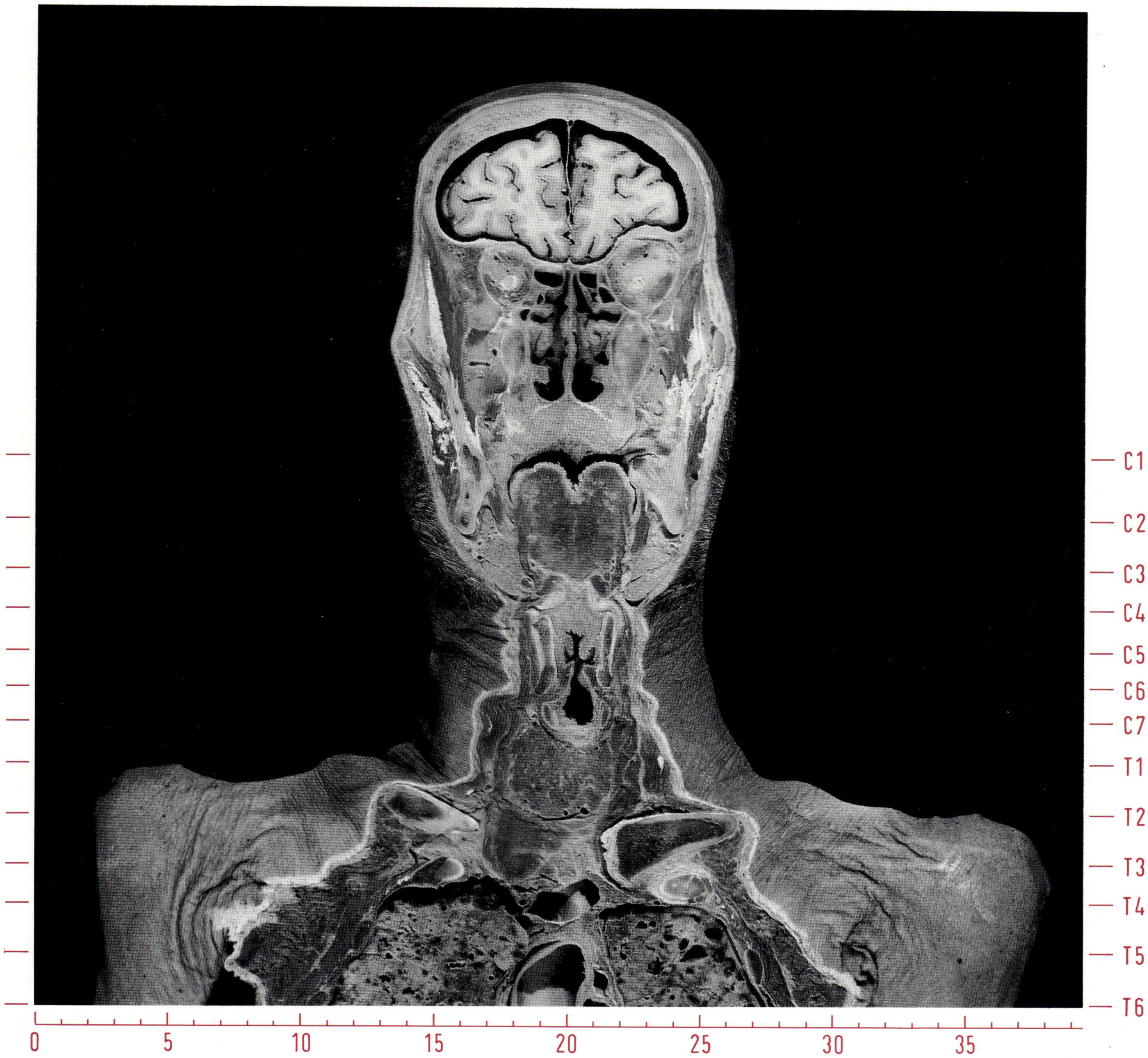

Fig. FC 5

SC

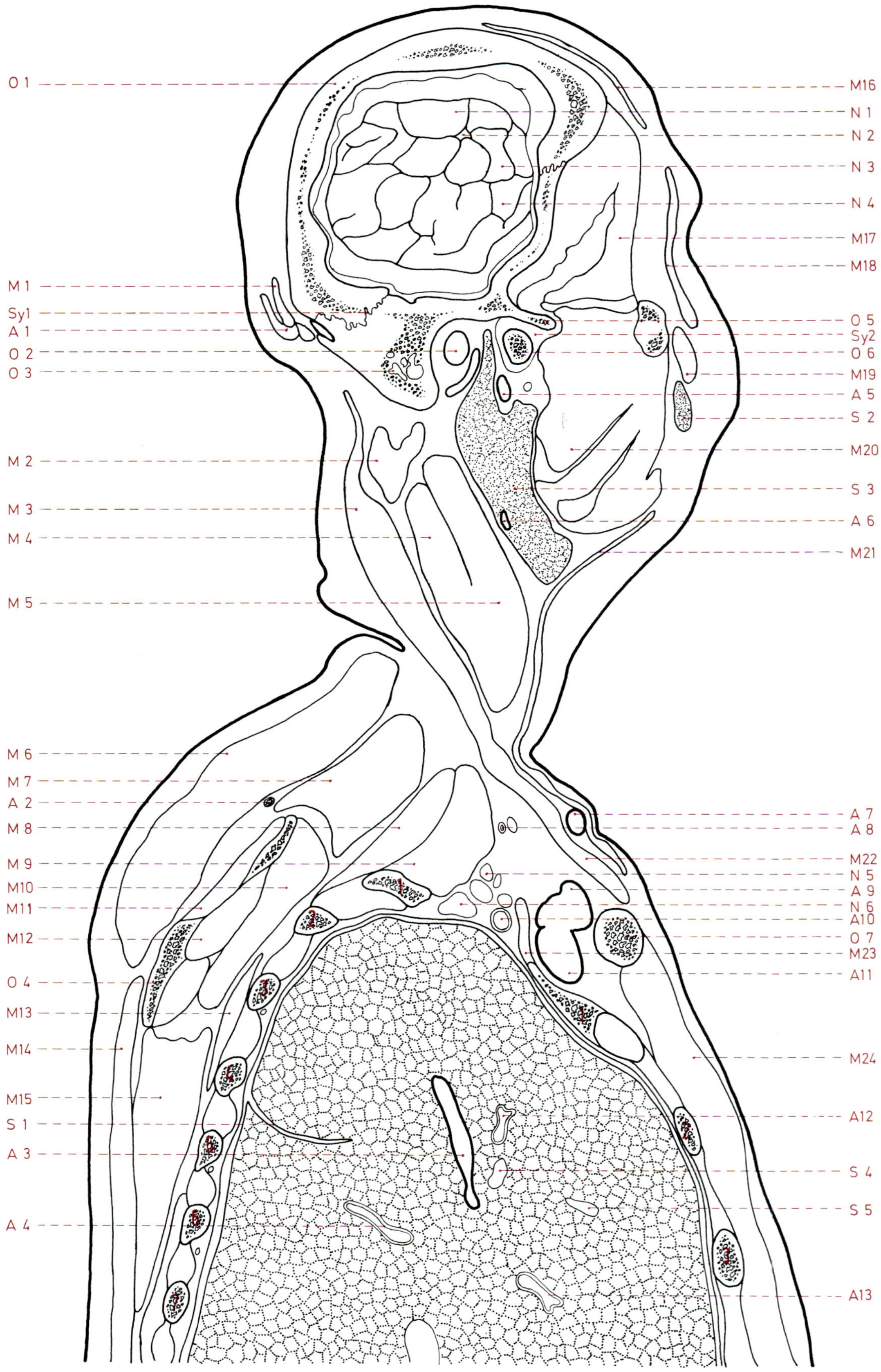
O 1
M 1
Sy1
A 1
O 2
O 3
M 2
M 3
M 4
M 5
M 6
M 7
A 2
M 8
M 9
M10
M11
M12
O 4
M13
M14
M15
S 1
A 3
A 4
M16
N 1
N 2
N 3
N 4
M17
M18
O 5
Sy2
O 6
M19
A 5
S 2
M20
S 3
A 6
M21
A 7
A 8
M22
N 5
A 9
N 6
A10
O 7
M23
A11
M24
A12
S 4
S 5
A13

Fig. SC 3D

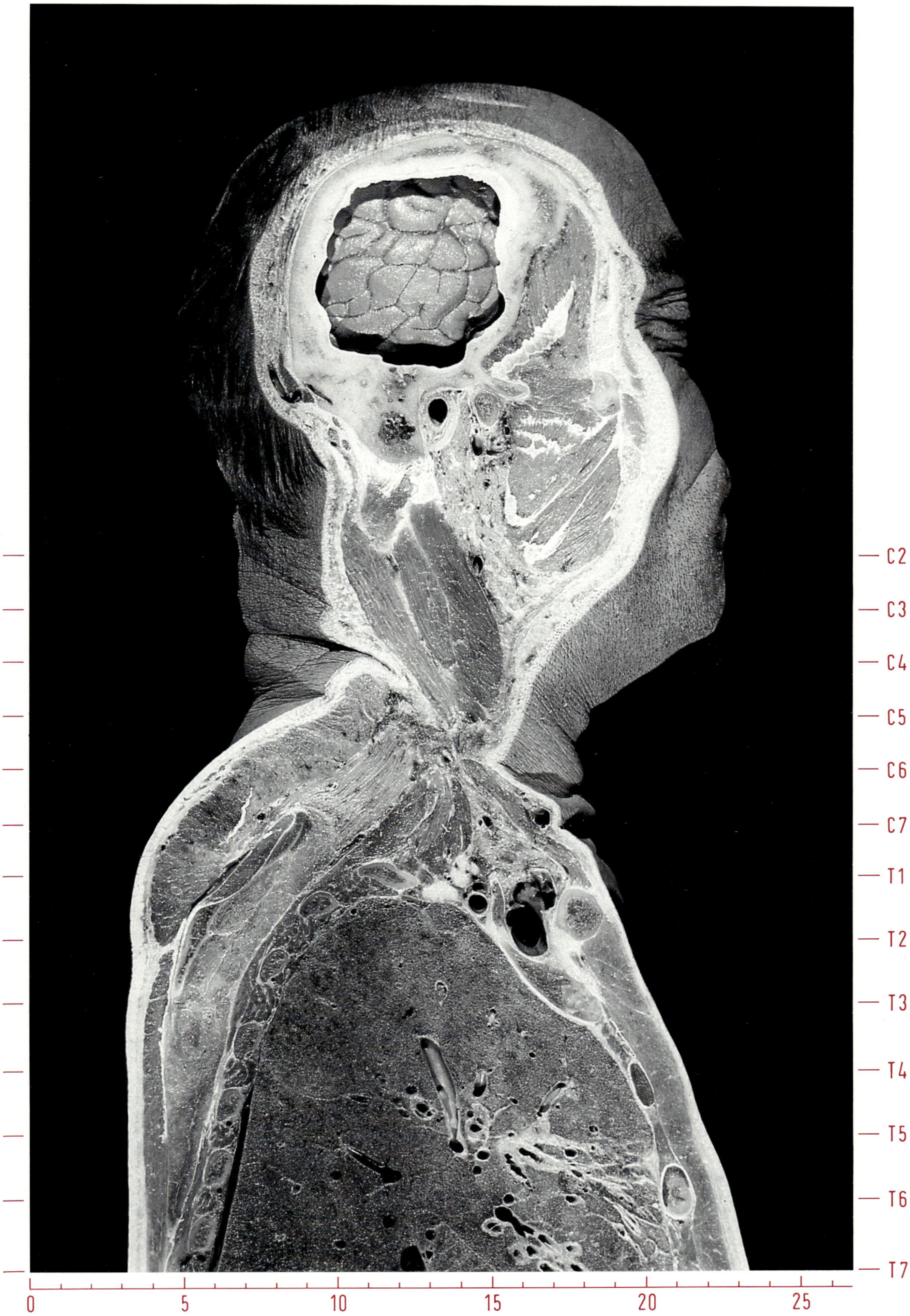

Fig. SC 3 D

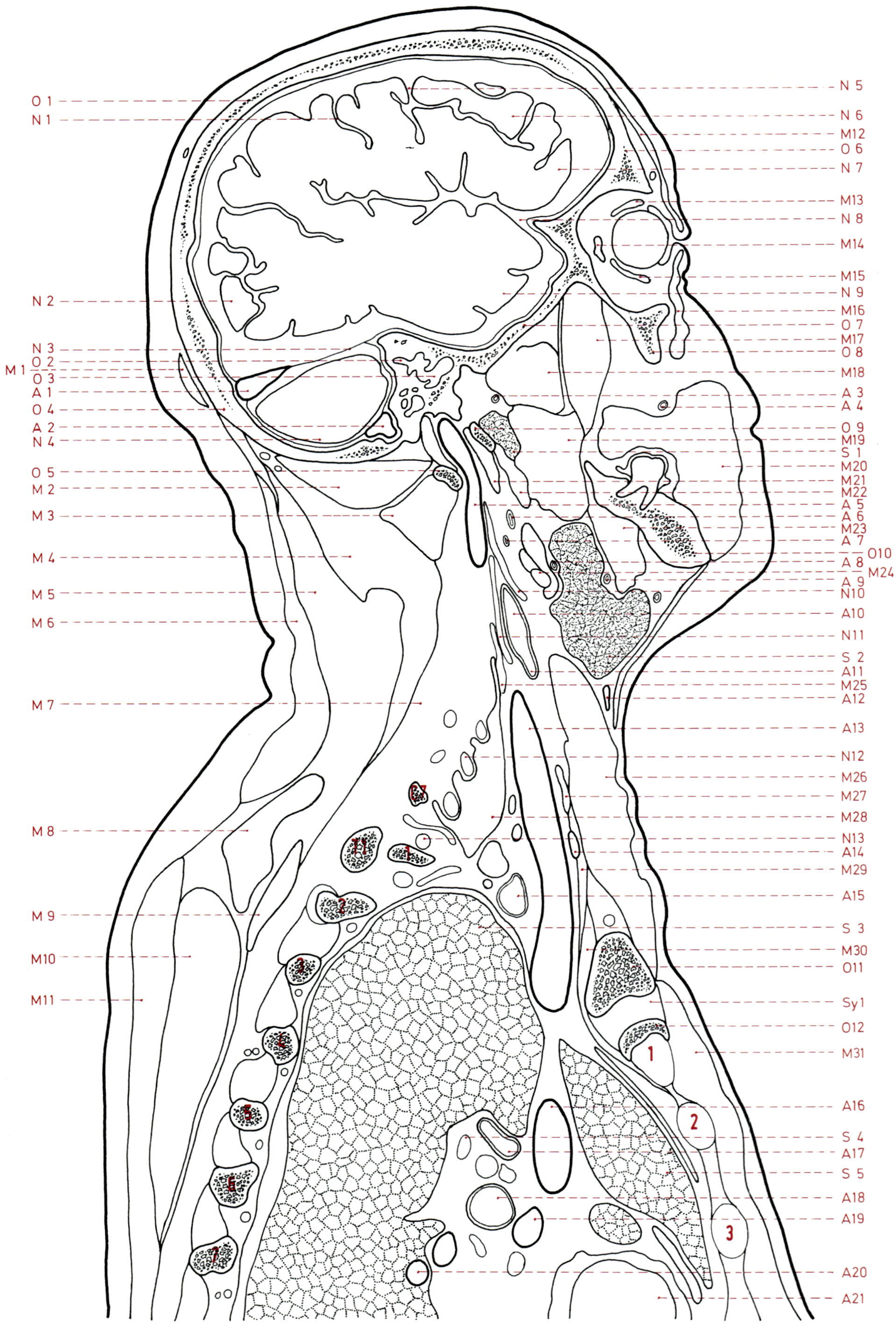
O 1
N 1
N 2
N 3
O 2
M 1
O 3
A 1
O 4
A 2
N 4
O 5
M 2
M 3
M 4
M 5
M 6
M 7
M 8
M 9
M10
M11
N 5
N 6
M12
O 6
N 7
M13
N 8
M14
M15
N 9
M16
O 7
M17
O 8
M18
A 3
A 4
O 9
M19
S 1
M20
M21
M22
A 5
A 6
M23
A 7
O10
A 8
M24
A 9
N10
A10
N11
S 2
A11
M25
A12
A13
N12
M26
M27
M28
N13
A14
M29
A15
S 3
M30
O11
Sy1
O12
M31
A16
S 4
A17
S 5
A18
A19
A20
A21
1
2
3

Fig. SC 2D

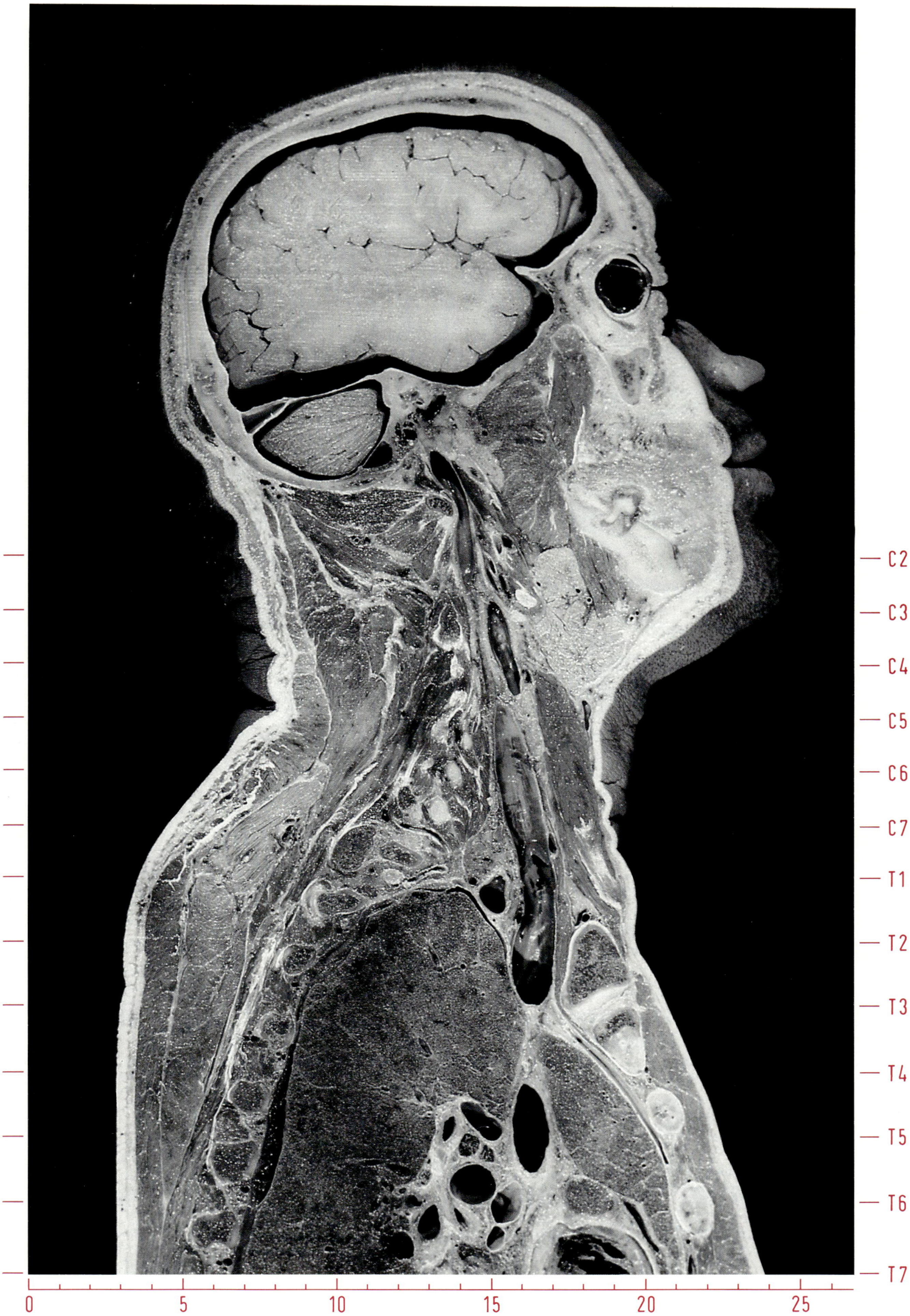

Fig. SC 2D

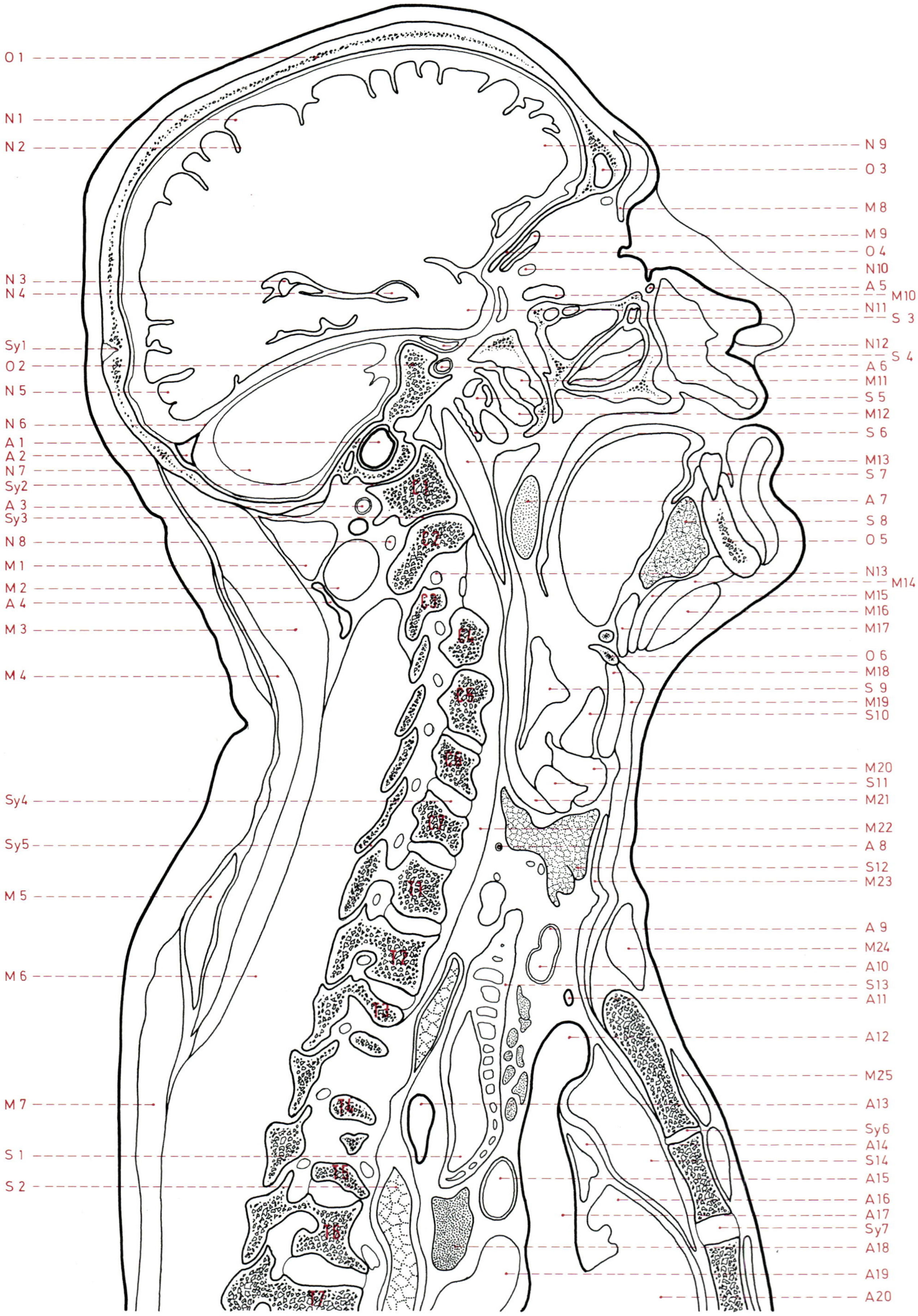
O 1
N 1
N 2
N 3
N 4
Sy1
O 2
N 5
N 6
A 1
A 2
N 7
Sy2
A 3
Sy3
N 8
M 1
M 2
A 4
M 3
M 4
Sy4
Sy5
M 5
M 6
M 7
S 1
S 2
N 9
O 3
M 8
M 9
O 4
N10
A 5
M10
N11
S 3
N12
S 4
A 6
M11
S 5
M12
S 6
M13
S 7
A 7
S 8
O 5
N13
M14
M15
M16
M17
O 6
M18
S 9
M19
S10
M20
S11
M21
M22
A 8
S12
M23
A 9
M24
A10
S13
A11
A12
M25
A13
Sy6
A14
S14
A15
A16
A17
Sy7
A18
A19
A20
C1
C2
C3
C4
C5
C6
C7
T1
T2
T3
T4
T5
T6
T7

Fig. SC 1D

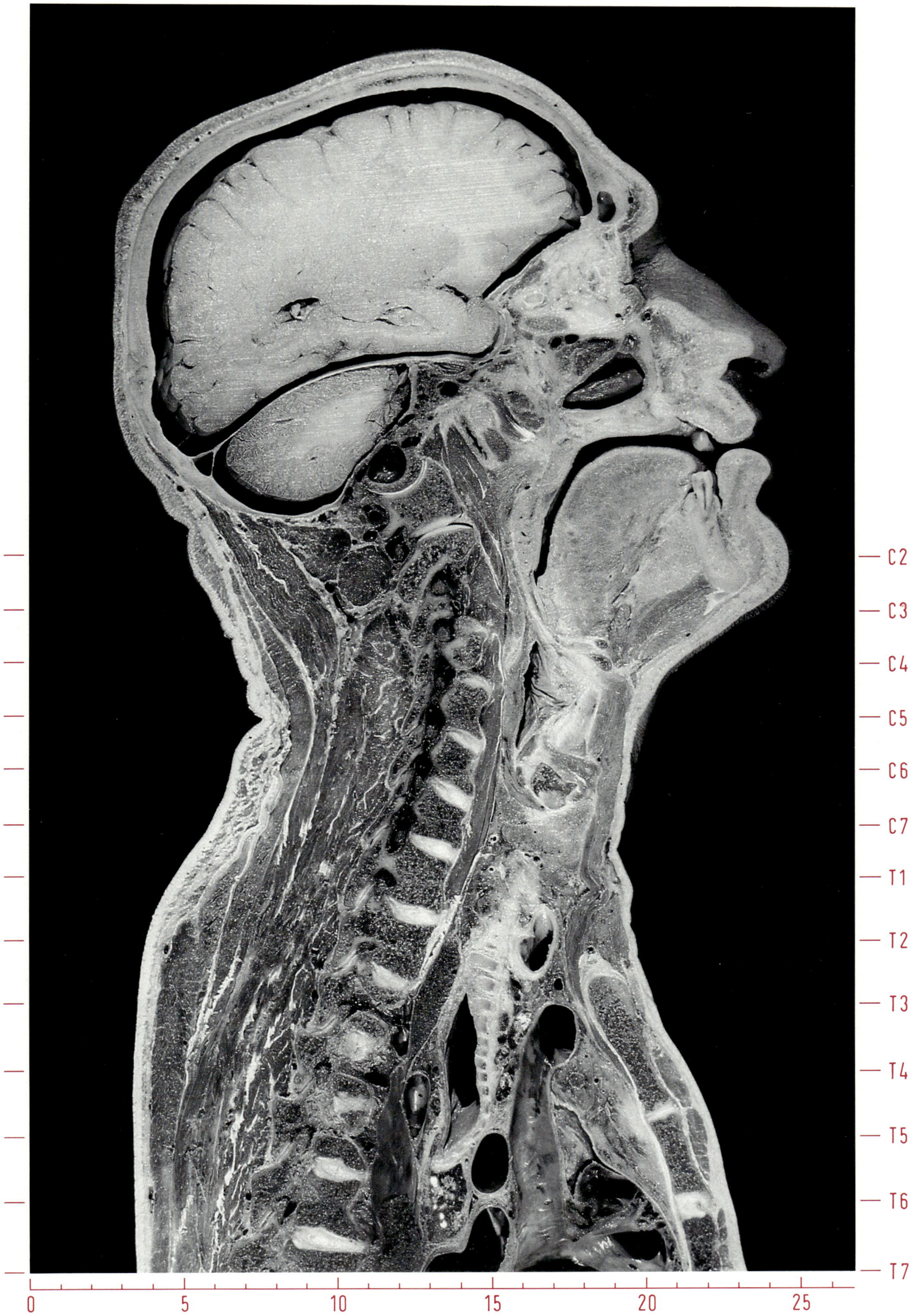

Fig. SC 1D

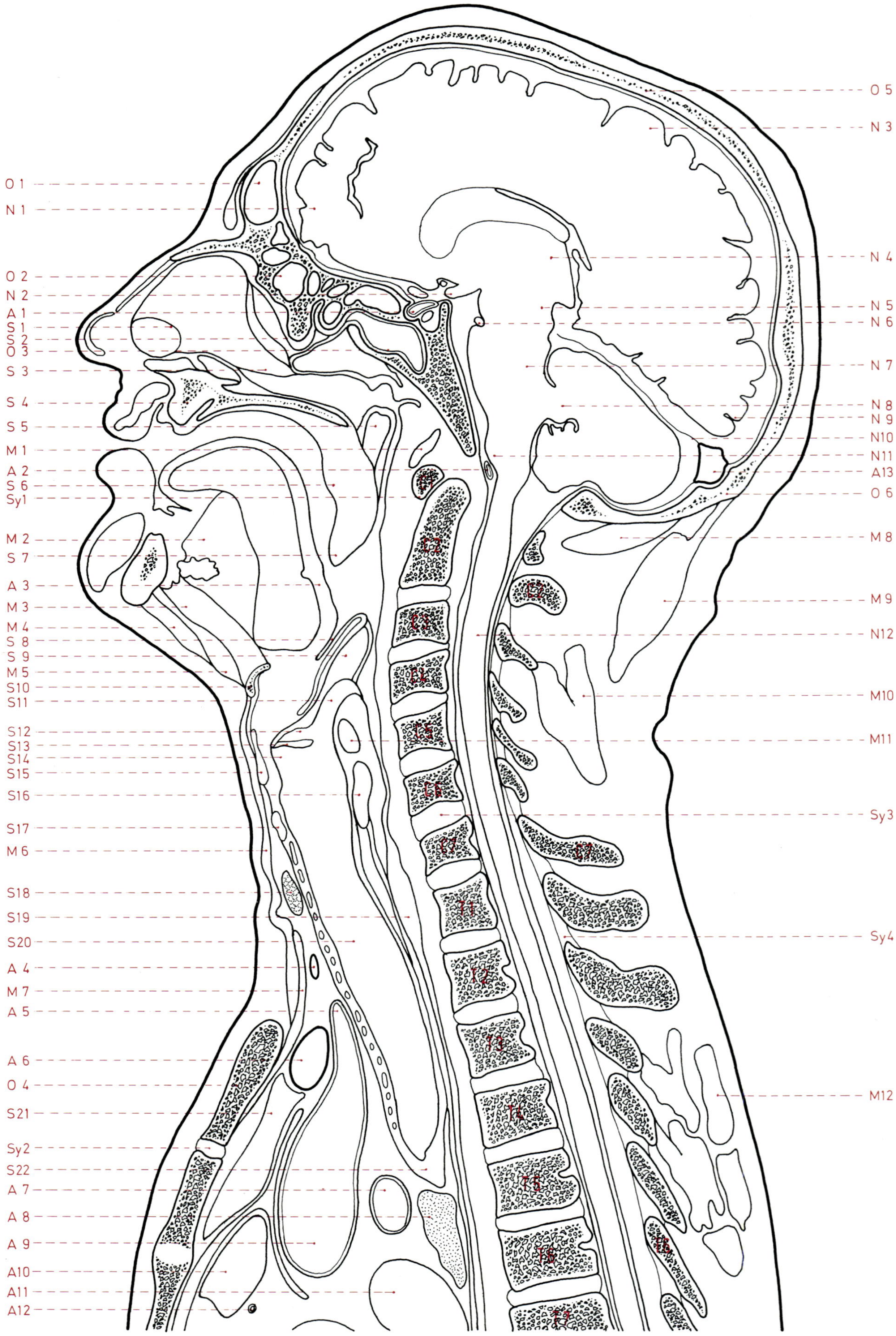

O 1
N 1
O 2
N 2
A 1
S 1
S 2
O 3
S 3
S 4
S 5
M 1
A 2
S 6
Sy1
M 2
S 7
A 3
M 3
M 4
S 8
S 9
M 5
S10
S11
S12
S13
S14
S15
S16
S17
M 6
S18
S19
S20
A 4
M 7
A 5
A 6
O 4
S21
Sy2
S22
A 7
A 8
A 9
A10
A11
A12
O 5
N 3
N 4
N 5
N 6
N 7
N 8
N 9
N10
N11
A13
O 6
M 8
M 9
N12
M10
M11
Sy3
Sy4
M12

Fig. SC MD

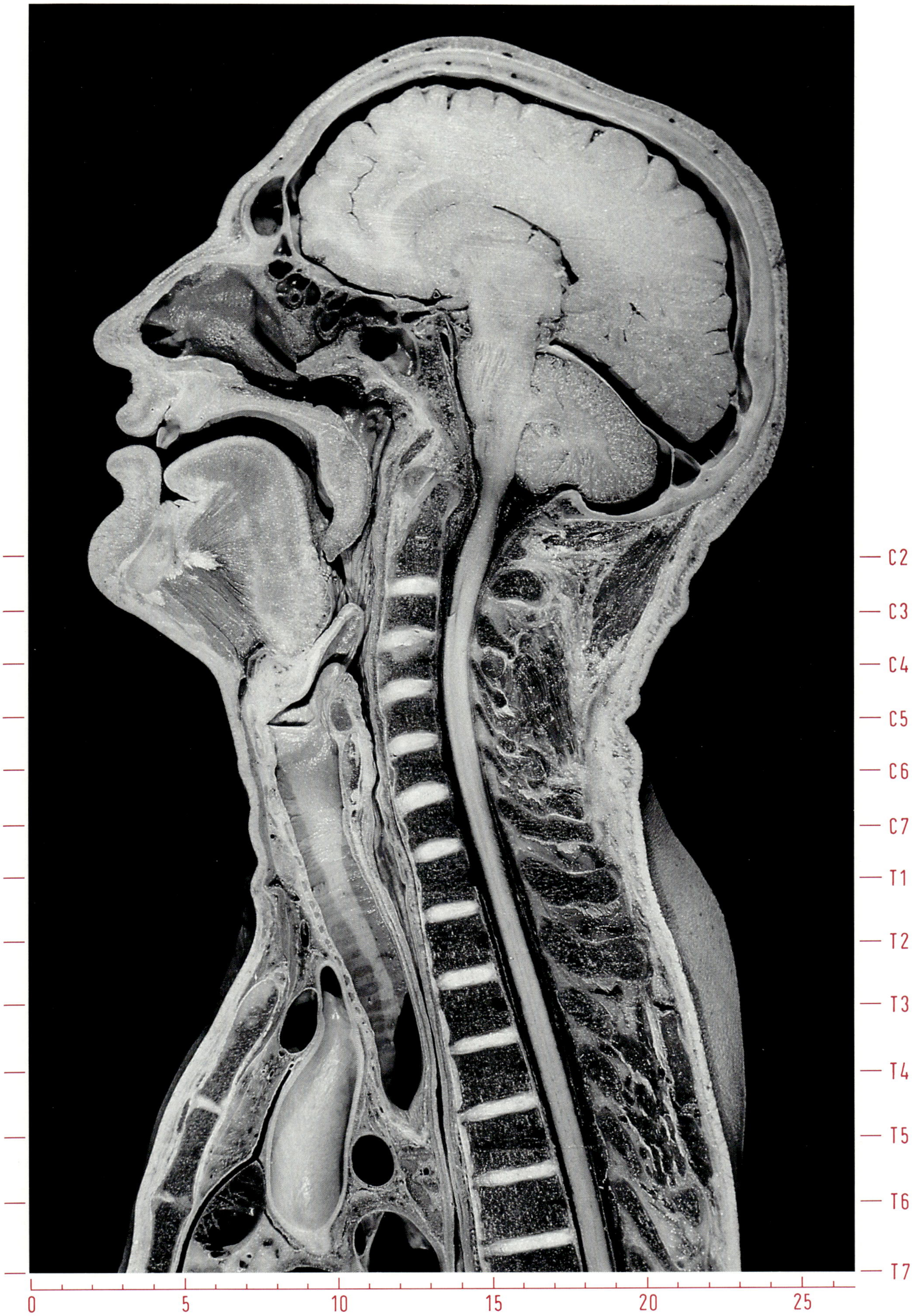

Fig. SC MD

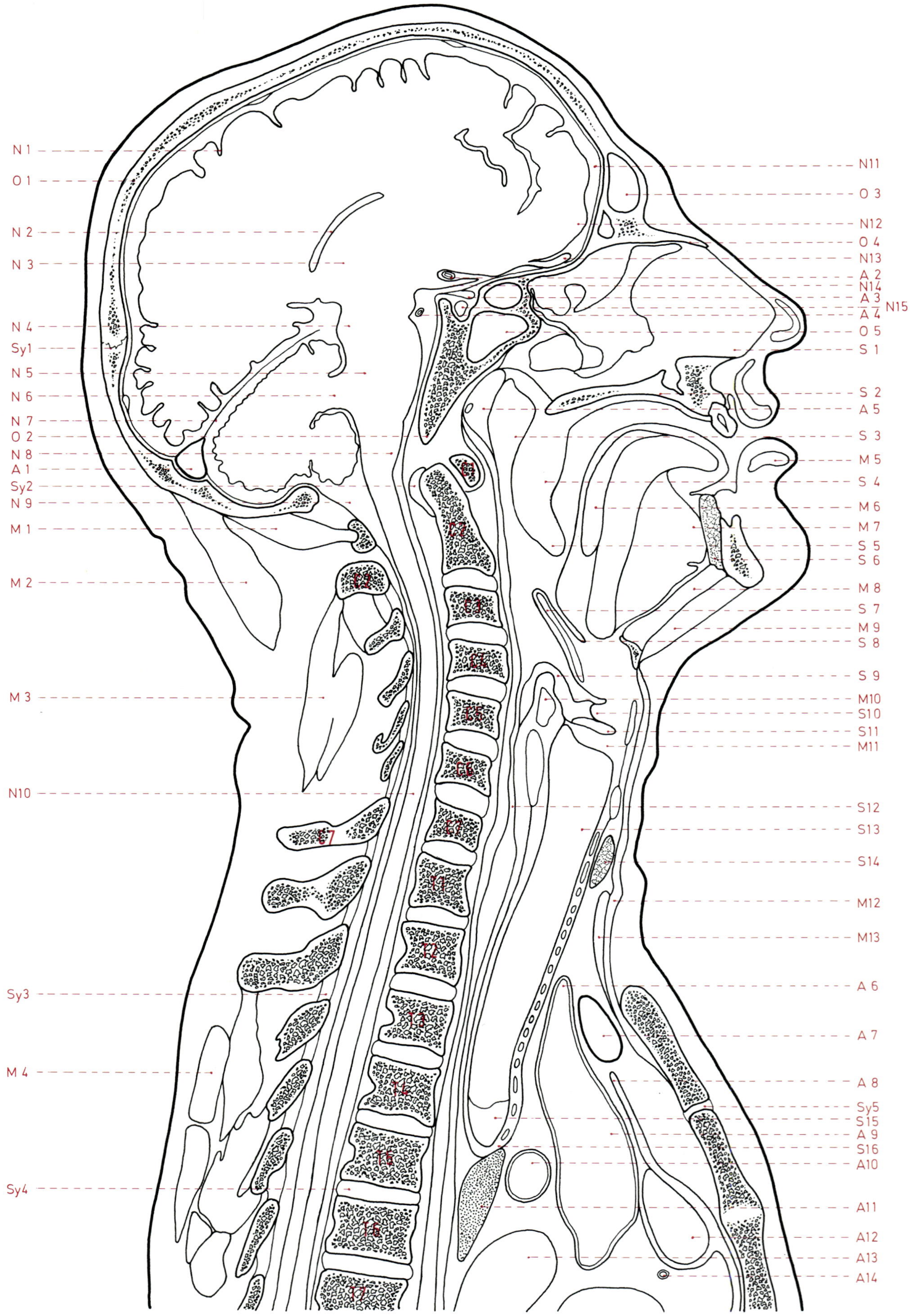
N 1
O 1
N 2
N 3
N 4
Sy1
N 5
N 6
N 7
O 2
N 8
A 1
Sy2
N 9
M 1
M 2
M 3
N10
Sy3
M 4
Sy4
N11
O 3
N12
O 4
N13
A 2
N14
A 3
N15
A 4
O 5
S 1
S 2
A 5
S 3
M 5
S 4
M 6
M 7
S 5
S 6
M 8
S 7
M 9
S 8
S 9
M10
S10
S11
M11
S12
S13
S14
M12
M13
A 6
A 7
A 8
Sy5
S15
A 9
S16
A10
A11
A12
A13
A14

Fig. SC MSi

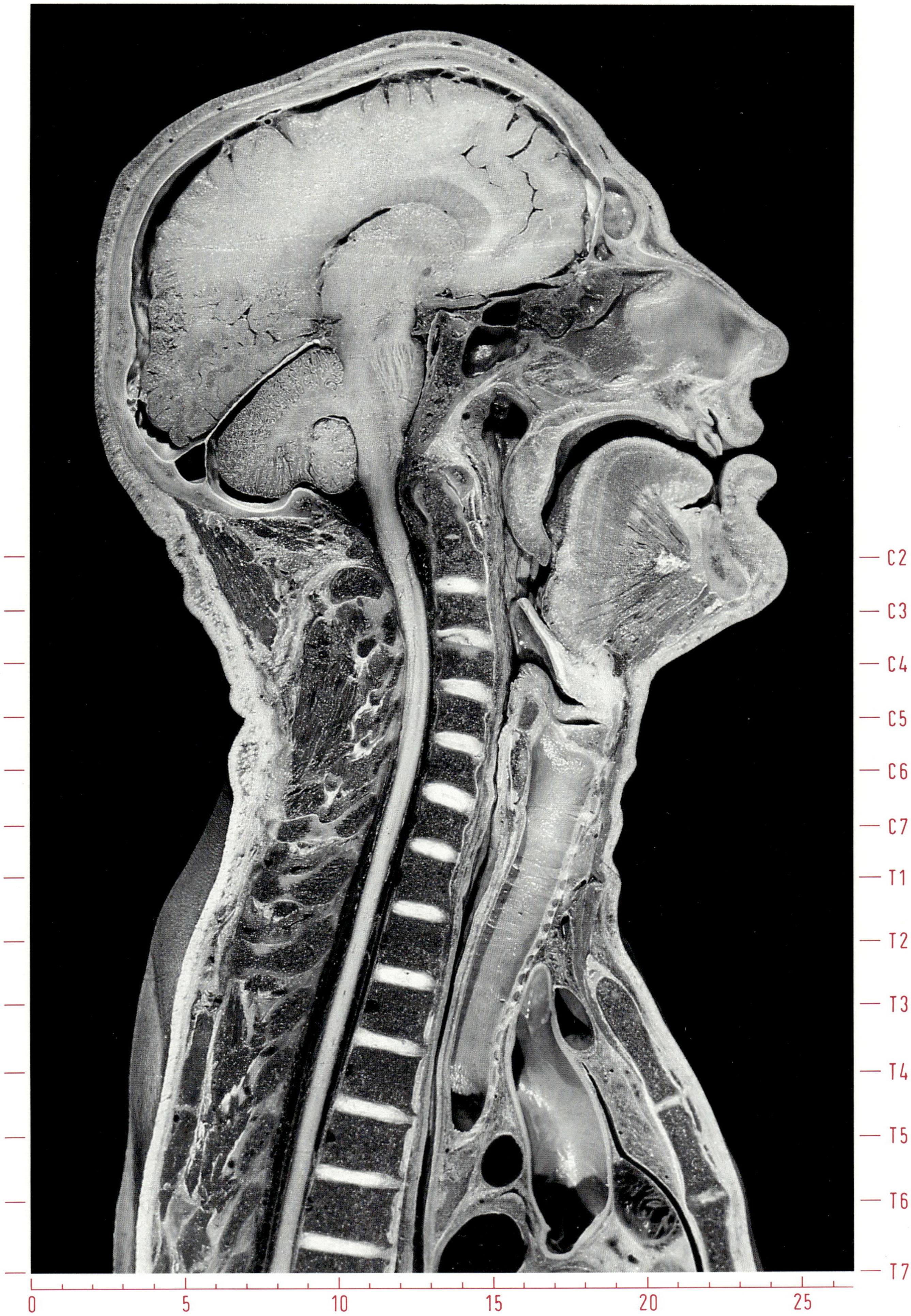

Fig. SC MSi

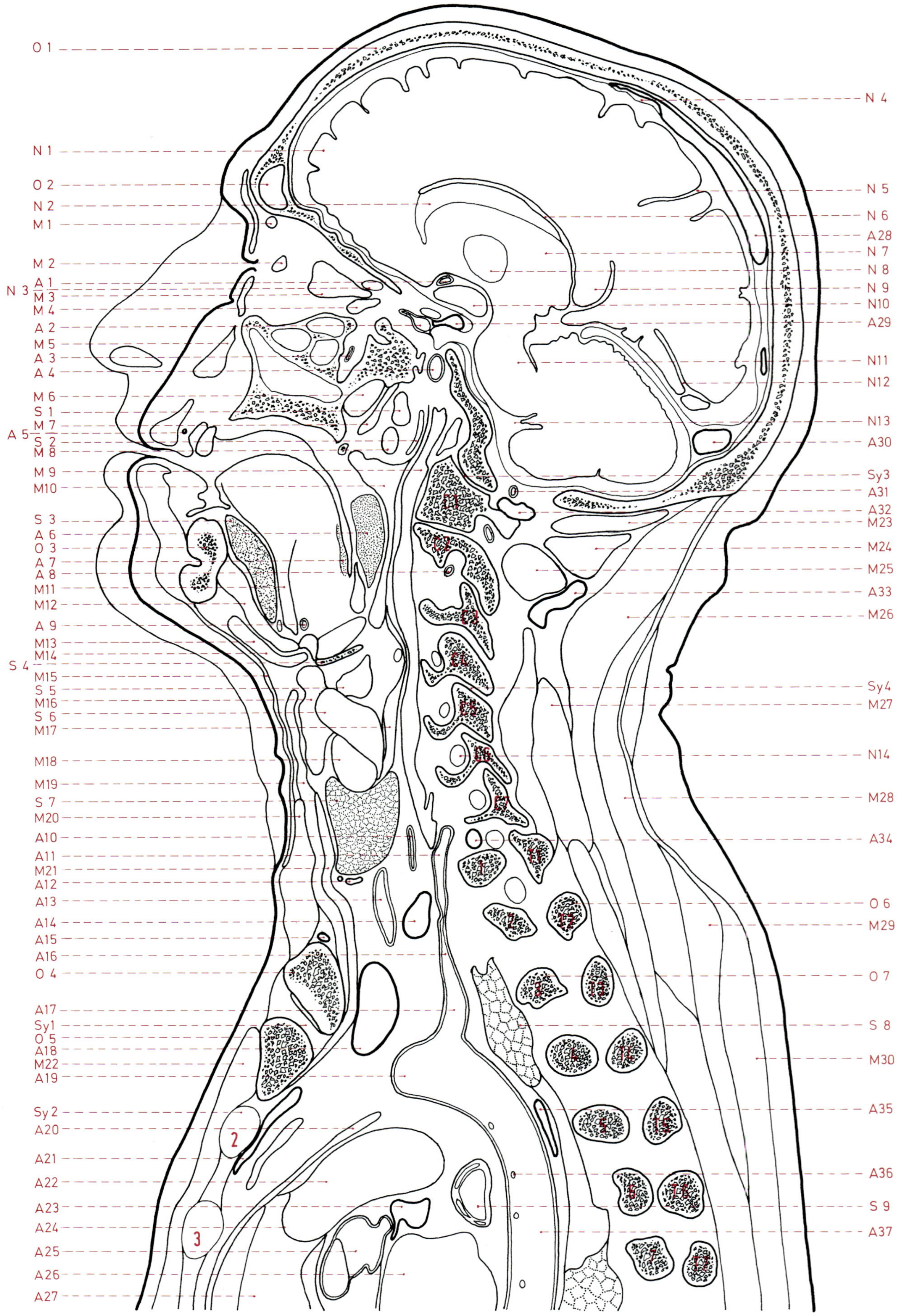
O 1
N 1
O 2
N 2
M 1
M 2
N 3
A 1
M 3
M 4
A 2
M 5
A 3
A 4
M 6
S 1
A 5
M 7
S 2
M 8
M 9
M10
S 3
A 6
O 3
A 7
A 8
M11
M12
A 9
M13
S 4
M14
M15
S 5
M16
S 6
M17
M18
M19
S 7
M20
A10
A11
M21
A12
A13
A14
A15
A16
O 4
A17
Sy1
O 5
A18
M22
A19
Sy 2
A20
A21
A22
A23
A24
A25
A26
A27
N 4
N 5
N 6
A28
N 7
N 8
N 9
N10
A29
N11
N12
N13
A30
Sy3
A31
A32
M23
M24
M25
A33
M26
Sy4
M27
N14
M28
A34
O 6
M29
O 7
S 8
M30
A35
A36
S 9
A37
2
3

Fig. SC 1 Si

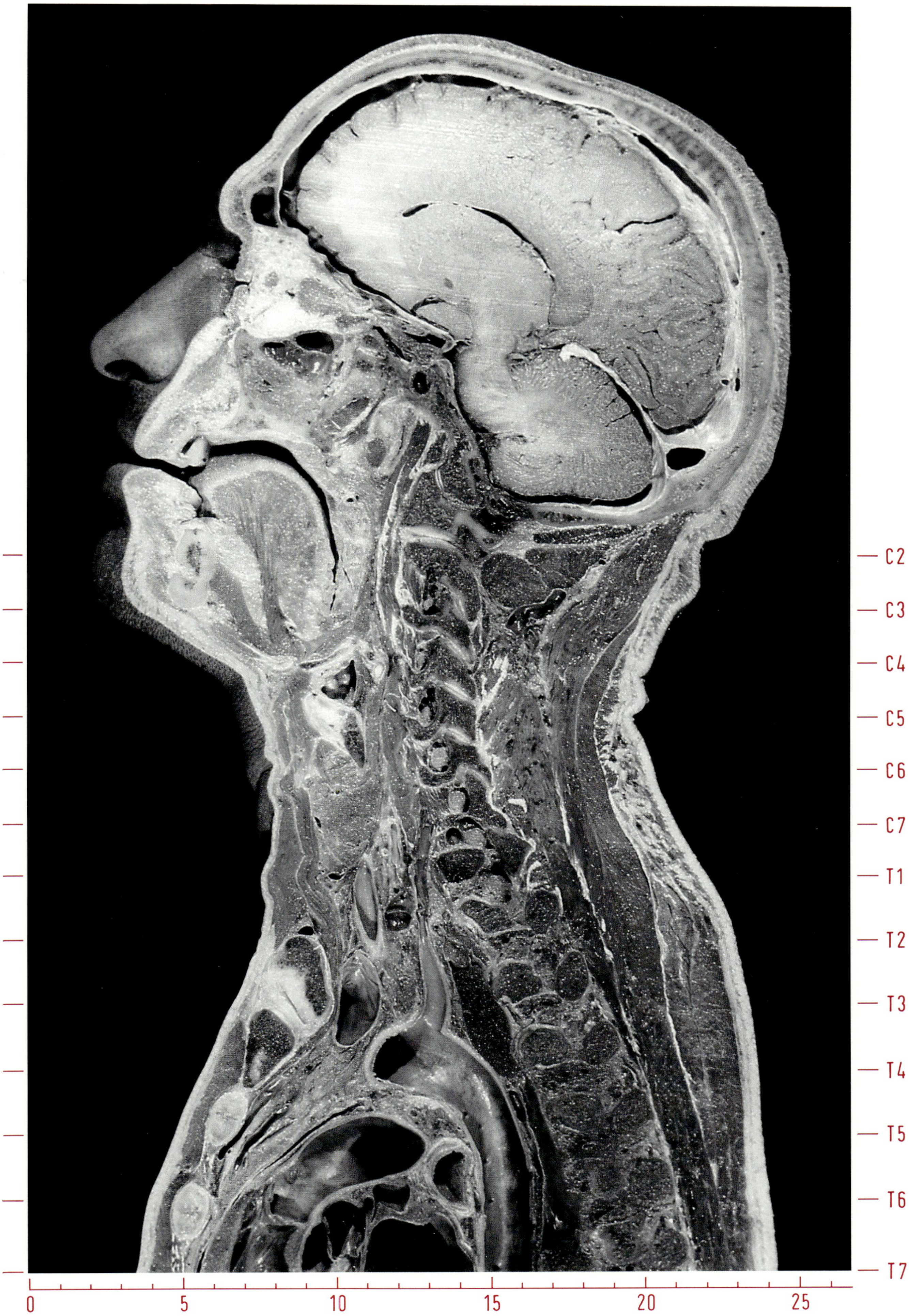

Fig. SC 1Si

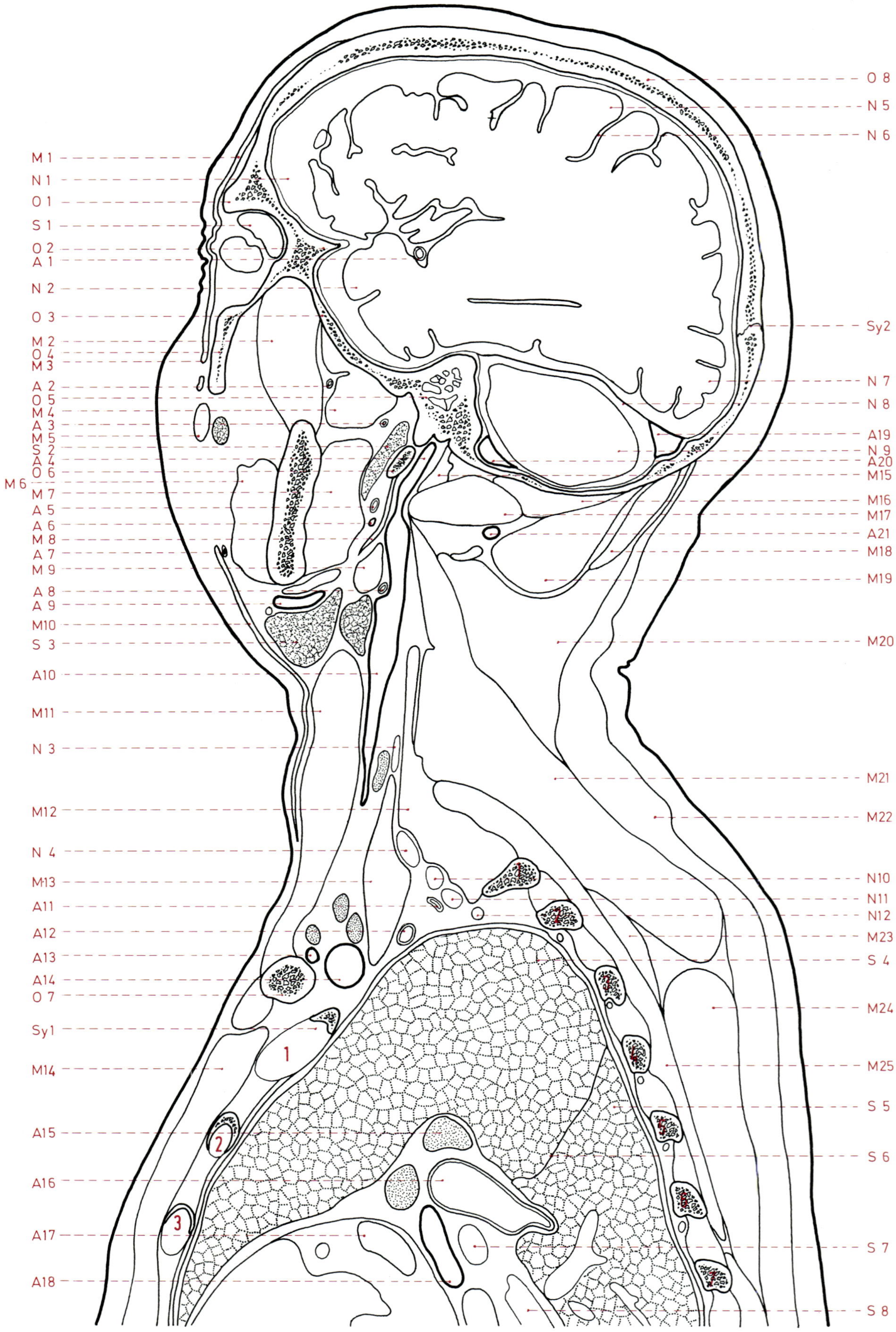

M 1
N 1
O 1
S 1
O 2
A 1
N 2
O 3
M 2
O 4
M 3
A 2
O 5
M 4
A 3
M 5
S 2
A 4
O 6
M 6
M 7
A 5
A 6
M 8
A 7
M 9
A 8
A 9
M10
S 3
A10
M11
N 3
M12
N 4
M13
A11
A12
A13
A14
O 7
Sy1
M14
A15
A16
A17
A18
O 8
N 5
N 6
Sy2
N 7
N 8
A19
N 9
A20
M15
M16
M17
A21
M18
M19
M20
M21
M22
N10
N11
N12
M23
S 4
M24
M25
S 5
S 6
S 7
S 8
1
2
3

Fig. SC 2 Si

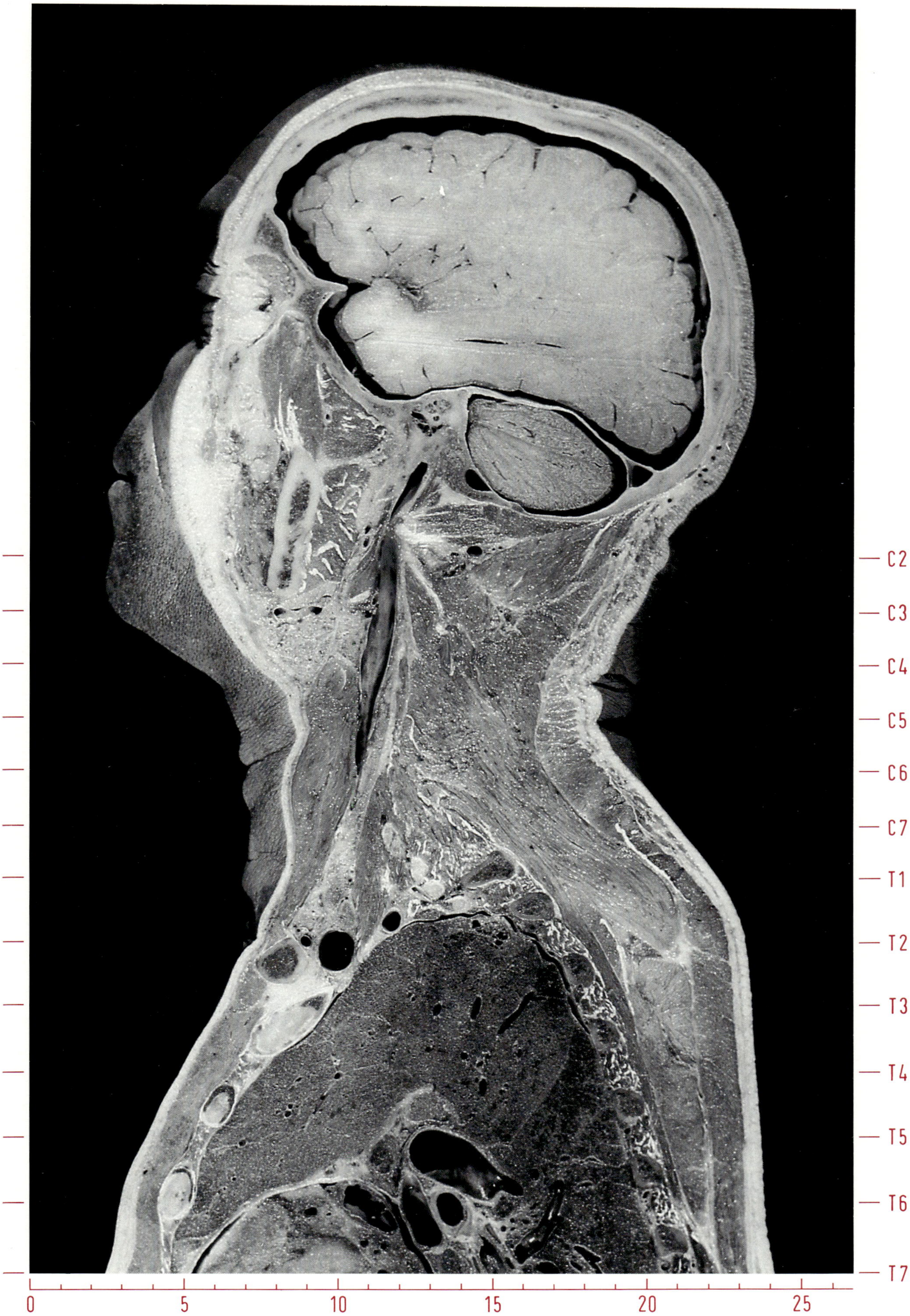

Fig. SC 2 Si

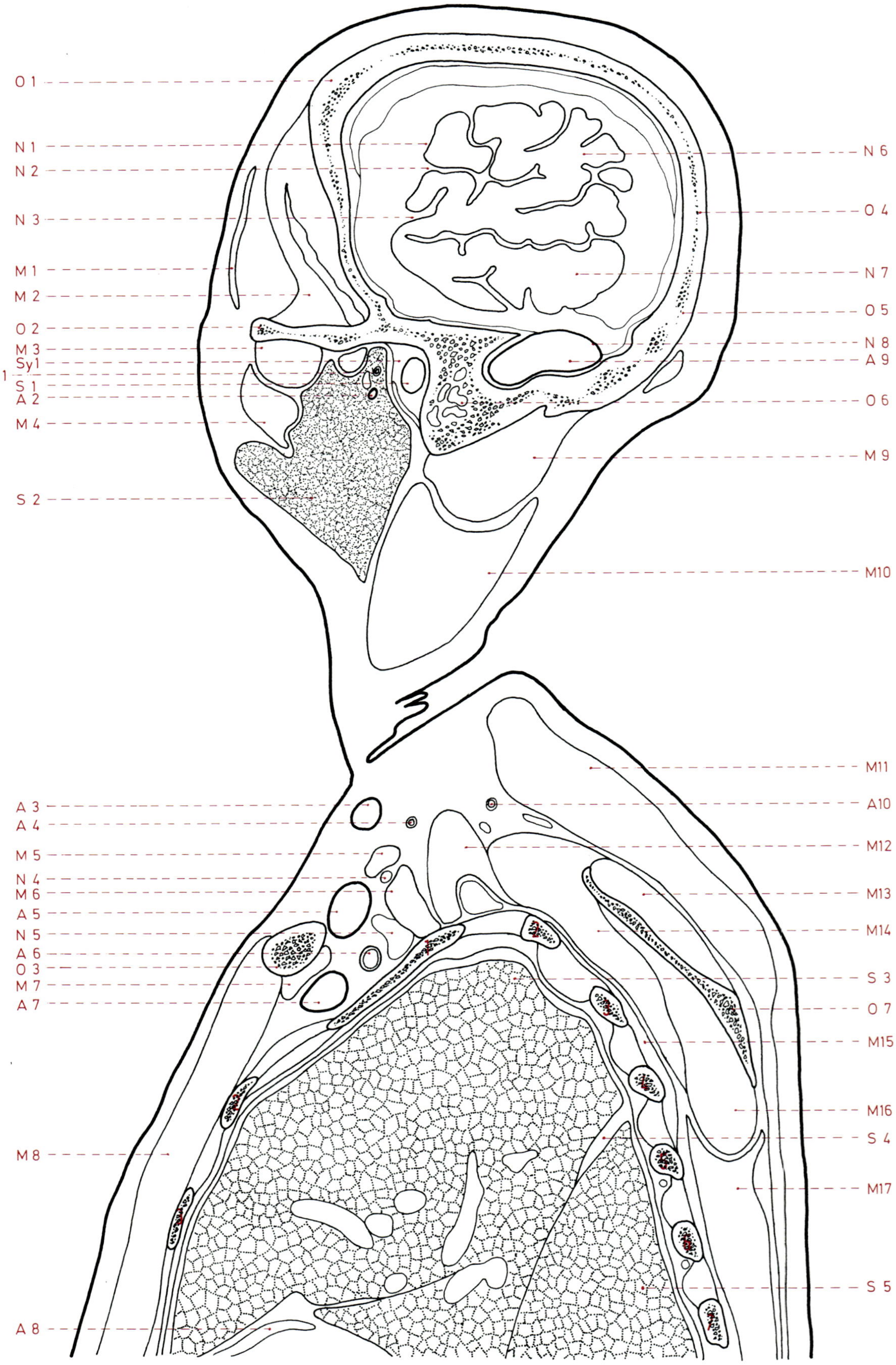
O 1
N 1
N 2
N 3
M 1
M 2
O 2
M 3
Sy1
A 1
S 1
A 2
M 4
S 2
N 6
O 4
N 7
O 5
N 8
A 9
O 6
M 9
M10
M11
A 3
A 4
M 5
N 4
M 6
A 5
N 5
A 6
O 3
M 7
A 7
M 8
A 8
A10
M12
M13
M14
S 3
O 7
M15
M16
S 4
M17
S 5

Fig. SC 3 Si

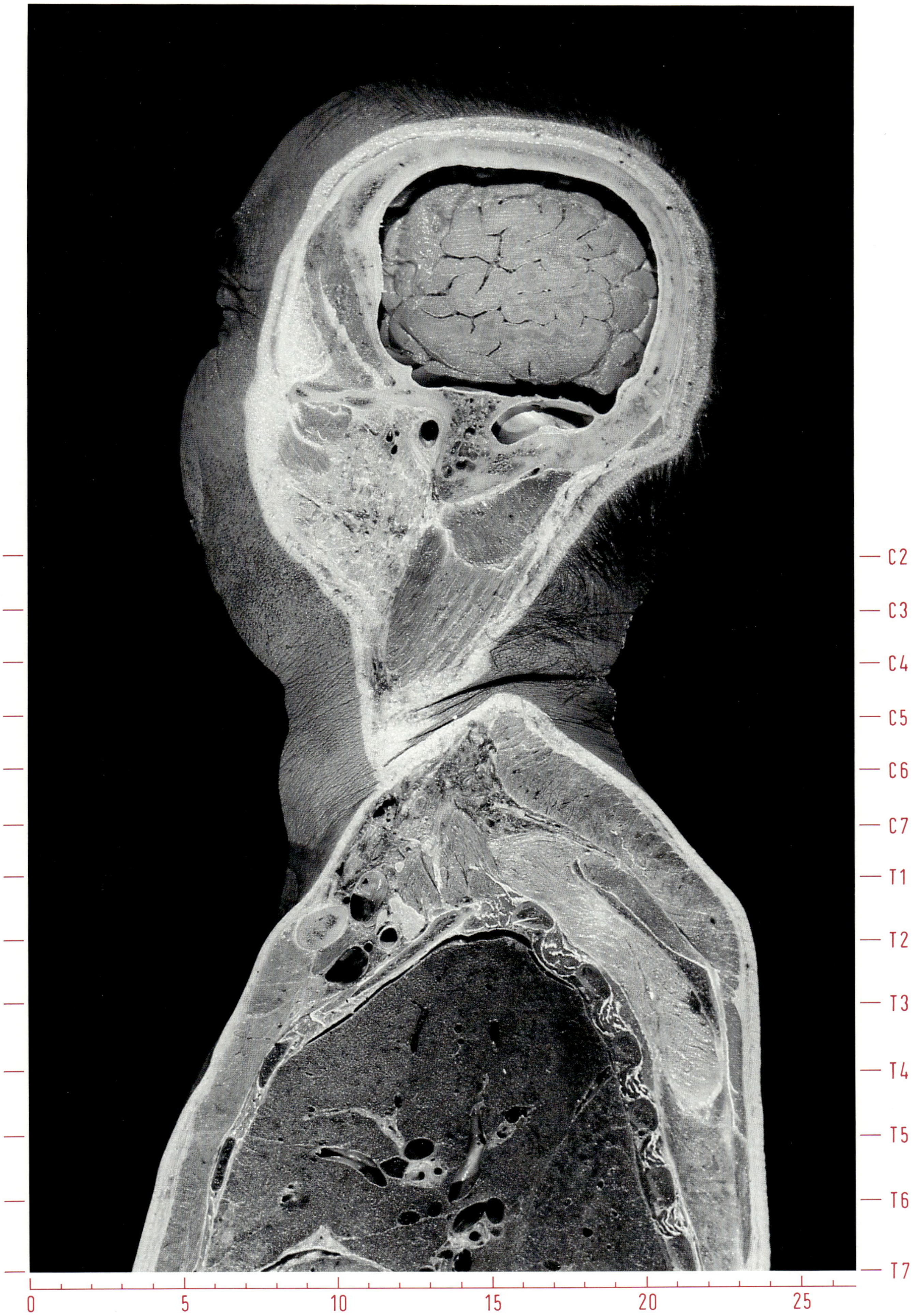

Fig. SC 3 Si

HC

D Si

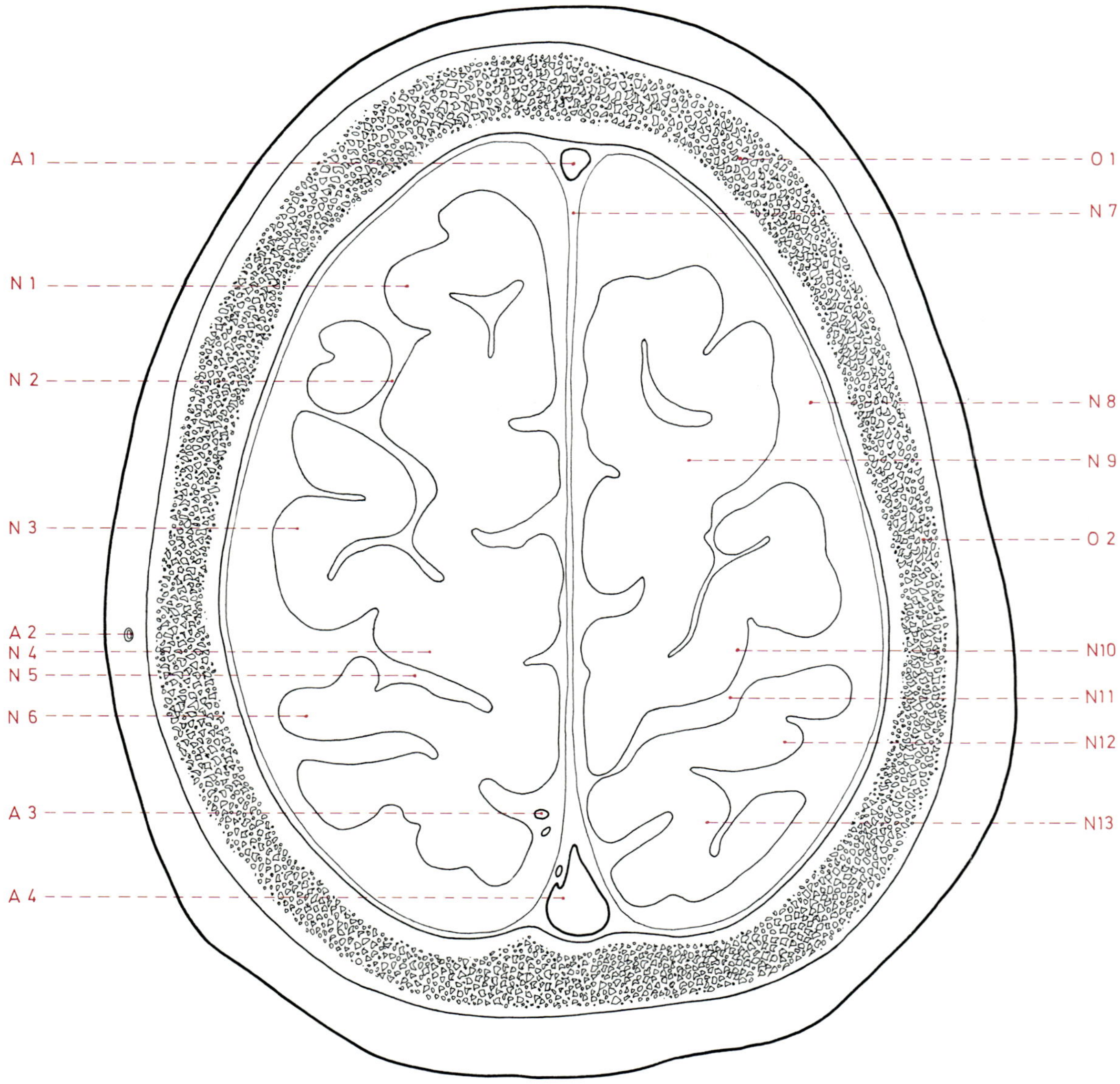

Fig. HC 1

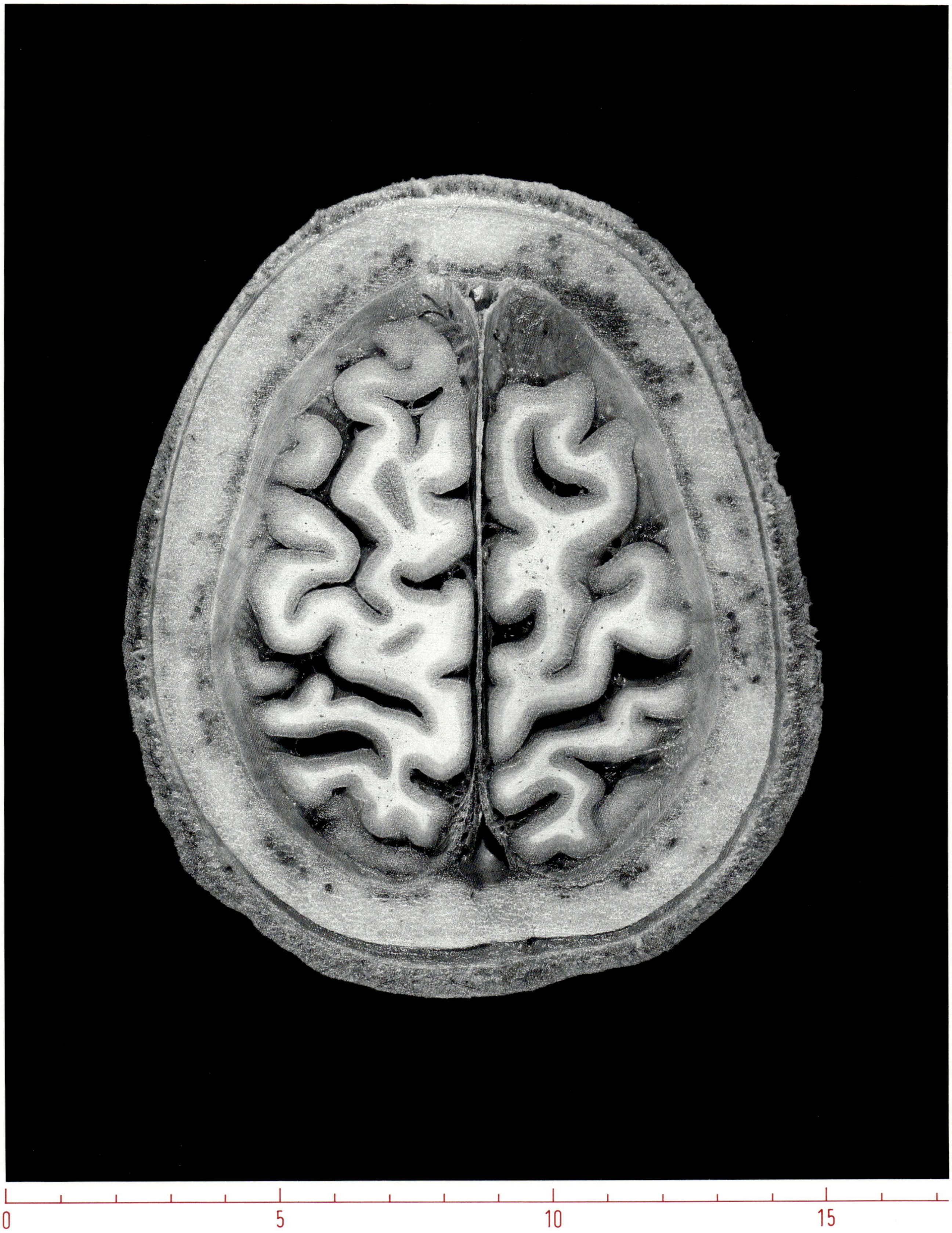

Fig. HC 1

Si D

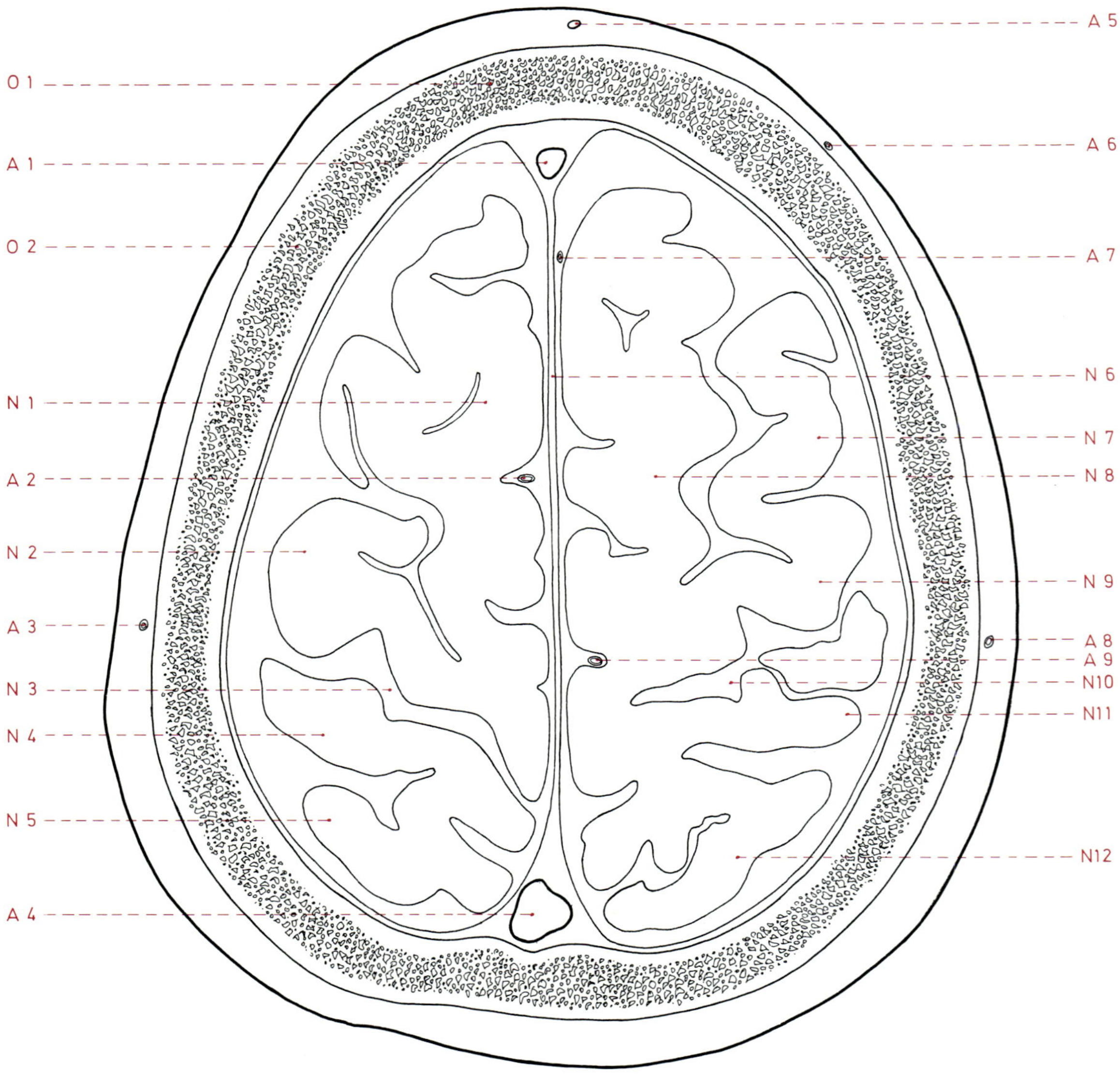

Fig. HC 2

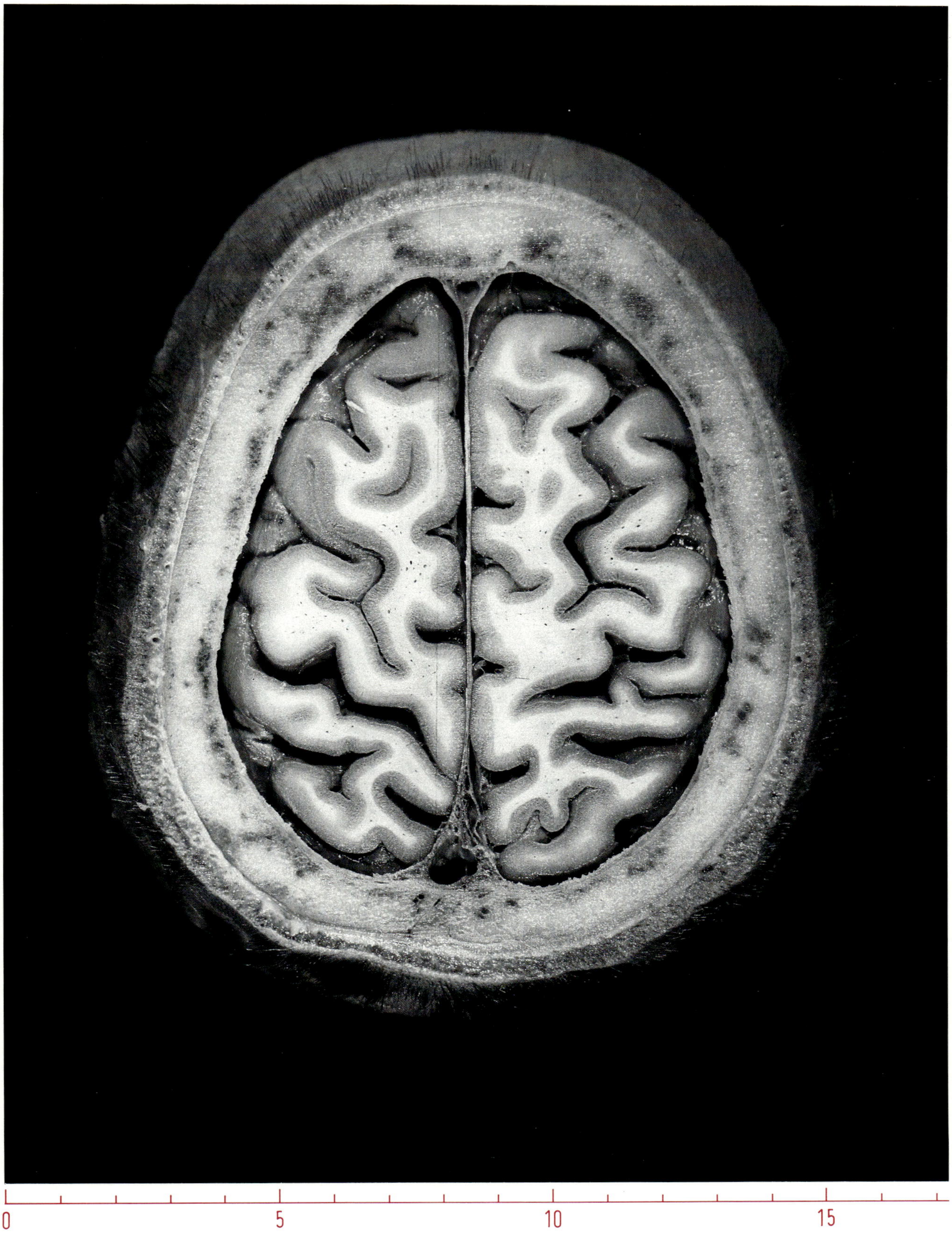

Fig. HC 2

D Si

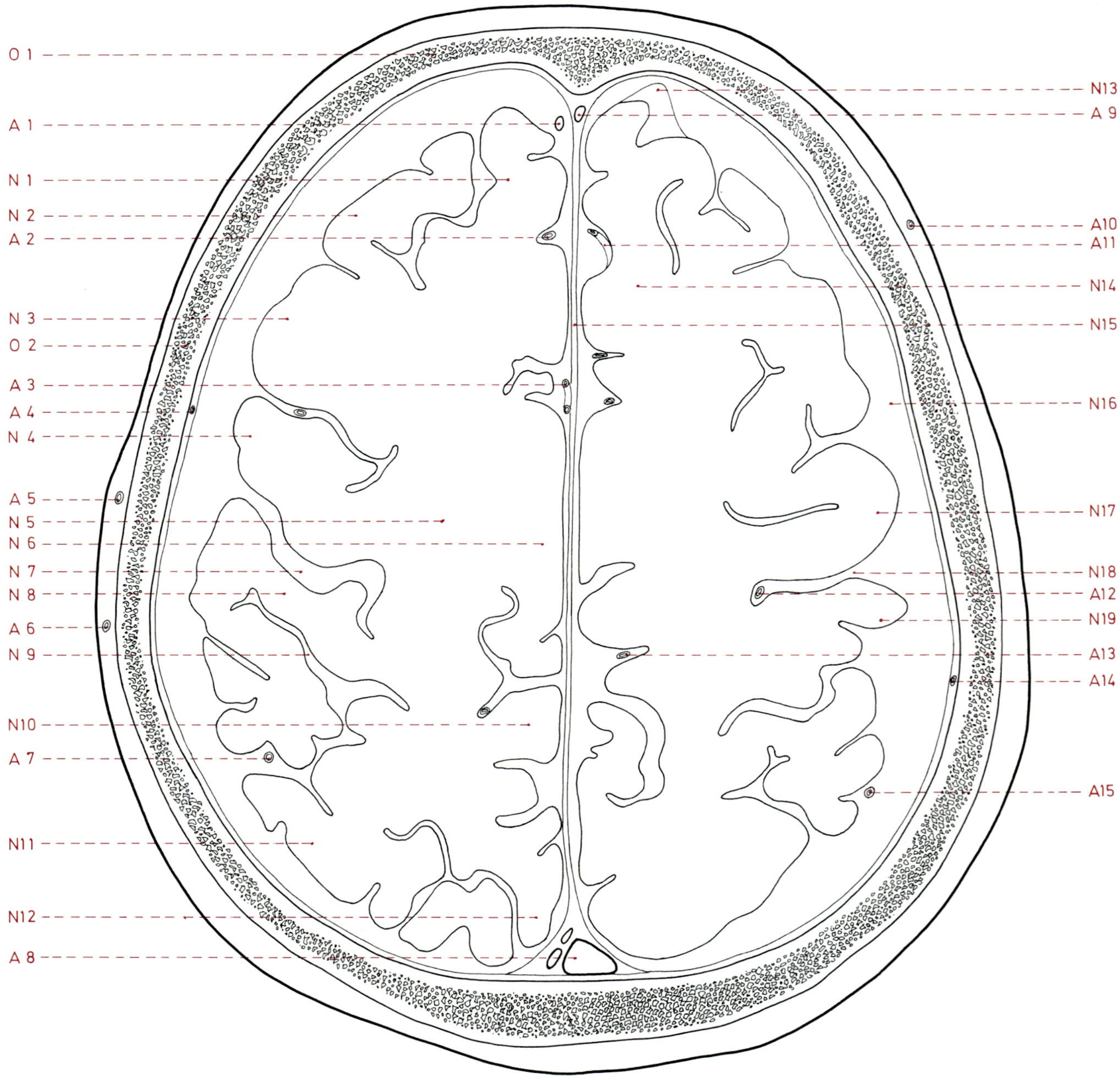

O 1
A 1
N 1
N 2
A 2
N 3
O 2
A 3
A 4
N 4
A 5
N 5
N 6
N 7
N 8
A 6
N 9
N10
A 7
N11
N12
A 8
N13
A 9
A10
A11
N14
N15
N16
N17
N18
A12
N19
A13
A14
A15

Fig. HC 3

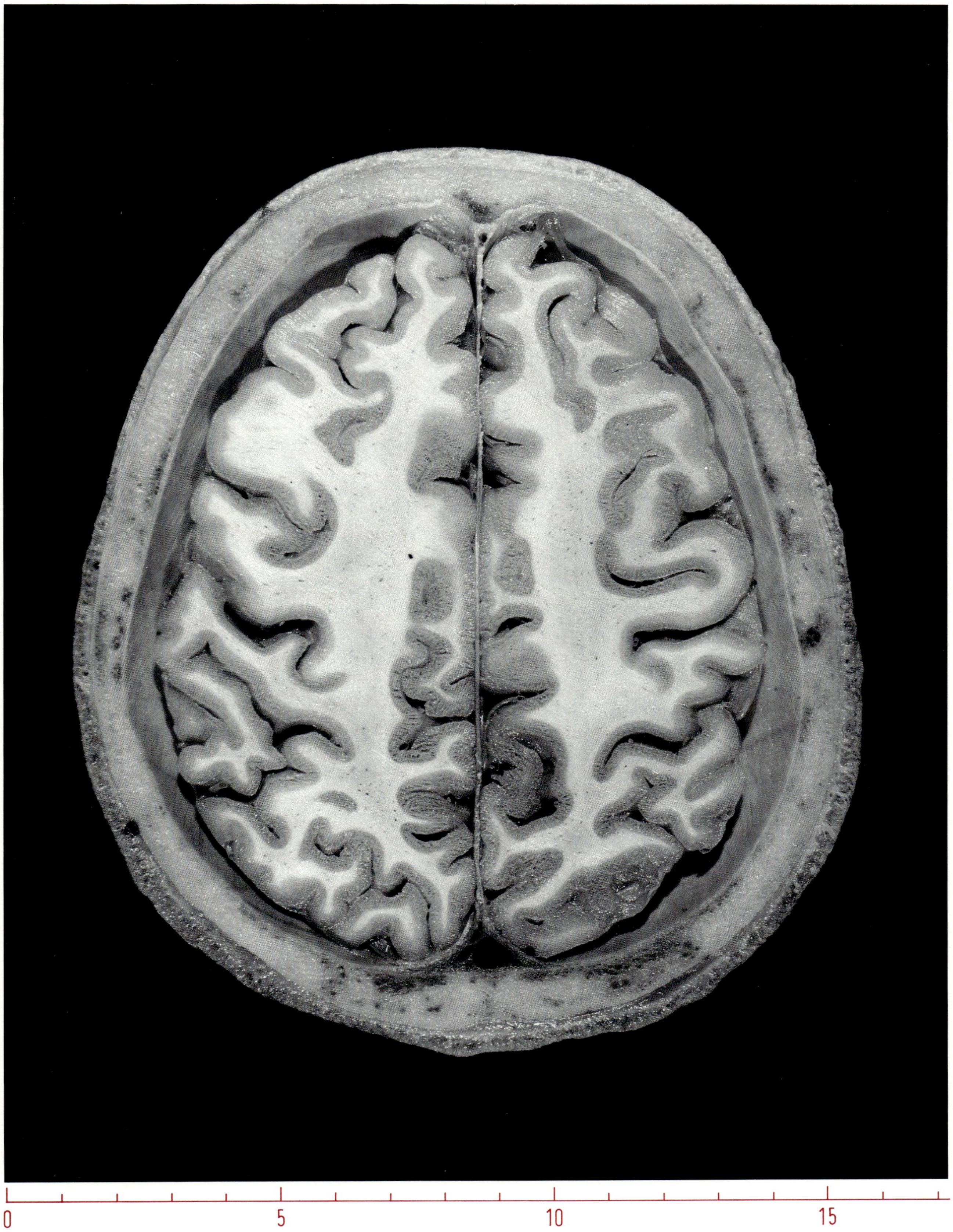

Fig. HC 3

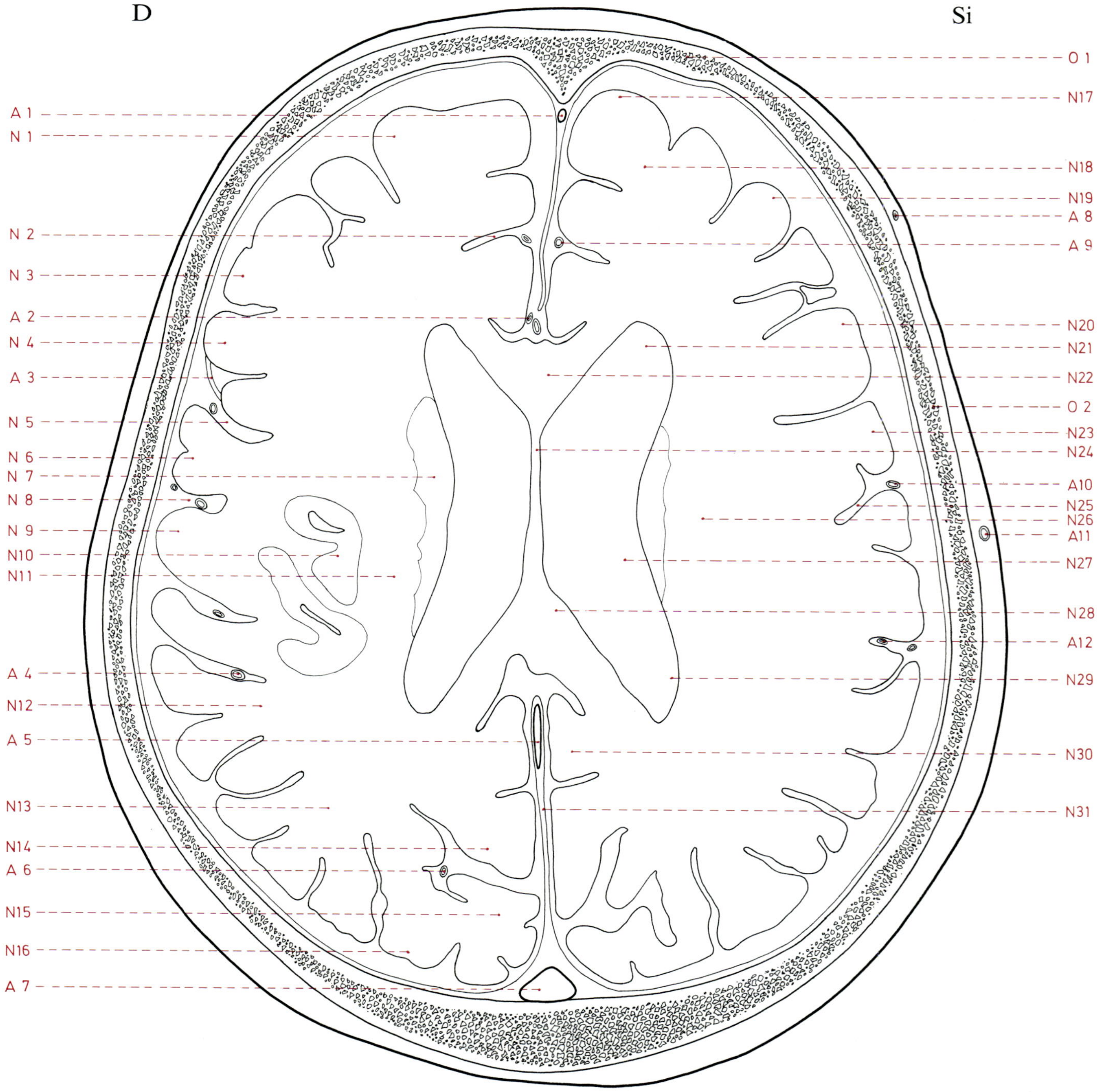
D
Si
A 1
N 1
N 2
N 3
A 2
N 4
A 3
N 5
N 6
N 7
N 8
N 9
N10
N11
A 4
N12
A 5
N13
N14
A 6
N15
N16
A 7
O 1
N17
N18
N19
A 8
A 9
N20
N21
N22
O 2
N23
N24
A10
N25
N26
A11
N27
N28
A12
N29
N30
N31

Fig. HC 4

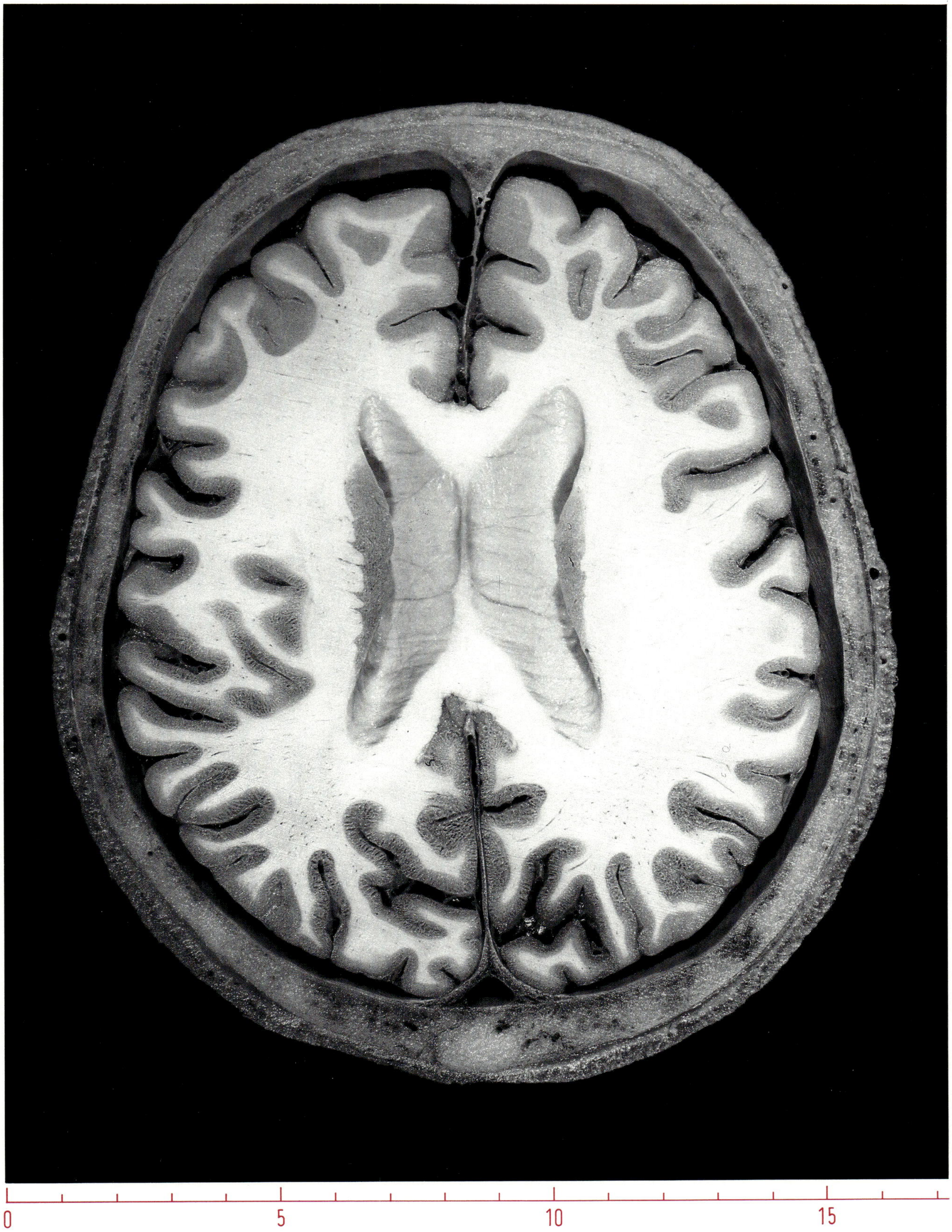

Fig. HC 4

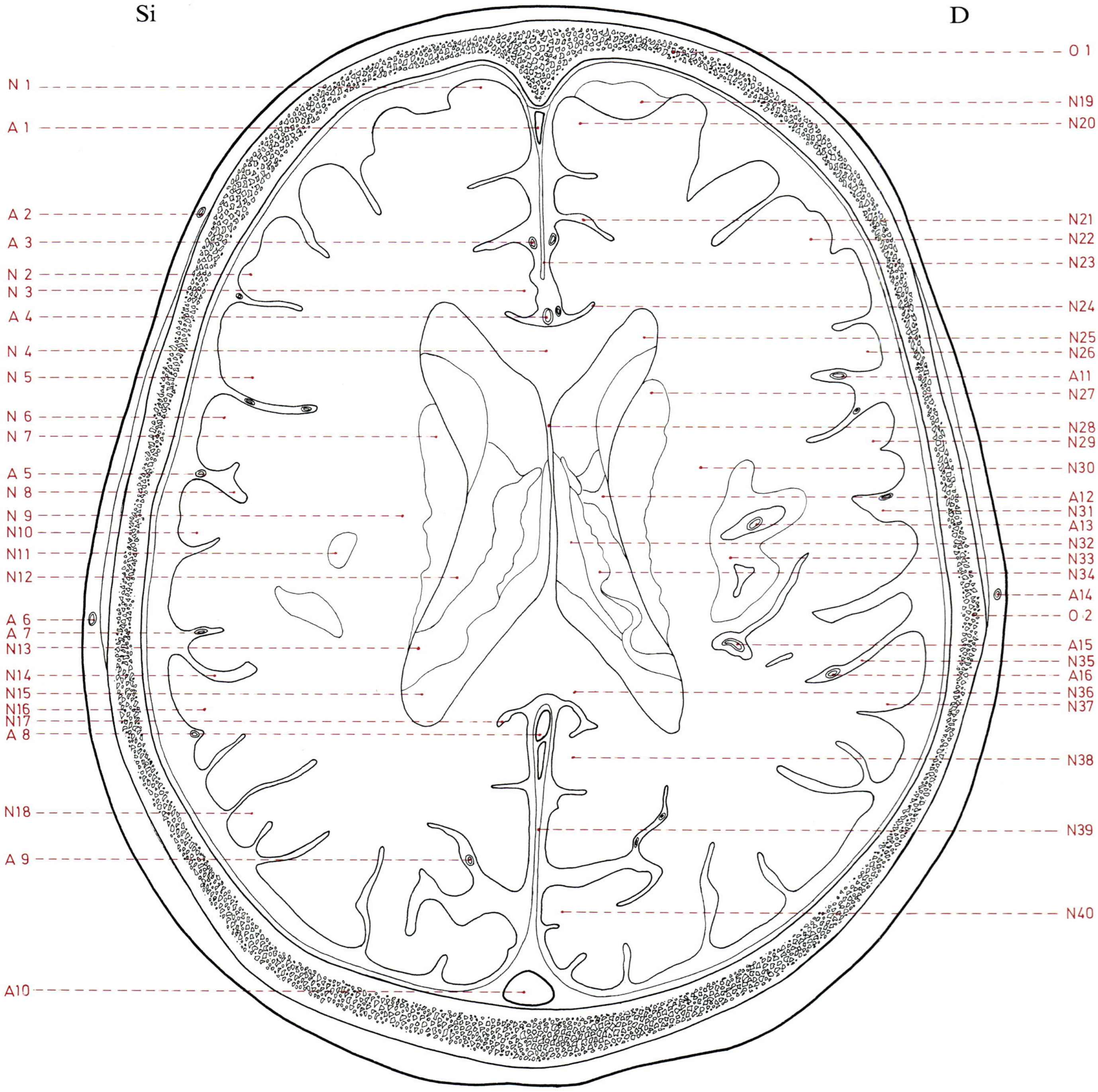

Si
D
N 1
A 1
A 2
A 3
N 2
N 3
A 4
N 4
N 5
N 6
N 7
A 5
N 8
N 9
N10
N11
N12
A 6
A 7
N13
N14
N15
N16
N17
A 8
N18
A 9
A10
O 1
N19
N20
N21
N22
N23
N24
N25
N26
A11
N27
N28
N29
N30
A12
N31
A13
N32
N33
N34
A14
O 2
A15
N35
A16
N36
N37
N38
N39
N40

Fig. HC 5

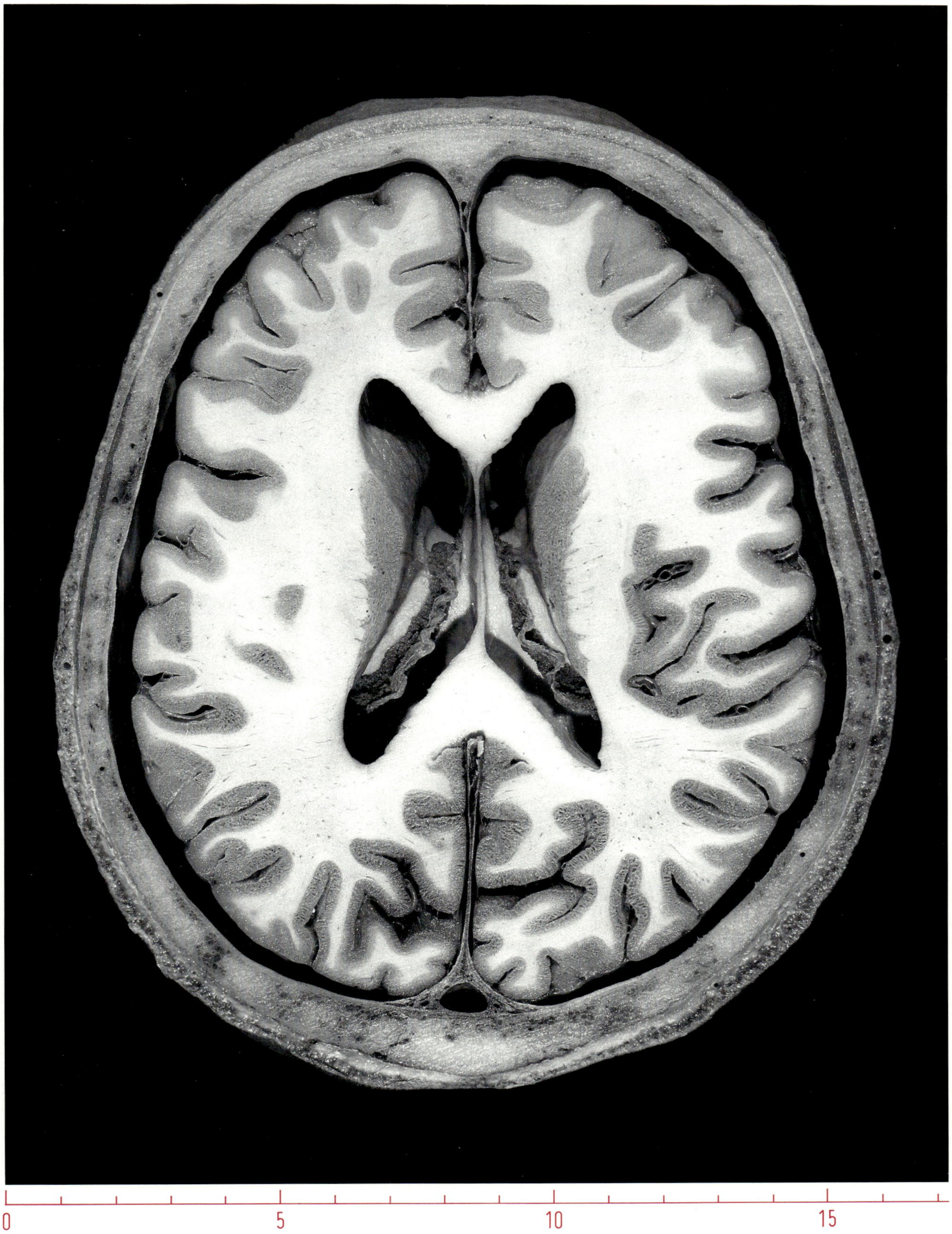

Fig. HC 5

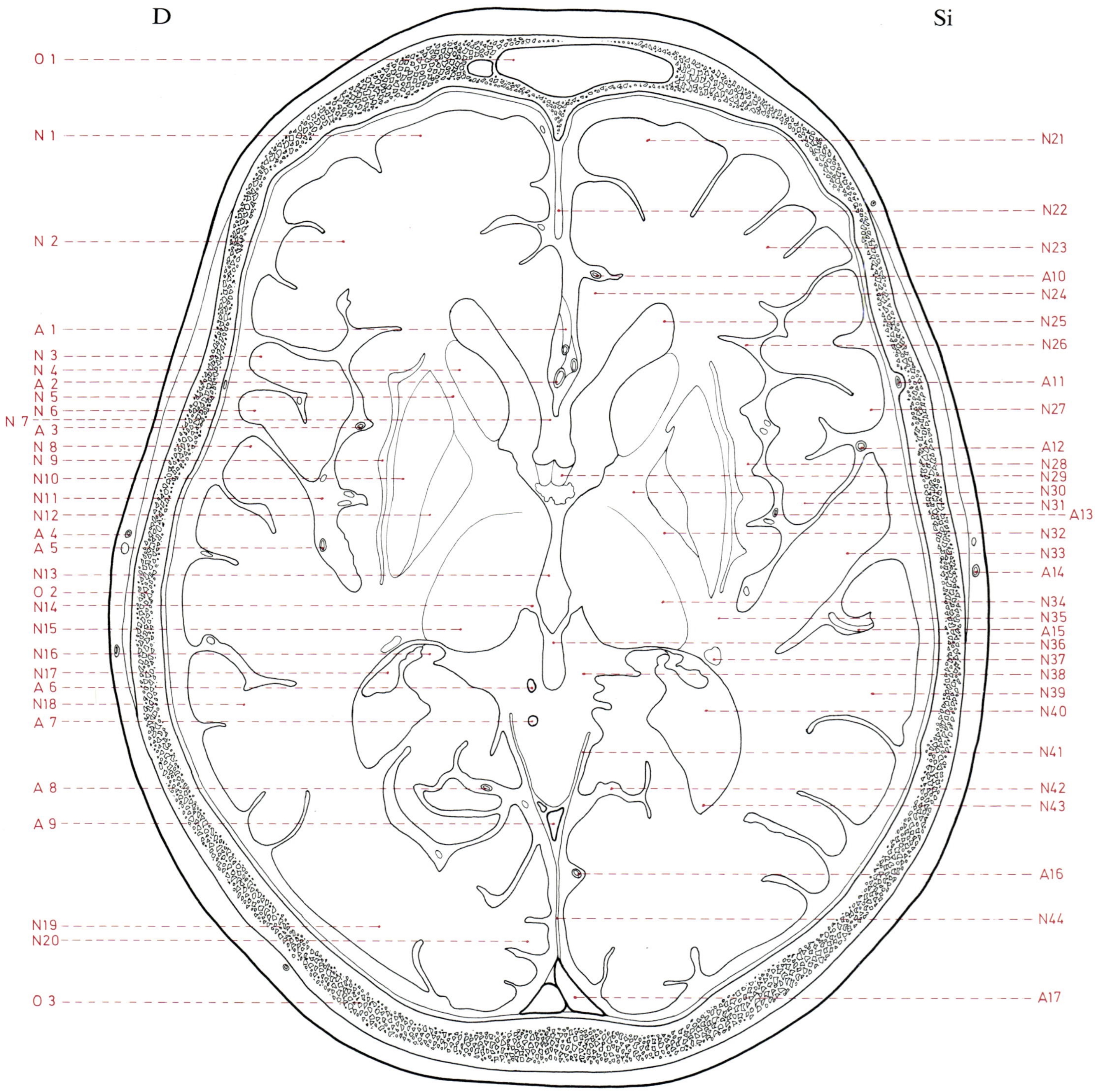
D
Si
O 1
N 1
N 2
A 1
N 3
N 4
A 2
N 5
N 6
N 7
A 3
N 8
N 9
N10
N11
N12
A 4
A 5
N13
O 2
N14
N15
N16
N17
A 6
N18
A 7
A 8
A 9
N19
N20
O 3
N21
N22
N23
A10
N24
N25
N26
A11
N27
A12
N28
N29
N30
N31
A13
N32
N33
A14
N34
N35
A15
N36
N37
N38
N39
N40
N41
N42
N43
A16
N44
A17

Fig. HC 6

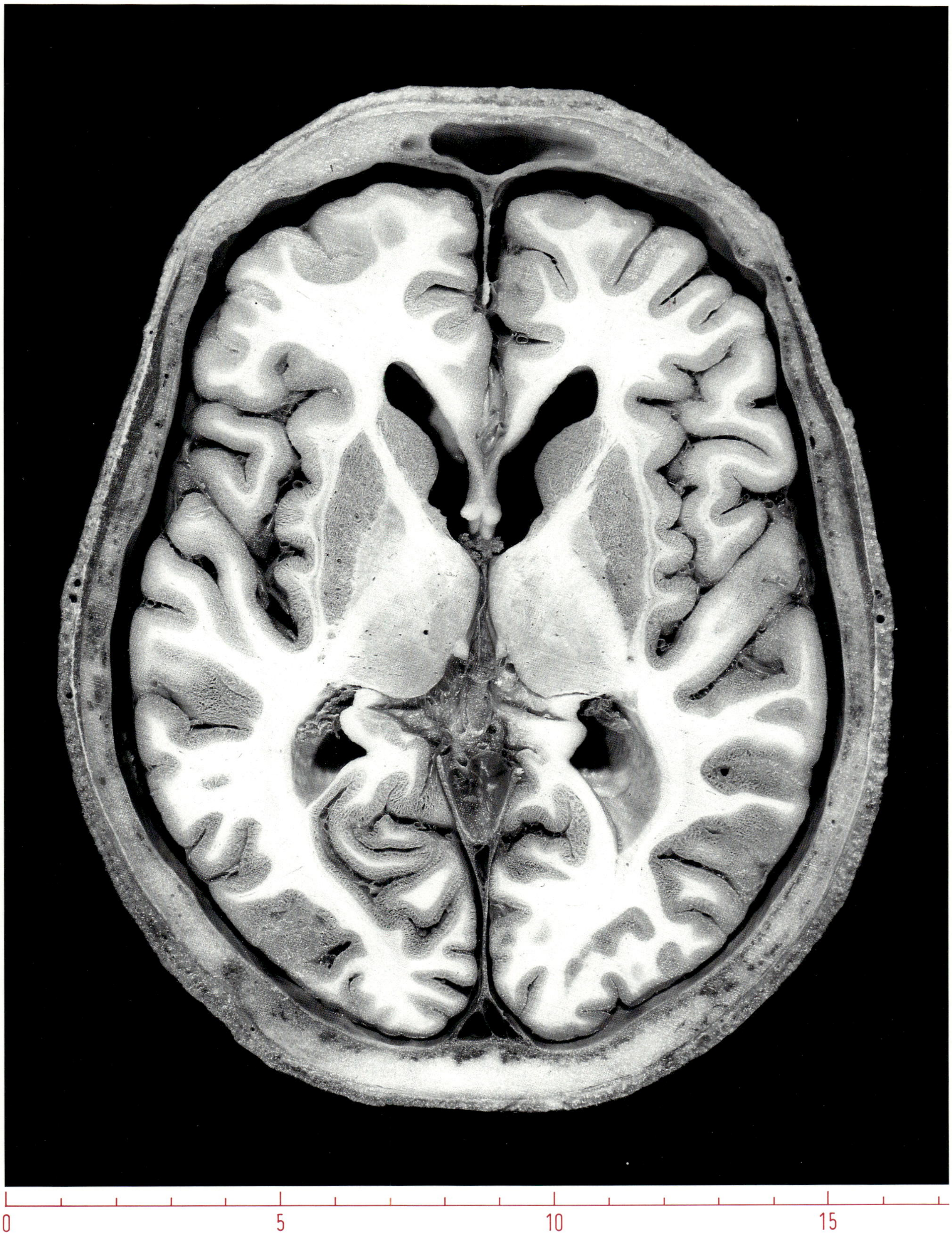

Fig. HC 6

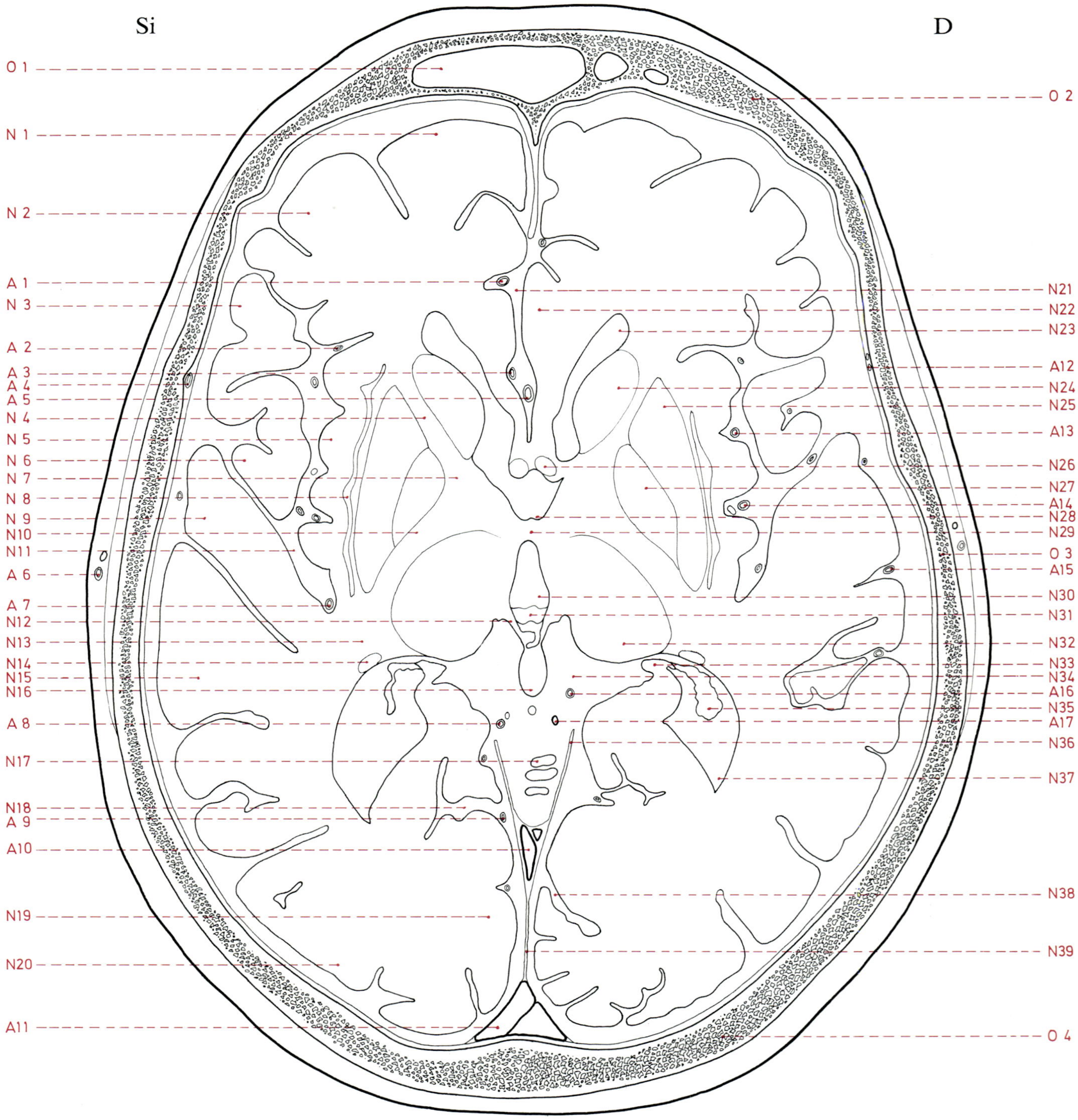

Si
D
O 1
N 1
N 2
A 1
N 3
A 2
A 3
A 4
A 5
N 4
N 5
N 6
N 7
N 8
N 9
N10
N11
A 6
A 7
N12
N13
N14
N15
N16
A 8
N17
N18
A 9
A10
N19
N20
A11
O 2
N21
N22
N23
A12
N24
N25
A13
N26
N27
A14
N28
N29
O 3
A15
N30
N31
N32
N33
N34
A16
N35
A17
N36
N37
N38
N39
O 4

Fig. HC 7

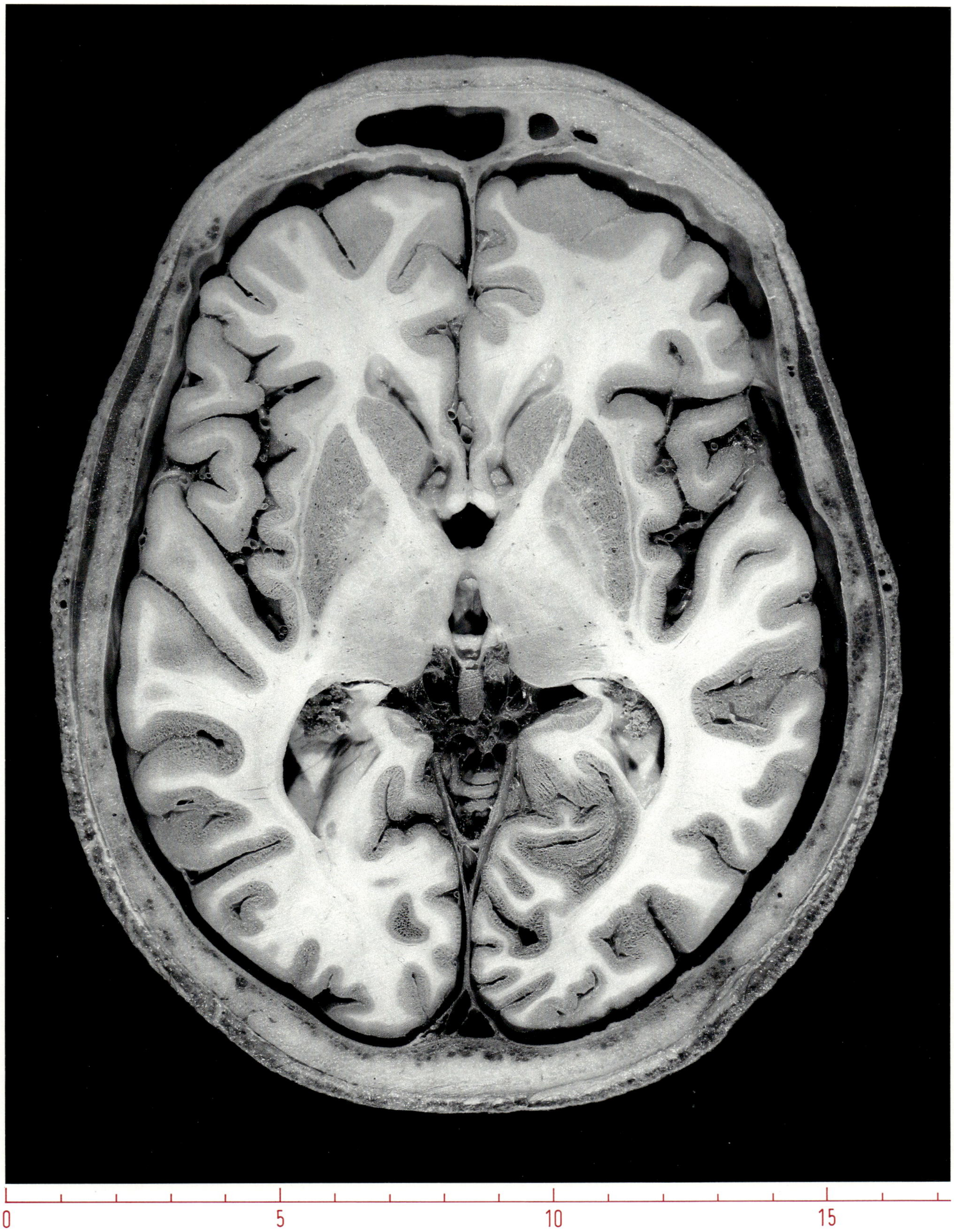

Fig. HC 7

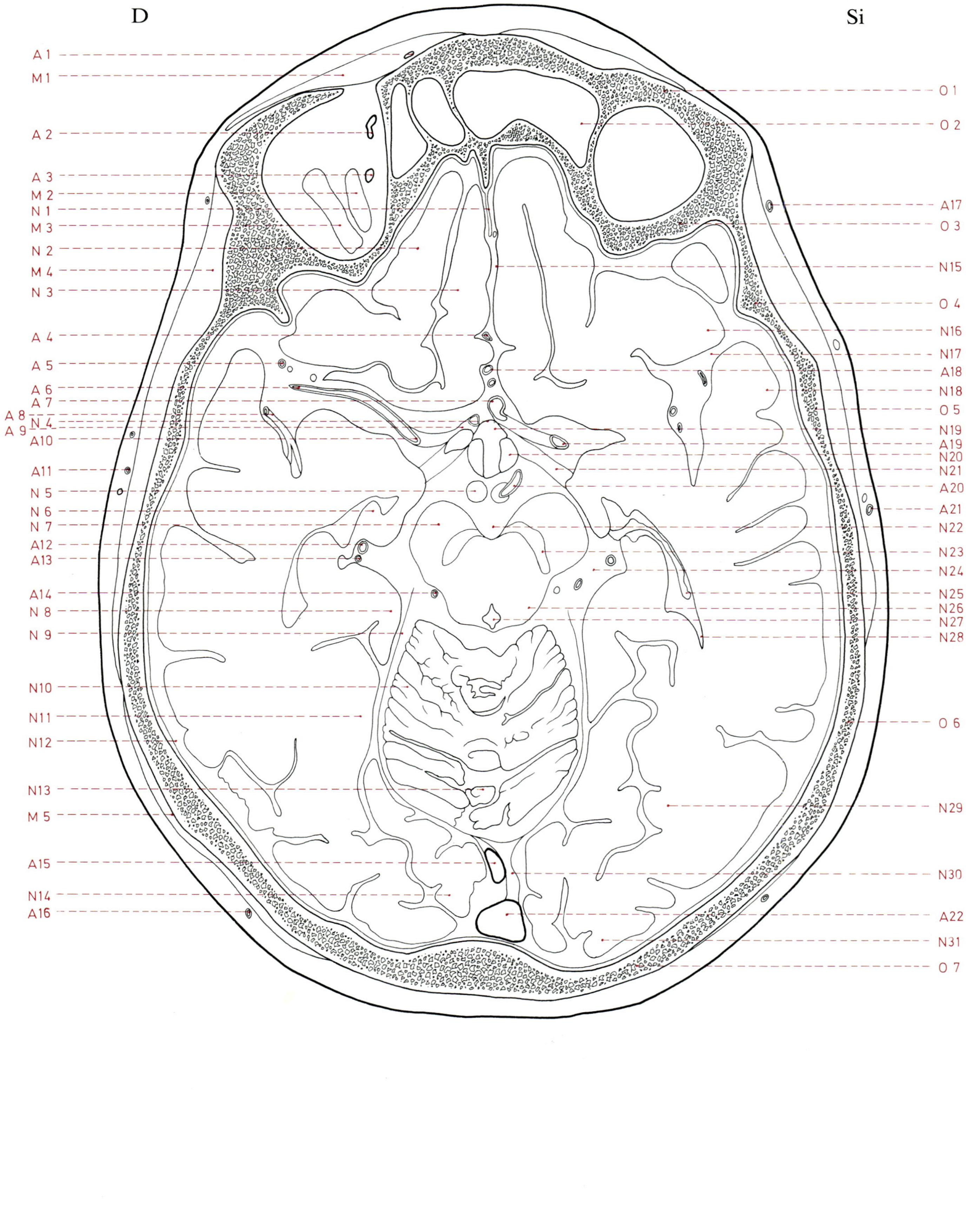

D
Si
A 1
M 1
A 2
A 3
M 2
N 1
M 3
N 2
M 4
N 3
A 4
A 5
A 6
A 7
A 8
N 4
A 9
A10
A11
N 5
N 6
N 7
A12
A13
A14
N 8
N 9
N10
N11
N12
N13
M 5
A15
N14
A16
O 1
O 2
A17
O 3
N15
O 4
N16
N17
A18
N18
O 5
N19
A19
N20
N21
A20
A21
N22
N23
N24
N25
N26
N27
N28
O 6
N29
N30
A22
N31
O 7

Fig. HC 8

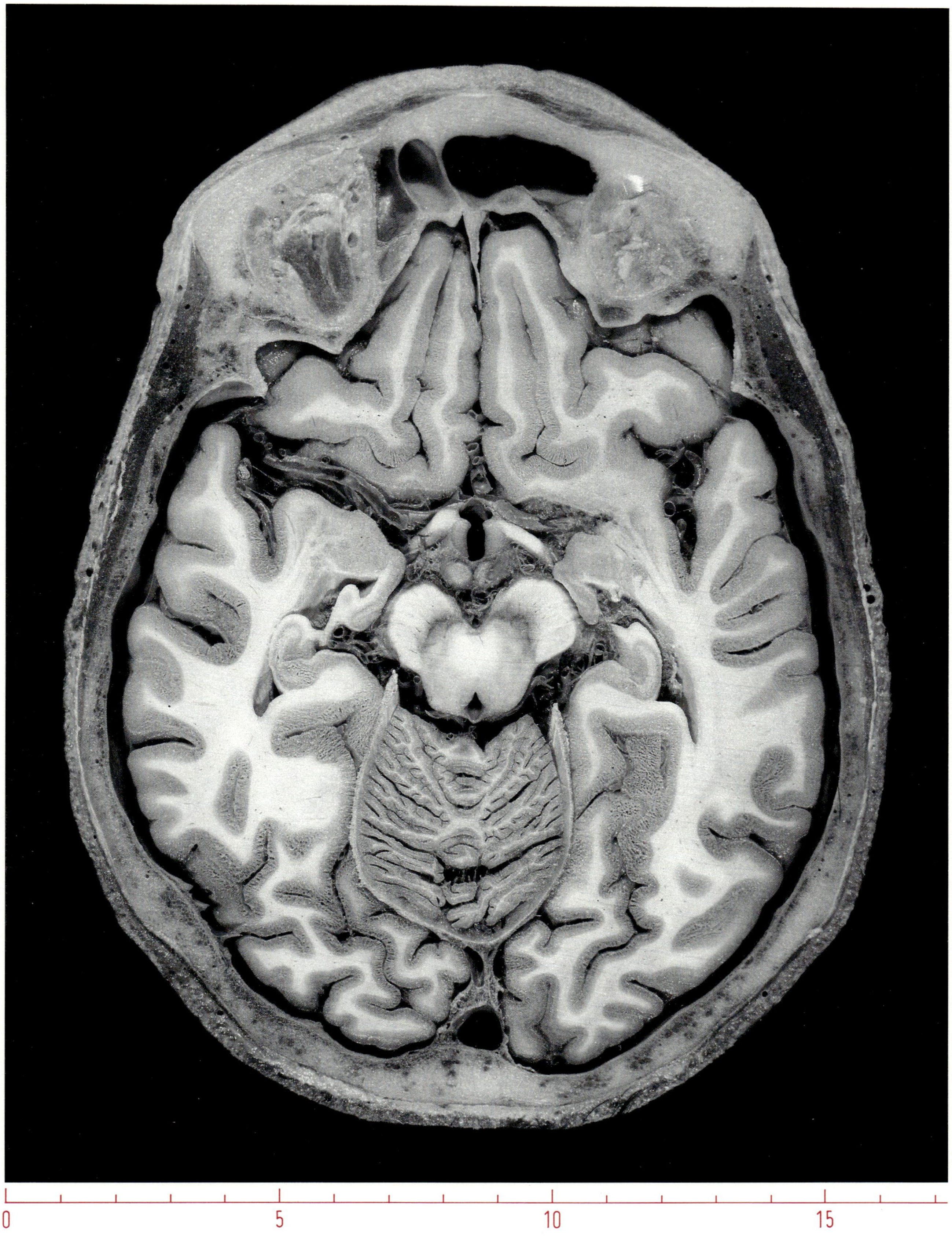

Fig. HC 8

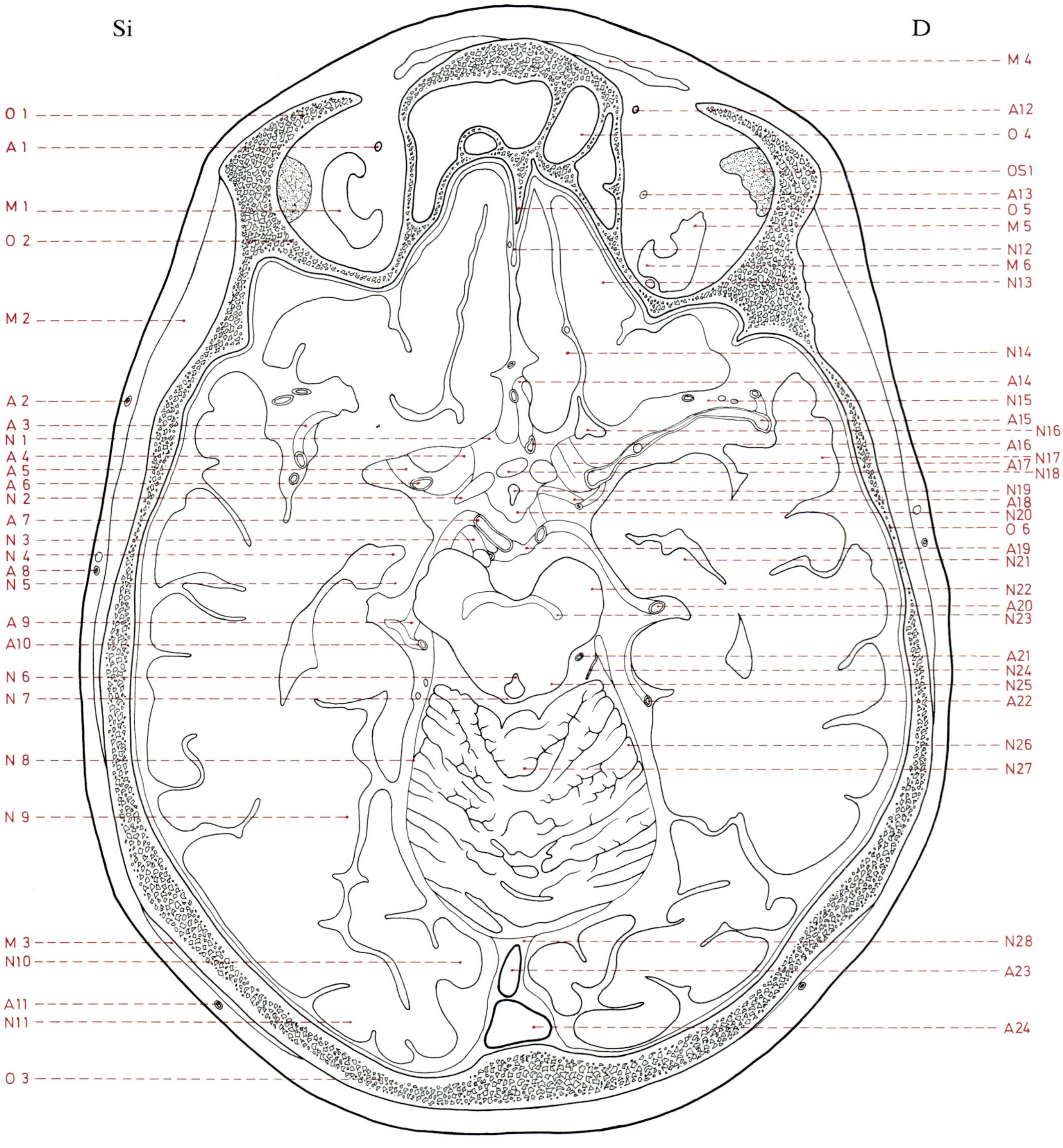
Si
D
O 1
A 1
M 1
O 2
M 2
A 2
A 3
N 1
A 4
A 5
A 6
N 2
A 7
N 3
N 4
A 8
N 5
A 9
A10
N 6
N 7
N 8
N 9
M 3
N10
A11
N11
O 3
M 4
A12
O 4
OS1
A13
O 5
M 5
N12
M 6
N13
N14
A14
N15
A15
N16
A16
N17
A17
N18
N19
A18
N20
O 6
A19
N21
N22
A20
N23
A21
N24
N25
A22
N26
N27
N28
A23
A24

Fig. HC 9

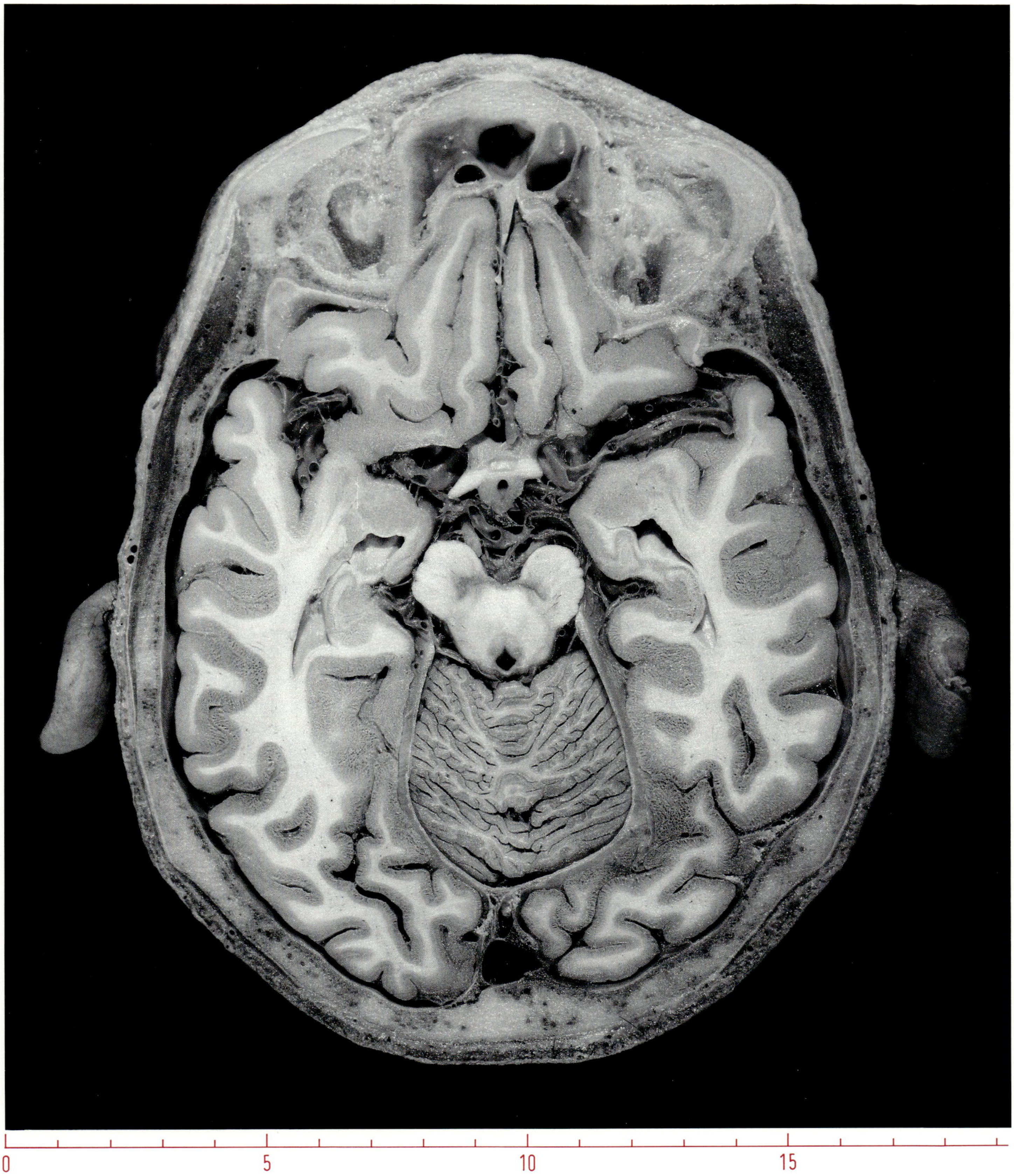

Fig. HC 9

D Si

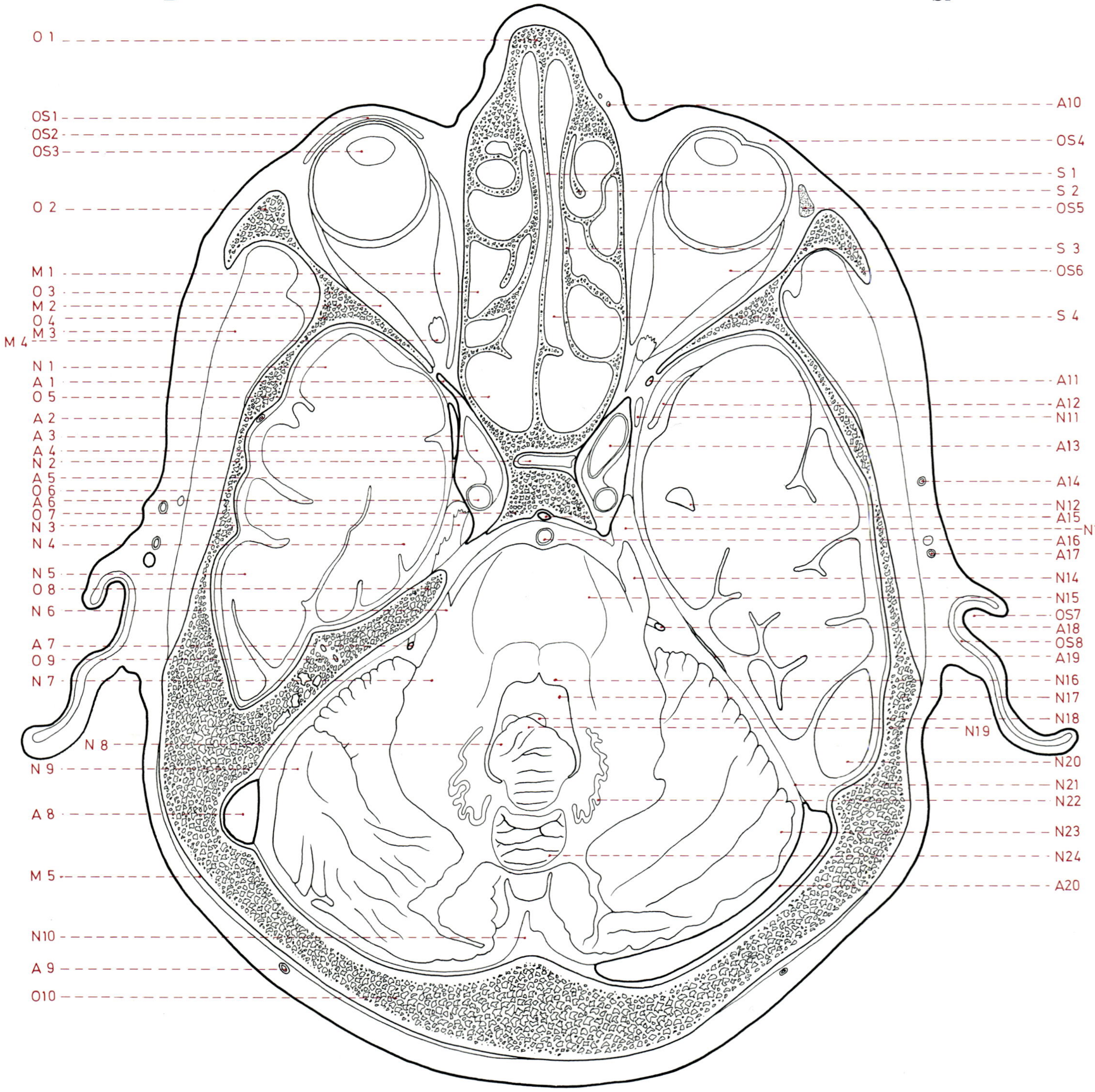
O 1
OS1
OS2
OS3
O 2
M 1
O 3
M 2
O 4
M 3
M 4
N 1
A 1
O 5
A 2
A 3
A 4
N 2
A 5
O 6
A 6
O 7
N 3
N 4
N 5
O 8
N 6
A 7
O 9
N 7
N 8
N 9
A 8
M 5
N10
A 9
O10
A10
OS4
S 1
S 2
OS5
S 3
OS6
S 4
A11
A12
N11
A13
A14
N12
A15
N13
A16
A17
N14
N15
OS7
A18
OS8
A19
N16
N17
N18
N19
N20
N21
N22
N23
N24
A20

Fig. HC 10

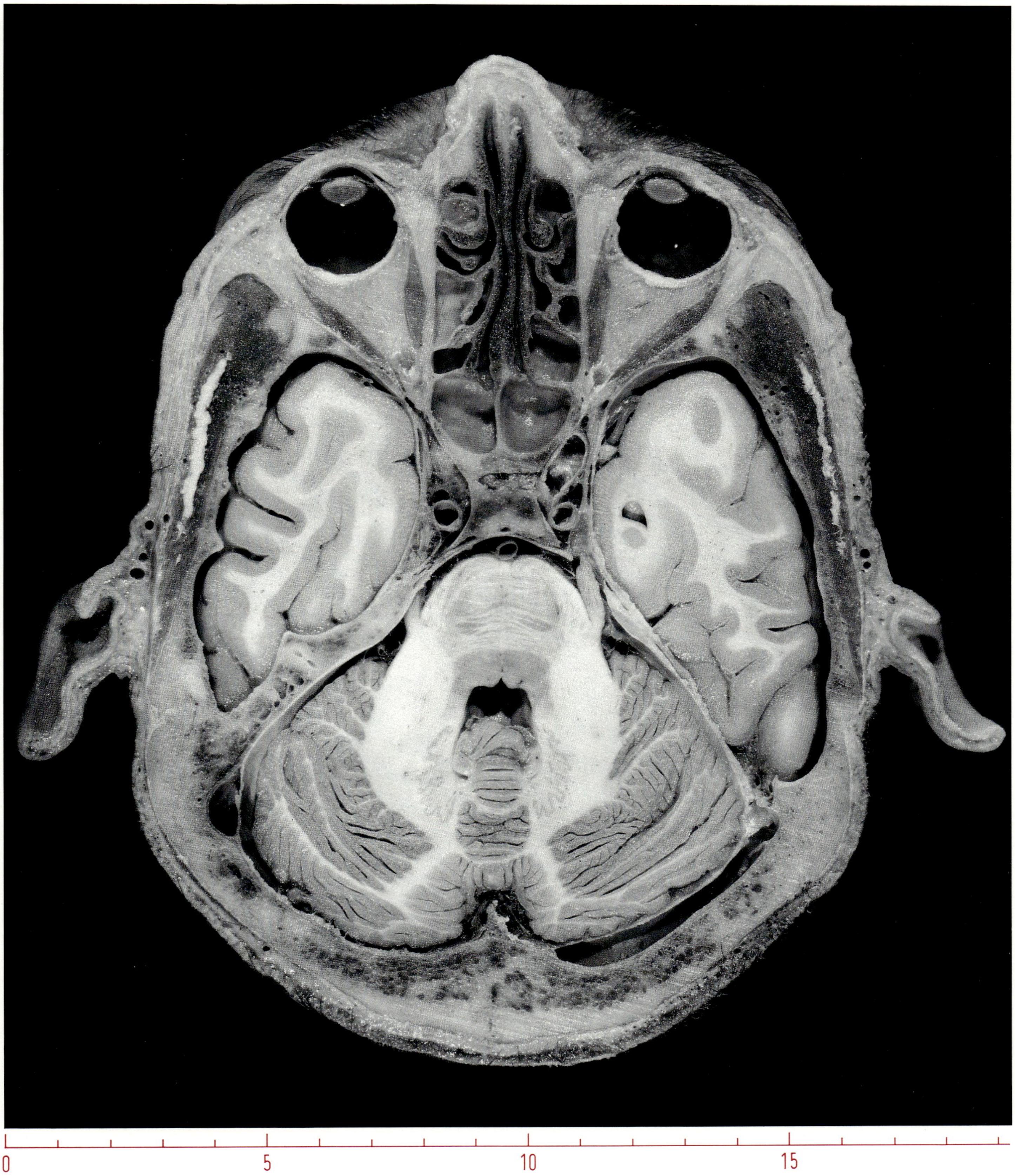

Fig. HC 10

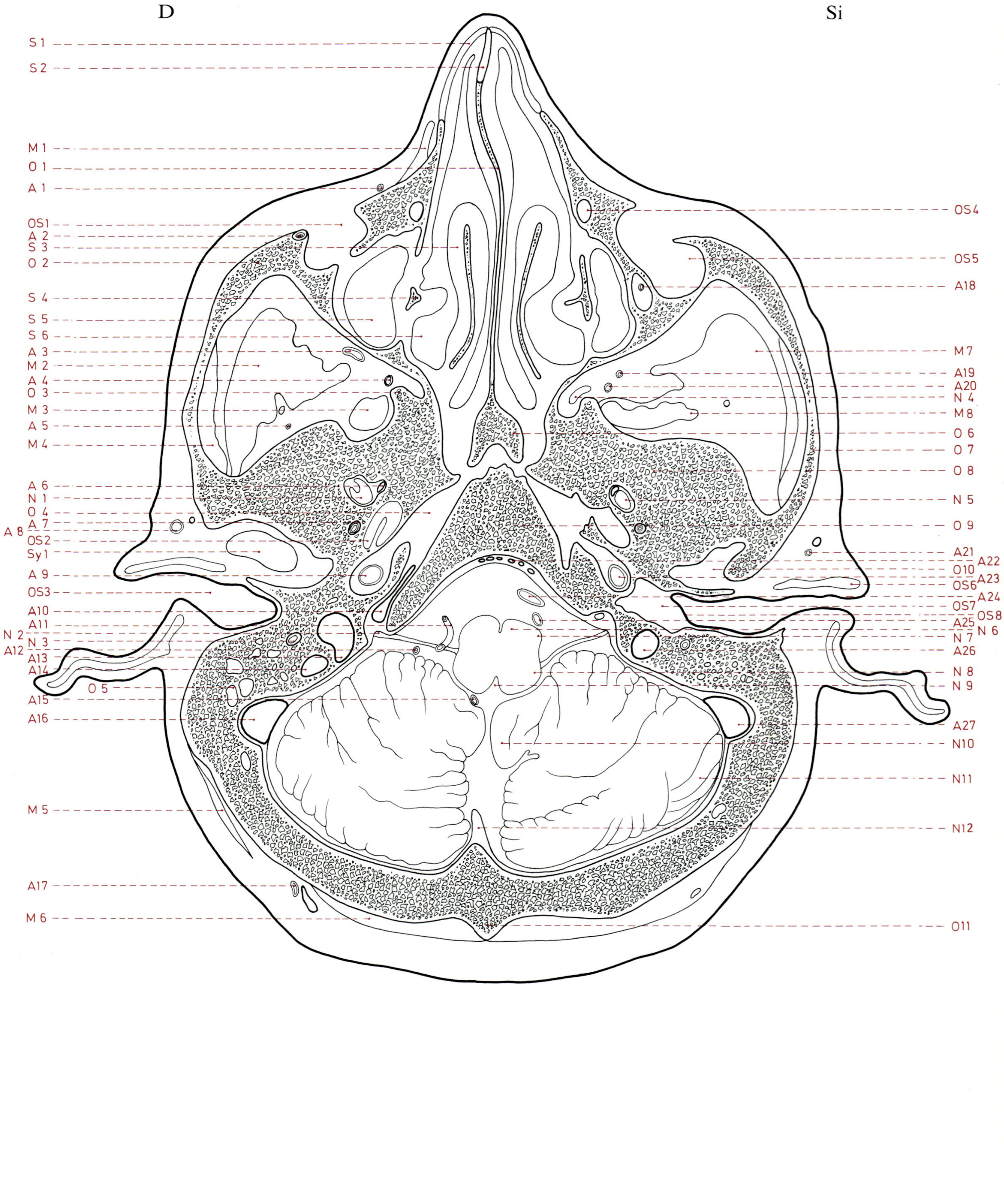
D
Si
S 1
S 2
M 1
O 1
A 1
OS1
A 2
S 3
O 2
S 4
S 5
S 6
A 3
M 2
A 4
O 3
M 3
A 5
M 4
A 6
N 1
O 4
A 7
A 8
OS2
Sy1
A 9
OS3
A10
A11
N 2
N 3
A12
A13
A14
O 5
A15
A16
M 5
A17
M 6
OS4
OS5
A18
M 7
A19
A20
N 4
M 8
O 6
O 7
O 8
N 5
O 9
A21
A22
O10
A23
OS6
A24
OS7
OS8
A25
N 6
N 7
A26
N 8
N 9
A27
N10
N11
N12
O11

Fig. HC 11

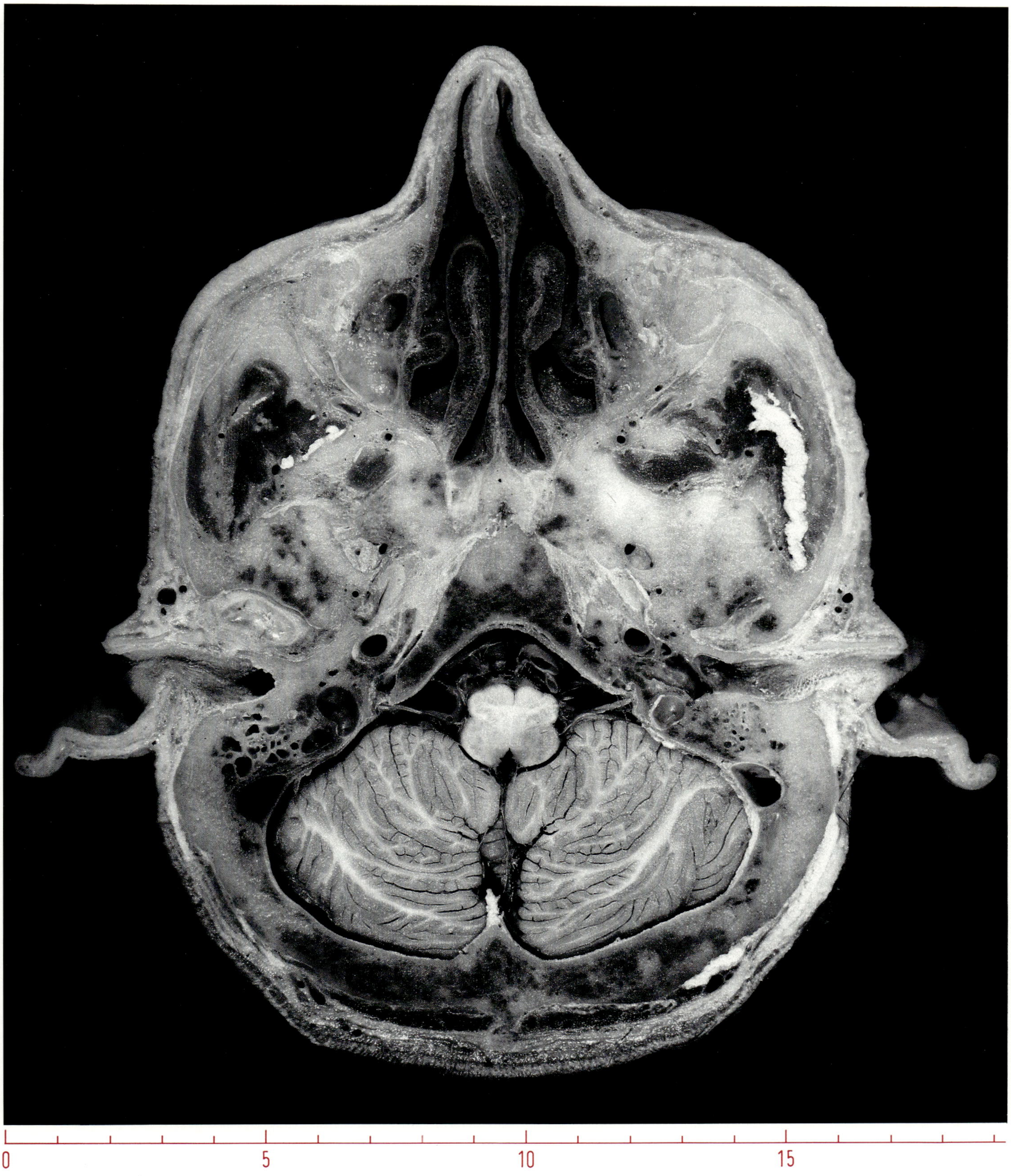

Fig. HC 11

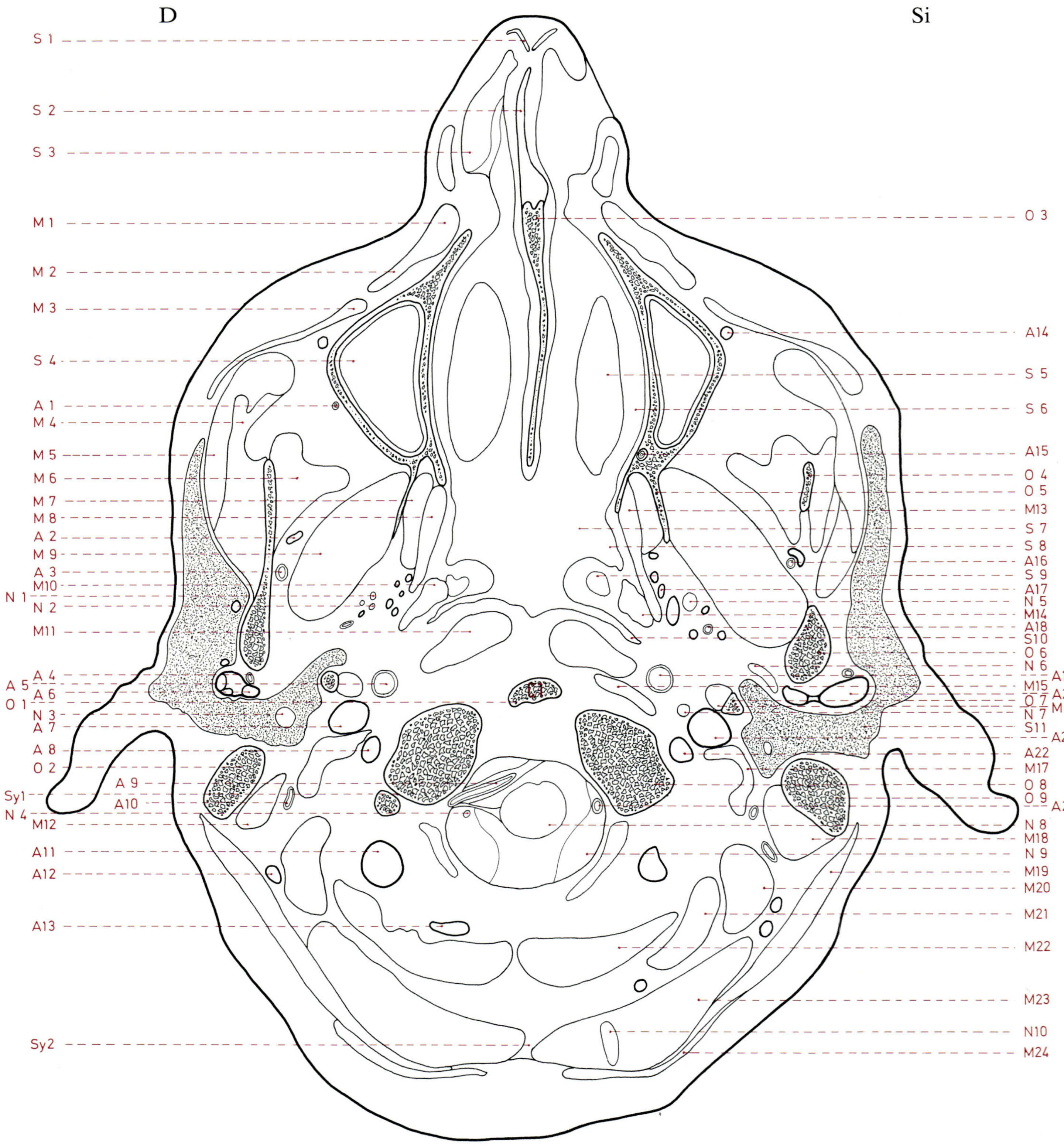
D
Si
S 1
S 2
S 3
M 1
M 2
M 3
S 4
A 1
M 4
M 5
M 6
M 7
M 8
A 2
M 9
A 3
M10
N 1
N 2
M11
A 4
A 5
A 6
O 1
N 3
A 7
A 8
O 2
A 9
Sy1
A10
N 4
M12
A11
A12
A13
Sy2
O 3
A14
S 5
S 6
A15
O 4
O 5
M13
S 7
S 8
A16
S 9
A17
N 5
M14
A18
S10
O 6
N 6
A19
M15
A20
O 7
M16
N 7
S11
A21
A22
M17
O 8
O 9
A23
N 8
M18
N 9
M19
M20
M21
M22
M23
N10
M24

Fig. HC 12

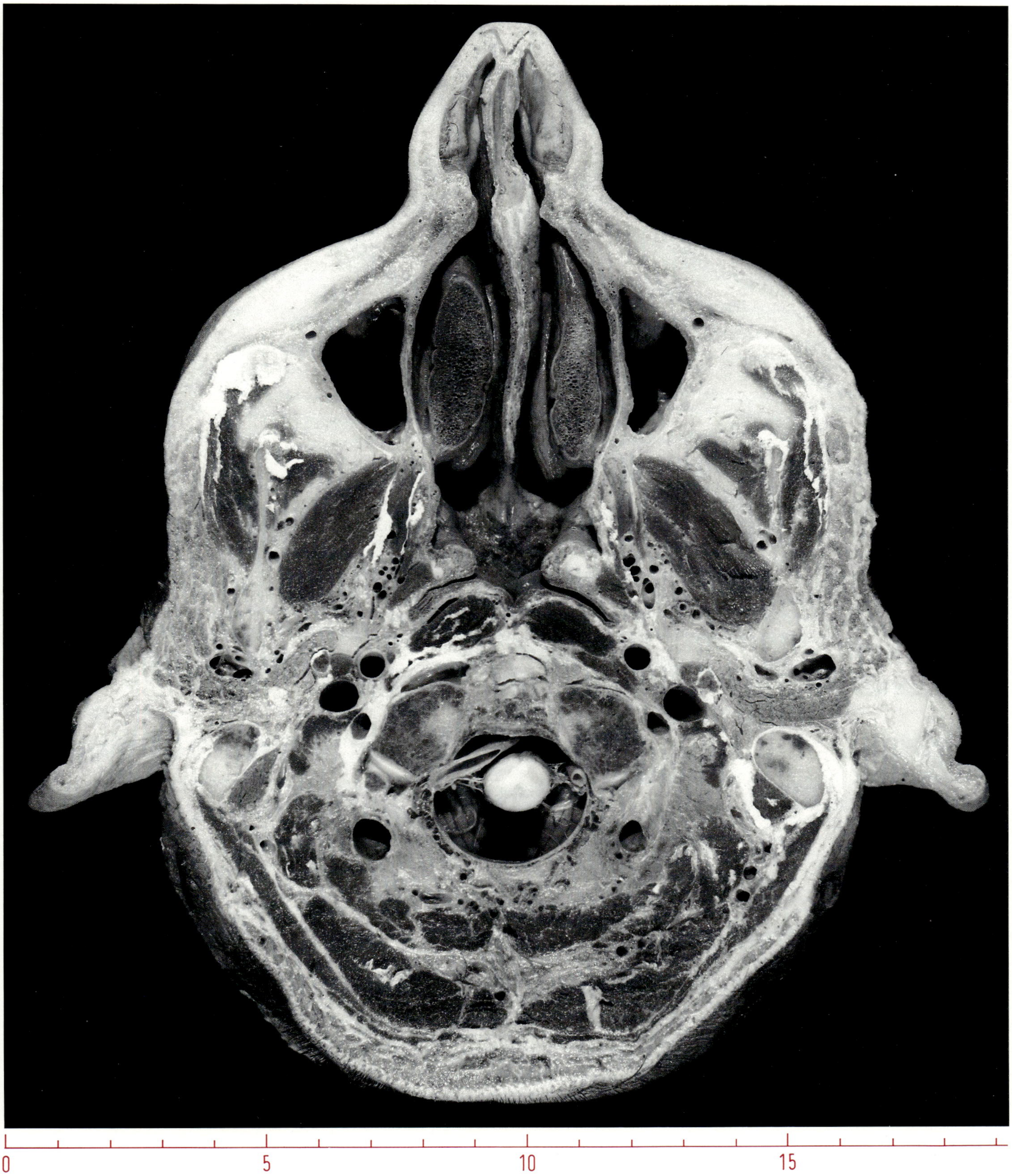

Fig. HC 12

D Si

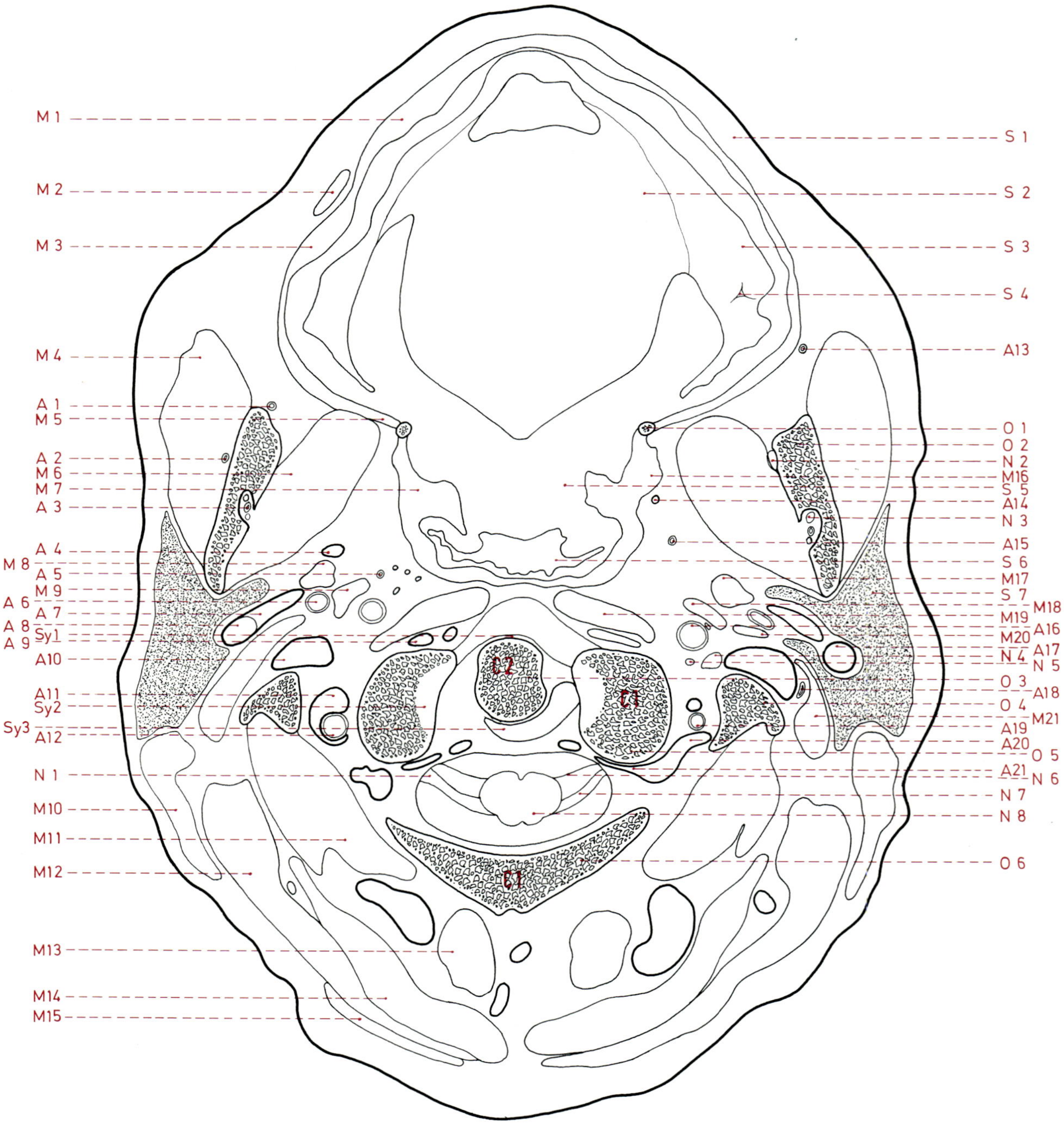
M 1
M 2
M 3
M 4
A 1
M 5
A 2
M 6
M 7
A 3
A 4
M 8
A 5
M 9
A 6
A 7
A 8
Sy1
A 9
A10
A11
Sy2
Sy3
A12
N 1
M10
M11
M12
M13
M14
M15
S 1
S 2
S 3
S 4
A13
O 1
O 2
N 2
M16
S 5
A14
N 3
A15
S 6
M17
S 7
M18
M19
A16
M20
A17
N 4
N 5
O 3
A18
O 4
M21
A19
A20
O 5
A21
N 6
N 7
N 8
O 6
C2
C1

Fig. HC 13

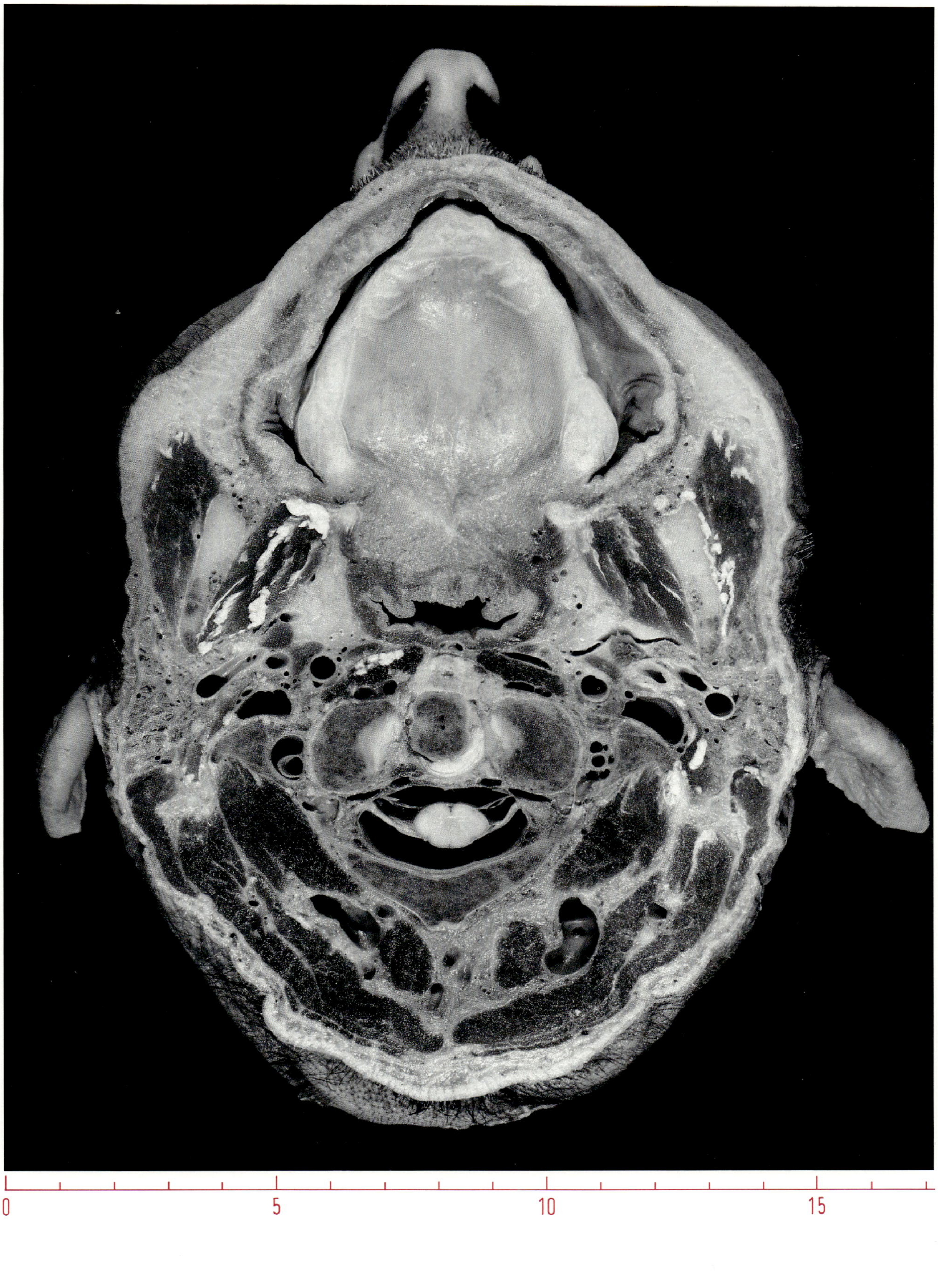

Fig. HC 13

D Si

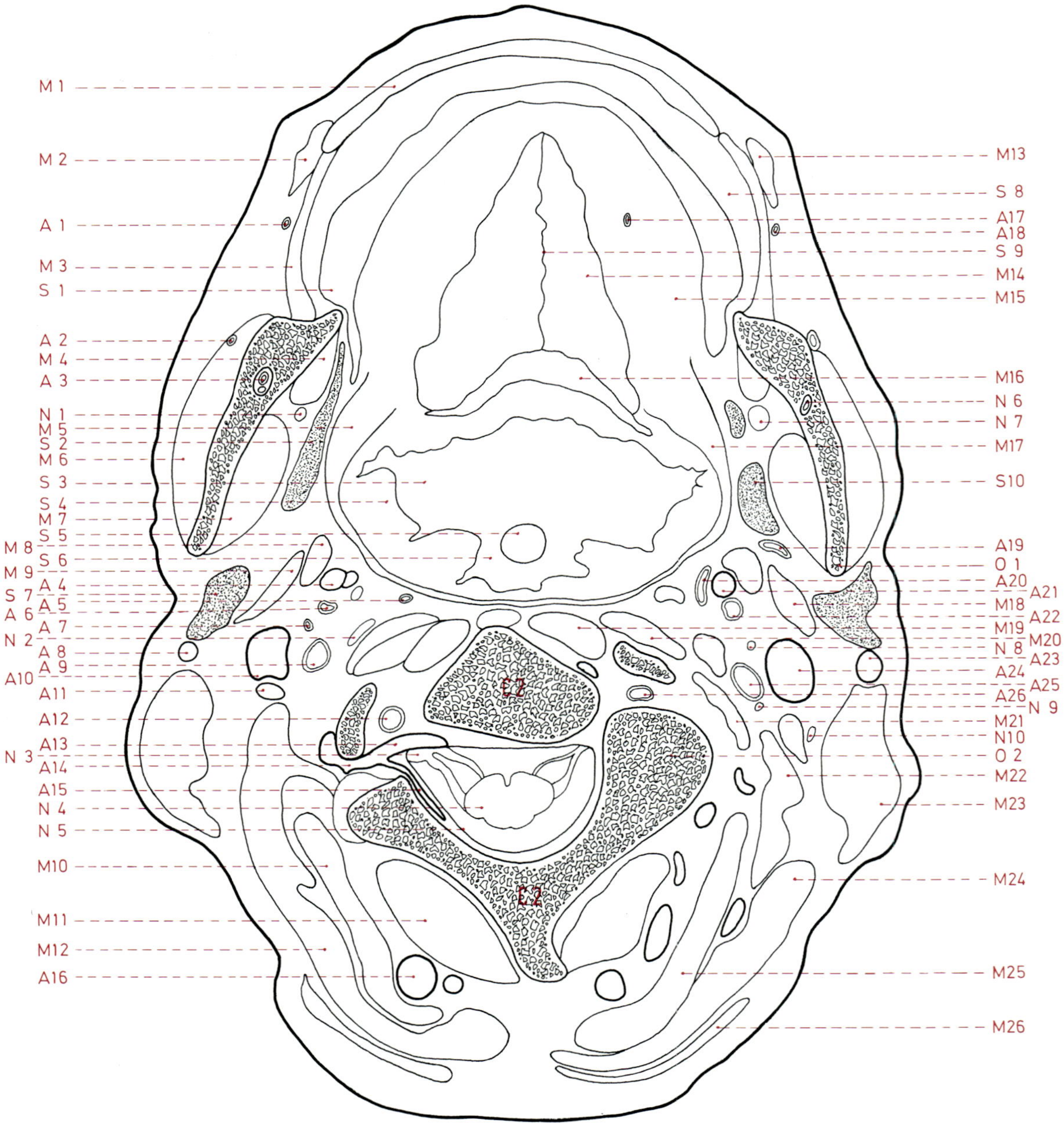
M 1
M 2
A 1
M 3
S 1
A 2
M 4
A 3
N 1
M 5
S 2
M 6
S 3
S 4
M 7
S 5
M 8
S 6
M 9
A 4
S 7
A 5
A 6
A 7
N 2
A 8
A 9
A10
A11
A12
A13
N 3
A14
A15
N 4
N 5
M10
M11
M12
A16
M13
S 8
A17
A18
S 9
M14
M15
M16
N 6
N 7
M17
S10
A19
O 1
A20
A21
M18
A22
M19
M20
N 8
A23
A24
A25
A26
N 9
M21
N10
O 2
M22
M23
M24
M25
M26
C2
C2

Fig. HC 14

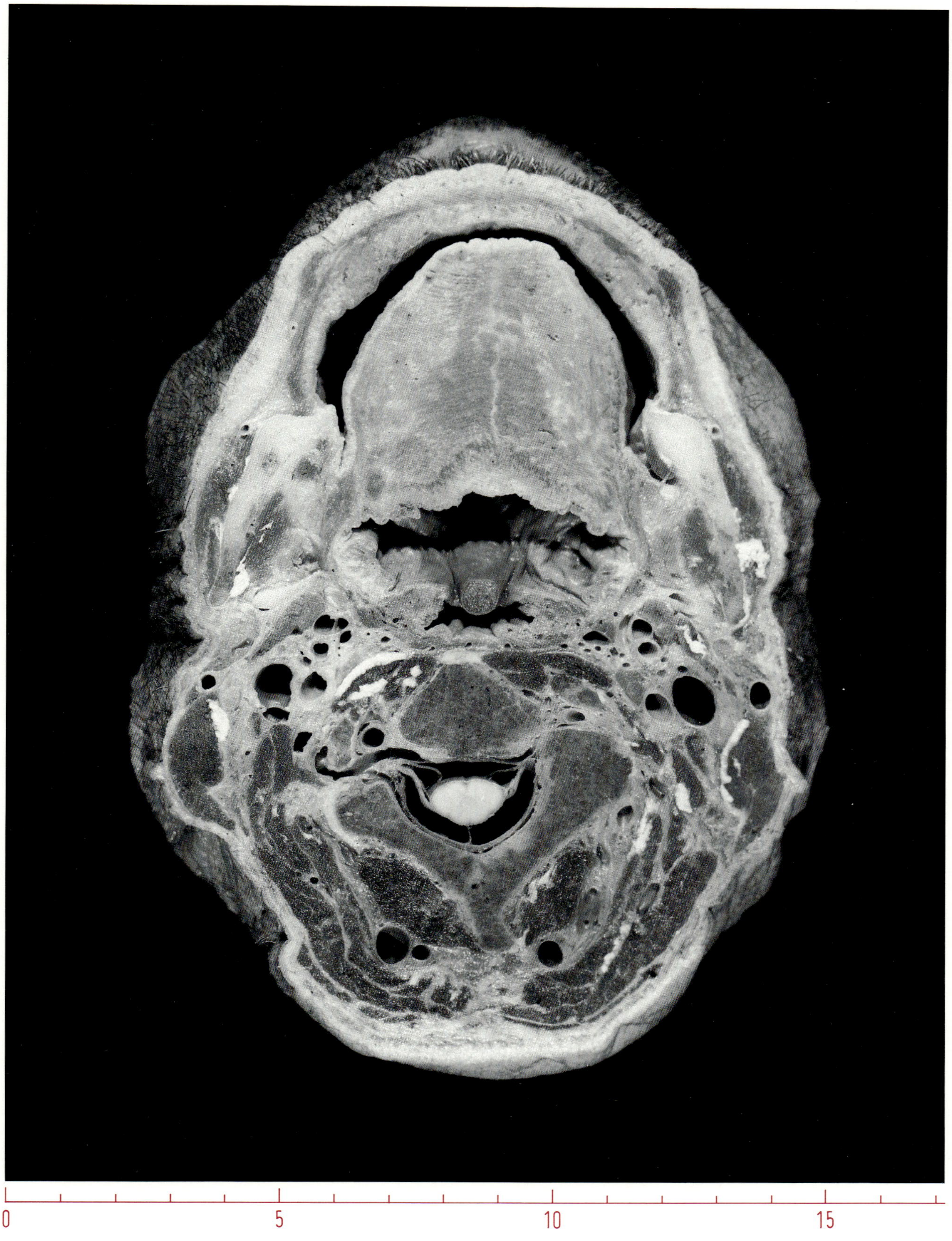

Fig. HC 14

D

Si

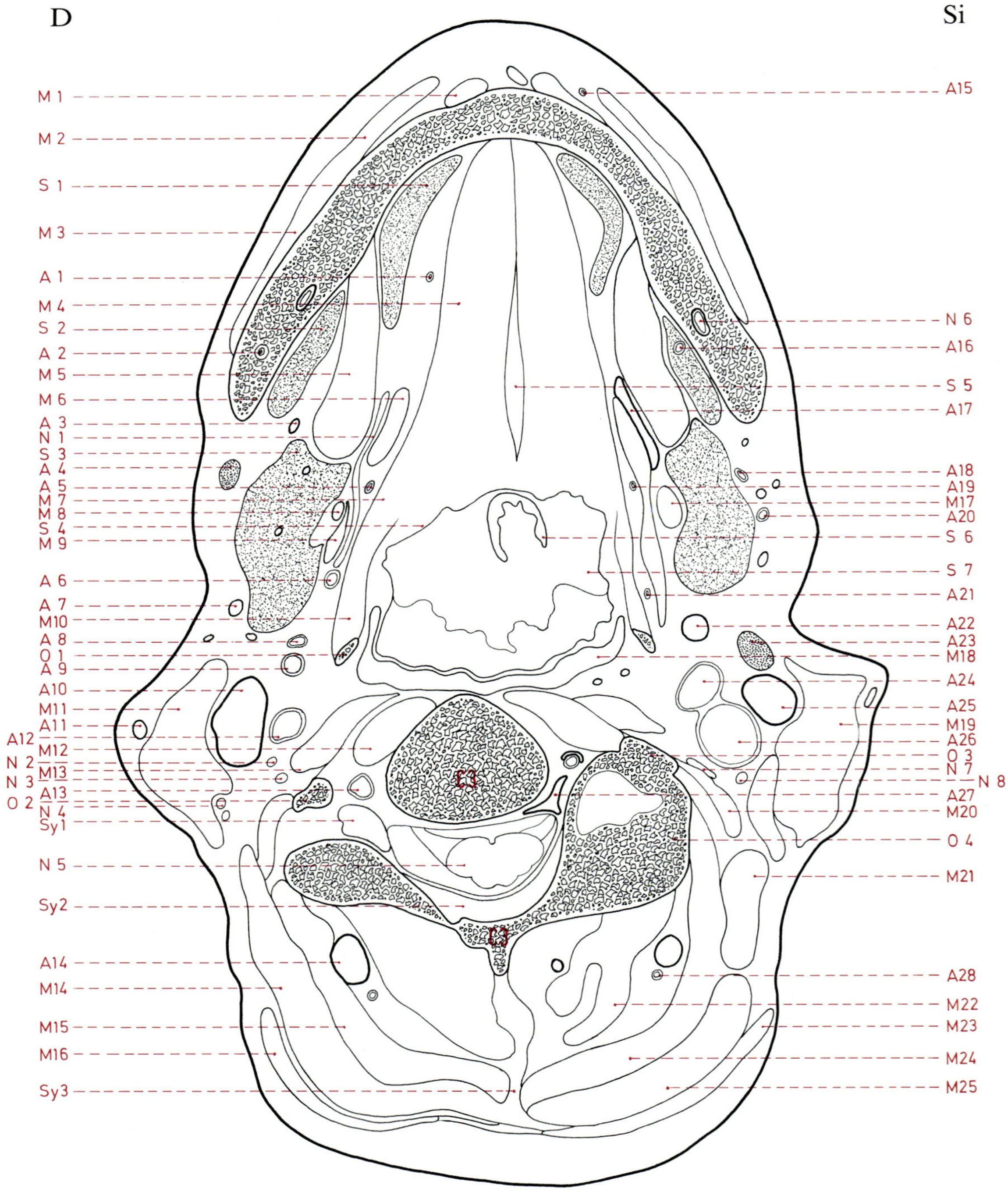

M 1
M 2
S 1
M 3
A 1
M 4
S 2
A 2
M 5
M 6
A 3
N 1
S 3
A 4
A 5
M 7
M 8
S 4
M 9
A 6
A 7
M10
A 8
O 1
A 9
A10
M11
A11
A12
M12
N 2
N 3
M13
O 2
A13
N 4
Sy1
N 5
Sy2
A14
M14
M15
M16
Sy3
A15
N 6
A16
S 5
A17
A18
A19
M17
A20
S 6
S 7
A21
A22
A23
M18
A24
A25
M19
A26
O 3
N 7
N 8
A27
M20
O 4
M21
A28
M22
M23
M24
M25

Fig. HC 15

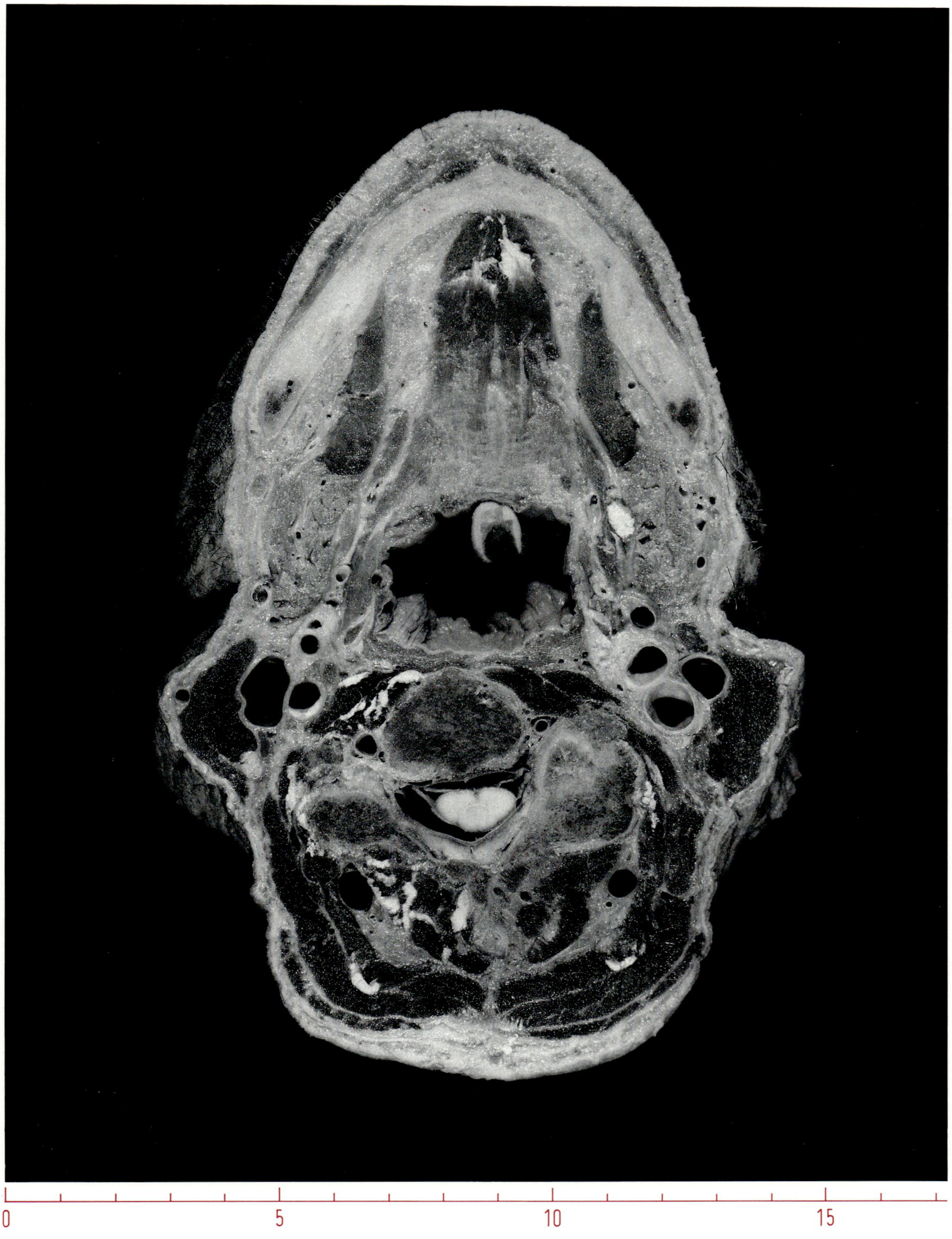

Fig. HC 15

D Si

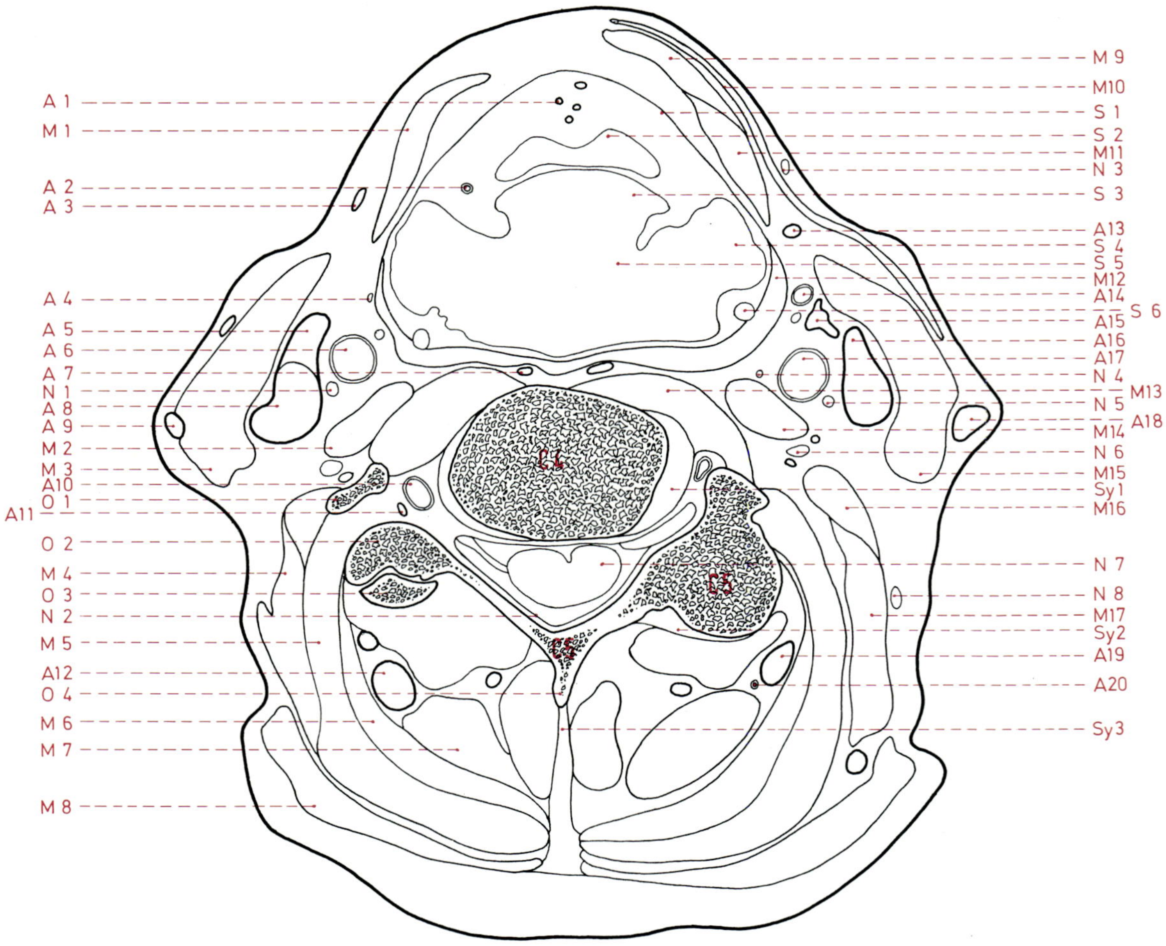
A 1
M 1
A 2
A 3
A 4
A 5
A 6
A 7
N 1
A 8
A 9
M 2
M 3
A10
O 1
A11
O 2
M 4
O 3
N 2
M 5
A12
O 4
M 6
M 7
M 8
M 9
M10
S 1
S 2
M11
N 3
S 3
A13
S 4
S 5
M12
A14
S 6
A15
A16
A17
N 4
M13
N 5
A18
M14
N 6
M15
Sy1
M16
N 7
N 8
M17
Sy2
A19
A20
Sy3

Fig. HC 16

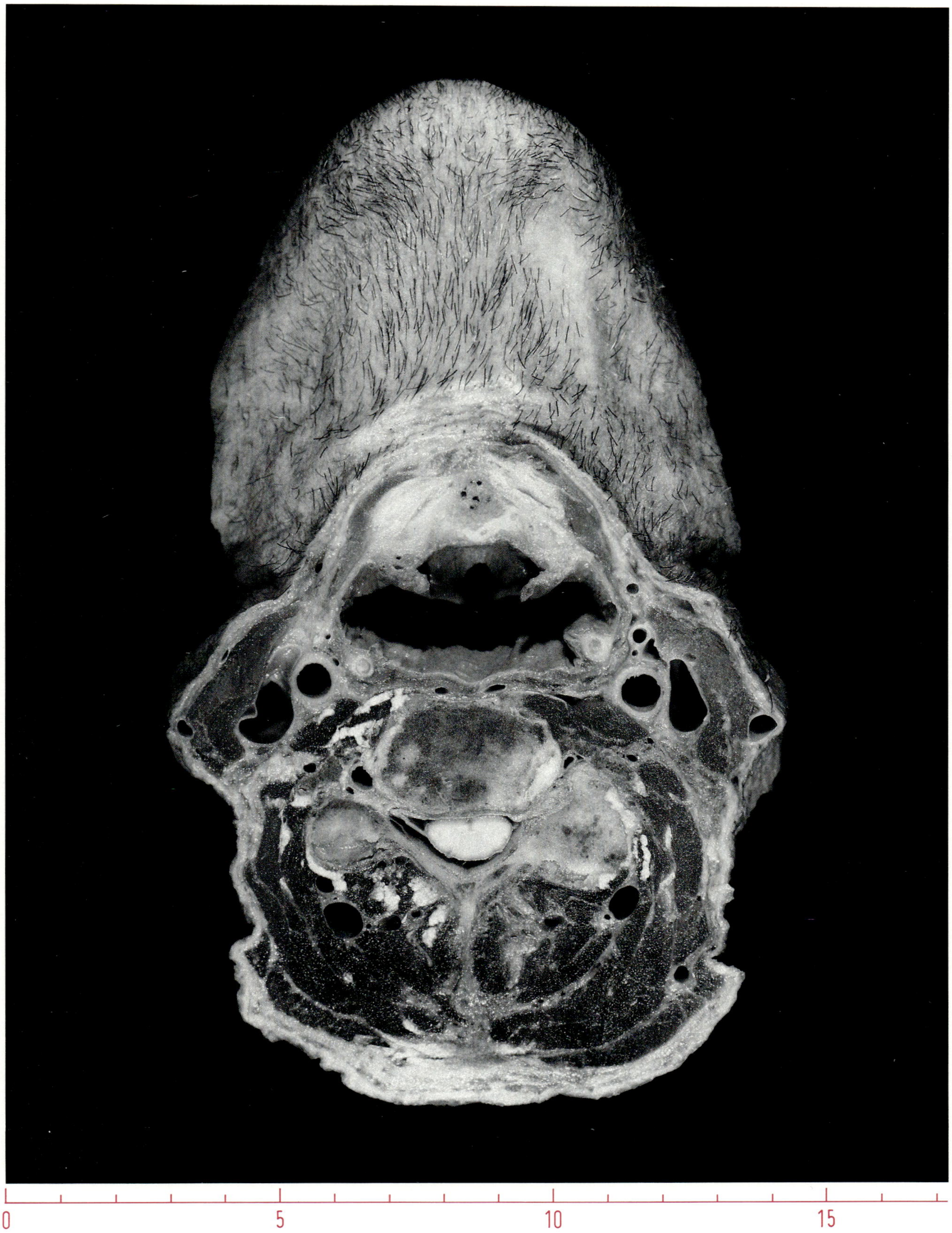

Fig. HC 16

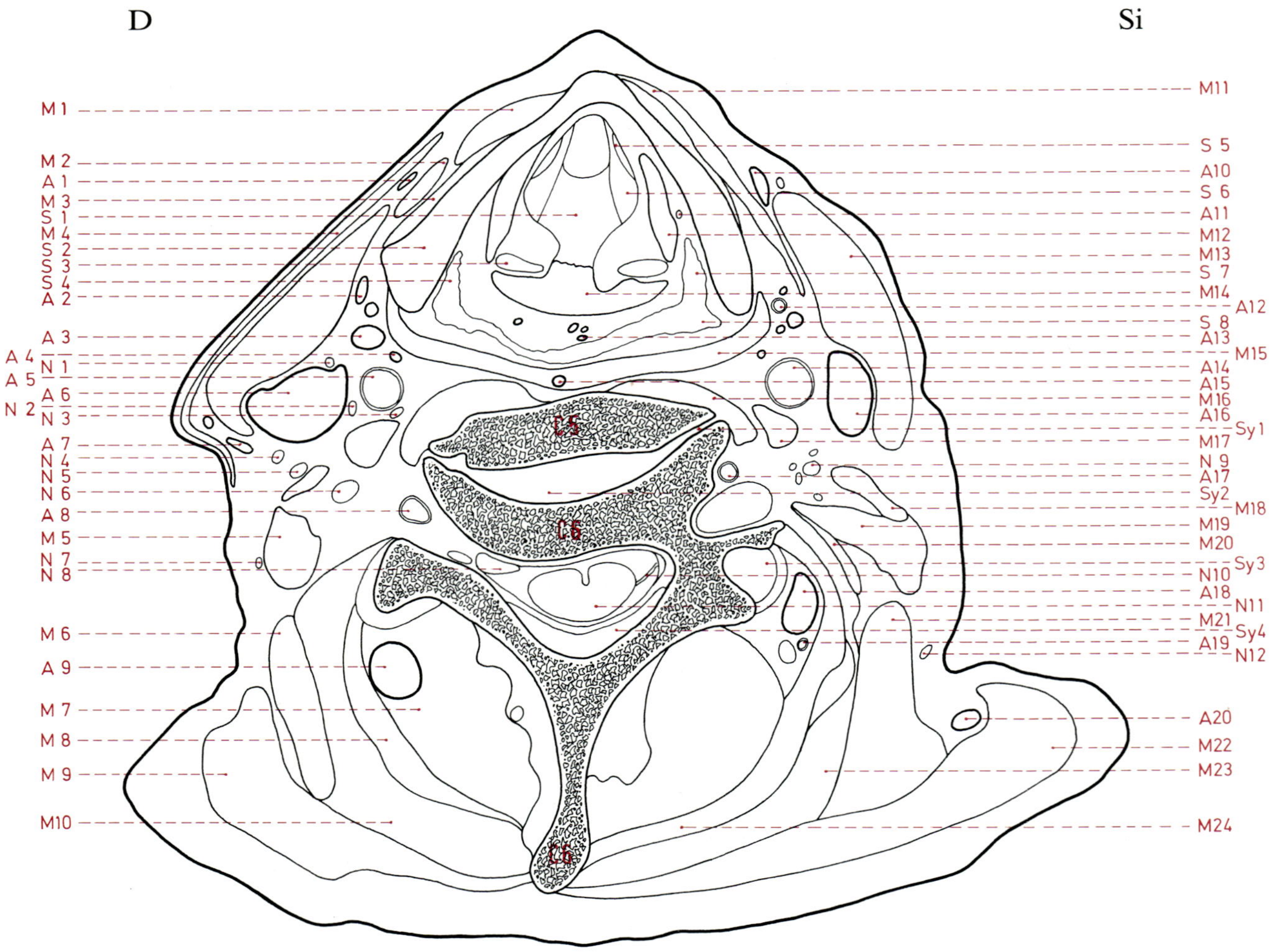
D
Si
M 1
M 2
A 1
M 3
S 1
M 4
S 2
S 3
S 4
A 2
A 3
A 4
N 1
A 5
A 6
N 2
N 3
A 7
N 4
N 5
N 6
A 8
M 5
N 7
N 8
M 6
A 9
M 7
M 8
M 9
M10
M11
S 5
A10
S 6
A11
M12
M13
S 7
M14
A12
S 8
A13
M15
A14
A15
M16
A16
Sy1
M17
N 9
A17
Sy2
M18
M19
M20
Sy3
N10
A18
N11
M21
Sy4
A19
N12
A20
M22
M23
M24
C5
C6
C6

Fig. HC 17

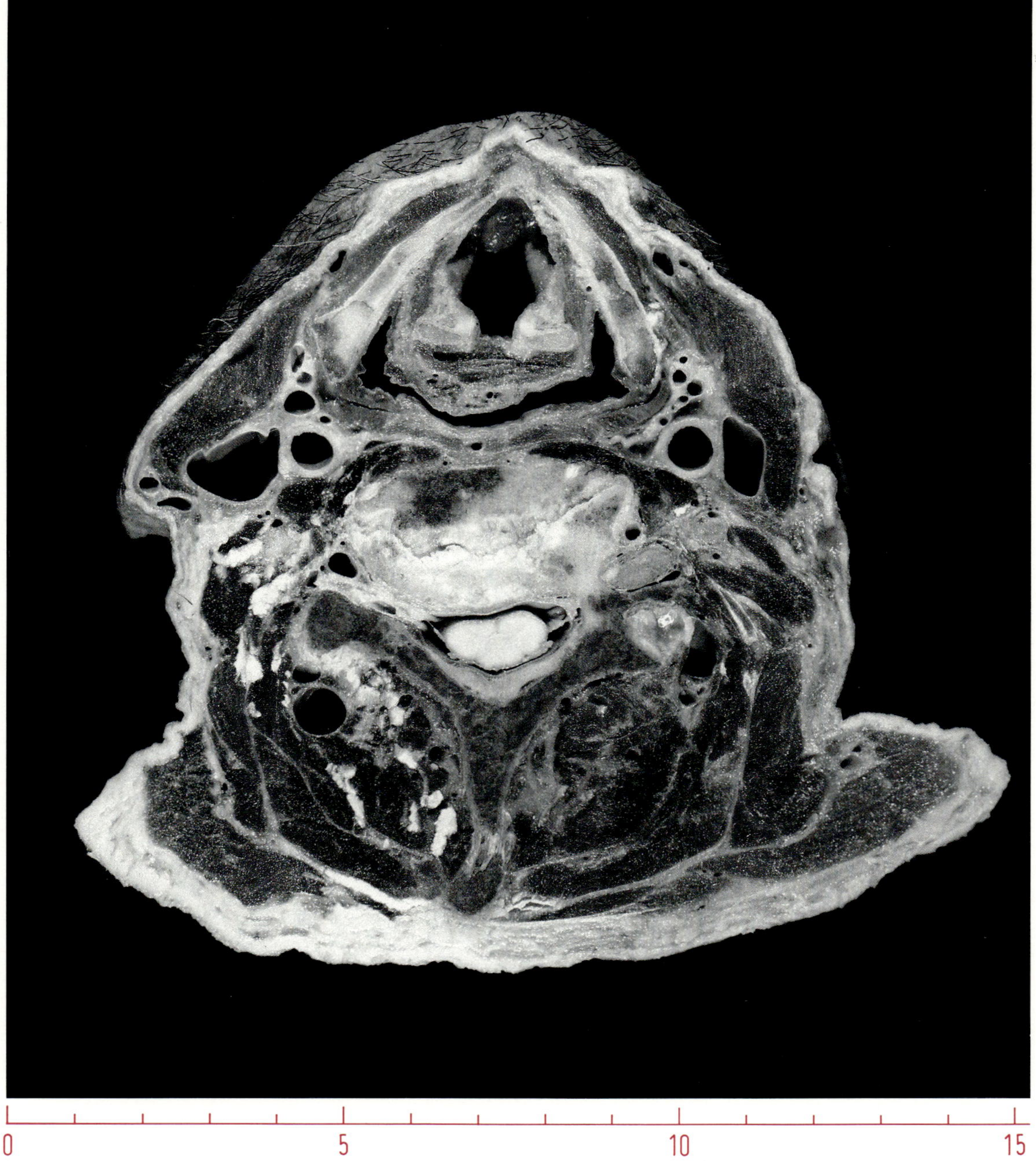

Fig. HC 17

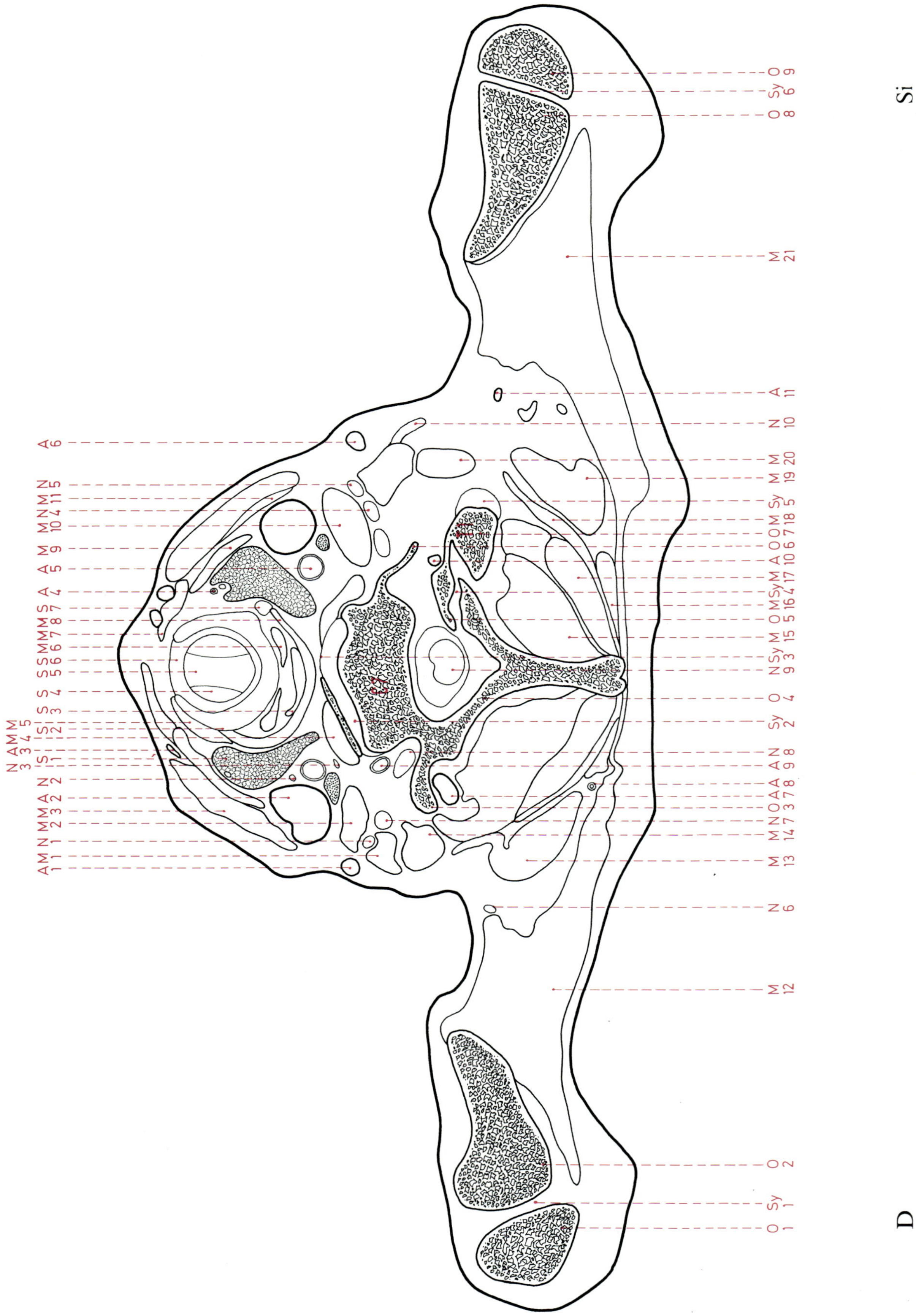

Fig. HC 18

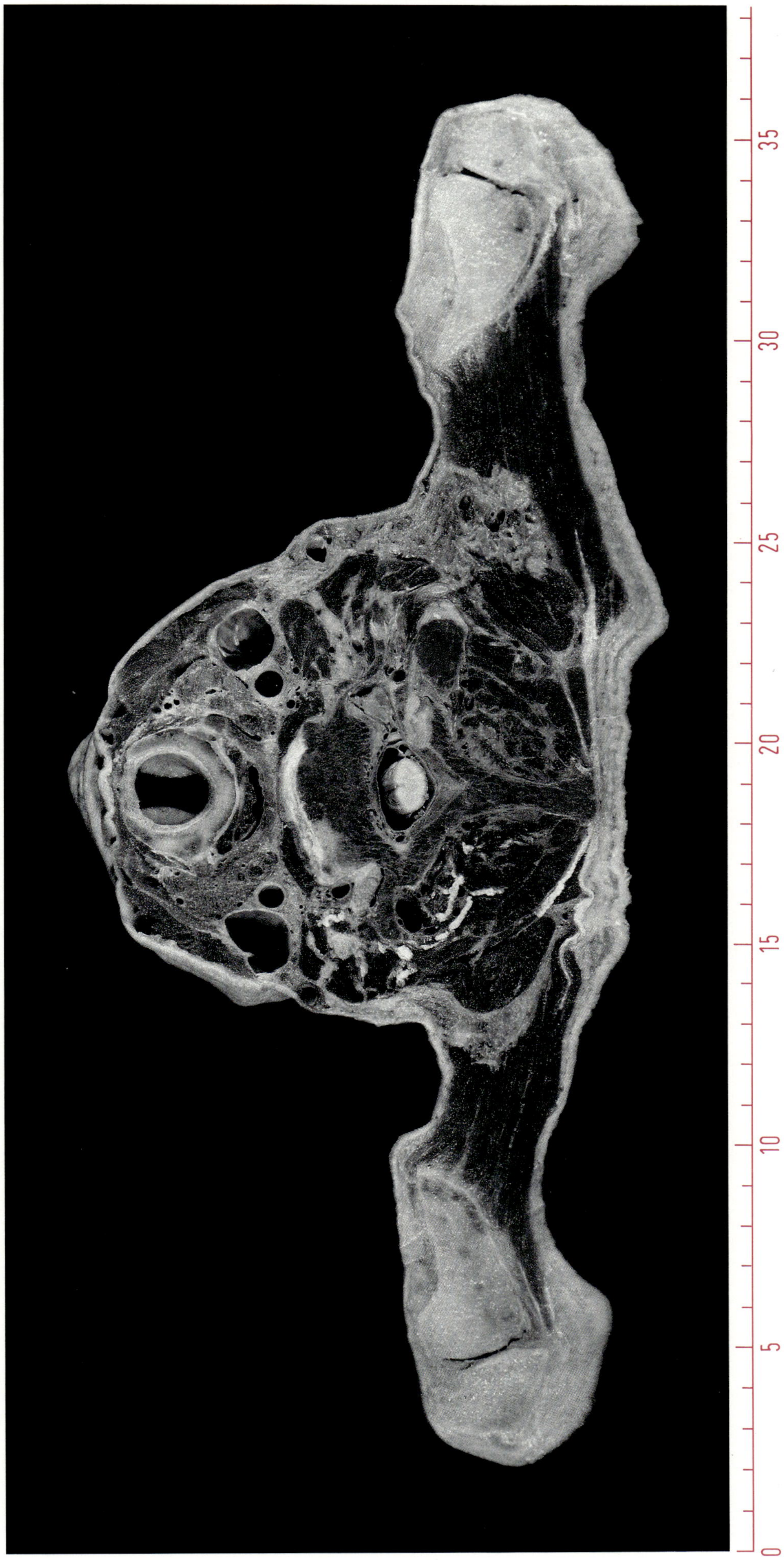

Fig. HC 18

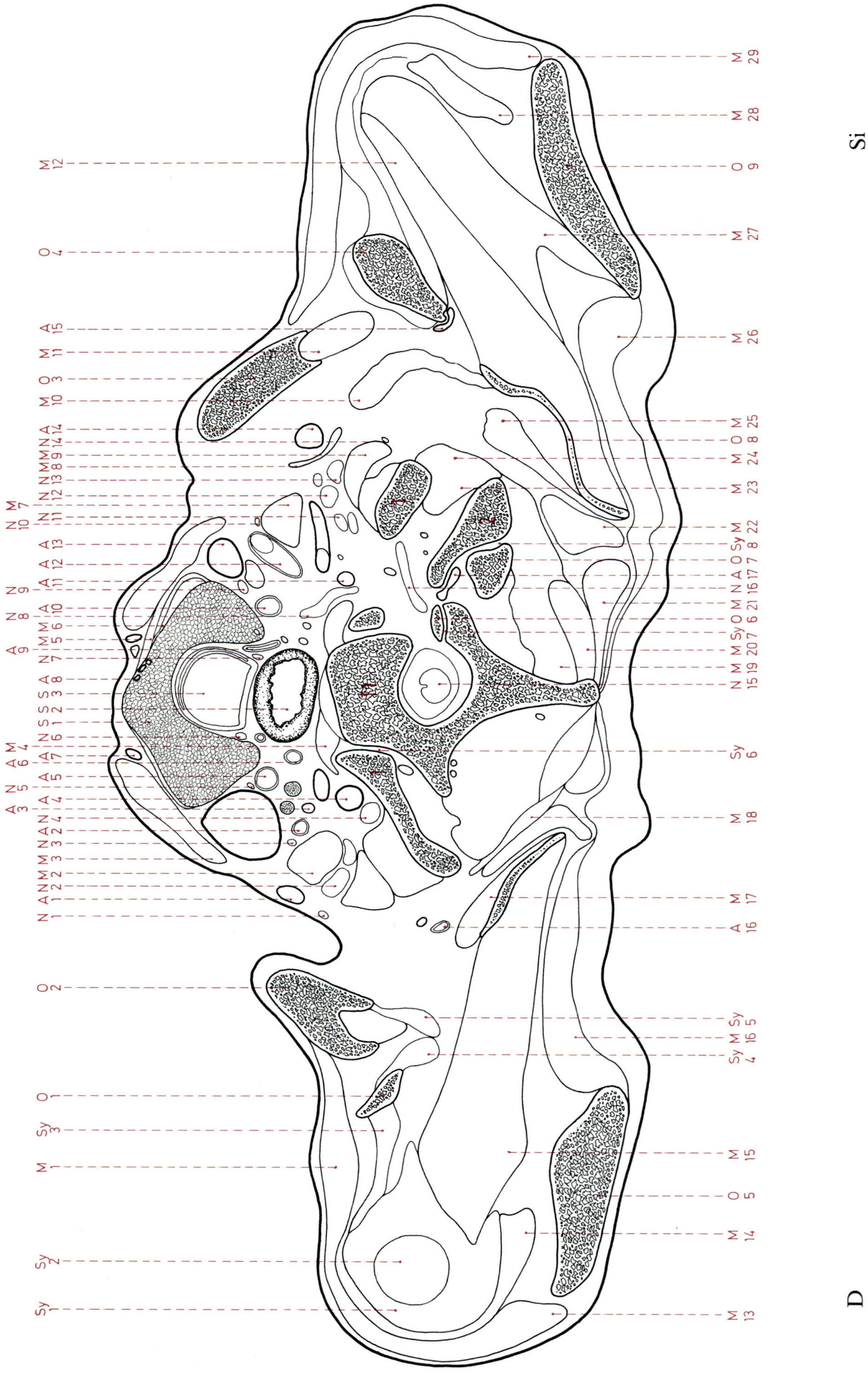

Fig. HC 19

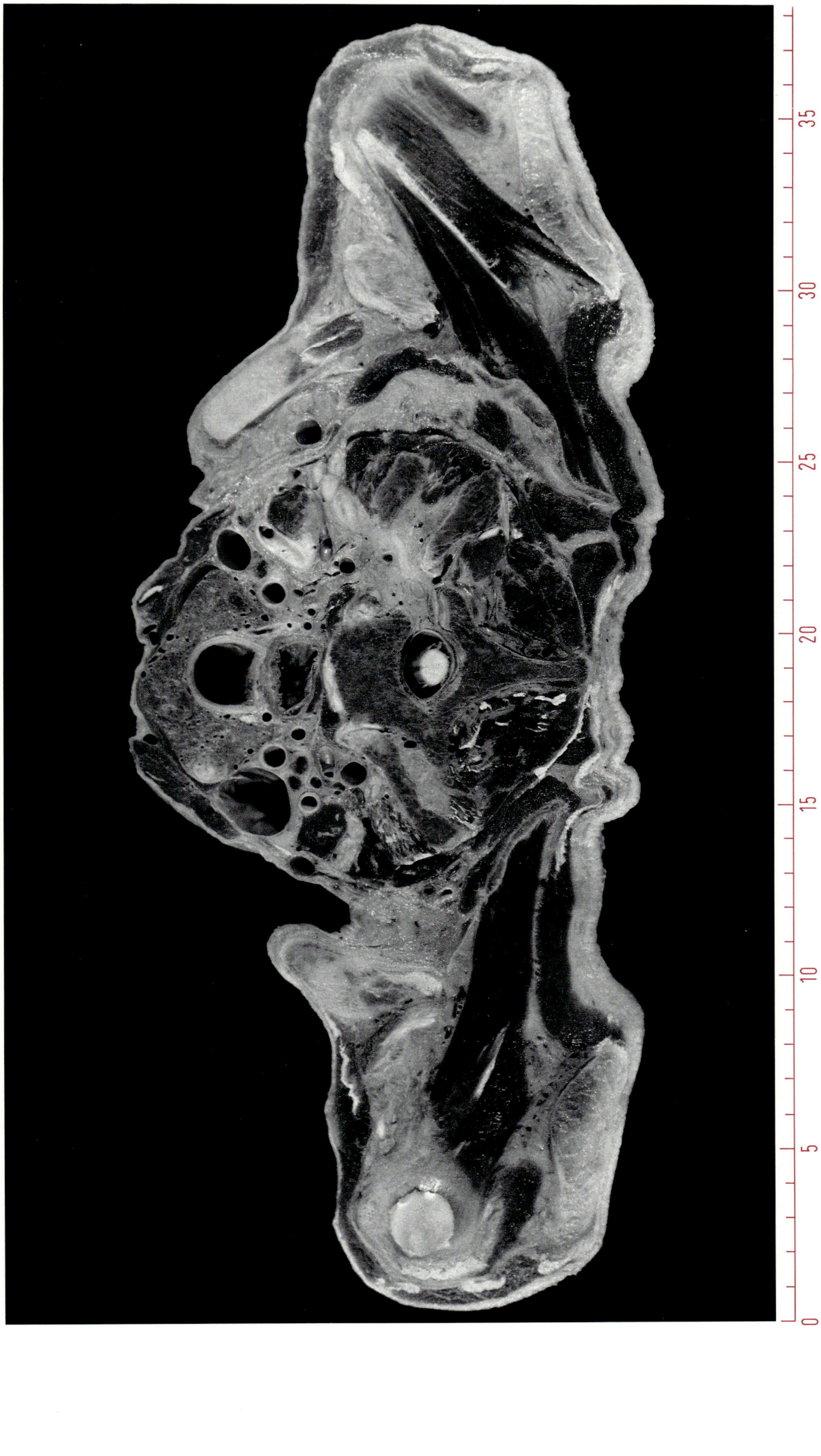

Fig. HC 19

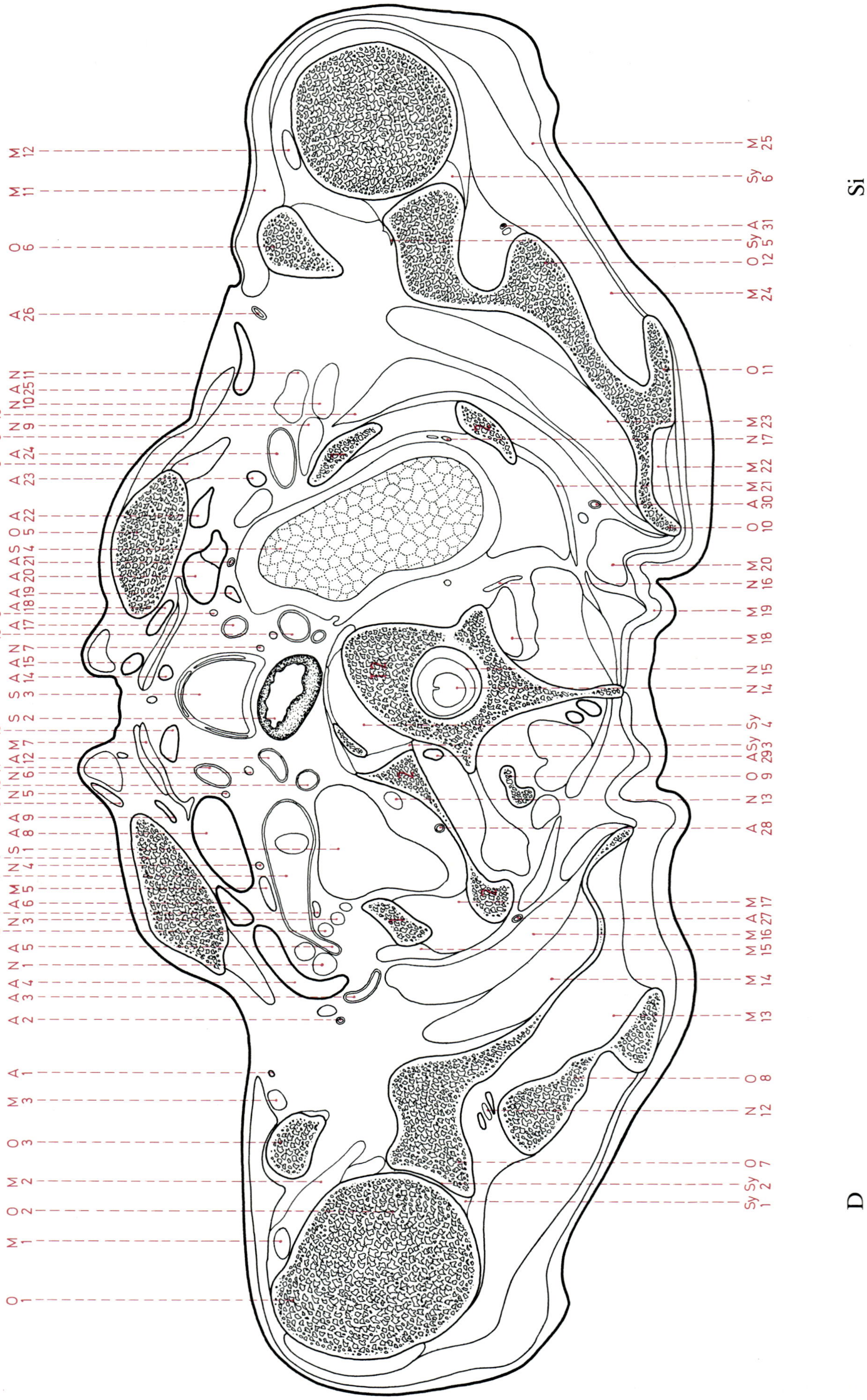

Fig. HC 20

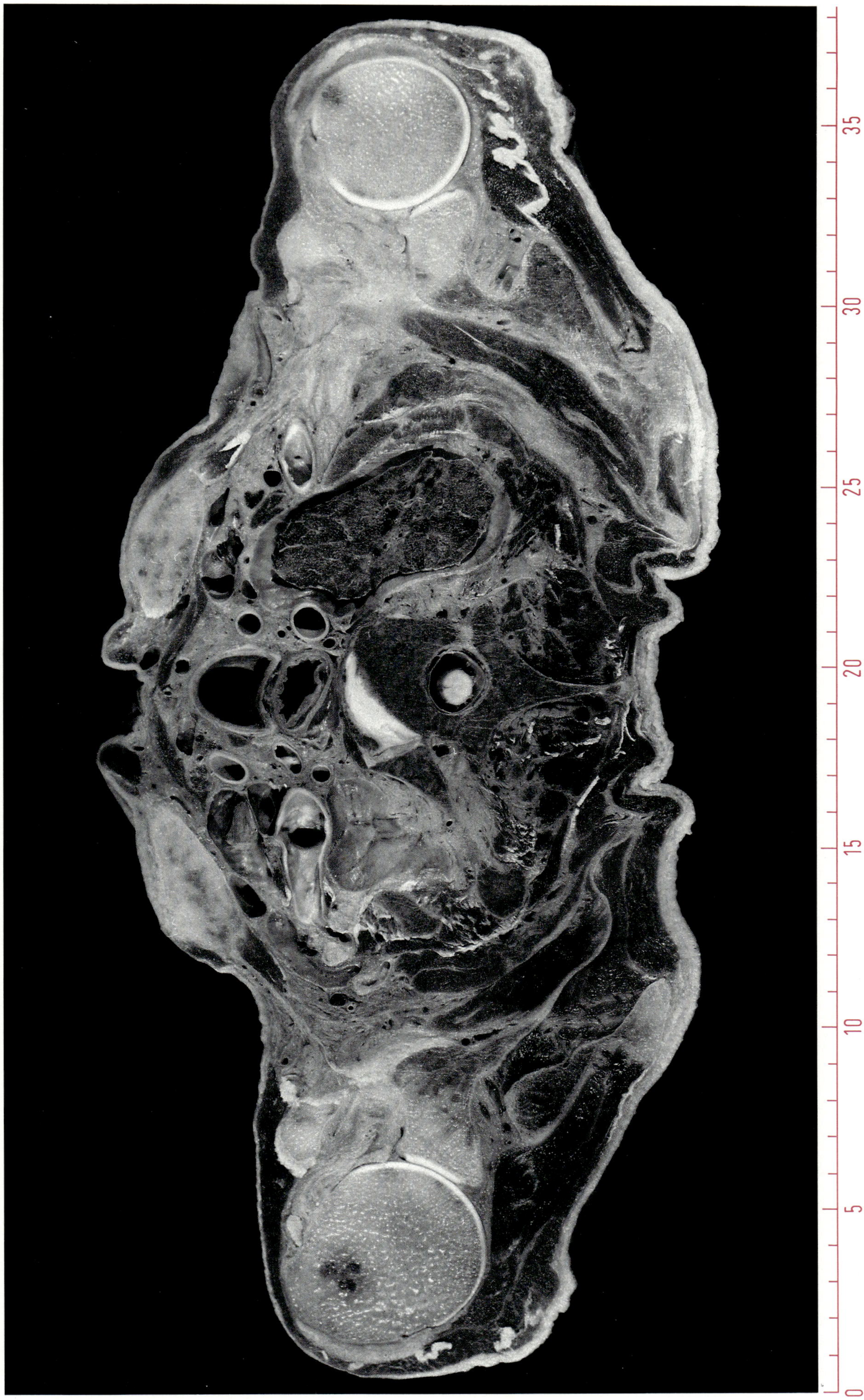

Fig. HC 20

FT

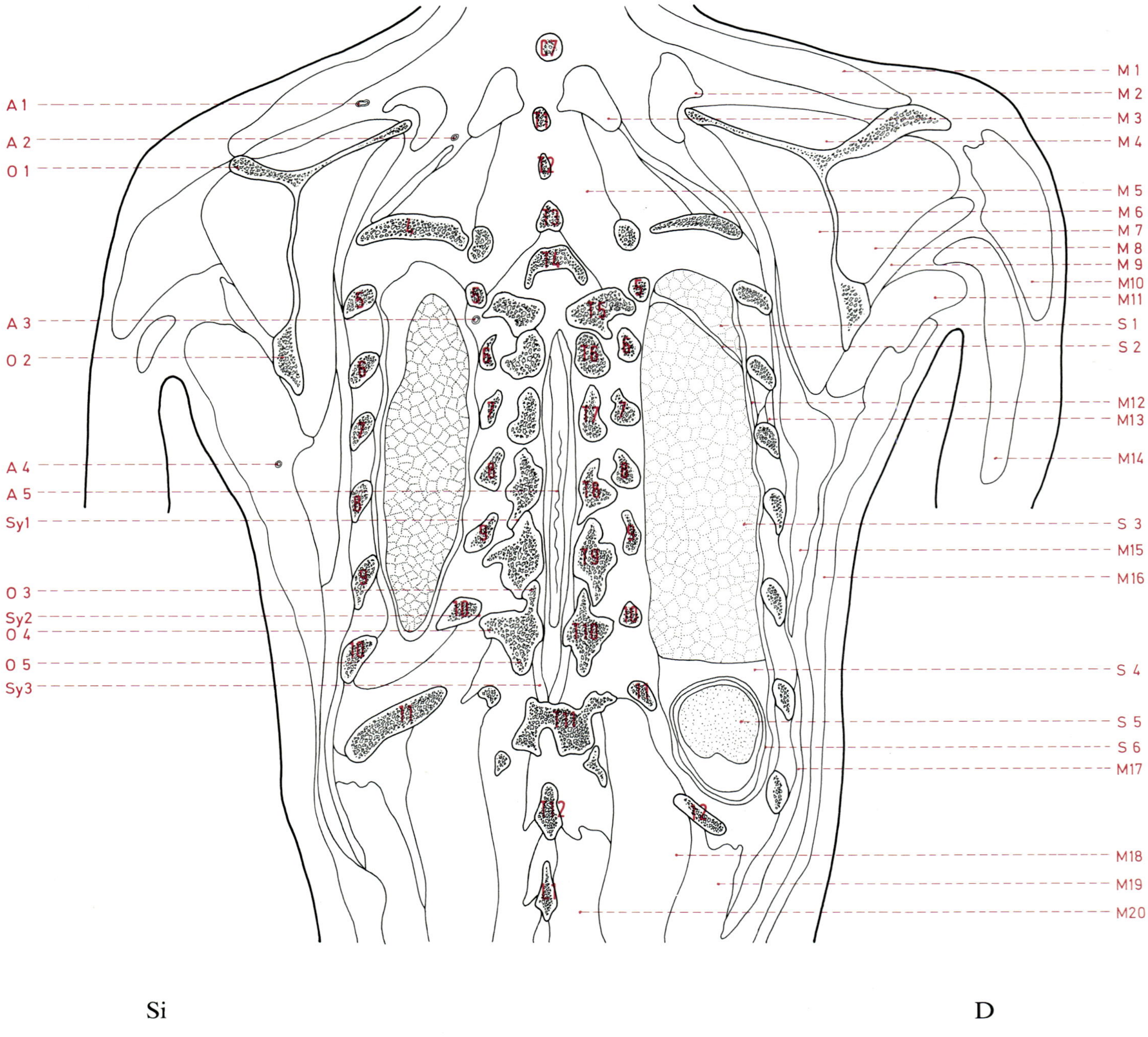

A 1
A 2
O 1
A 3
O 2
A 4
A 5
Sy1
O 3
Sy2
O 4
O 5
Sy3
M 1
M 2
M 3
M 4
M 5
M 6
M 7
M 8
M 9
M10
M11
S 1
S 2
M12
M13
M14
S 3
M15
M16
S 4
S 5
S 6
M17
M18
M19
M20
Si
D

Fig. FT 1

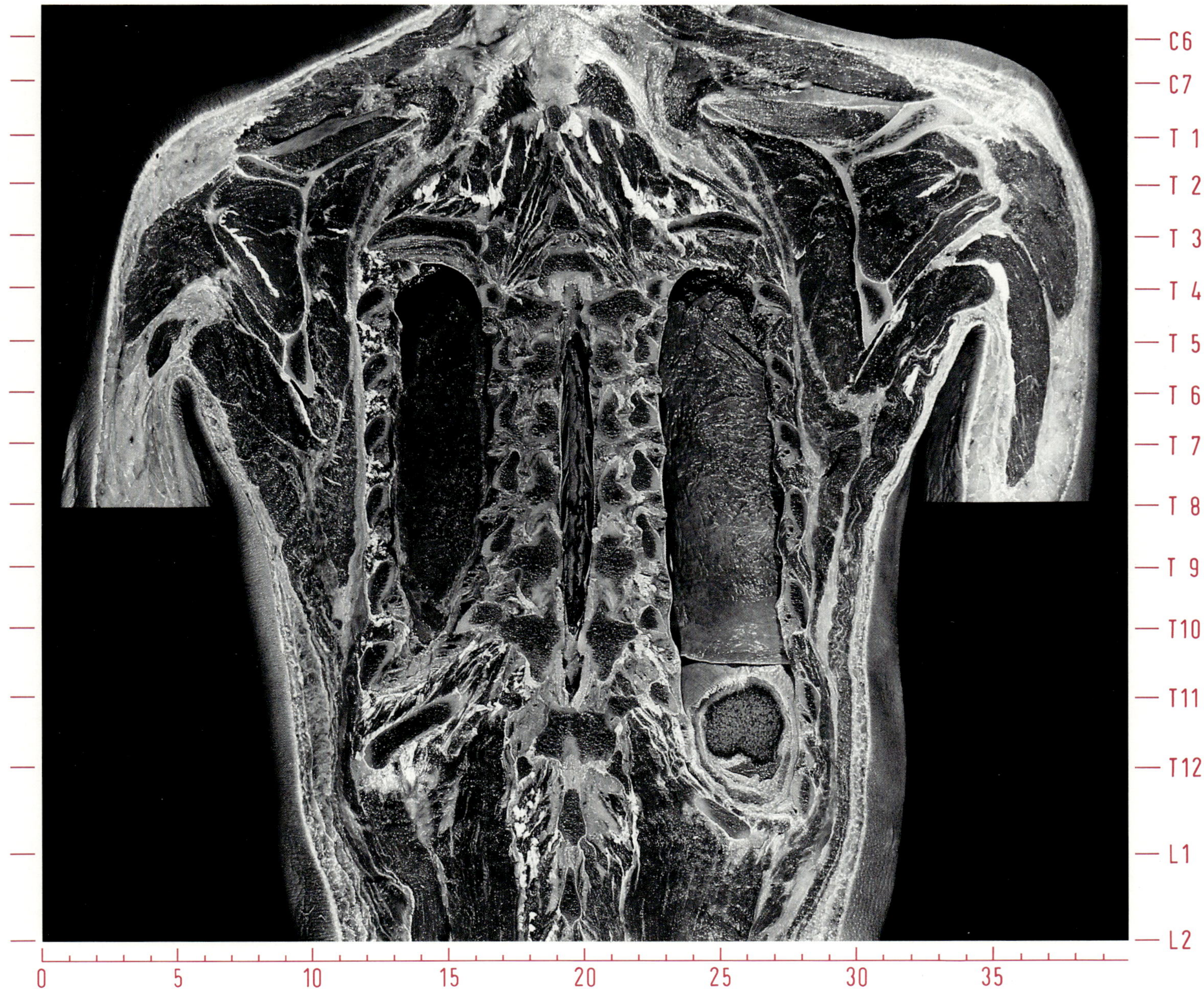

Fig. FT 1

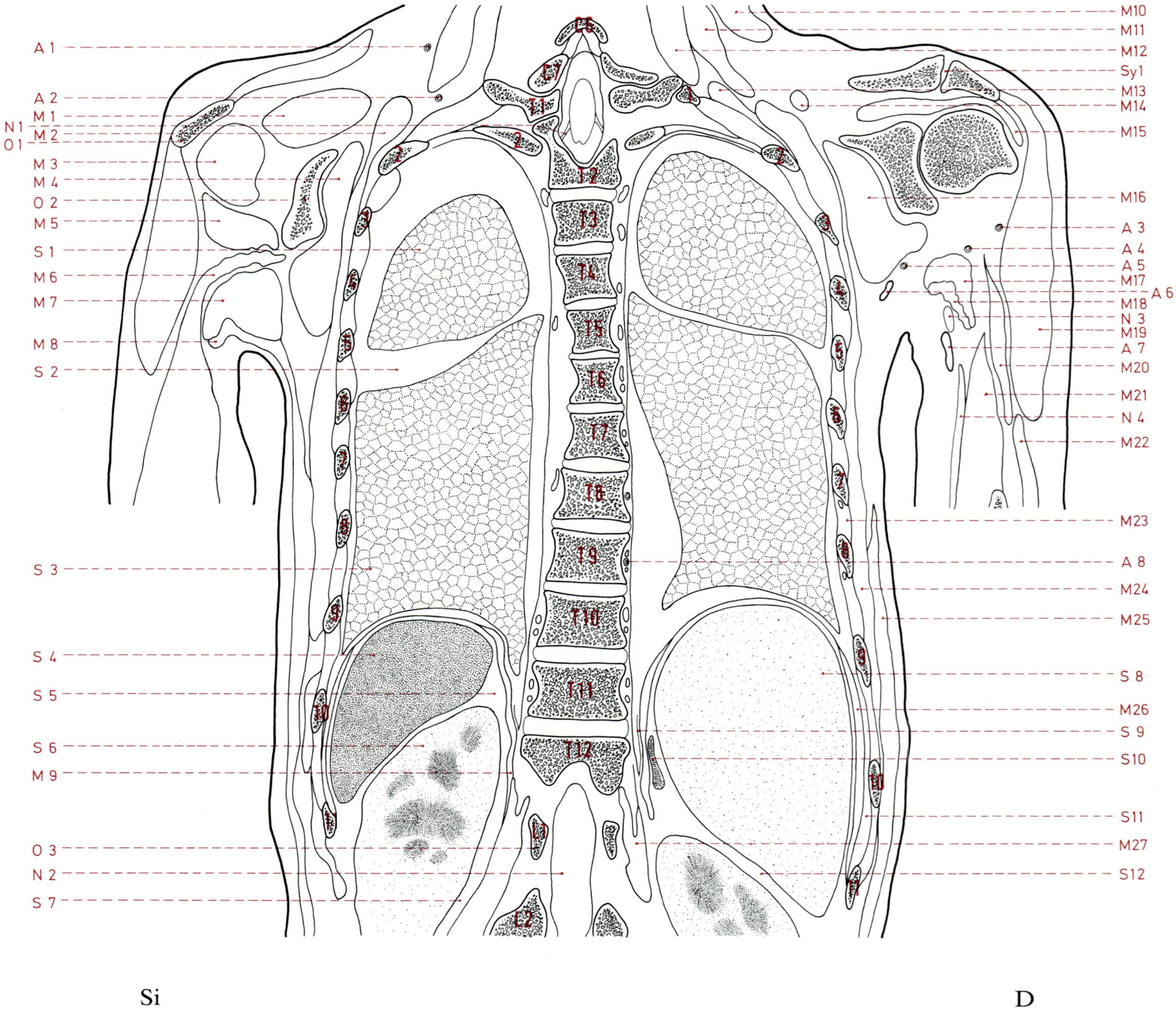
A 1
A 2
M 1
N1
M 2
O1
M 3
M 4
O 2
M 5
S 1
M 6
M 7
M 8
S 2
S 3
S 4
S 5
S 6
M 9
O 3
N 2
S 7
M10
M11
M12
Sy1
M13
M14
M15
M16
A 3
A 4
A 5
M17
A 6
M18
N 3
M19
A 7
M20
M21
N 4
M22
M23
A 8
M24
M25
S 8
M26
S 9
S10
S11
M27
S12
C6
C7
T1
T2
T3
T4
T5
T6
T7
T8
T9
T10
T11
T12
L1
L2
Si
D

Fig. FT 2

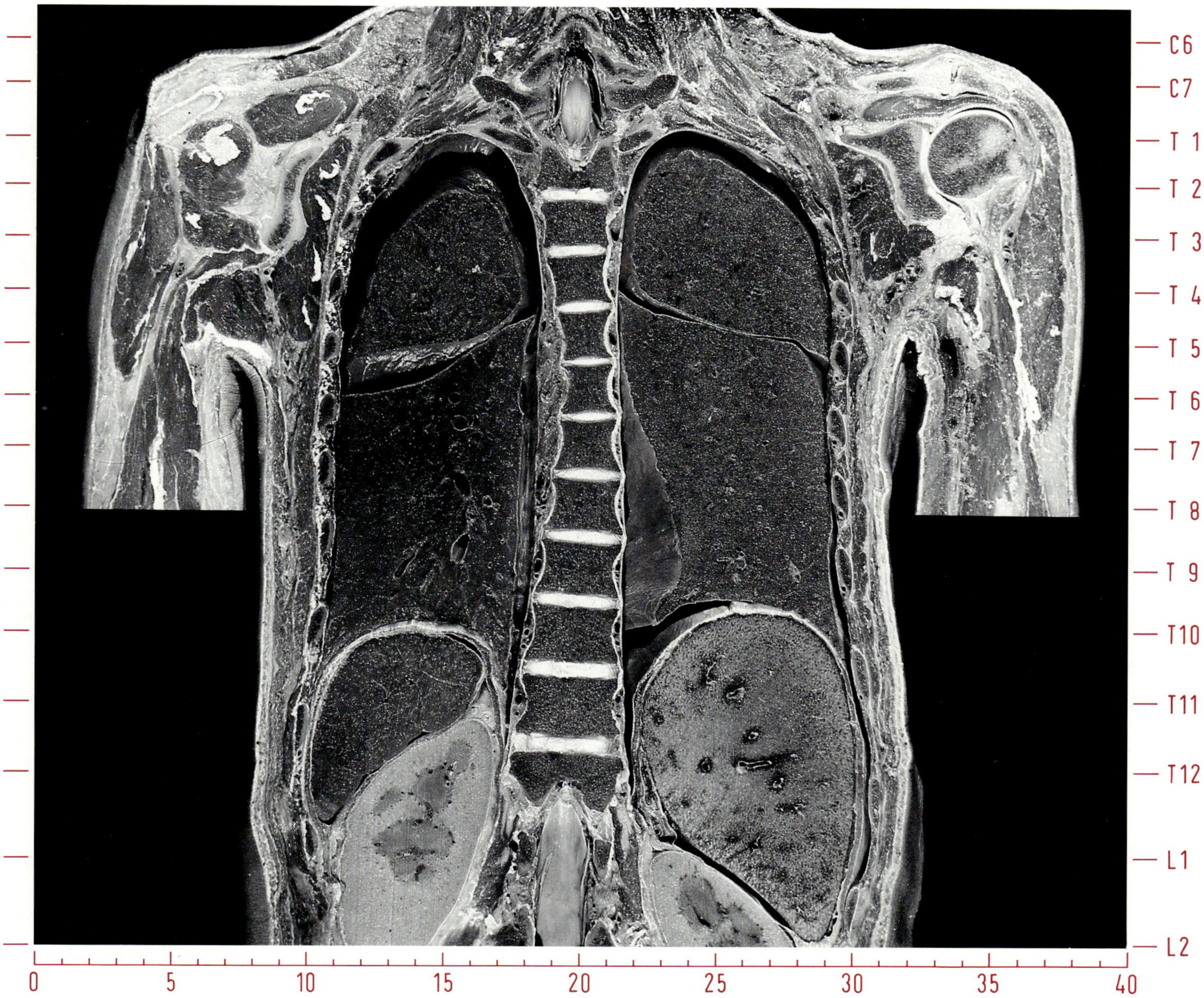
C6
C7
T 1
T 2
T 3
T 4
T 5
T 6
T 7
T 8
T 9
T10
T11
T12
L1
L2
0
5
10
15
20
25
30
35
40

Fig. FT 2

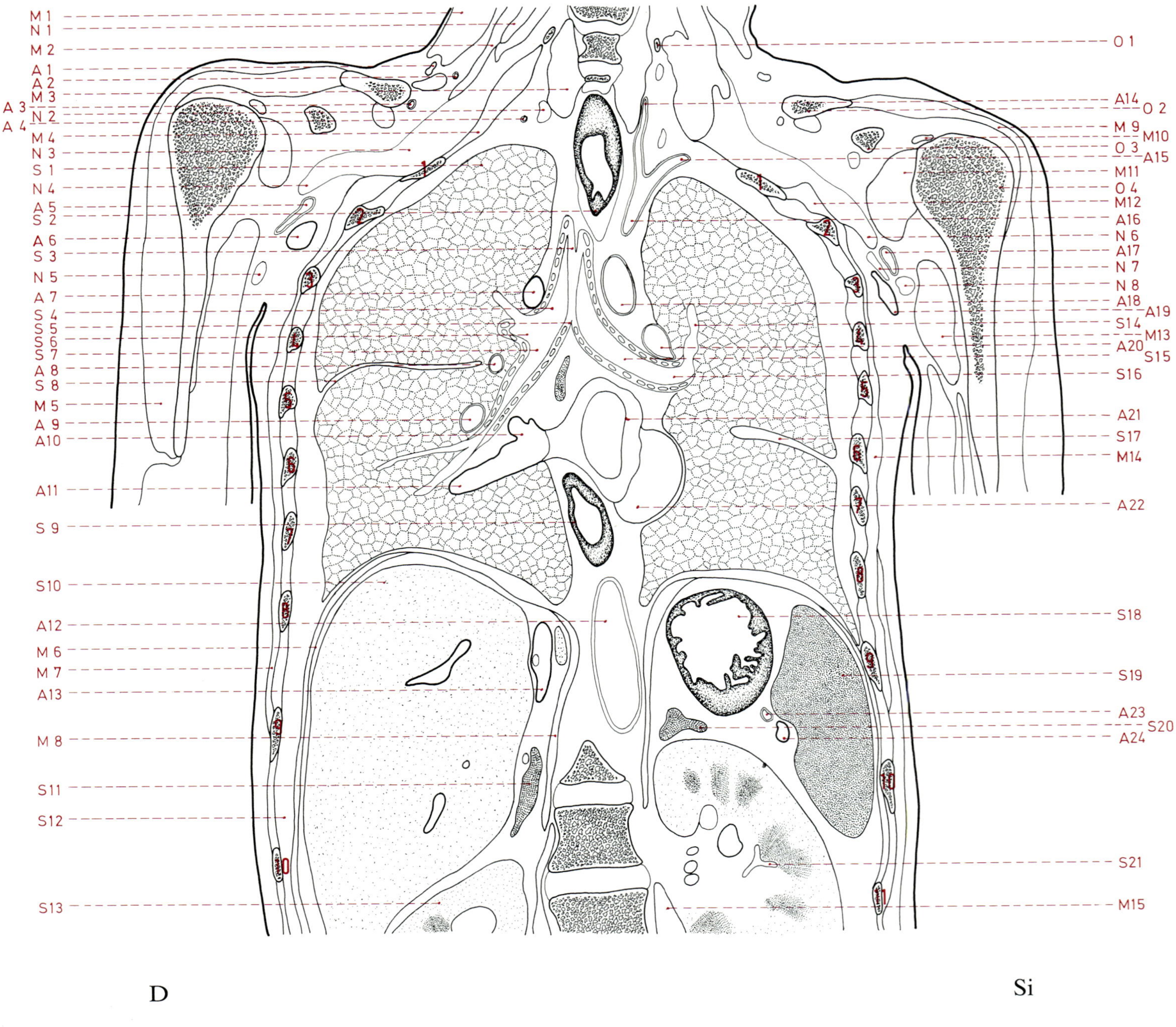
M 1
N 1
M 2
A 1
A 2
M 3
A 3
N 2
A 4
M 4
N 3
S 1
N 4
A 5
S 2
A 6
S 3
N 5
A 7
S 4
S 5
S 6
S 7
A 8
S 8
M 5
A 9
A10
A11
S 9
S10
A12
M 6
M 7
A13
M 8
S11
S12
S13
O 1
A14
O 2
M 9
M10
O 3
A15
M11
O 4
M12
A16
N 6
A17
N 7
N 8
A18
A19
S14
M13
A20
S15
S16
A21
S17
M14
A22
S18
S19
A23
S20
A24
S21
M15
D
Si

Fig. FT 3 a

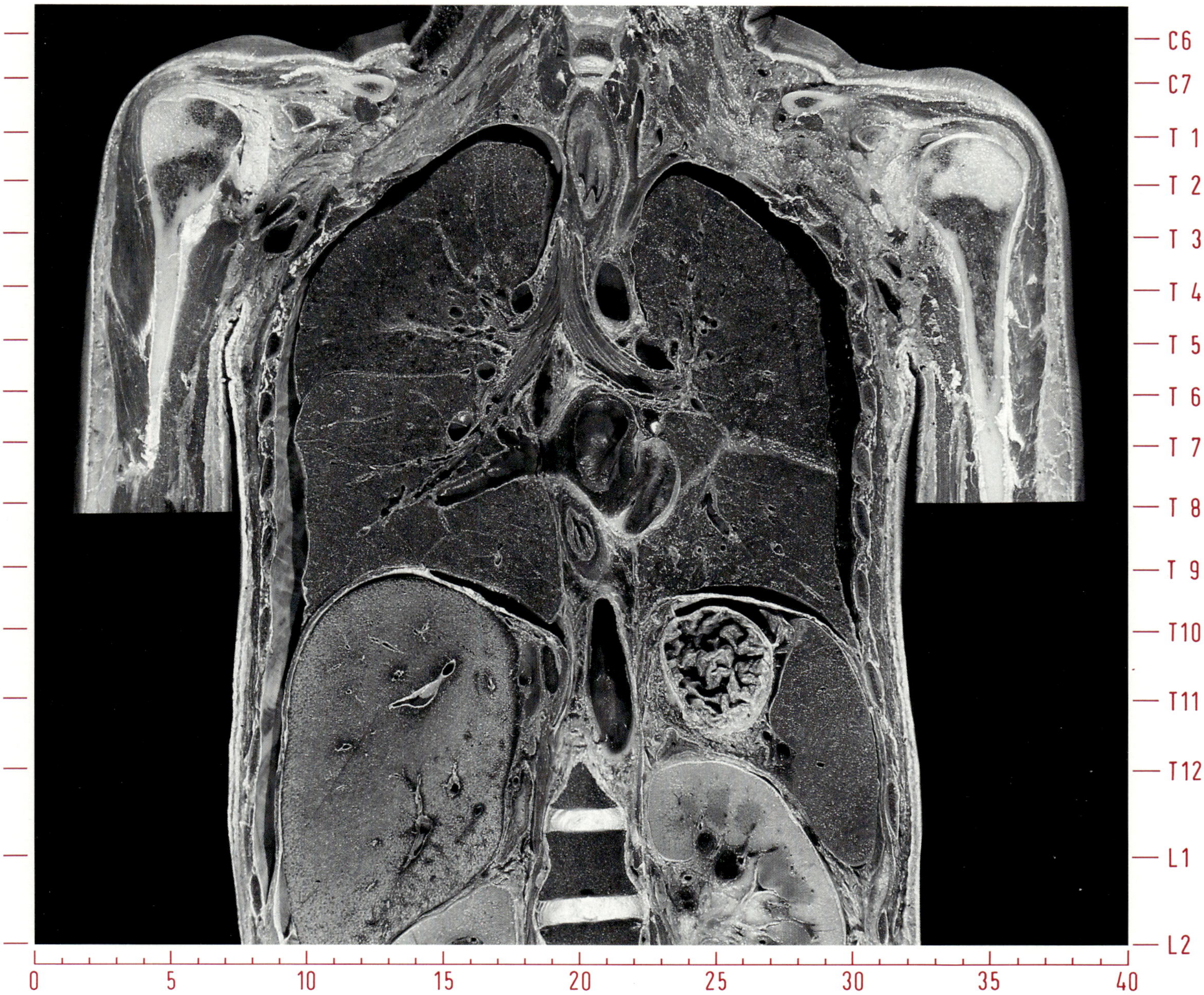
C6
C7
T 1
T 2
T 3
T 4
T 5
T 6
T 7
T 8
T 9
T10
T11
T12
L1
L2
0
5
10
15
20
25
30
35
40

Fig. FT 3a

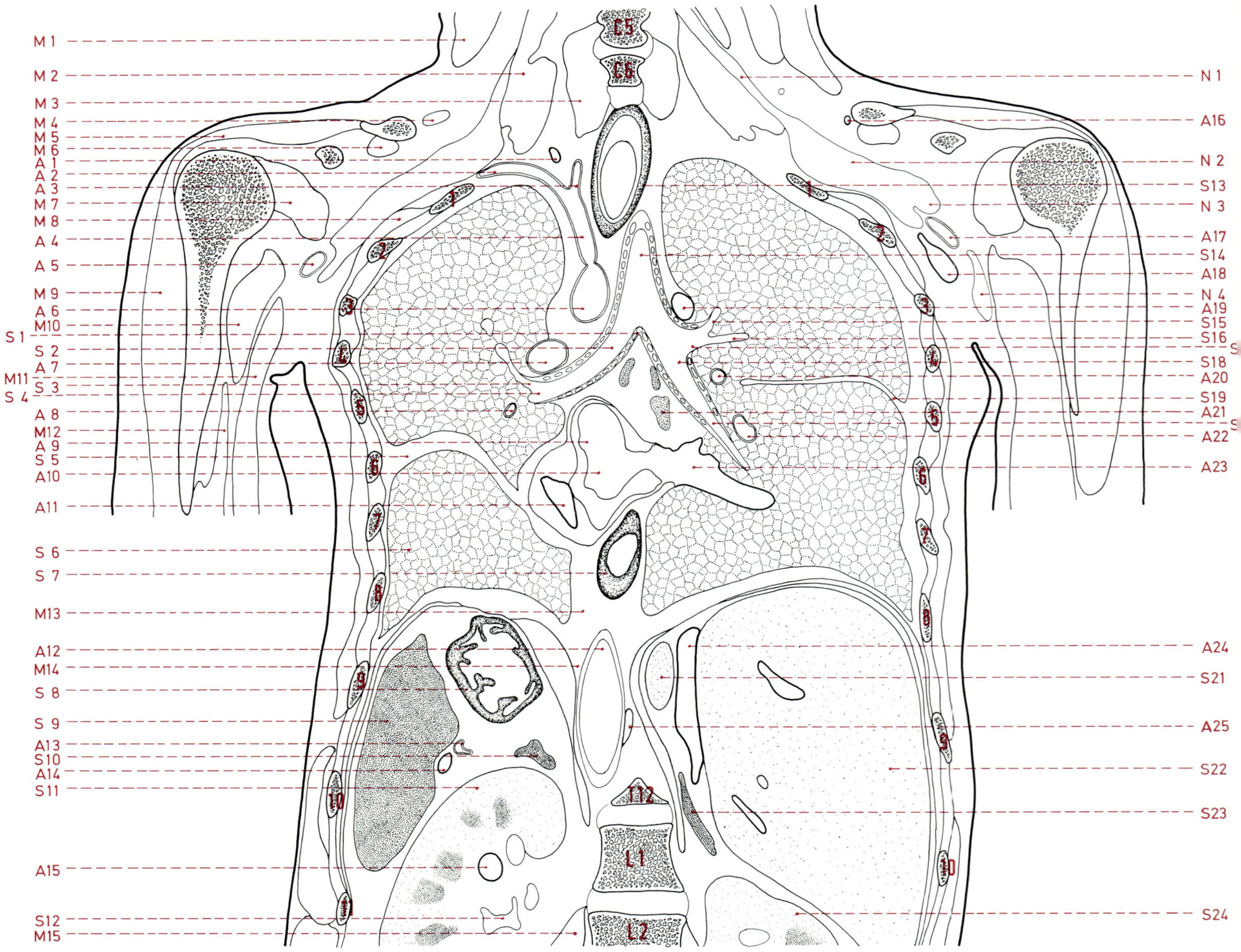
M 1
M 2
M 3
M 4
M 5
M 6
A 1
A 2
A 3
M 7
M 8
A 4
A 5
M 9
A 6
M10
S 1
S 2
A 7
M11
S 4
S 3
A 8
M12
A 9
S 5
A10
A11
S 6
S 7
M13
A12
M14
S 8
S 9
A13
S10
A14
S11
A15
S12
M15
N 1
A16
N 2
S13
N 3
A17
S14
A18
N 4
A19
S15
S16
S18
A20
S19
A21
A22
A23
A24
S21
A25
S22
S23
S24
C5
C6
T12
L1
L2

Si

D

Fig. FT 3p

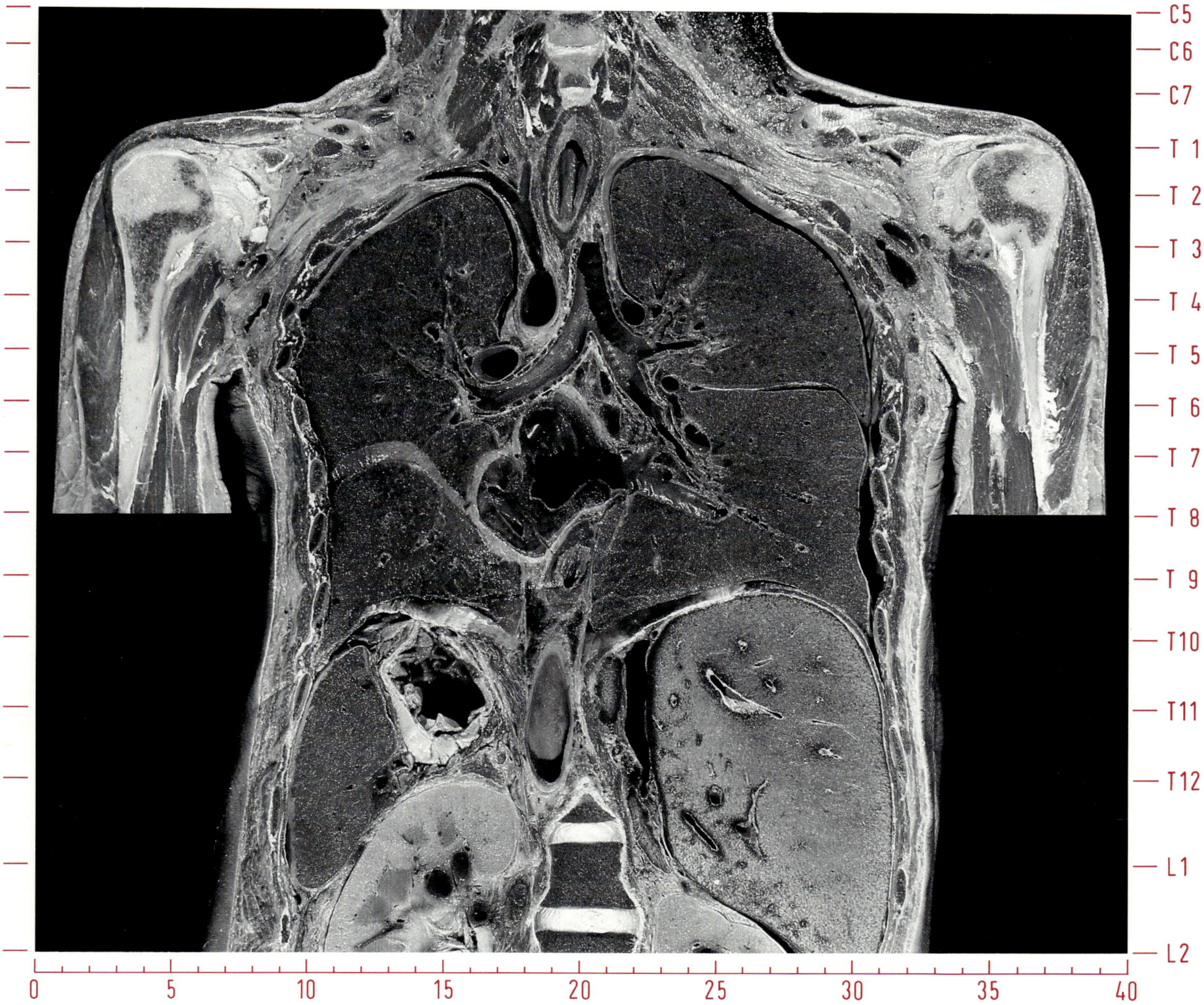
C5
C6
C7
T 1
T 2
T 3
T 4
T 5
T 6
T 7
T 8
T 9
T10
T11
T12
L1
L2
0
5
10
15
20
25
30
35
40

Fig. FT 3p

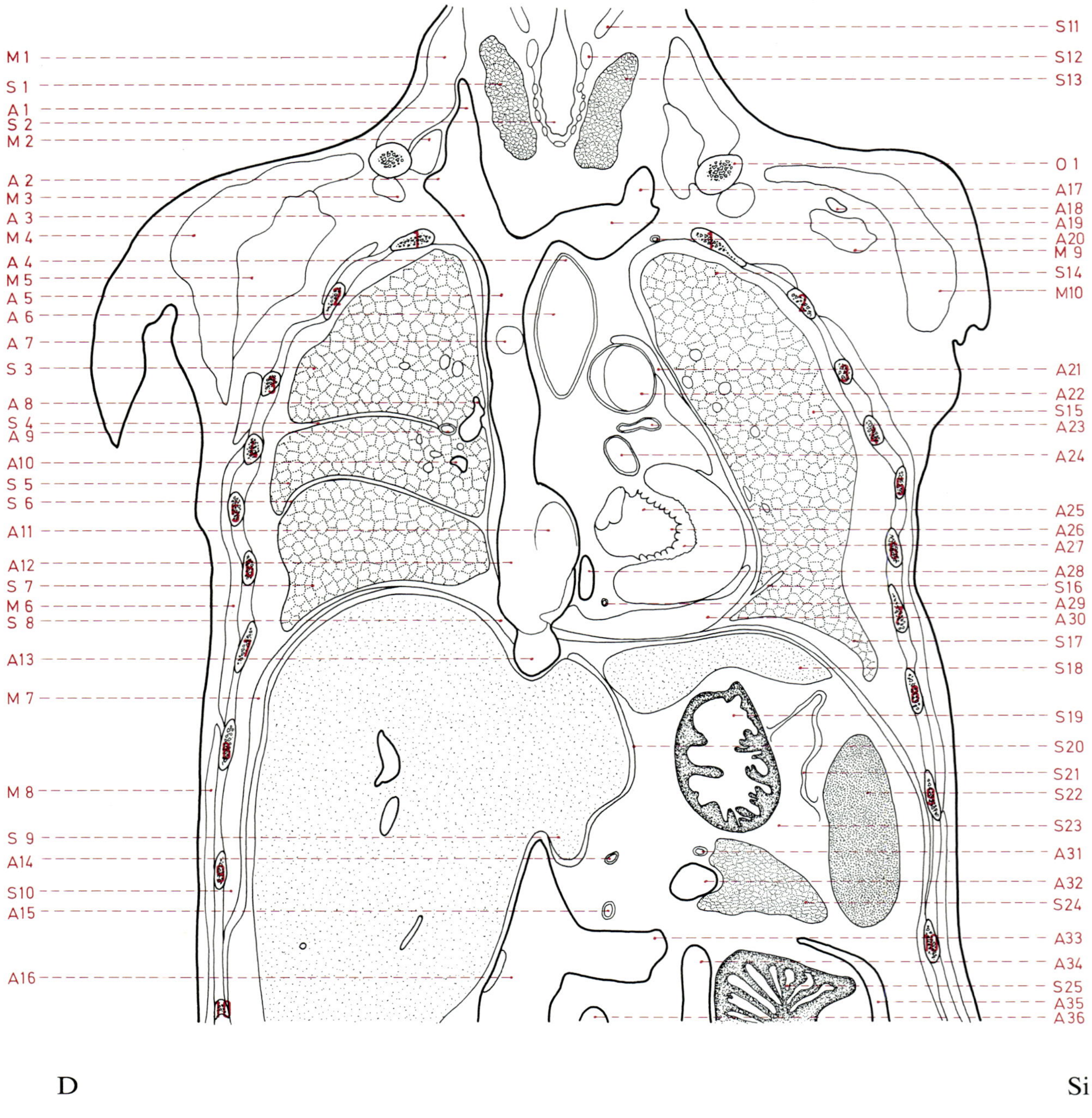
M 1
S 1
A 1
S 2
M 2
A 2
M 3
A 3
M 4
A 4
M 5
A 5
A 6
A 7
S 3
A 8
S 4
A 9
A10
S 5
S 6
A11
A12
S 7
M 6
S 8
A13
M 7
M 8
S 9
A14
S10
A15
A16
S11
S12
S13
O 1
A17
A18
A19
A20
M 9
S14
M10
A21
A22
S15
A23
A24
A25
A26
A27
A28
S16
A29
A30
S17
S18
S19
S20
S21
S22
S23
A31
A32
S24
A33
A34
S25
A35
A36
D
Si

Fig. FT 4

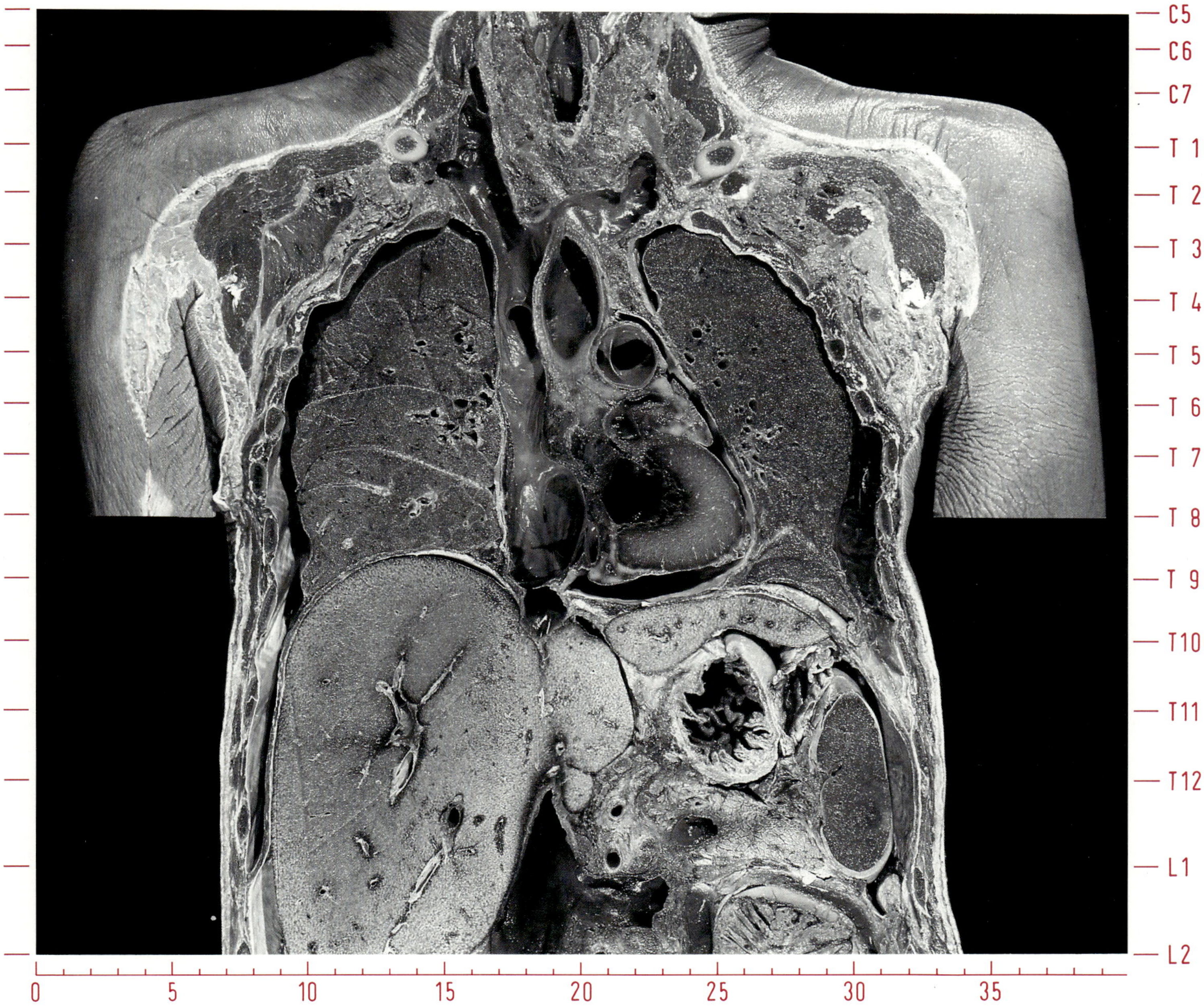
C5
C6
C7
T 1
T 2
T 3
T 4
T 5
T 6
T 7
T 8
T 9
T10
T11
T12
L1
L2
0
5
10
15
20
25
30
35

Fig. FT 4

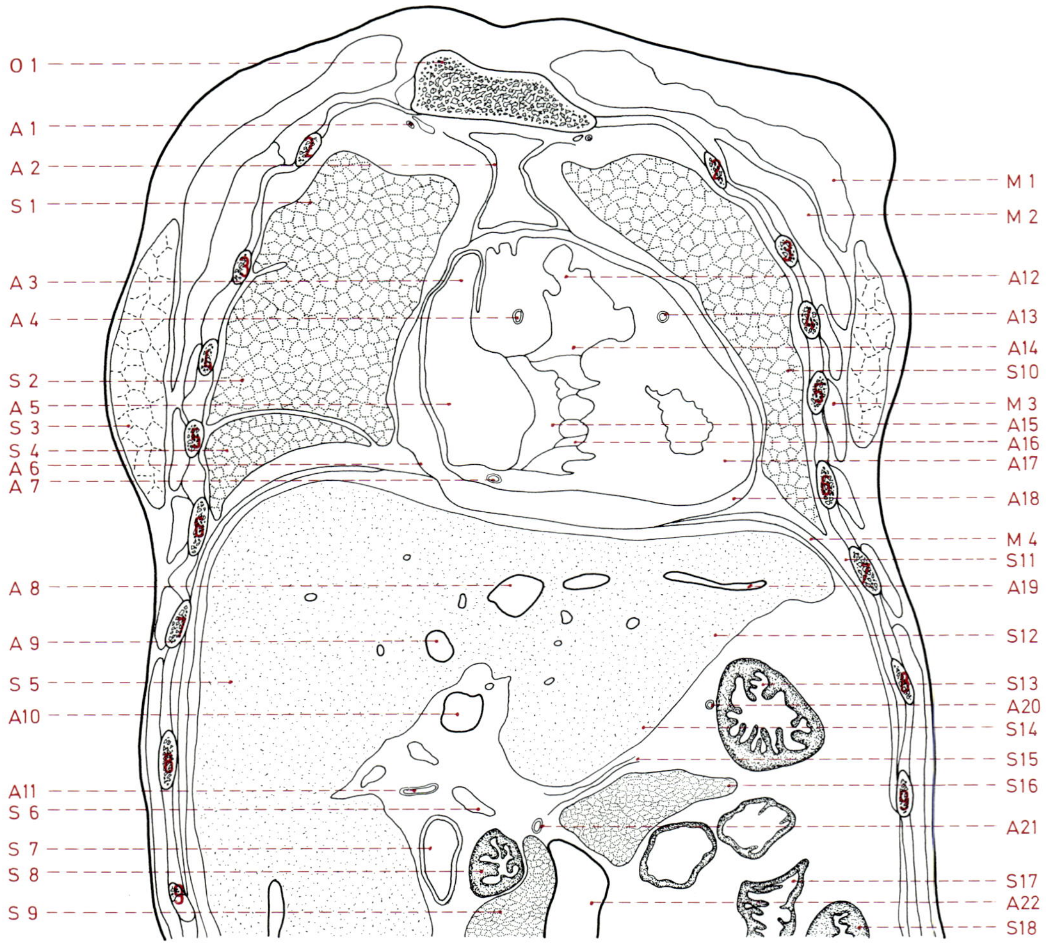

O 1
A 1
A 2
S 1
A 3
A 4
S 2
A 5
S 3
S 4
A 6
A 7
A 8
A 9
S 5
A10
A11
S 6
S 7
S 8
S 9
M 1
M 2
A12
A13
A14
S10
M 3
A15
A16
A17
A18
M 4
S11
A19
S12
S13
A20
S14
S15
S16
A21
S17
A22
S18
D
Si

Fig. FT 5

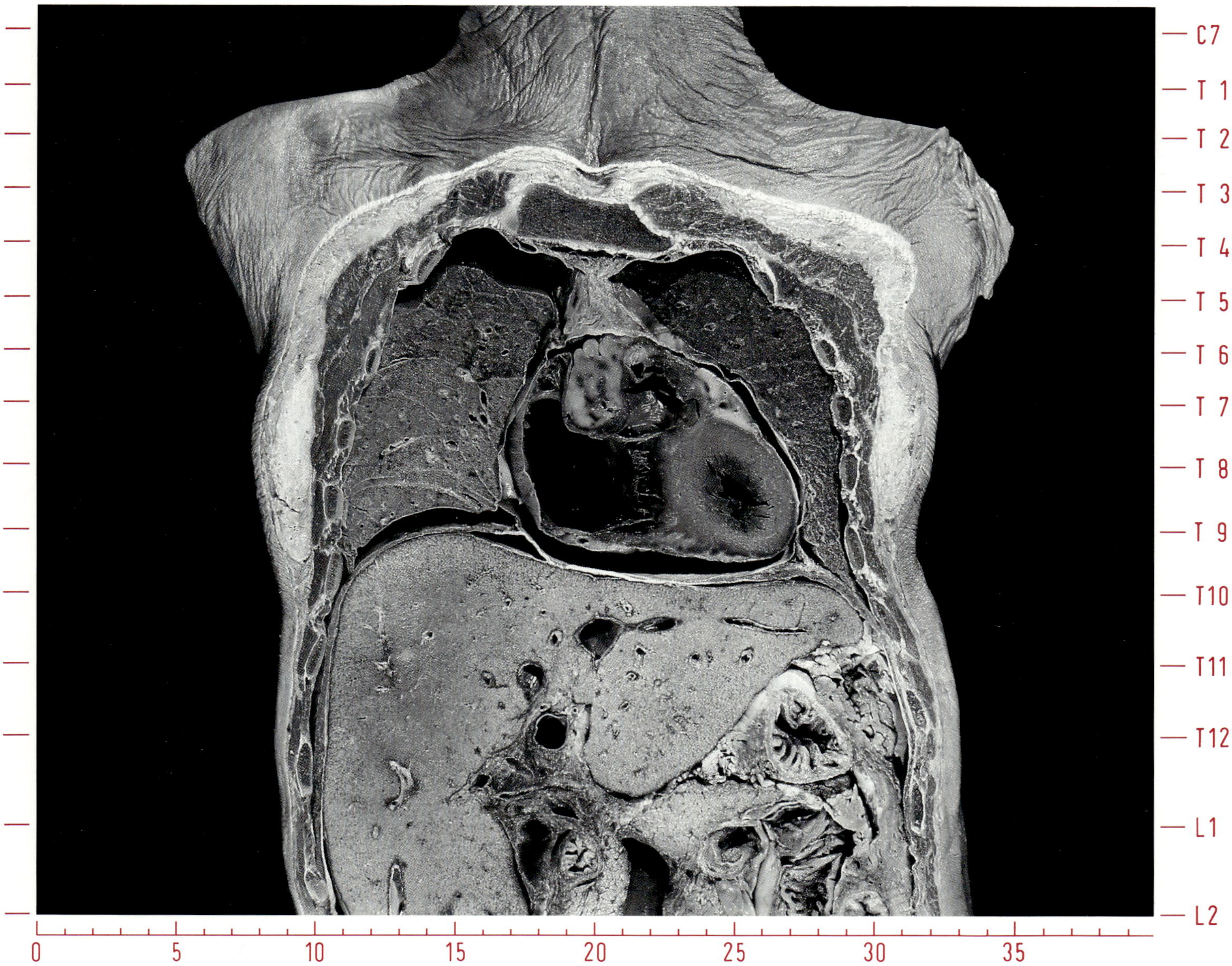

Fig. FT 5

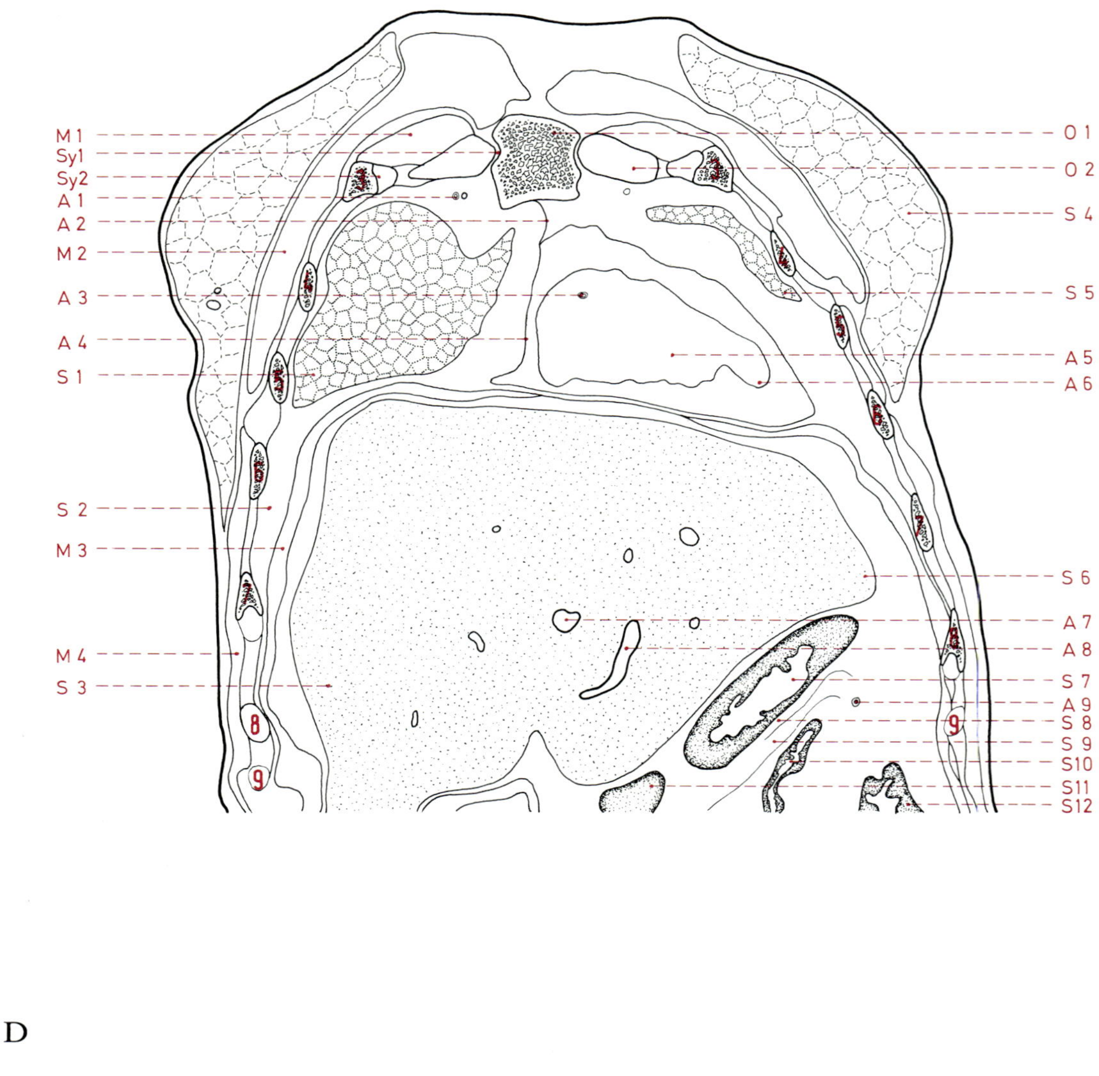
M 1
Sy1
Sy2
A 1
A 2
M 2
A 3
A 4
S 1
S 2
M 3
M 4
S 3
O 1
O 2
S 4
S 5
A 5
A 6
S 6
A 7
A 8
S 7
A 9
S 8
S 9
S10
S11
S12
8
9
D
Si

Fig. FT 6

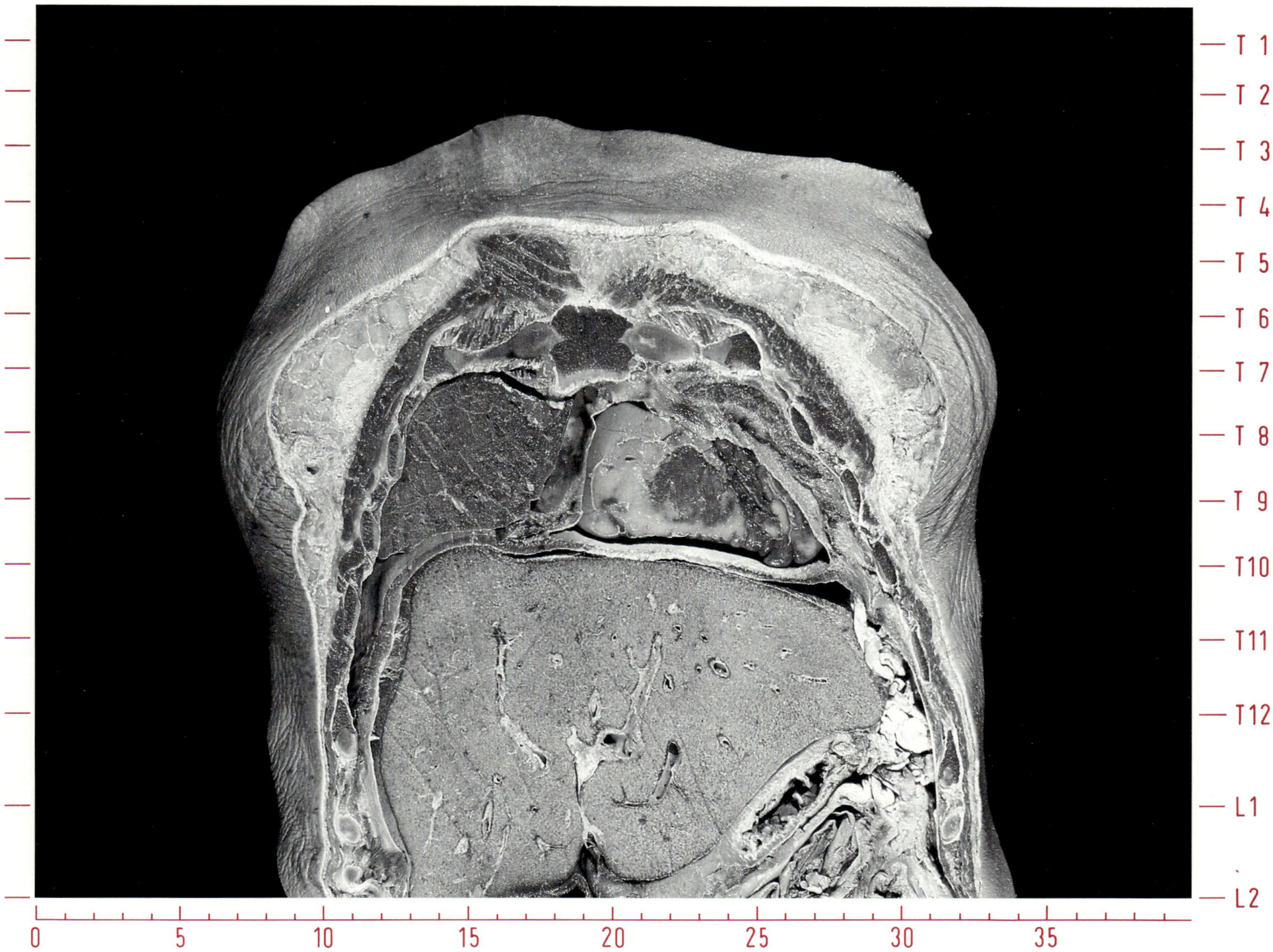

Fig. FT 6

ST

♂

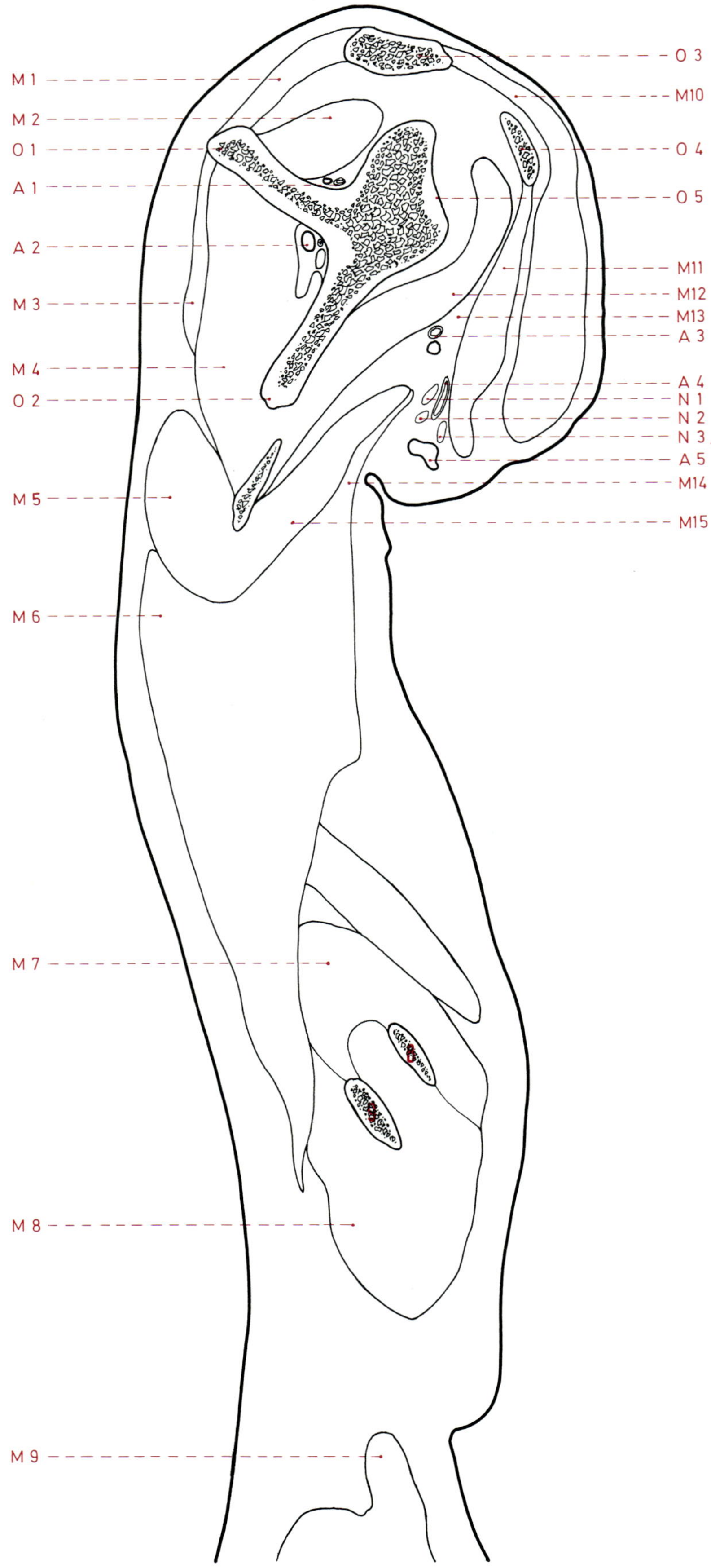

M 1
M 2
O 1
A 1
A 2
M 3
M 4
O 2
M 5
M 6
M 7
M 8
M 9
O 3
M10
O 4
O 5
M11
M12
M13
A 3
A 4
N 1
N 2
N 3
A 5
M14
M15

Fig. ST♂ 7D

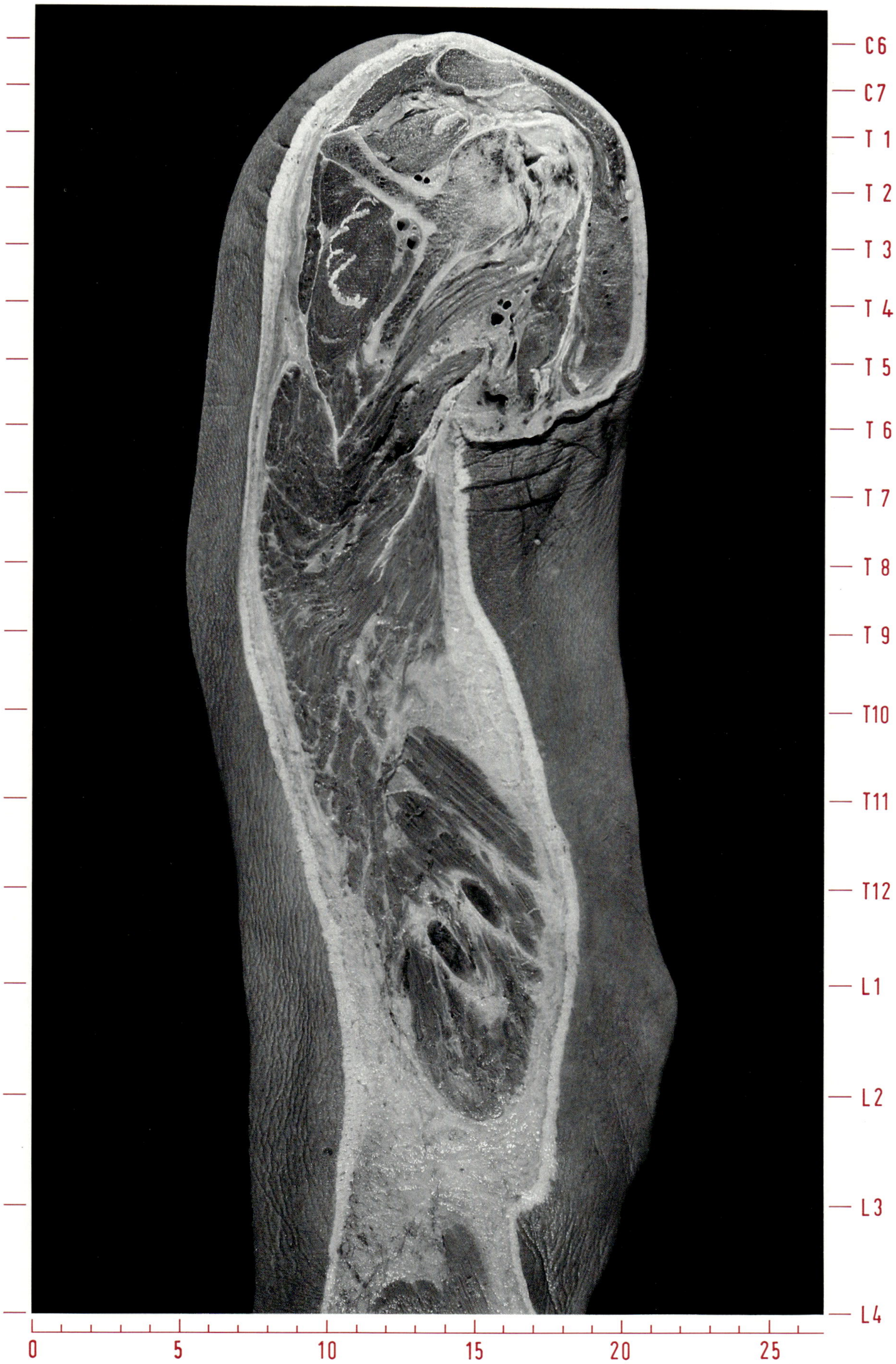

Fig. ST♂ 7D

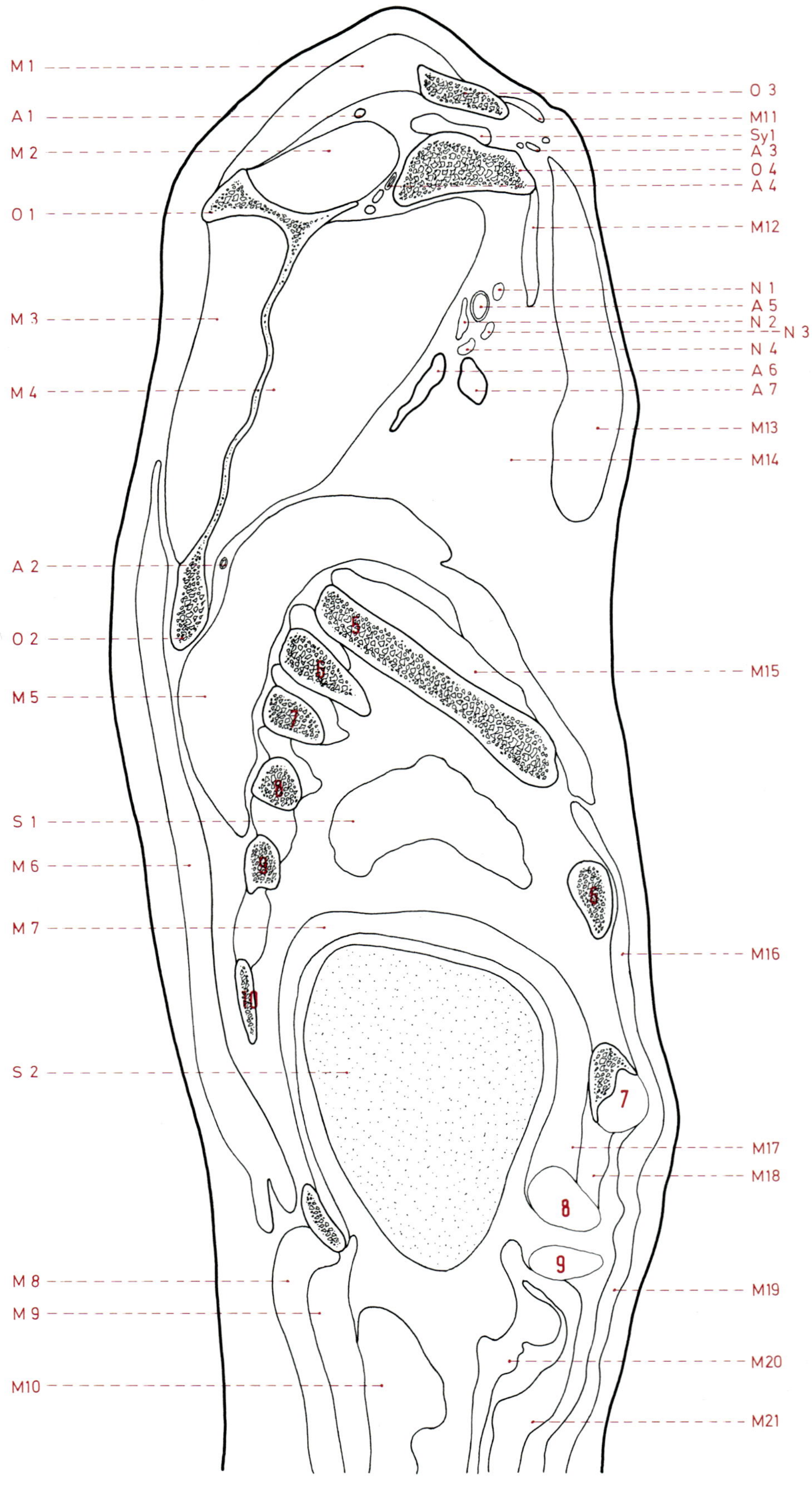

M 1
A 1
M 2
O 1
M 3
M 4
A 2
O 2
M 5
S 1
M 6
M 7
S 2
M 8
M 9
M10
O 3
M11
Sy1
A 3
O 4
A 4
M12
N 1
A 5
N 2
N 3
N 4
A 6
A 7
M13
M14
M15
M16
M17
M18
M19
M20
M21
5
6
7
8
9
10
6
7
8
9

Fig. ST♂ 6D

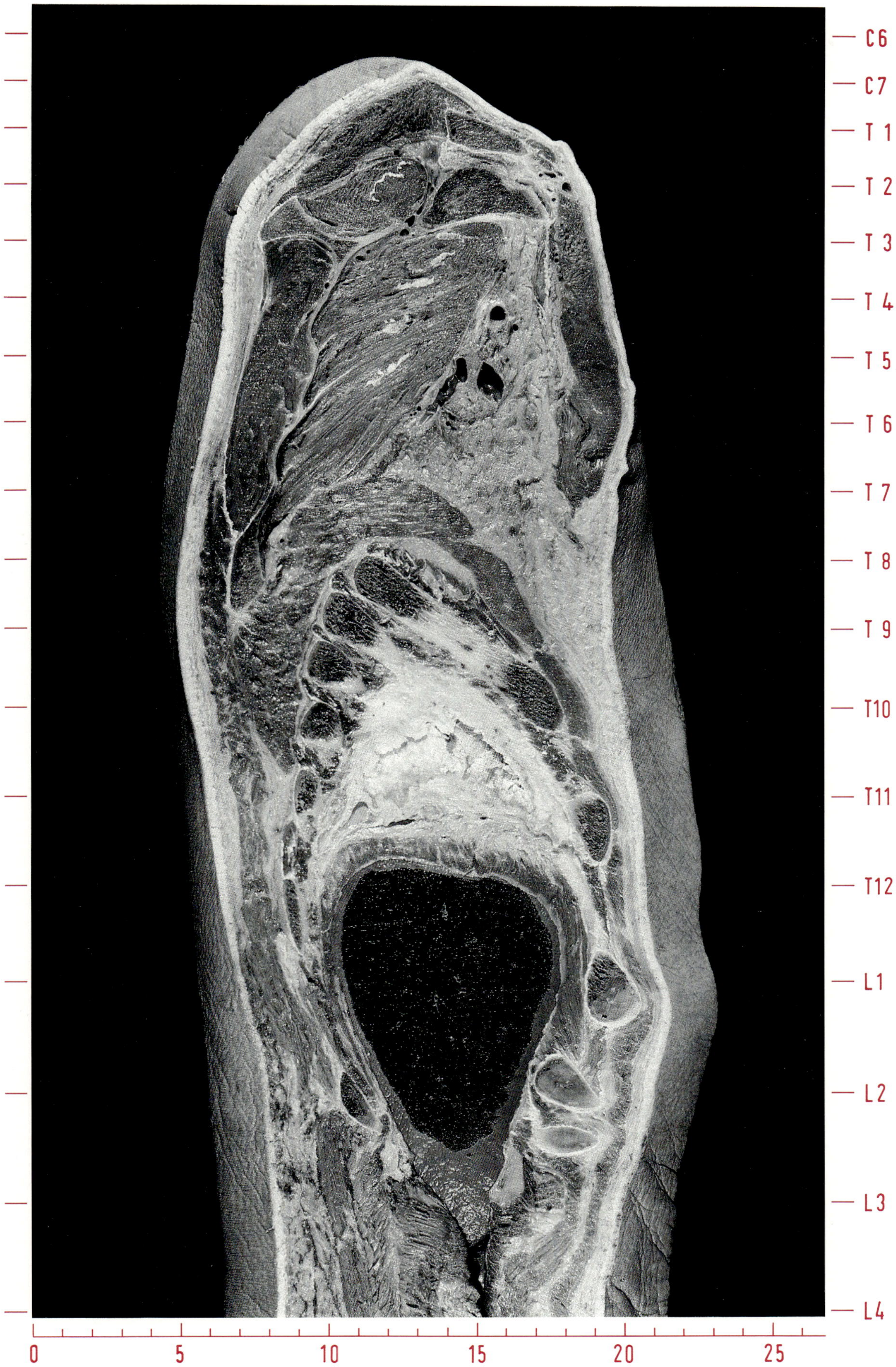

Fig. ST♂ 6D

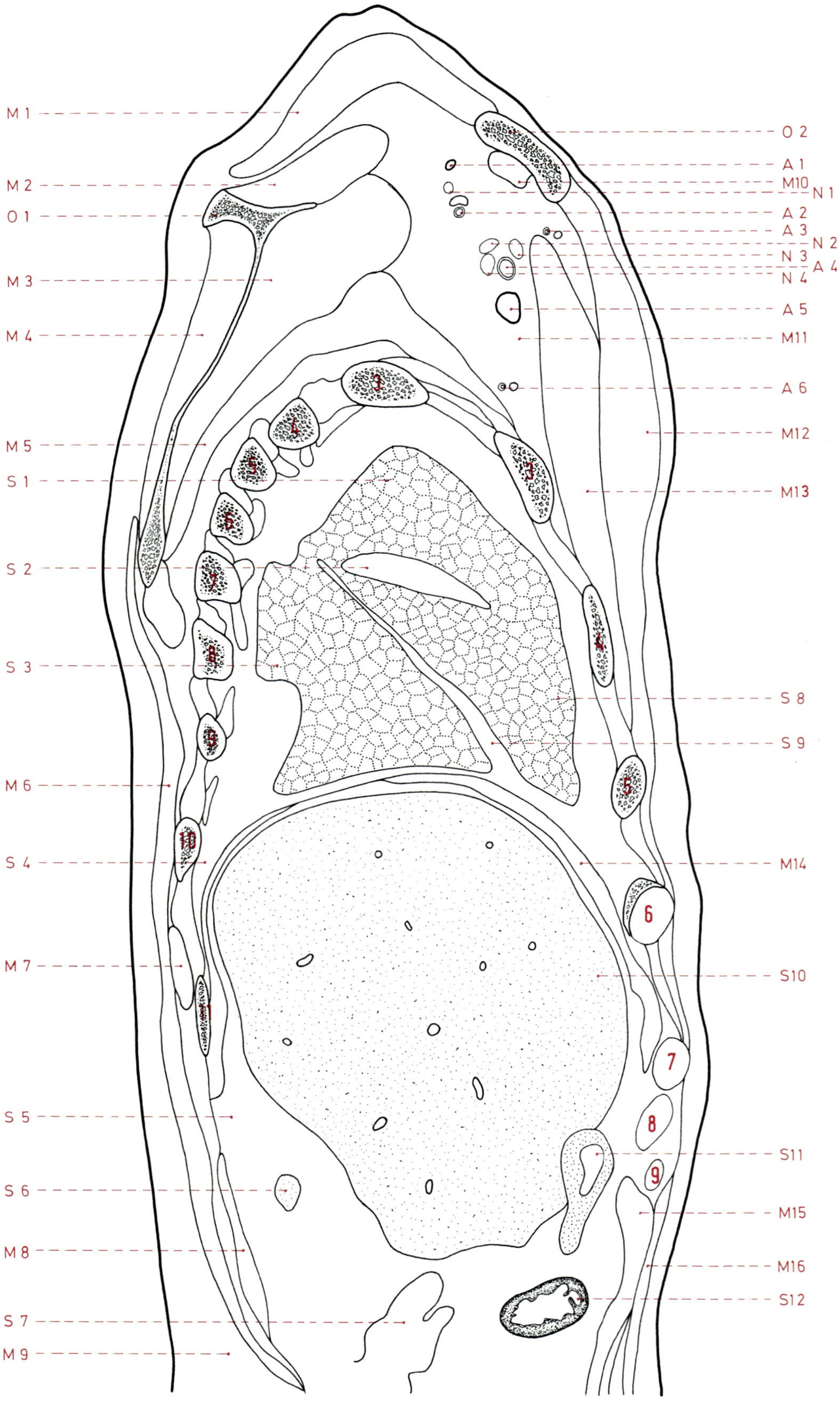

M 1
M 2
O 1
M 3
M 4
M 5
S 1
S 2
S 3
M 6
S 4
M 7
S 5
S 6
M 8
S 7
M 9
O 2
A 1
M10
N 1
A 2
A 3
N 2
N 3
A 4
N 4
A 5
M11
A 6
M12
M13
S 8
S 9
M14
S10
S11
M15
M16
S12

Fig. ST♂ 5D

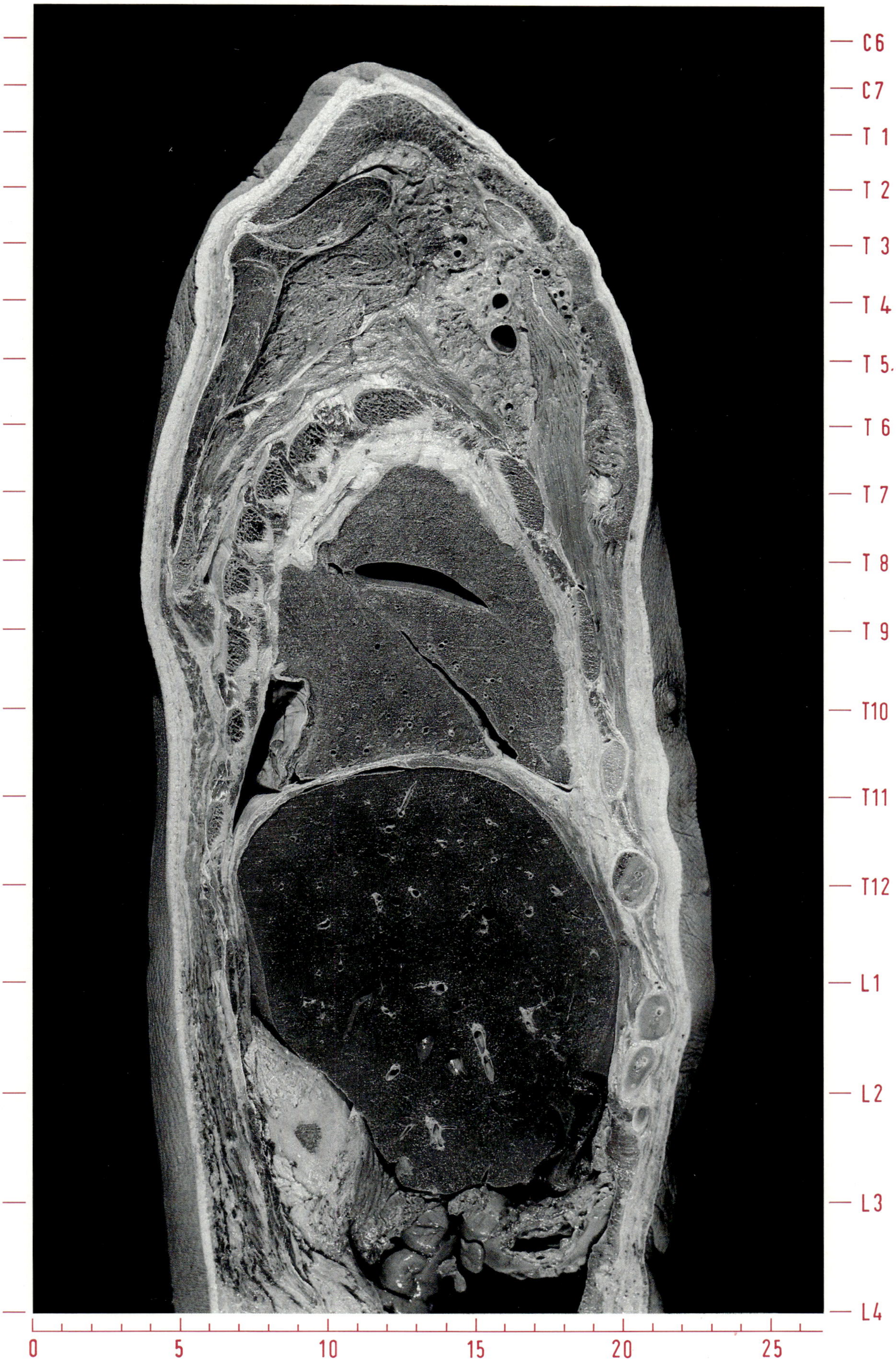

Fig. ST♂ 5D

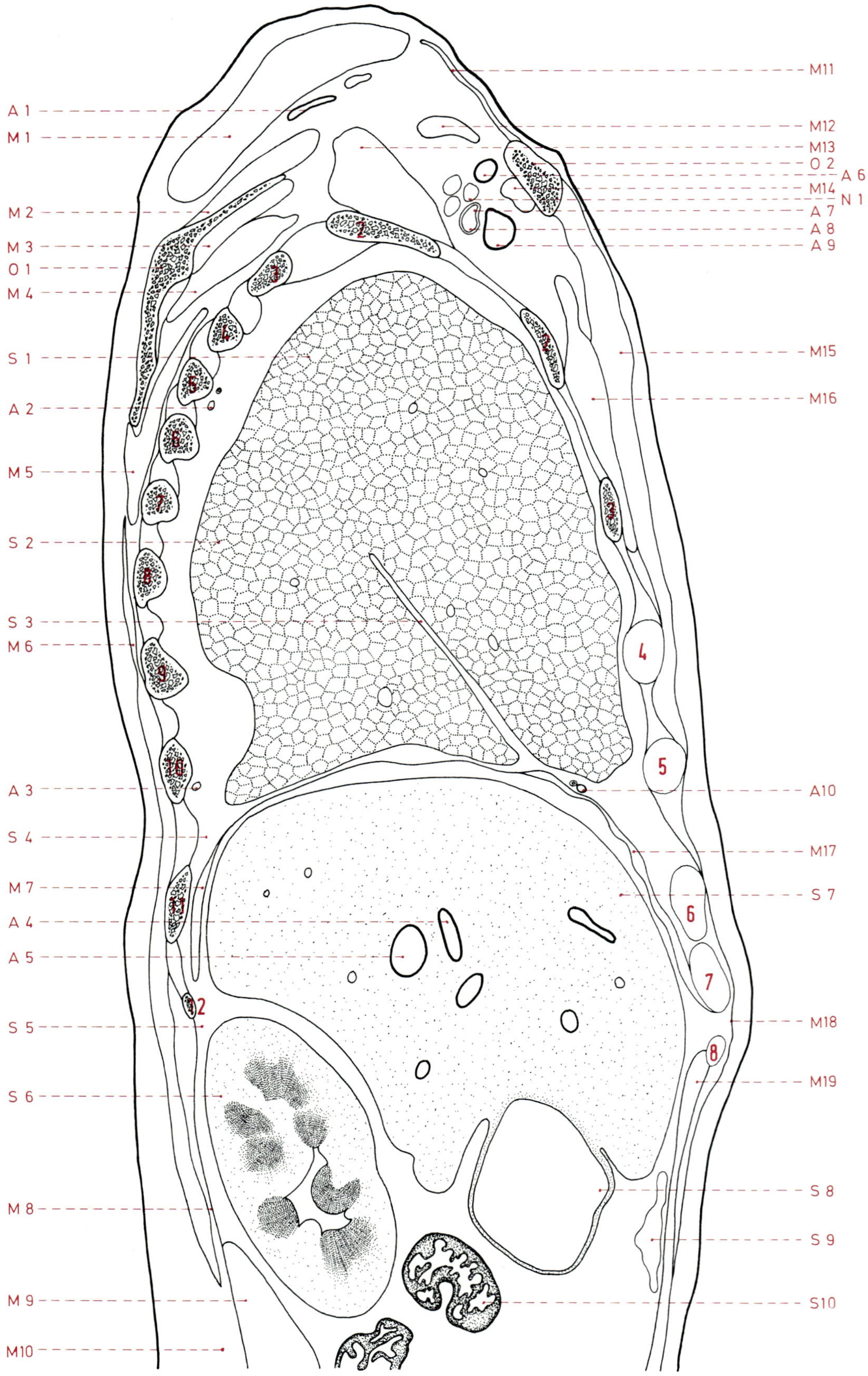
A 1
M 1
M 2
M 3
O 1
M 4
S 1
A 2
M 5
S 2
S 3
M 6
A 3
S 4
M 7
A 4
A 5
S 5
S 6
M 8
M 9
M10
M11
M12
M13
O 2
A 6
M14
N 1
A 7
A 8
A 9
M15
M16
A10
M17
S 7
M18
M19
S 8
S 9
S10

Fig. ST♂ 4D

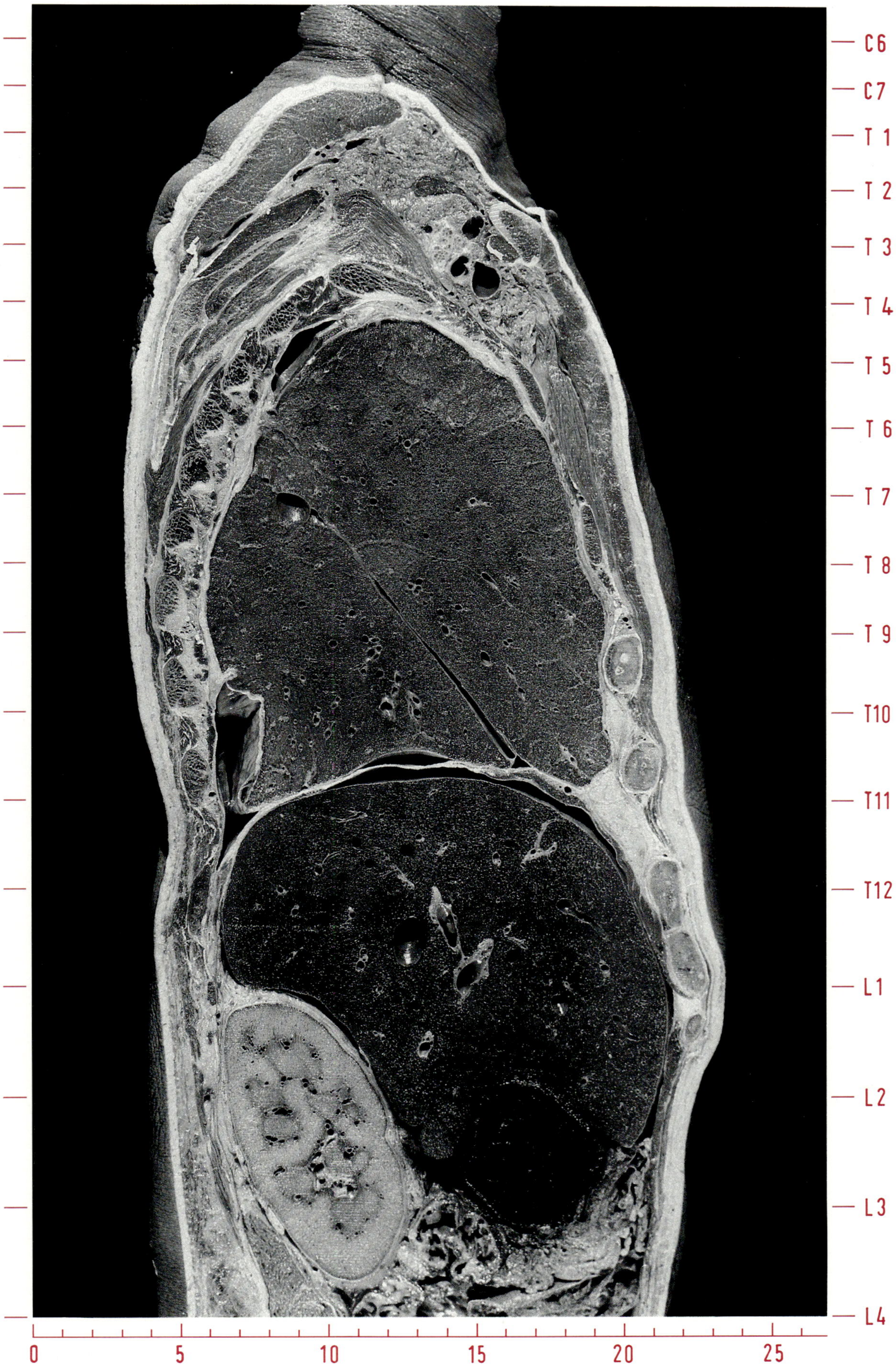

Fig. ST♂ 4D

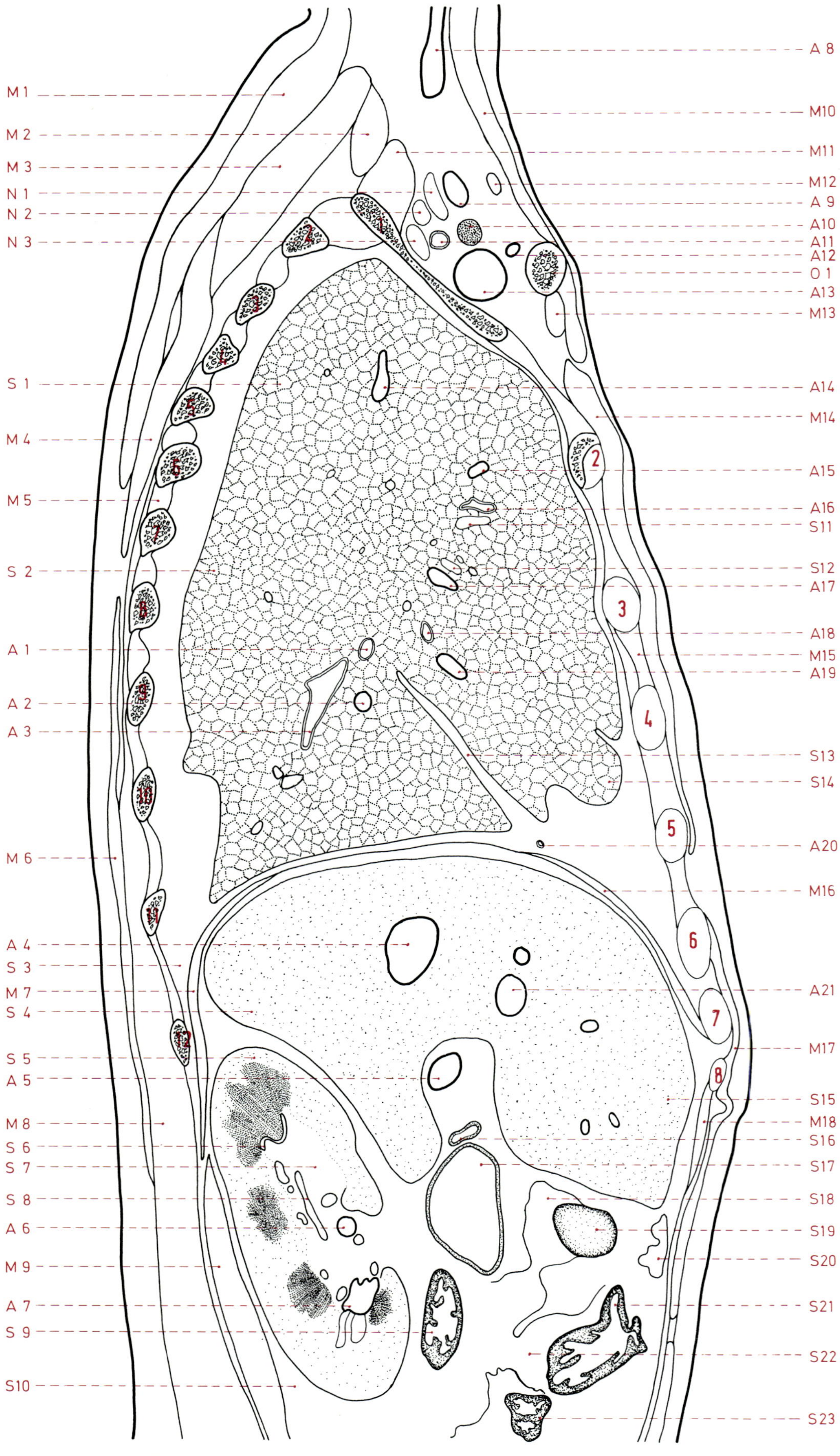

M 1
M 2
M 3
N 1
N 2
N 3
S 1
M 4
M 5
S 2
A 1
A 2
A 3
M 6
A 4
S 3
M 7
S 4
S 5
A 5
M 8
S 6
S 7
S 8
A 6
M 9
A 7
S 9
S10
A 8
M10
M11
M12
A 9
A10
A11
A12
O 1
A13
M13
A14
M14
A15
A16
S11
S12
A17
A18
M15
A19
S13
S14
A20
M16
A21
M17
S15
M18
S16
S17
S18
S19
S20
S21
S22
S23

Fig. ST♂ 3D

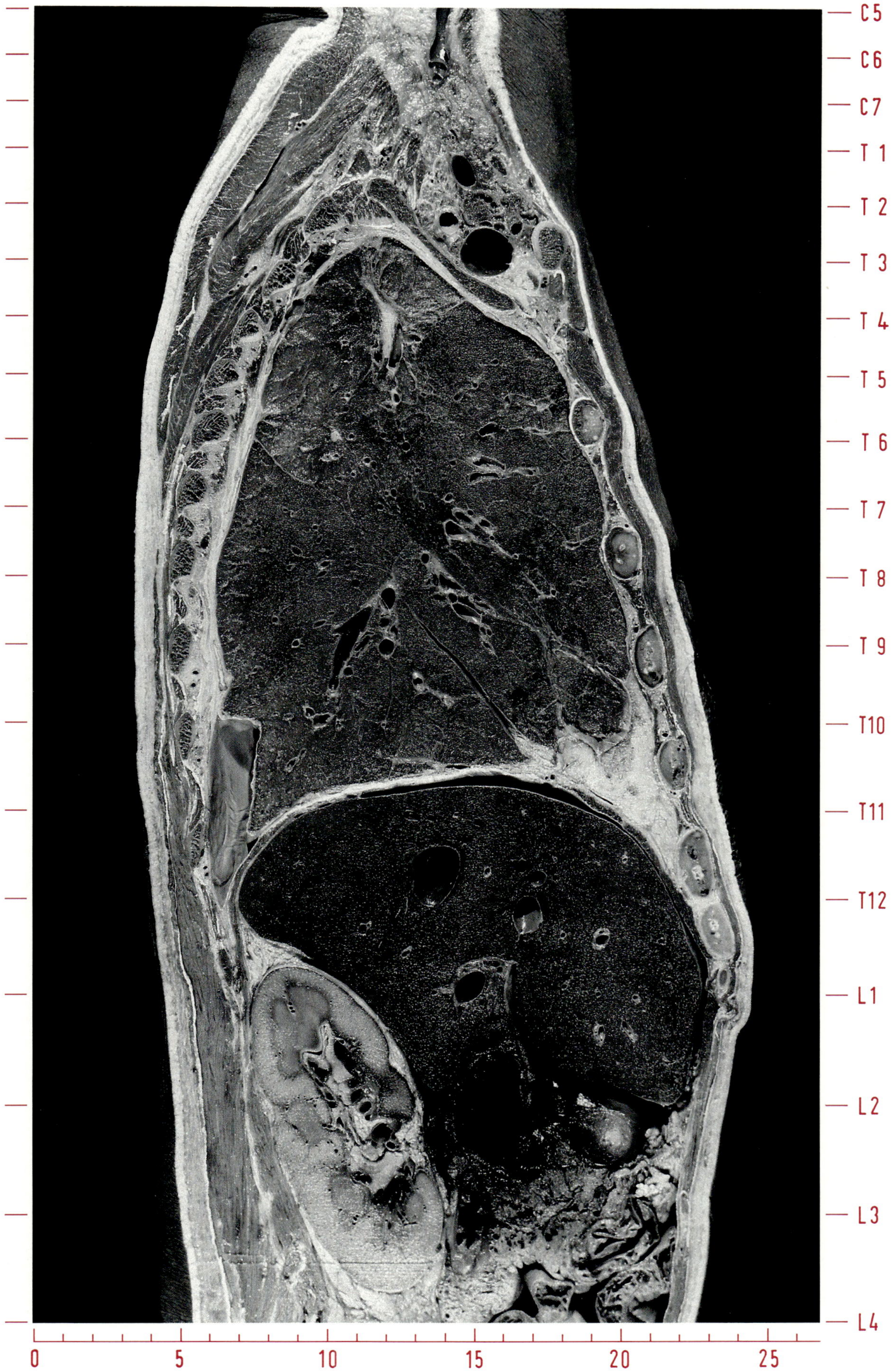
C5
C6
C7
T 1
T 2
T 3
T 4
T 5
T 6
T 7
T 8
T 9
T10
T11
T12
L1
L2
L3
L4
0
5
10
15
20
25

Fig. ST♂ 3D

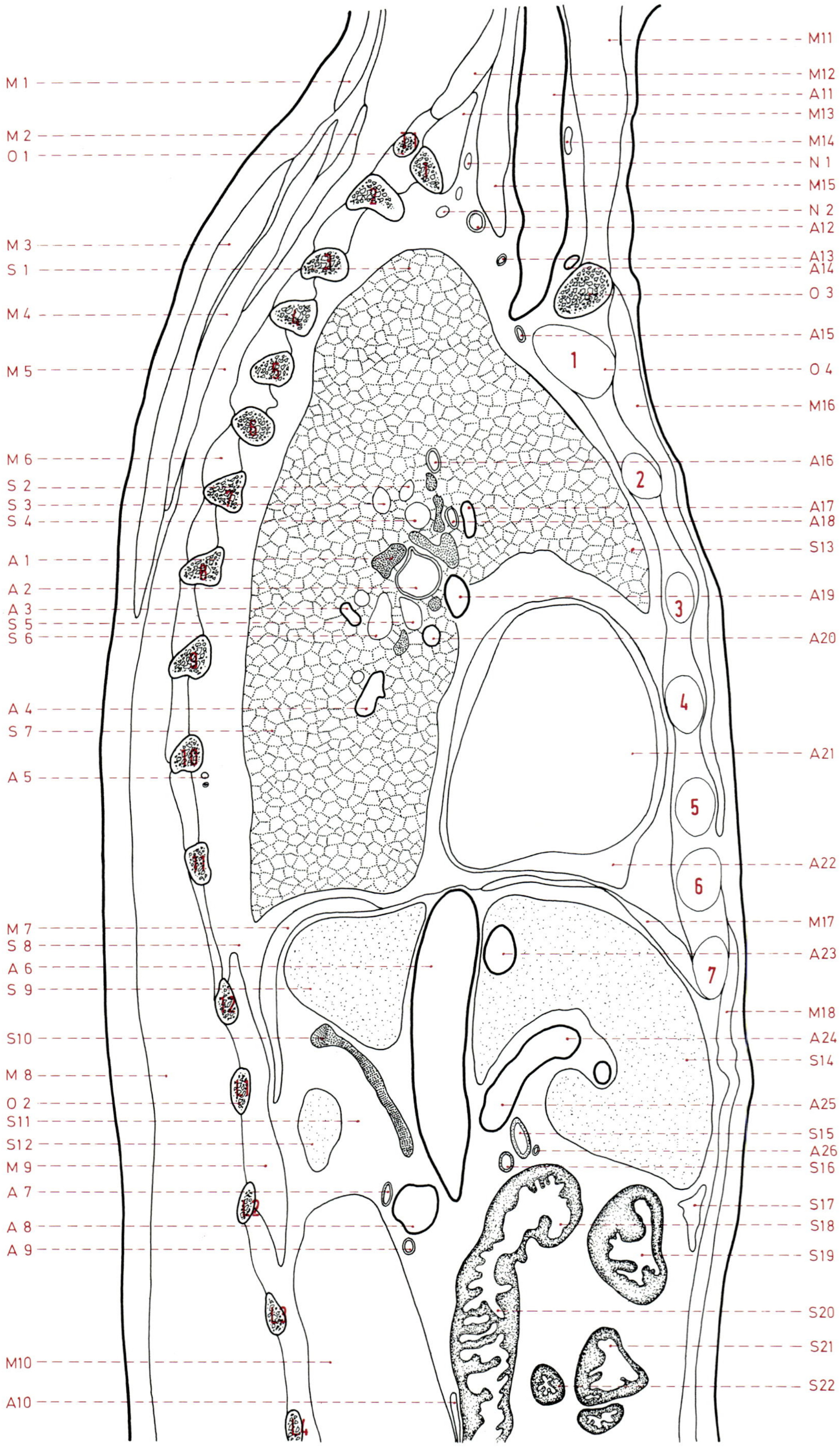
M 1
M 2
O 1
M 3
S 1
M 4
M 5
M 6
S 2
S 3
S 4
A 1
A 2
A 3
S 5
S 6
A 4
S 7
A 5
M 7
S 8
A 6
S 9
S10
M 8
O 2
S11
S12
M 9
A 7
A 8
A 9
M10
A10
M11
M12
A11
M13
M14
N 1
M15
N 2
A12
A13
A14
O 3
A15
O 4
M16
A16
A17
A18
S13
A19
A20
A21
A22
M17
A23
M18
A24
S14
A25
S15
A26
S16
S17
S18
S19
S20
S21
S22

Fig. ST♂ 2D

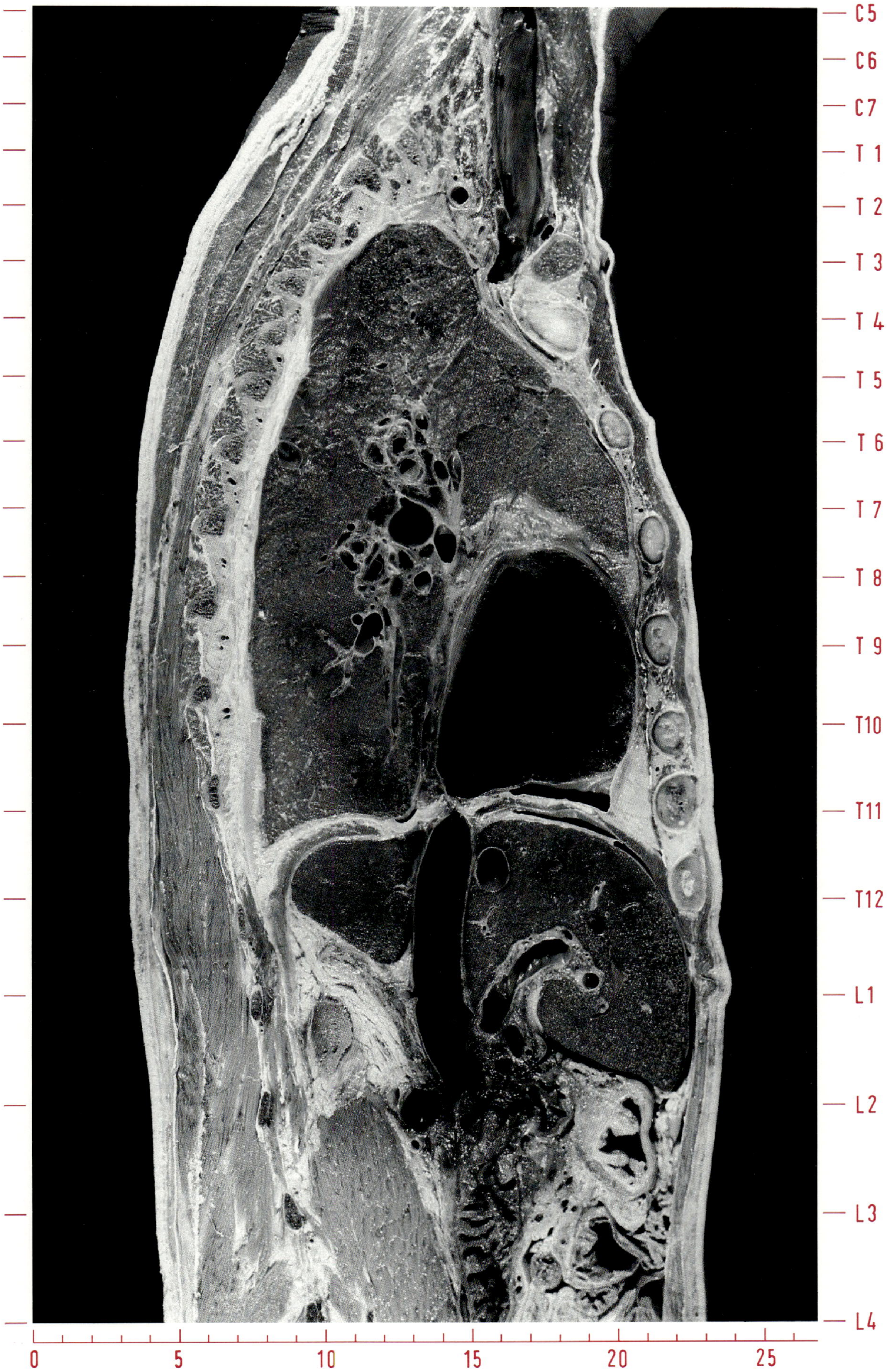
C5
C6
C7
T 1
T 2
T 3
T 4
T 5
T 6
T 7
T 8
T 9
T10
T11
T12
L1
L2
L3
L4
0
5
10
15
20
25

Fig. ST♂ 2D

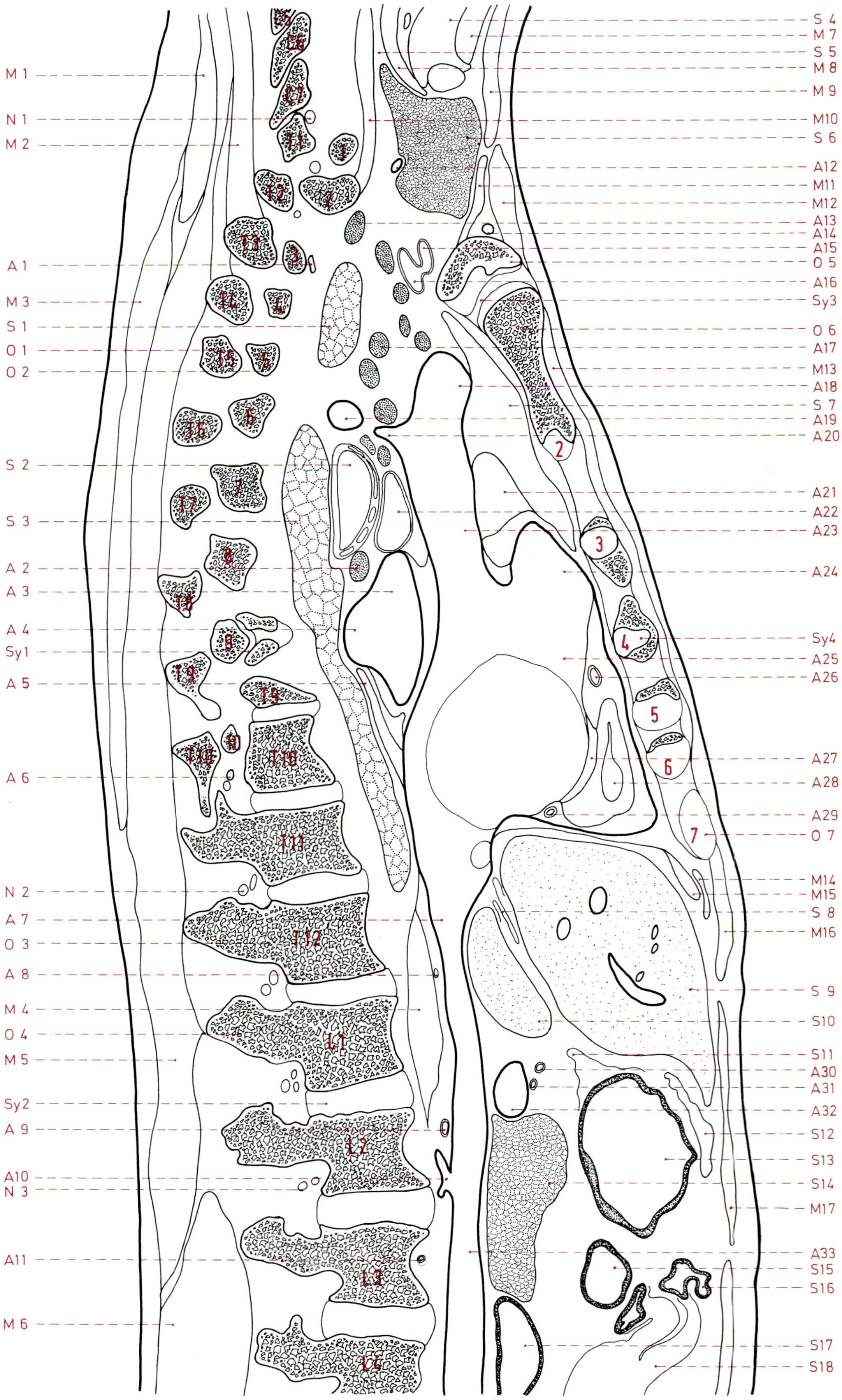
M 1
N 1
M 2
A 1
M 3
S 1
O 1
O 2
S 2
S 3
A 2
A 3
A 4
Sy1
A 5
A 6
N 2
A 7
O 3
A 8
M 4
O 4
M 5
Sy2
A 9
A10
N 3
A11
M 6
S 4
M 7
S 5
M 8
M 9
M10
S 6
A12
M11
M12
A13
A14
A15
O 5
A16
Sy3
O 6
A17
M13
A18
S 7
A19
A20
A21
A22
A23
A24
Sy4
A25
A26
A27
A28
A29
O 7
M14
M15
S 8
M16
S 9
S10
S11
A30
A31
A32
S12
S13
S14
M17
A33
S15
S16
S17
S18

Fig. ST♂ 1D

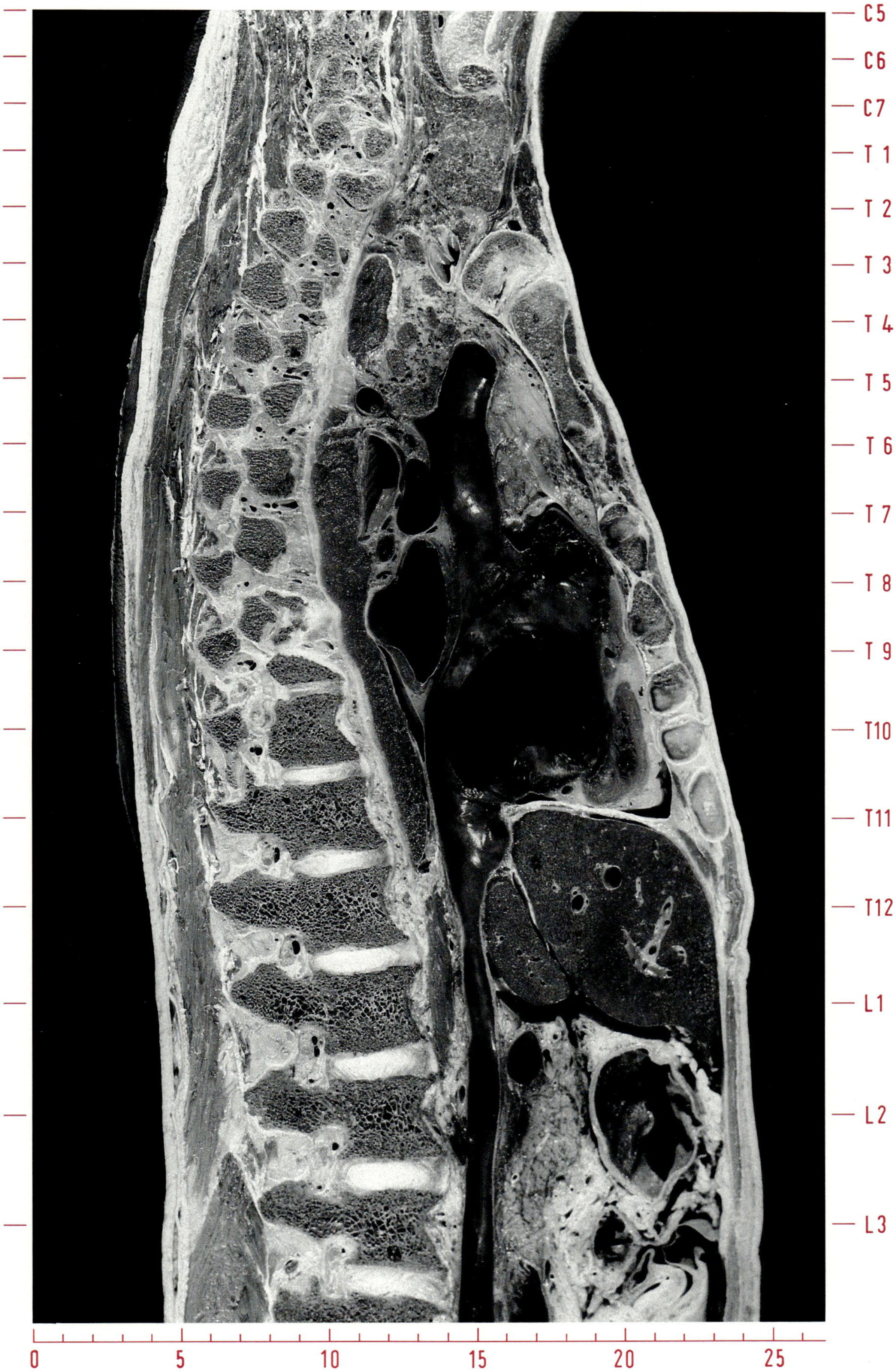

Fig. ST♂ 1D

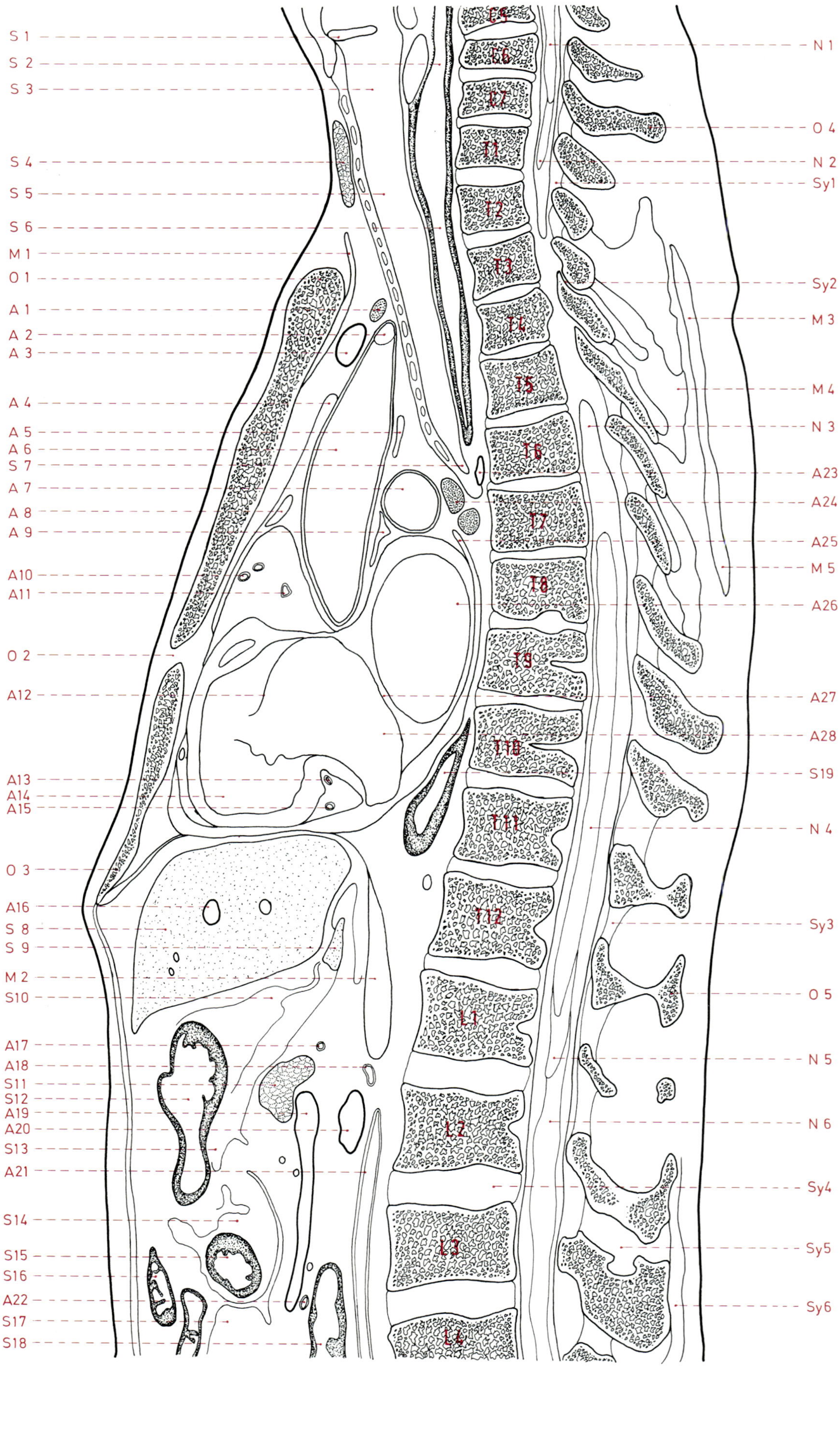
S 1
S 2
S 3
S 4
S 5
S 6
M 1
O 1
A 1
A 2
A 3
A 4
A 5
A 6
S 7
A 7
A 8
A 9
A10
A11
O 2
A12
A13
A14
A15
O 3
A16
S 8
S 9
M 2
S10
A17
A18
S11
S12
A19
A20
S13
A21
S14
S15
S16
A22
S17
S18
N 1
O 4
N 2
Sy1
Sy2
M 3
M 4
N 3
A23
A24
A25
M 5
A26
A27
A28
S19
N 4
Sy3
O 5
N 5
N 6
Sy4
Sy5
Sy6
C5
C6
C7
T1
T2
T3
T4
T5
T6
T7
T8
T9
T10
T11
T12
L1
L2
L3
L4

Fig. ST♂ MD

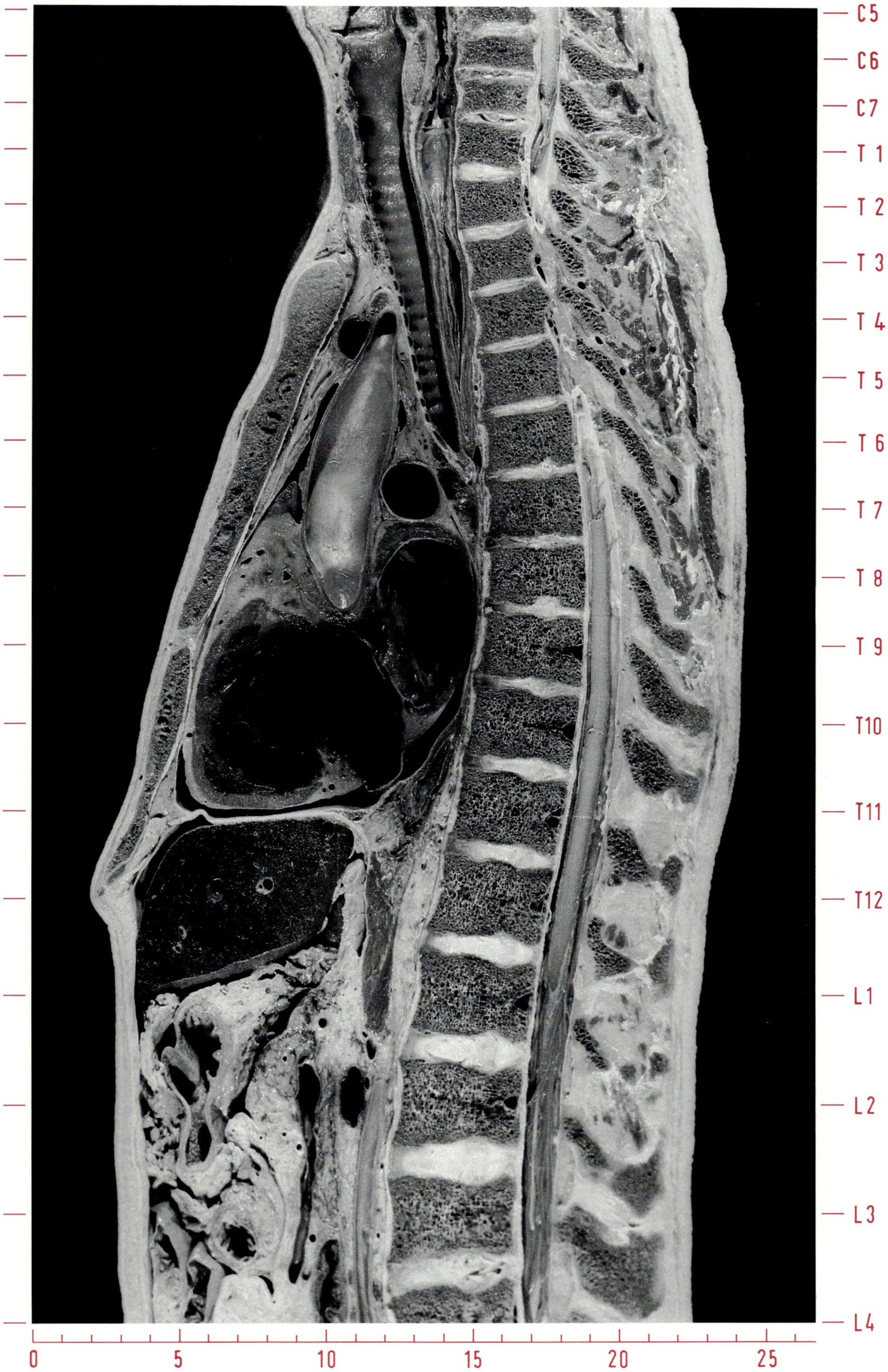
C5
C6
C7
T 1
T 2
T 3
T 4
T 5
T 6
T 7
T 8
T 9
T10
T11
T12
L1
L2
L3
L4
0
5
10
15
20
25

Fig. ST♂ MD

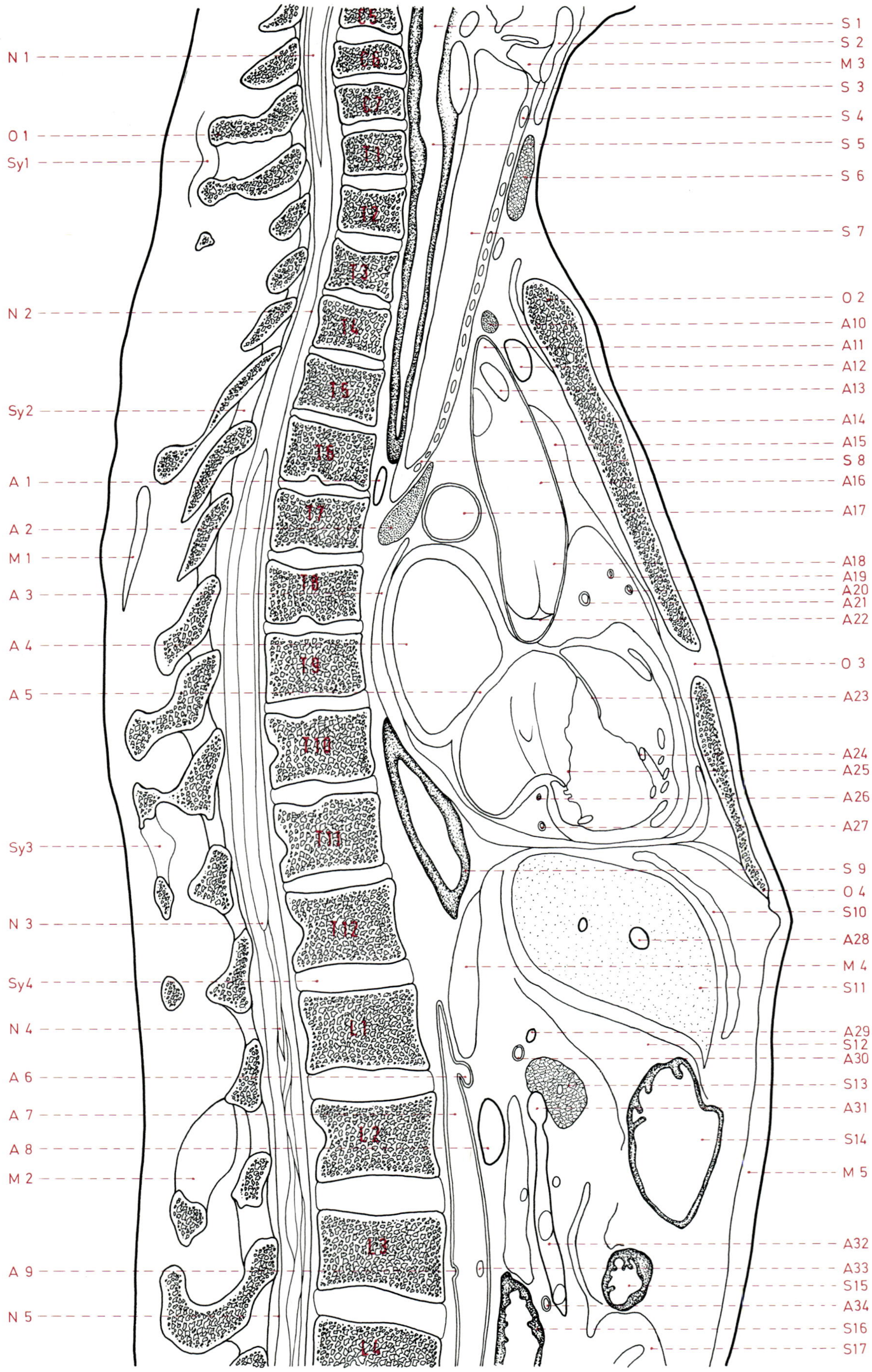
N 1
O 1
Sy1
N 2
Sy2
A 1
A 2
M 1
A 3
A 4
A 5
Sy3
N 3
Sy4
N 4
A 6
A 7
A 8
M 2
A 9
N 5
S 1
S 2
M 3
S 3
S 4
S 5
S 6
S 7
O 2
A10
A11
A12
A13
A14
A15
S 8
A16
A17
A18
A19
A20
A21
A22
O 3
A23
A24
A25
A26
A27
S 9
O 4
S10
A28
M 4
S11
A29
S12
A30
S13
A31
S14
M 5
A32
A33
S15
A34
S16
S17

Fig. ST♂ MSi

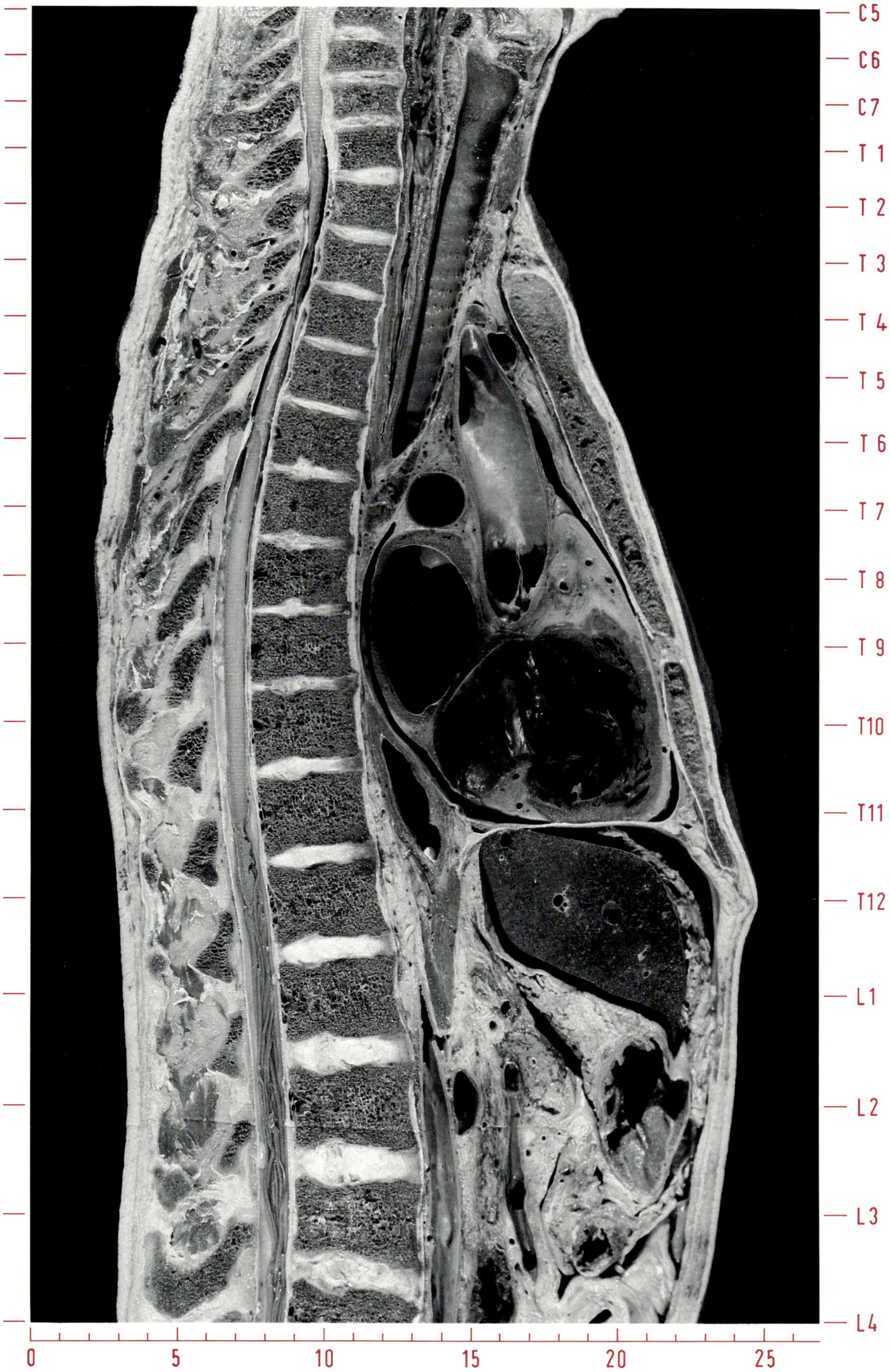
C5
C6
C7
T 1
T 2
T 3
T 4
T 5
T 6
T 7
T 8
T 9
T10
T11
T12
L1
L2
L3
L4
0
5
10
15
20
25

Fig. ST♂ MSi

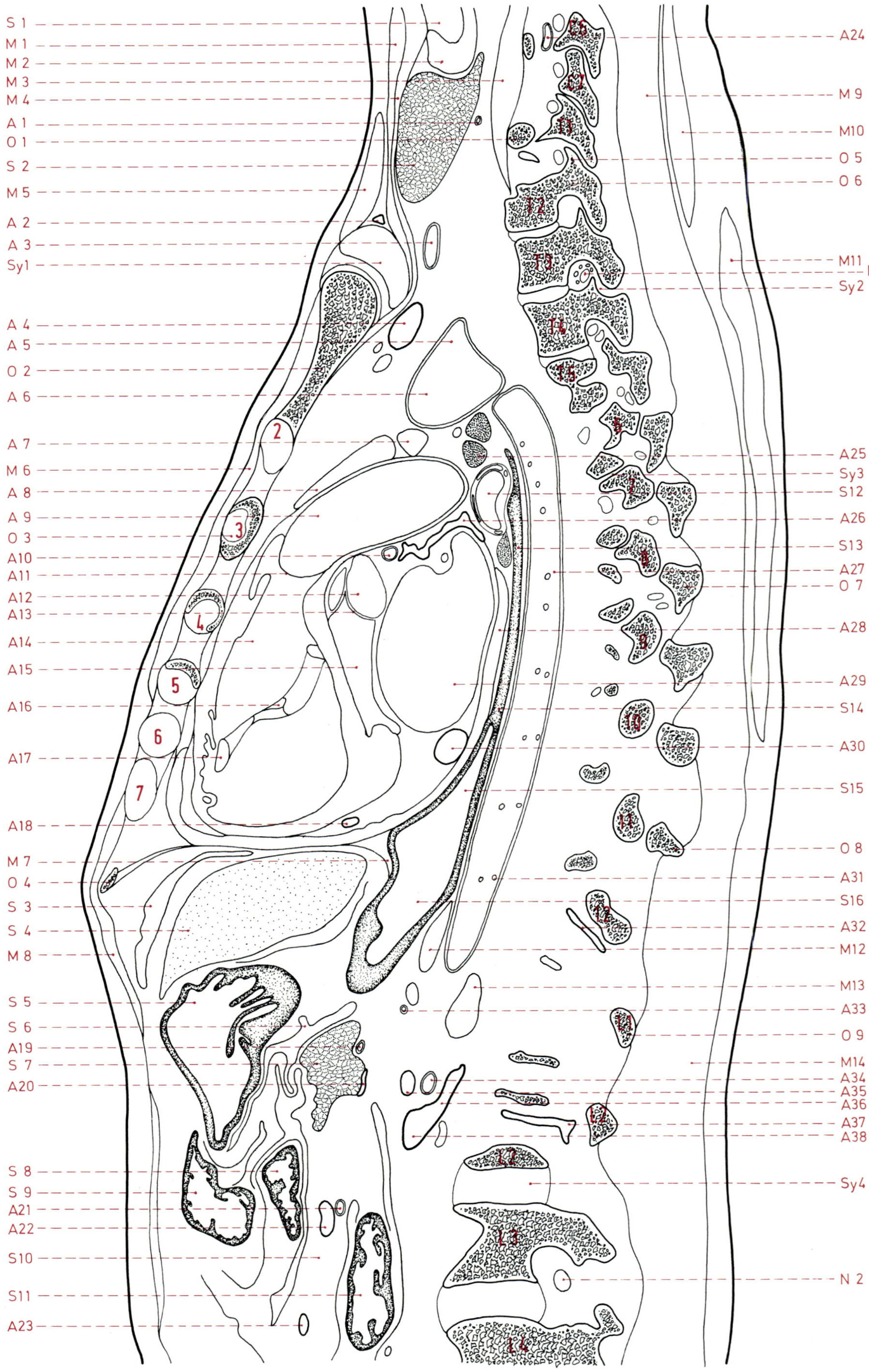
S 1
M 1
M 2
M 3
M 4
A 1
O 1
S 2
M 5
A 2
A 3
Sy1
A 4
A 5
O 2
A 6
A 7
M 6
A 8
A 9
O 3
A10
A11
A12
A13
A14
A15
A16
A17
A18
M 7
O 4
S 3
S 4
M 8
S 5
S 6
A19
S 7
A20
S 8
S 9
A21
A22
S10
S11
A23
2
3
4
5
6
7
A24
M 9
M10
O 5
O 6
M11
N1
Sy2
A25
Sy3
S12
A26
S13
A27
O 7
A28
A29
S14
A30
S15
O 8
A31
S16
A32
M12
M13
A33
O 9
M14
A34
A35
A36
A37
A38
Sy4
N 2

Fig. ST♂ 1 Si

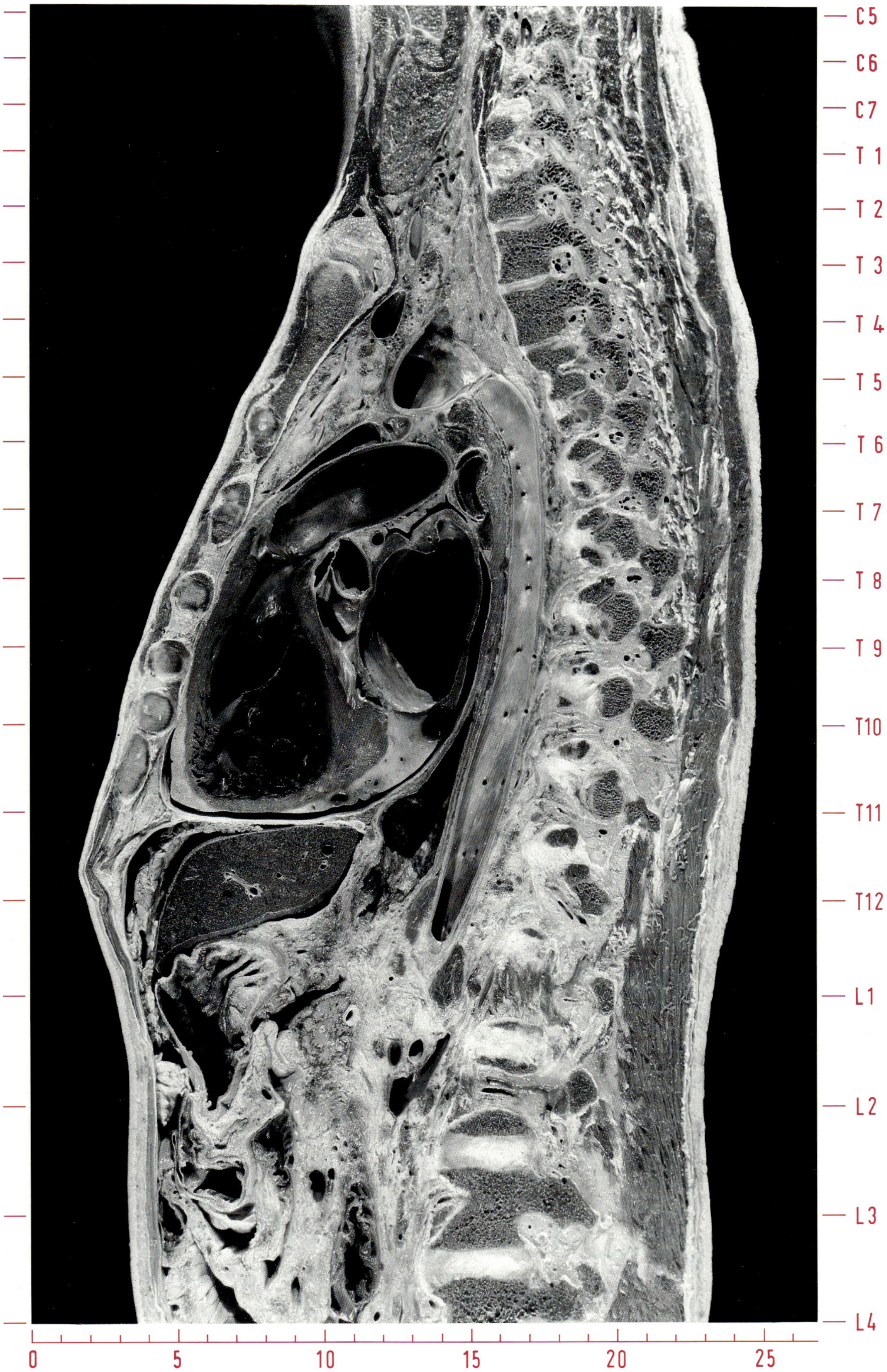
C5
C6
C7
T 1
T 2
T 3
T 4
T 5
T 6
T 7
T 8
T 9
T10
T11
T12
L1
L2
L3
L4
0
5
10
15
20
25

Fig. ST♂ 1 Si

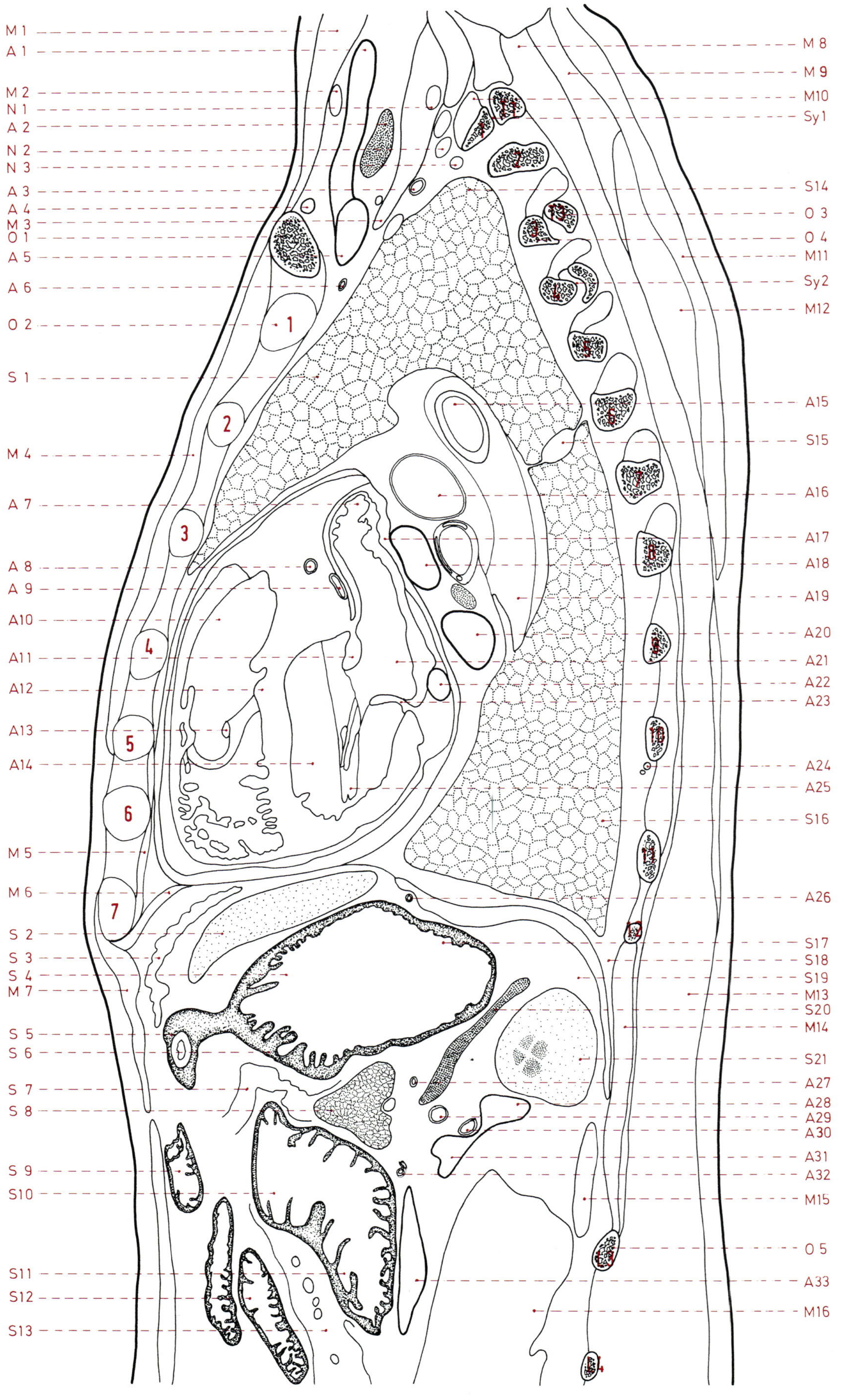

M 1
A 1
M 2
N 1
A 2
N 2
N 3
A 3
A 4
M 3
O 1
A 5
A 6
O 2
S 1
M 4
A 7
A 8
A 9
A10
A11
A12
A13
A14
M 5
M 6
S 2
S 3
S 4
M 7
S 5
S 6
S 7
S 8
S 9
S10
S11
S12
S13
M 8
M 9
M10
Sy1
S14
O 3
O 4
M11
Sy2
M12
A15
S15
A16
A17
A18
A19
A20
A21
A22
A23
A24
A25
S16
A26
S17
S18
S19
M13
S20
M14
S21
A27
A28
A29
A30
A31
A32
M15
O 5
A33
M16
1
2
3
4
5
6
7

Fig. ST♂ 2 Si

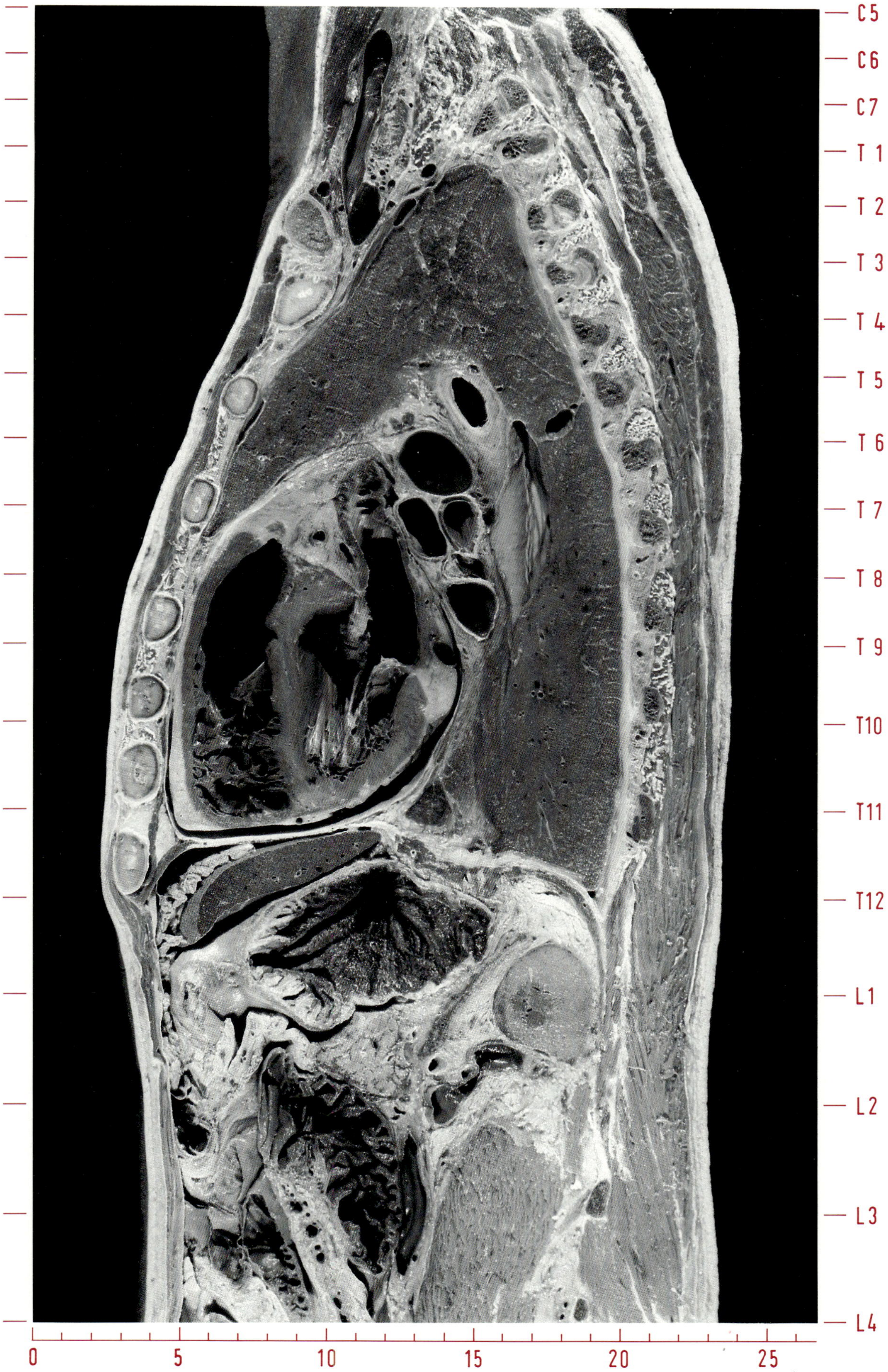
C5
C6
C7
T 1
T 2
T 3
T 4
T 5
T 6
T 7
T 8
T 9
T10
T11
T12
L1
L2
L3
L4
0
5
10
15
20
25

Fig. ST♂ 2 Si

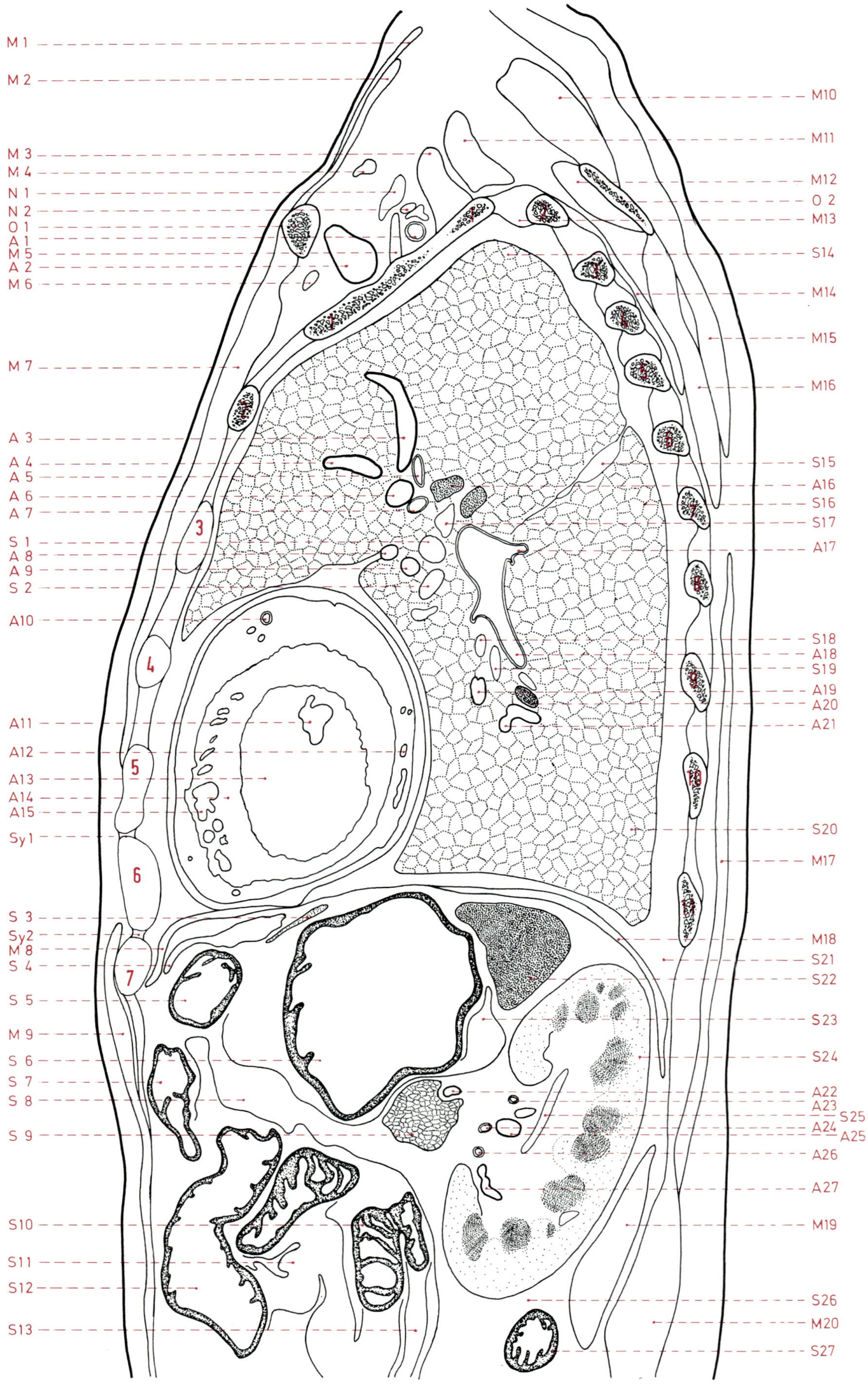
M 1
M 2
M 3
M 4
N 1
N 2
O 1
A 1
M 5
A 2
M 6
M 7
A 3
A 4
A 5
A 6
A 7
S 1
A 8
A 9
S 2
A10
A11
A12
A13
A14
A15
Sy1
S 3
Sy2
M 8
S 4
S 5
M 9
S 6
S 7
S 8
S 9
S10
S11
S12
S13
M10
M11
M12
O 2
M13
S14
M14
M15
M16
S15
A16
S16
S17
A17
S18
A18
S19
A19
A20
A21
S20
M17
M18
S21
S22
S23
S24
A22
A23
S25
A24
A25
A26
A27
M19
S26
M20
S27
1
2
3
4
5
6
7
8
9
10
11

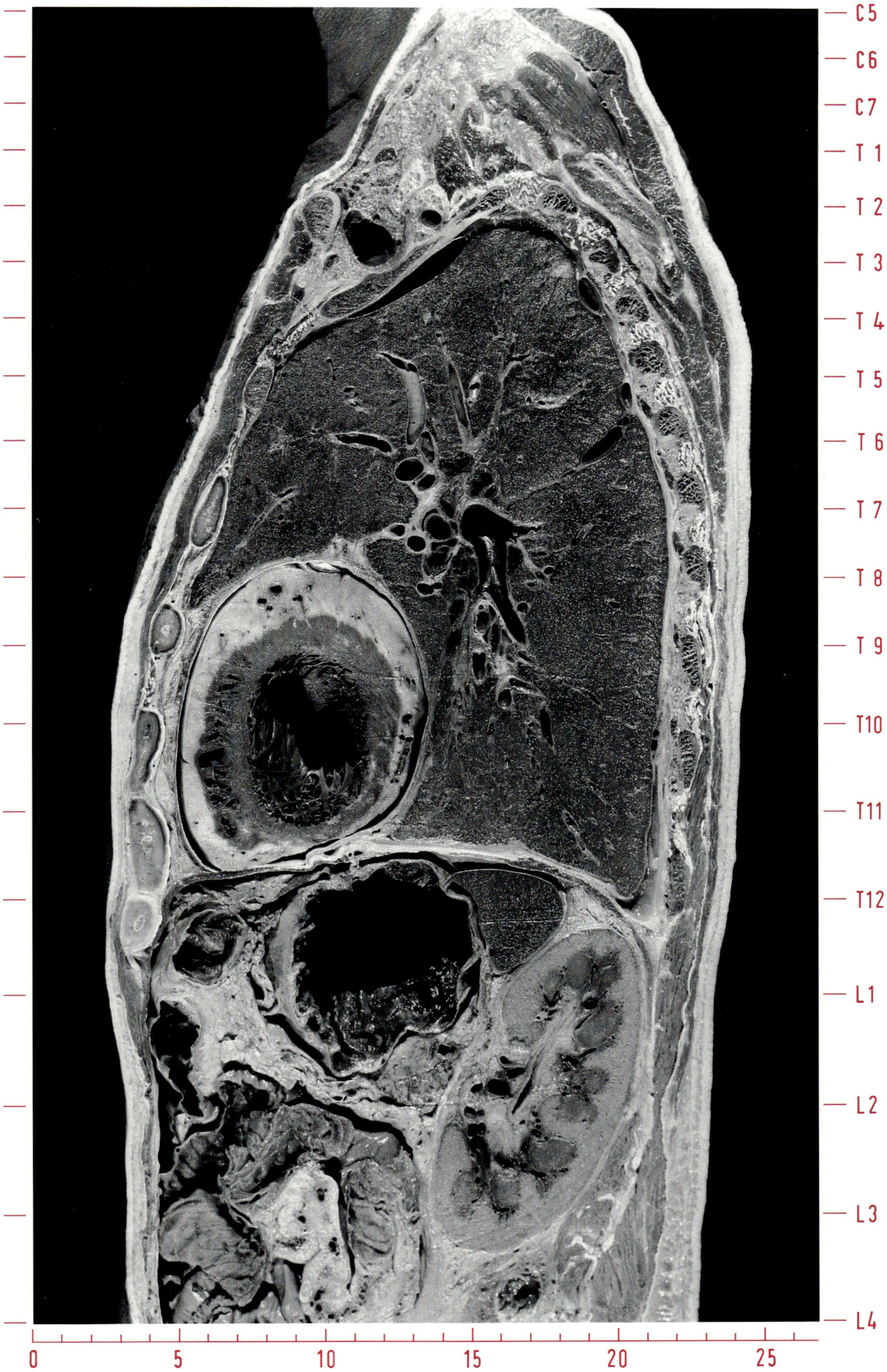
C5
C6
C7
T 1
T 2
T 3
T 4
T 5
T 6
T 7
T 8
T 9
T10
T11
T12
L1
L2
L3
L4
0
5
10
15
20
25

Fig. ST♂ 3 Si

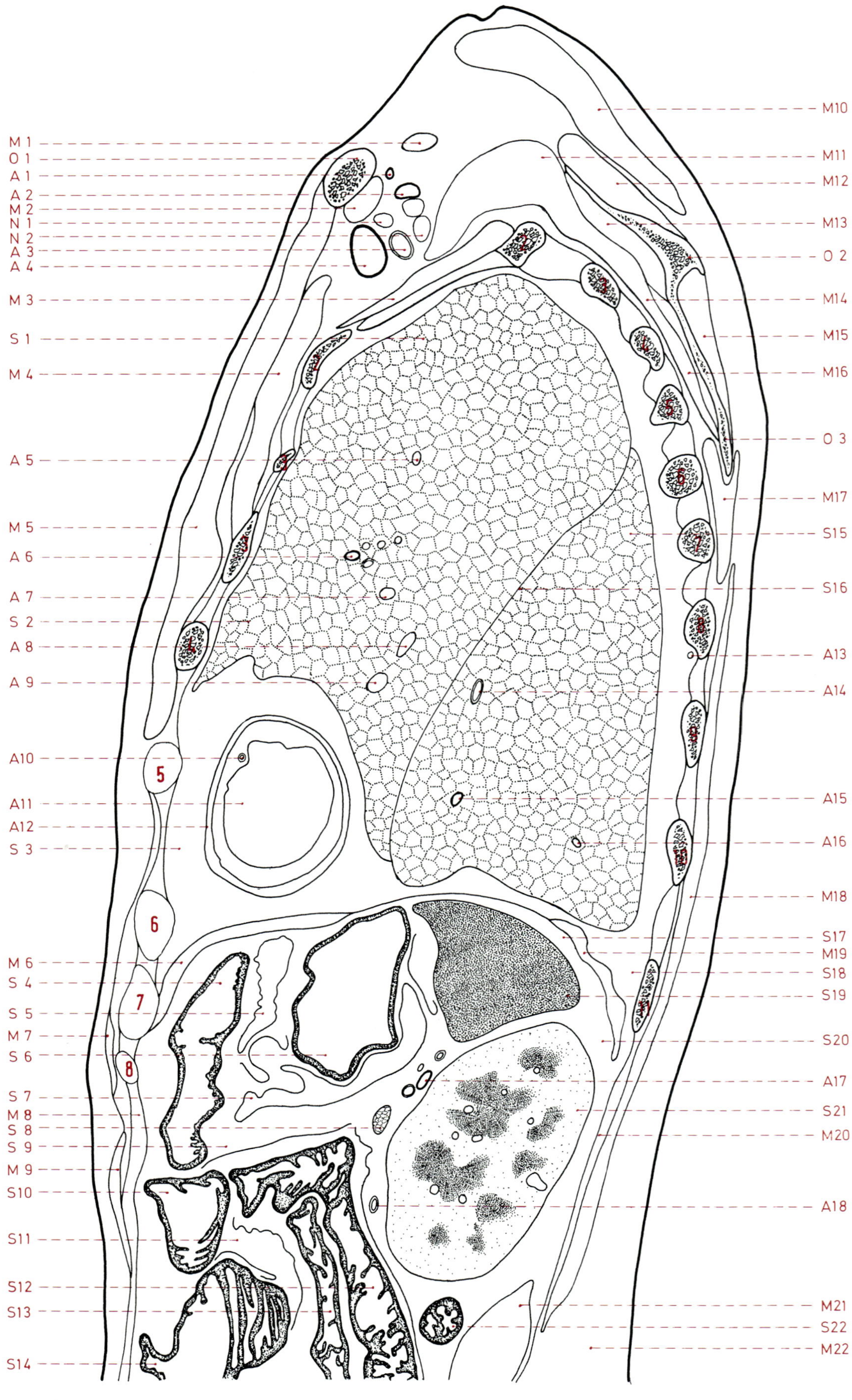
M 1
O 1
A 1
A 2
M 2
N 1
N 2
A 3
A 4
M 3
S 1
M 4
A 5
M 5
A 6
A 7
S 2
A 8
A 9
A10
A11
A12
S 3
M 6
S 4
S 5
M 7
S 6
S 7
M 8
S 8
S 9
M 9
S10
S11
S12
S13
S14
M10
M11
M12
M13
O 2
M14
M15
M16
O 3
M17
S15
S16
A13
A14
A15
A16
M18
S17
M19
S18
S19
S20
A17
S21
M20
A18
M21
S22
M22
5
6
7
8

Fig. ST♂ 4 Si

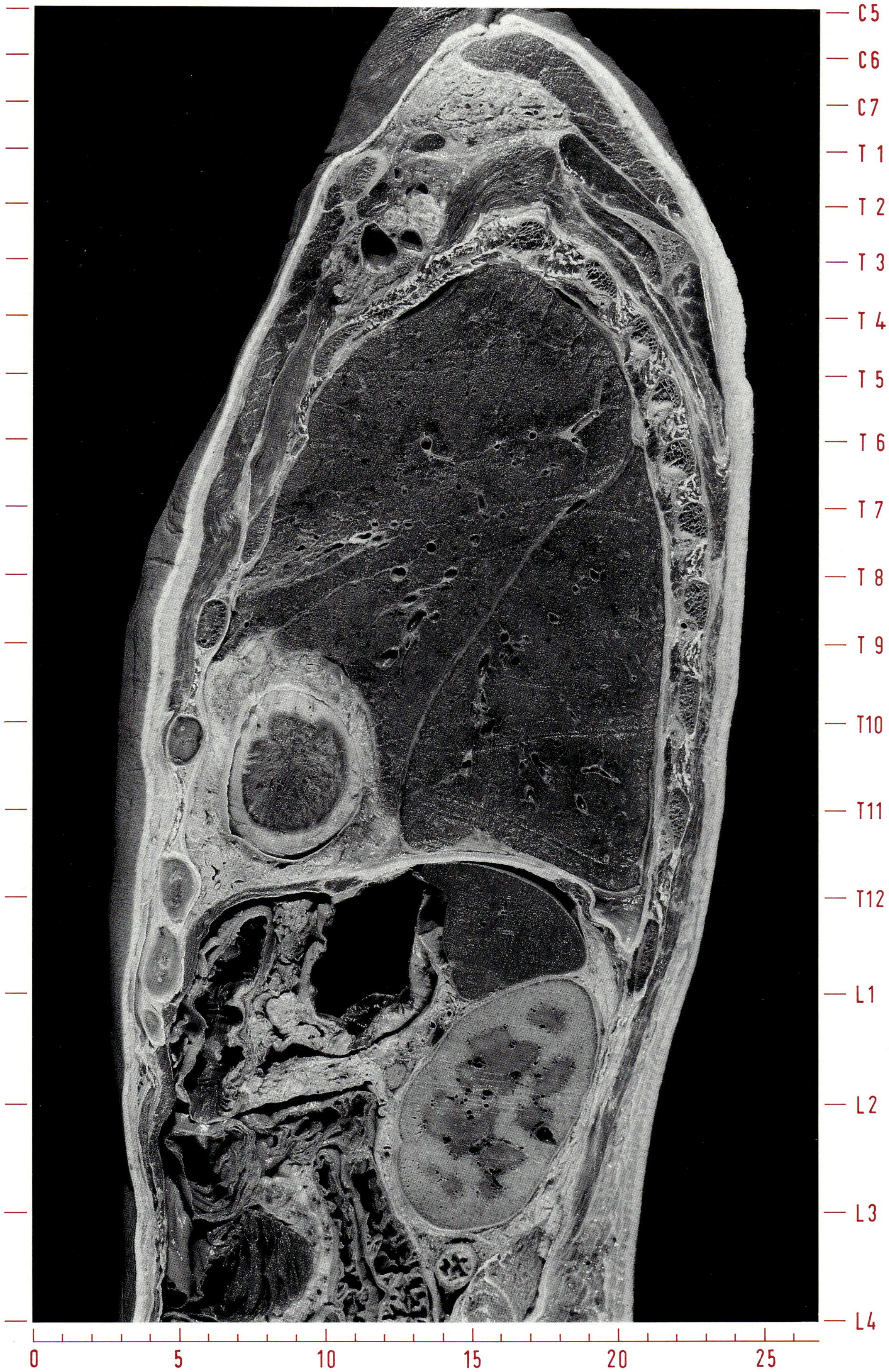
C5
C6
C7
T 1
T 2
T 3
T 4
T 5
T 6
T 7
T 8
T 9
T10
T11
T12
L1
L2
L3
L4
0
5
10
15
20
25

Fig. ST♂ 4Si

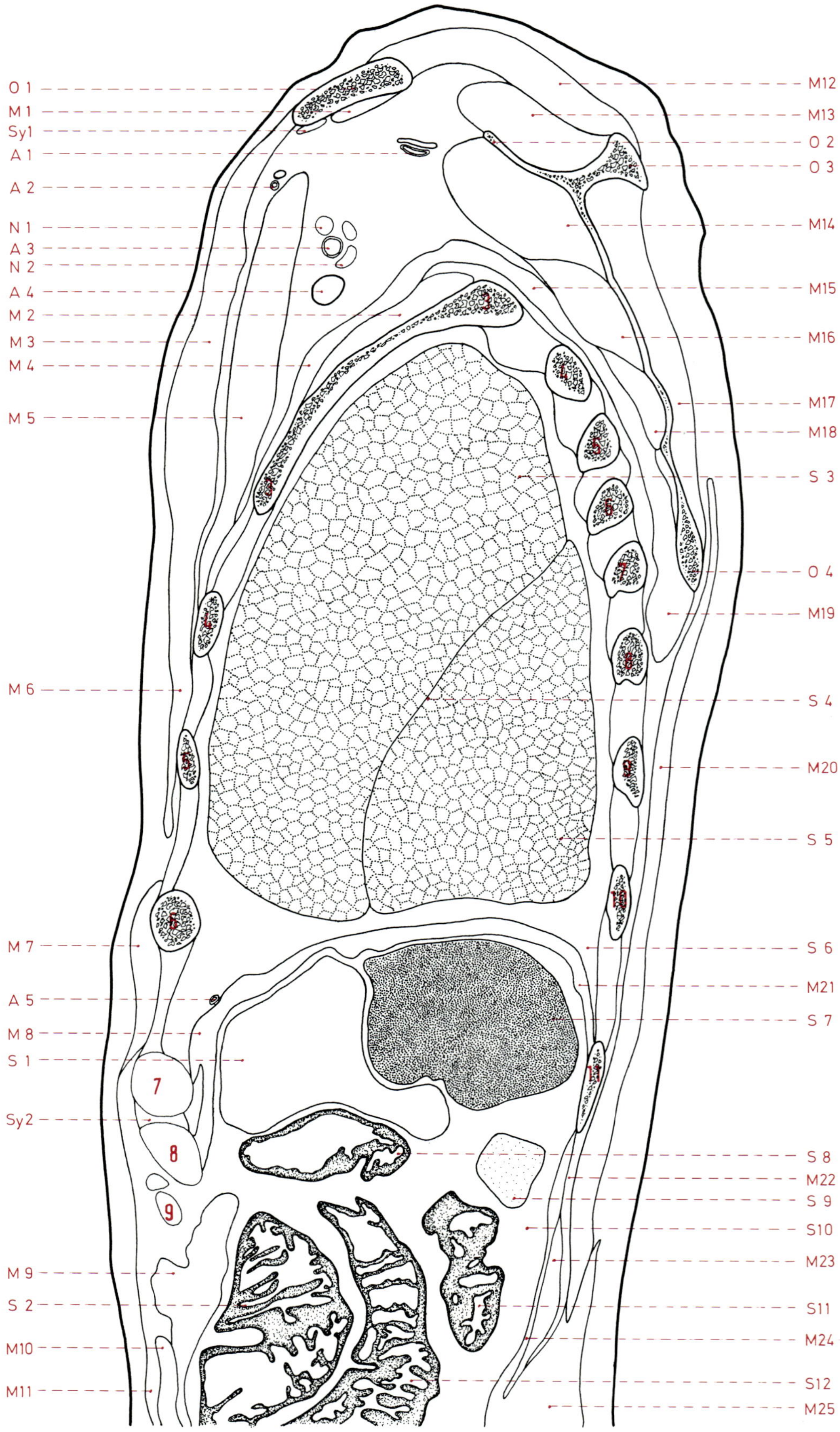
O 1
M 1
Sy1
A 1
A 2
N 1
A 3
N 2
A 4
M 2
M 3
M 4
M 5
M 6
M 7
A 5
M 8
S 1
Sy2
M 9
S 2
M10
M11
M12
M13
O 2
O 3
M14
M15
M16
M17
M18
S 3
O 4
M19
S 4
M20
S 5
S 6
M21
S 7
S 8
M22
S 9
S10
M23
S11
M24
S12
M25
7
8
9

Fig. ST♂ 5 Si

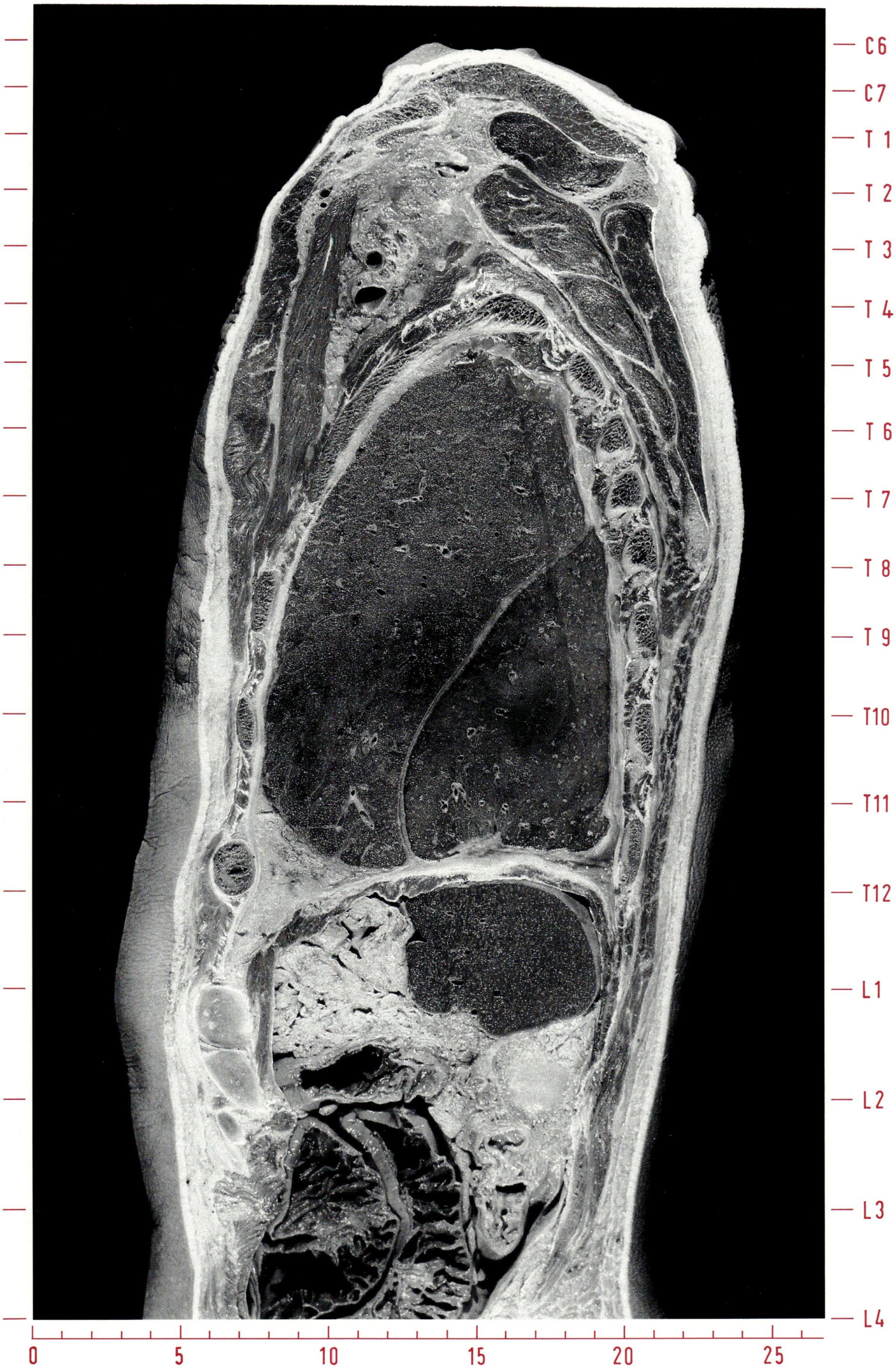

Fig. ST♂ 5 Si

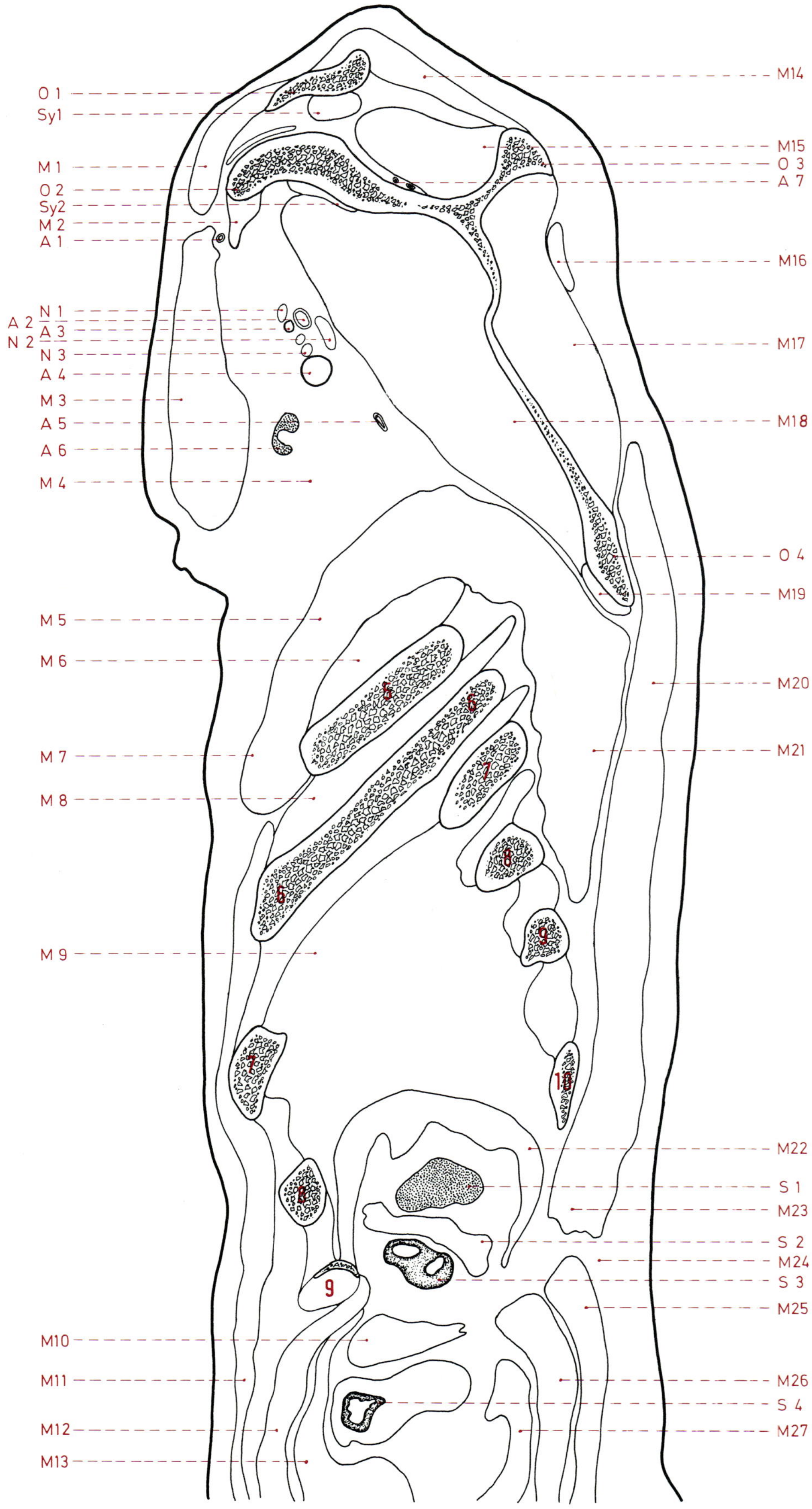
O 1
Sy1
M 1
O 2
Sy2
M 2
A 1
N 1
A 2
A 3
N 2
N 3
A 4
M 3
A 5
A 6
M 4
M 5
M 6
M 7
M 8
M 9
M10
M11
M12
M13
M14
M15
O 3
A 7
M16
M17
M18
O 4
M19
M20
M21
M22
S 1
M23
S 2
M24
S 3
M25
M26
S 4
M27
5
6
7
8
9
10
6
7
8
9

Fig. ST♂ 6 Si

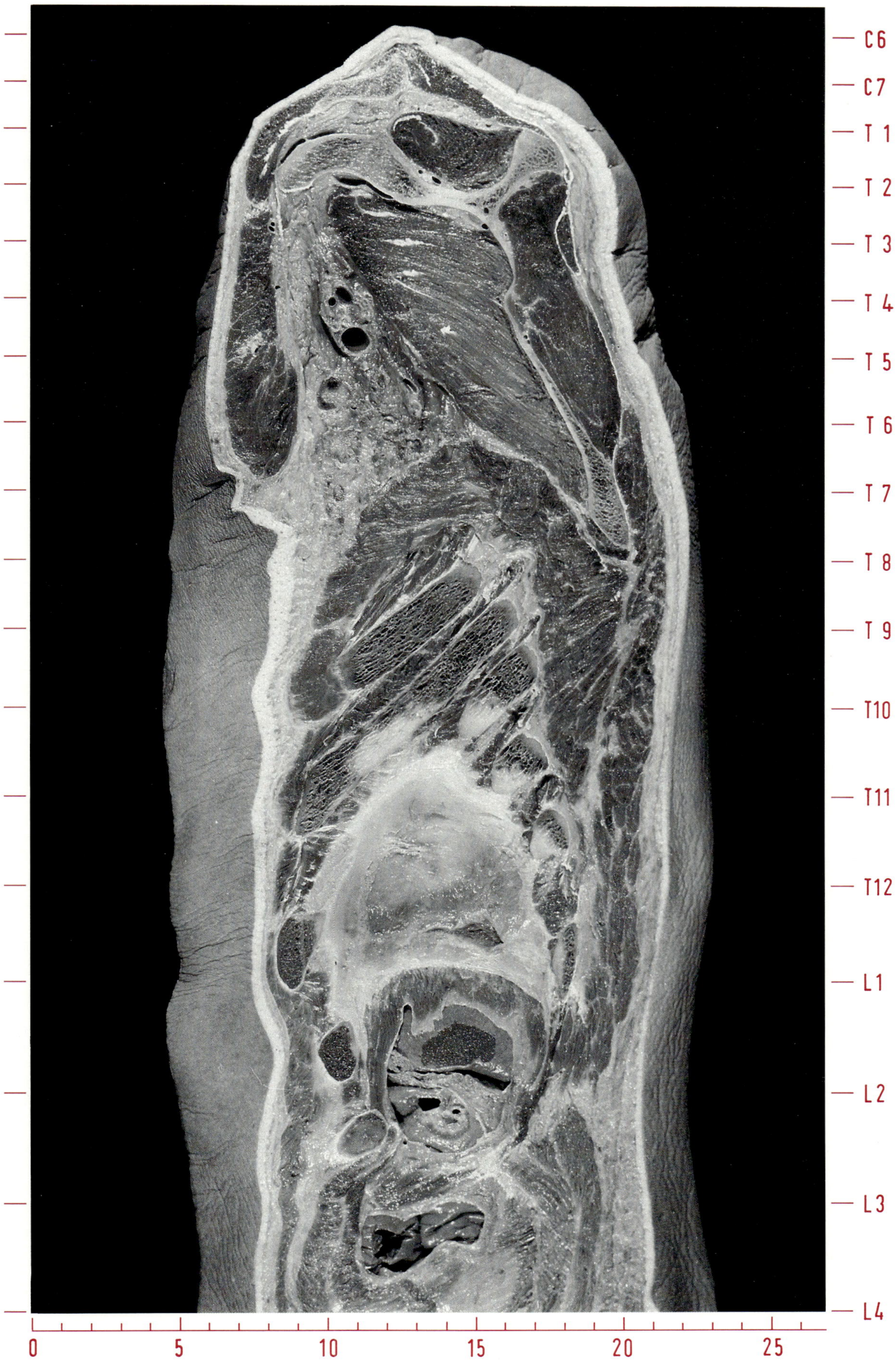

Fig. ST♂ 6 Si

ST
♀

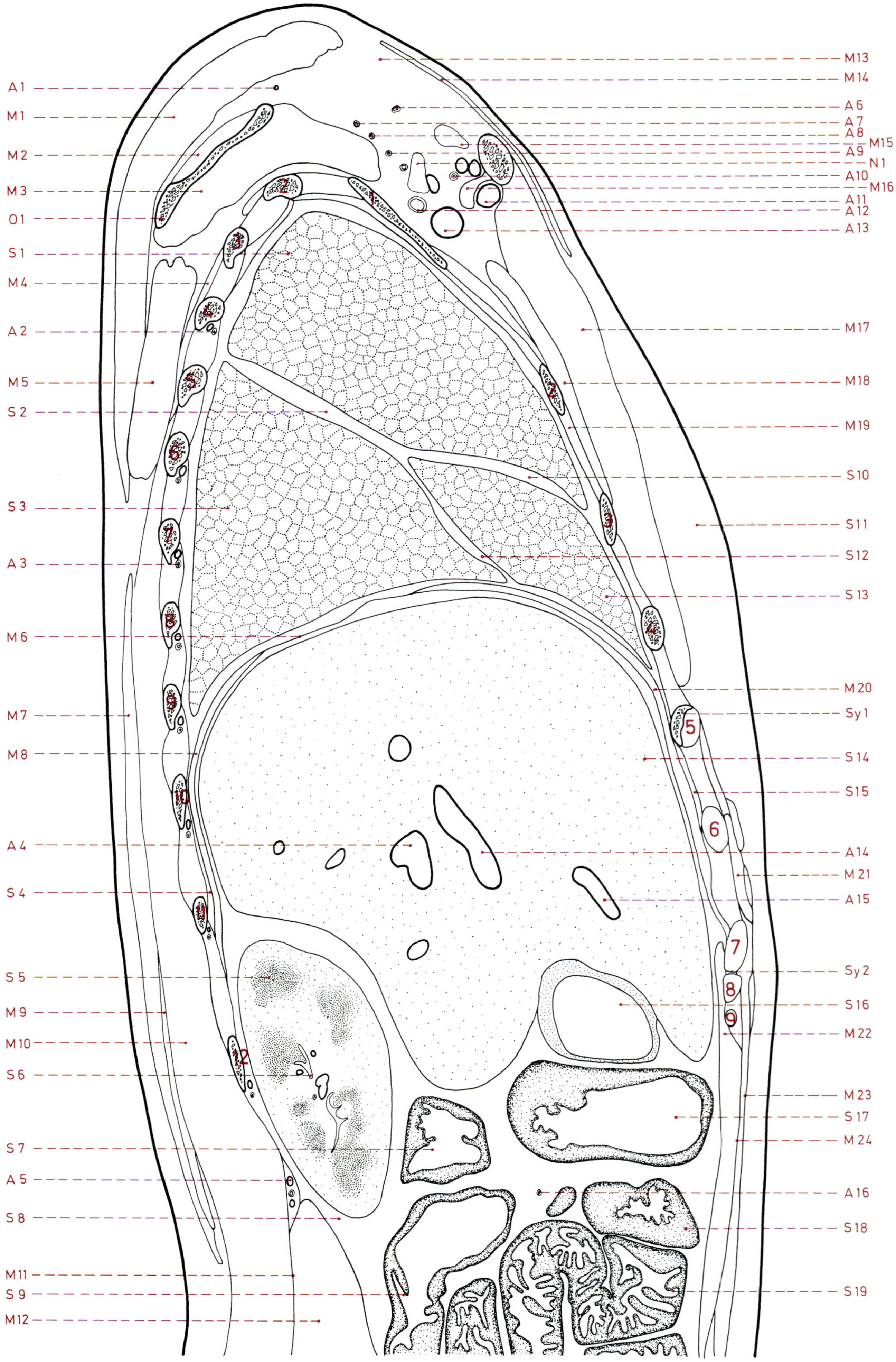
A1
M1
M2
M3
O1
S1
M4
A2
M5
S2
S3
A3
M6
M7
M8
A4
S4
S5
M9
M10
S6
S7
A5
S8
M11
S9
M12
M13
M14
A6
A7
A8
A9
M15
N1
A10
M16
A11
A12
A13
M17
M18
M19
S10
S11
S12
S13
M20
Sy1
S14
S15
A14
M21
A15
Sy2
S16
M22
M23
S17
M24
A16
S18
S19

Fig. ST♀ 1D

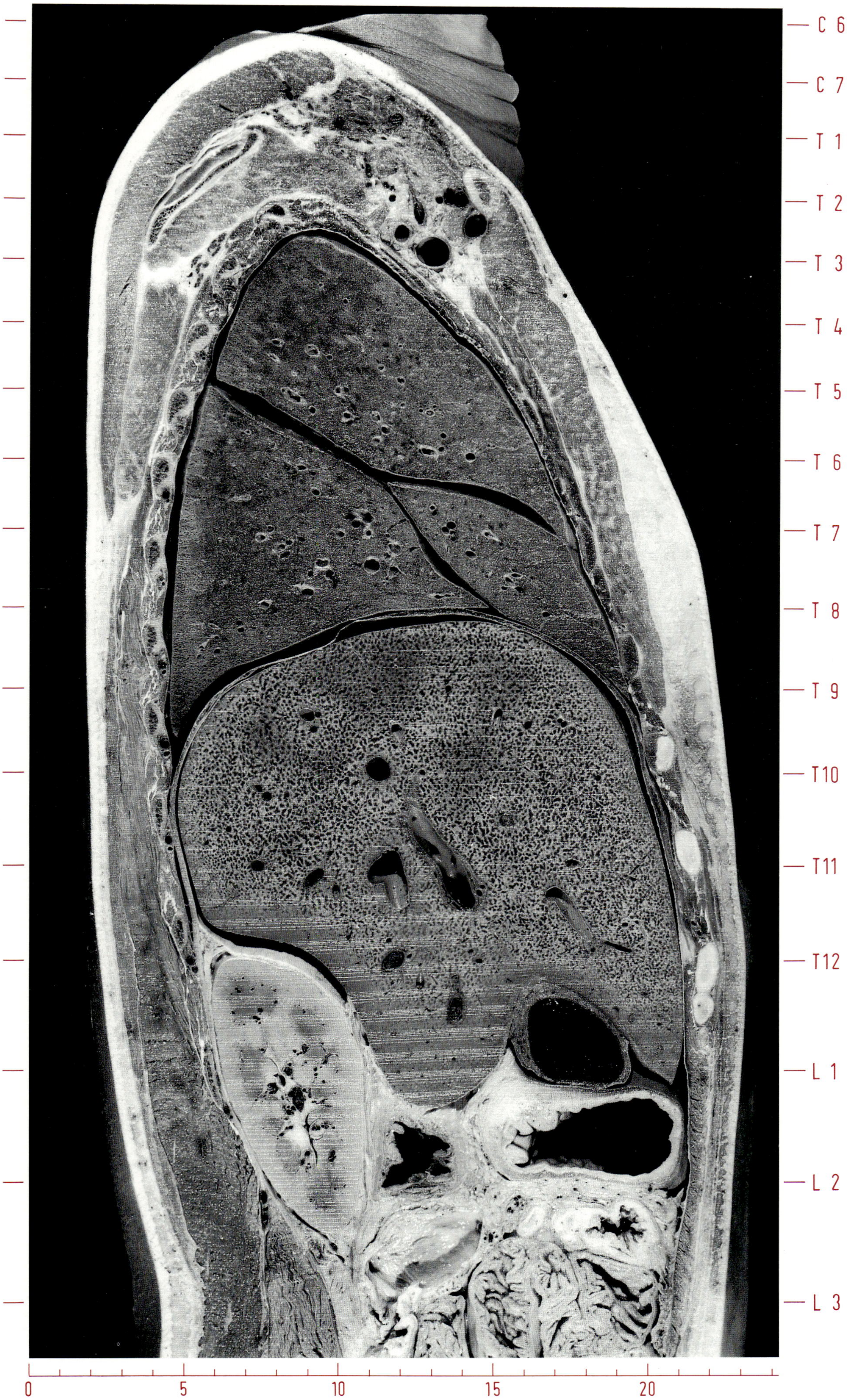

Fig. ST♀ 1D

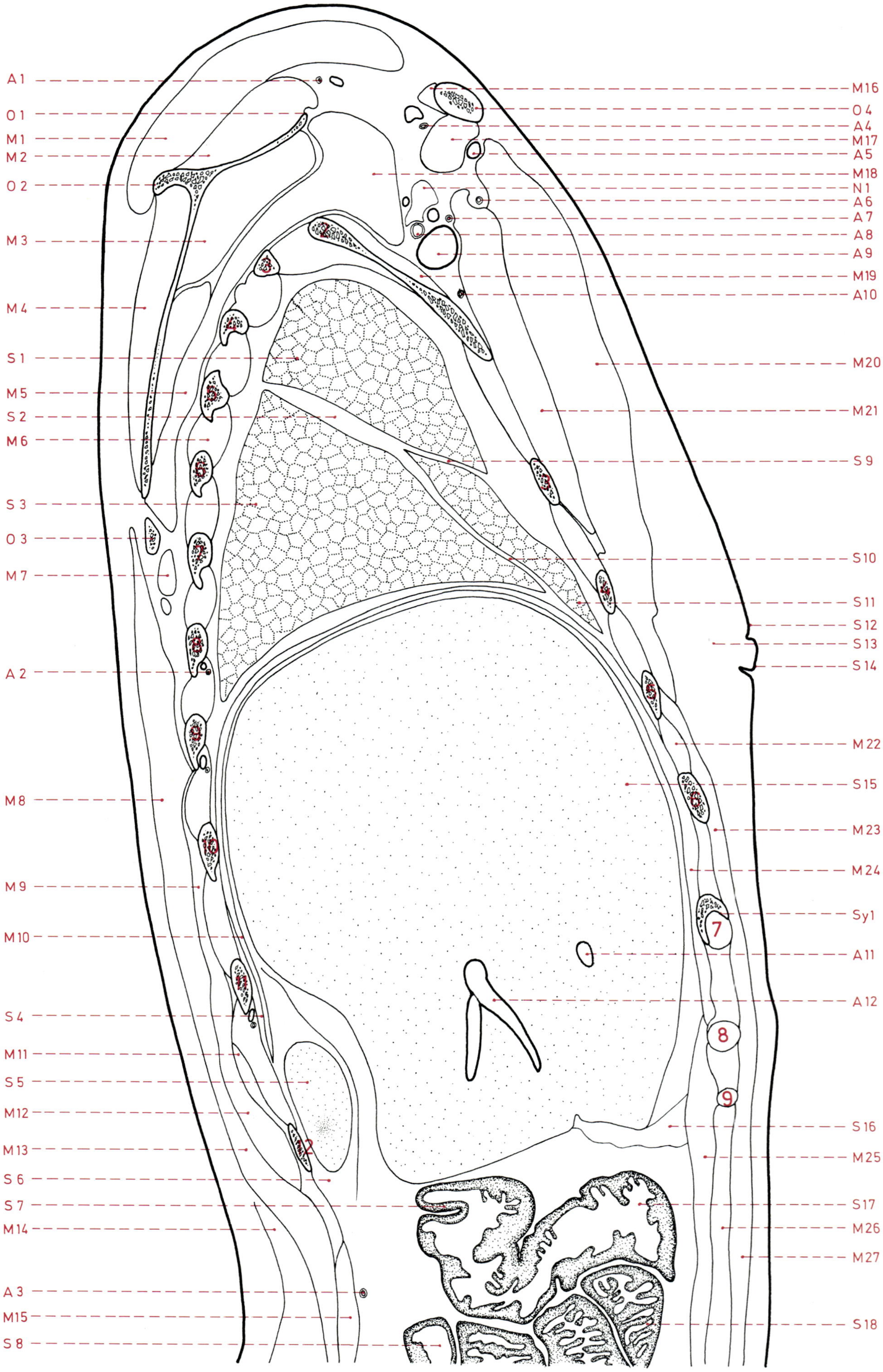

A1
O1
M1
M2
O2
M3
M4
S1
M5
S2
M6
S3
O3
M7
A2
M8
M9
M10
S4
M11
S5
M12
M13
S6
S7
M14
A3
M15
S8
M16
O4
A4
M17
A5
M18
N1
A6
A7
A8
A9
M19
A10
M20
M21
S9
S10
S11
S12
S13
S14
M22
S15
M23
M24
Sy1
A11
A12
S16
M25
S17
M26
M27
S18

Fig. ST♀ 2D

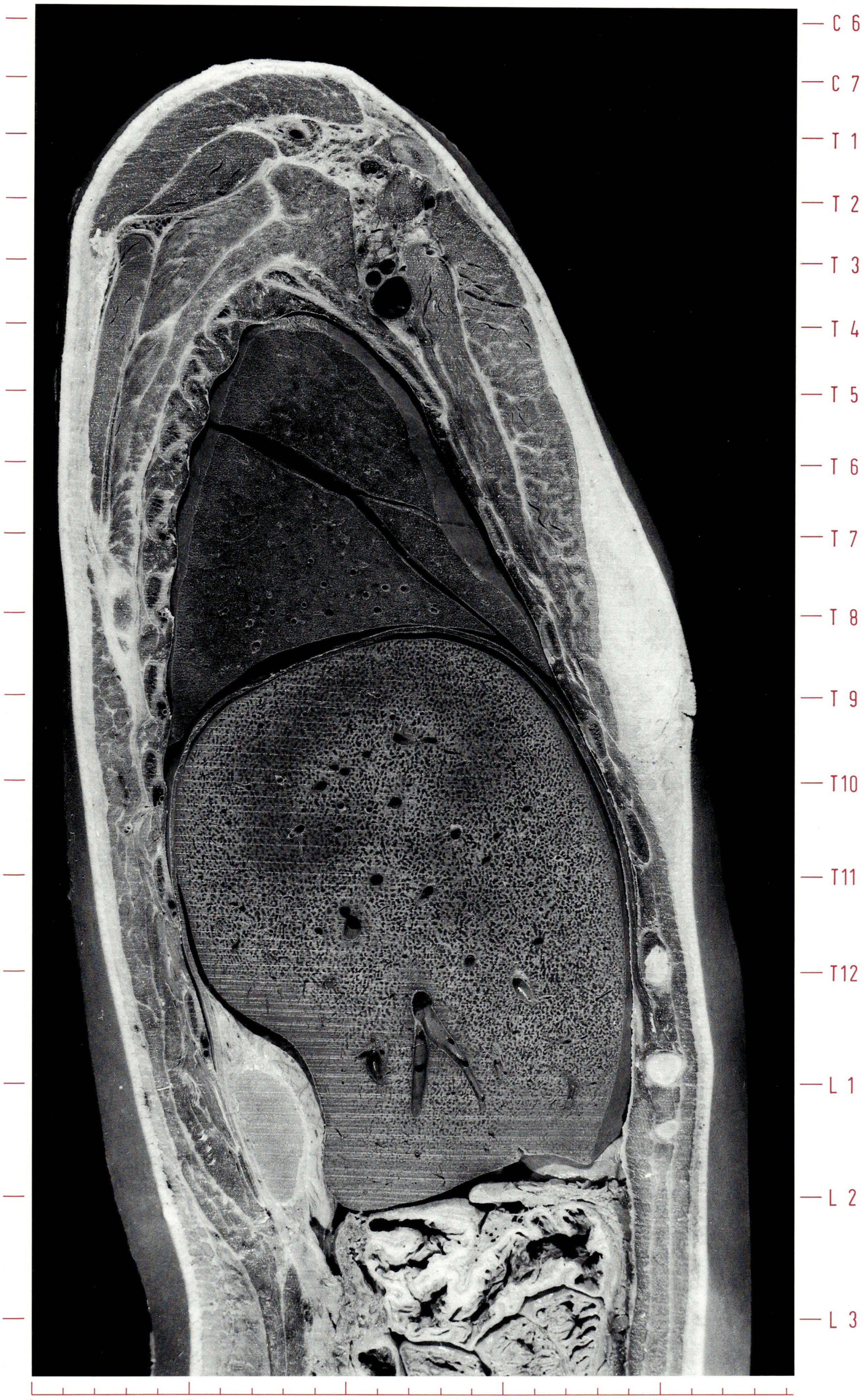

Fig. ST♀ 2D

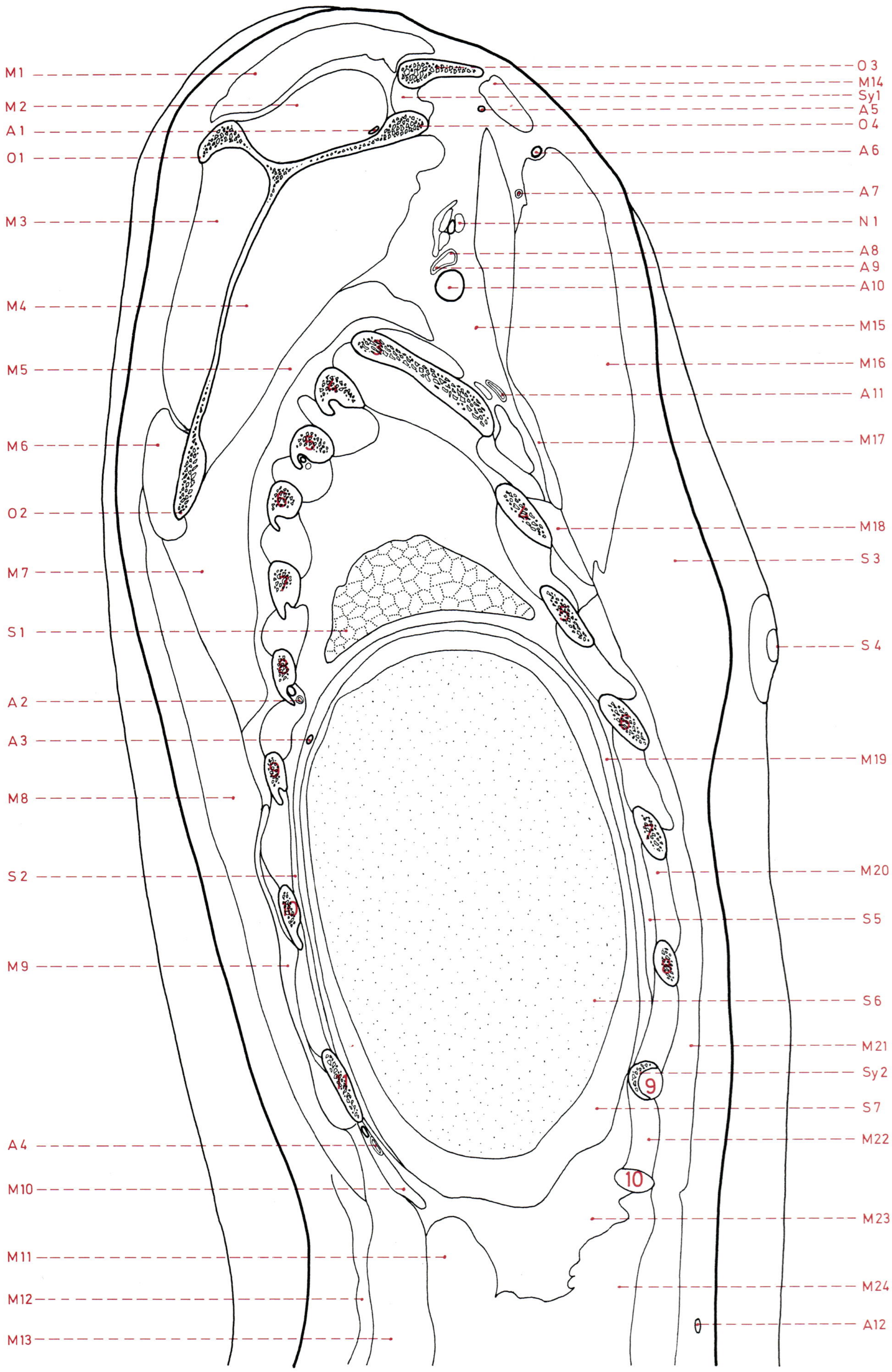
M1
M2
A1
O1
M3
M4
M5
M6
O2
M7
S1
A2
A3
M8
S2
M9
A4
M10
M11
M12
M13
O3
M14
Sy1
A5
O4
A6
A7
N1
A8
A9
A10
M15
M16
A11
M17
M18
S3
S4
M19
M20
S5
S6
M21
Sy2
S7
M22
M23
M24
A12
9
10

Fig. ST♀ 3D

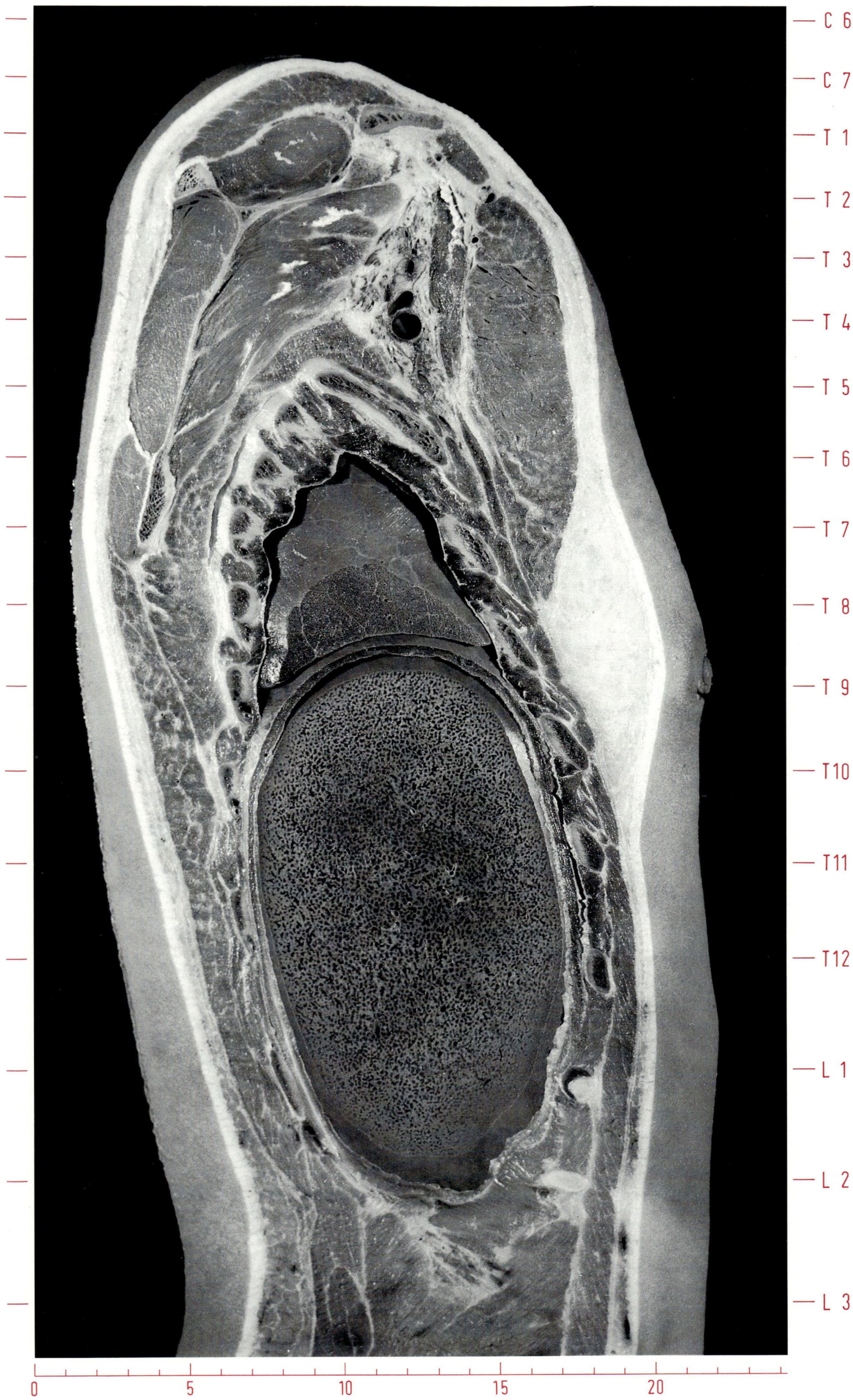

Fig. ST♀ 3D

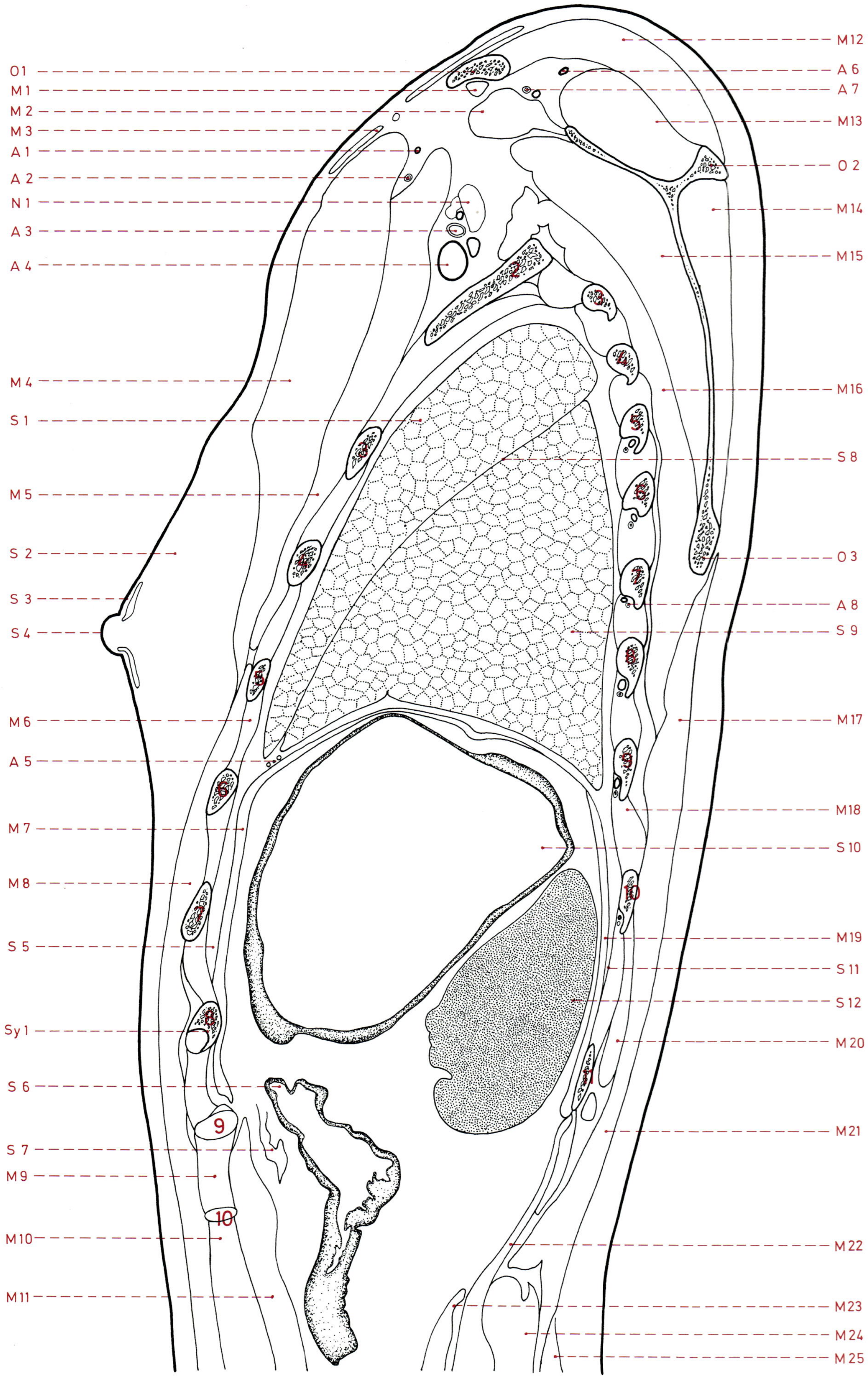

O1
M1
M2
M3
A1
A2
N1
A3
A4
M4
S1
M5
S2
S3
S4
M6
A5
M7
M8
S5
Sy1
S6
S7
M9
M10
M11
M12
A6
A7
M13
O2
M14
M15
M16
S8
O3
A8
S9
M17
M18
S10
M19
S11
S12
M20
M21
M22
M23
M24
M25

Fig. ST♀ Si

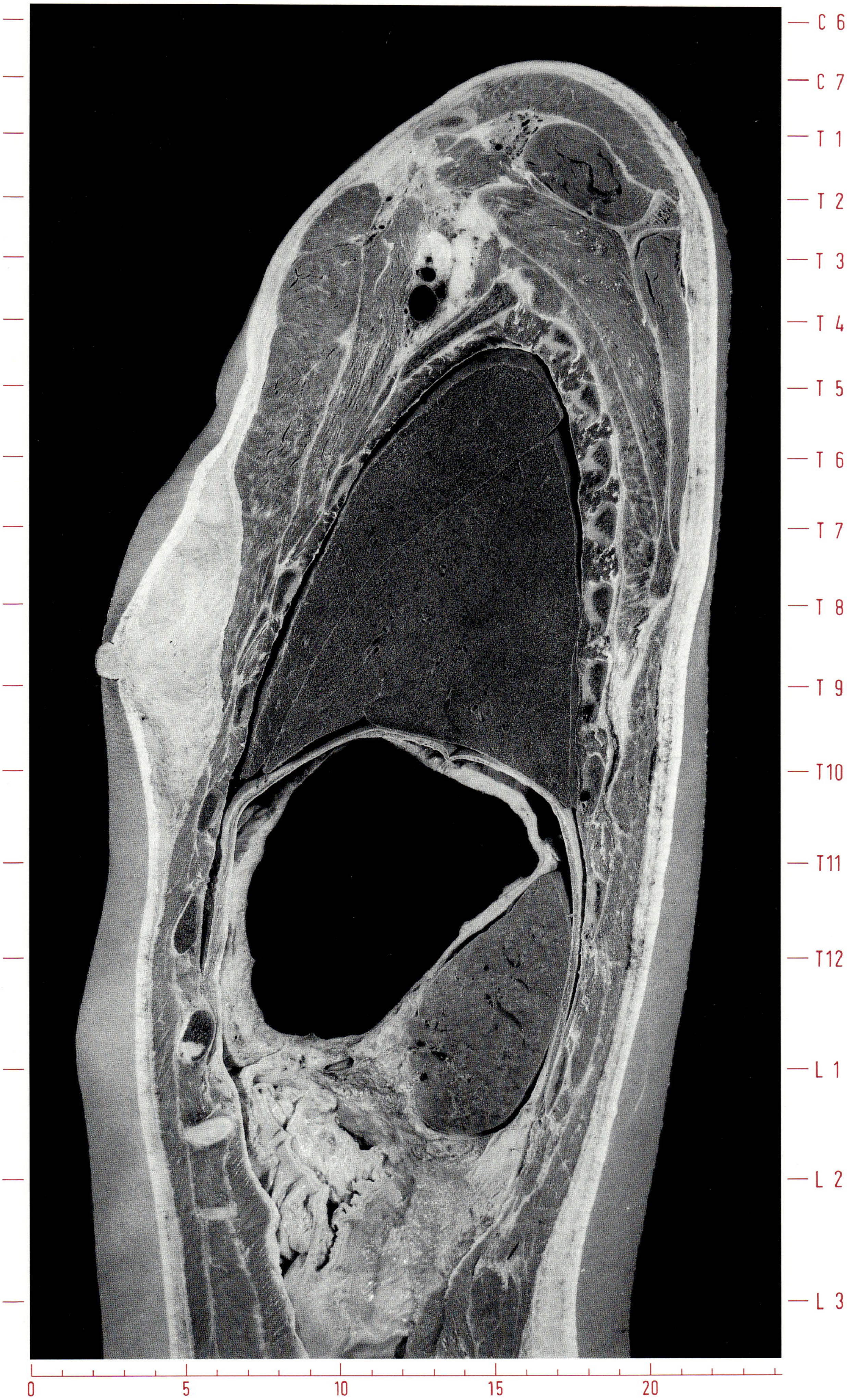

Fig. ST♀ Si

HT

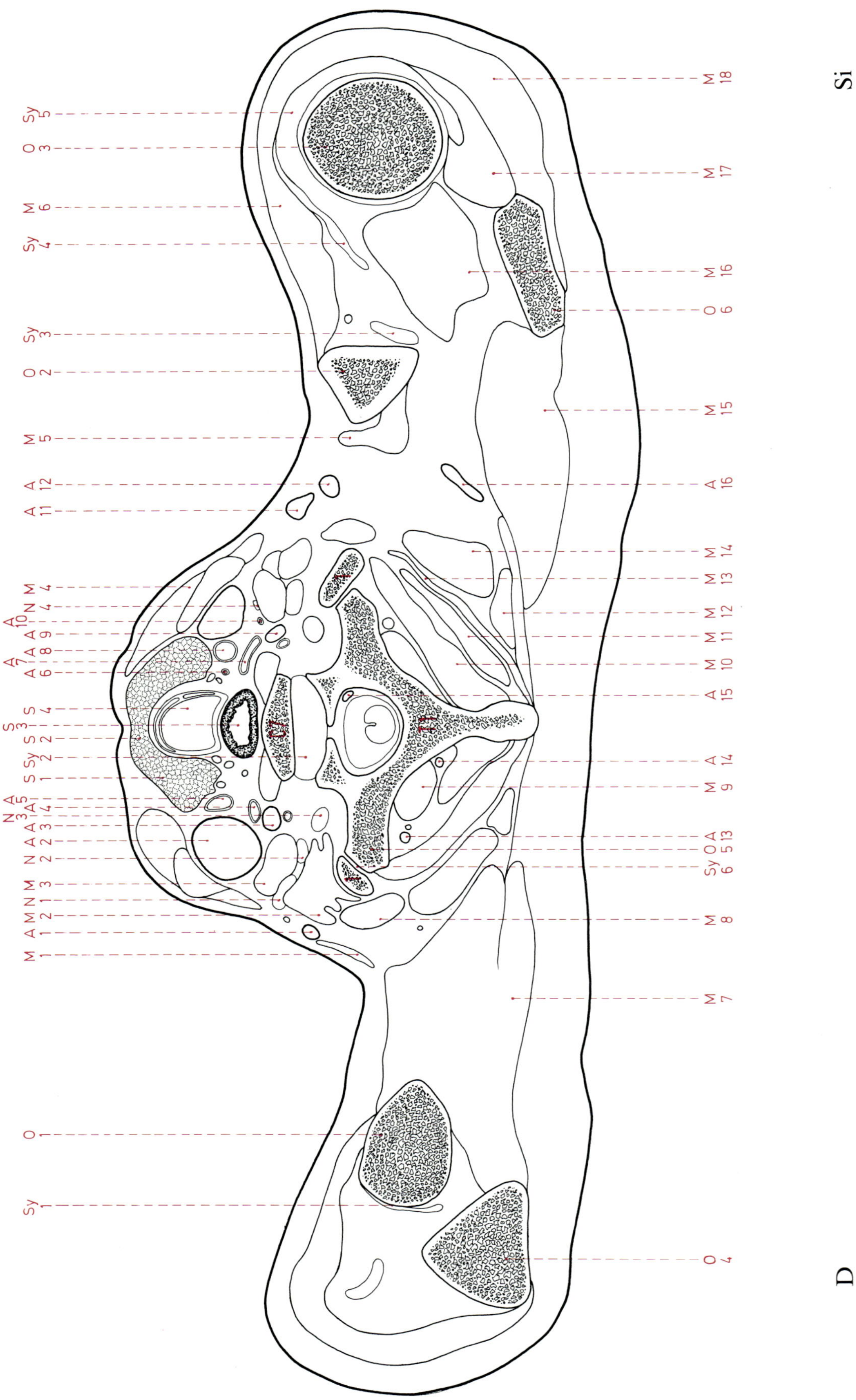
Si
D
M 18
M 17
M 16
O 6
M 15
A 16
M 14
M 13
M 12
M 11
M 10
A 15
A 14
M 9
A 13
O 5
Sy 6
M 8
M 7
O 4
Sy 5
O 3
M 6
Sy 4
Sy 3
O 2
M 5
A 12
A 11
M 4
N 4
M 3
N 2
M 2
N 1
A 1
M 1
O 1
Sy 1

Fig. HT 1

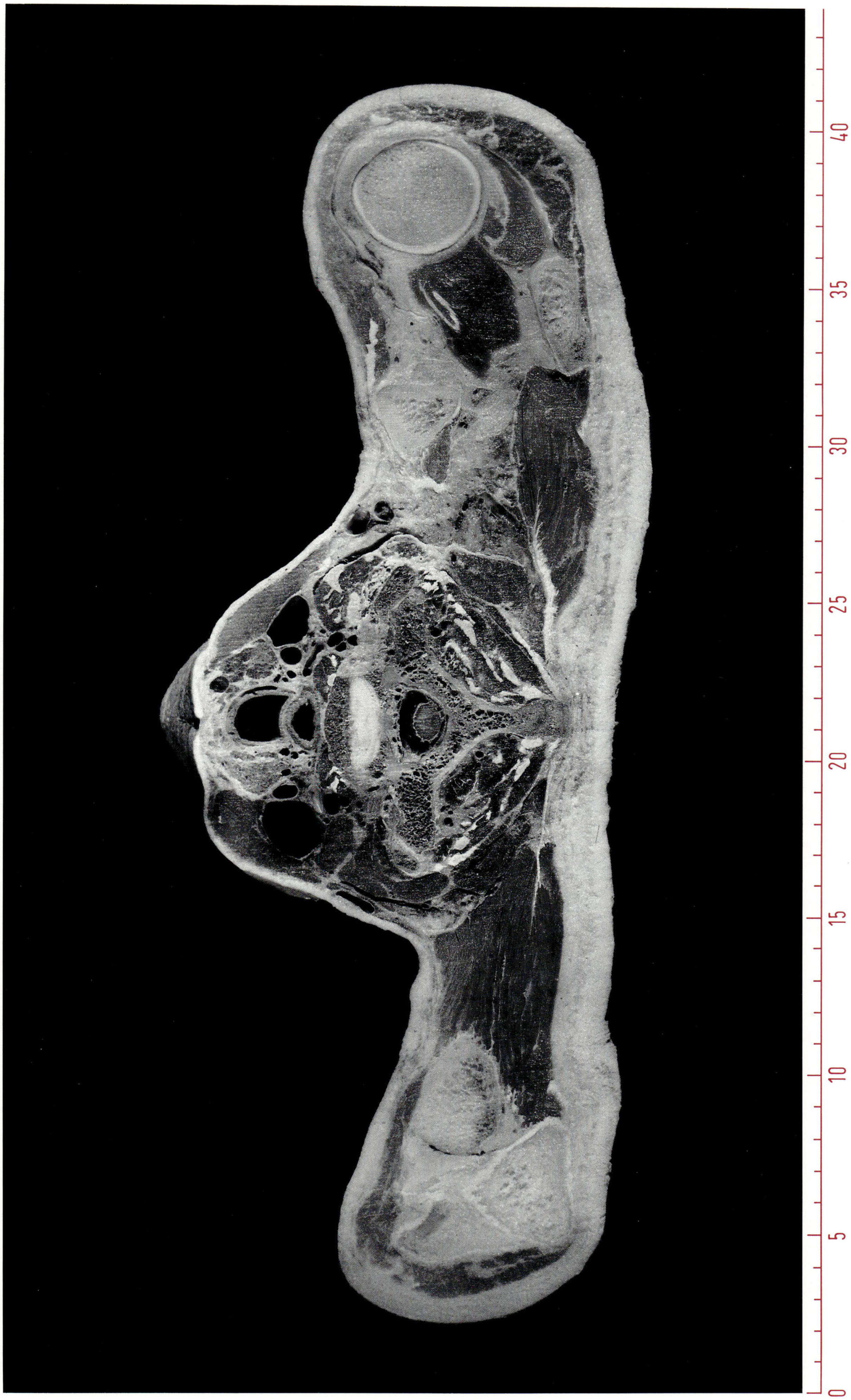

Fig. HT 1

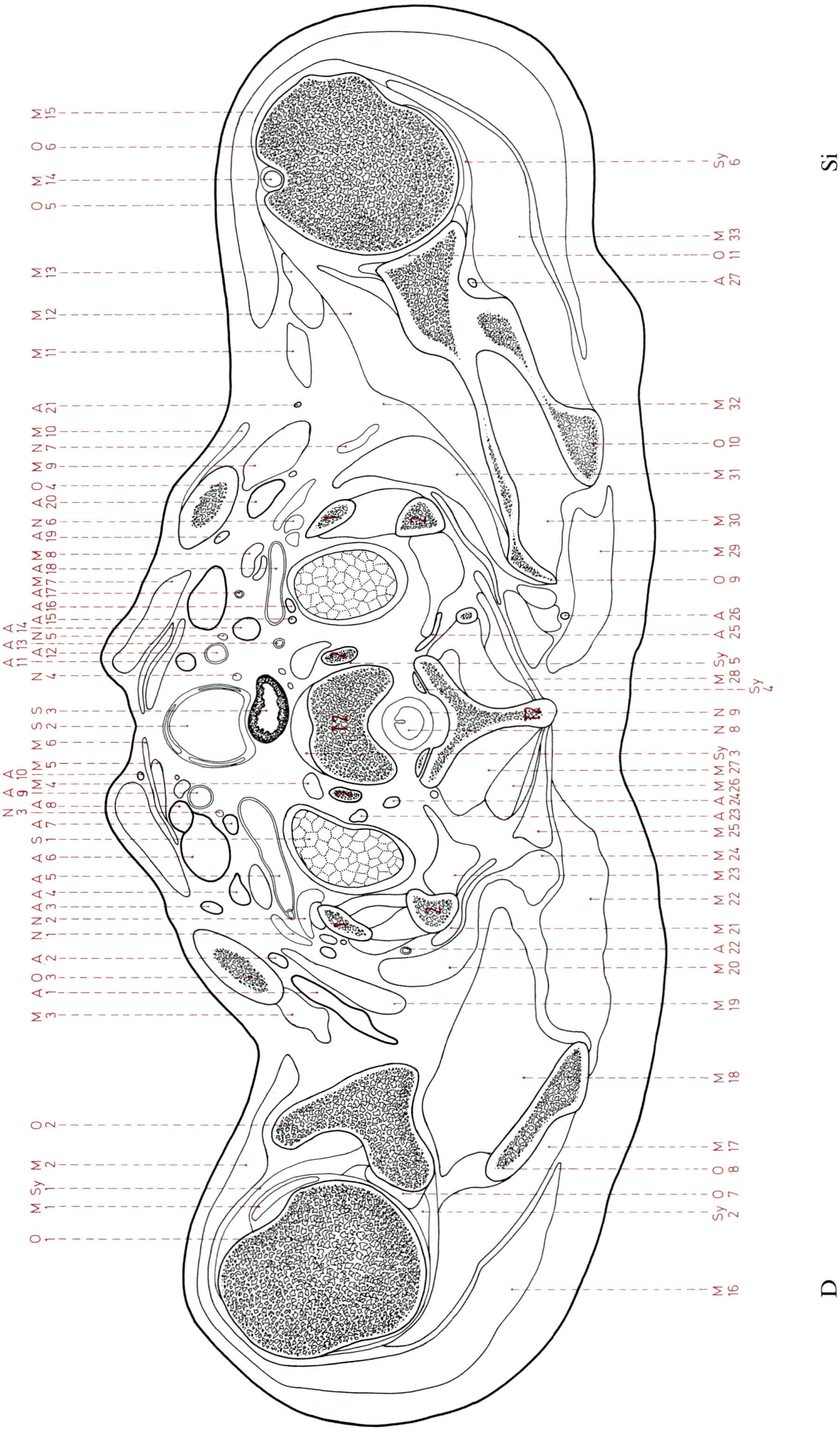

Fig. HT 2

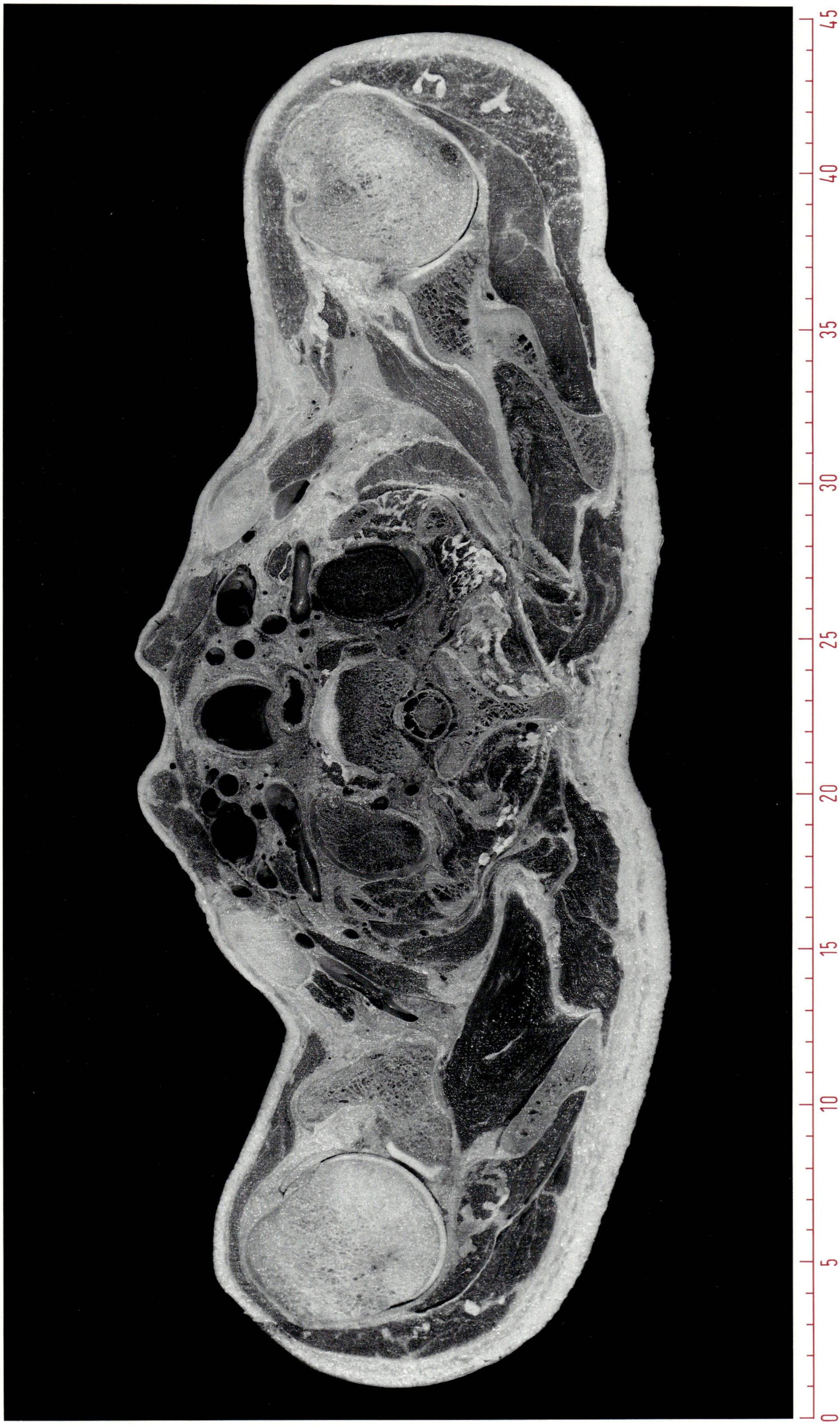

Fig. HT 2

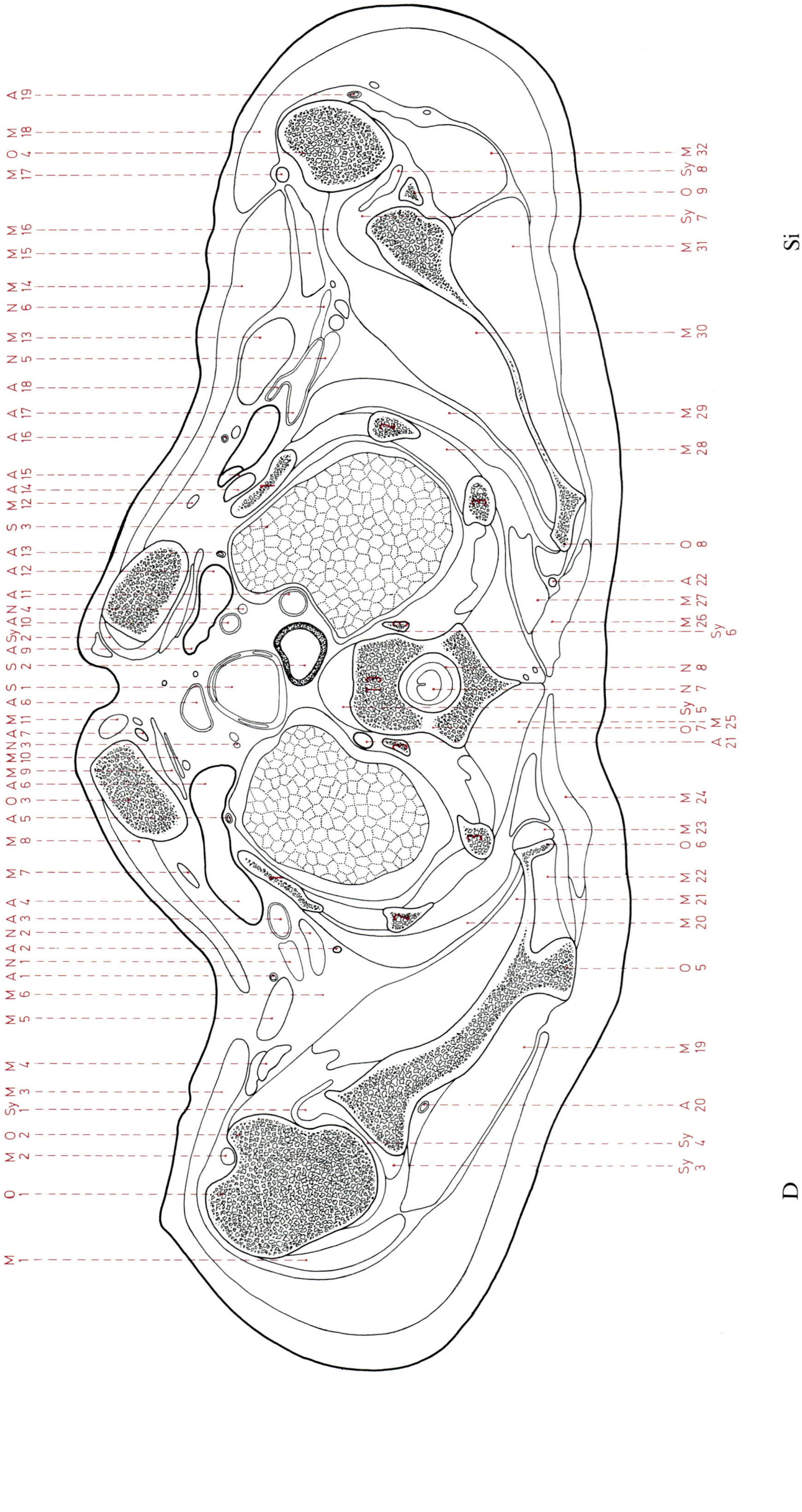

Fig. HT 3

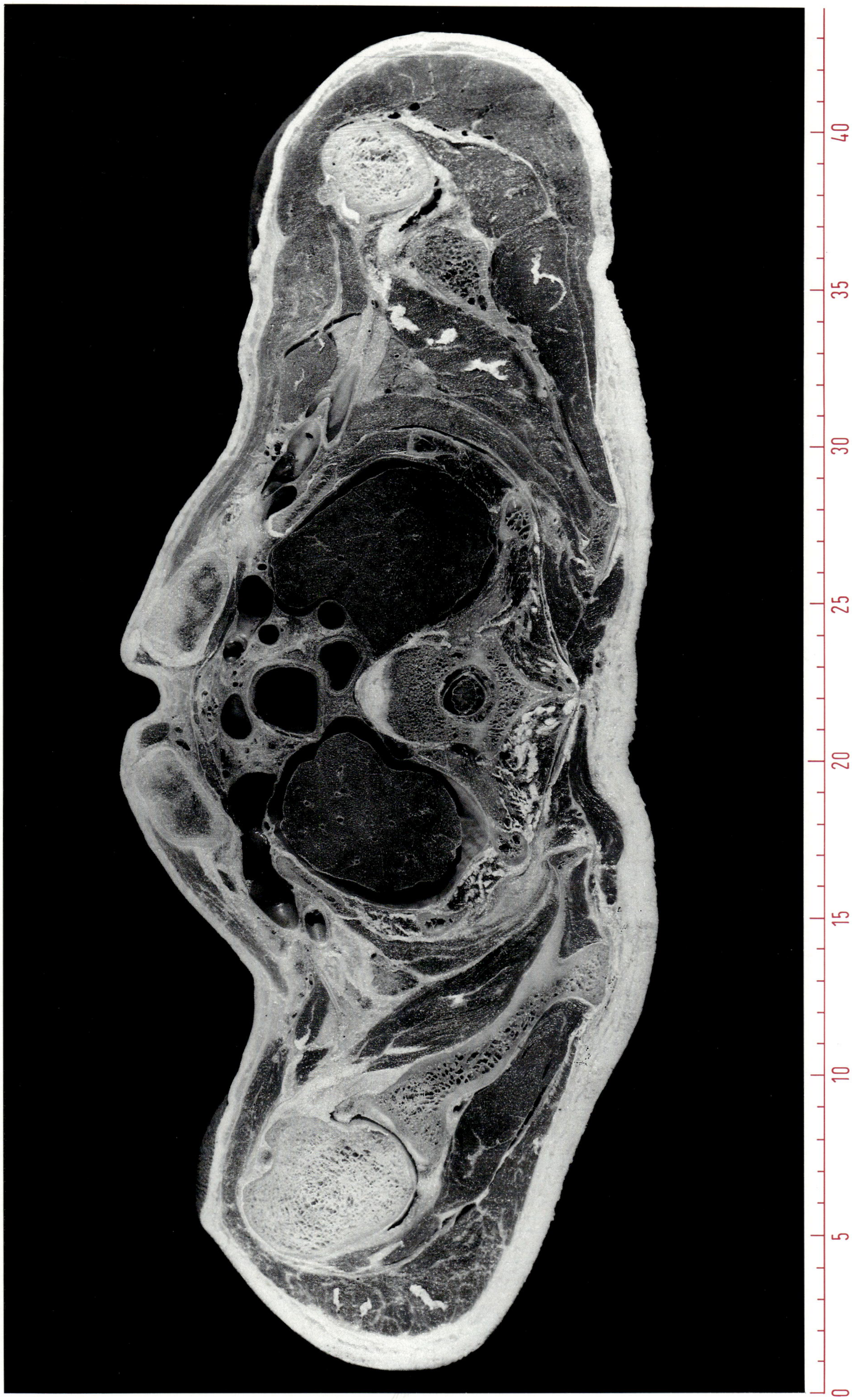

Fig. HT 3

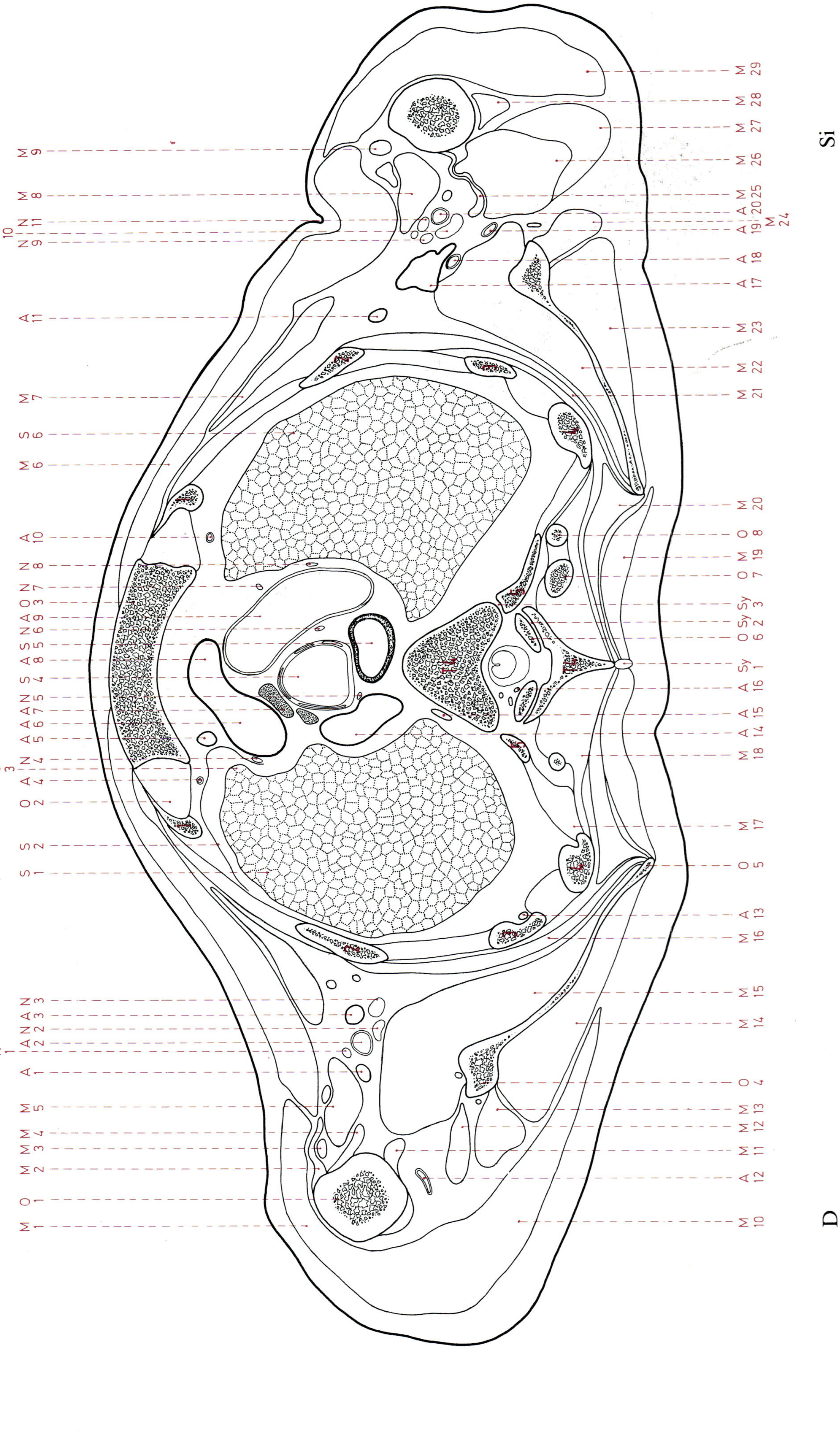
M 9
M 8
N 11
N 10
N 9
A 11
M 7
S 6
M 6
A 10
N 8
N 7
O 3
A 9
N 6
S 5
A 8
S 4
N 5
A 7
A 6
A 5
N 4
S 3
A 4
O 2
S 2
S 1
N 3
A 3
N 2
A 2
N 1
A 1
M 5
M 4
M 3
M 2
O 1
M 1
M 29
M 28
M 27
M 26
M 25
A 20
A 19
M 24
A 18
A 17
M 23
M 22
M 21
M 20
O 8
M 19
O 7
Sy 3
Sy 2
O 6
Sy 1
A 16
A 15
A 14
M 18
M 17
O 5
A 13
M 16
M 15
M 14
O 4
M 13
M 12
M 11
A 12
M 10
Si
D

Fig. HT 4

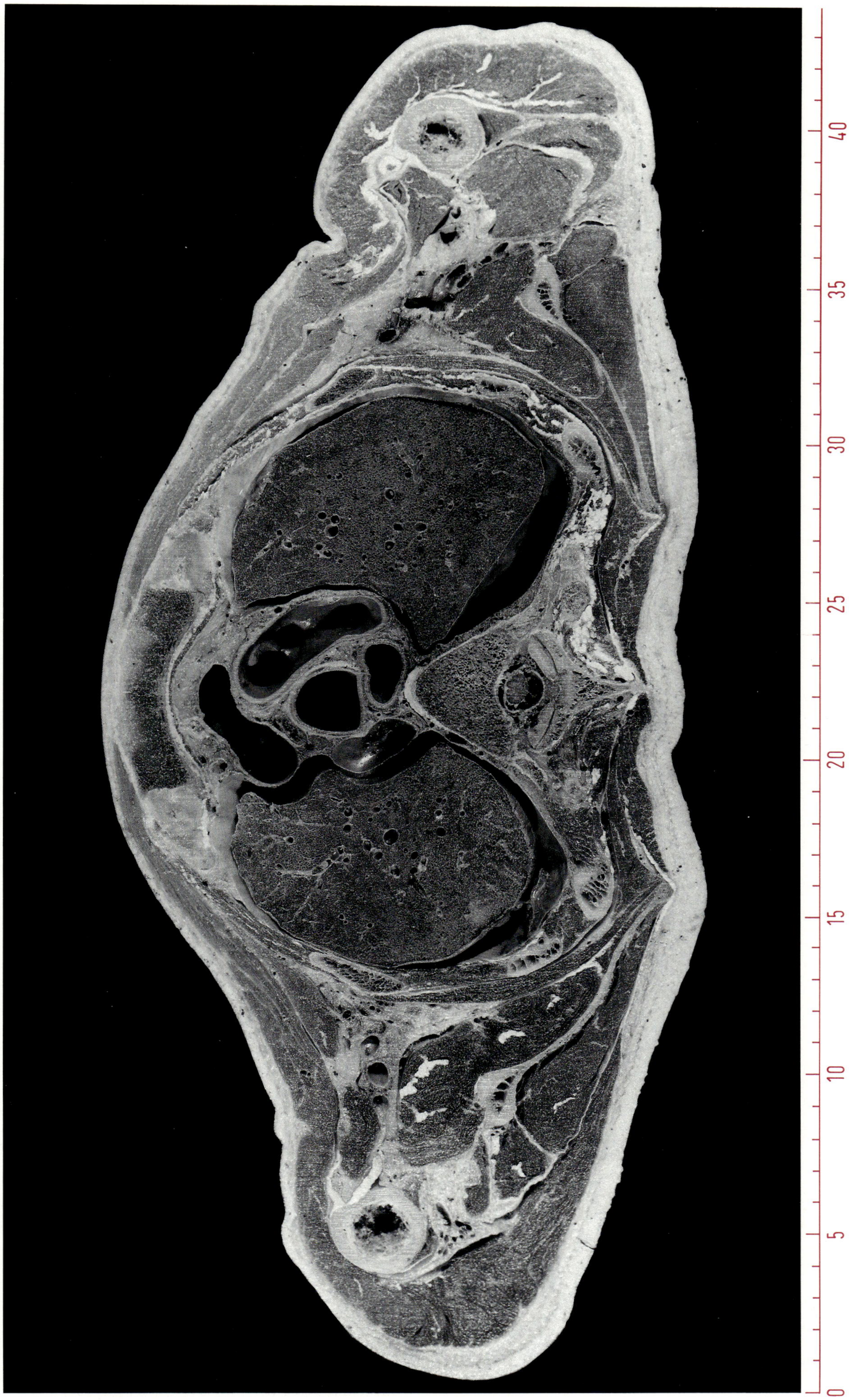

Fig. HT 4

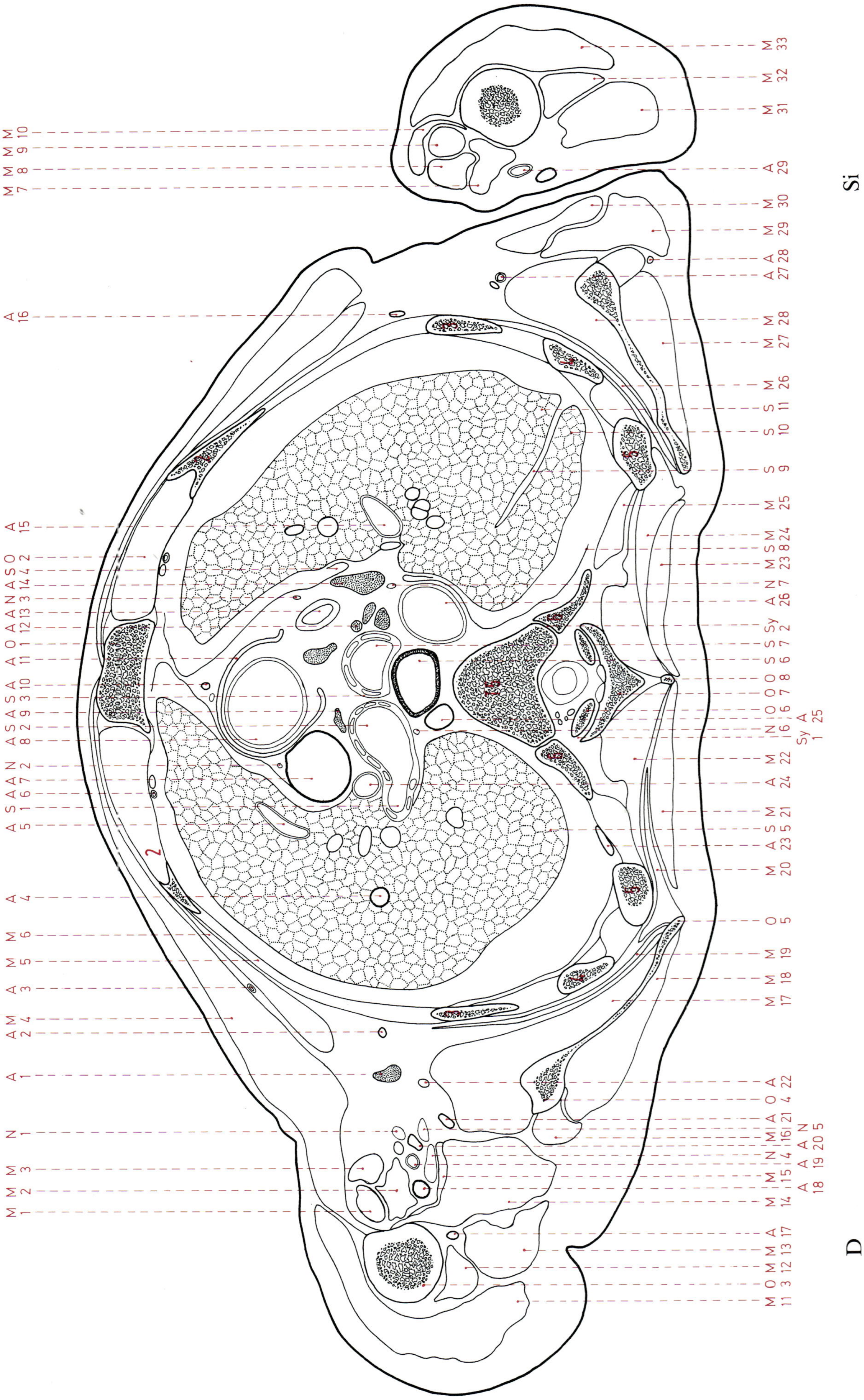

Fig. HT 5

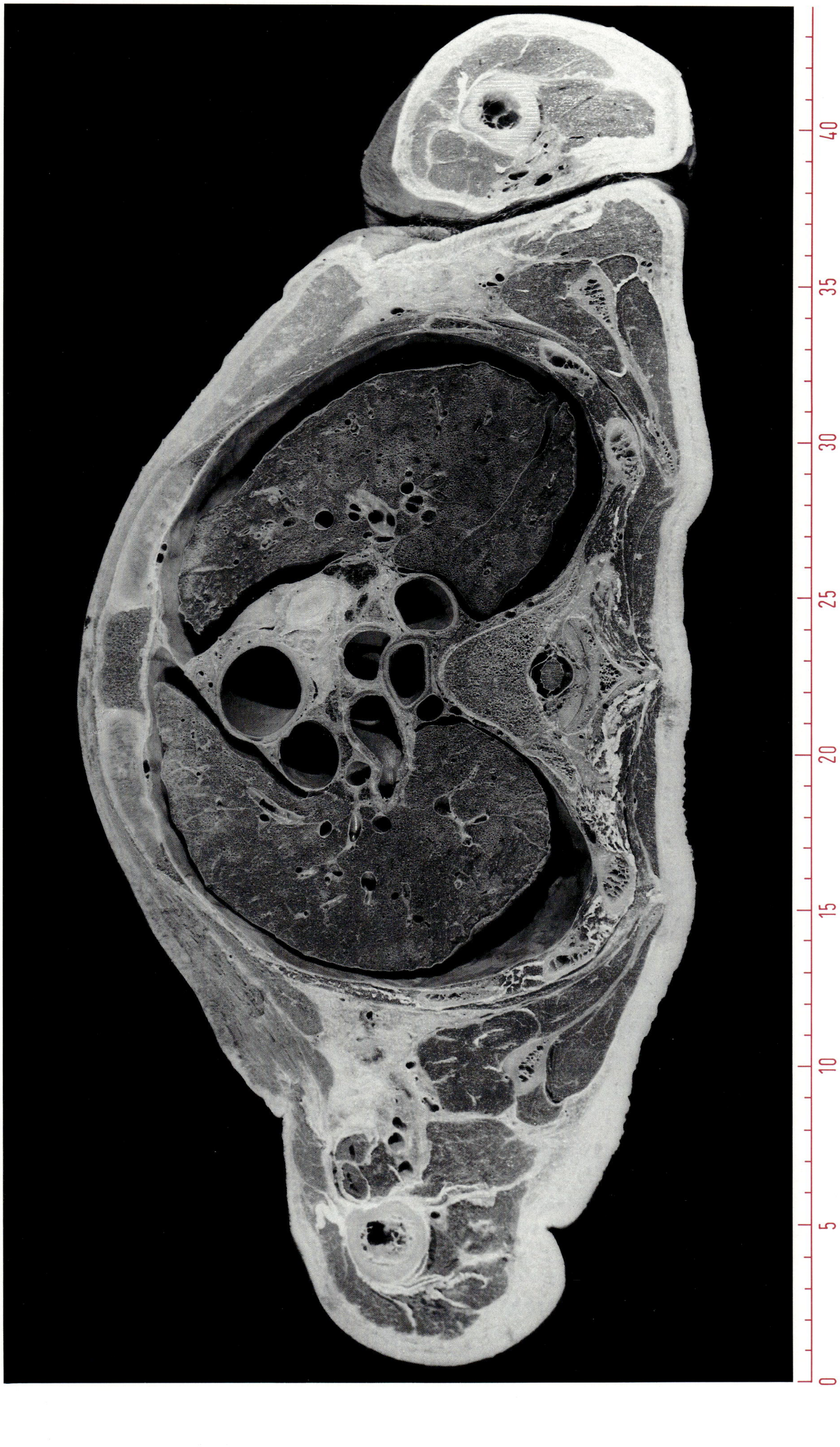

Fig. HT 5

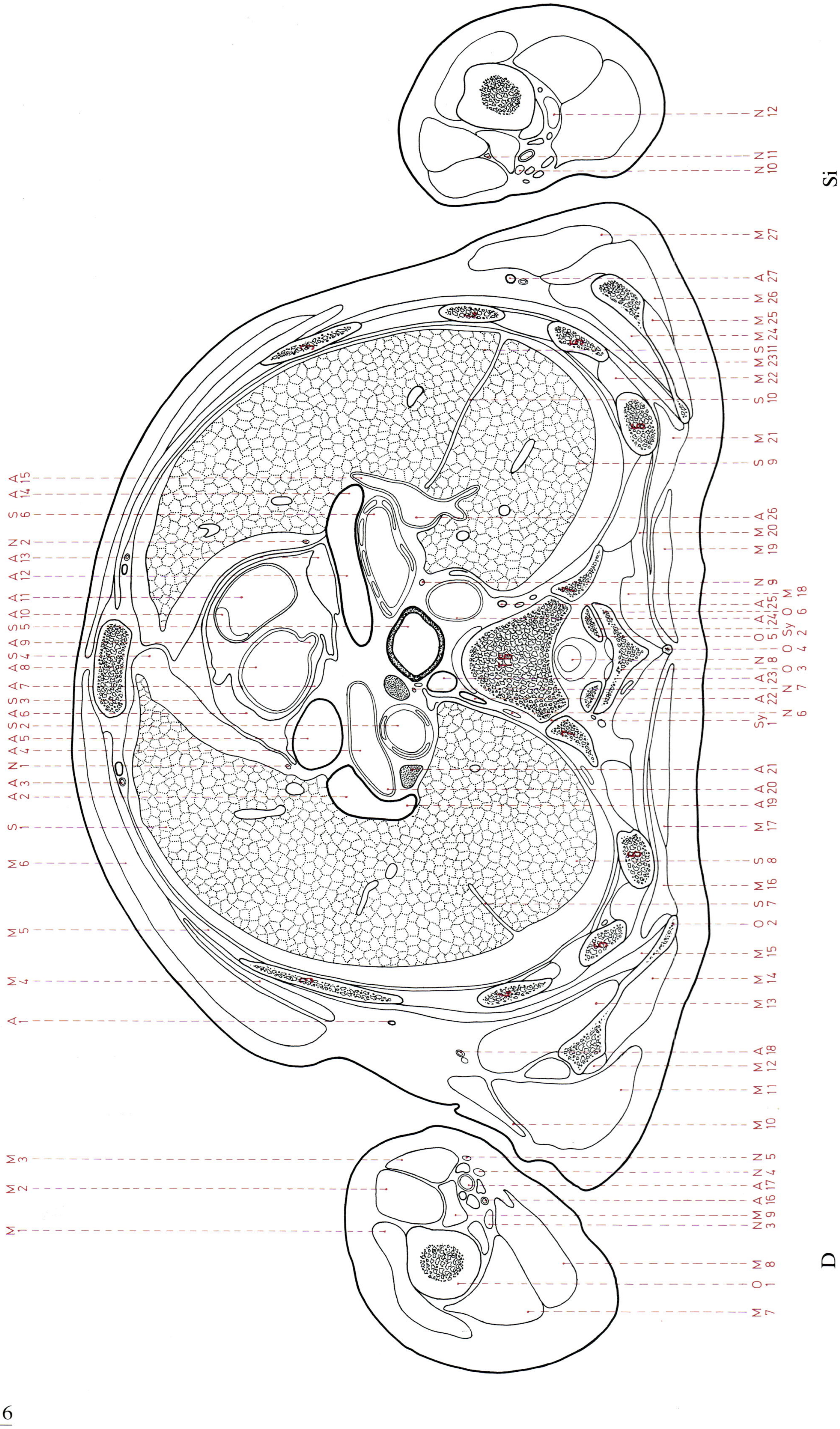

Fig. HT 6

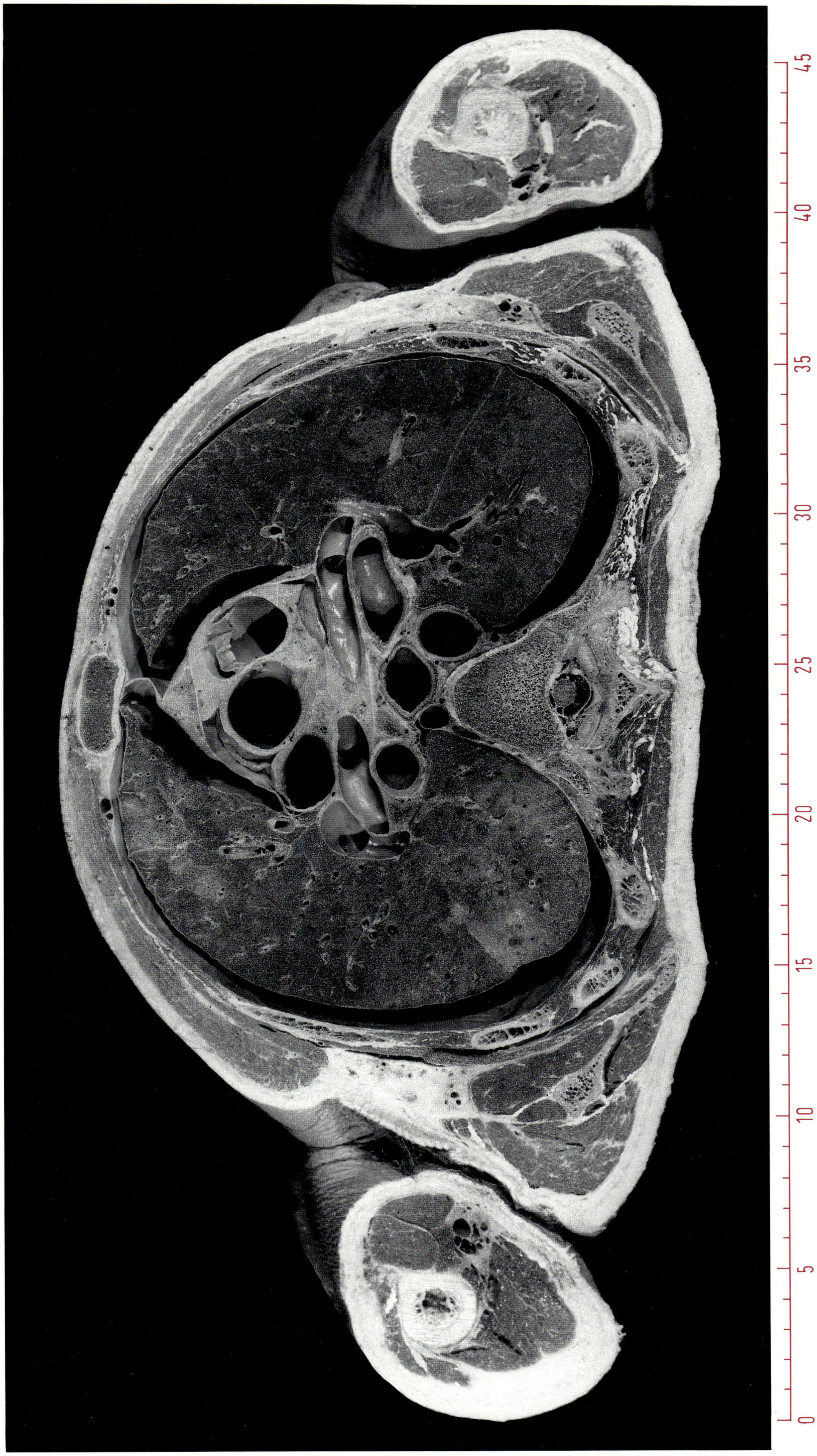

Fig. HT 6

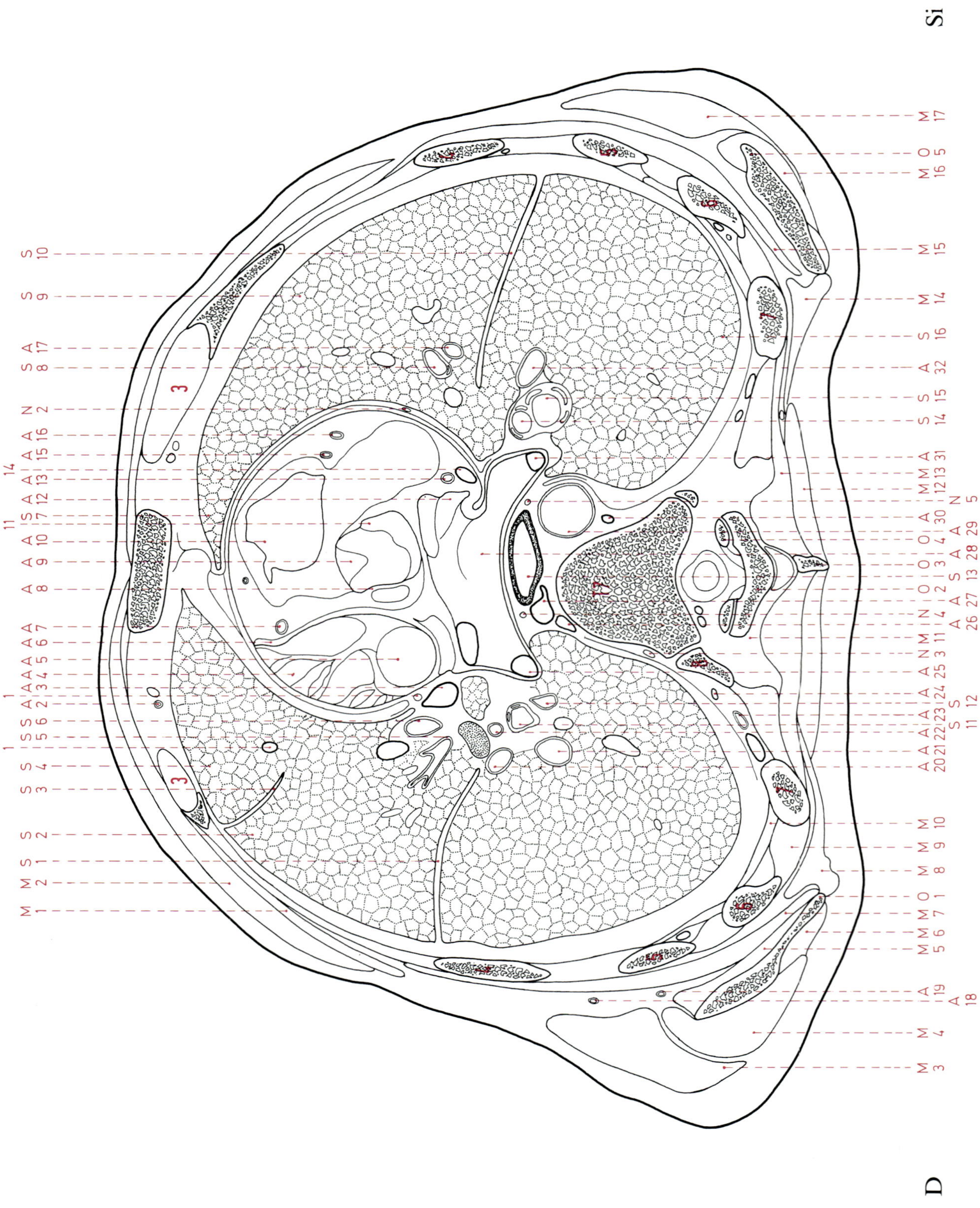

Fig. HT 7

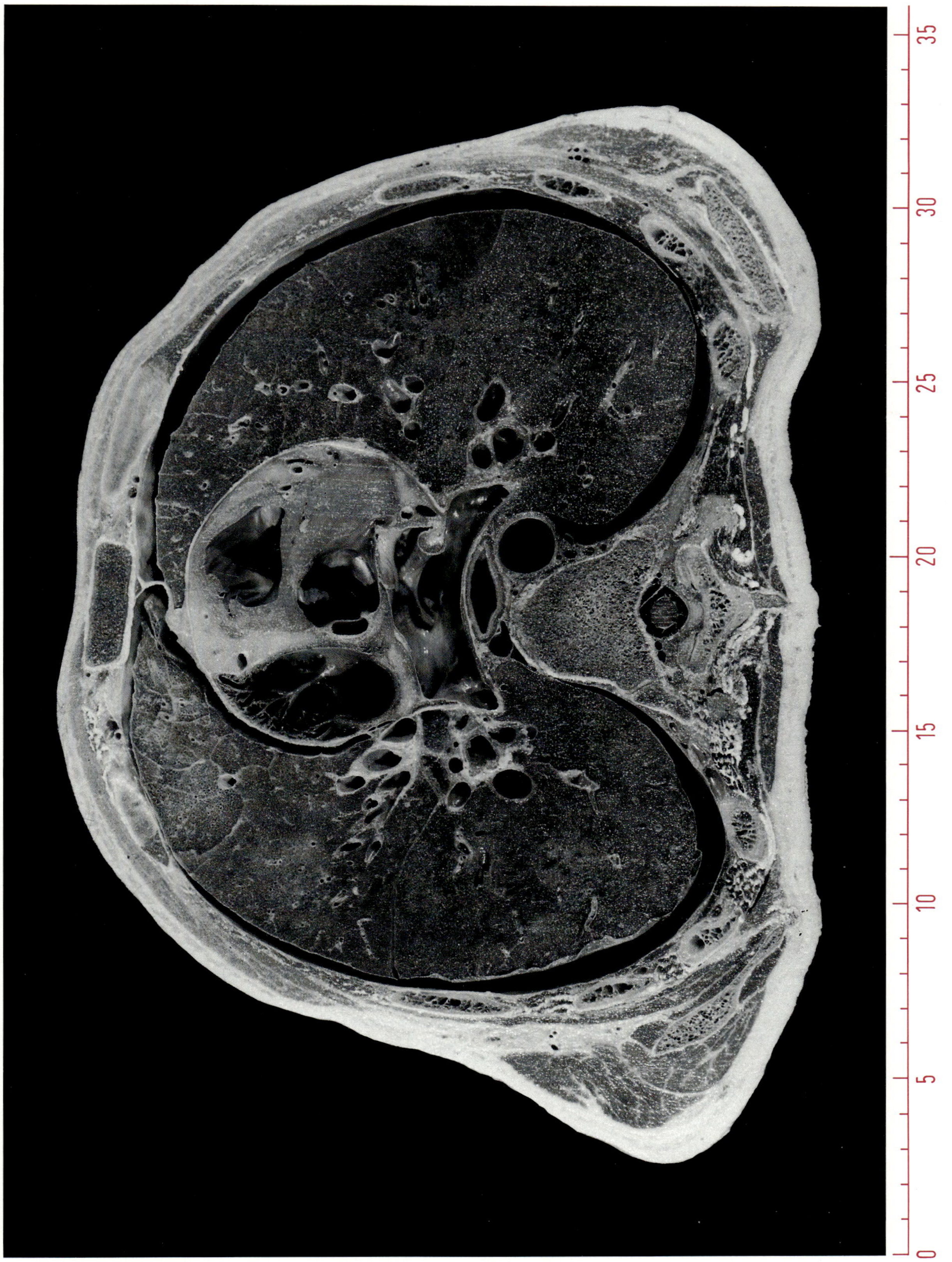

Fig. HT 7

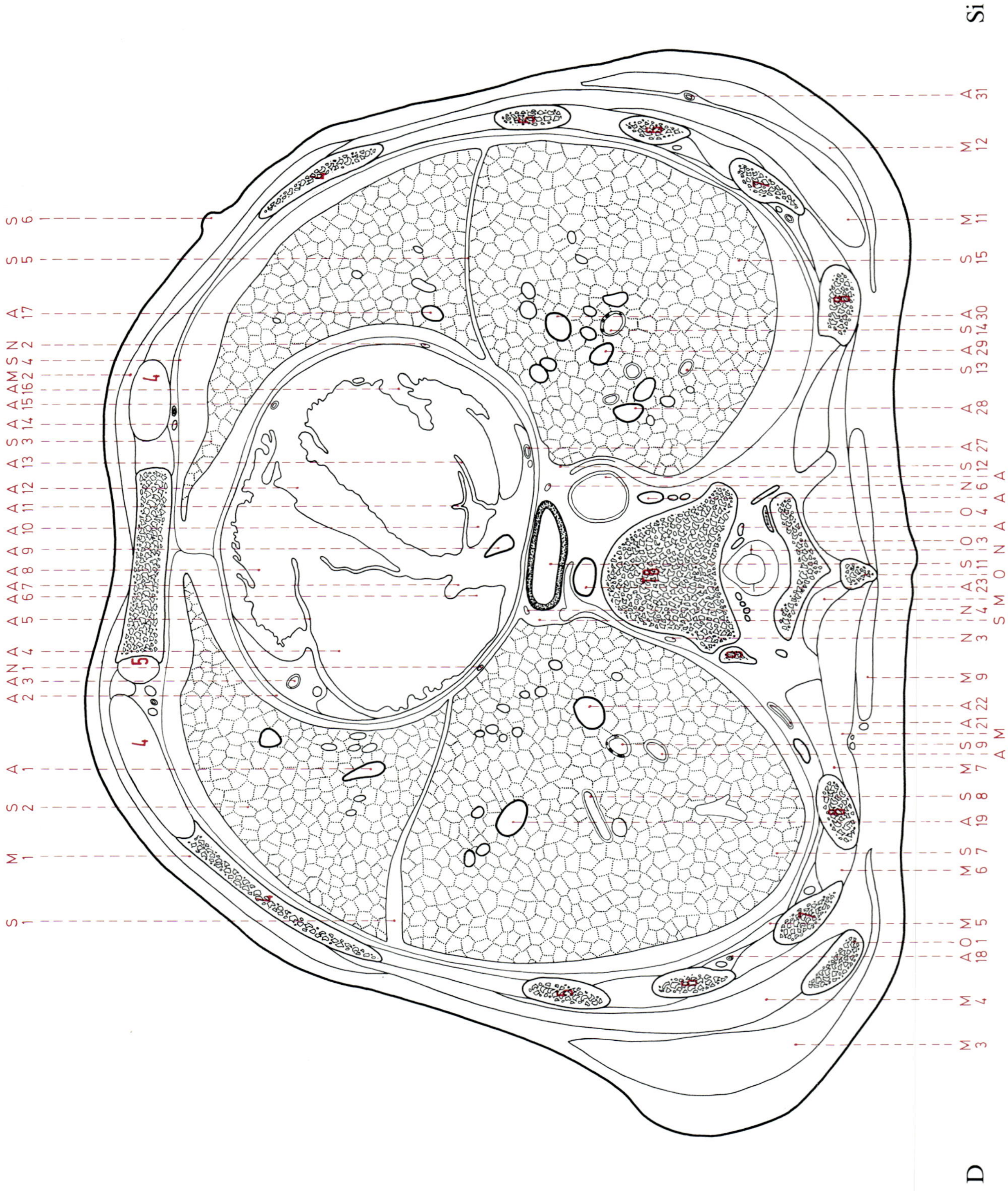

Fig. HT 8

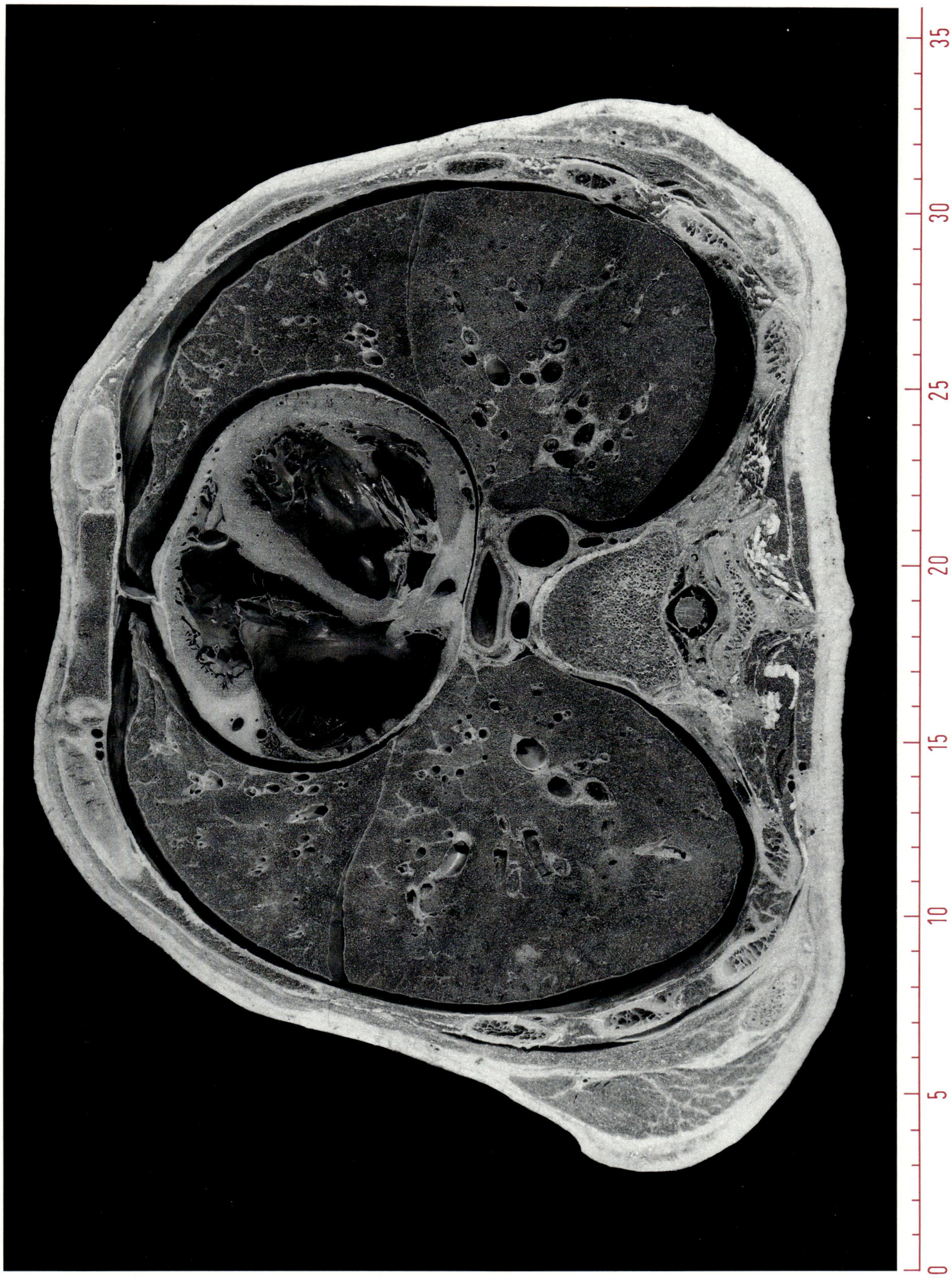

Fig. HT 8

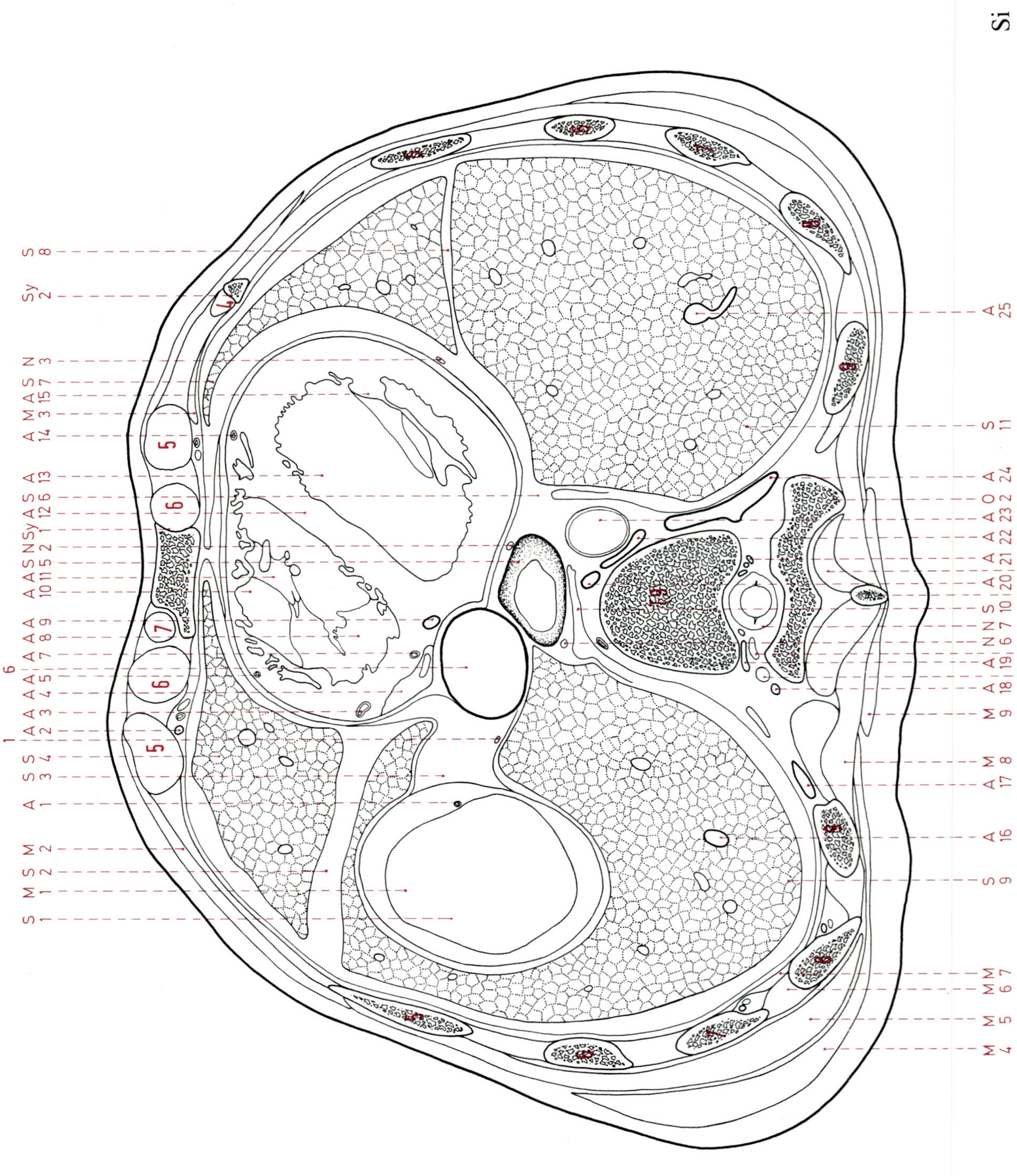

Fig. HT 9

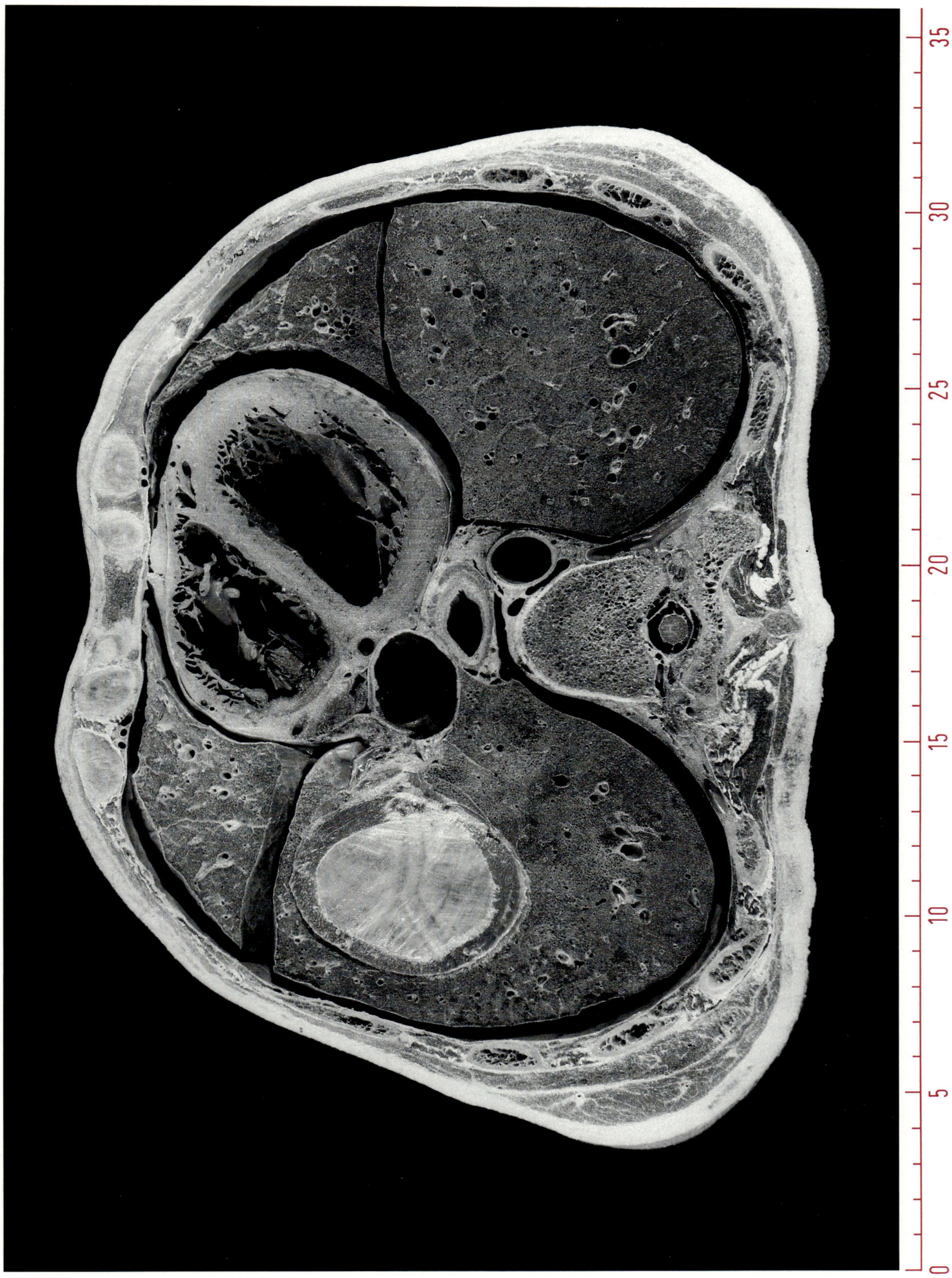

Fig. HT 9

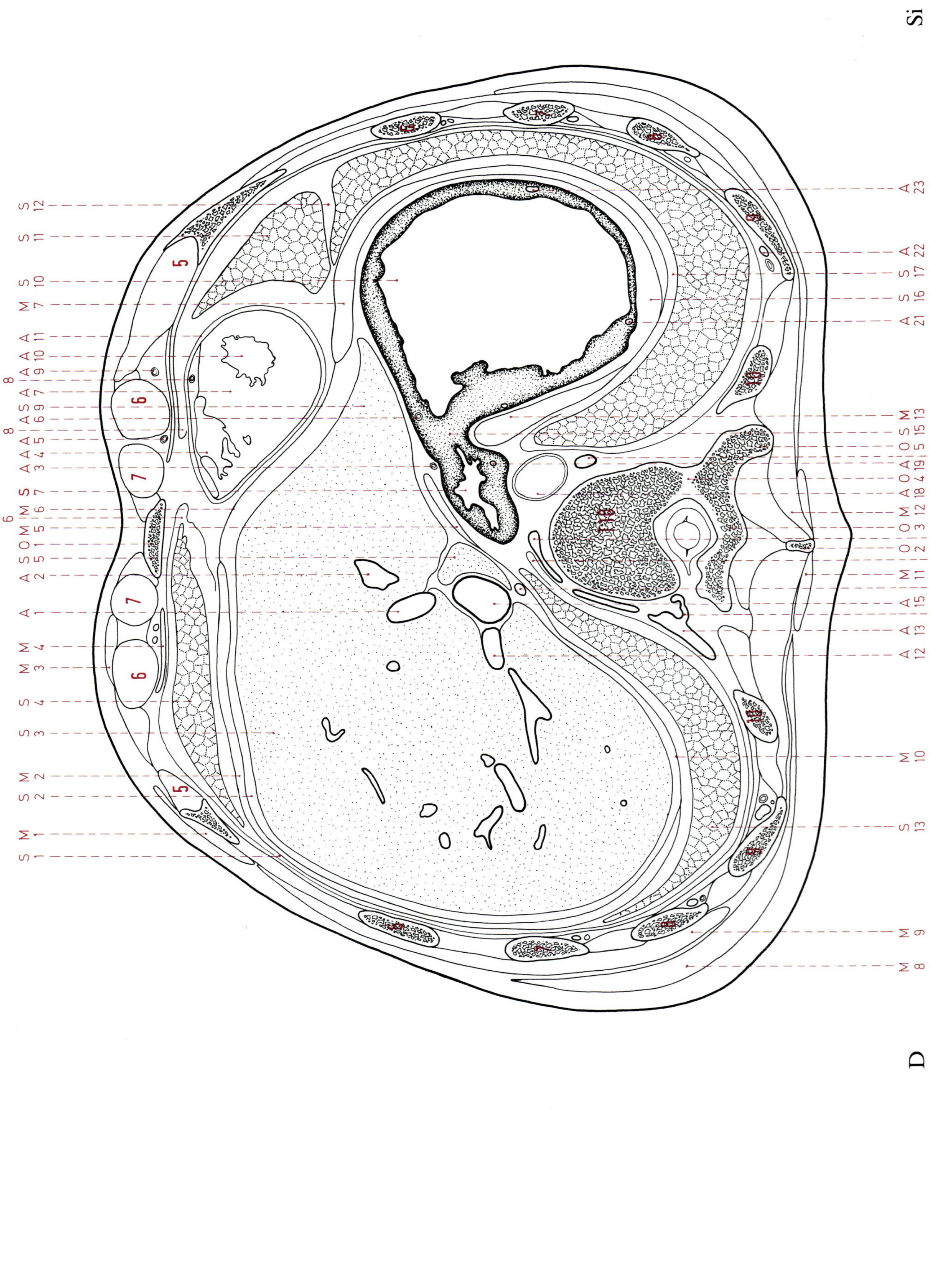

Fig. HT 10

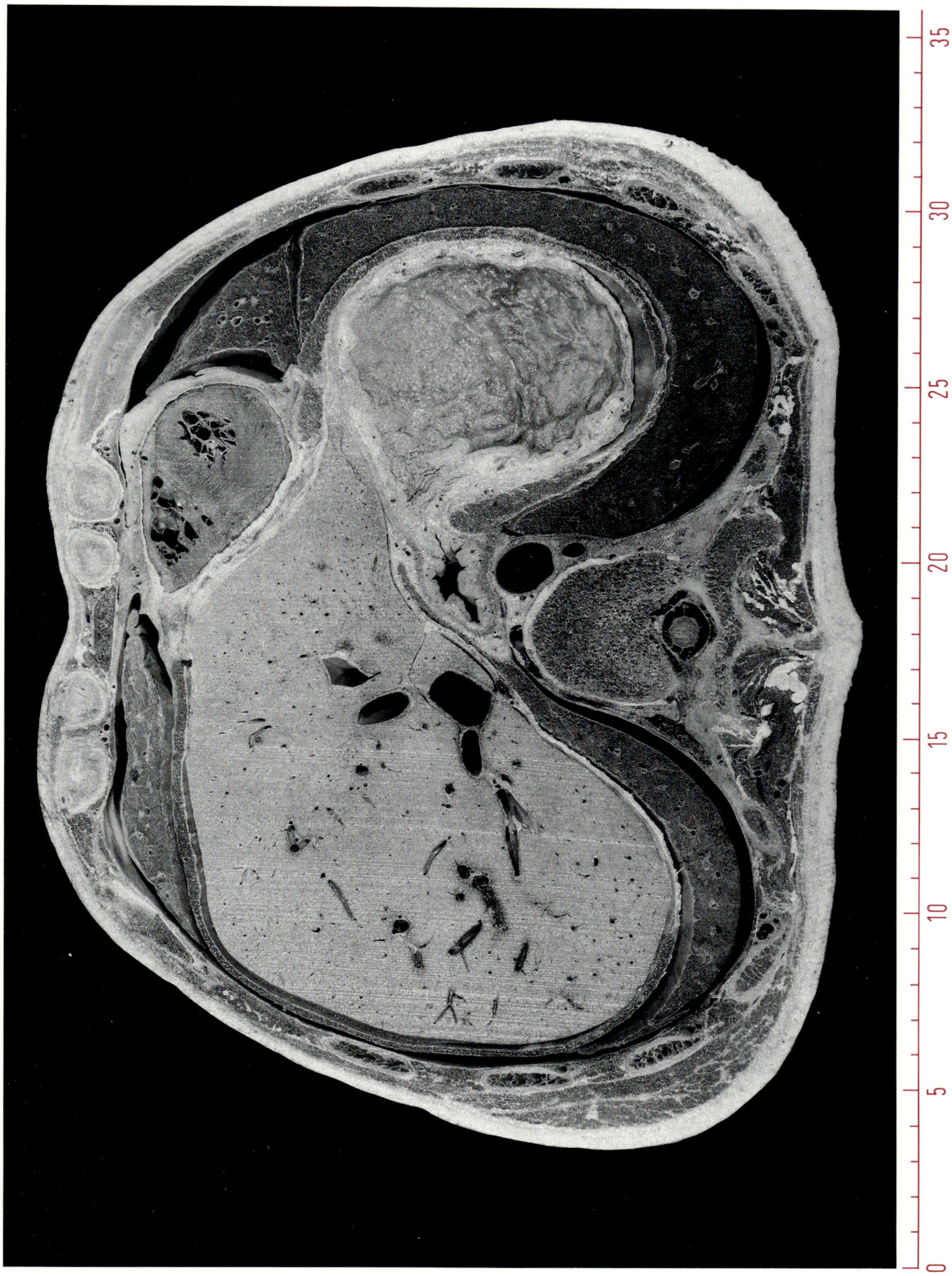

Fig. HT 10

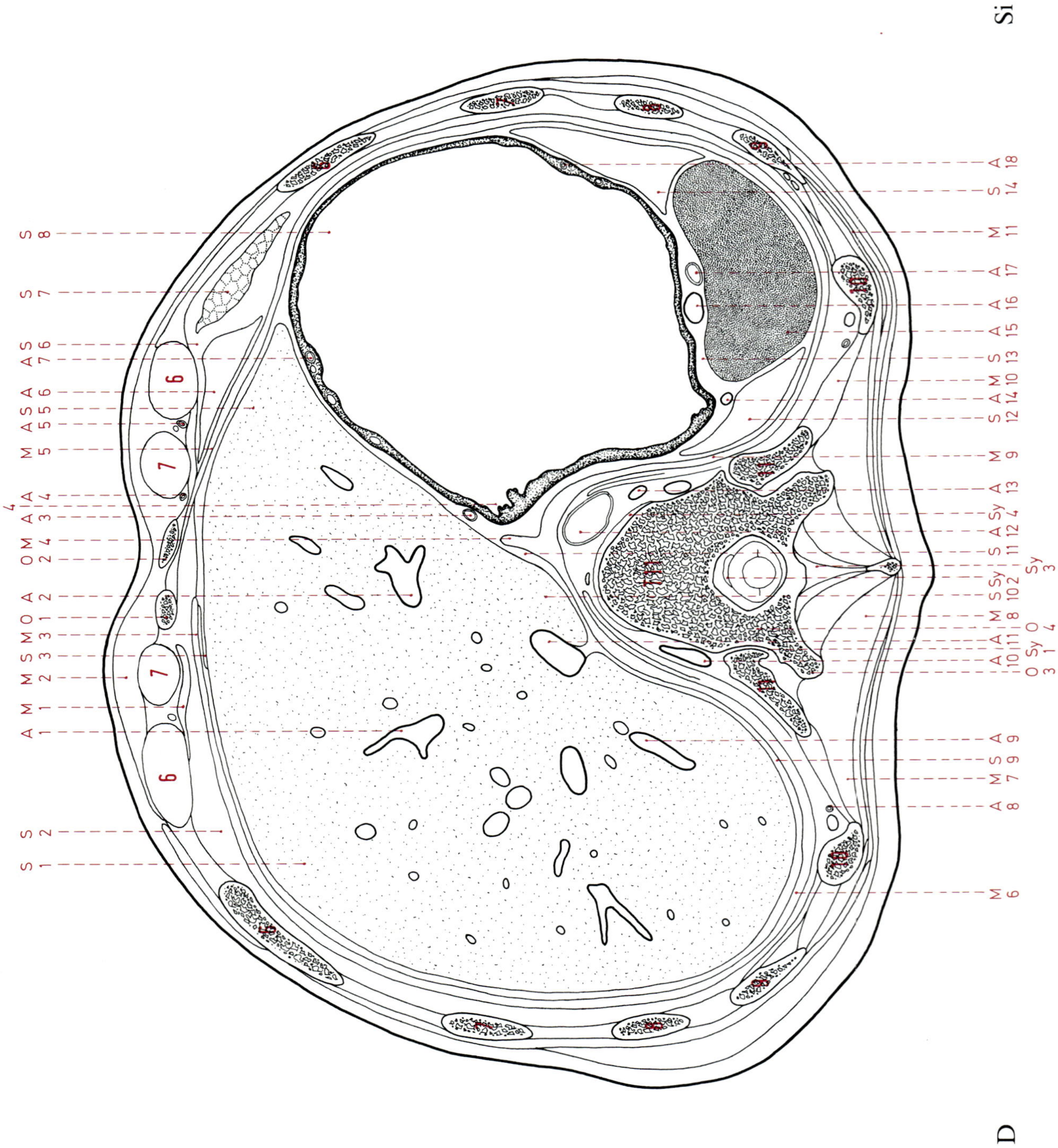

Fig. HT 11

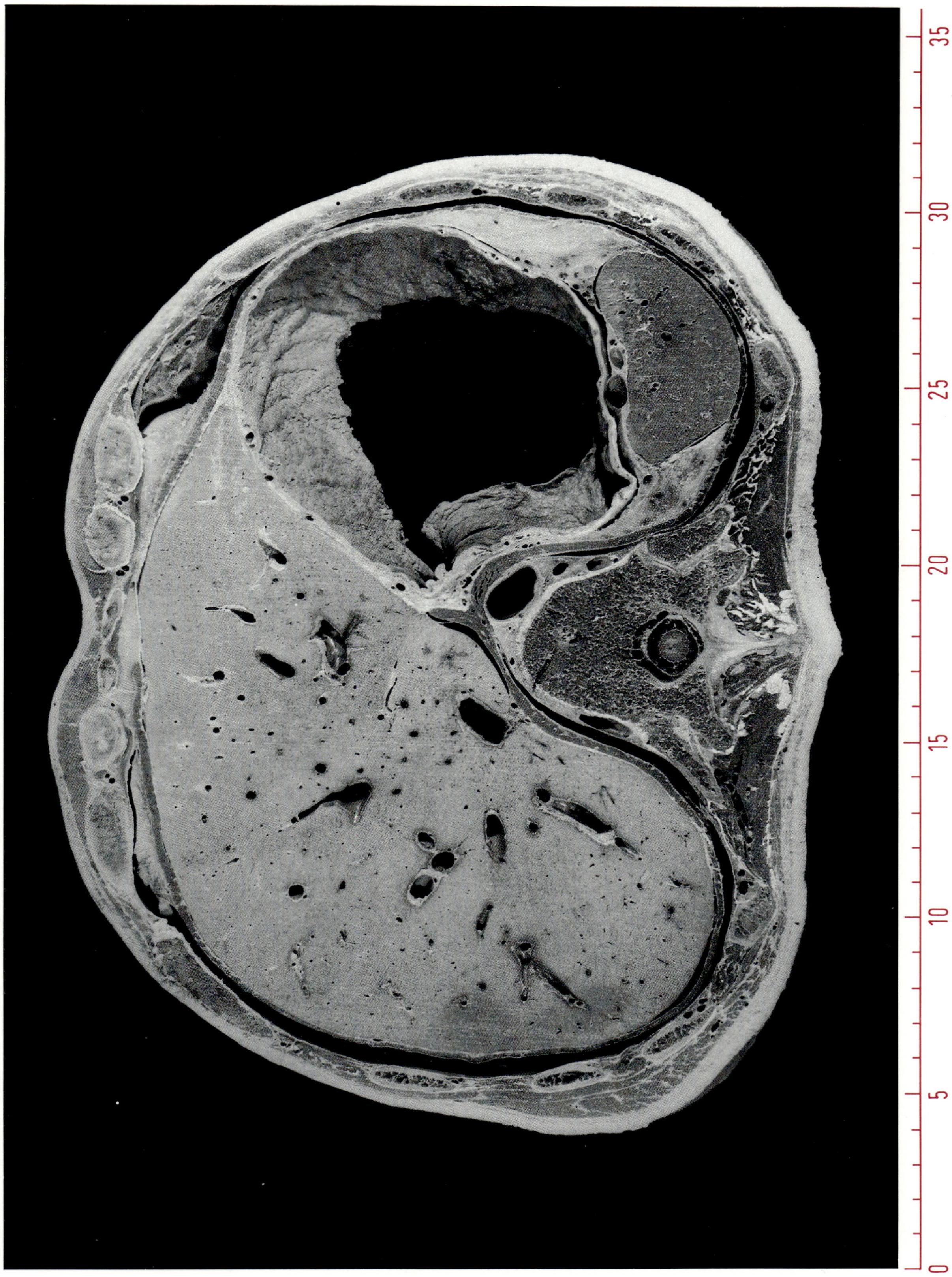

Fig. HT 11

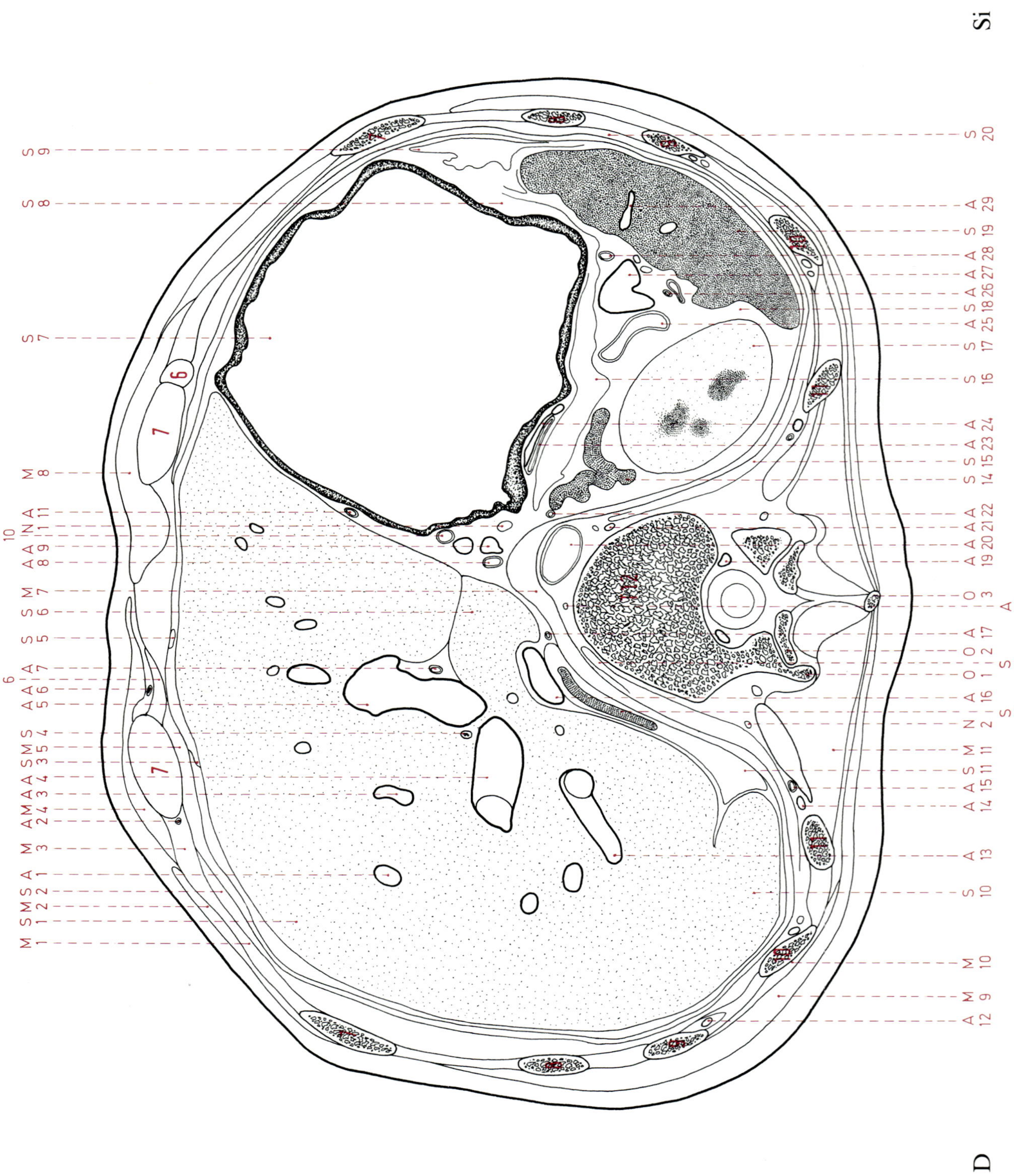

Fig. HT 12

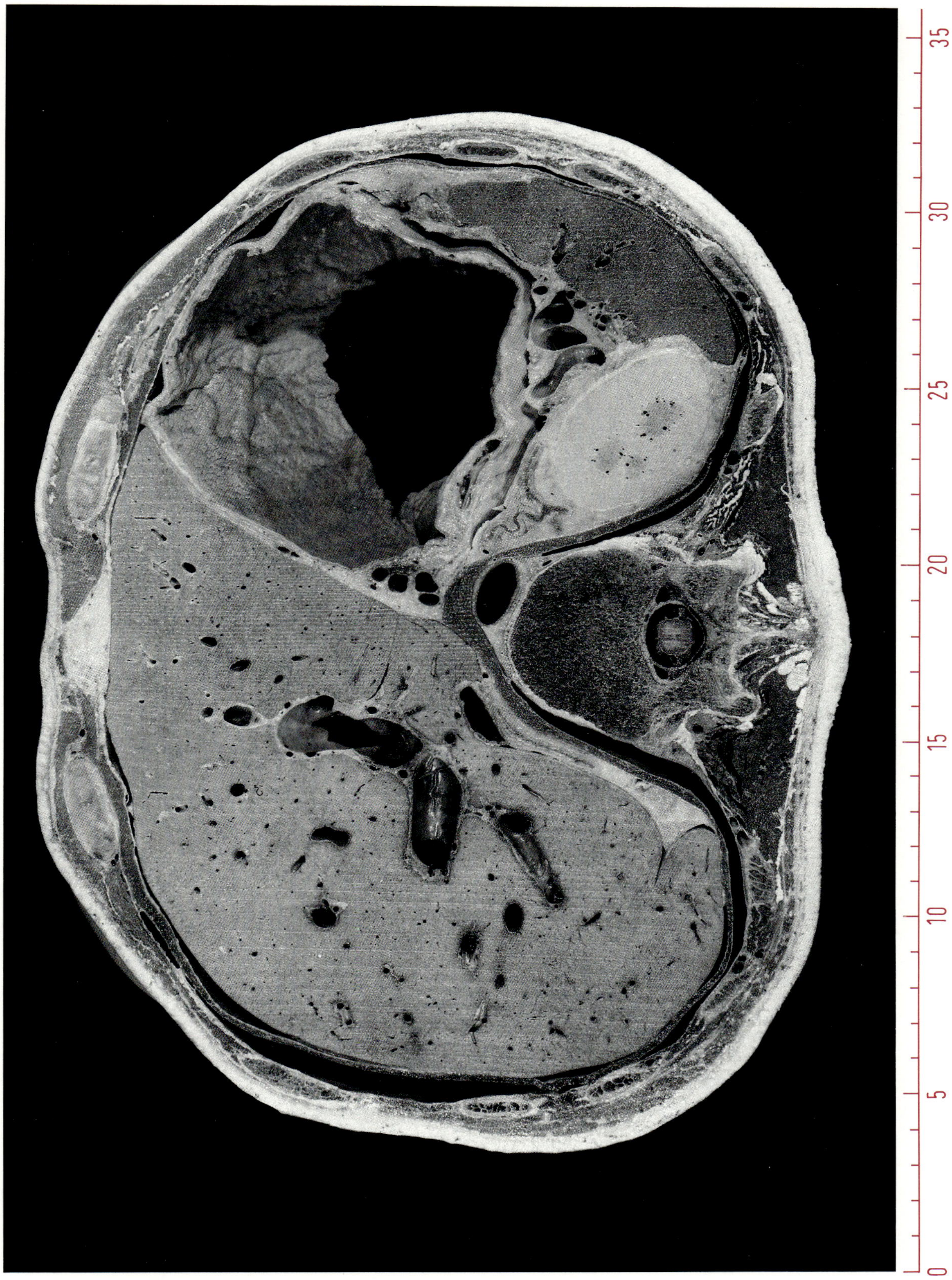

Fig. HT 12

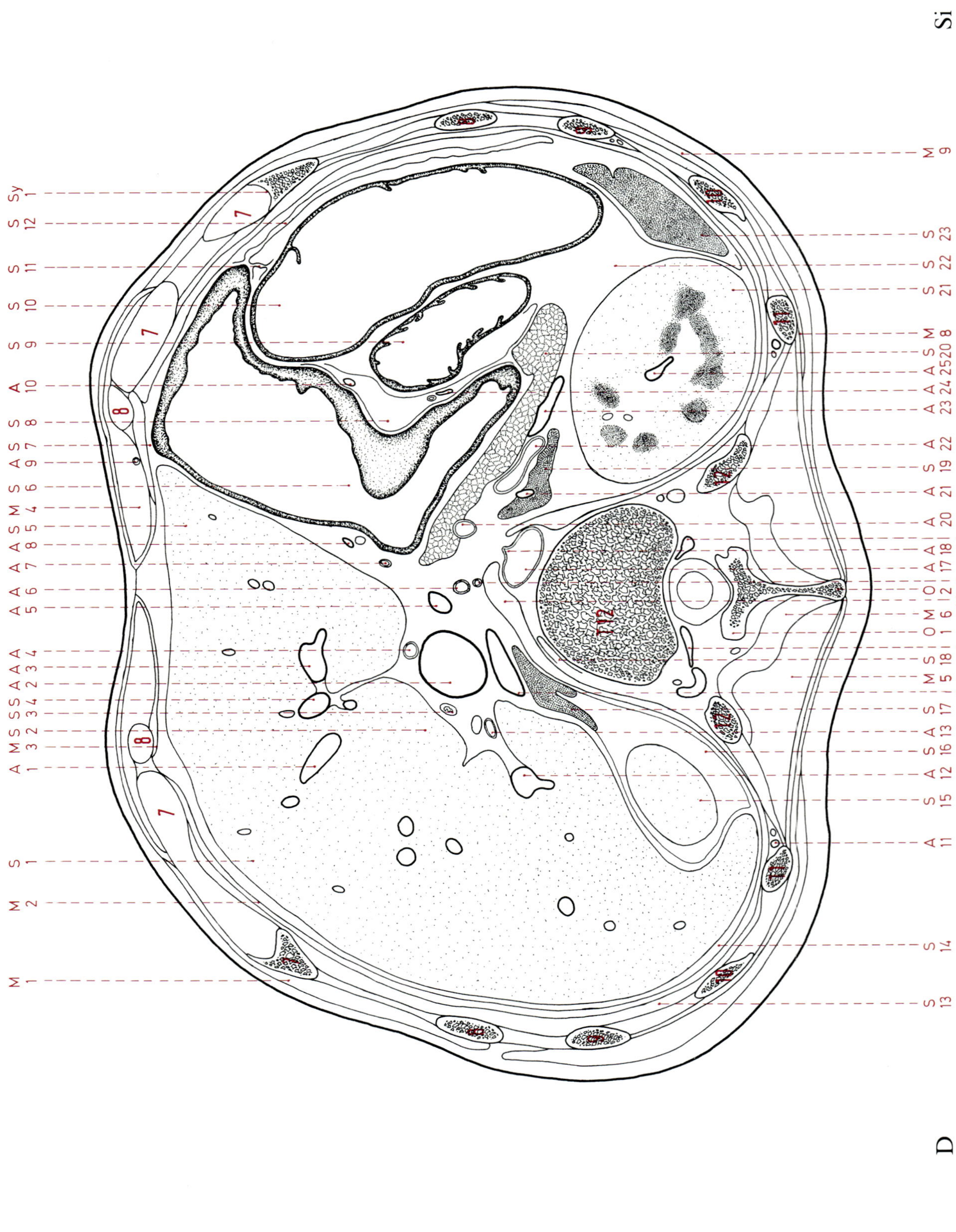

Fig. HT 13

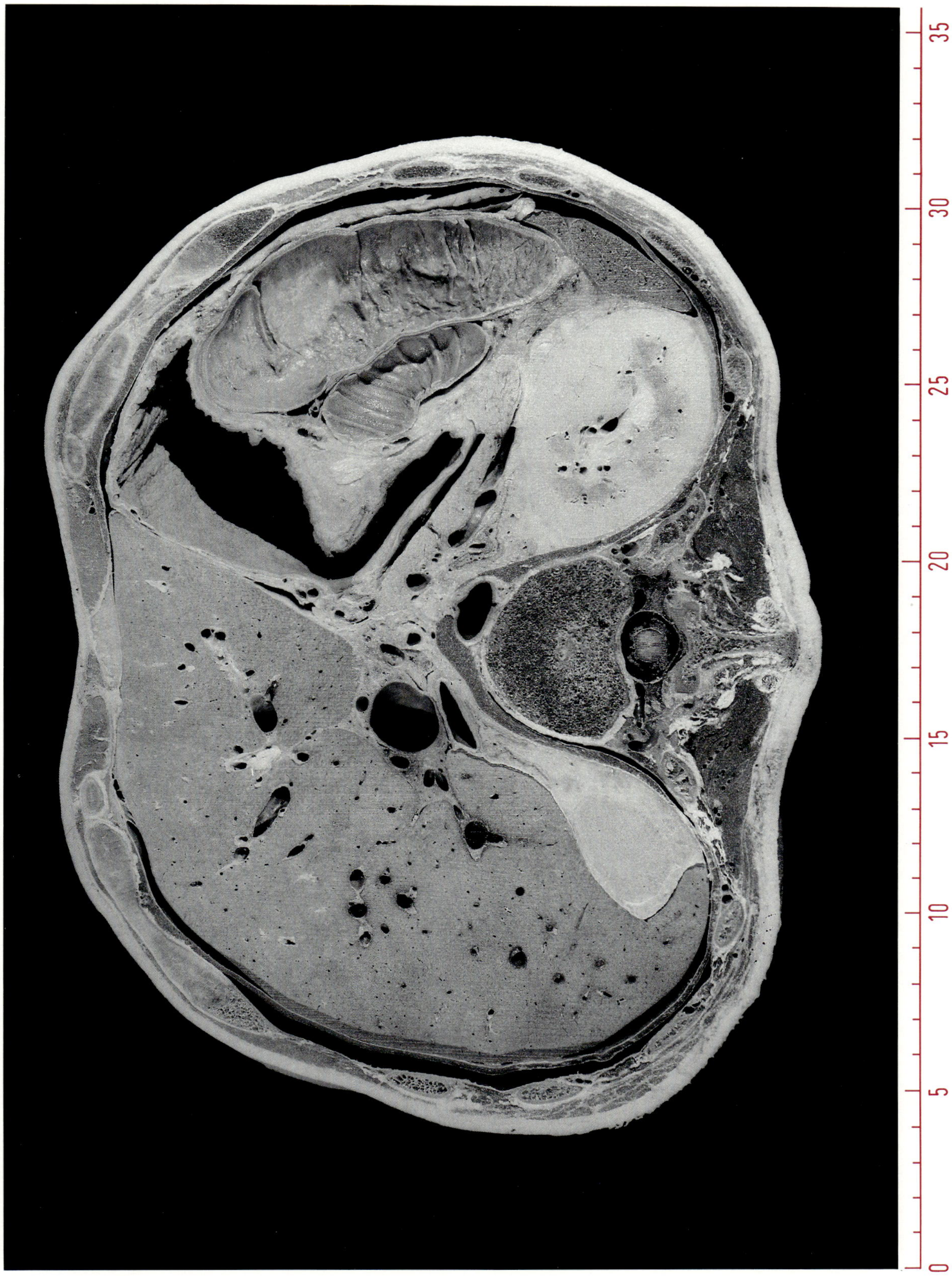

Fig. HT 13

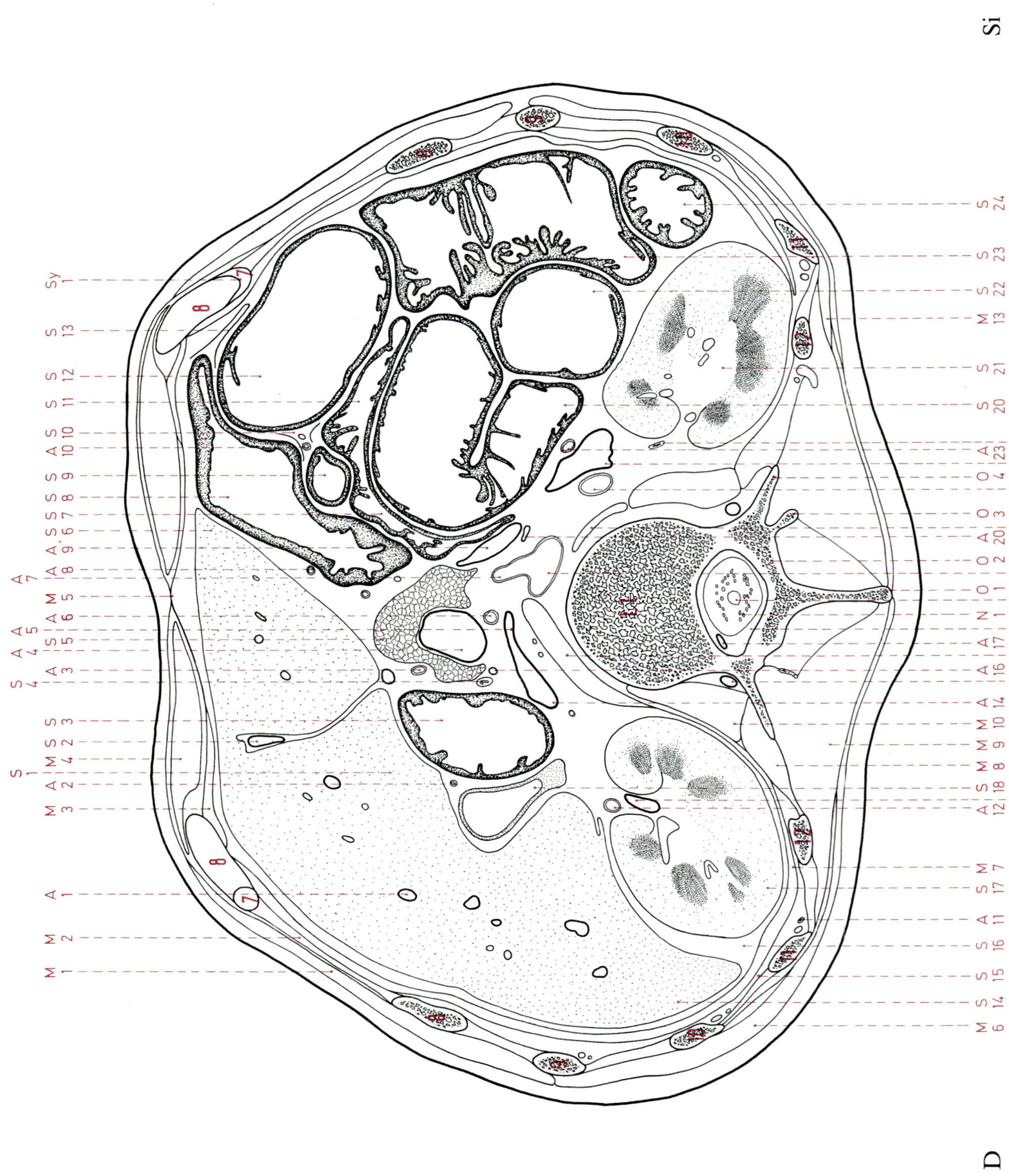

Fig. HT 14

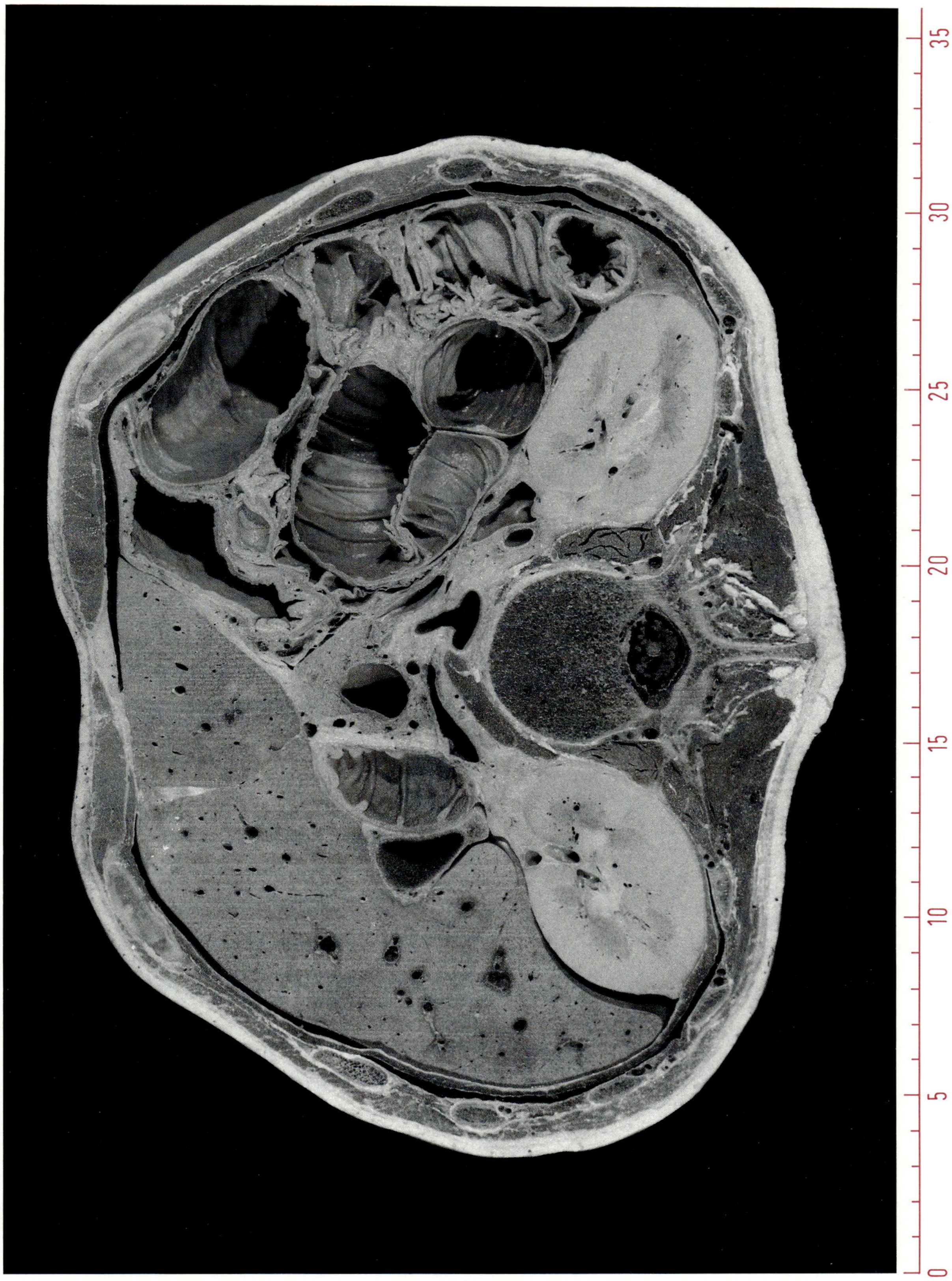

Fig. HT 14

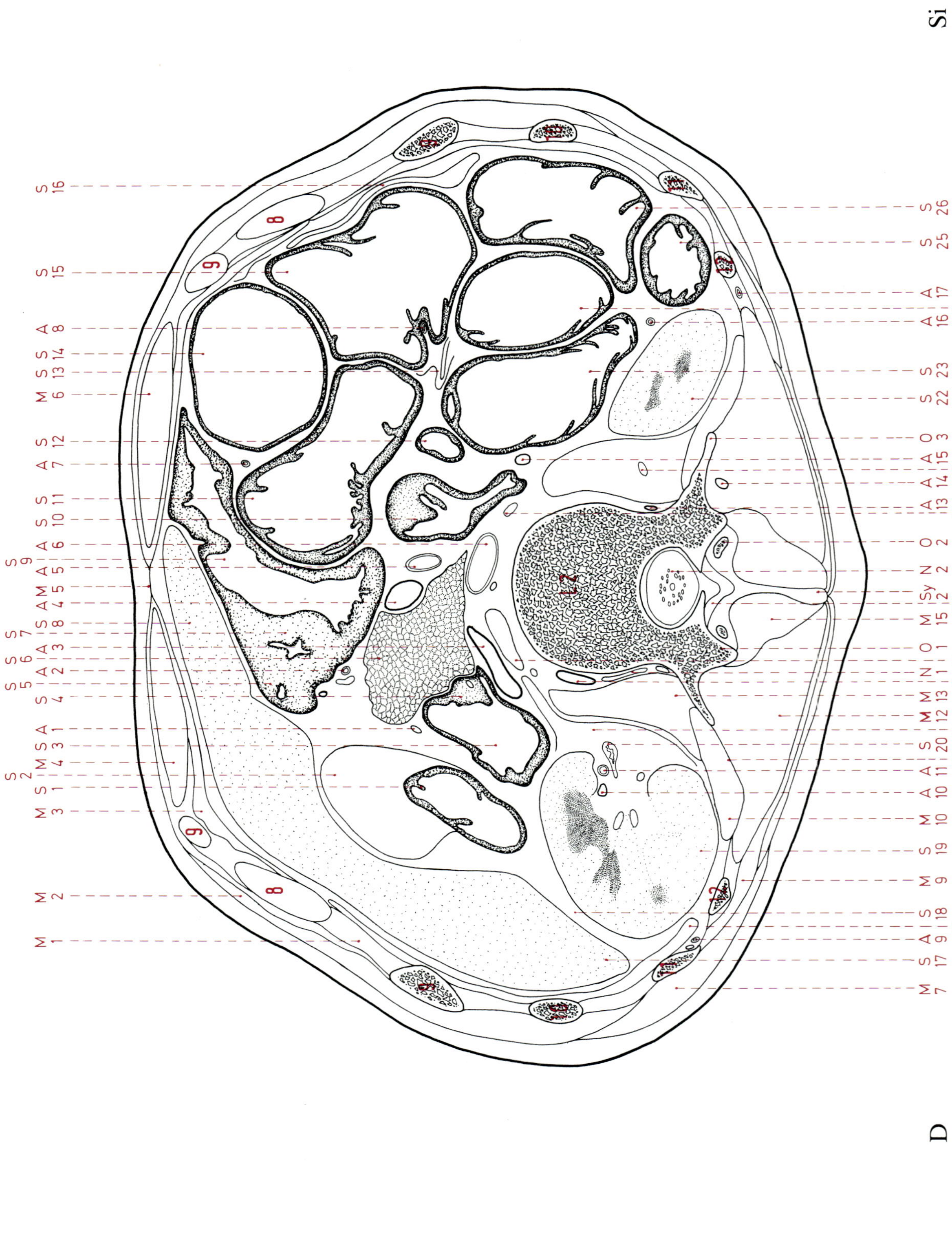

Fig. HT 15

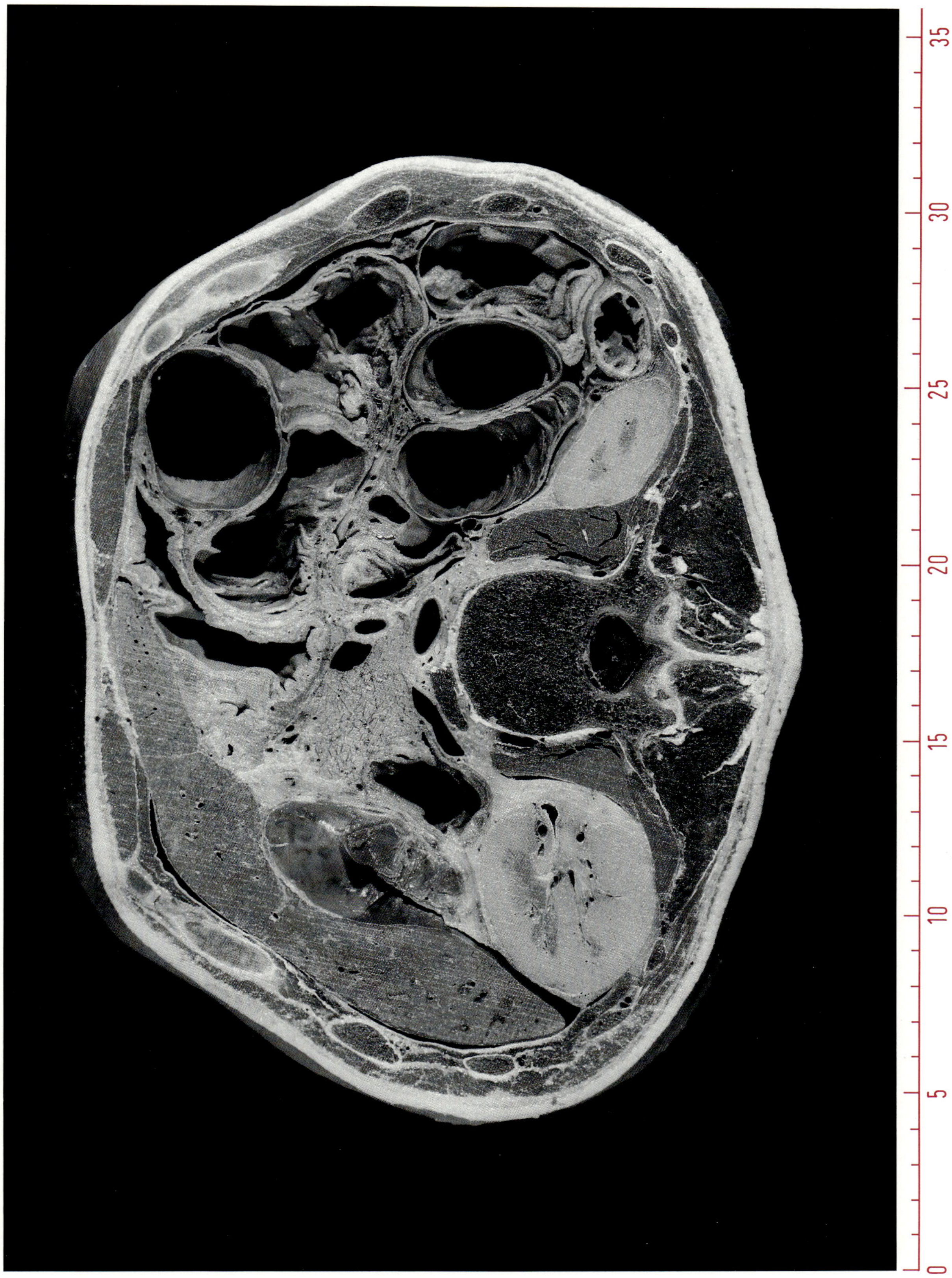

Fig. HT 15

Koritké/Sick
Atlas of Sectional Human Anatomy
Vol. 2: Abdomen, Pelvis

J. G. Koritké and H. Sick

Atlas of Sectional Human Anatomy

Frontal, Sagittal, and Horizontal Planes

Vol. 2: Abdomen, Pelvis

Drawings in collaboration with B. Lafleuriel
Photographs by R. Becker

138 Photographs and Drawings

Urban & Schwarzenberg · Baltimore – Munich 1983

Authors' addresses:

Jean Georges Koritké
Chairman
Department of Anatomy
Faculty of Medicine
Louis Pasteur University
4, rue Kirschleger
F-67085 Strasbourg CEDEX
France

Henri Sick
Professor of Anatomy
Department of Anatomy
Faculty of Medicine
Louis Pasteur University
4, rue Kirschleger
F-67085 Strasbourg CEDEX
France

This atlas consists of two seperate volumes:

Vol. 1: Head, Neck, Thorax (incl. Index)

Vol. 2: Abdomen, Pelvis (incl. Index)

Printed in Germany by Kösel, Kempten

Library of Congress Cataloging in Publication Data

Koritké, Jean Georges, 1928 –
Atlas of sectional human anatomy.

Bibliography: v. 1, p. 185; v. 2, p. 172
Includes indexes.
Contents: v. 1. Head, neck, and thorax –
v. 2. Abdomen and pelvis.
1. Anatomy, Human–Atlases. 2. Anatomy, Surgical and topographical–Atlases. I. Sick, Henri. II. Title.
[DNLM: 1. Anatomy, Regional–Atlases. QS 17 K84a]
QM25.K67 1983 611′.0022′2 82-13558
ISBN 0-8067-1030-6 (U.S. : set)
ISBN 0-8067-1031-4 (U.S. : v. 1)
ISBN 0-8067-1032-2 (U.S. : v. 2)

Vol. 2: ISBN 0-8067-1032-2 (Baltimore)
ISBN 3-541-71032-2 (Munich)
Complete set: ISBN 0-8067-1030-6 (Baltimore)
ISBN 3-541-71030-6 (Munich)

Preface

Illustrations of anatomical sections have existed for a very long time. Some of these sections which are considered characteristic have been presented in papers for essentially didactic purposes. Various parts of the body, in particular the head and the thorax, have been studied using series of sections which have usually been made in one plane in space. Only in exceptional cases have these been made in the three major planes.

New methods of radiologic and tomographic exploration have given a new thrust to the topographical anatomy of sections. It is no longer only the anatomists who deal with pictures of sections of the human body on a daily basis. Today, clinicians in ever-increasing numbers must do the same.

Topographical anatomy has been a part of the Institute of Anatomy at the Department of Medicine in Strasbourg for a long time. It was here that G. Joessel left the most remarkable reproductions of serial sections ever published in their entirety. It was also here that Ph. Bellocq was a lecturer. Here he published his doctoral thesis "Medical-Surgical Anatomy" which actually contained several sections identical to those prepared for this publication.

This atlas could not have been completed without the great help received from the staff at the Institute of Anatomy. We are also grateful to Mr. Robert Becker, the photographer at the Institute of Pathological Anatomy (Director: Prof. Y. Legal). Mr. Bernard Lafleuriel, the illustrator at the Department of Medicine in Strasbourg helped to prepare the diagrams. The control radiographs were made possible through the help of Prof. Francois Kempf.

The publishing house, Urban & Schwarzenberg, has done everything in its power to make the presentation clear and the quality of the print excellent. We are extremely grateful to them, and to Prof. Dr. H. J. Clemens especially, for the insight making it possible to publish an atlas of such high quality.

Strasbourg, April 1982

J. G. Koritké
H. Sick

Table of Contents

Text

Illustrations

Abdomen (Abdomen and pelvis)

Pelvis

Table of Contents Vol. 1

(Condensation)

Introduction

(H. J. Clemens)

An atlas of anatomy normally presents illustrations of a specific anatomical region or organs that are superimposed within that region. Only rarely does it present horizontal sections of the trunk or the extremities, and other sections are even more rare. With the advent of transverse computerized tomography and ultrasonography (echography) and their increasingly common use in clinical examinations, it has become necessary to show serial sections of the various planes of the body. These two diagnostic imaging methods provide the clinician with representations of sections of the human body; interpreting them, however, raises a number of problems, due in part to the lack of literature on normal cross-sectional human anatomy.

Tomographic sections have been prepared in radiology since 1921 (Bocage). These present layers of the body shown by blurring the adjacent zones. The principle involved concerns moving either the X-ray tube and the sensitive plate with respect to the stationary patient or moving the patient and the sensitive plate with respect to the stationary source of radiation. Pictures of sections can be obtained from orientation in all three dimensions. Without minimizing the contribution of this diagnostic method, it should be mentioned that tomography cannot be used to present an exact reproduction of the anatomical substratum.

There has been considerable improvement in the imaging of sections of the human body resulting from computerized tomography (CT). Unlike the methods that were used before this development, the X-rays pass through only the layer to be studied, and yield an exact anatomical reproduction without any displacement. In conventional radiology, the absorption of the X-rays by the body is shown by the various degrees of blackening of the film or by the differences in the luminous intensity of a fluorescent screen. In CT, this absorption is recorded by highly sensitive detectors (scintillation crystals). The source of radiation as well as the system for detecting and measuring the absorption is located behind the patient, and moves horizontally around the body (360° in 2–5 seconds). The impulses detected by the system are amplified electronically and transformed into potential differences. These are sent to a computer where the difference in density is determined for each point of the picture. This information is then sent to a monitor showing the differences in light intensity. The resolving power for two adjacent structures is approximately 2 mm. The tomographic picture displayed on the monitor can be photographed and thus preserved. CT can show differences in absorption of approximately 0.5%, whereas X-ray film only shows differences of 4%. New computer programs will make it possible to produce frontal, sagittal, and oblique sections. In addition they will reduce exposure time and thus artefacts resulting from movement of the heart, the lungs, or other organs.

Two different devices are being used at present: the cranial computer tomograph and the whole body tomograph. The sections of the encephalon are generally made in the axial direction, but depending on the problems presented, the head is sometimes retroflexed (to examine the eye socket) or anteflexed (to examine the posterior fossa). The orbito-meatal line is used as the reference plane in standard examinations. Ten to twelve sections with a thickness of 1 cm are needed to examine the contents of the braincase.

Whole body CT presents transverse sections of the trunk viewed from the inferior side. The section is generally 1 cm thick and can be made at any height of the trunk with the patient supine.

Cranial CT has been of great benefit in medicine. CT of the trunk, however, is only used diagnostically in certain cases since its application overlaps with both ultrasonography and nuclear medicine. In addition to its great importance diagnostically, CT is also used to plan and monitor the progress of therapeutic measures.

An ultrasonic examination, like CT, also makes it possible to present slices of the body. This is based on a completely different physical principle: the reflection of an ultrasonic wave.

Ultrasonic waves emitted by a sender reach the surface of tissues that have varying degrees of permeability. The waves are thus reflected at different intensities. The reflected waves return to the sender, which serves as a receiver between emissions, and there are transformed electronically to luminous points of varying brightness

displayed on a screen. Unlike radiology or scintigraphy, this method makes it possible to obtain pictures of sections from any plane of the body. In general, the acoustic borders of the organs are identical to the anatomical borders. The quality of the ultrasonic picture as well as its usefulness depend on the experience and skill of the operator.
The penetration of an ultrasonic wave into organs depends on the degree to which these organs absorb certain frequencies. Generally, the higher the frequency emitted, the lower the penetration. Because bone almost completely reflects ultrasonic waves and air almost completely absorbs them, organs and tissues located behind bone or air cannot be examined using this method. Bone itself cannot be examined, air becomes an obstacle in the lungs, and intestinal gases block that region. Unlike CT, differences in organ density cannot readily be determined by ultrasonography. Liquids can be distinguished from solids with a high degree of certainty, but resolving power is far less that that of computed tomography.
The advantages of this method should be stressed in addition to the drawbacks:
It is a noninvasive method; neither the patient nor the operator is exposed to any dangerous radiation. The procedure is not expensive and can be repeated an unlimited number of times to follow the course of any process. In the same examination, several organs can be observed from various angles depending on the clinical problems involved and the anatomical factors in the region being examined. Less patient compliance is required than with CT (immobility, length of examination). Like classical radiology and CT, ultrasonic methods also make it possible to observe the dynamic phenomena in the human organism and to present views of topographical anatomy.
The "real time scan" is an orienting examination method that can produce series of sonograms at a rate of 16 to 30 per second. The examination can be performed ventrally, dorsally, or from the flanks, and the probe is moved continuously so that any dynamic changes appearing on the screen can be observed and interpreted without interruption. The equipment required is relatively small, easy to handle, and thus portable. This procedure can be compared with radioscopy.
The "storage scan" stores the information obtained by ultrasonography on transverse or longitudinal sections in a computer. This procedure has an advantage over the real time scan in that transverse, longitudinal, and oblique sections can be obtained from the entire epigastric region. These sections are similar to those obtained with CT and are easier to decipher than those obtained with the real time scan. The power of resolution is greater since the borders of the organs are analyzed from several sides.
The A scan presents a graphic reproduction and makes it possible, among other things, to recognize various parenchymatous organs and to make a distinction between tumors and cysts. Its main application is in echoencephalography – a one-dimensional ultrasonic method in which the amplitudes displayed on the screen do not depend on the intensity of the echos. New devices have been developed recently that make it possible to store a picture displayed during a real time scan without interrupting the examination. This picture can be photographed or entered into a computer memory and recalled whenever required. When the picture is being taken, the contrast, brightness, and positive or negative representation are adjusted automatically according to a preset program. Distances of 1 mm can be measured between two sliding marks on this picture. The result appears immediately when the desired function is called.
It is important that the dynamic phenomena can be recorded on video tape and that the static images from the real time scan can be reproduced on standard photographic equipment. Computerized tomography and ultrasonography can supply us with pictures of sections from nearly every part of the body. The sections can vary in thickness, but are generally 1 cm thick with CT, and a few millimeters thick with the ultrasonic method. There are great differences in the power of resolution for both methods, and improvement is a constant goal. Physical limitations are such that sonograms are more difficult to interpret anatomically than are CT scans.
Before the use of ultrasonics and CT became common, physicians had no noninvasive means of obtaining pictures of sections of the human body. For this reason, they did not have to study such sections in the dissecting room or the hospital. Problems involved in presenting these sections are becoming greater now that it is possible to show them in three dimensions.
It is important to note that the clinician should be wary of many CT atlases in which tomograms are placed next to sections of normal or pathological anatomy. The CT pictures are often not of high quality since they may have been taken with first or second generation equipment. They often give nothing more than a relatively rough orientation since details are lacking and the sections are spaced too far apart. As far as a comparison of sections of pathological anatomy and CT pictures is con-

cerned, knowledge concerning sections of normal anatomy is required before the pathological pictures encountered daily in clinical practice can be detected and then interpreted. As is the case in pathology, it is impossible to present one volume with all the forms, varieties, anomalies, and characteristics depending on age and physiological state. There are also varying degrees of modification due to the preparation methods.

This atlas, like other publications including anatomical illustrations, is a collection of "normal" pictures. Artefacts and pathological lesions have been omitted to the greatest possible extent.

Materials and Methods

We were able to progressively improve our techniques over the long period of time involved in the preparation of this atlas. If new series of sections had to be prepared, the report on the experimentation, which has also been adopted for one part of the sections shown in this atlas, could be defined as follows:

Selection and Preservation of the Material

The material must be normal, *i.e.,* free of any lesions, malformations, or deviations as compared to macroscopic standards. Our subjects were preferably mediolineal young adults whose history and cause of death allowed us to assume that the anatomical structures of the regions to be studied were complete. Approximately 30 subjects meeting these requirements were used in this study. The series of sections presented here were supplied by only 11 of them.

As soon as the cadaver is brought to the Institute of Anatomy it is embalmed by injecting a preserving fluid (consisting mainly of alcohol, formaldehyde solution, and glycerine) into one of the femoral arteries. The nerve centers are not sufficiently fixed by this method, so they are injected with 100 cc of a 20% formaldehyde solution into the leptomeningeal spaces. The cadaver is then immersed in a 30% solution of ordinary alcohol where it lies perfectly horizontally for three months. This length of fixation appears to be the most appropriate. If it is shorter, fixation is incomplete, resulting in soft tissues which are difficult to work with and preserve. If it is longer, the colors and contrasts between neighboring tissues are altered. The sections are also preserved in a 30% solution of alcohol.

Production and Preparation of Sections

After removing the cadaver from the fixing fluid, it is placed in the horizontal position in the freezing room. The body is placed and maintained in the anatomical position for the complete freezing period (one week at −20° C). After this period, the different hardened tissues present a uniform resistance to the progression of a saw blade. The low temperature of preparation and the great length of the saw blade reduce the heat generated, and thus prevent any alteration of the slices.

The body, frozen *in toto,* is cut into blocks along the transverse or horizontal planes perpendicular to the median sagittal plane or the frontal plane upon which the subject is lying. These cuts are made at a certain distance from the areas to be studied. Thus, the head and neck block includes the upper part of the thorax; the thorax block extends from the middle of the cervical column to the middle of the lumbar column and the abdomen block from the region above the diaphragm to the middle third of the thigh.

Each block is cut into sections along one of the following planes of reference.

The plane of reference for frontal sections is the plane upon which the subject is lying. The eye-ear plane (Frankfurt, horizontal, orbito-meatal line) is thus perpendicular to this frontal plane of reference.

The plane of reference for sagittal sections is the median plane. It is difficult to locate and follow since the spinal column presents lateral curves, pronounced to varying degrees, in nearly all cases.

The plane of reference for horizontal sections depends on the region to be studied. It is represented by the eye-ear plane (Frankfurt horizontal, orbito-meatal line) for the head and neck and by the plane passing through the two antero-superior iliac spines for the thorax, abdomen, and pelvis.

The sections made from the plane of reference are normally of uniform thickness for a given block. The thickness varies, however, for the different blocks. It ranges from 1.5 cm for horizontal sections from the cephalo-cervical region to 2.5 cm for frontal sections. The very large size of the frontal sections made it necessary to increase their thickness. The block is placed on an adjustable guiding plane parallel to the saw blade in order to ensure that the section is of uniform thickness.

Immediately after sectioning, each specimen is placed on a plate and covered with plastic material permitting easy handling. The accessible surface is placed on a similar plate by simply turning it over. The surfaces of the sections are washed with running water and a sponge. The contents of the natural cavities are thus removed while the specimen is still in the frozen state. While these simple operations are being performed, the remarkable appearance of the shape and colors of the organs can be observed.

The washed sections are covered with plastic sheet material to prevent dehydration. They are then stored in the refrigerating room. Before photographing, they are studied separately to determine degree of normality, to touch up an organ of primary importance located at a very slight distance from the cutting plane, and to regularize certain contours.

Photography

The photographic equipment consists of a 9×12 Technika Linhof camera with a 4.5/150 mm Xenar F objective. Normal exposure is ½ sec with the aperture set at 16. Two 500 W lamps are placed 1.5 m away from the specimen.

Pictures are taken of both sides of each specimen next to a graduated rule. Enlargements are made in order to present horizontal sections of the trunk in their natural size, to present horizontal sections of the head and neck as 1.5 times their natural size, and to present all sagittal and frontal sections as 0.75 times their natural size. A ruler with the conversion data is contained in the back of this volume.

The photographs have been chosen to a large extent on the basis of their clinical importance. Thus, horizontal sections are viewed from the bottom side; when elements of primary importance only appear on the top side, however, this side has been added to complete the information. All frontal sections are viewed from the front. Some posterior views considered absolutely necessary complete each series. All sagittal sections are viewed from the exterior.

The frontal sections are numbered starting from the posterior plane of reference upon which the subject is lying. The letter "a" indicates that the specimen is viewed from the front and "p" that it is viewed from the back. The sagittal sections are numbered starting from the median plane and moving to the left and the right. The median plane divides the specimen into a right segment viewed from the left (MD) and a left segment viewed from the right (MSi). The horizontal sections are numbered from top to bottom within the same block and are marked with respect to a plane of reference.

Thus, each section is defined as follows: by a first letter indicating the general orientation (F=frontal; S=sagittal; H=horizontal), by a second letter qualifying the overall region to which it belongs (C=cephalo-cervical; T=thoracic; A=abdominal; P=pelvic), by a symbol, if required, indicating the sex (♀ or ♂), by a number marking it within the block, and perhaps by an indication as to the position of the observer with respect to the plane of the section (superior, inferior, right, left, anterior, posterior).

Analysis of Sections

During specimen preparation, the major objective is to keep the organs in their original position.

During the process of identification, however, the organs can be treated individually. A sufficient order must be retained, although, to allow for later verification.

First the sections are radiographed. Each radiologic negative is photographed. The photographic enlarger used to project the radiologic image of the skeleton makes it possible to set up a grid displaying the height of the center of the vertebral bodies. This grid is applied to each side of the photographs of frontal and sagittal sections.

The organ parts separated by the saw cut are sutured to each other and then to the specimen box wall. After all the elements of the section have been joined, it can be immersed in the fixing fluid and handled without being damaged.

Tracing paper is placed on each photograph, which is then compared with the original specimen. The radiologic negative makes it easier to recognize the skeletal elements, but quite frequently the entire anatomical region must be reconstructed. The muscle or aponeurotic layers are separated from each other. Each vascular lumen in all sections is injected systematically to determine the origin and area of the vessels. The full path of the nerves is followed from one section to the next. In some instances we had to revert to histological techniques to support, with certainty, some of our findings.

Drawings

Each drawing, traced from a photograph, is a true reflection of the actual anatomy. Only those elements that are considered absolutely necessary to interpret the photograph are added to the drawing using a conventional code. The photograph thus serves as the primary reference. The drawing directs the observer and stimulates his research, but should in no way limit his field of investigation. Because each photograph contains a vast amount of interesting elements, we were often faced with difficult decisions regarding the representation and references to the labels. Regrettably the multitude of direction lines made it impossible to place the nomenclature in a good readable way at the lines. To achieve a better clearness a code system has been employed whose abbreviations should be understandable internationally.

The references to labels are based on a letter and a number permitting rapid access to the terminology by referring to a system (A=angiology; M=myology; N=nervous system; O=osteology; OS=sense organs; S=splanchnology; Sy=syndesmology). Each reference is explained in the "Index of Labels" section.

The choice of nomenclature is always a difficult problem. To make the terms as accessible as possible, all labels have been reviewed for their appropriateness to the English-speaking clinician and strict adherence to Nomina Anatomica has been abandoned in favor of practical terminology.

Directions for Use

The most important document is the photograph, a true reflection of the actual anatomy. From the very beginning, the observer is able to record a large amount of information, with the progression of the analysis indicated by the diagram and then the labels.

Photographs

The photographs reproduced on the right-hand pages convey different types of information: general appearance of the section, shape of its surfaces, positions, proportions, impressions of the density of the organs. The entire section or each of its elements can be situated within the body by referring to the following: the letters characterizing each section with respect to the planes of reference; the height of the vertebra shown on the osseous section as soon as it appears; the height of the center of the vertebral body indicated on both sides of all photographs of frontal and sagittal sections; the level of the ribs; and the scale shown at the bottom of each photograph.

Drawings

Each diagram, an interpretation of the actual anatomy, brings out the essential characteristic elements of the section. The thickness of the features and the character of the grid representing the organs indicate to which system the organs shown belong. References are also made using keys. Symmetric organs are not necessarily presented in a bilateral manner. This makes the drawing clearer. It was not always possible to superimpose extremely thin planes. A single feature can give an outline of several structures. Cuts of vertebrae and ribs are designated in any case.

Labels

The labels supply further information and help to check the identification of the photograph and the interpretation of the drawing.

Abbreviations have been used for the identification and orientation of the section, the classification of organs referred to by large systems, and finally in the labels. The first examination of the photograph, the interpretation of the drawing, and then reading the labels should logically lead the observer to re-examine the photograph in order to record the numerous elements that could not be referred to in the space allowed. It should then be possible to imagine and compare the image that will be supplied by CT, ultrasound, or any other new technique technology brings to us.

Abbreviations

1. Identification and Orientation of the Sections

a: anterior – anterior
p: posterior – posterior
D: right – dexter
Si: left – sinister
F: frontal – frontalis
H: horizontal – horizontalis
S: sagittal – sagittalis
C: head, neck – caput, collum
T: thorax – thorax
A: abdomen – abdomen
P: pelvis – pelvis
♀: female – femininum
♂: male – masculinum

The vertebrae are indicated by the letters C (cervical – cervicalis), T (thoracic – thoracicus), L (lumbar – lumbalis), S (sacral – sacralis) and Co (coccygeal – coccygea). The number that corresponds to their level is added to the letter. The ribs are marked by a number indicating their level ("3" refers to the third rib).

2. Association of the Organ to a Large System

A: Angiology – Angiologia
M: Myology – Myologia
N: Nervous System – Systema Nervosum
O: Osteology – Osteologia
OS: Sense Organs – Organa Sensuum
S: Splanchnology – Splanchnologia
Sy: Syndesmology, Arthrology – Syndesmologia

3. Abbreviations used in the Labels

a.: artery – arteria
art.: articulation – articulatio
b.: bursa – bursa
C.: cervical (vertebra) – cervicalis (vertebra)
Co.: coccyx – os coccygis
gl.: gland – glandula
L.: lumbar (vertebra) – lumbalis (vertebra)
lig.: ligament – ligamentum
m.: muscle – musculus
n.: nerve – nervus
no.: node (lymphatic) – nodus (lyymphaticus)
r.: ramus – ramus
S.: sacral (vertebra) – sacralis (vertebra)
v.: vein – vena
( ): specifies the preceding term

References

Adachi, B.: Das Arteriensystem der Japaner. Verlag der Kaiserlich-Japanischen Universität zu Kyoto, Kyoto 1928.

Alfidi, R., J. Haaga, M. Weinstein, J. de Groot: Computed tomography of the human body. An atlas of normal anatomy. C. V. Mosby Co., St. Louis, 1977.

Aubaniac, R., J. Porot: Radio-anatomie générale de la tête. Masson, Paris, 1955.

Baert, A., L. Jeanmart, A. Wackenheim: Clinical computor tomography, Springer, Berlin, Heidelberg, New York, 1978.

Baert, A., L. Jeanmart, A. Wackenheim: Atlas of computor tomography, Springer, Berlin, Heidelberg, New York, 1980.

Bellocq, Ph.: Anatomie médico-chirurgicale. Anatomie des formes extérieures et Anatomie des régions. Masson, Paris 1925–1957, I–XI

Bo, W., I. Meschan, W. Krueger: Basic atlas of cross-sectional anatomy. W. B. Saunders Co., Philadelphia, London, Toronto, 1980.

Bocage, A. E. M.: Procédé et dispositifs de radiographie sur plaque en mouvement. Brevet n^0 536464 (1921/22).

Braune, W.: Topographisch–anatomischer Atlas nach Durchschnitten an gefrorenen Cadavern. Veit u. Comp., Leipzig, 1875.

Carter, B., J. Morehead, S. Wolpert, S. Hammerschlag, H. Griffiths, P. Kahn: Cross-Sectional anatomy. Computed tomography and ultrasound correlation. Appleton-Century-Crofts, New York 1977.

Delmas, A., B. Pertuiset: Topométrie crânio-encéphalique chez l'Homme. Masson, Paris/Thomas, Springfield, 1959.

Eycleshymer, A., D. Schoemaker: A cross-sectional anatomy. Appleton-Century-Crofts, New York 1970.

Feifenbaum, H.: Echography. Lea & Febiger, Philadelphia, 1972.

Friedemann, G., U. Mödder: Leistungsfähigkeit der Ganzkörpertomographie. Deutsches Ärzteblatt, 48, 2891–2899 (1978).

Gambarelli, J., G. Guerinel, L. Chevrot, M. Mattei: Coupes sériées du corps humain. Anatomie – Radiologie – Scanner. Springer, Berlin, Heidelberg, New York/Masson, Paris, New York, Barcelone, Milan, 1977

Gebauer, A., U. Scherer: Differentialindikation für die Computertomographie und die Ultraschalluntersuchung. Internist, 19, 568–578 (1978).

Gonzalesz, C., C. Grossman, E. Palacios: Computed brain and orbital tomography. Technique and interpretation. Wiley & Sons, New York, London, Sydney, Toronto, 1976

Hounsfield, G. H.: Computerized transverse axial scanning (tomography). I. Description of system. Br. J. Radiol., 46, 1016 (1973).

Hovelacque, A., O. Monod, H. Evrard: Treize coupes horizontales du thorax. Maloine, Paris, 1938.

Joessel, G.: Lehrbuch der topographisch-chirurgischen Anatomie. Max Cohen, Bonn, 1884.

Lanz, T. v., W. Wachsmuth: Praktische Anatomie. Hals. Springer, Berlin, Göttingen, Heidelberg, 1955.

Lanz, T. v., W. Wachsmuth: Praktische Anatomie. Kopf. Springer, Berlin, New York, Heidelberg, 1979.

Lissner, J., W. Schwarz: Technik und Apparaturen der Ganzkörper-Computertomographie. Internist, 19, 559–563 (1978).

Luschka, H. v.: Die Lage der Bauchorgane des Menschen. C. F. Müllersche Hofbuchhandlung, Carlsruhe, 1873.

Lyons, E. A.: A color atlas of sectional anatomy chest, abdomen and pelvis. C. V. Mosby Co., St. Louis, 1978.

Matsui, T., A. Hirano: An atlas of the human brain for computerized tomography. Fischer, Stuttgart, New York, 1978.

Nieuwenhuys, R., J. Voogd, C. V. Huijzen: The human central nervous system. Springer, Berlin, Heidelberg, New York, 1978.

Pernkopf, E.: Topographische Anatomie des Menschen. Urban & Schwarzenberg, Berlin, Wien, 1937–1960 I–IV.

Pernkopf, E.: Atlas der topographischen und angewandten Anatomie des Menschen. Urban & Schwarzenberg, München, Wien, Baltimore, 1980, I. u. II.

Rettenmeier, G.: Sonographischer Oberbauchstatus. Aussagefähigkeit und Indikationen der Ultraschall-Schnittbilduntersuchung des Oberbauches. Internist, 17, 549 (1976).

Roy-Camille, R.: Coupes horizontales du tronc. Masson, Paris, 1959.

Sarazin, R., J. Champetier et Ch. Contamin: Quatorze coupes sagittales du médiastin. Maloine, Paris, 1965.

Schaltenbrand, G.: Einführung in die stereotaktischen Operationen mit einem Atlas des menschlichen Gehirns. Thieme, Stuttgart, 1959.

Scherer, U.: Computertomographische Untersuchungen der Oberbauchorgane. Internist, 19, 579–585 (1978).

Sobotta, J.: Atlas d'anatomie humaine. Urban & Schwarzenberg, München, Wien, Baltimore/ Maloine, Paris, 1977.

Struve, C.: Ultraschalltomographie des Abdomens. Urban & Schwarzenberg, München, Wien, Baltimore, 2. Aufl., 1981.

Testut, L., O. Jacob: Traité d'anatomie topographique avec applications médico-chirurgicales. Doin, Paris, 1905.

Thompson, J. R. and A. N. Hasso: Correlative sectional anatomy of the head and neck. C. V. Mosby Co., St. Louis, 1980.

Töndury, G.: Angewandte und topographische Anatomie. Thieme, Stuttgart, 4. Aufl., 1970.

Wyman, A., T. Lawson, L. Goodman: Transverse anatomy of the human thorax, abdomen and pelvis. An atlas of anatomic, radiologic, computed tomographic and ultrasonic correlation. Little Brown, Boston, 1978.

FA

♂

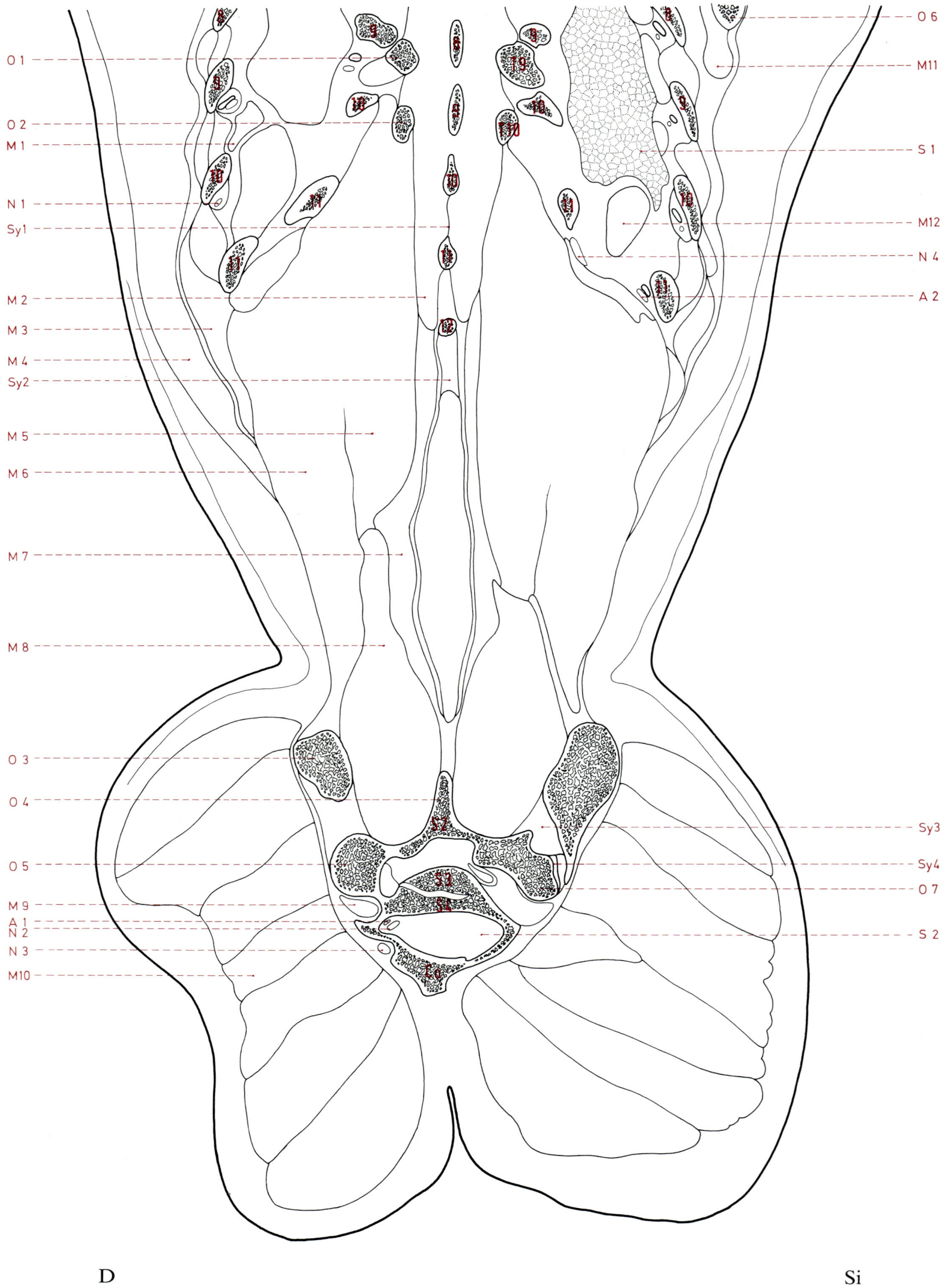
O 1
O 2
M 1
N 1
Sy1
M 2
M 3
M 4
Sy2
M 5
M 6
M 7
M 8
O 3
O 4
O 5
M 9
A 1
N 2
N 3
M10
O 6
M11
S 1
M12
N 4
A 2
Sy3
Sy4
O 7
S 2
D
Si

Fig. FA♂ 1

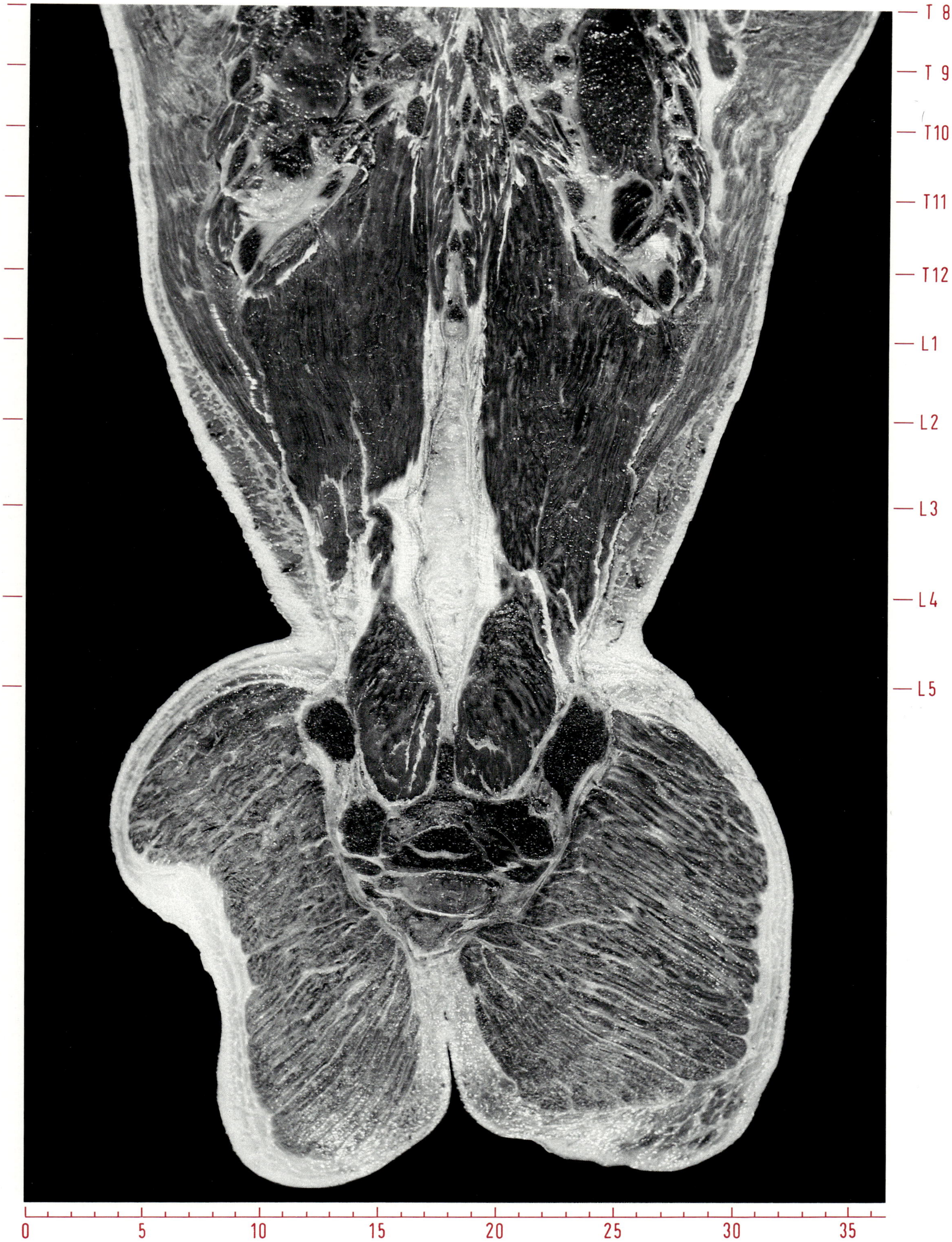

Fig. FA♂ 1

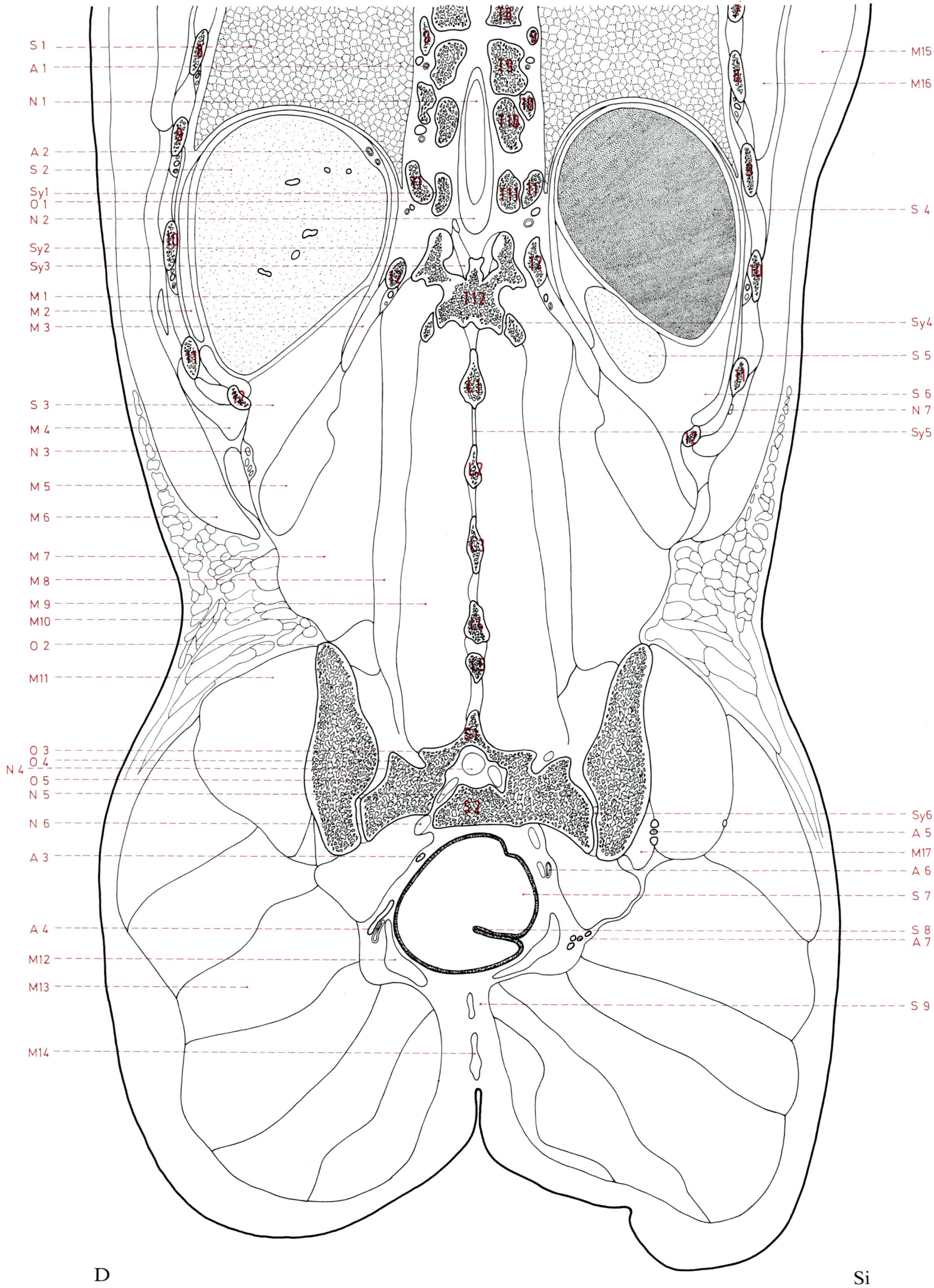
S 1
A 1
N 1
A 2
S 2
Sy1
O 1
N 2
Sy2
Sy3
M 1
M 2
M 3
S 3
M 4
N 3
M 5
M 6
M 7
M 8
M 9
M10
O 2
M11
O 3
O 4
N 4
O 5
N 5
N 6
A 3
A 4
M12
M13
M14
M15
M16
S 4
Sy4
S 5
S 6
N 7
Sy5
Sy6
A 5
M17
A 6
S 7
S 8
A 7
S 9
T12
L1
L2
L3
L4
L5
S1
S2
D
Si

Fig. FA♂ 2

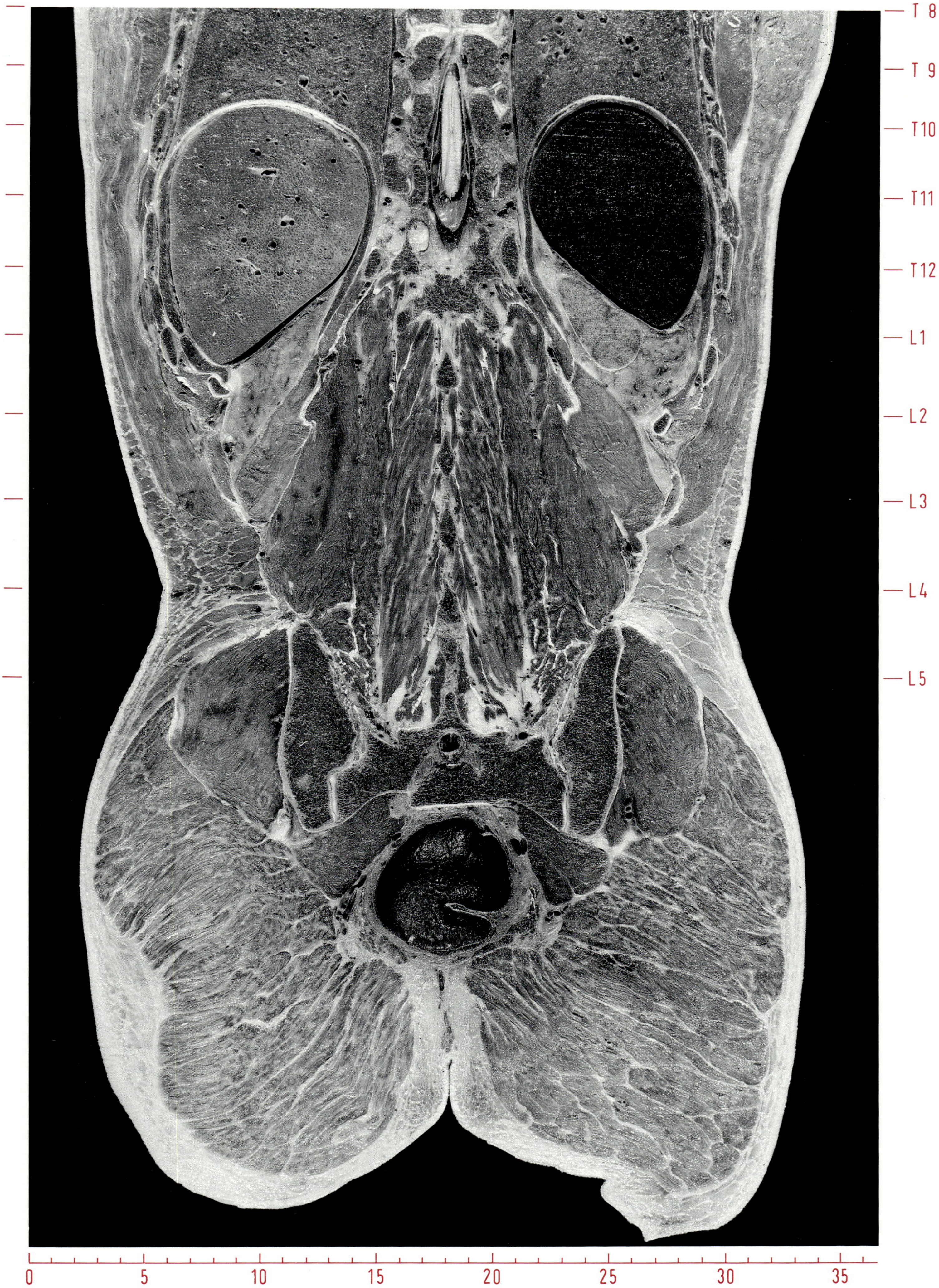

Fig. FA♂ 2

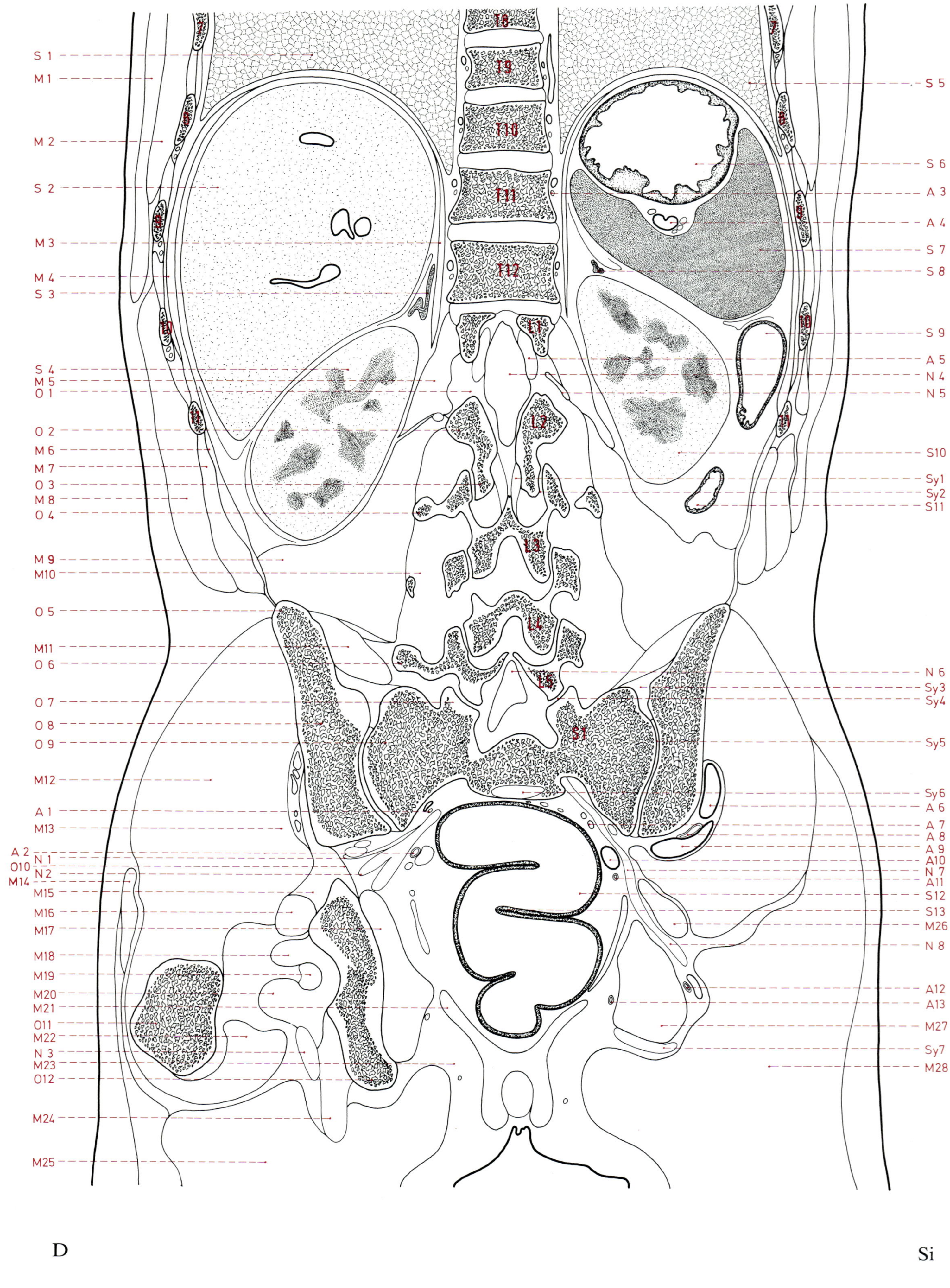
S 1
M 1
M 2
S 2
M 3
M 4
S 3
S 4
M 5
O 1
O 2
M 6
M 7
O 3
M 8
O 4
M 9
M10
O 5
M11
O 6
O 7
O 8
O 9
M12
A 1
M13
A 2
N 1
O10
N 2
M14
M15
M16
M17
M18
M19
M20
M21
O11
M22
N 3
M23
O12
M24
M25
S 5
S 6
A 3
A 4
S 7
S 8
S 9
A 5
N 4
N 5
S10
Sy1
Sy2
S11
N 6
Sy3
Sy4
Sy5
Sy6
A 6
A 7
A 8
A 9
A10
N 7
A11
S12
S13
M26
N 8
A12
A13
M27
Sy7
M28
T8
T9
T10
T11
T12
L1
L2
L3
L4
L5
S1
D
Si

Fig. FA♂ 3

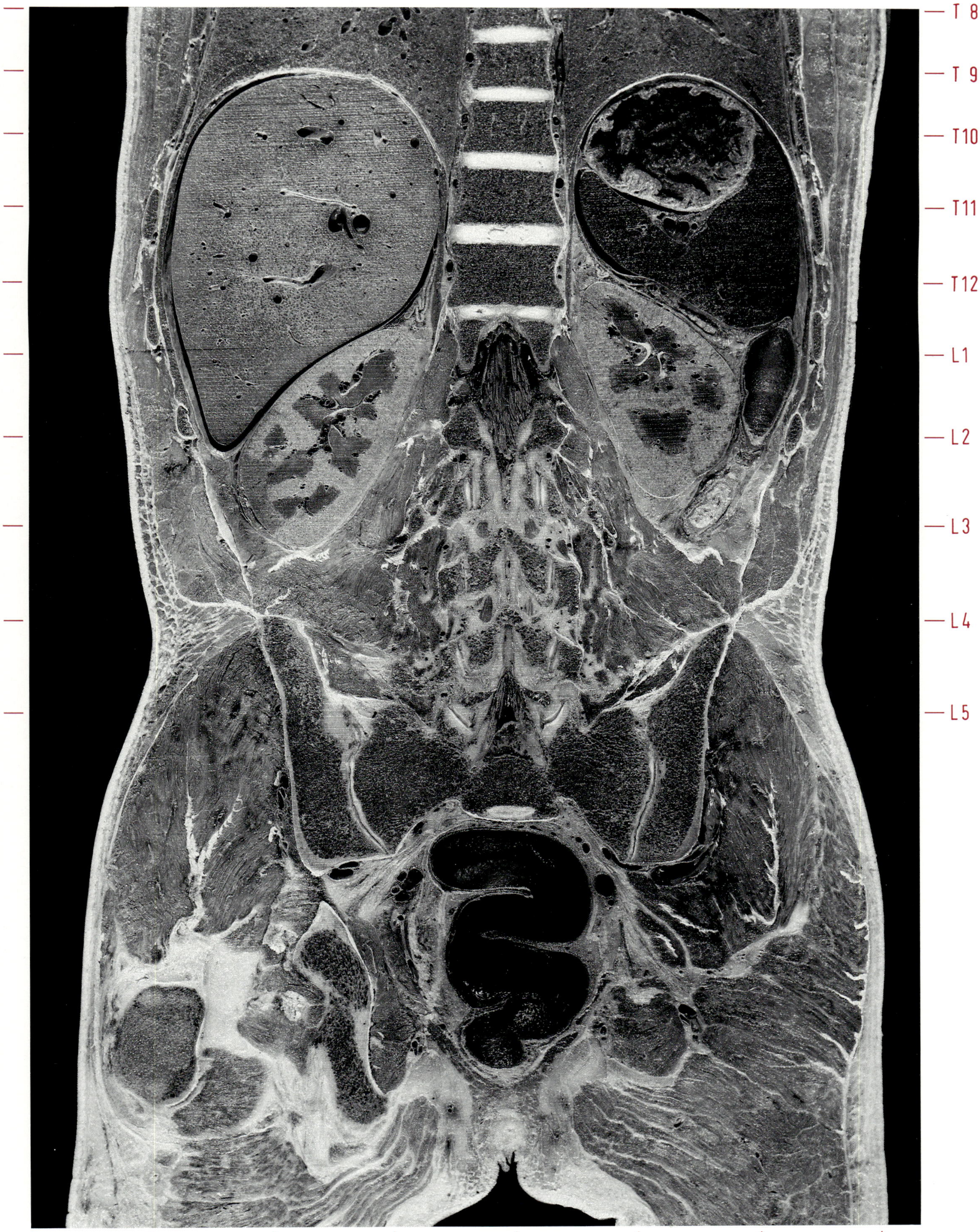

Fig. FA♂ 3

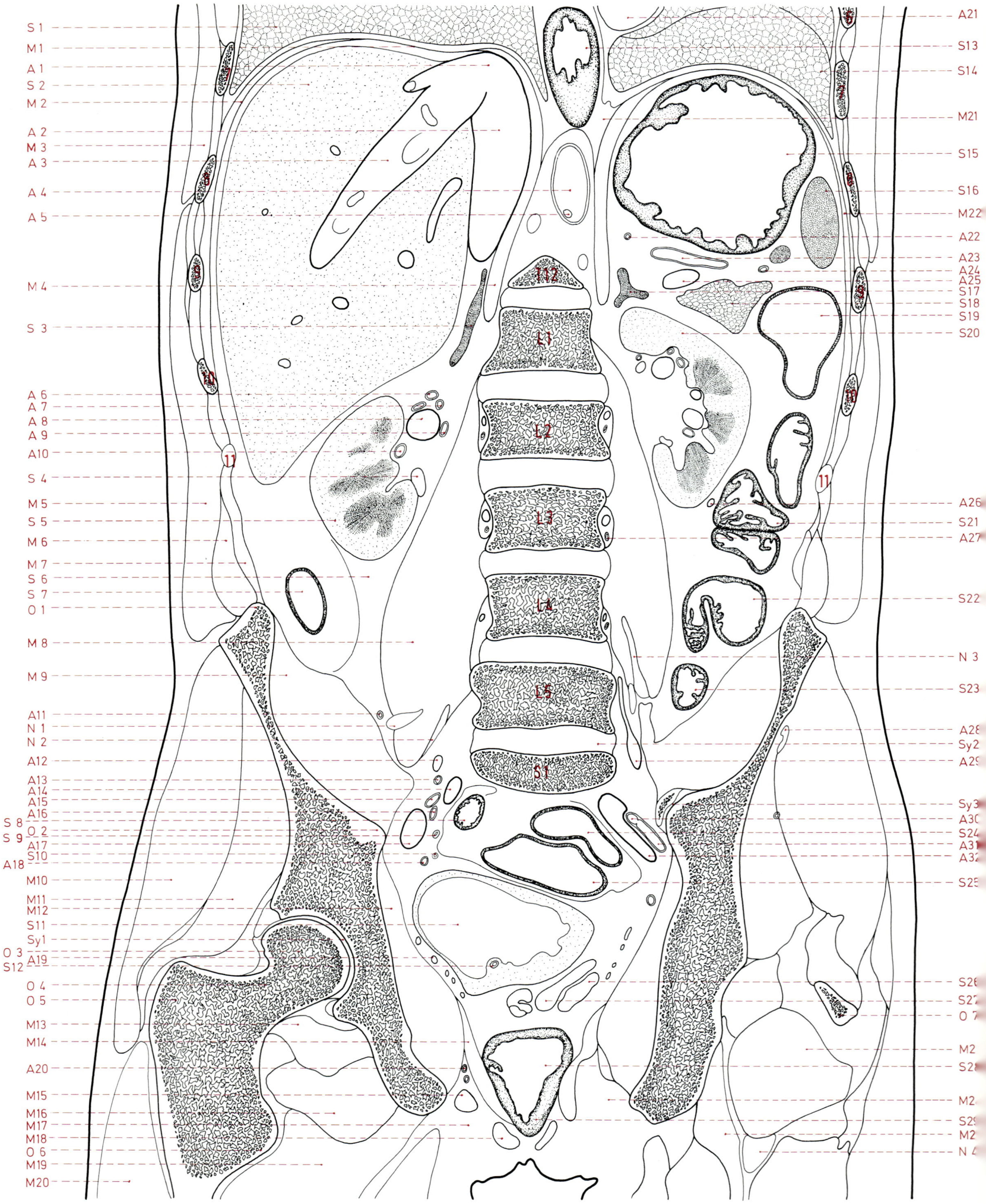

S 1
M 1
A 1
S 2
M 2
A 2
M 3
A 3
A 4
A 5
M 4
S 3
A 6
A 7
A 8
A 9
A10
S 4
M 5
S 5
M 6
M 7
S 6
S 7
O 1
M 8
M 9
A11
N 1
N 2
A12
A13
A14
A15
A16
S 8
O 2
S 9
A17
S10
A18
M10
M11
M12
S11
Sy1
O 3
A19
S12
O 4
O 5
M13
M14
A20
M15
M16
M17
M18
O 6
M19
M20
A21
S13
S14
M21
S15
S16
M22
A22
A23
A24
A25
S17
S18
S19
S20
A26
S21
A27
S22
N 3
S23
A28
Sy2
S25
T12
L1
L2
L3
L4
L5
S1
6
7
8
9
10
11

D

Si

Fig. FA♂ 4

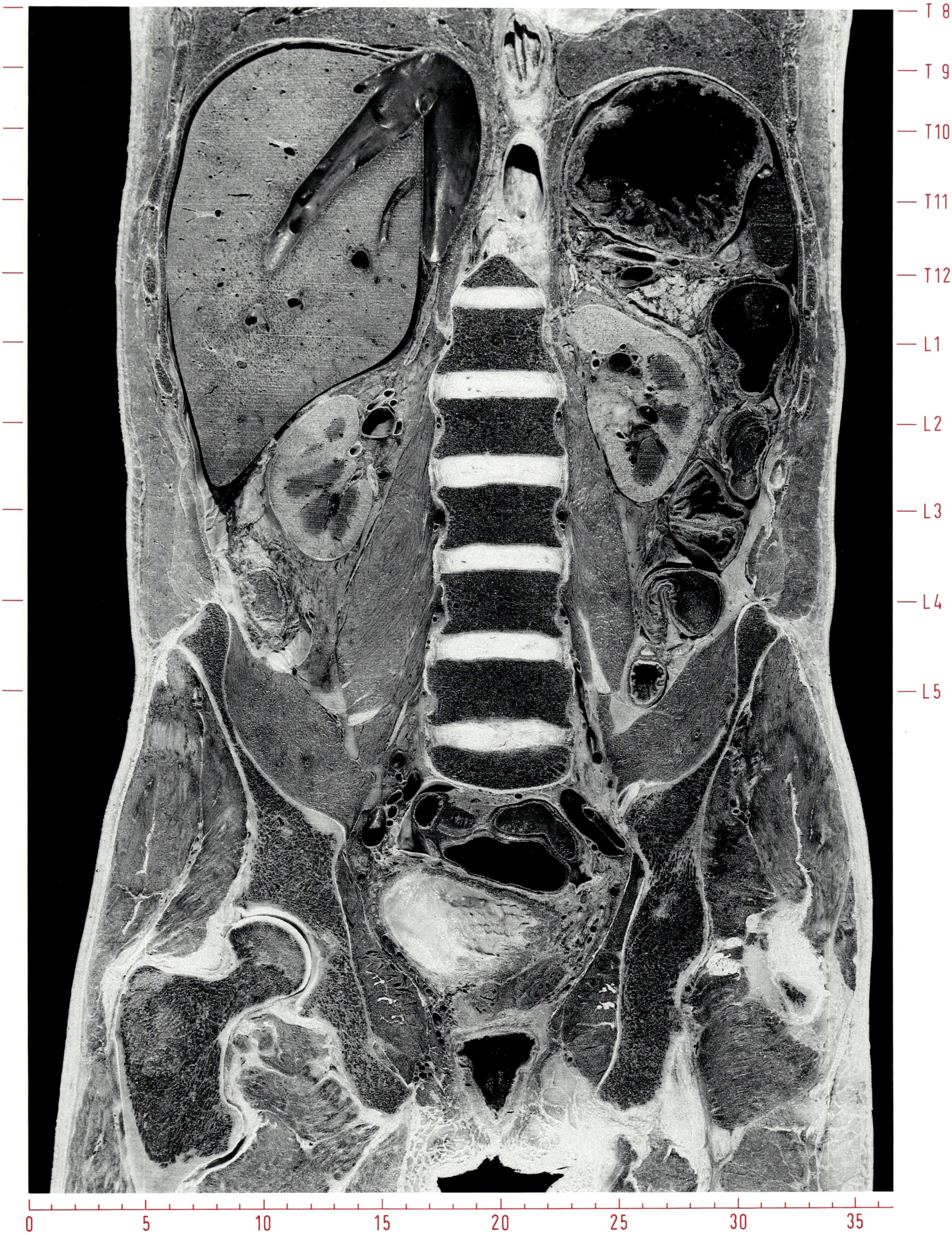

Fig. FA♂ 4

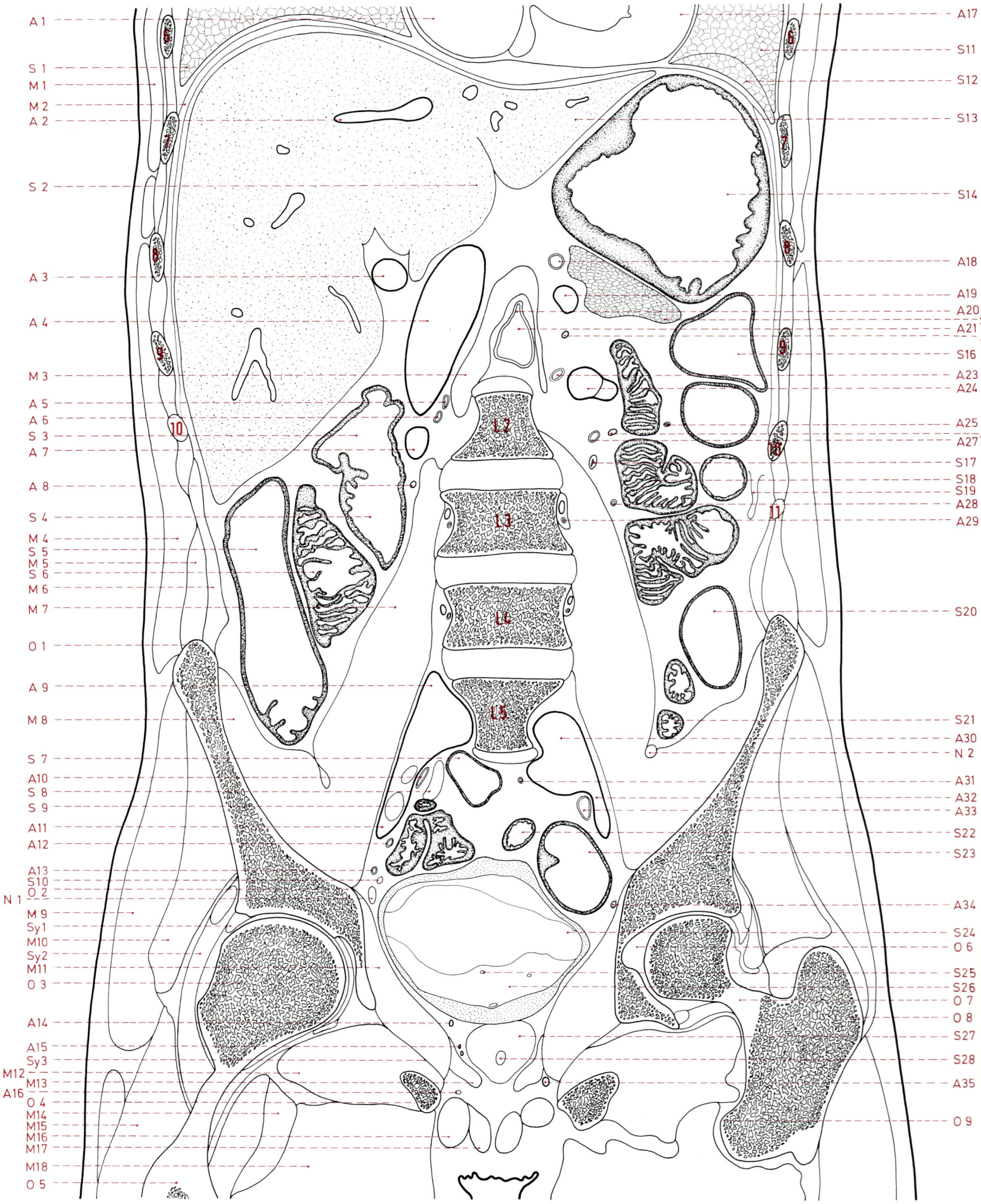
A 1
S 1
M 1
M 2
A 2
S 2
A 3
A 4
M 3
A 5
A 6
S 3
A 7
A 8
S 4
M 4
S 5
M 5
S 6
M 6
M 7
O 1
A 9
M 8
S 7
A10
S 8
S 9
A11
A12
A13
S10
O 2
N 1
M 9
Sy1
M10
Sy2
M11
O 3
A14
A15
Sy3
M12
M13
A16
O 4
M14
M15
M16
M17
M18
O 5
A17
S11
S12
S13
S14
A18
A19
A20
A21
S16
A23
A24
A25
A27
S17
S18
S19
A28
A29
S20
S21
A30
N 2
A31
A32
A33
S22
S23
A34
S24
O 6
S25
S26
O 7
O 8
S27
S28
A35
O 9
L2
L3
L5
10
11

D

Si

Fig. FA♂ 5

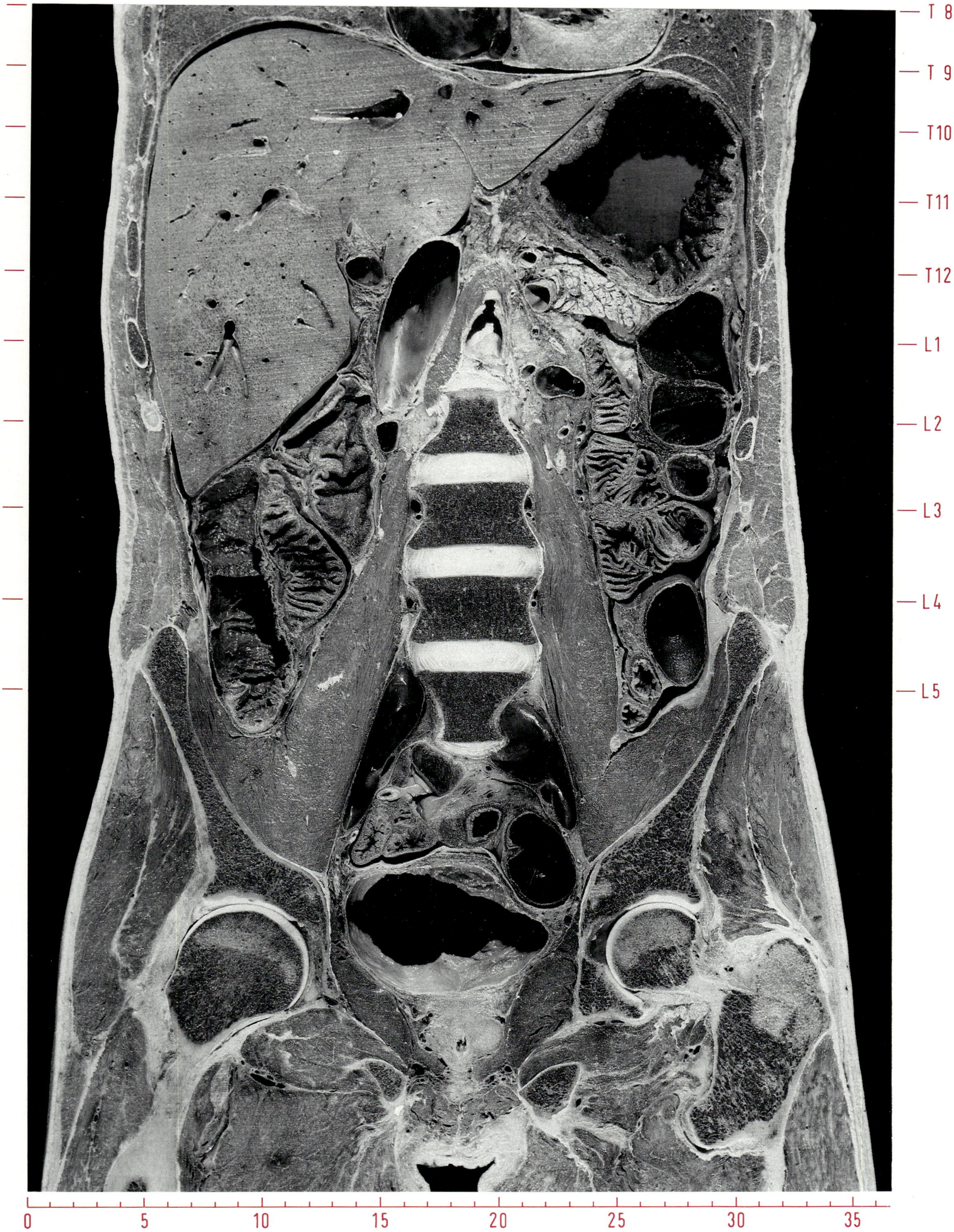
T 8
T 9
T10
T11
T12
L1
L2
L3
L4
L5
0
5
10
15
20
25
30
35

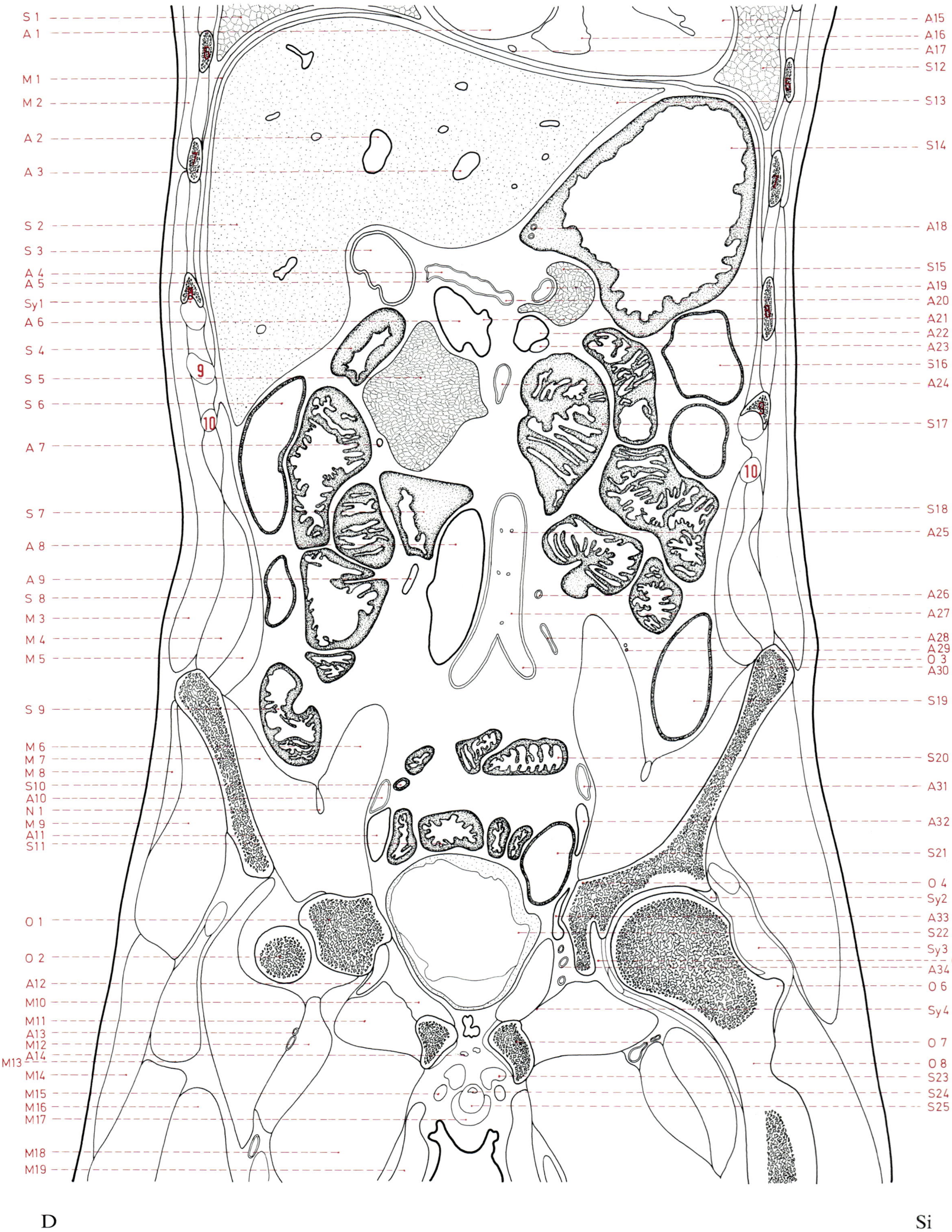
S 1
A 1
M 1
M 2
A 2
A 3
S 2
S 3
A 4
A 5
Sy1
A 6
S 4
S 5
S 6
A 7
S 7
A 8
A 9
S 8
M 3
M 4
M 5
S 9
M 6
M 7
M 8
S10
A10
N 1
M 9
A11
S11
O 1
O 2
A12
M10
M11
A13
M12
A14
M13
M14
M15
M16
M17
M18
M19
A15
A16
A17
S12
S13
S14
A18
S15
A19
A20
A21
A22
A23
S16
A24
S17
S18
A25
A26
A27
A28
A29
O 3
A30
S19
S20
A31
A32
S21
O 4
Sy2
A33
S22
Sy3
A34
O 6
Sy4
O 7
O 8
S23
S24
S25
9
10
10
D
Si

Fig. FA♂ 6

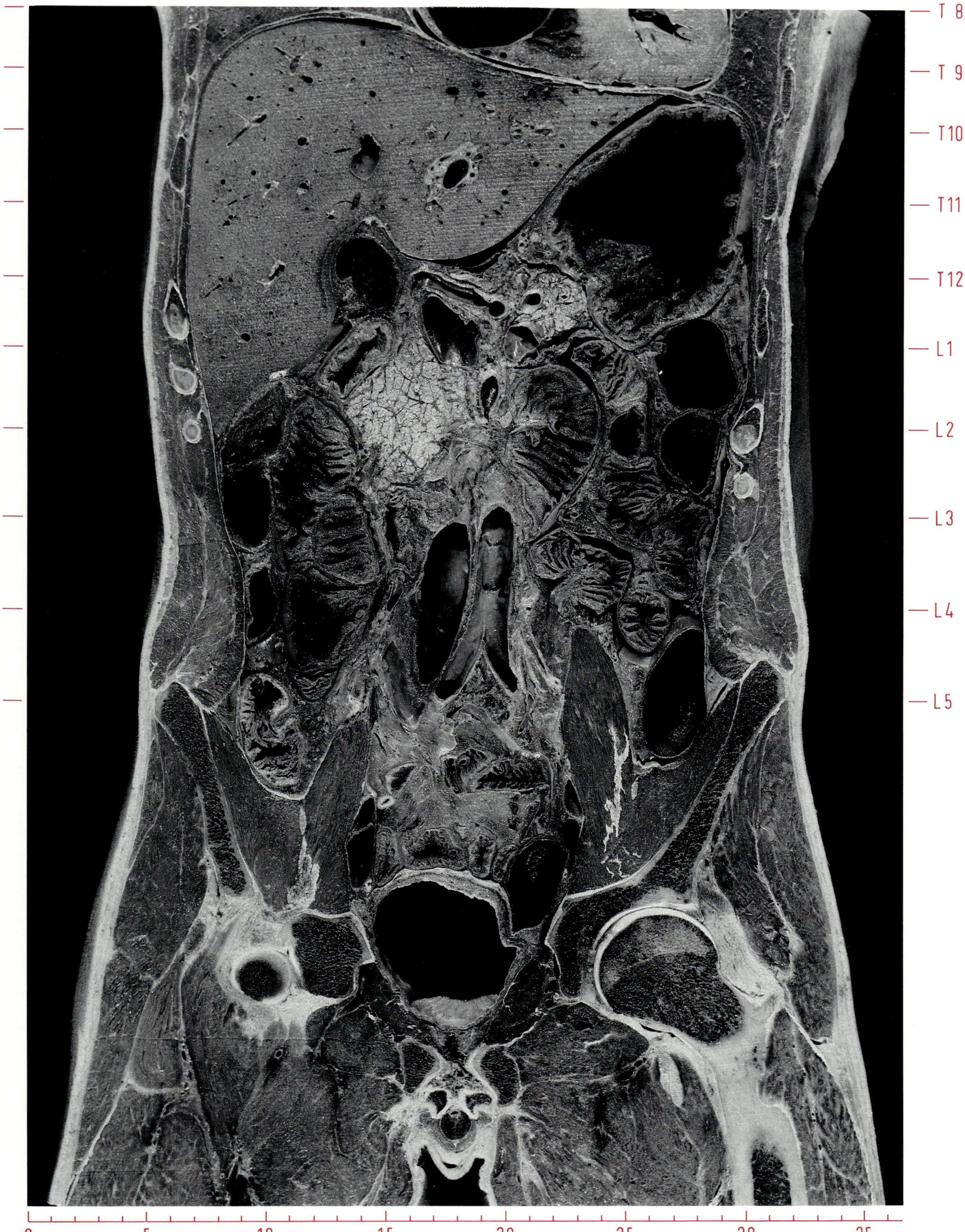

Fig. FA♂ 6

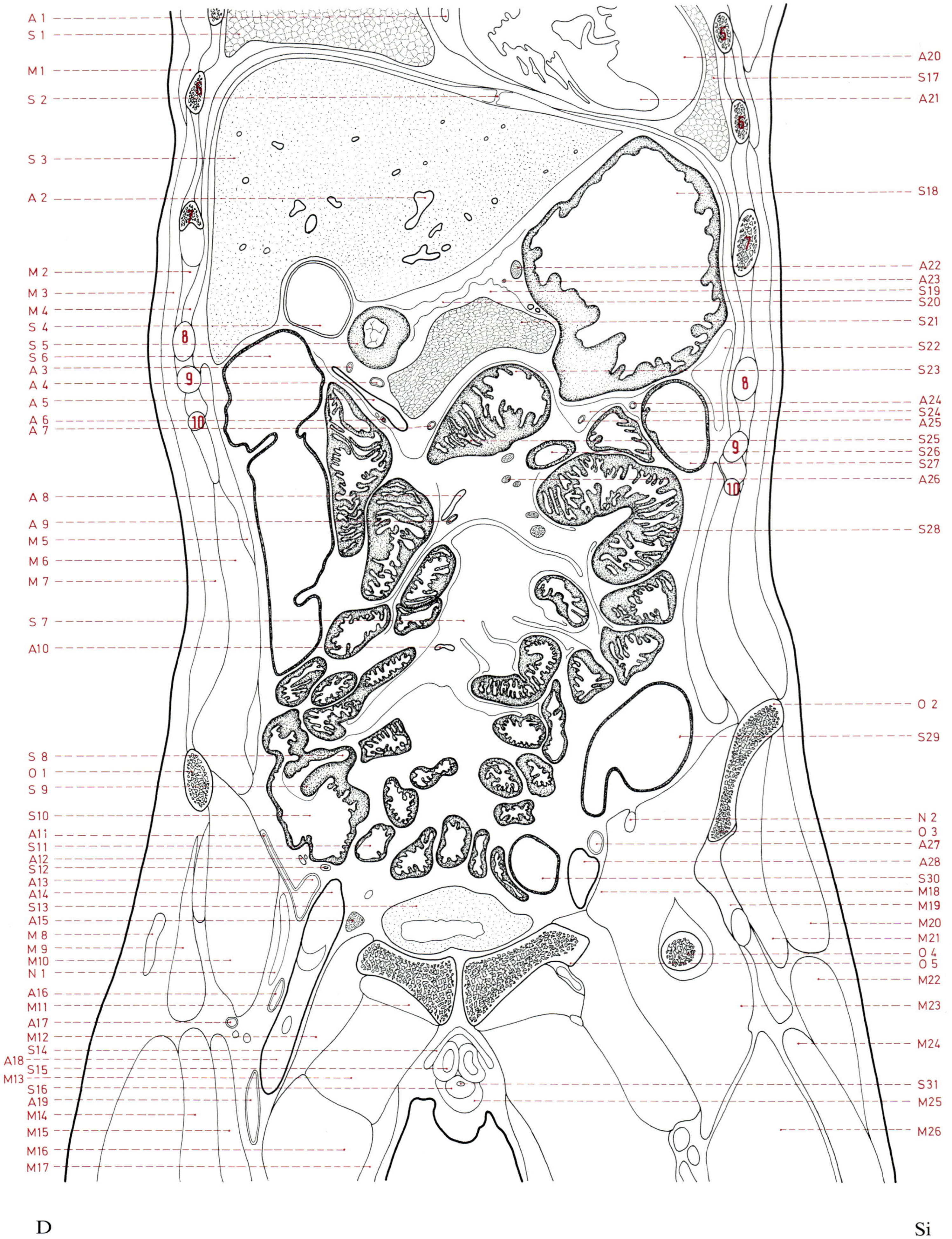
A 1
S 1
M 1
S 2
S 3
A 2
M 2
M 3
M 4
S 4
S 5
S 6
A 3
A 4
A 5
A 6
A 7
A 8
A 9
M 5
M 6
M 7
S 7
A10
S 8
O 1
S 9
S10
A11
S11
A12
S12
A13
A14
S13
A15
M 8
M 9
M10
N 1
A16
M11
A17
M12
S14
A18
S15
M13
S16
A19
M14
M15
M16
M17
A20
S17
A21
S18
A22
A23
S19
S20
S21
S22
S23
A24
S24
A25
S25
S26
S27
A26
S28
O 2
S29
N 2
O 3
A27
A28
S30
M18
M19
M20
M21
O 4
O 5
M22
M23
M24
S31
M25
M26
5
6
7
8
9
10
D
Si

Fig. FA♂ 7

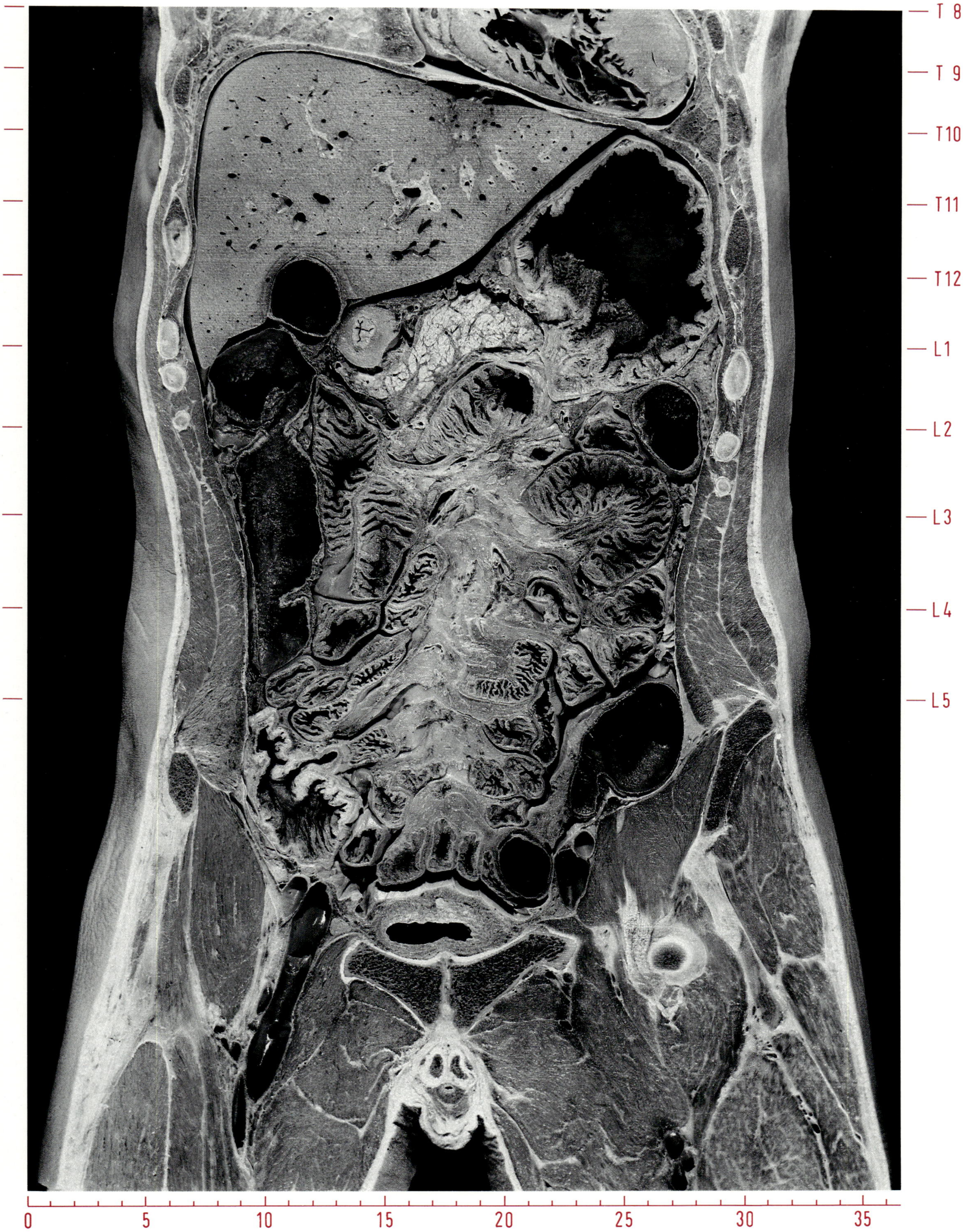

Fig. FA♂ 7

FA

♀

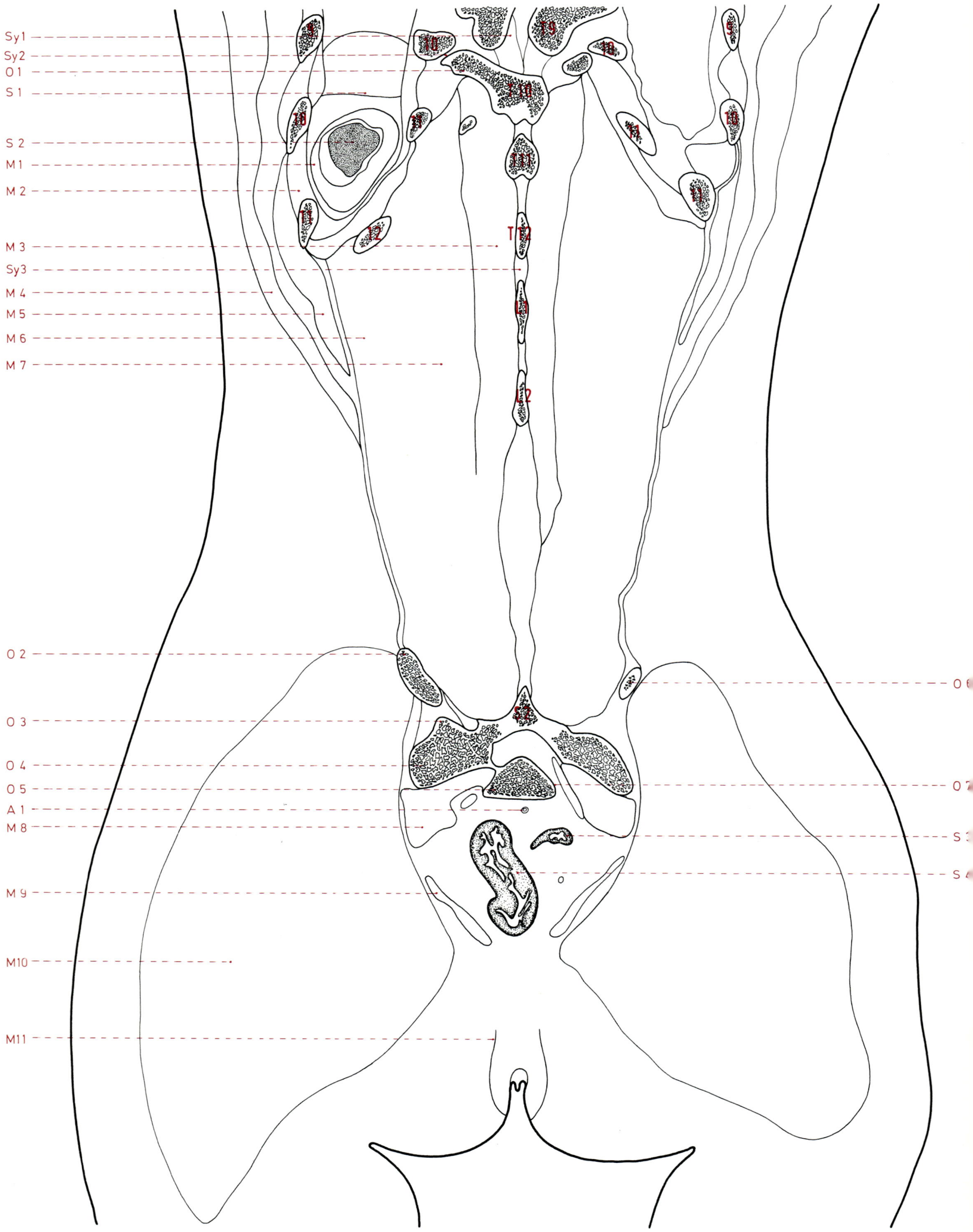
Sy1
Sy2
O 1
S 1
S 2
M 1
M 2
M 3
Sy3
M 4
M 5
M 6
M 7
O 2
O 3
O 4
O 5
A 1
M 8
M 9
M10
M11
T9
T10
T11
T12

Si D

Fig. FA♀ 1

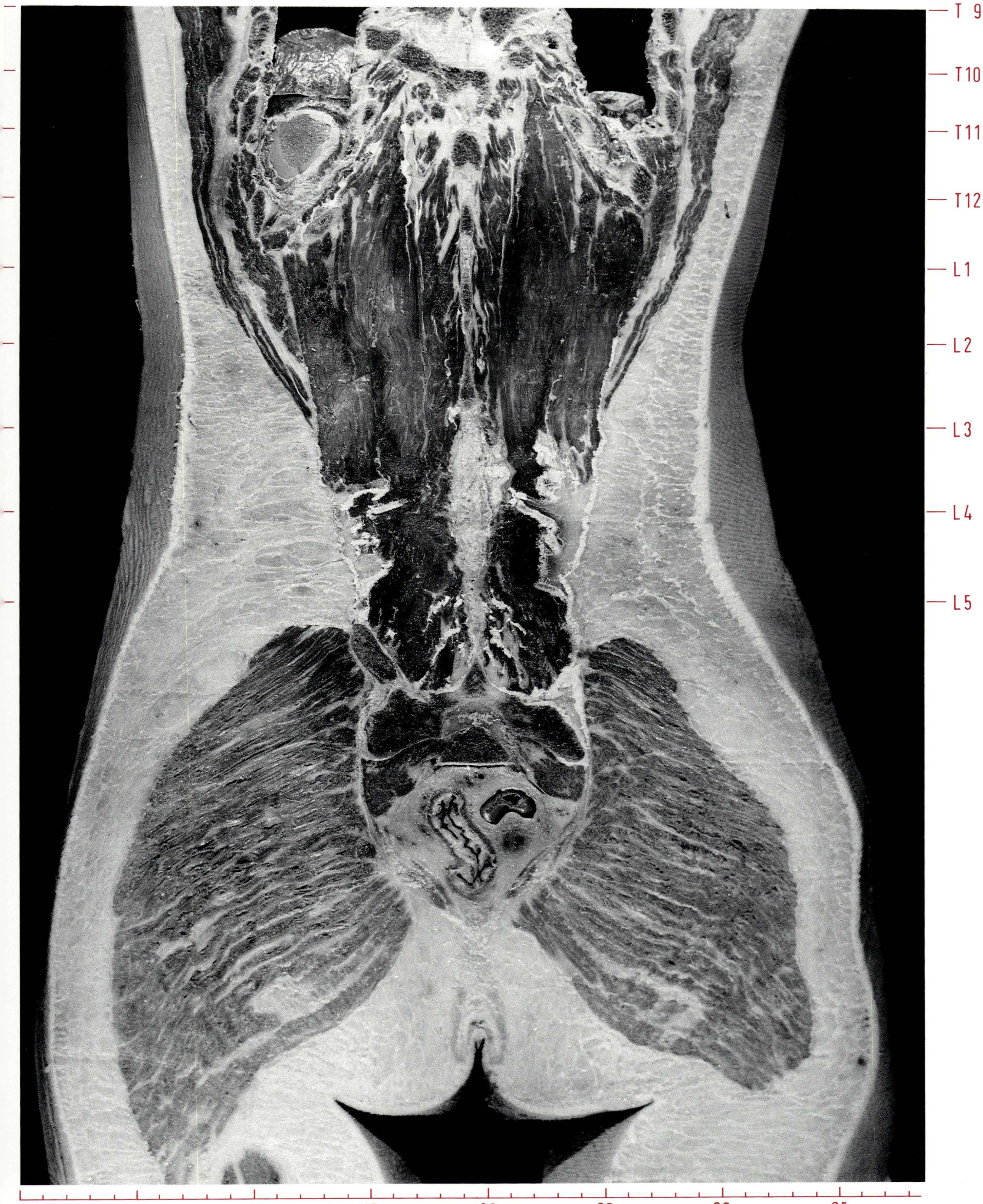

Fig. FA♀ 1

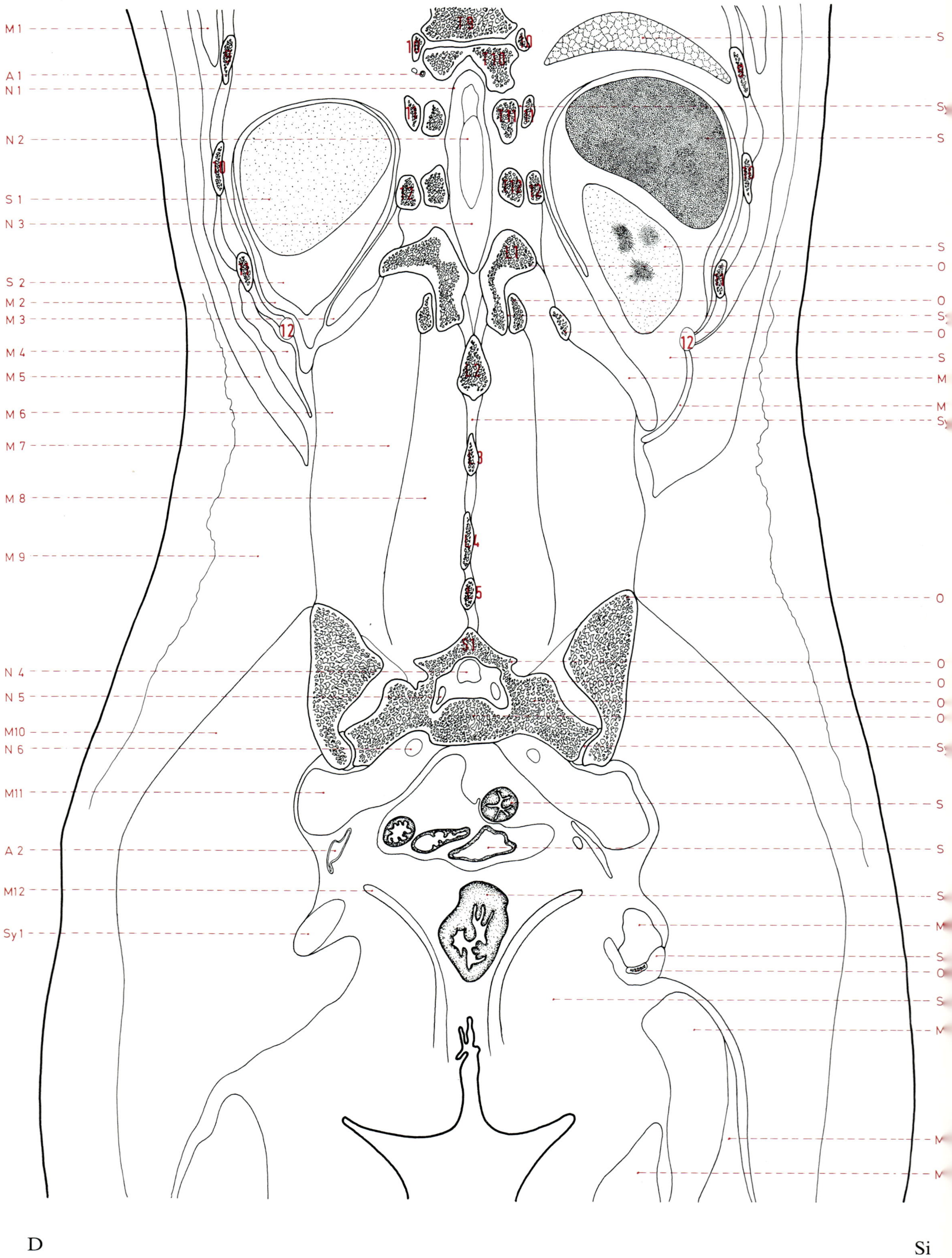
M1
A1
N1
N2
S1
N3
S2
M2
M3
M4
M5
M6
M7
M8
M9
N4
N5
M10
N6
M11
A2
M12
Sy1
D
Si

<u>Fig. FA♀ 2</u>

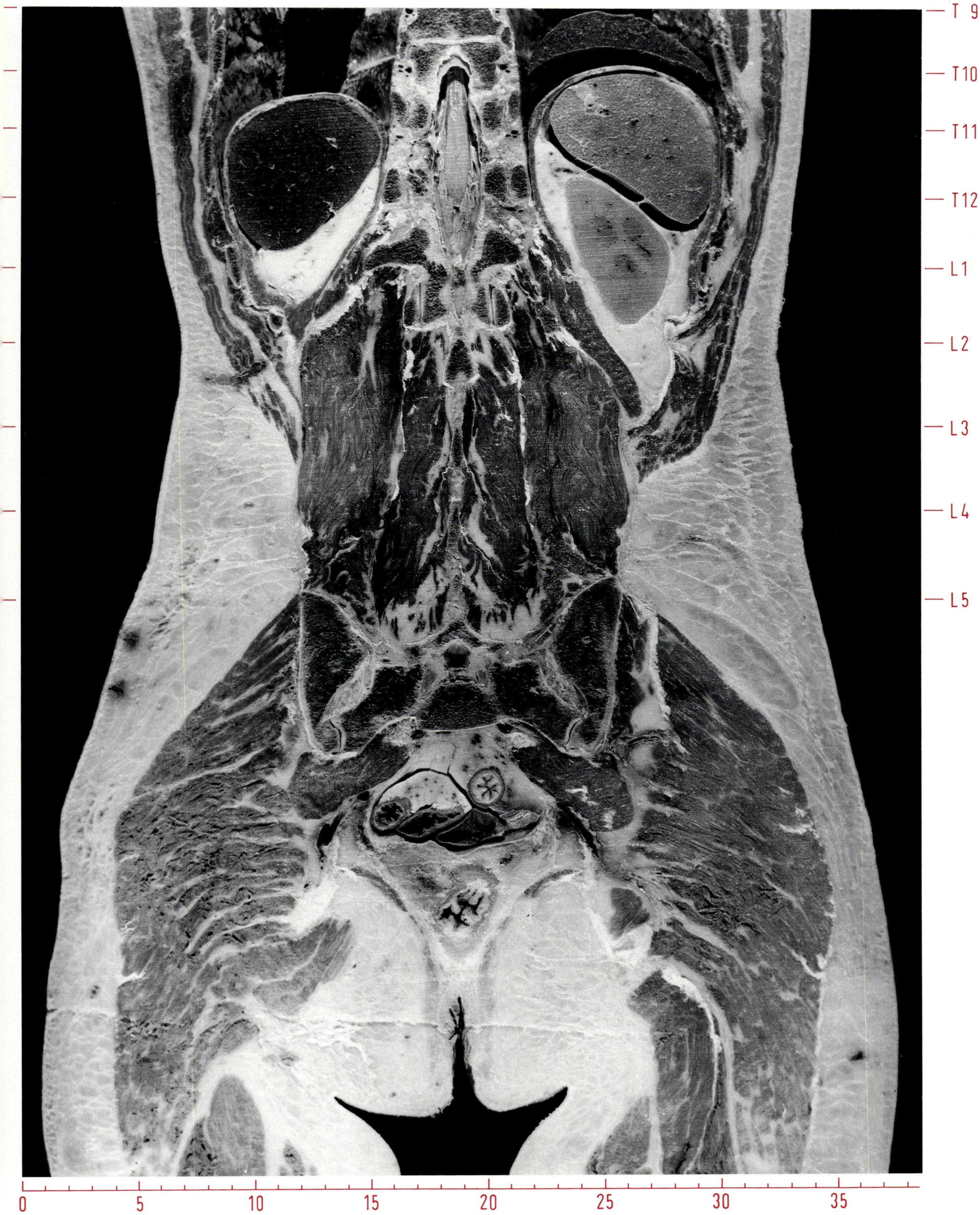

Fig. FA♀ 2

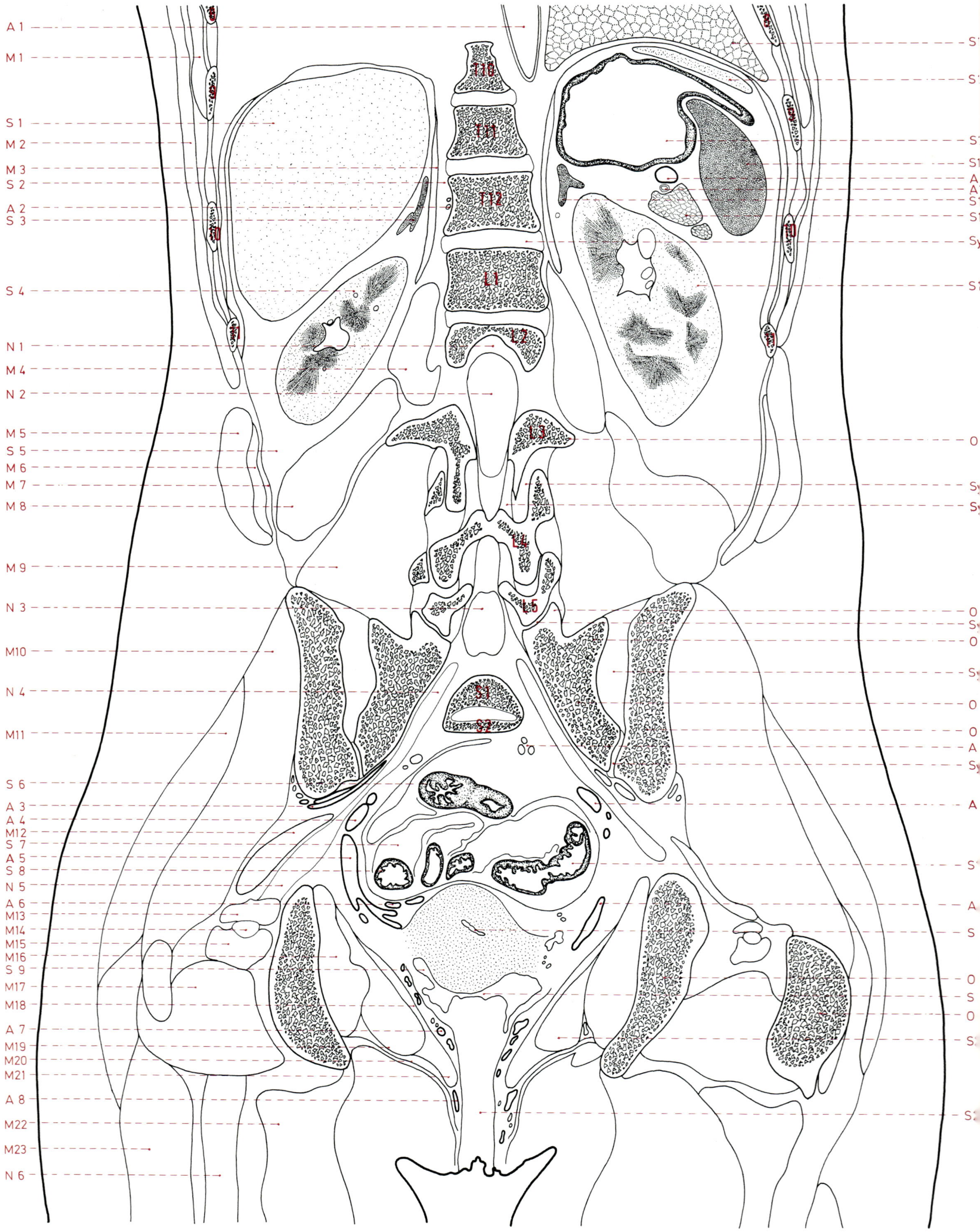
A 1
M 1
S 1
M 2
M 3
S 2
A 2
S 3
S 4
N 1
M 4
N 2
M 5
S 5
M 6
M 7
M 8
M 9
N 3
M10
N 4
M11
S 6
A 3
A 4
M12
S 7
A 5
S 8
N 5
A 6
M13
M14
M15
M16
S 9
M17
M18
A 7
M19
M20
M21
A 8
M22
M23
N 6
T10
T11
T12
L1
L2
L3
L4
L5
S1
S2

D Si

Fig. FA♀ 3

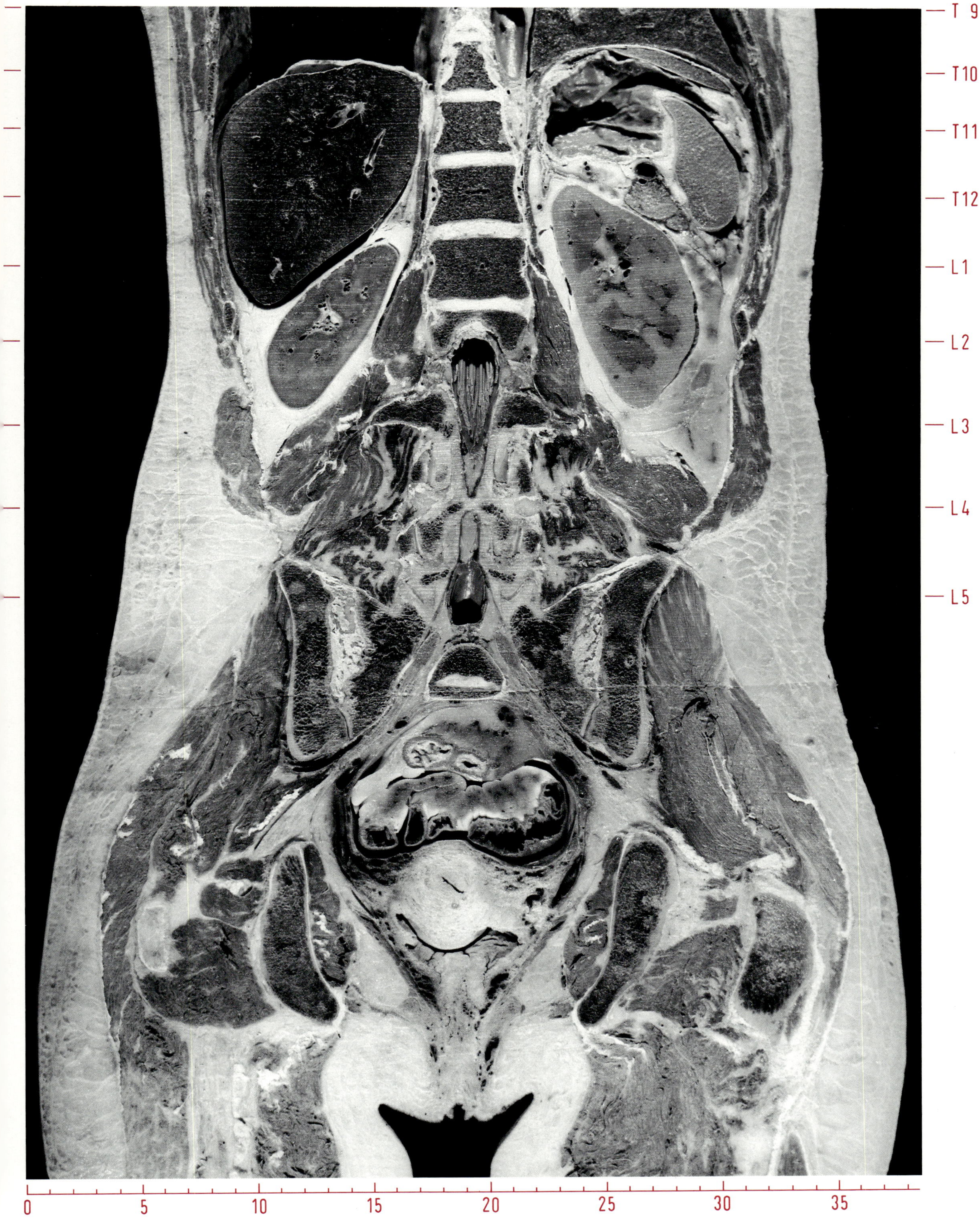

Fig. FA♀ 3

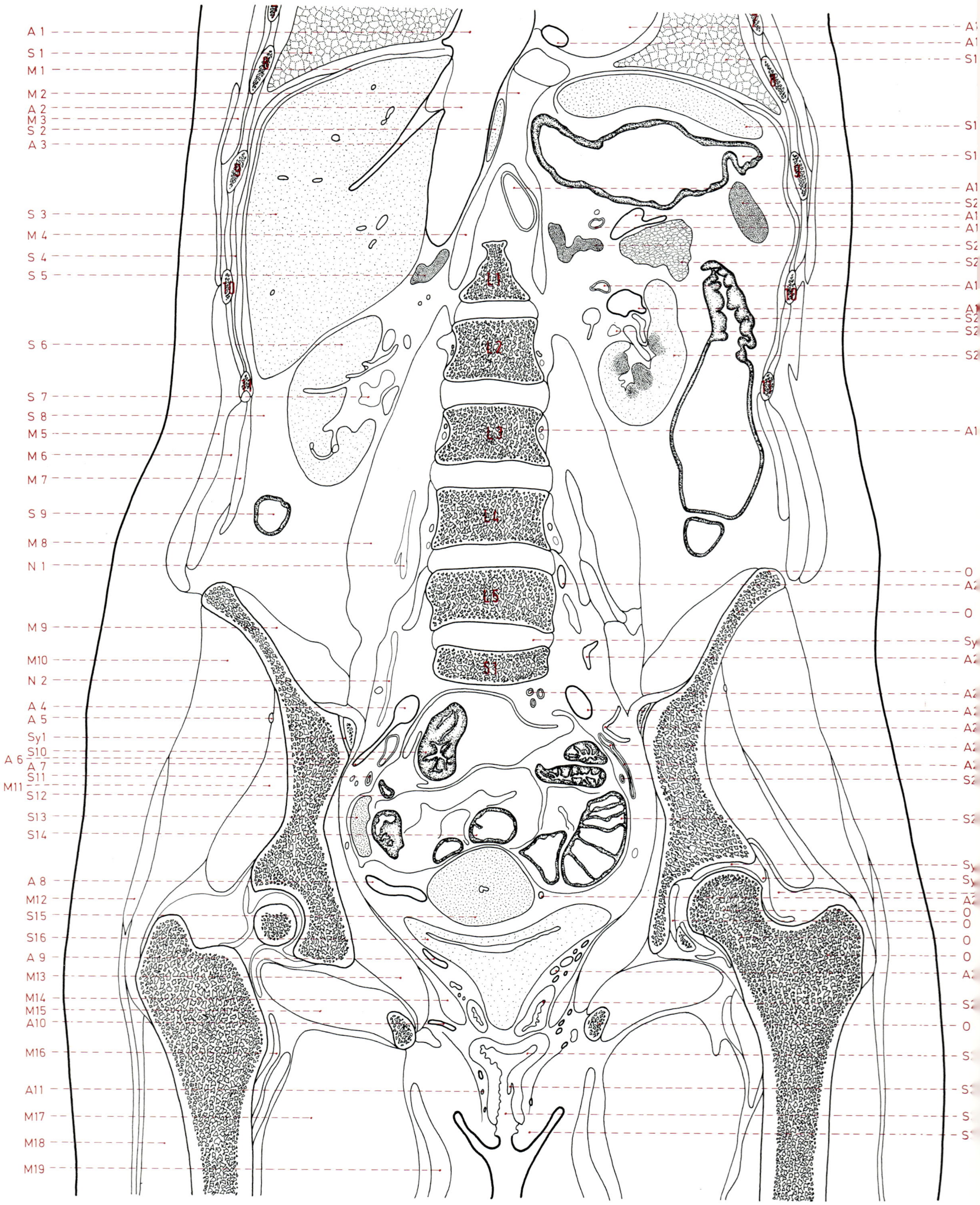

A 1
S 1
M 1
M 2
A 2
M 3
S 2
A 3
S 3
M 4
S 4
S 5
S 6
S 7
S 8
M 5
M 6
M 7
S 9
M 8
N 1
M 9
M10
N 2
A 4
A 5
Sy1
S10
A 6
A 7
S11
M11
S12
S13
S14
A 8
M12
S15
S16
A 9
M13
M14
M15
A10
M16
A11
M17
M18
M19
L1
L2
L3
L4
L5
S1

D

Si

Fig. FA♀ 4

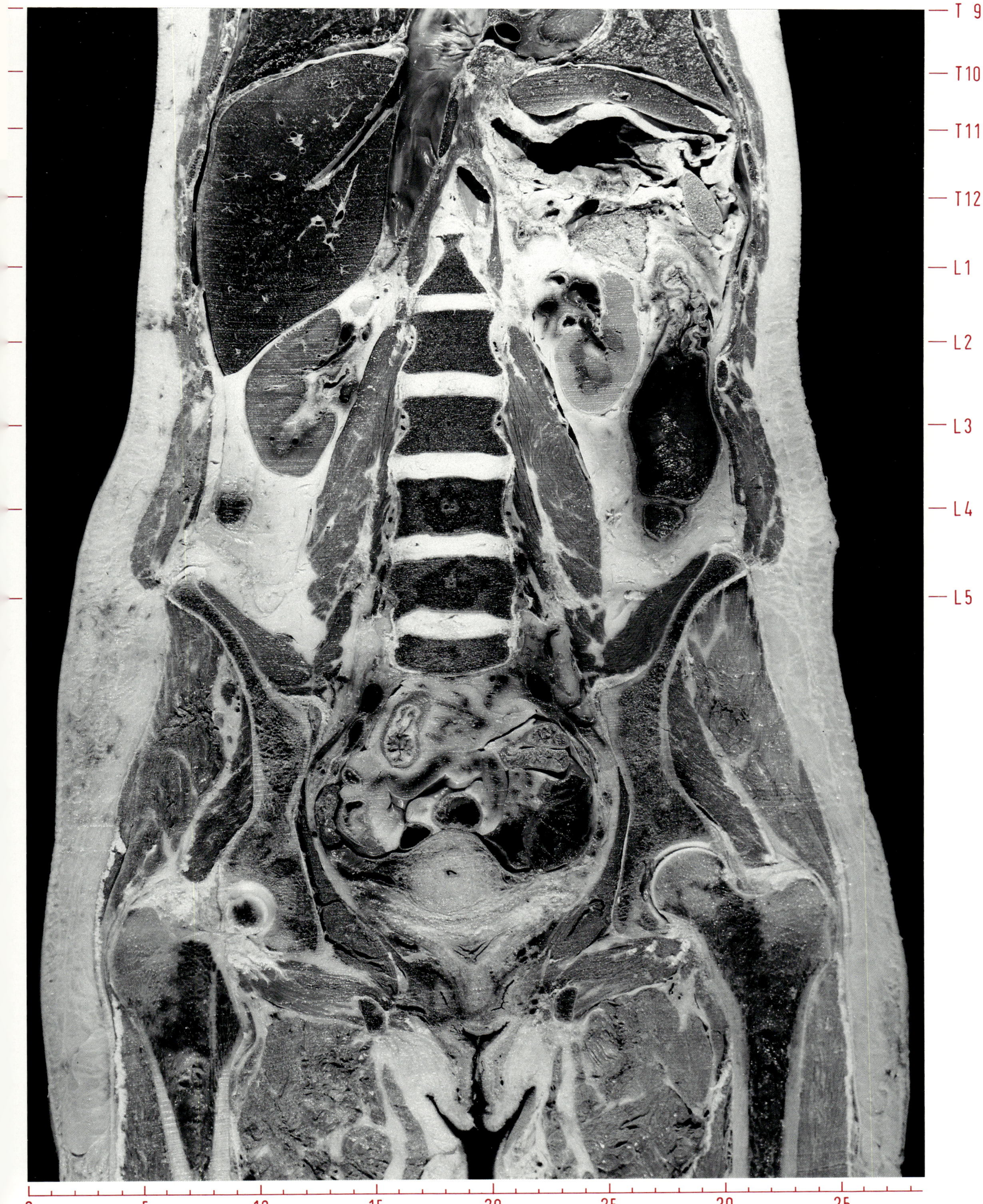

Fig. FA♀ 4

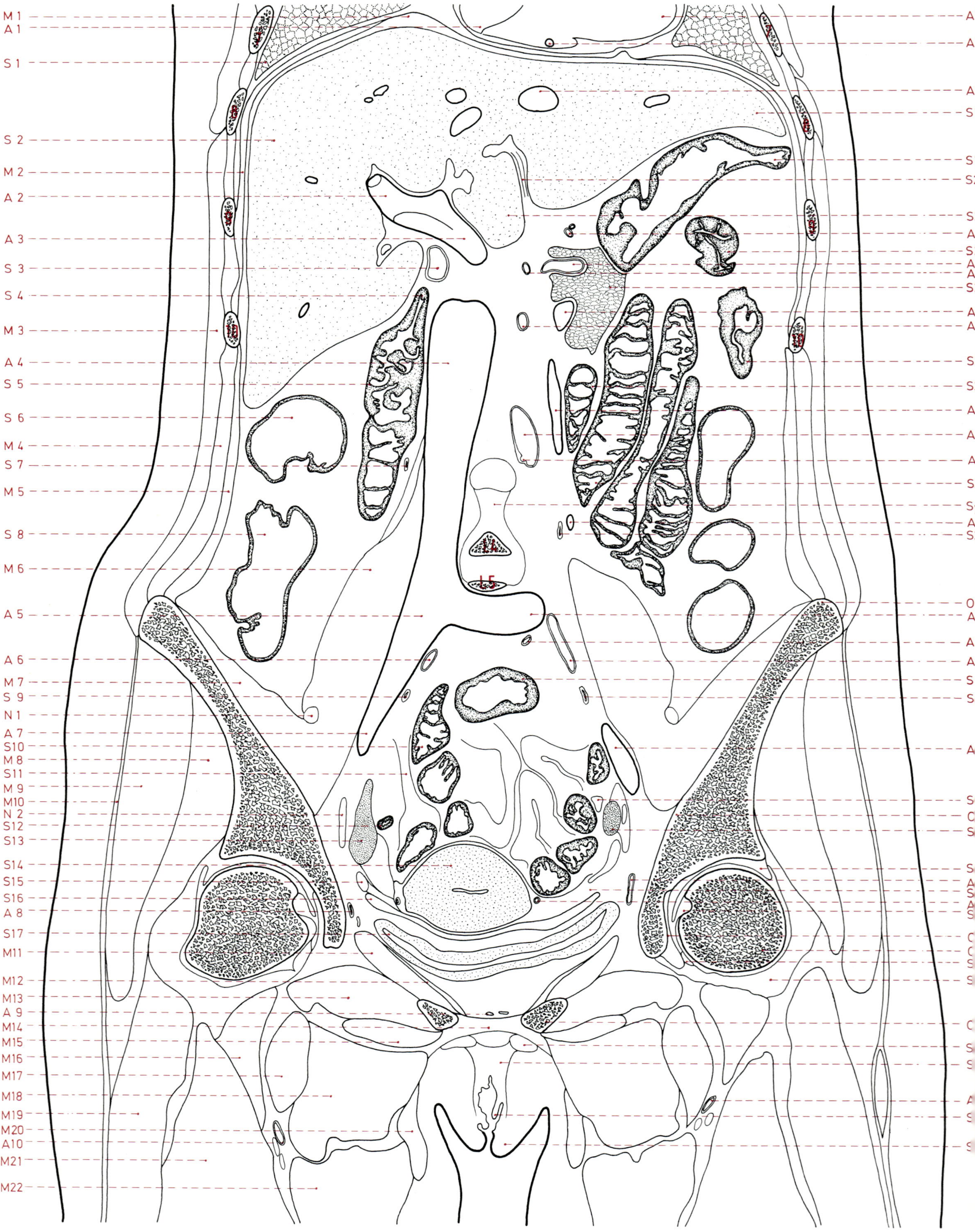
M 1
A 1
S 1
S 2
M 2
A 2
A 3
S 3
S 4
M 3
A 4
S 5
S 6
M 4
S 7
M 5
S 8
M 6
A 5
A 6
M 7
S 9
N 1
A 7
S10
M 8
S11
M 9
M10
N 2
S12
S13
S14
S15
S16
A 8
S17
M11
M12
M13
A 9
M14
M15
M16
M17
M18
M19
M20
A10
M21
M22
L4
L5

D

Si

Fig. FA♀ 5a

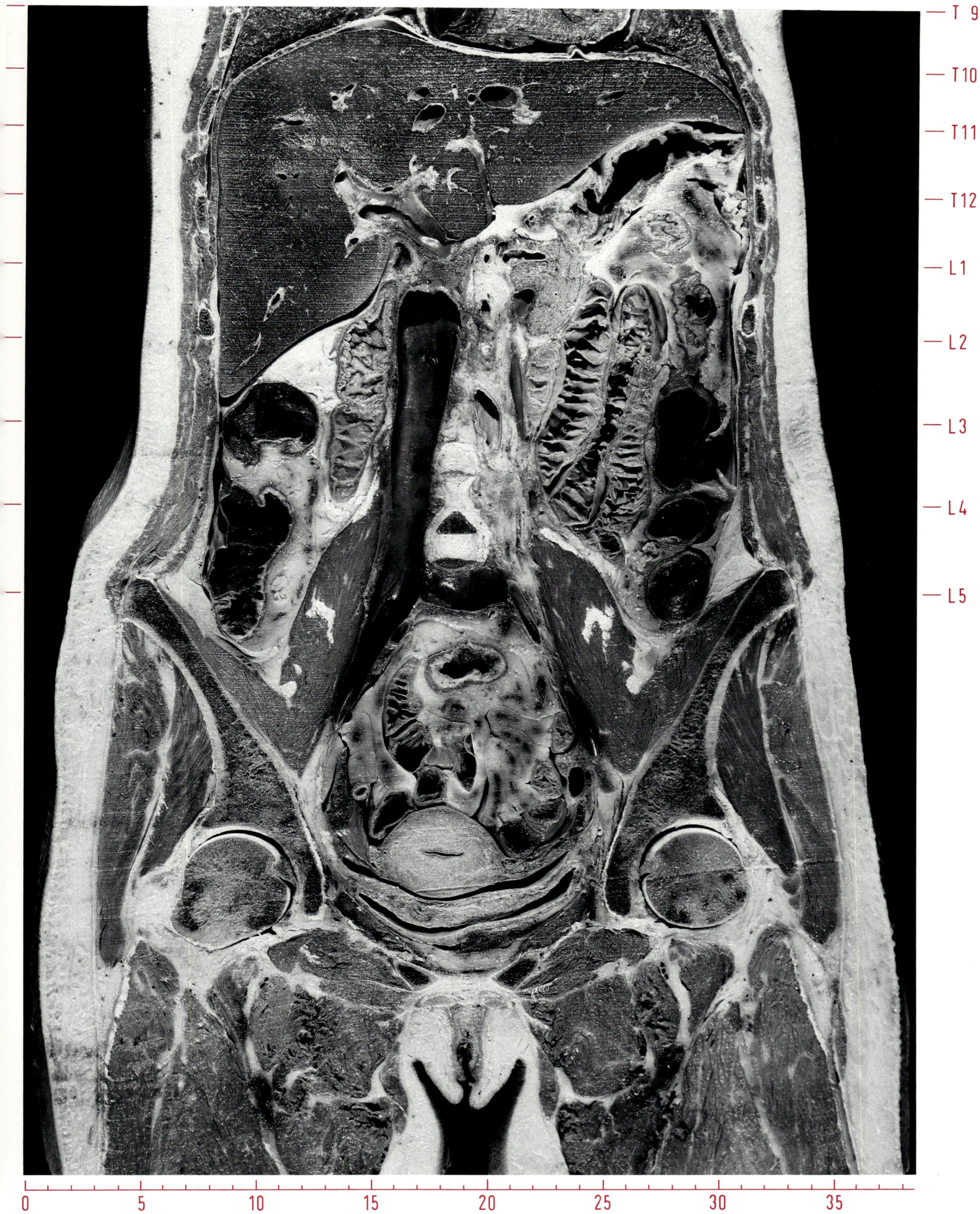

Fig. FA♀ 5a

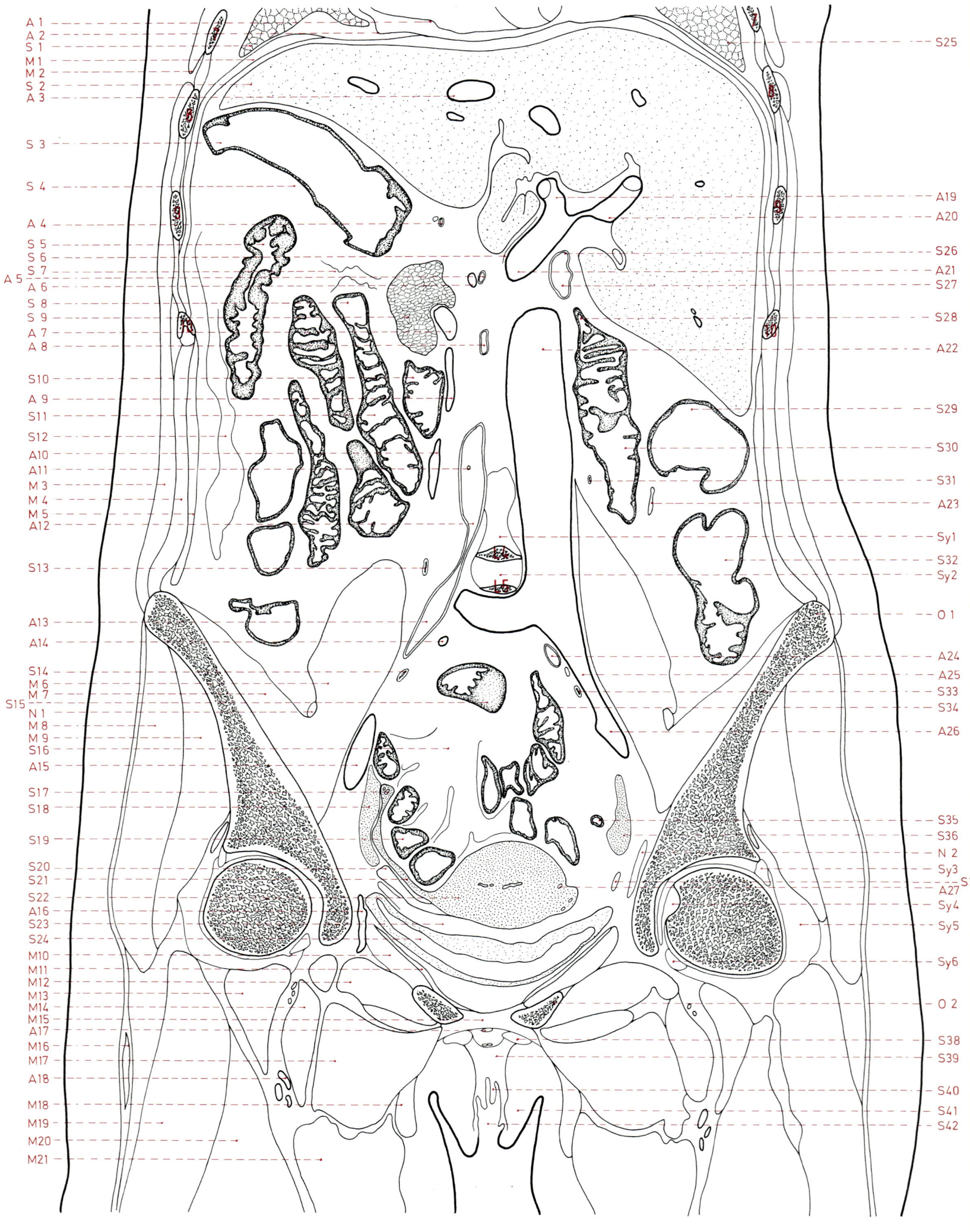
A 1
A 2
S 1
M 1
M 2
S 2
A 3
S 3
S 4
A 4
S 5
S 6
S 7
A 5
A 6
S 8
S 9
A 7
A 8
S10
A 9
S11
S12
A10
A11
M 3
M 4
M 5
A12
S13
A13
A14
S14
M 6
M 7
S15
N 1
M 8
M 9
S16
A15
S17
S18
S19
S20
S21
S22
A16
S23
S24
M10
M11
M12
M13
M14
M15
A17
M16
M17
A18
M18
M19
M20
M21
S25
A19
A20
S26
A21
S27
S28
A22
S29
S30
S31
A23
Sy1
S32
Sy2
O 1
A24
A25
S33
S34
A26
S35
S36
N 2
Sy3
A27
Sy4
Sy5
Sy6
O 2
S38
S39
S40
S41
S42
L4
L5
Si
D

Fig. FA♀5p

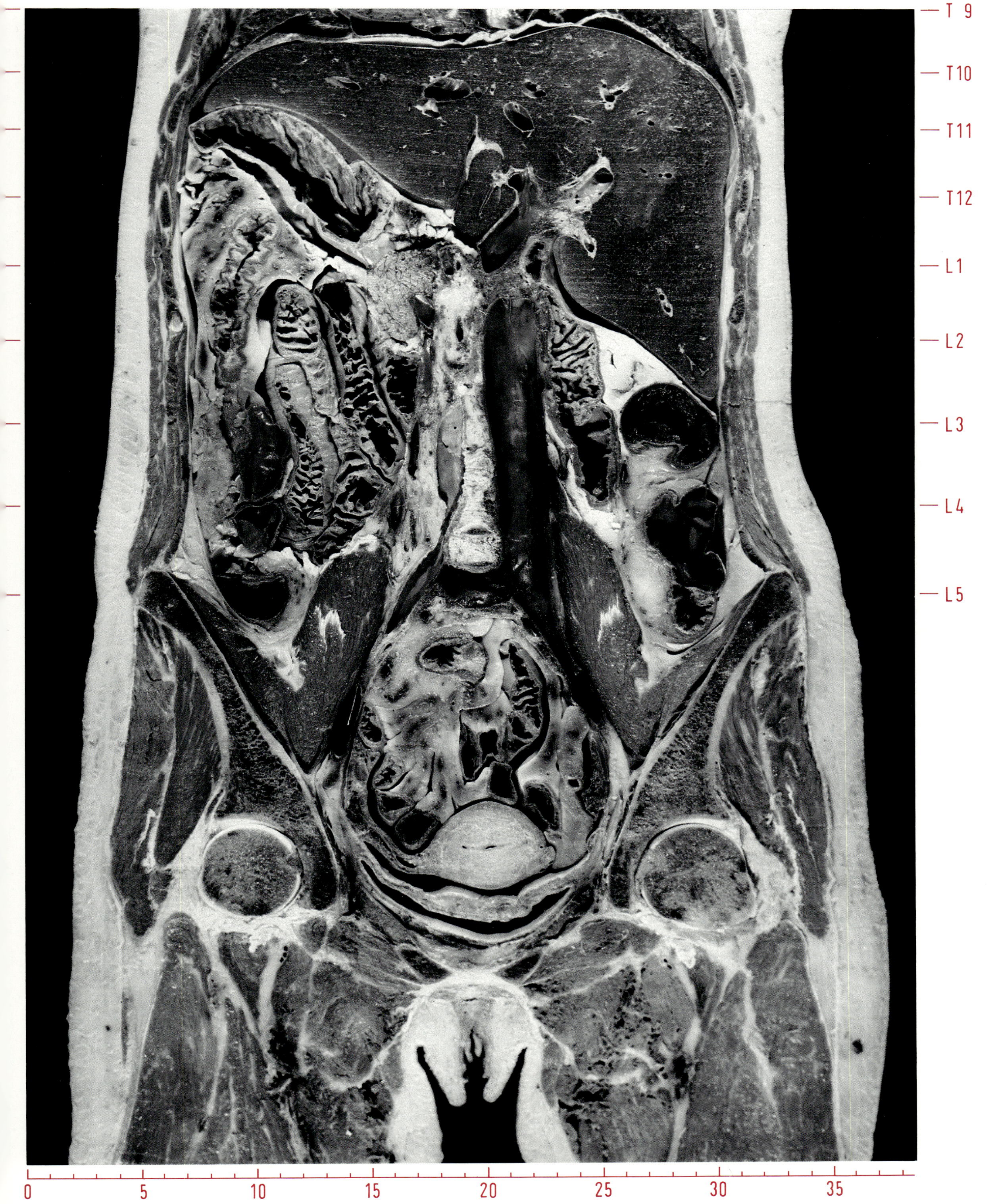

Fig. FA ♀ 5 p

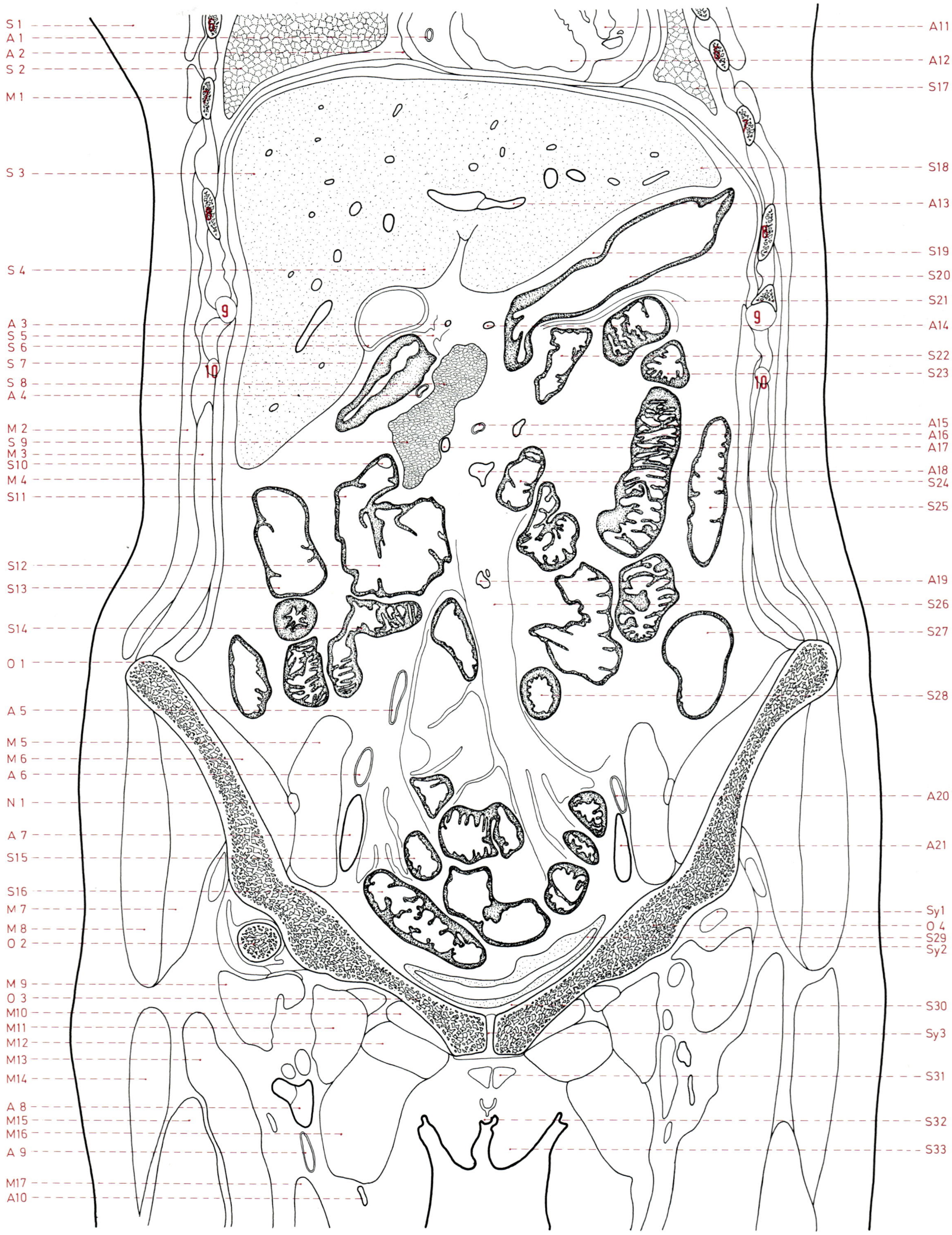
S 1
A 1
A 2
S 2
M 1
S 3
S 4
A 3
S 5
S 6
S 7
S 8
A 4
M 2
S 9
M 3
S10
M 4
S11
S12
S13
S14
O 1
A 5
M 5
M 6
A 6
N 1
A 7
S15
S16
M 7
M 8
O 2
M 9
O 3
M10
M11
M12
M13
M14
A 8
M15
M16
A 9
M17
A10
A11
A12
S17
S18
A13
S19
S20
S21
A14
S22
S23
A15
A16
A17
A18
S24
S25
A19
S26
S27
S28
A20
A21
Sy1
O 4
S29
Sy2
S30
Sy3
S31
S32
S33
6
7
8
9
10
D
Si

Fig. FA♀ 6

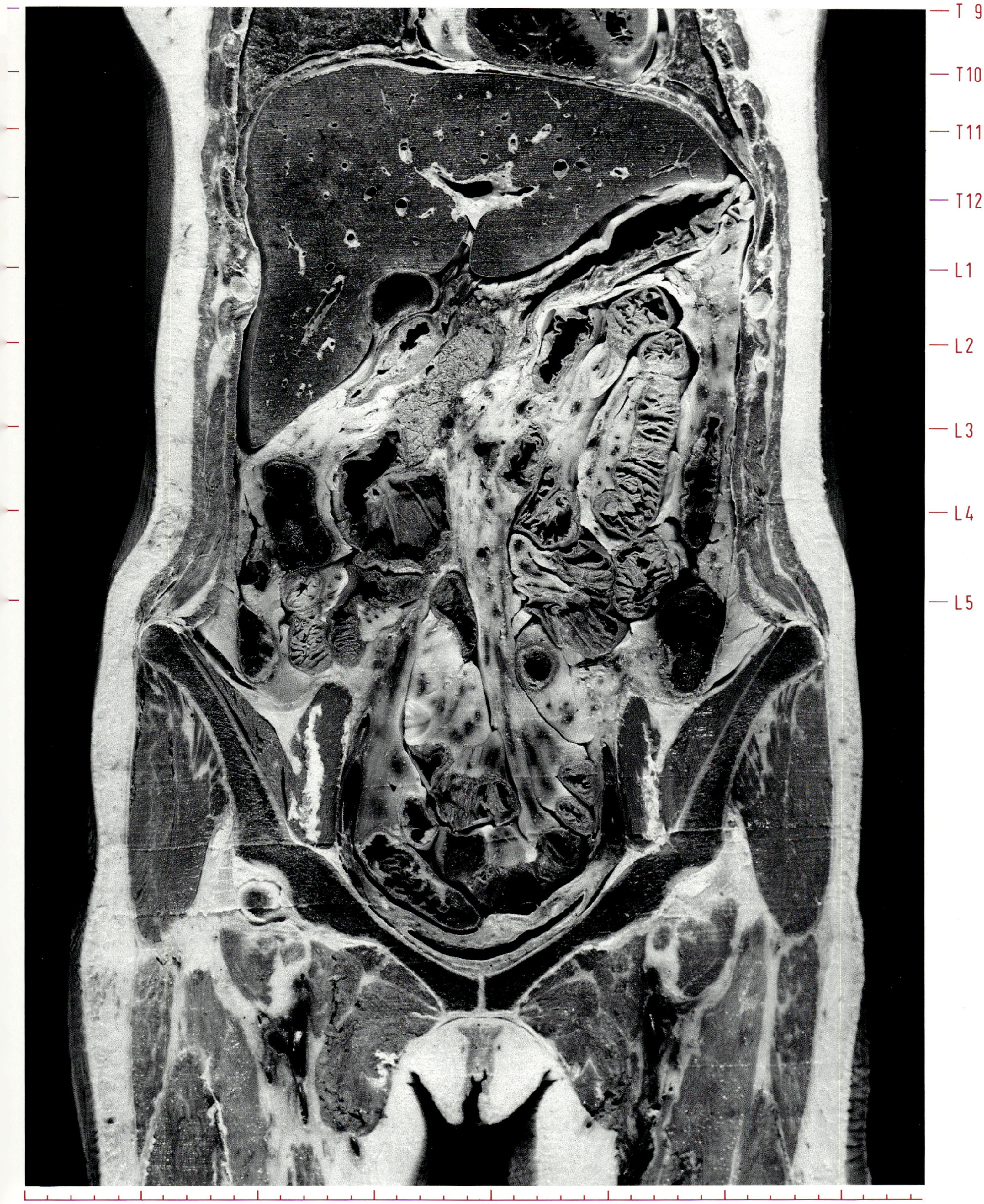

Fig. FA♀ 6

SA

♂

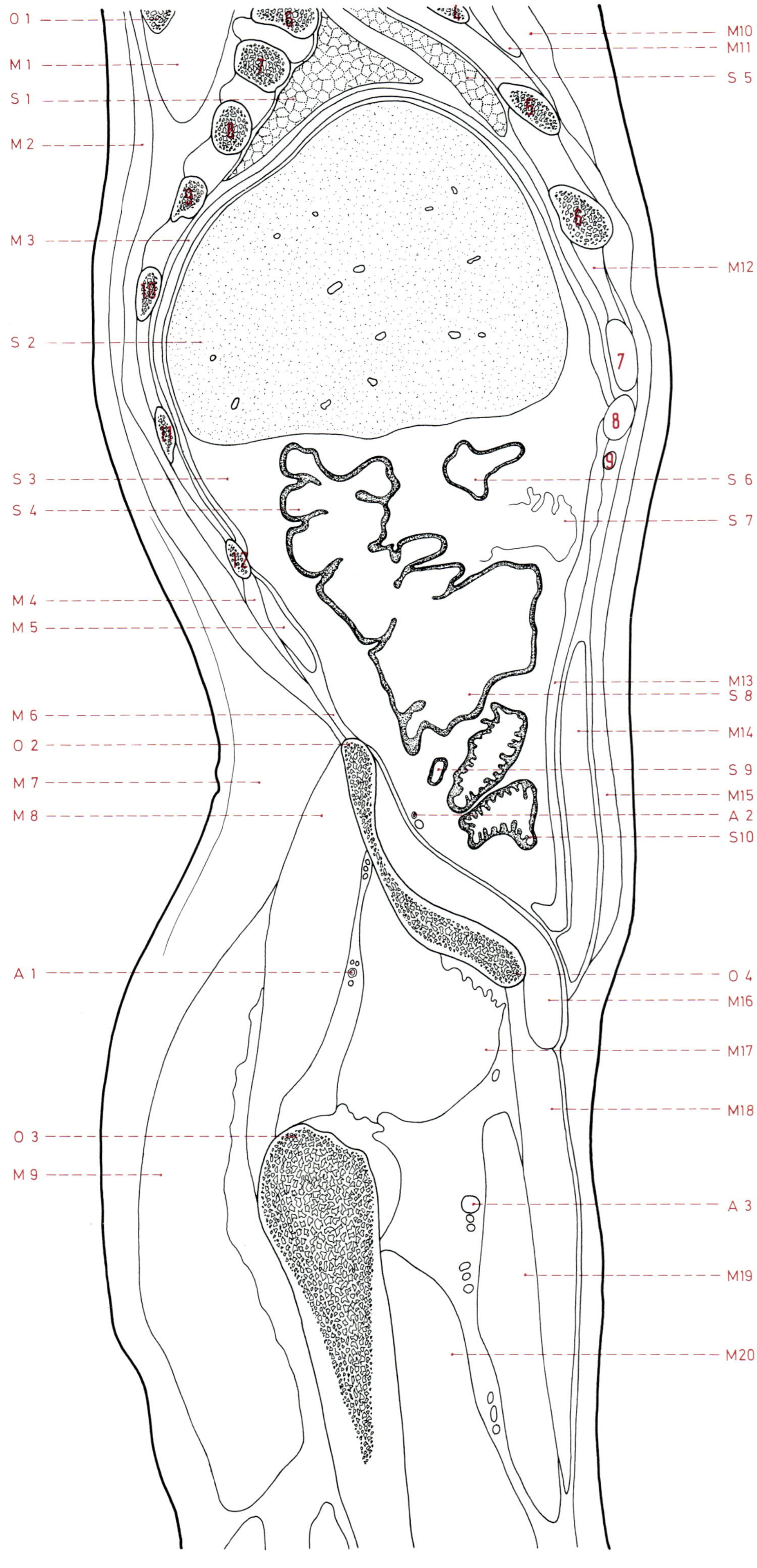
O 1
M 1
S 1
M 2
M 3
S 2
S 3
S 4
M 4
M 5
M 6
O 2
M 7
M 8
A 1
O 3
M 9
M10
M11
S 5
M12
S 6
S 7
M13
S 8
M14
S 9
M15
A 2
S10
O 4
M16
M17
M18
A 3
M19
M20
1
2
3
4
5
6
7
8
9
10
11
12

<u>Fig. SA♂ 6D</u>

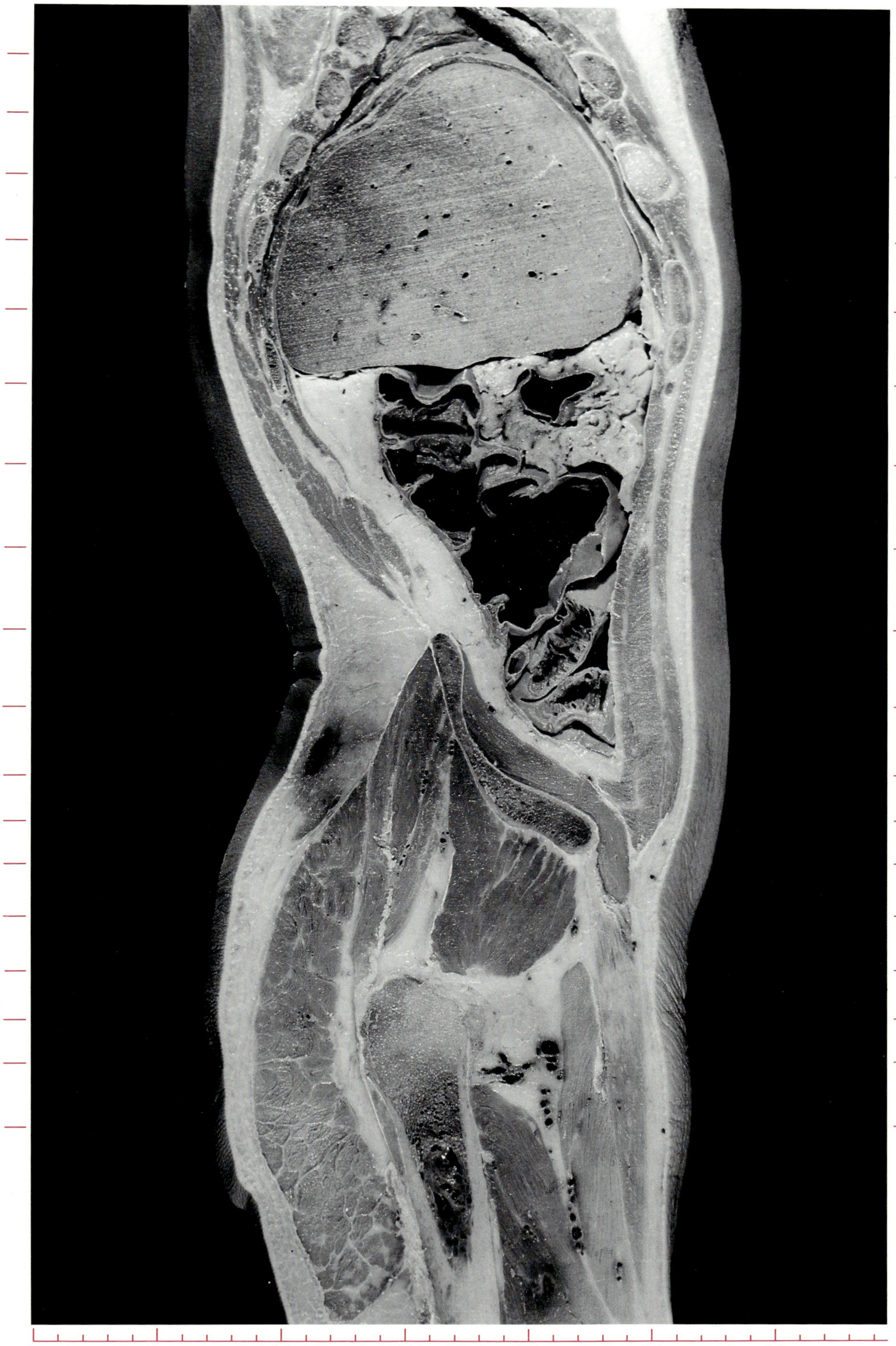
Th 8
Th 9
Th10
Th11
Th12
L1
L2
L3
L4
L5
S1
S2
S3
S4
S5
Co1
Co2
APEX
0
5
10
15
20
25
30

Fig. SA♂ 6D

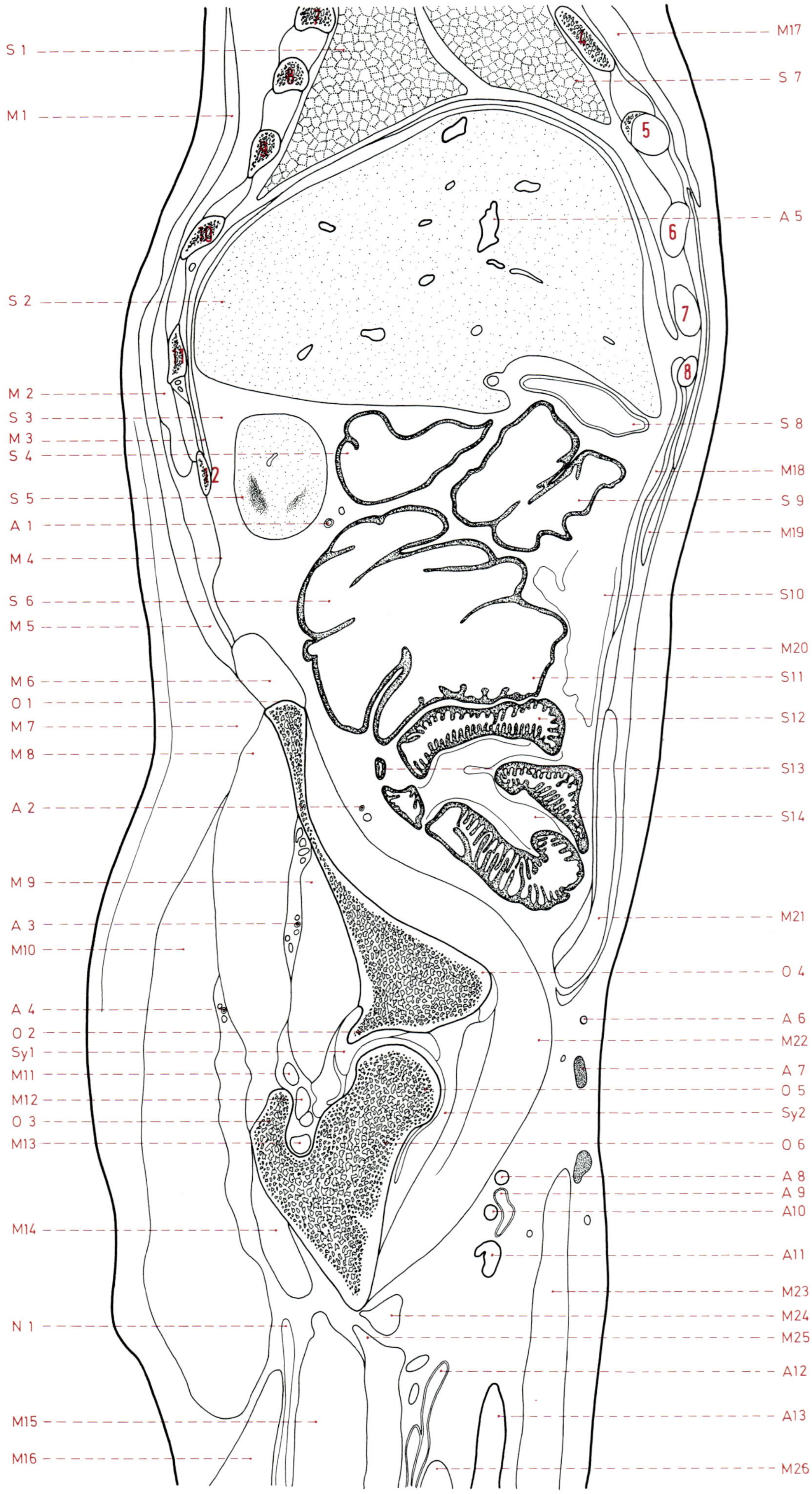
S 1
M 1
S 2
M 2
S 3
M 3
S 4
S 5
A 1
M 4
S 6
M 5
M 6
O 1
M 7
M 8
A 2
M 9
A 3
M10
A 4
O 2
Sy1
M11
M12
O 3
M13
M14
N 1
M15
M16
M17
S 7
A 5
S 8
M18
S 9
M19
S10
M20
S11
S12
S13
S14
M21
O 4
A 6
M22
A 7
O 5
Sy2
O 6
A 8
A 9
A10
A11
M23
M24
M25
A12
A13
M26

Fig. SA♂ 5D

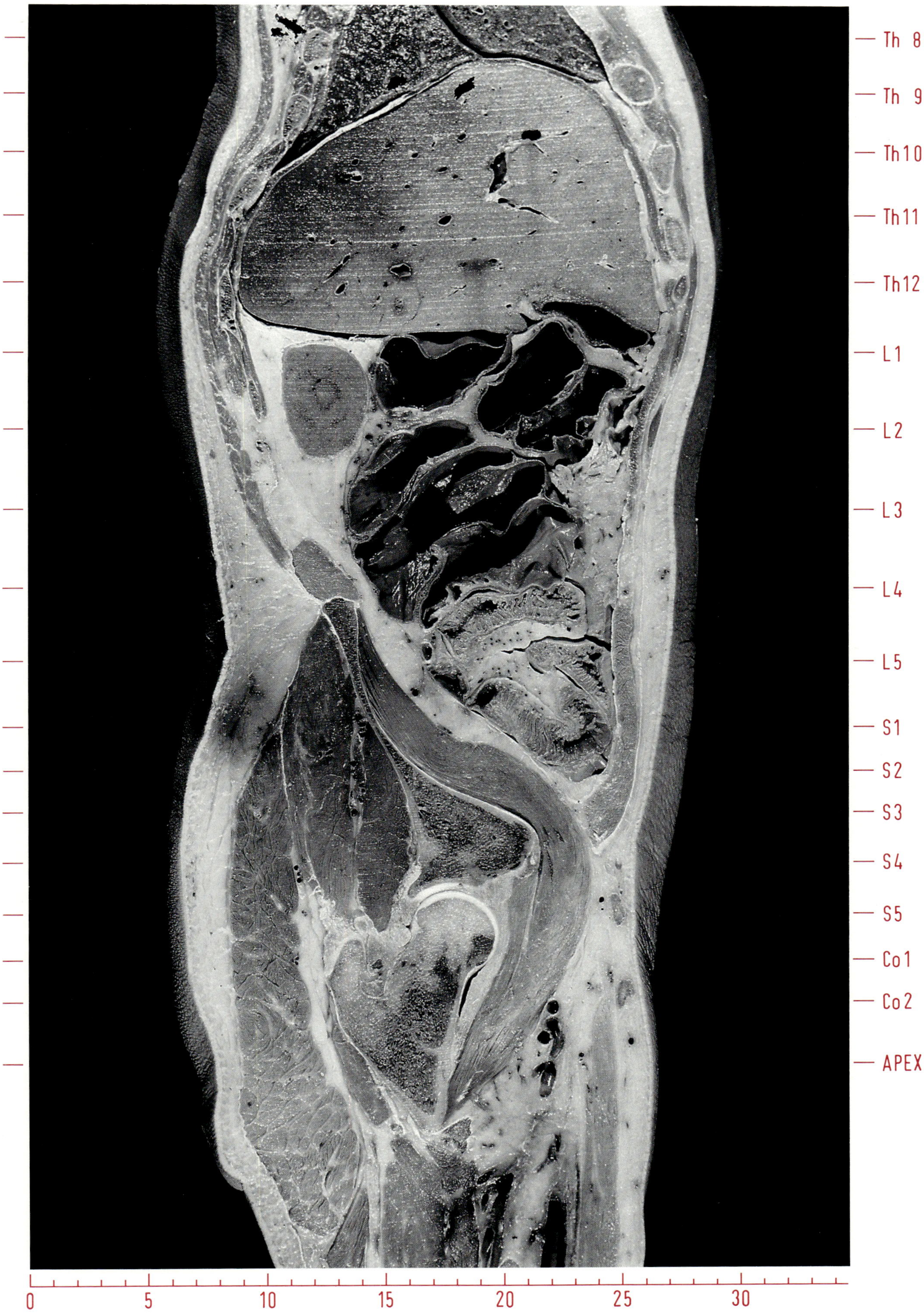
Th 8
Th 9
Th10
Th11
Th12
L1
L2
L3
L4
L5
S1
S2
S3
S4
S5
Co1
Co2
APEX
0
5
10
15
20
25
30

Fig. SA♂ 5D

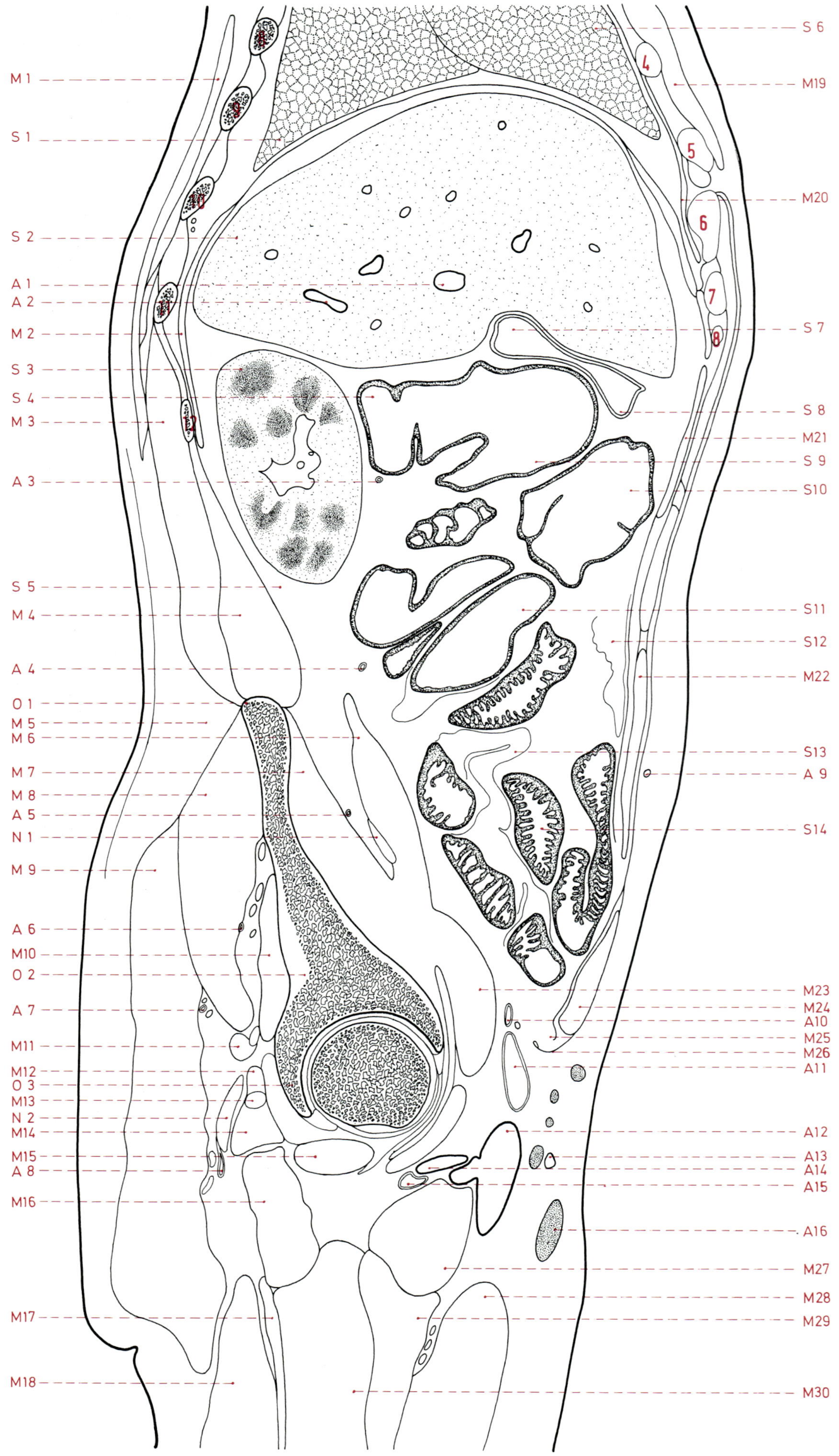
M1
S1
S2
A1
A2
M2
S3
S4
M3
A3
S5
M4
A4
O1
M5
M6
M7
M8
A5
N1
M9
A6
M10
O2
A7
M11
M12
O3
M13
N2
M14
M15
A8
M16
M17
M18
S6
M19
M20
S7
S8
M21
S9
S10
S11
S12
M22
S13
A9
S14
M23
M24
A10
M25
M26
A11
A12
A13
A14
A15
A16
M27
M28
M29
M30

Fig. SA♂ 4D

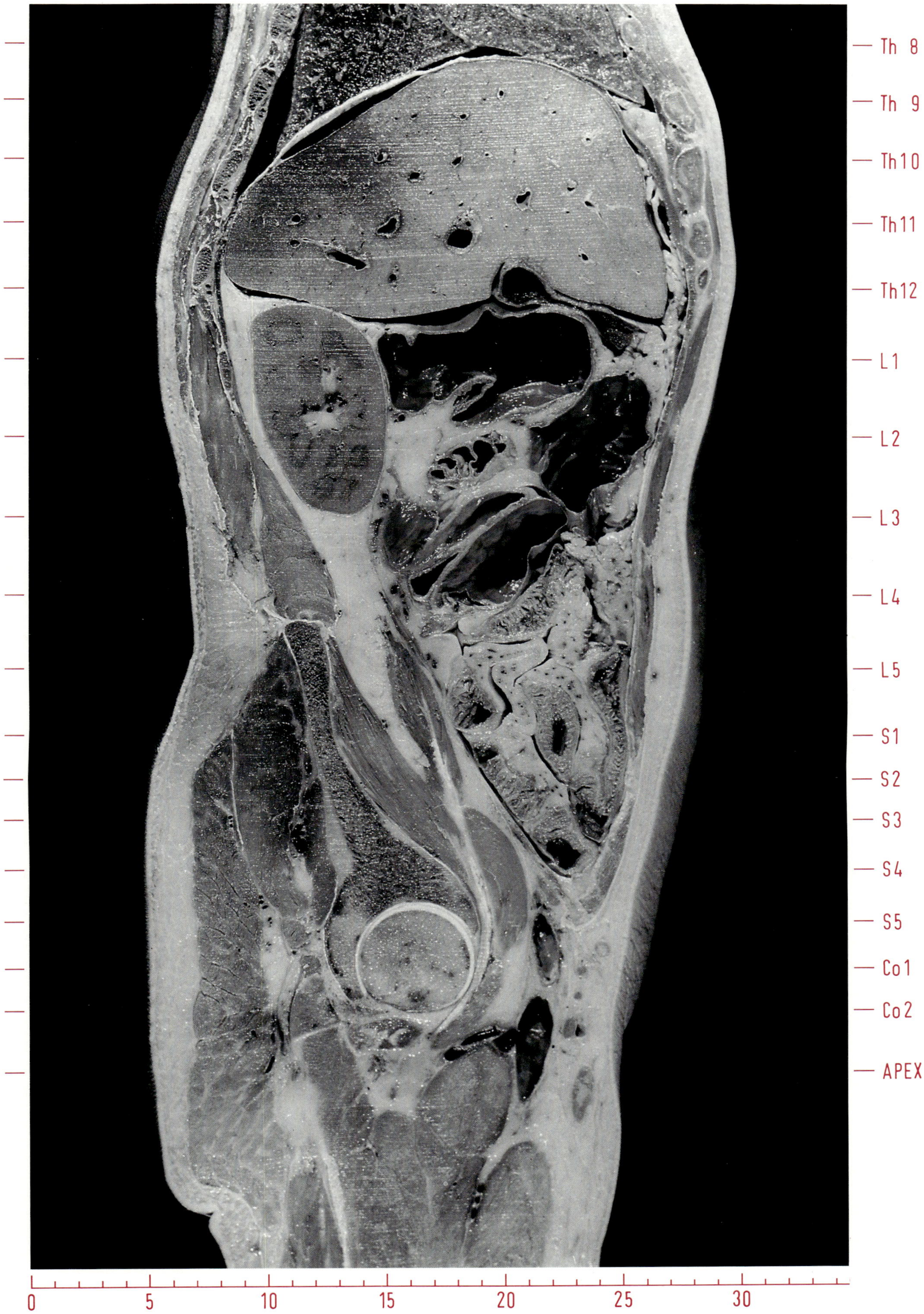

Th 8
Th 9
Th10
Th11
Th12
L1
L2
L3
L4
L5
S1
S2
S3
S4
S5
Co1
Co2
APEX
0
5
10
15
20
25
30

Fig. SA♂ 4D

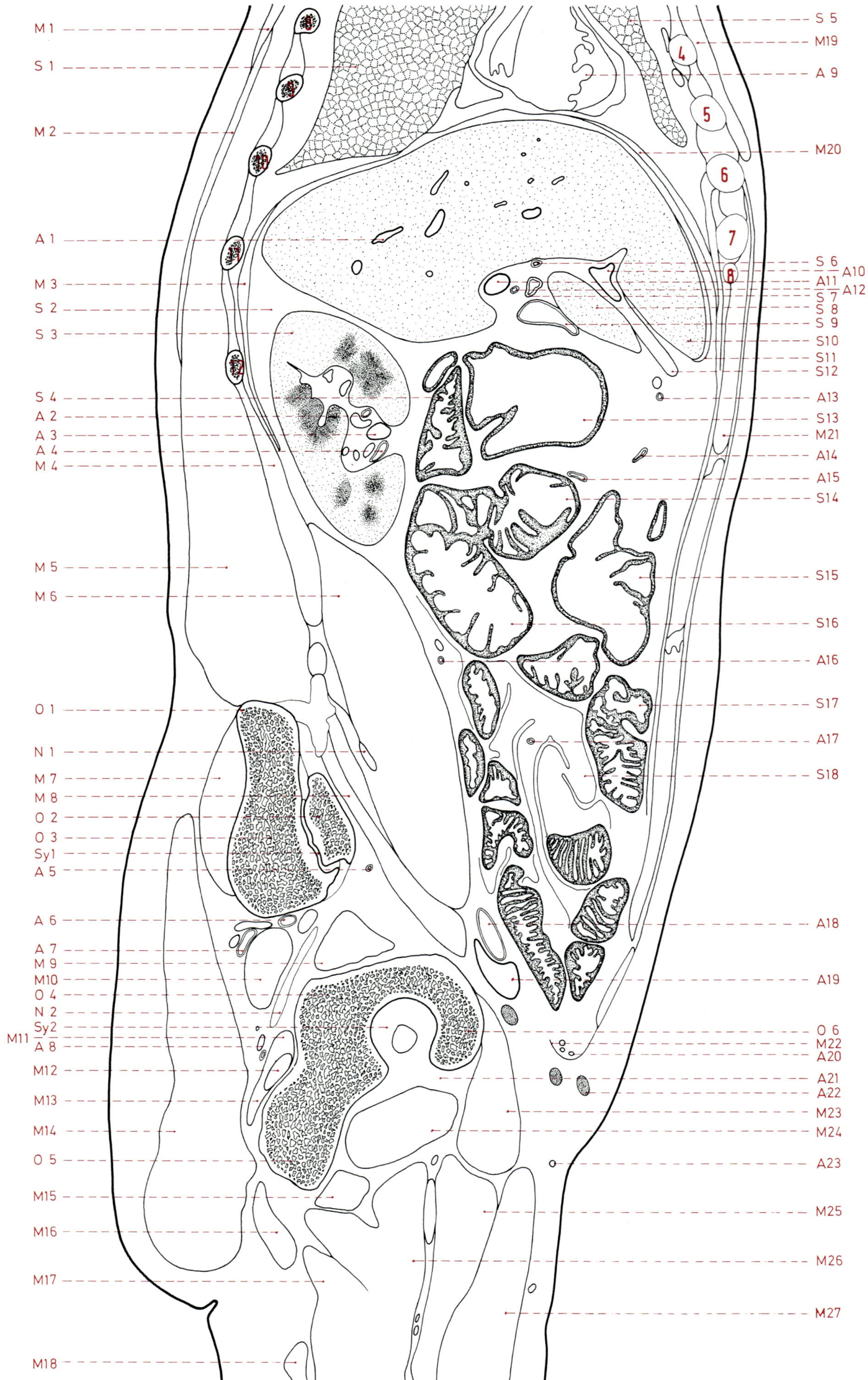
M 1
S 1
M 2
A 1
M 3
S 2
S 3
S 4
A 2
A 3
A 4
M 4
M 5
M 6
O 1
N 1
M 7
M 8
O 2
O 3
Sy1
A 5
A 6
A 7
M 9
M10
O 4
N 2
Sy2
M11
A 8
M12
M13
M14
O 5
M15
M16
M17
M18
S 5
M19
A 9
M20
S 6
A10
A11
A12
S 7
S 8
S 9
S10
S11
S12
A13
S13
M21
A14
A15
S14
S15
S16
A16
S17
A17
S18
A18
A19
O 6
M22
A20
A21
A22
M23
M24
A23
M25
M26
M27
4
5
6
7
8

Fig. SA♂ 3D

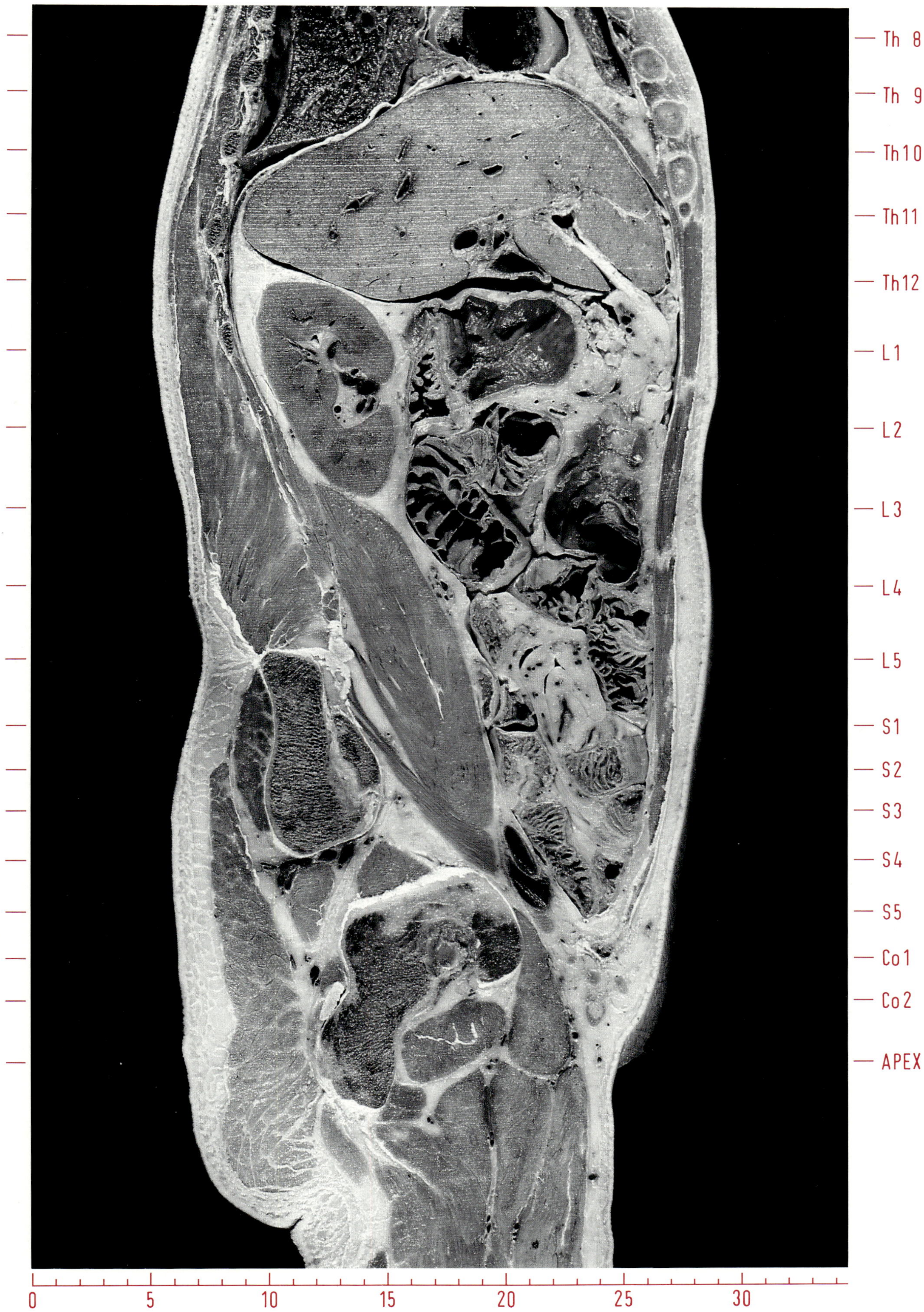
Th 8
Th 9
Th10
Th11
Th12
L1
L2
L3
L4
L5
S1
S2
S3
S4
S5
Co1
Co2
APEX
0
5
10
15
20
25
30

Fig. SA♂ 3D

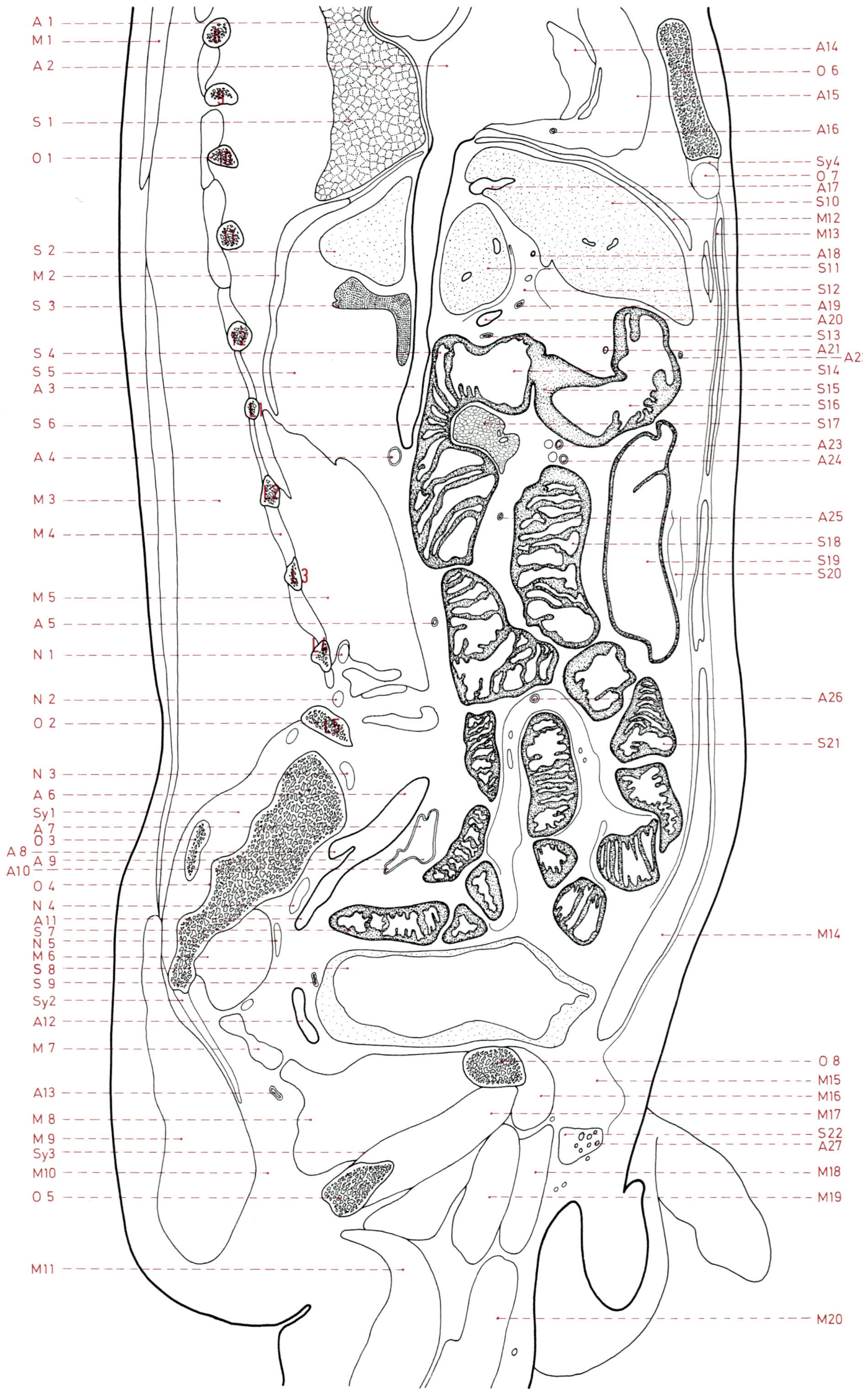

A 1
M 1
A 2
S 1
O 1
S 2
M 2
S 3
S 4
S 5
A 3
S 6
A 4
M 3
M 4
M 5
A 5
N 1
N 2
O 2
N 3
A 6
Sy1
A 7
O 3
A 8
A 9
A10
O 4
N 4
A11
S 7
N 5
M 6
S 8
S 9
Sy2
A12
M 7
A13
M 8
M 9
Sy3
M10
O 5
M11
A14
O 6
A15
A16
Sy4
O 7
A17
S10
M12
M13
A18
S11
S12
A19
A20
S13
A21
A22
S14
S15
S16
S17
A23
A24
A25
S18
S19
S20
A26
S21
M14
O 8
M15
M16
M17
S22
A27
M18
M19
M20

Fig. SA♂ 2D

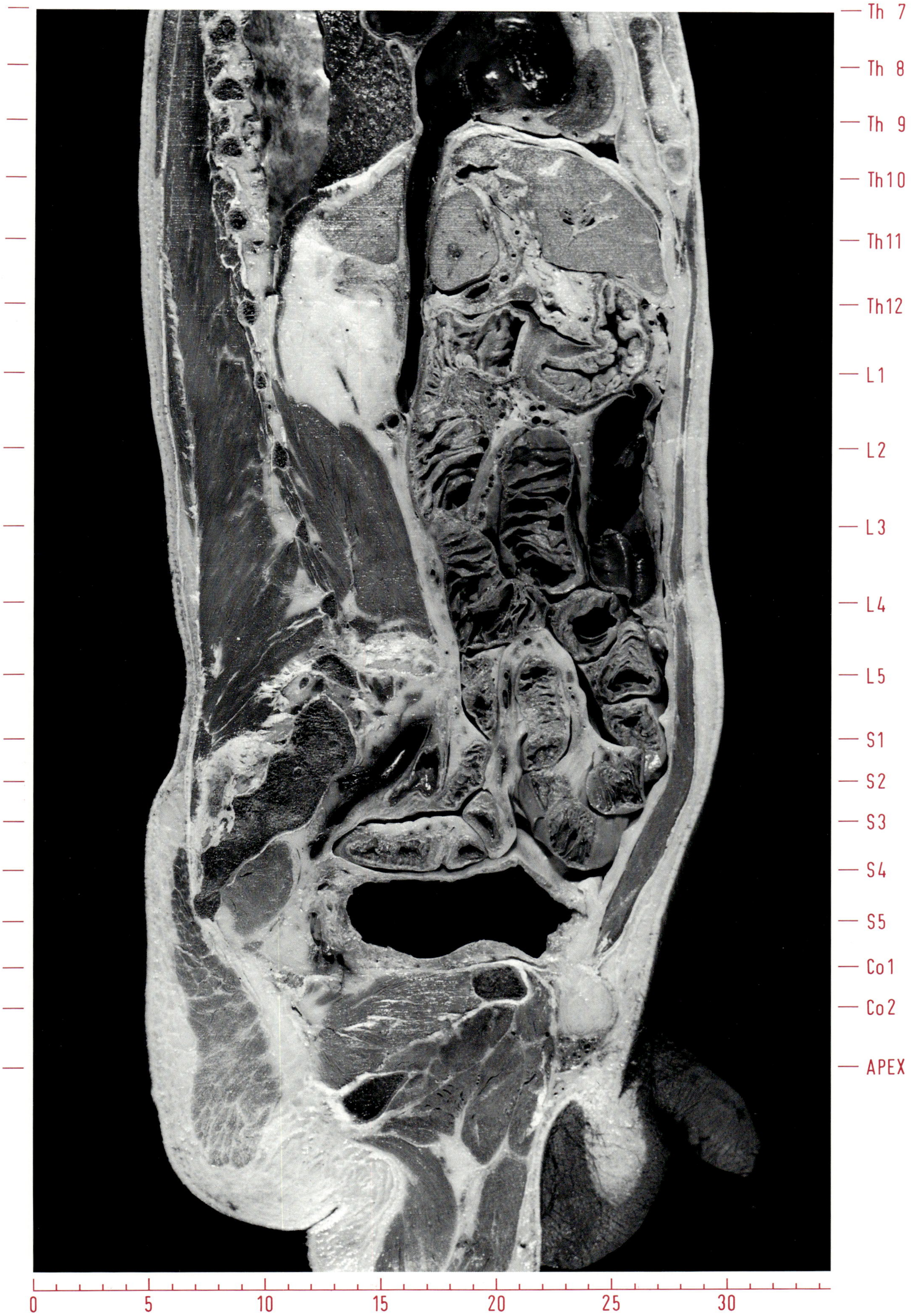
Th 7
Th 8
Th 9
Th10
Th11
Th12
L1
L2
L3
L4
L5
S1
S2
S3
S4
S5
Co1
Co2
APEX
0
5
10
15
20
25
30

Fig. SA♂ 2D

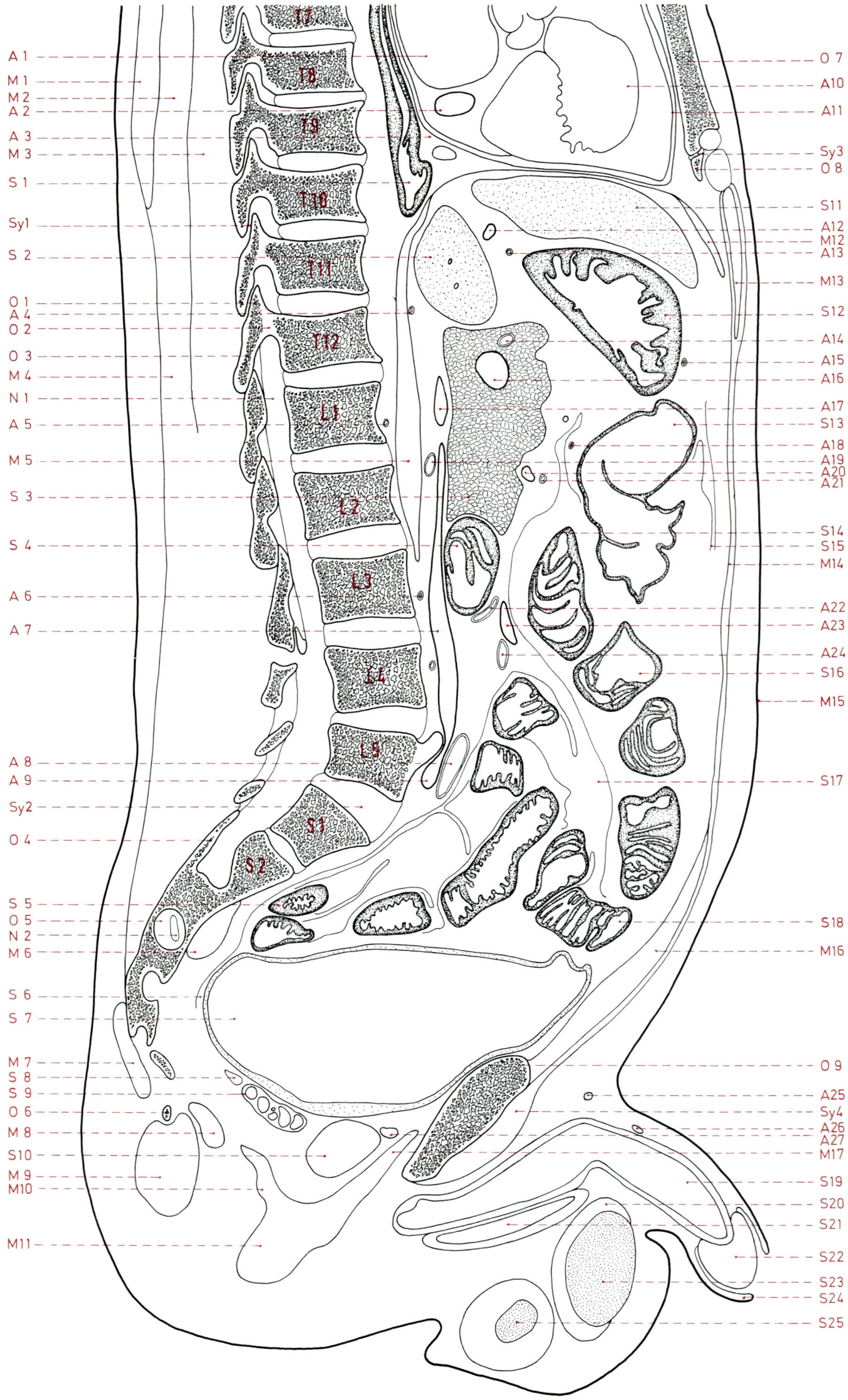
A 1
M 1
M 2
A 2
A 3
M 3
S 1
Sy1
S 2
O 1
A 4
O 2
O 3
M 4
N 1
A 5
M 5
S 3
S 4
A 6
A 7
A 8
A 9
Sy2
O 4
S 5
O 5
N 2
M 6
S 6
S 7
M 7
S 8
S 9
O 6
M 8
S10
M 9
M10
M11
T7
T8
T9
T10
T11
T12
L1
L2
L3
L4
L5
S1
S2
O 7
A10
A11
Sy3
O 8
S11
A12
M12
A13
M13
S12
A14
A15
A16
A17
S13
A18
A19
A20
A21
S14
S15
M14
A22
A23
A24
S16
M15
S17
S18
M16
O 9
A25
Sy4
A26
A27
M17
S19
S20
S21
S22
S23
S24
S25

Fig. SA♂ 1D

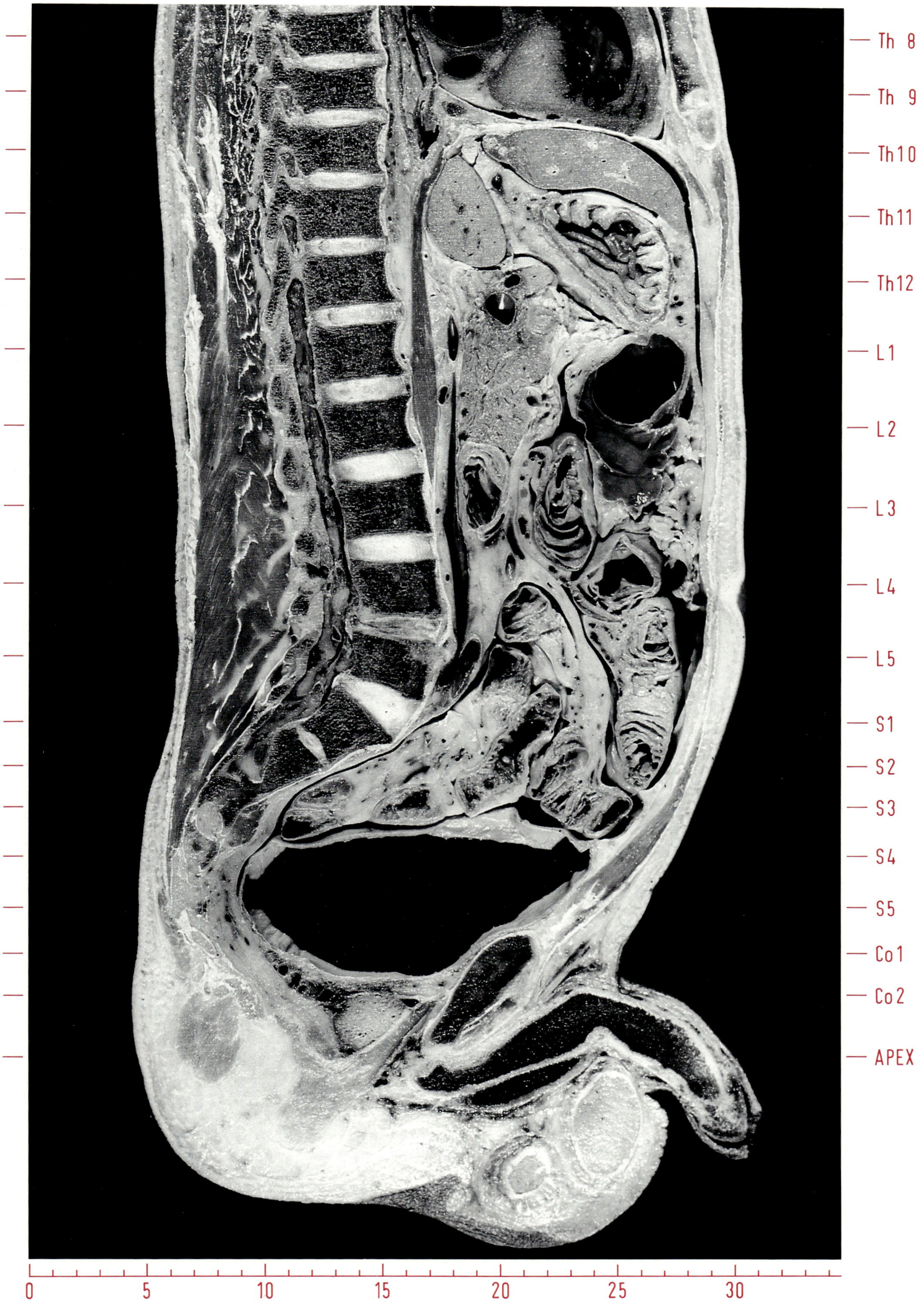
Th 8
Th 9
Th10
Th11
Th12
L1
L2
L3
L4
L5
S1
S2
S3
S4
S5
Co1
Co2
APEX
0
5
10
15
20
25
30

Fig. SA♂ 1D

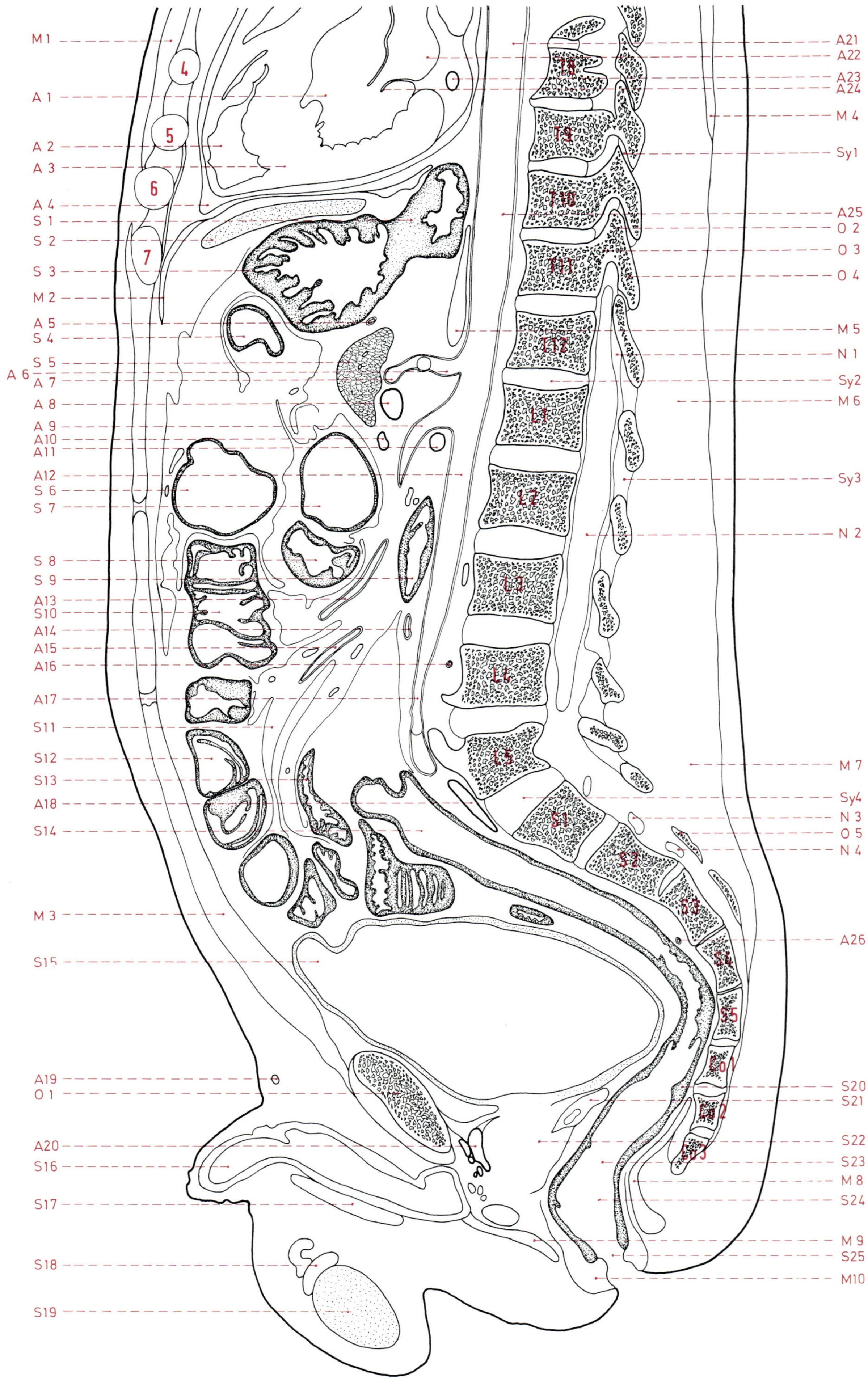
M1
A1
A2
A3
A4
S1
S2
S3
M2
A5
S4
S5
A6
A7
A8
A9
A10
A11
A12
S6
S7
S8
S9
A13
S10
A14
A15
A16
A17
S11
S12
S13
A18
S14
M3
S15
A19
O1
A20
S16
S17
S18
S19
A21
A22
A23
A24
M4
Sy1
A25
O2
O3
O4
M5
N1
Sy2
M6
Sy3
N2
M7
Sy4
N3
O5
N4
A26
S20
S21
S22
S23
M8
S24
M9
S25
M10
4
5
6
7
T8
T9
T10
T11
T12
L1
L2
L3
L4
L5
S1
S2
S3
S4
S5
Co1
Co2
Co3

Fig. SA♂ 1 Si

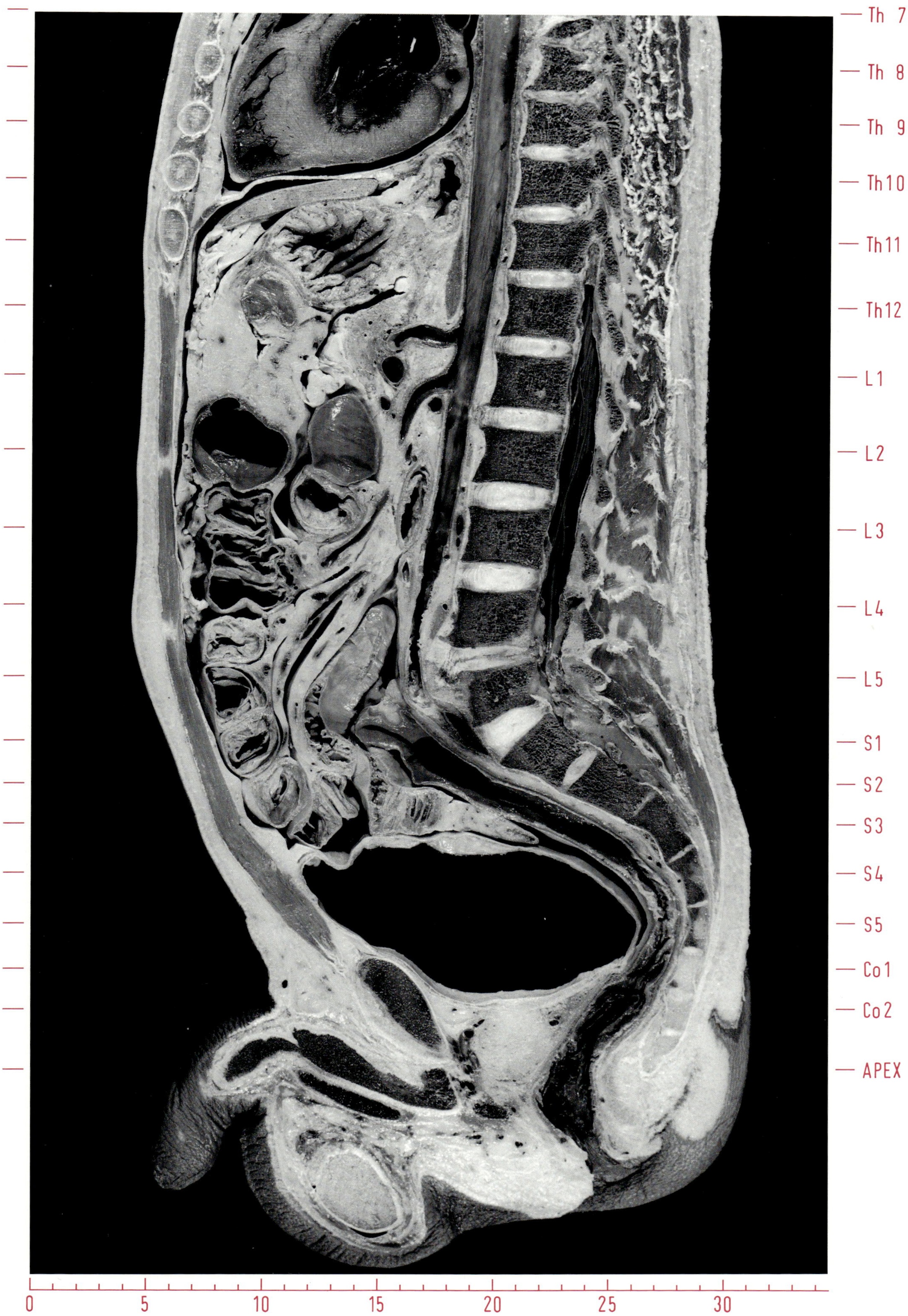

Th 7
Th 8
Th 9
Th10
Th11
Th12
L1
L2
L3
L4
L5
S1
S2
S3
S4
S5
Co1
Co2
APEX
0
5
10
15
20
25
30

Fig. SA♂ 1Si

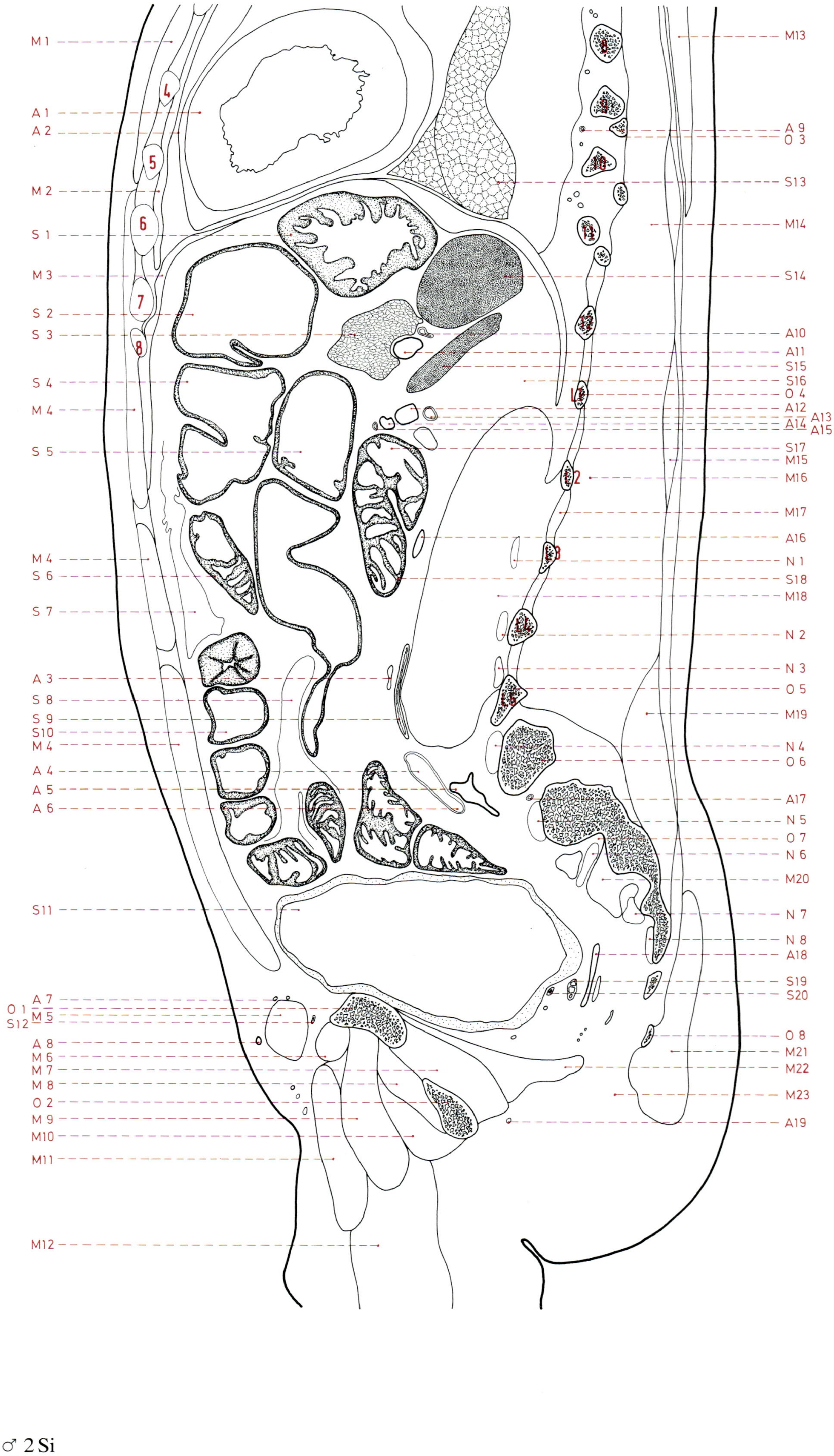
M 1
A 1
A 2
M 2
S 1
M 3
S 2
S 3
S 4
M 4
S 5
M 4
S 6
S 7
A 3
S 8
S 9
S10
M 4
A 4
A 5
A 6
S11
A 7
O 1
M 5
S12
A 8
M 6
M 7
M 8
O 2
M 9
M10
M11
M12
M13
A 9
O 3
S13
M14
S14
A10
A11
S15
S16
O 4
A12
A13
A14
A15
S17
M15
M16
M17
A16
N 1
S18
M18
N 2
N 3
O 5
M19
N 4
O 6
A17
N 5
O 7
N 6
M20
N 7
N 8
A18
S19
S20
O 8
M21
M22
M23
A19

Fig. SA♂ 2 Si

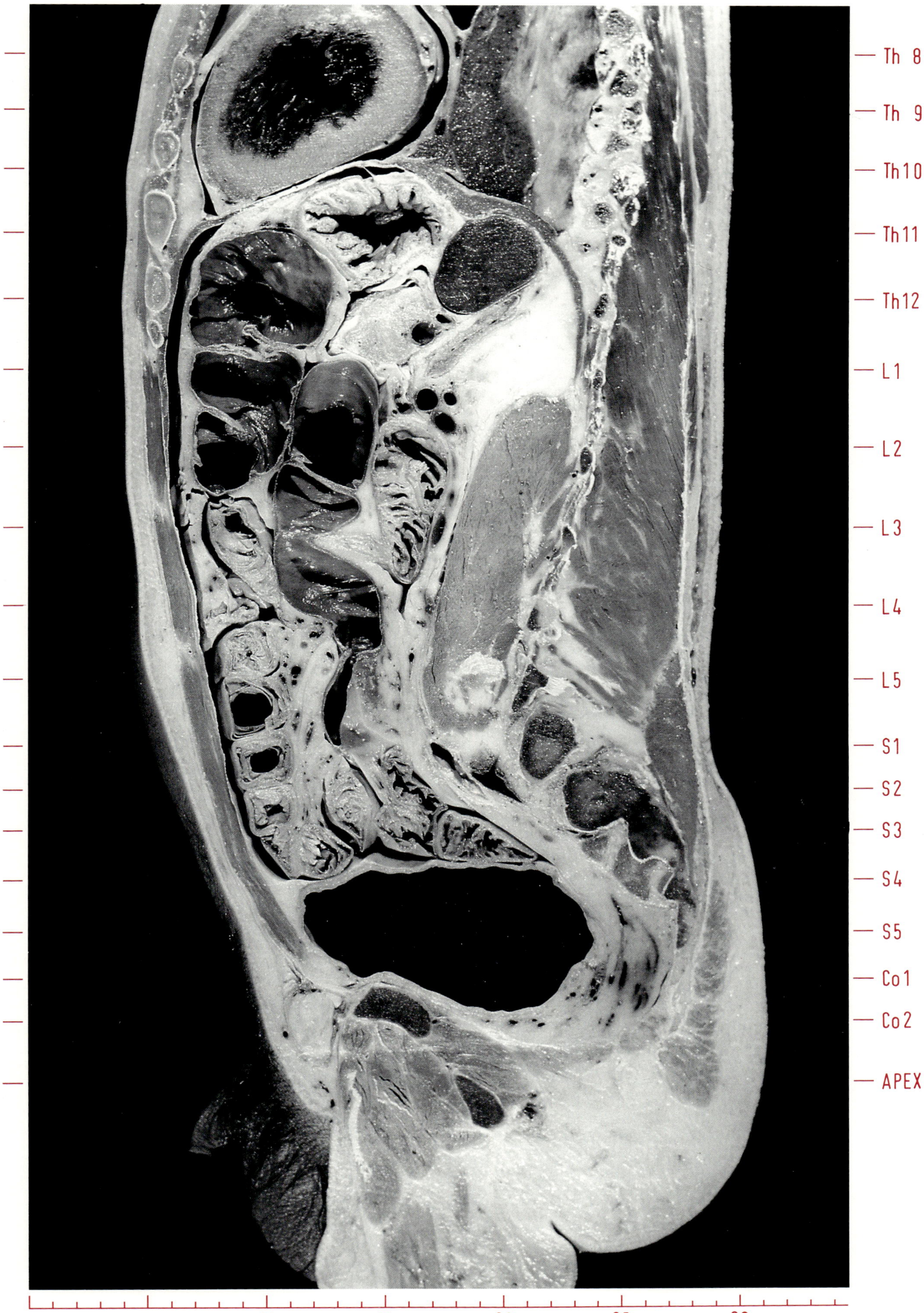
Th 8
Th 9
Th10
Th11
Th12
L1
L2
L3
L4
L5
S1
S2
S3
S4
S5
Co1
Co2
APEX
0
5
10
15
20
25
30

Fig. SA♂ 2Si

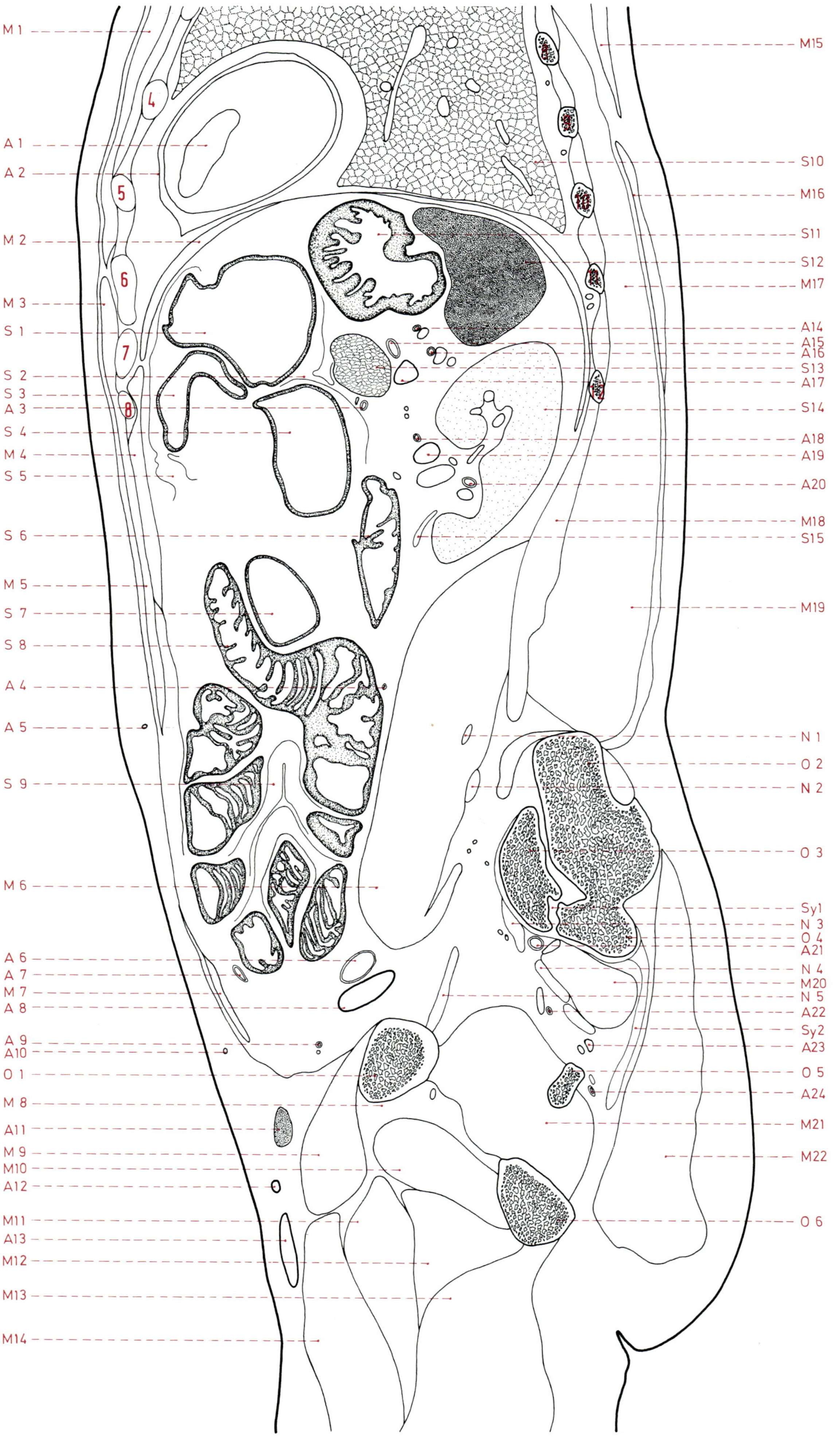

M 1
A 1
A 2
M 2
M 3
S 1
S 2
S 3
A 3
S 4
M 4
S 5
S 6
M 5
S 7
S 8
A 4
A 5
S 9
M 6
A 6
A 7
M 7
A 8
A 9
A10
O 1
M 8
A11
M 9
M10
A12
M11
A13
M12
M13
M14
M15
S10
M16
S11
S12
M17
A14
A15
A16
S13
A17
S14
A18
A19
A20
M18
S15
M19
N 1
O 2
N 2
O 3
Sy1
N 3
O 4
A21
N 4
M20
N 5
A22
Sy2
A23
O 5
A24
M21
M22
O 6
4
5
6
7
8
10

Fig. SA♂ 3 Si

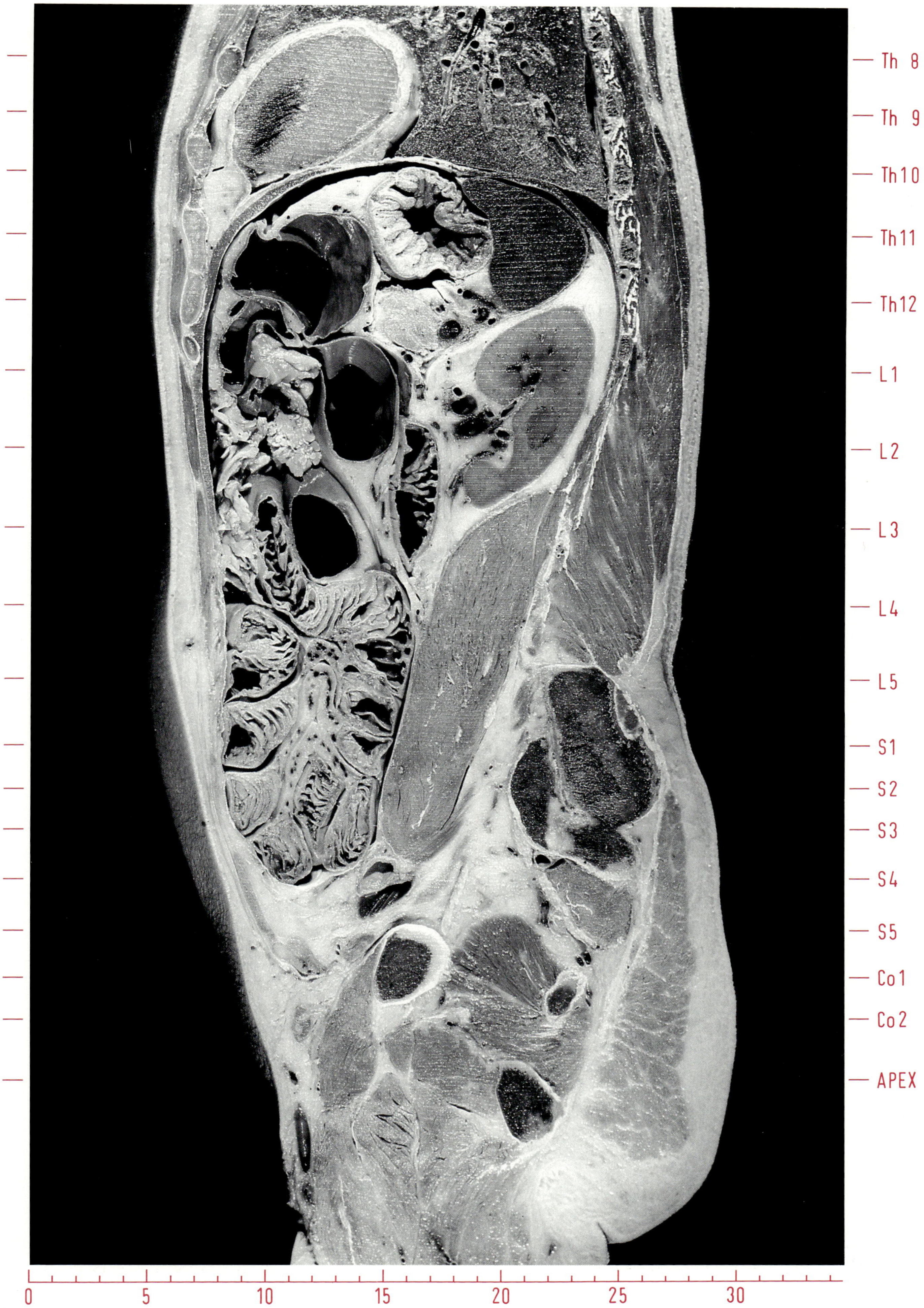
Th 8
Th 9
Th10
Th11
Th12
L1
L2
L3
L4
L5
S1
S2
S3
S4
S5
Co1
Co2
APEX
0
5
10
15
20
25
30

Fig. SA♂ 3Si

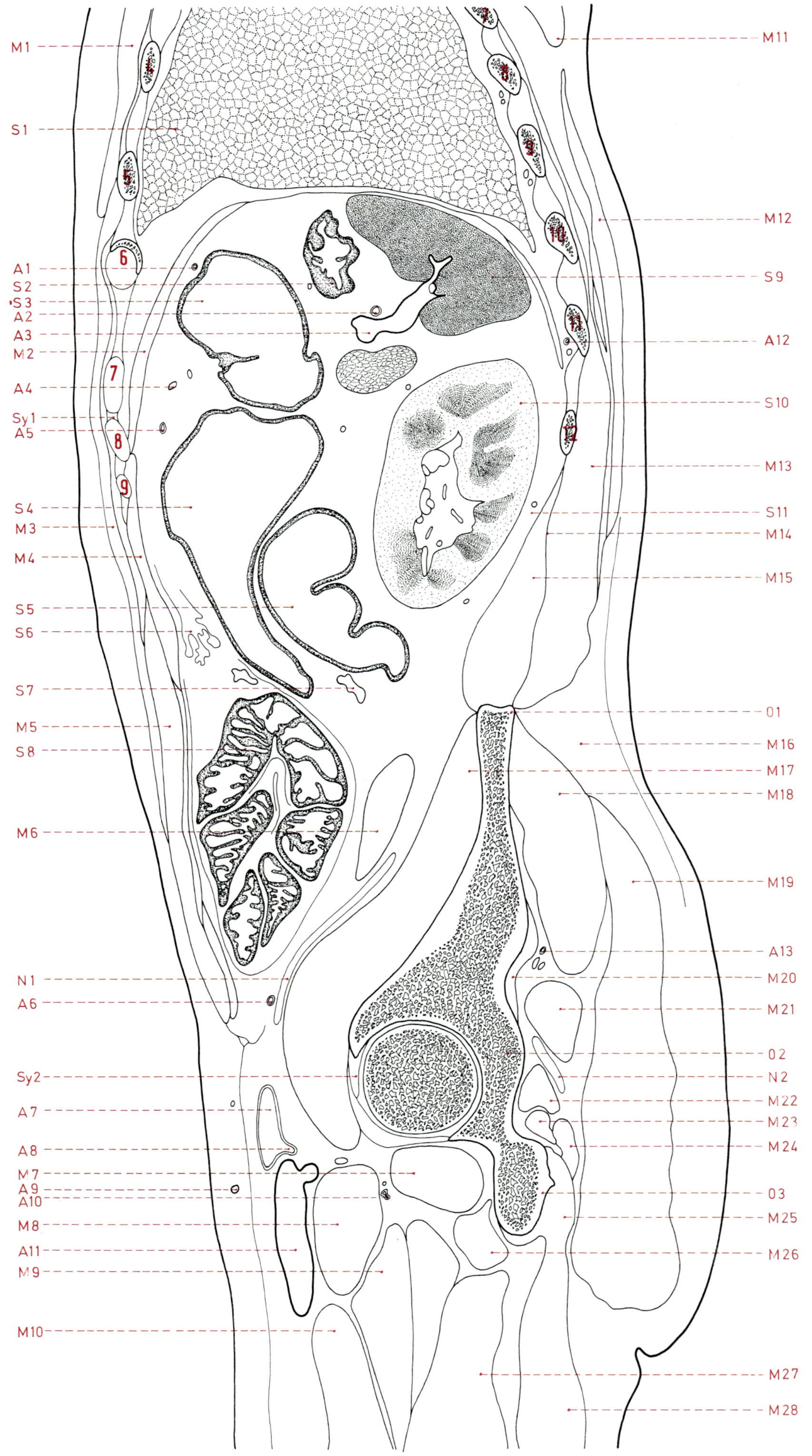

M1
S1
A1
S2
S3
A2
A3
M2
A4
Sy1
A5
S4
M3
M4
S5
S6
S7
M5
S8
M6
N1
A6
Sy2
A7
A8
M7
A9
A10
M8
A11
M9
M10
M11
M12
S9
A12
S10
M13
S11
M14
M15
01
M16
M17
M18
M19
A13
M20
M21
02
N2
M22
M23
M24
03
M25
M26
M27
M28

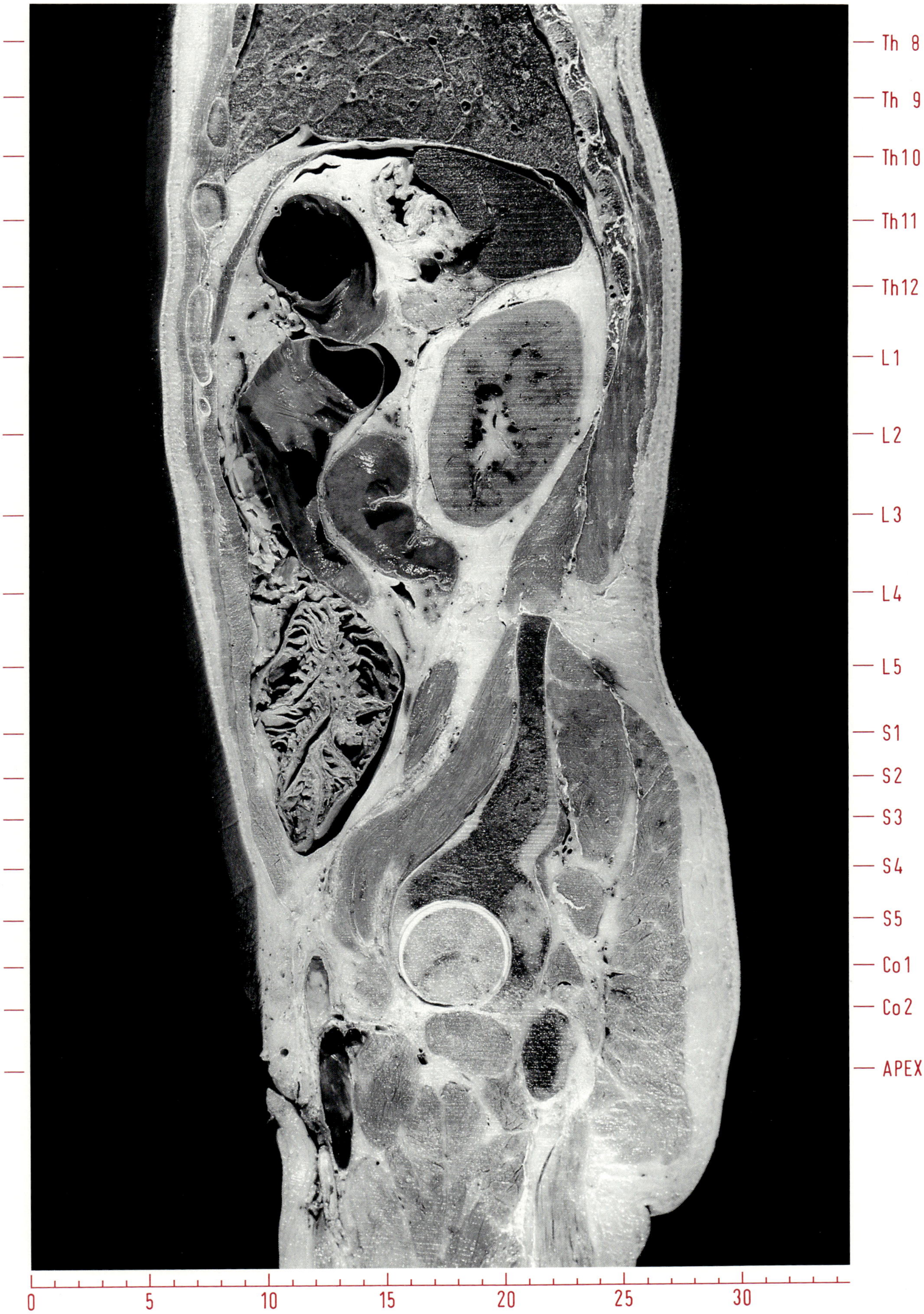
Th 8
Th 9
Th10
Th11
Th12
L1
L2
L3
L4
L5
S1
S2
S3
S4
S5
Co1
Co2
APEX
0
5
10
15
20
25
30

Fig. SA♂ 4Si

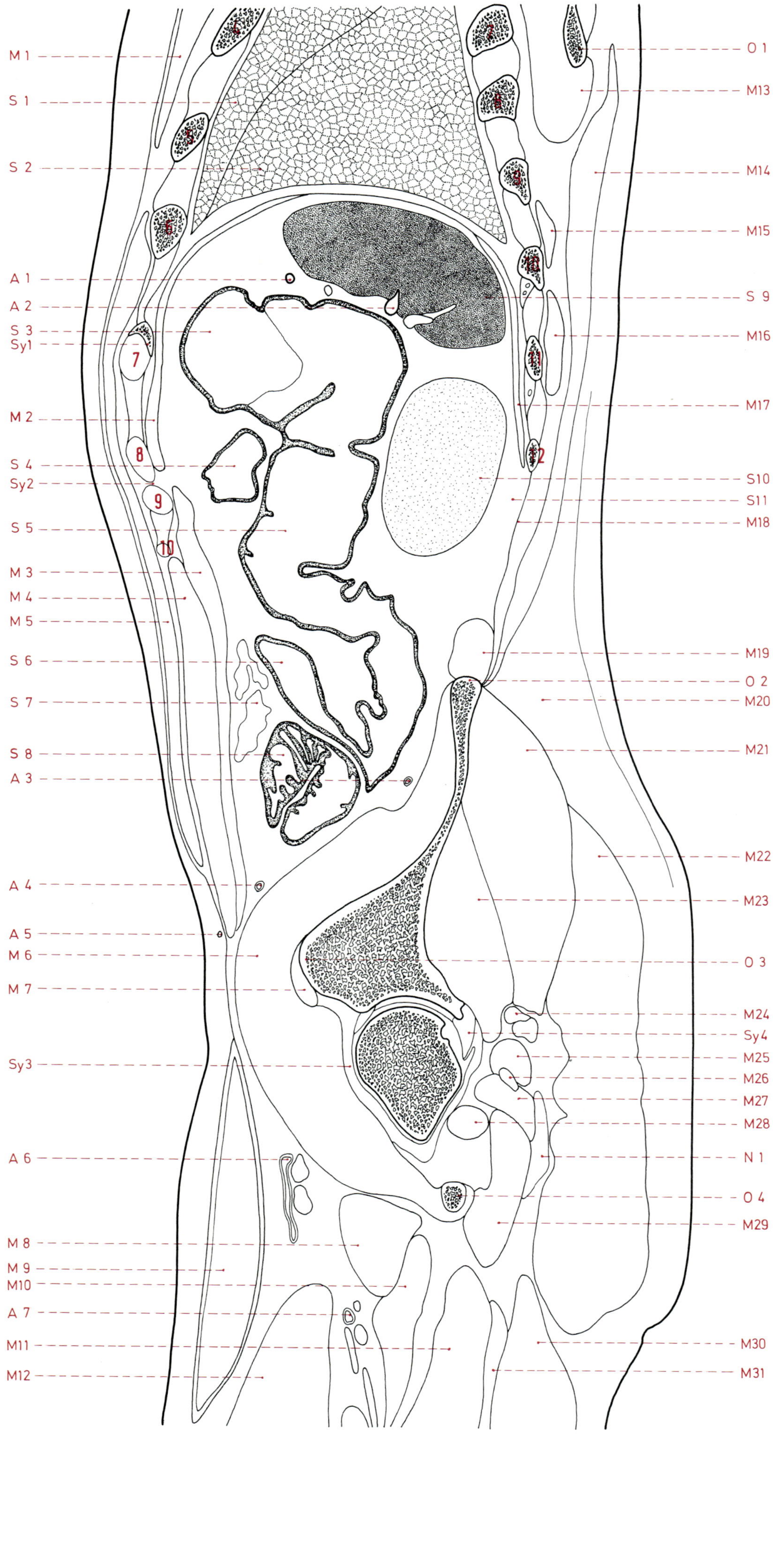
M 1
S 1
S 2
A 1
A 2
S 3
Sy1
M 2
S 4
Sy2
S 5
M 3
M 4
M 5
S 6
S 7
S 8
A 3
A 4
A 5
M 6
M 7
Sy3
A 6
M 8
M 9
M10
A 7
M11
M12
O 1
M13
M14
M15
S 9
M16
M17
S10
S11
M18
M19
O 2
M20
M21
M22
M23
O 3
M24
Sy4
M25
M26
M27
M28
N 1
O 4
M29
M30
M31
1
2
3
4
5
6
7
8
9
10
11
12

Fig. SA♂ 5 Si

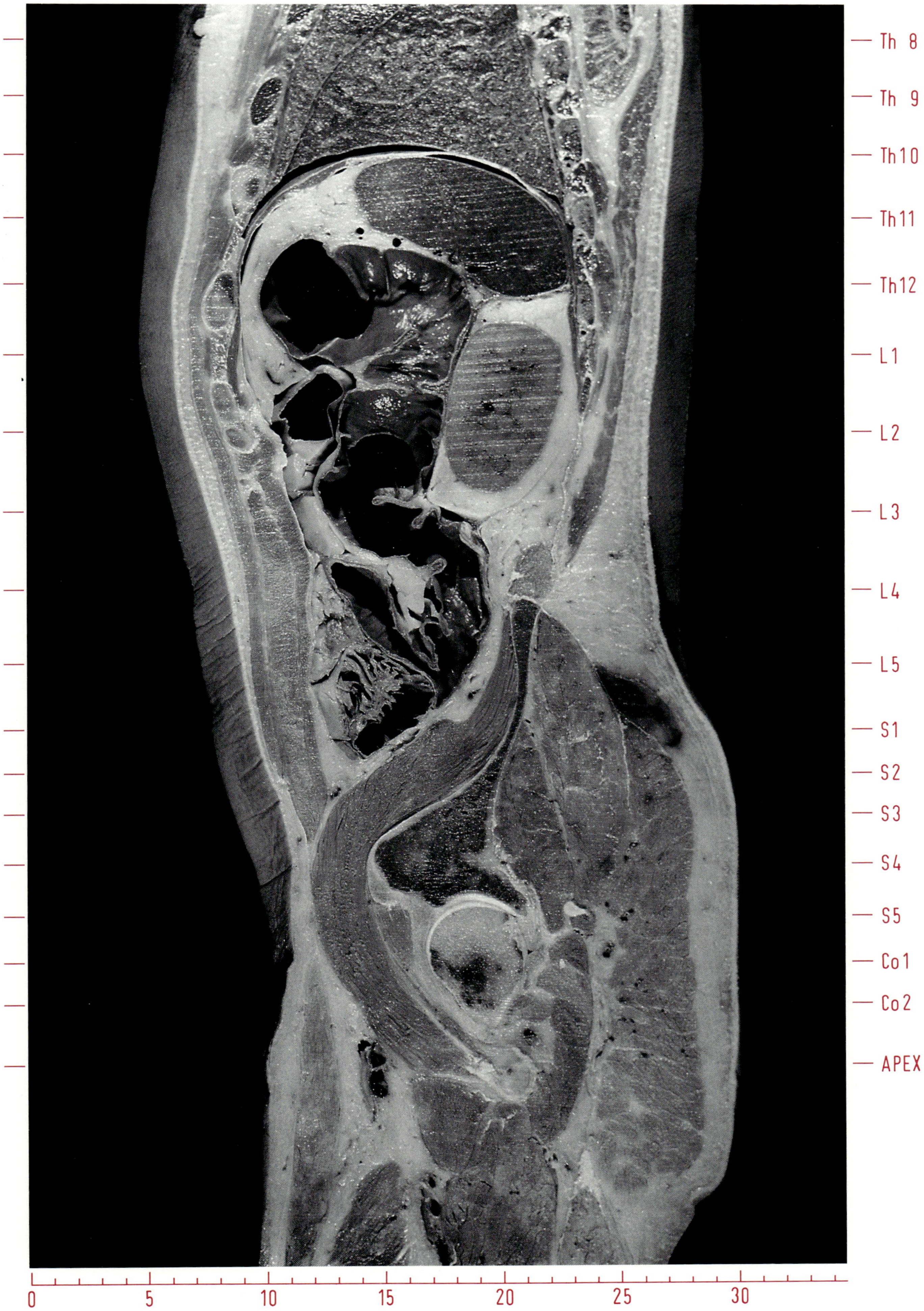
Th 8
Th 9
Th10
Th11
Th12
L1
L2
L3
L4
L5
S1
S2
S3
S4
S5
Co1
Co2
APEX
0
5
10
15
20
25
30

Fig. SA♂ 5Si

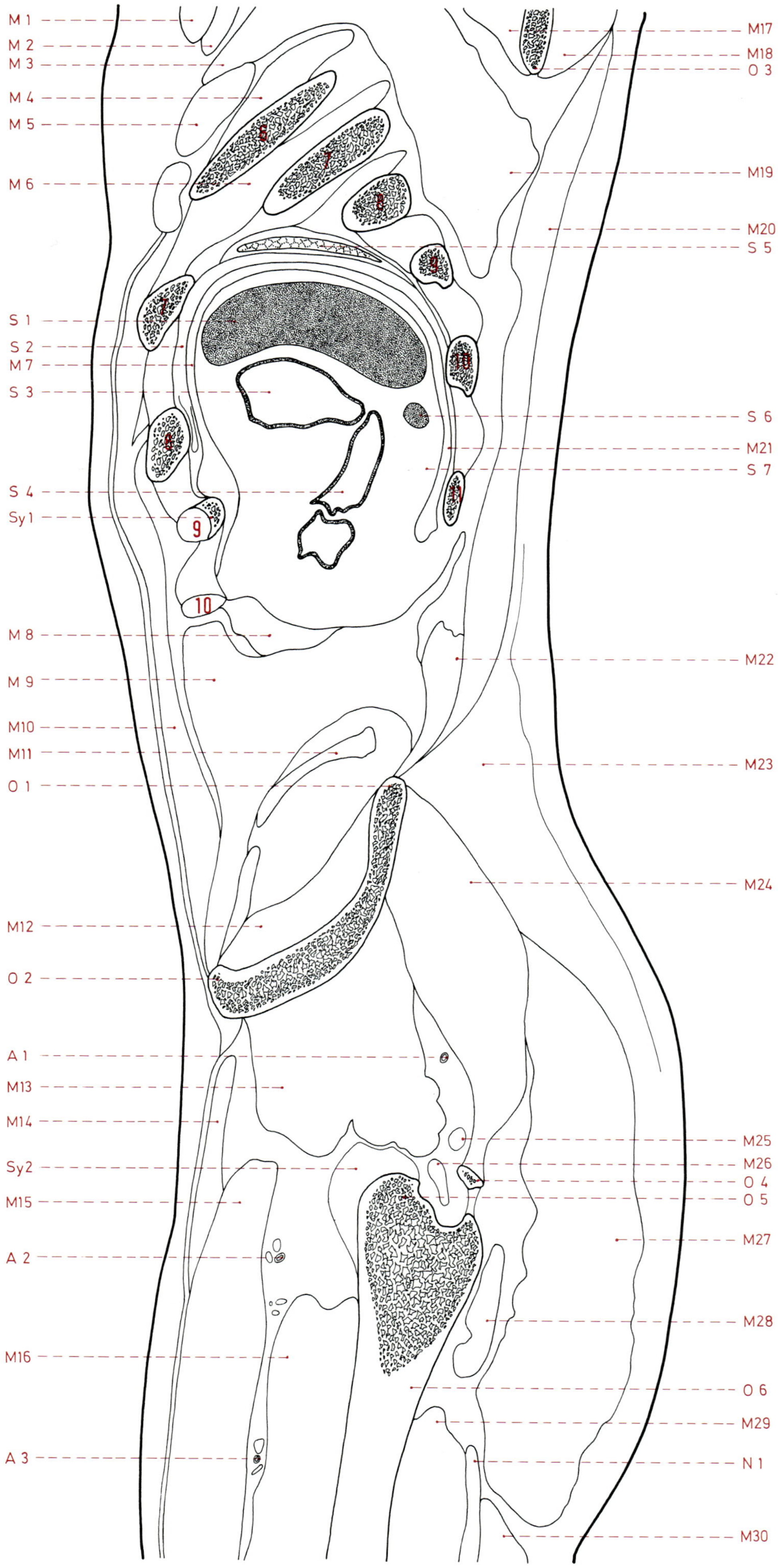
M 1
M 2
M 3
M 4
M 5
M 6
S 1
S 2
M 7
S 3
S 4
Sy 1
M 8
M 9
M10
M11
O 1
M12
O 2
A 1
M13
M14
Sy2
M15
A 2
M16
A 3
M17
M18
O 3
M19
M20
S 5
S 6
M21
S 7
M22
M23
M24
M25
M26
O 4
O 5
M27
M28
O 6
M29
N 1
M30
6
7
8
9
10
11
9
10

Fig. SA♂ 6 Si

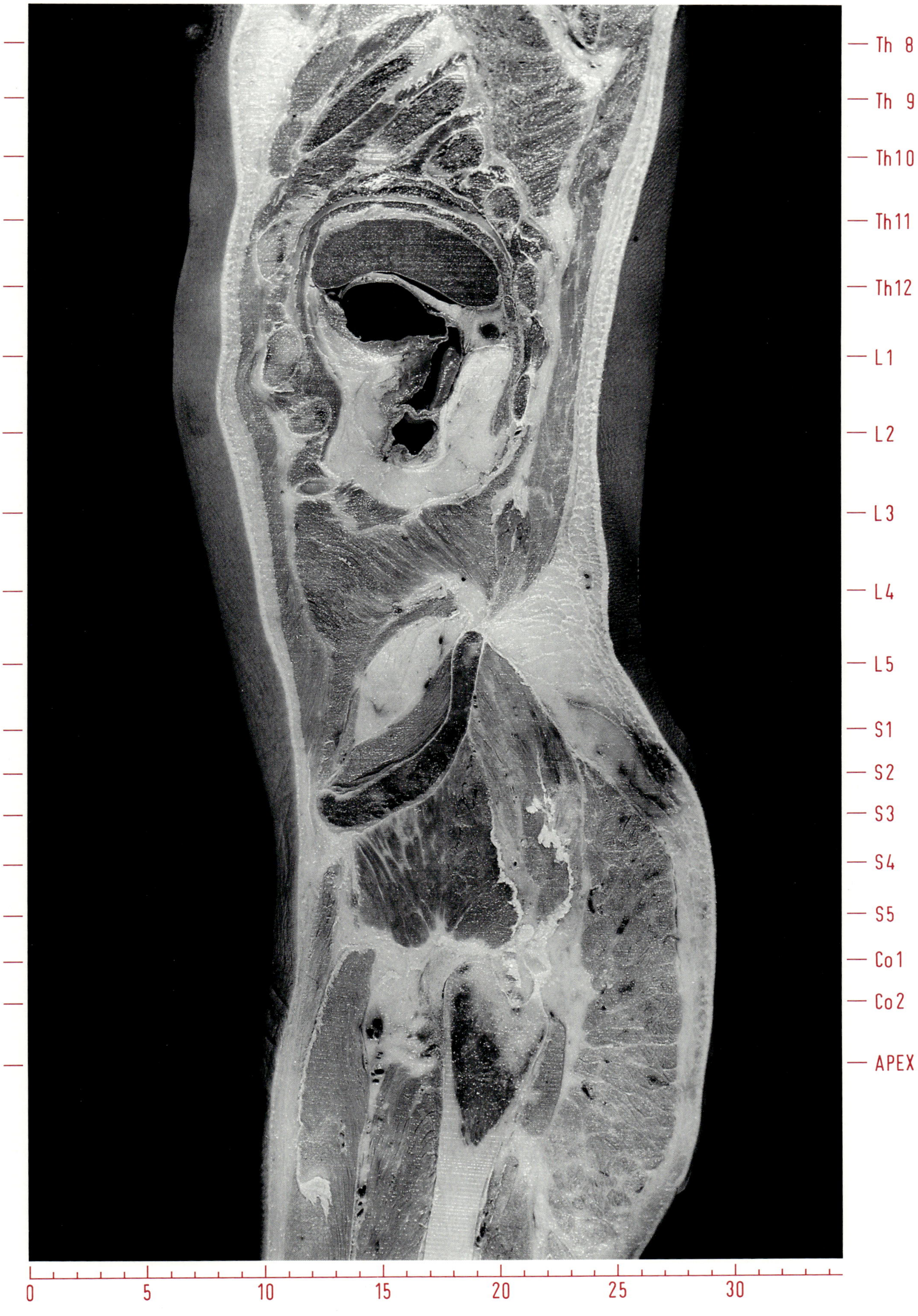
Th 8
Th 9
Th10
Th11
Th12
L1
L2
L3
L4
L5
S1
S2
S3
S4
S5
Co1
Co2
APEX
0
5
10
15
20
25
30

Fig. SA♂ 6Si

SA

♀

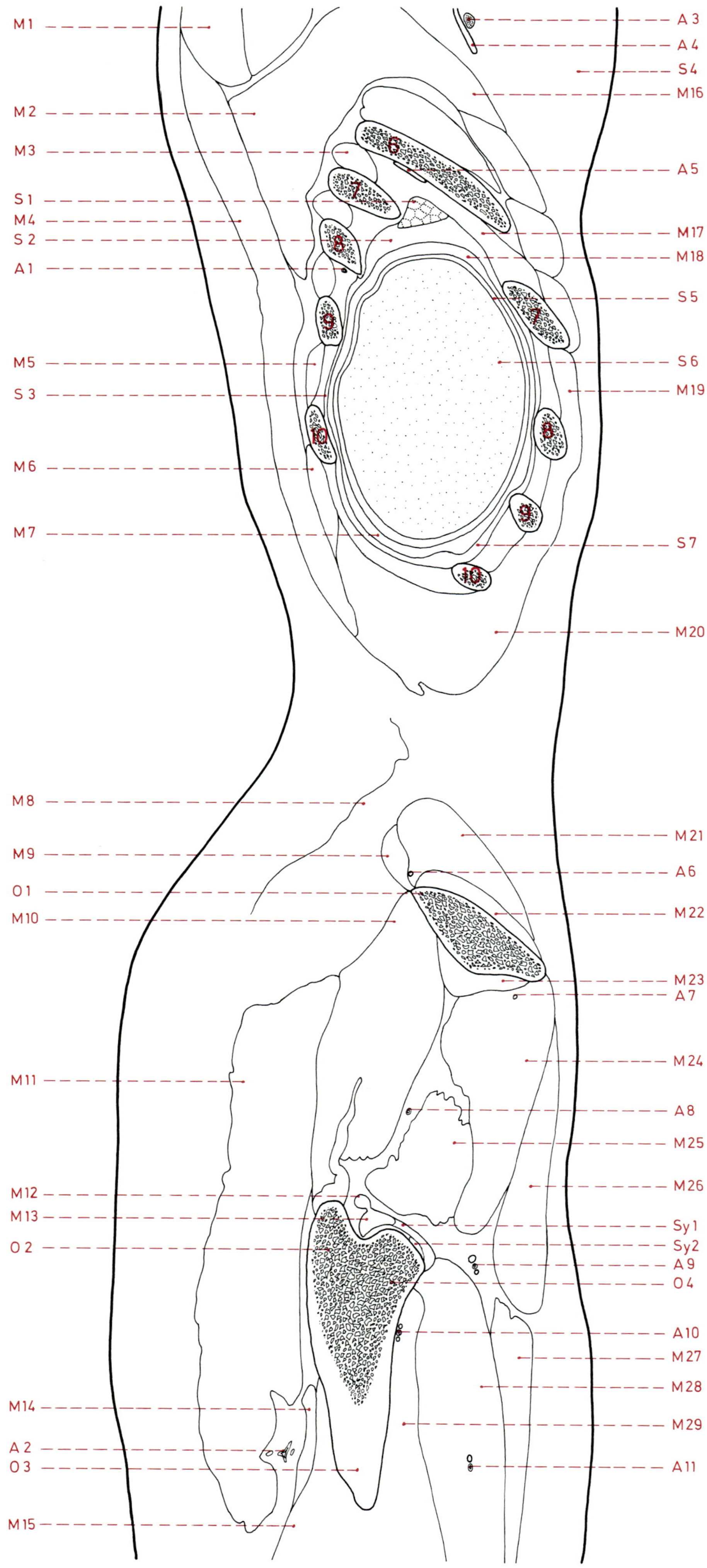
M1
M2
M3
S1
M4
S2
A1
M5
S3
M6
M7
M8
M9
O1
M10
M11
M12
M13
O2
M14
A2
O3
M15
A3
A4
S4
M16
A5
M17
M18
S5
S6
M19
S7
M20
M21
A6
M22
M23
A7
M24
A8
M25
M26
Sy1
Sy2
A9
O4
A10
M27
M28
M29
A11
6
7
8
9
10
7
8
9
10

Fig. SA♀ 6D

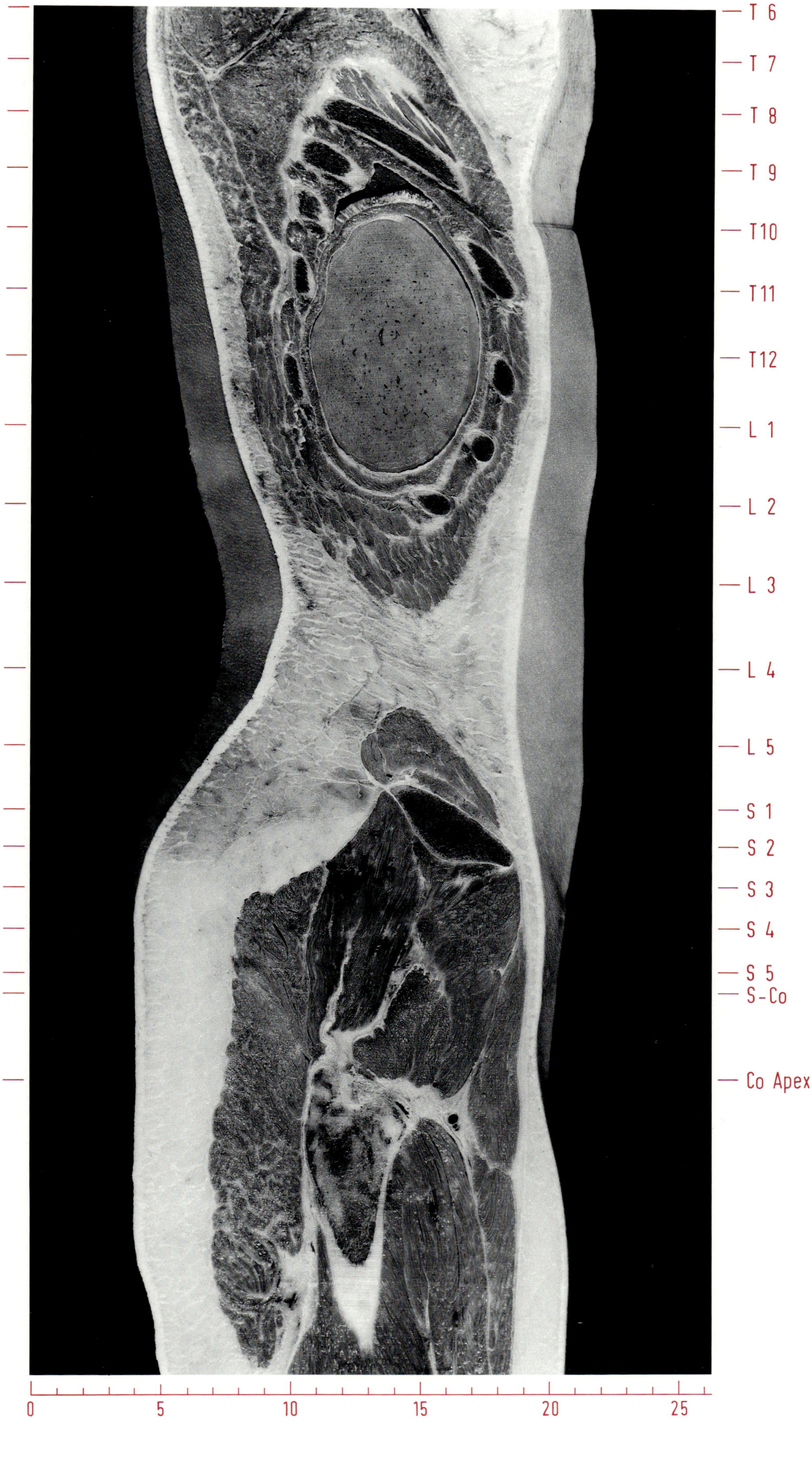
T 6
T 7
T 8
T 9
T10
T11
T12
L 1
L 2
L 3
L 4
L 5
S 1
S 2
S 3
S 4
S 5
S-Co
Co Apex
0
5
10
15
20
25

Fig. SA♀ 6D

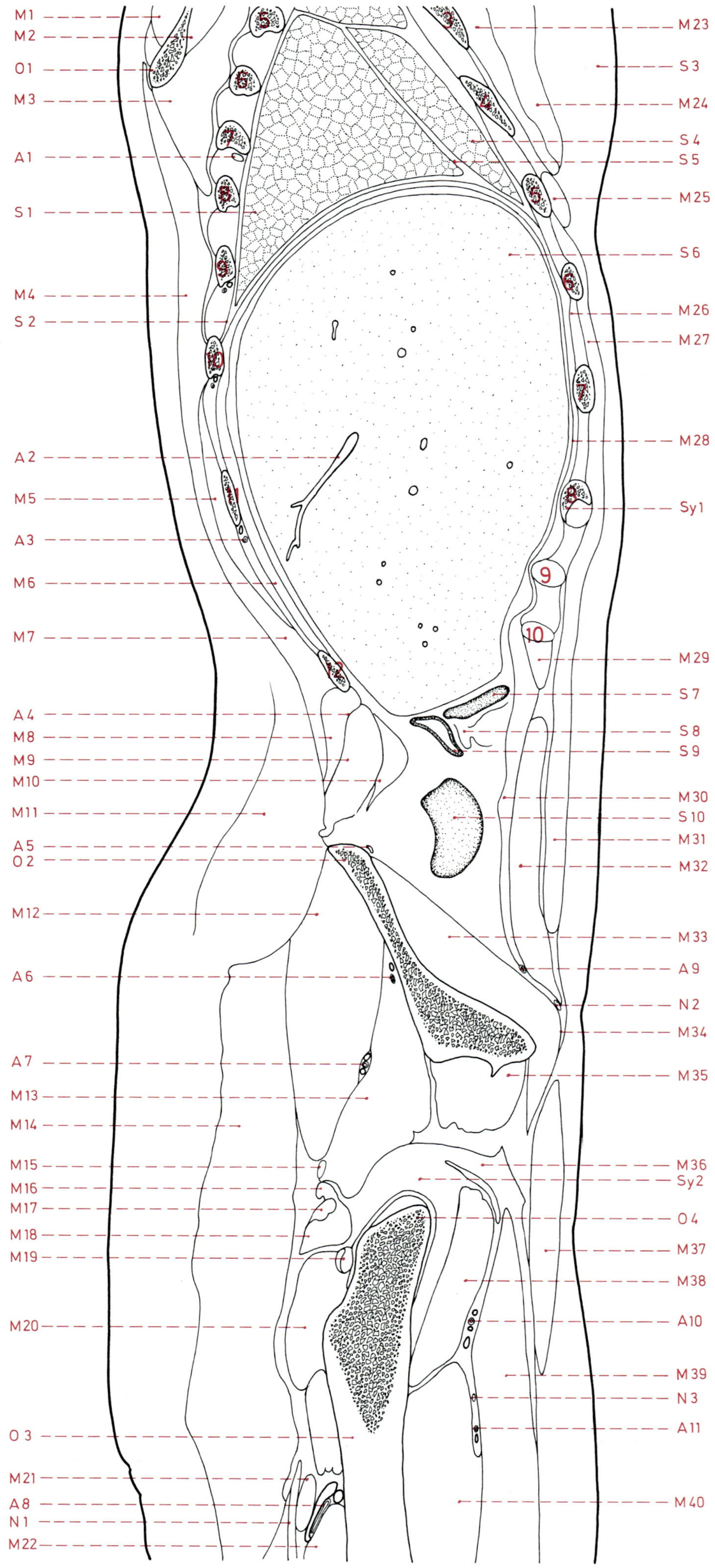
M1
M2
O1
M3
A1
S1
M4
S2
A2
M5
A3
M6
M7
A4
M8
M9
M10
M11
A5
O2
M12
A6
A7
M13
M14
M15
M16
M17
M18
M19
M20
O3
M21
A8
N1
M22
M23
S3
M24
S4
S5
M25
S6
M26
M27
M28
Sy1
M29
S7
S8
S9
M30
S10
M31
M32
M33
A9
N2
M34
M35
M36
Sy2
O4
M37
M38
A10
M39
N3
A11
M40

Fig. SA♀ 5D

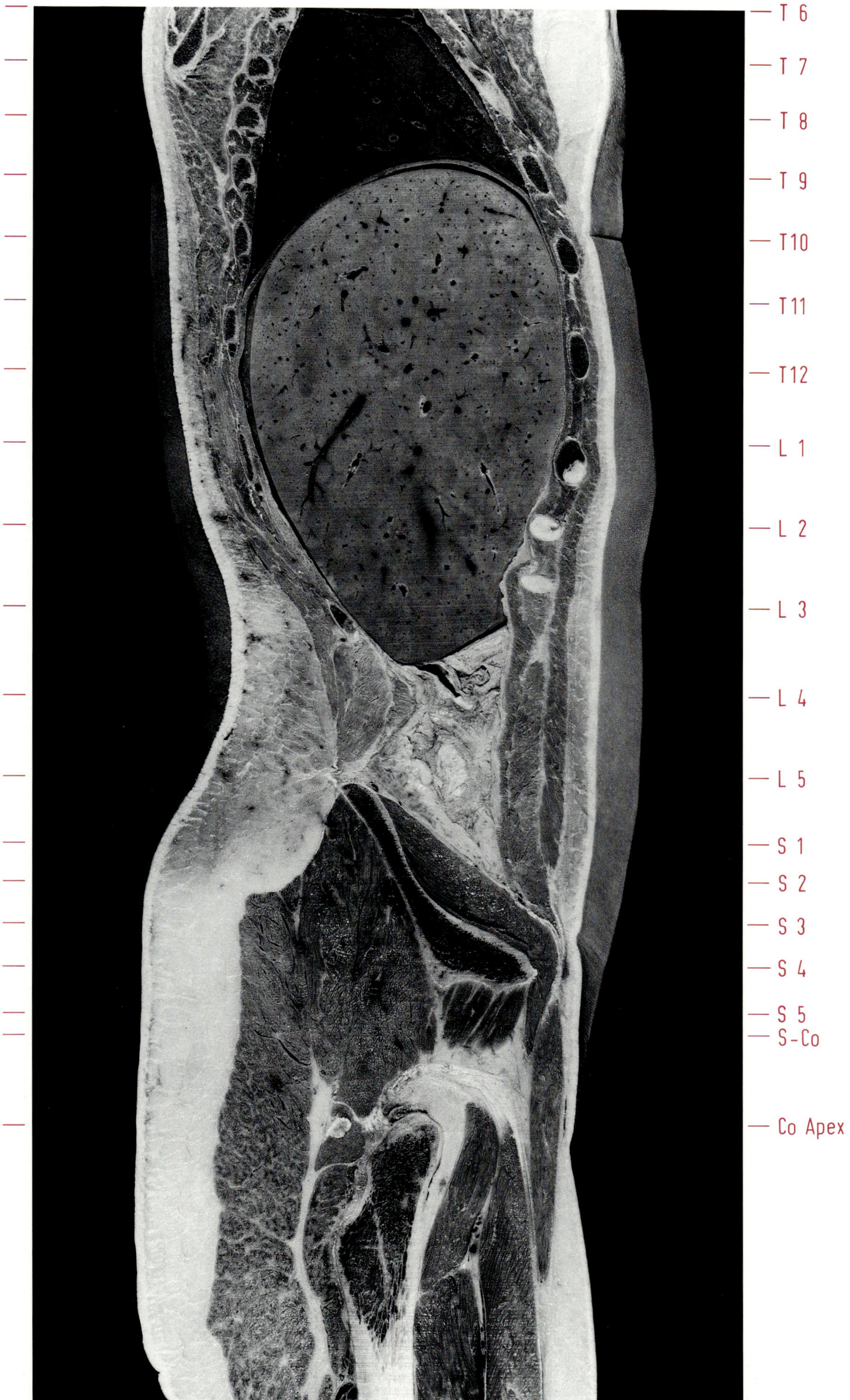
T 6
T 7
T 8
T 9
T10
T11
T12
L 1
L 2
L 3
L 4
L 5
S 1
S 2
S 3
S 4
S 5
S-Co
Co Apex
0
5
10
15
20
25

Fig. SA♀ 5D

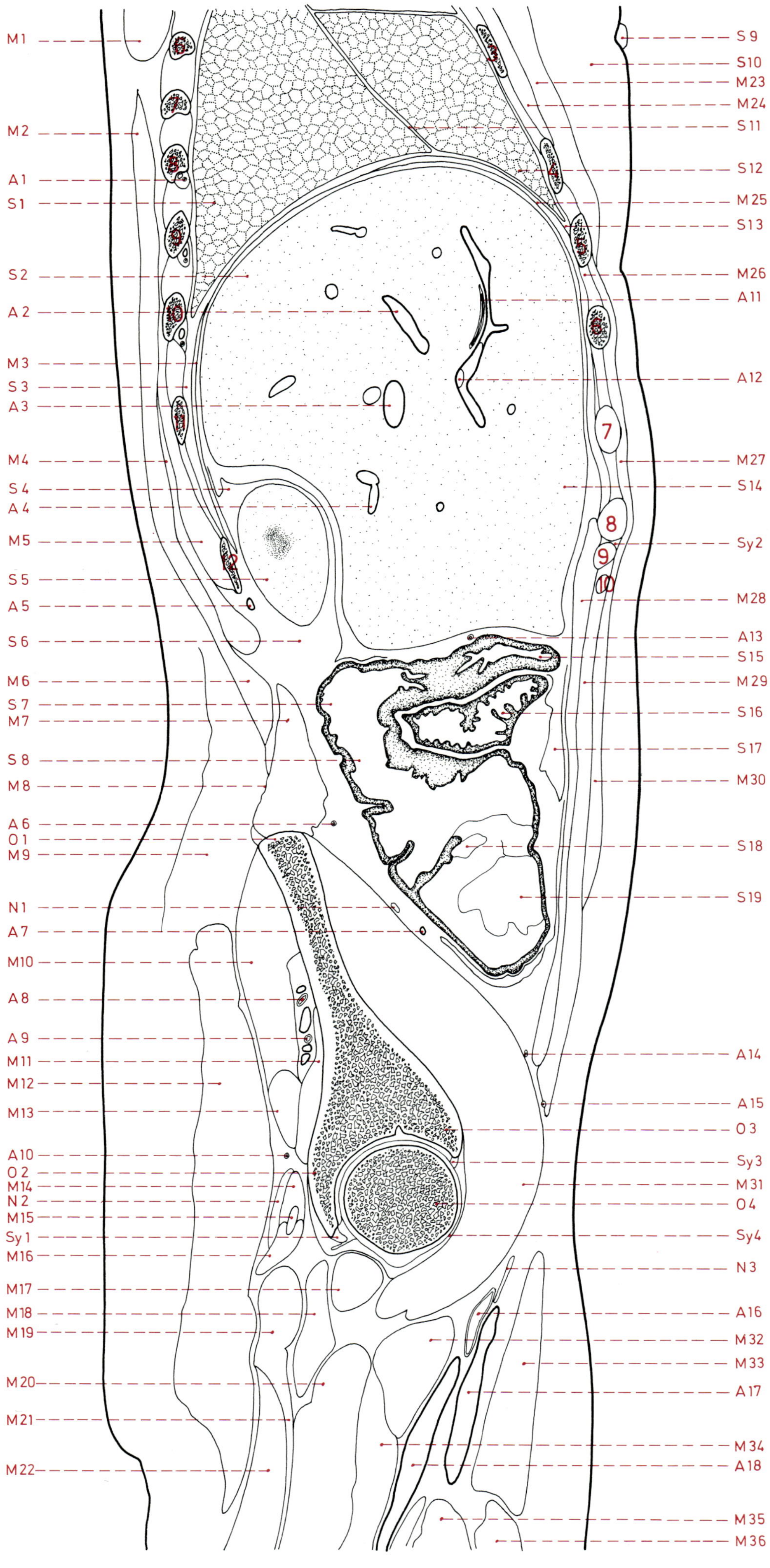
M1
M2
A1
S1
S2
A2
M3
S3
A3
M4
S4
A4
M5
S5
A5
S6
M6
S7
M7
S8
M8
A6
O1
M9
N1
A7
M10
A8
A9
M11
M12
M13
A10
O2
M14
N2
M15
Sy1
M16
M17
M18
M19
M20
M21
M22
S9
S10
M23
M24
S11
S12
M25
S13
M26
A11
A12
M27
S14
Sy2
M28
A13
S15
M29
S16
S17
M30
S18
S19
A14
A15
O3
Sy3
M31
O4
Sy4
N3
A16
M32
M33
A17
M34
A18
M35
M36

Fig. SA♀ 4D

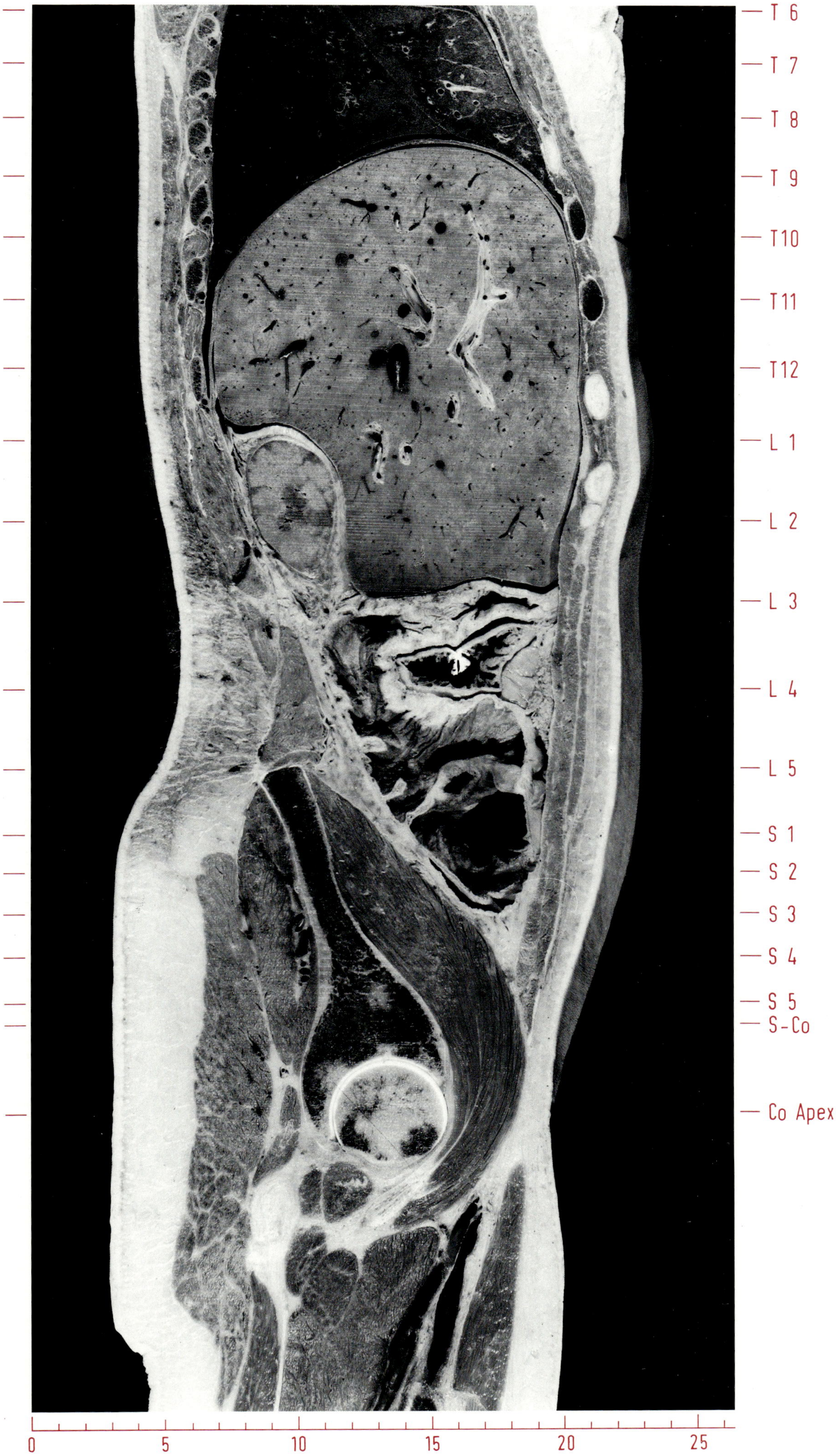
T 6
T 7
T 8
T 9
T10
T11
T12
L 1
L 2
L 3
L 4
L 5
S 1
S 2
S 3
S 4
S 5
S-Co
Co Apex
0
5
10
15
20
25

Fig. SA♀ 4D

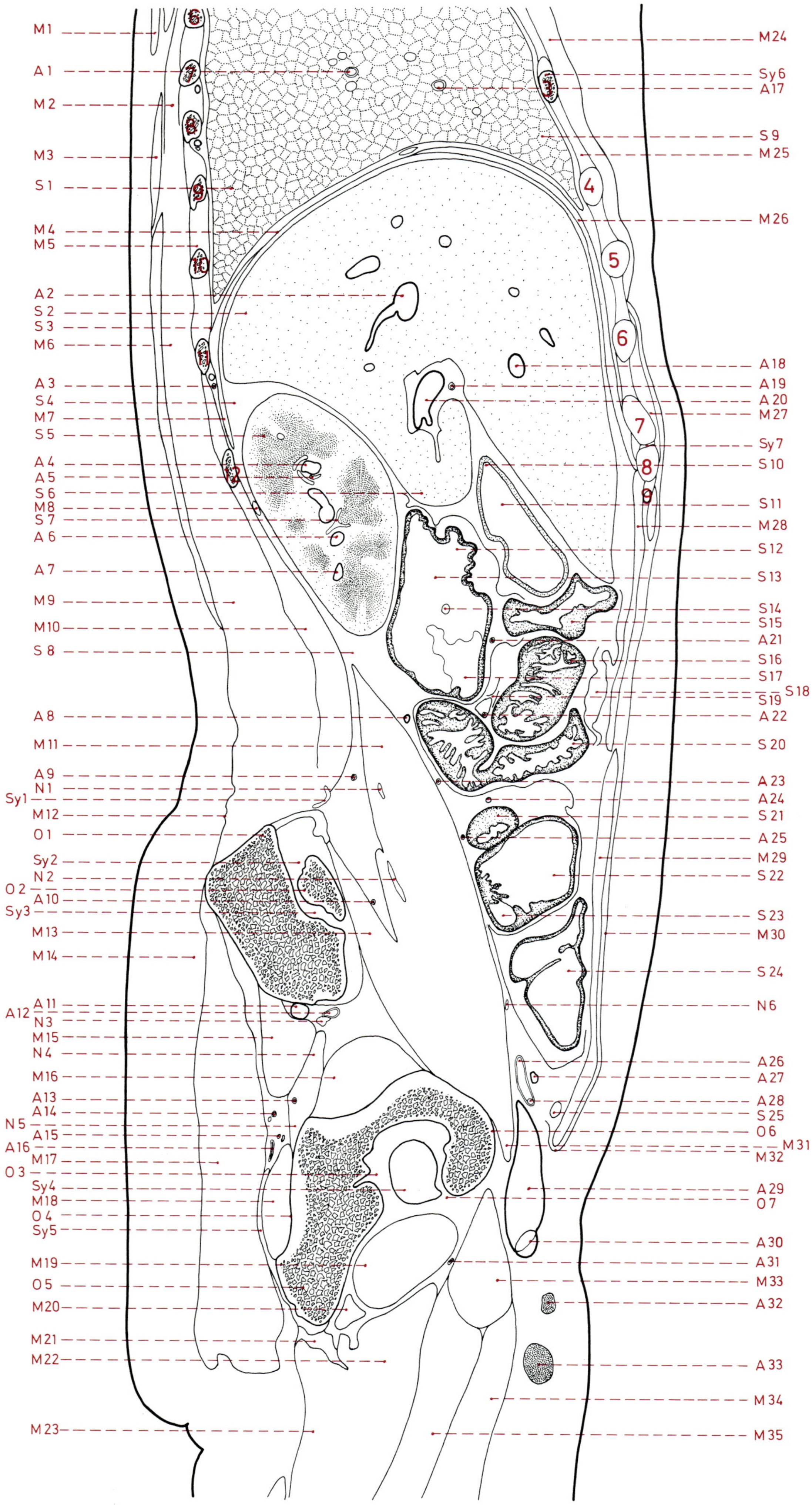
M1
A1
M2
M3
S1
M4
M5
A2
S2
S3
M6
A3
S4
M7
S5
A4
A5
S6
M8
S7
A6
A7
M9
M10
S8
A8
M11
A9
N1
Sy1
M12
O1
Sy2
N2
O2
A10
Sy3
M13
M14
A11
A12
N3
M15
N4
M16
A13
A14
N5
A15
A16
M17
O3
Sy4
M18
O4
Sy5
M19
O5
M20
M21
M22
M23
M24
Sy6
A17
S9
M25
M26
A18
A19
A20
M27
Sy7
S10
S11
M28
S12
S13
S14
S15
A21
S16
S17
S18
S19
A22
S20
A23
A24
S21
A25
M29
S22
S23
M30
S24
N6
A26
A27
A28
S25
O6
M31
M32
A29
O7
A30
A31
M33
A32
A33
M34
M35

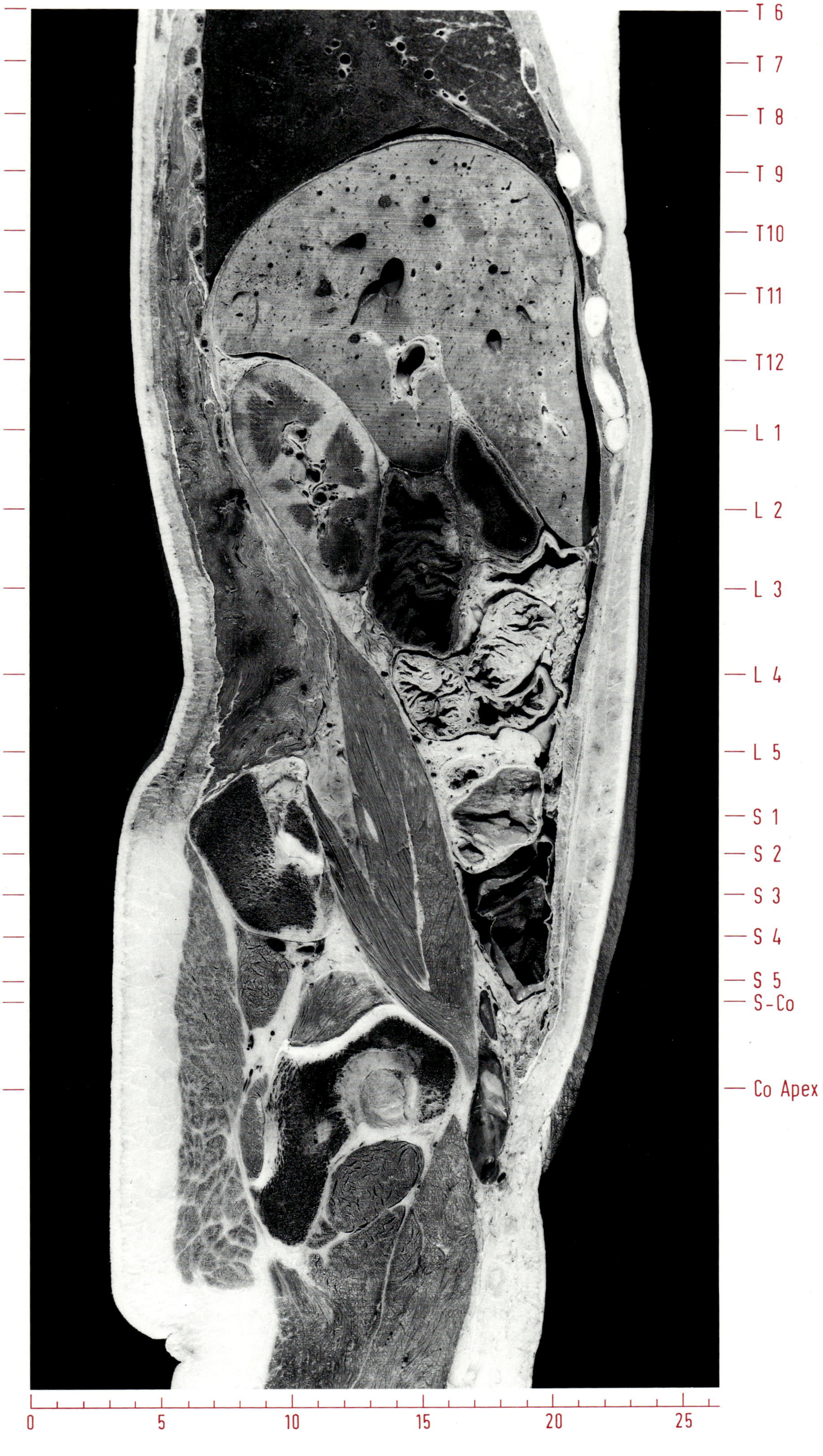
T 6
T 7
T 8
T 9
T10
T11
T12
L 1
L 2
L 3
L 4
L 5
S 1
S 2
S 3
S 4
S 5
S-Co
Co Apex
0
5
10
15
20
25

Fig. SA♀ 3D

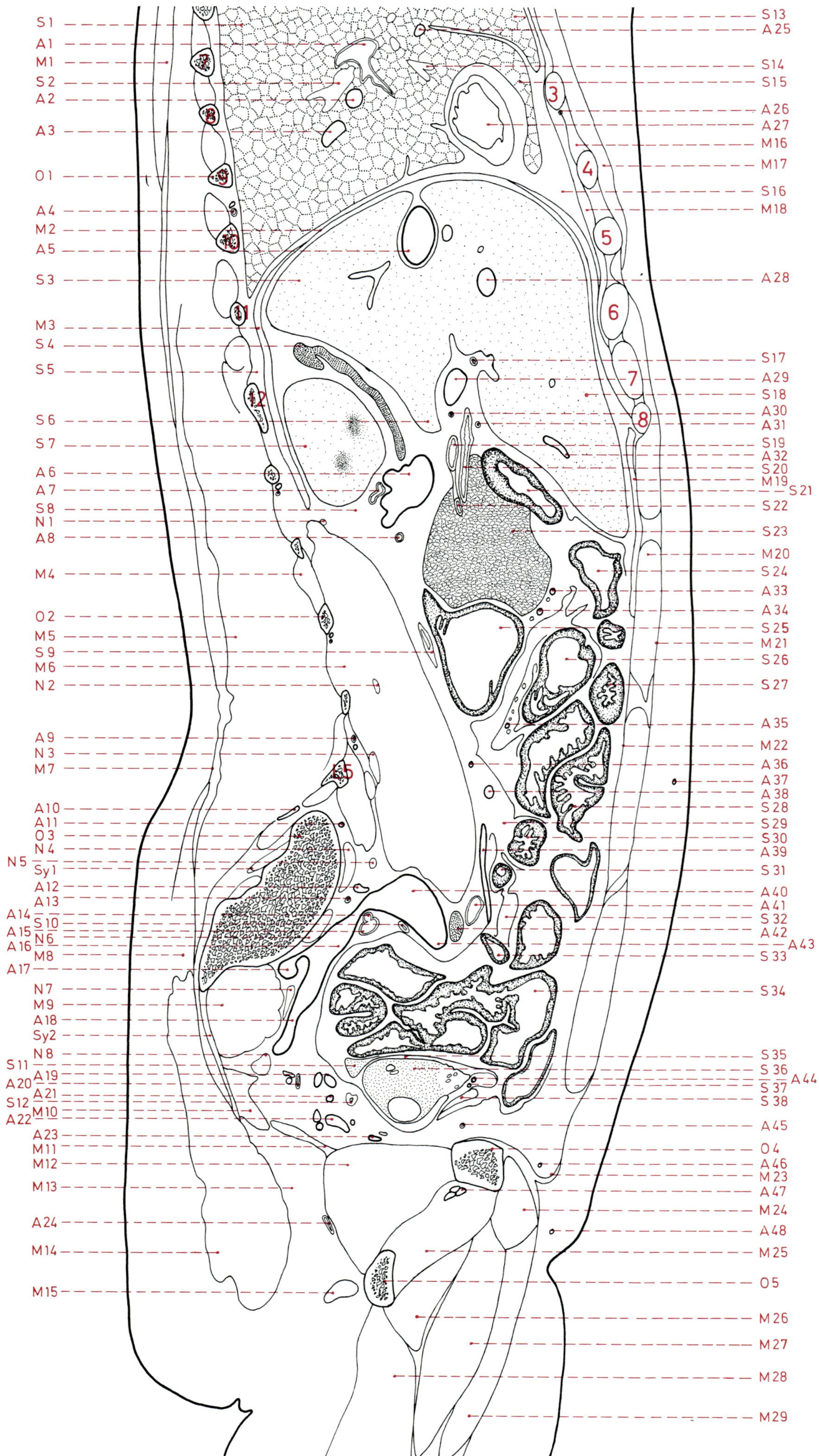
S1
A1
M1
S2
A2
A3
O1
A4
M2
A5
S3
M3
S4
S5
S6
S7
A6
A7
S8
N1
A8
M4
O2
M5
S9
M6
N2
A9
N3
M7
A10
A11
O3
N4
N5
Sy1
A12
A13
A14
S10
A15
N6
A16
M8
A17
N7
M9
A18
Sy2
N8
S11
A19
A20
A21
S12
M10
A22
A23
M11
M12
M13
A24
M14
M15
S13
A25
S14
S15
A26
A27
M16
M17
S16
M18
A28
S17
A29
S18
A30
A31
S19
A32
S20
M19
S21
S22
S23
M20
S24
A33
A34
S25
M21
S26
S27
A35
M22
A36
A37
A38
S28
S29
S30
A39
S31
A40
A41
S32
A42
A43
S33
S34
S35
S36
A44
S37
S38
A45
O4
A46
M23
A47
M24
A48
M25
O5
M26
M27
M28
M29
3
4
5
6
7
8
11
12
15

Fig. SA♀ 2D

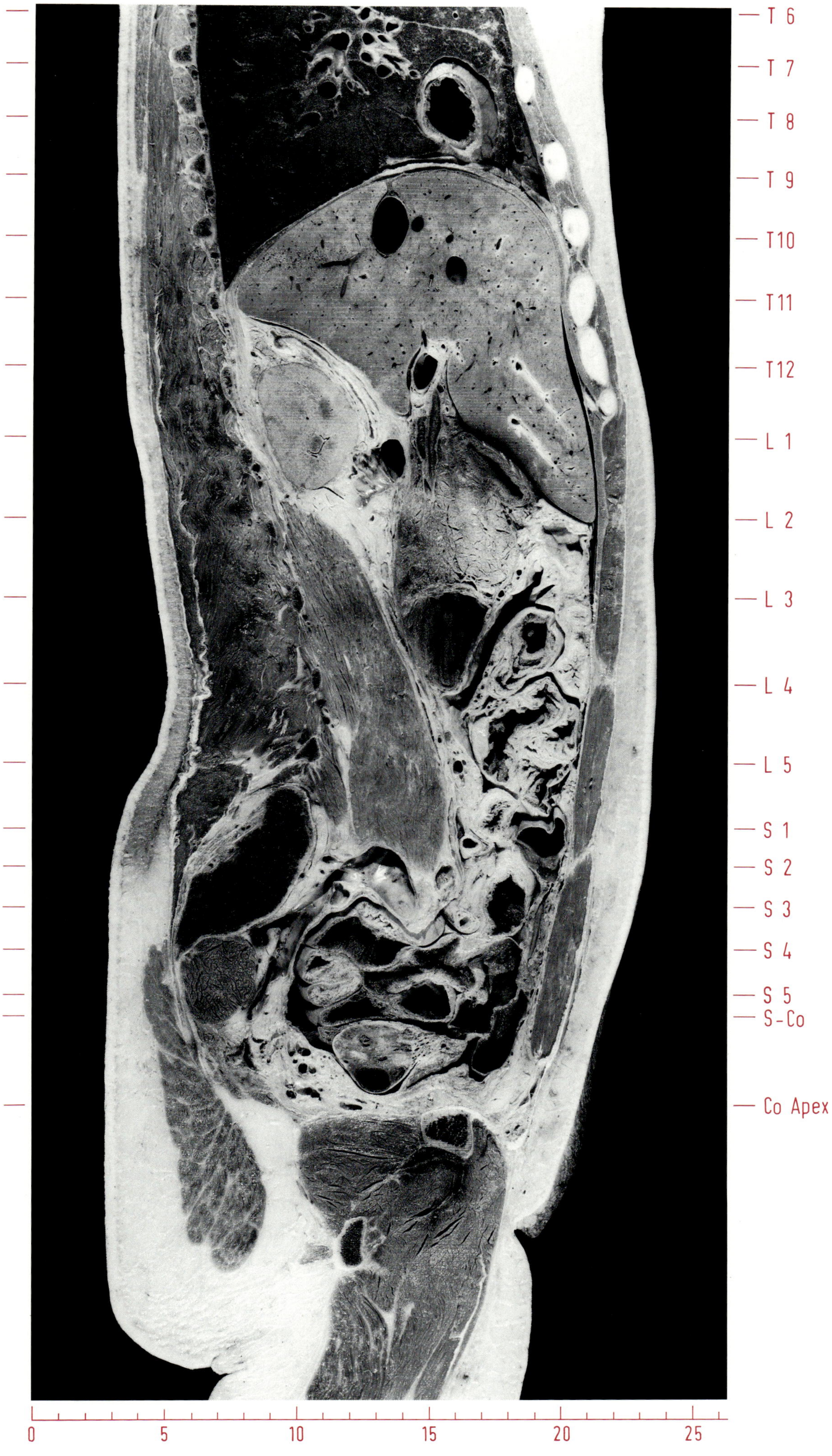
T 6
T 7
T 8
T 9
T10
T11
T12
L 1
L 2
L 3
L 4
L 5
S 1
S 2
S 3
S 4
S 5
S-Co
Co Apex
0
5
10
15
20
25

Fig. SA♀ 2D

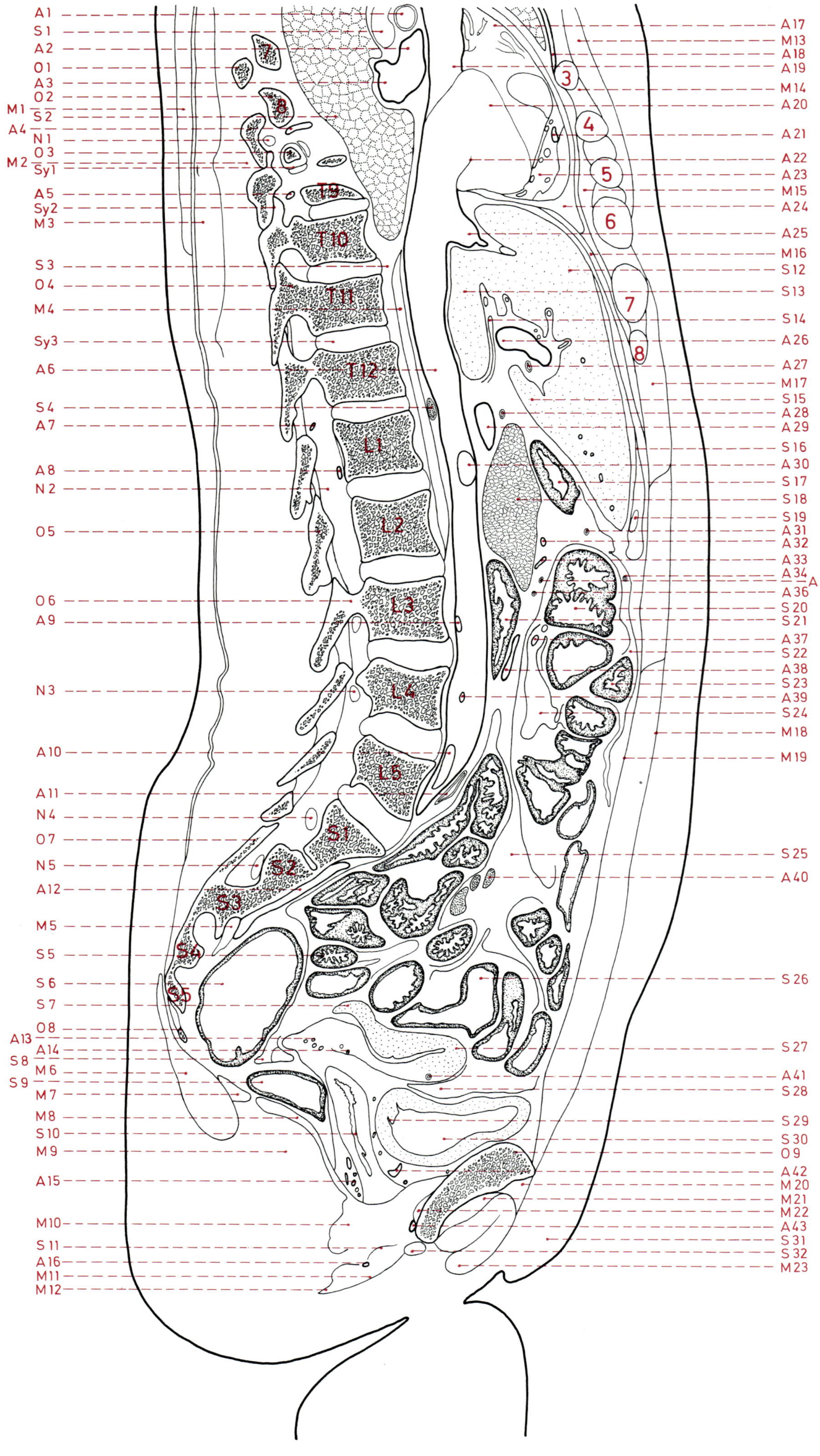
A1
S1
A2
O1
A3
O2
M1
S2
A4
N1
O3
M2
Sy1
A5
Sy2
M3
S3
O4
M4
Sy3
A6
S4
A7
A8
N2
O5
O6
A9
N3
A10
A11
N4
O7
N5
A12
M5
S5
S6
S7
O8
A13
A14
S8
M6
S9
M7
M8
S10
M9
A15
M10
S11
A16
M11
M12
A17
M13
A18
A19
M14
A20
A21
A22
A23
M15
A24
A25
M16
S12
S13
S14
A26
A27
M17
S15
A28
A29
S16
A30
S17
S18
S19
A31
A32
A33
A34
A35
A36
S20
S21
A37
S22
A38
S23
A39
S24
M18
M19
S25
A40
S26
S27
A41
S28
S29
S30
O9
A42
M20
M21
M22
A43
S31
S32
M23
T9
T10
T11
T12
L1
L2
L3
L4
L5
S1
S2
S3
S4
S5
3
4
5
6
7
8
7
8

Fig. SA♀ 1D

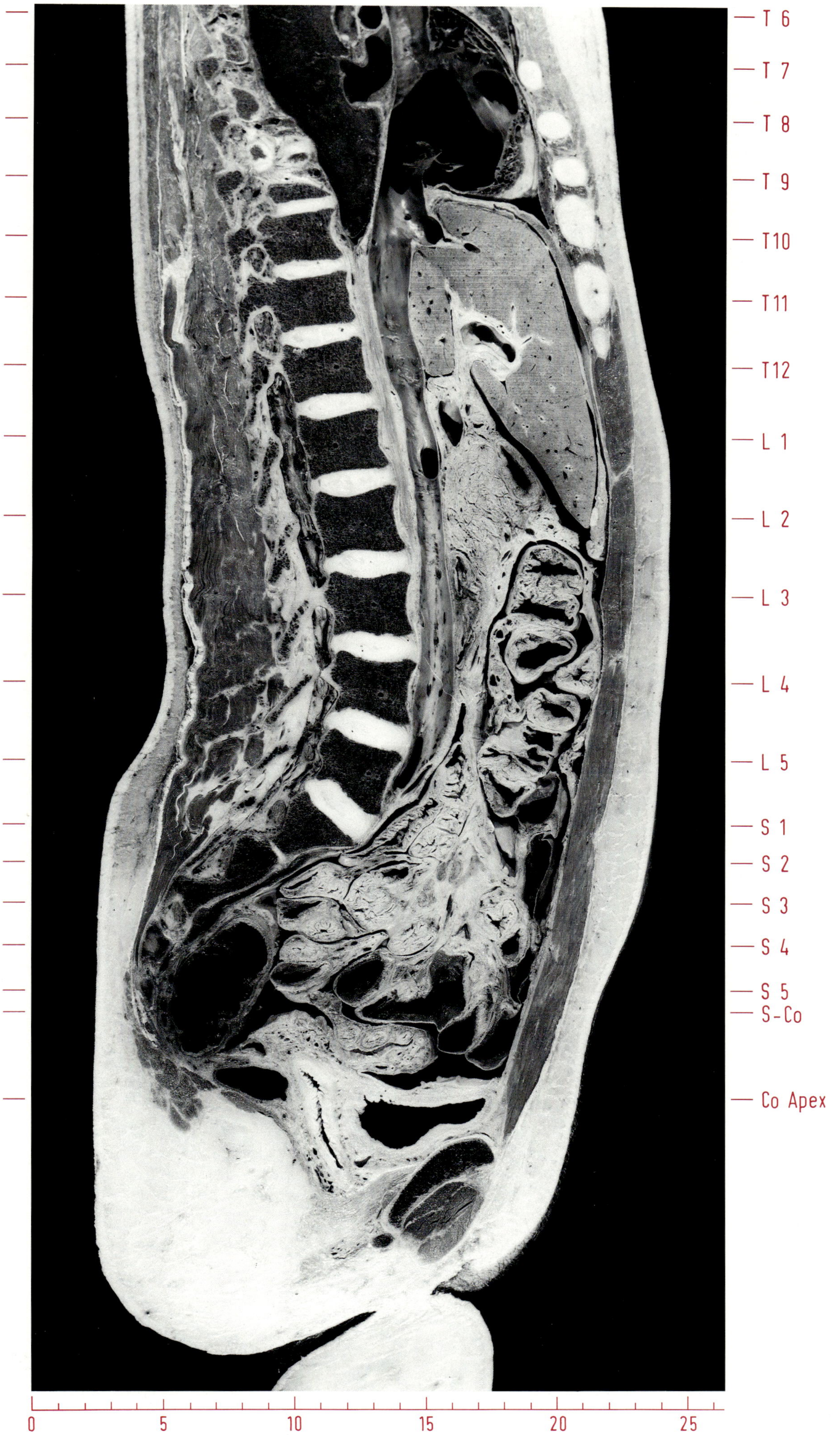
T 6
T 7
T 8
T 9
T10
T11
T12
L 1
L 2
L 3
L 4
L 5
S 1
S 2
S 3
S 4
S 5
S-Co
Co Apex
0
5
10
15
20
25

Fig. SA♀ 1D

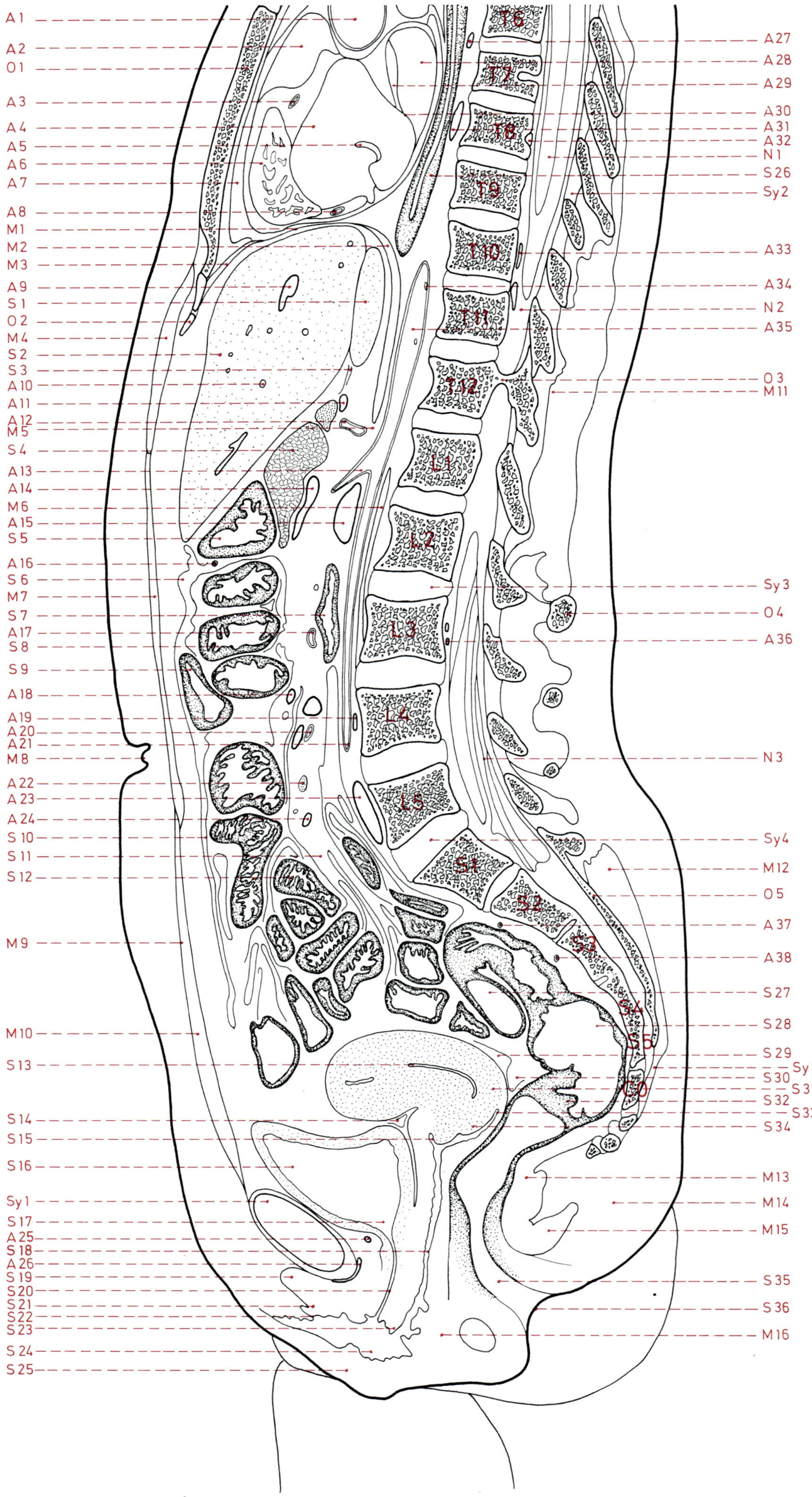
A1
A2
O1
A3
A4
A5
A6
A7
A8
M1
M2
M3
A9
S1
O2
M4
S2
S3
A10
A11
A12
M5
S4
A13
A14
M6
A15
S5
A16
S6
M7
S7
A17
S8
S9
A18
A19
A20
A21
M8
A22
A23
A24
S10
S11
S12
M9
M10
S13
S14
S15
S16
Sy1
S17
A25
S18
A26
S19
S20
S21
S22
S23
S24
S25
T6
T7
T8
T9
T10
T11
T12
L1
L2
L3
L4
L5
S1
S2
S3
S4
S5
Co
A27
A28
A29
A30
A31
A32
N1
S26
Sy2
A33
A34
N2
A35
O3
M11
Sy3
O4
A36
N3
Sy4
M12
O5
A37
A38
S27
S28
S29
S30
Sy5
S31
S32
S33
S34
M13
M14
M15
S35
S36
M16

Fig. SA♀ MD

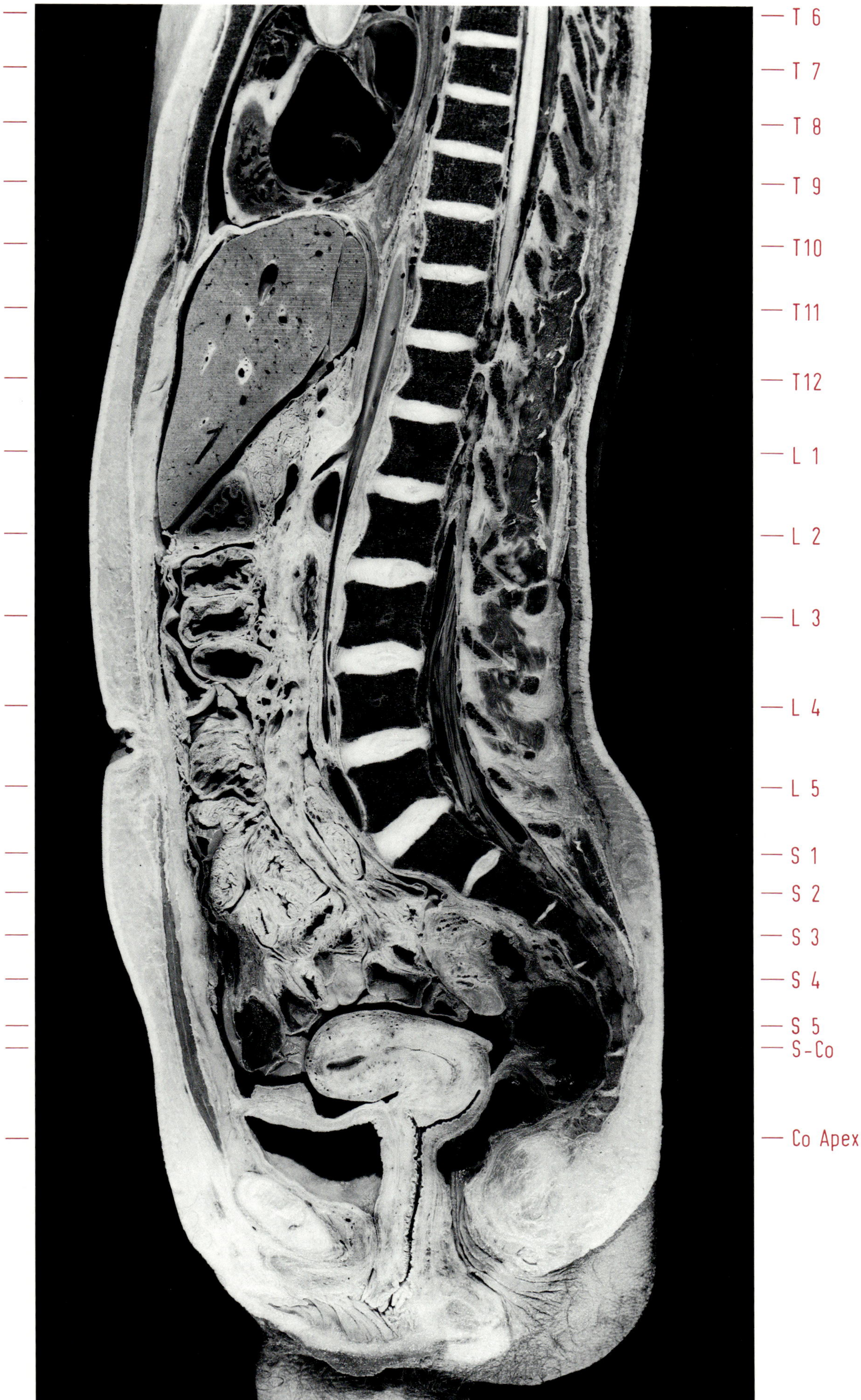
T 6
T 7
T 8
T 9
T10
T11
T12
L 1
L 2
L 3
L 4
L 5
S 1
S 2
S 3
S 4
S 5
S-Co
Co Apex
0
5
10
15
20
25

Fig. SA♀ MD

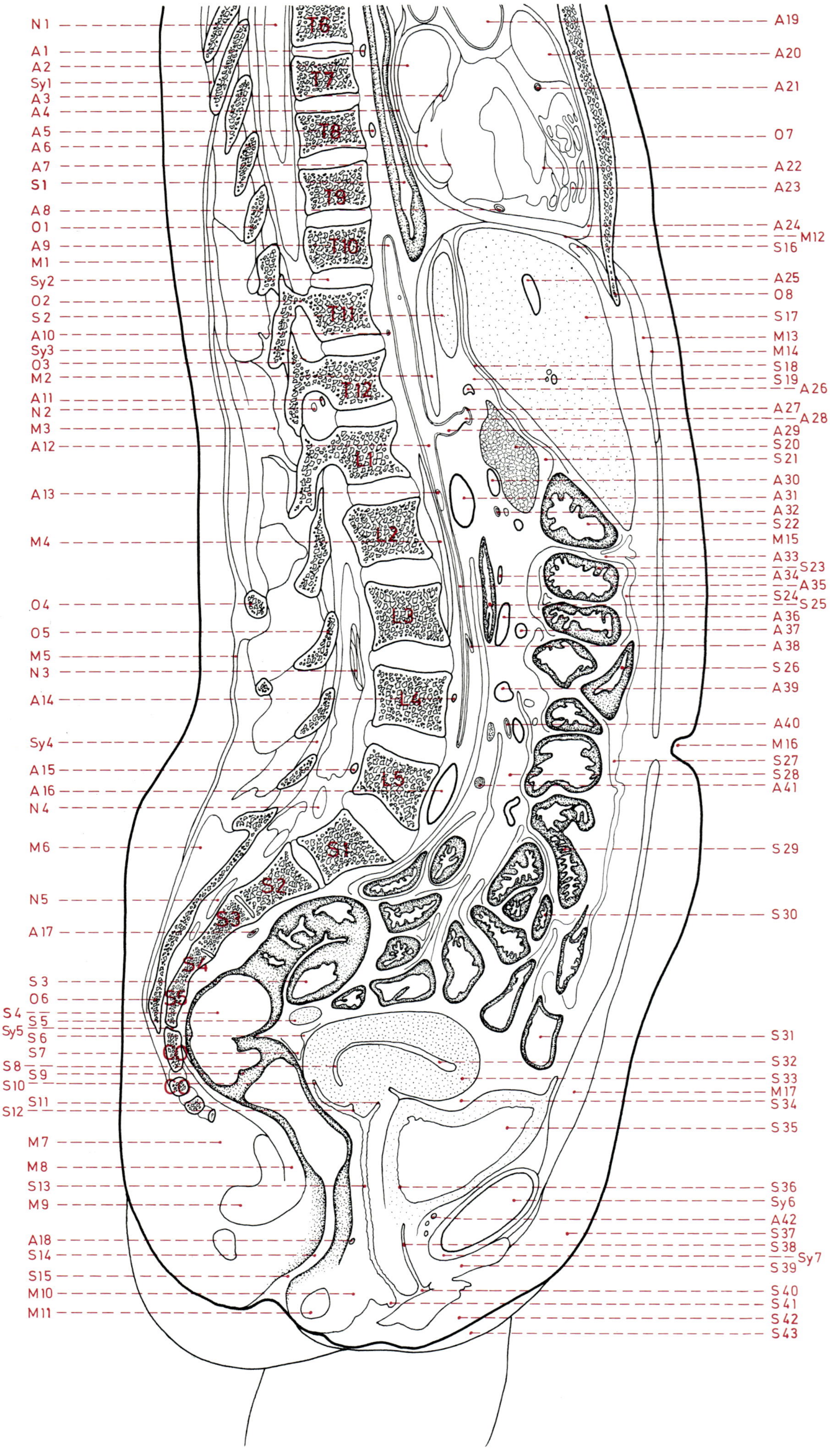
N1
A1
A2
Sy1
A3
A4
A5
A6
A7
S1
A8
O1
A9
M1
Sy2
O2
S2
A10
Sy3
O3
M2
A11
N2
M3
A12
A13
M4
O4
O5
M5
N3
A14
Sy4
A15
A16
N4
M6
N5
A17
S3
O6
S4
S5
Sy5
S6
S7
S8
S9
S10
S11
S12
M7
M8
S13
M9
A18
S14
S15
M10
M11
T6
T7
T8
T9
T10
T11
T12
L1
L2
L3
L4
L5
S1
S2
S3
S4
S5
A19
A20
A21
O7
A22
A23
A24
M12
S16
A25
O8
S17
M13
M14
S18
S19
A26
A27
A28
A29
S20
S21
A30
A31
A32
S22
M15
A33
S23
A34
A35
S24
S25
A36
A37
A38
S26
A39
A40
M16
S27
S28
A41
S29
S30
S31
S32
S33
M17
S34
S35
S36
Sy6
A42
S37
S38
Sy7
S39
S40
S41
S42
S43

Fig. SA♀ MSi

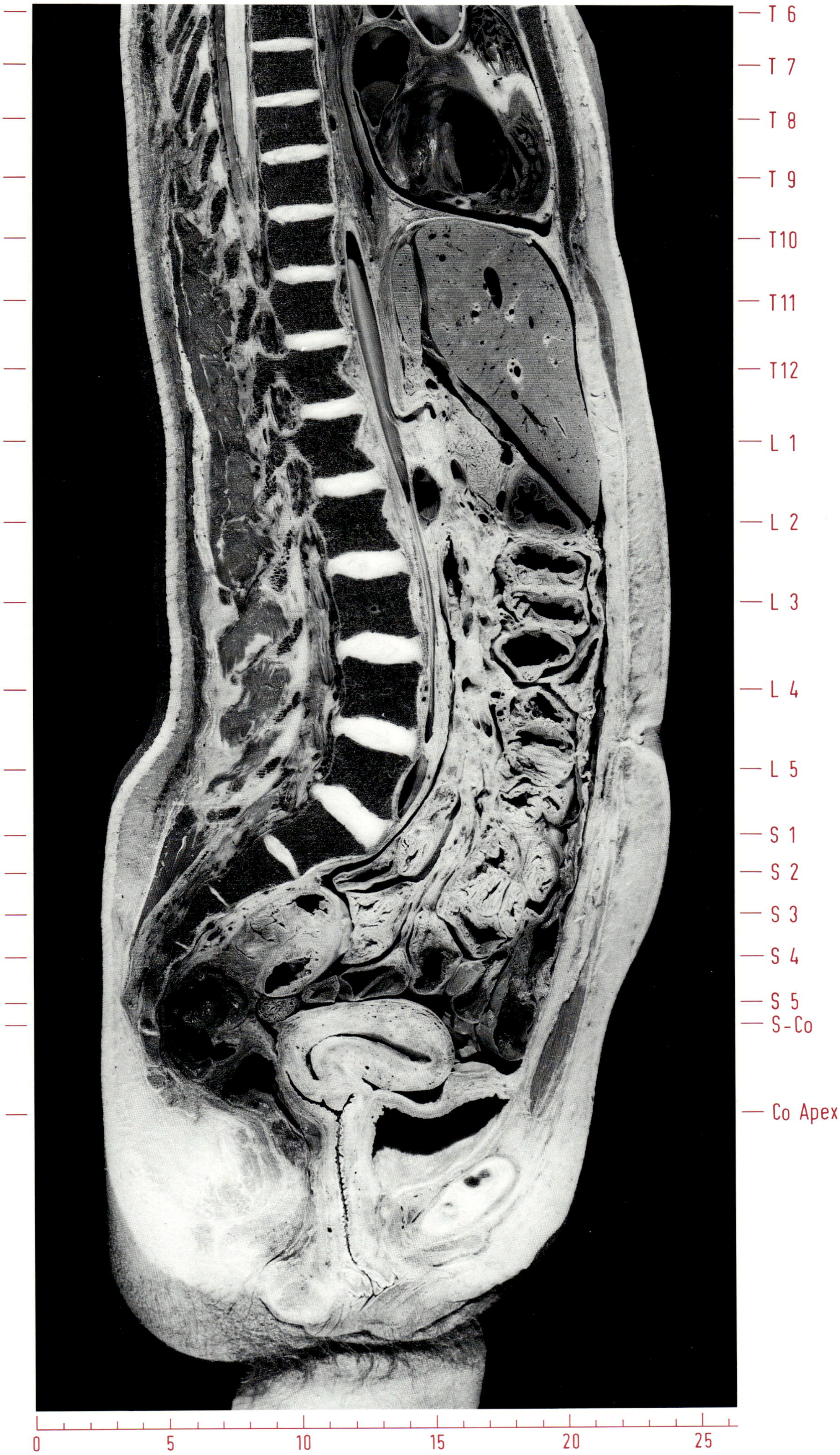
T 6
T 7
T 8
T 9
T10
T11
T12
L 1
L 2
L 3
L 4
L 5
S 1
S 2
S 3
S 4
S 5
S-Co
Co Apex
0
5
10
15
20
25

Fig. SA♀ MSi

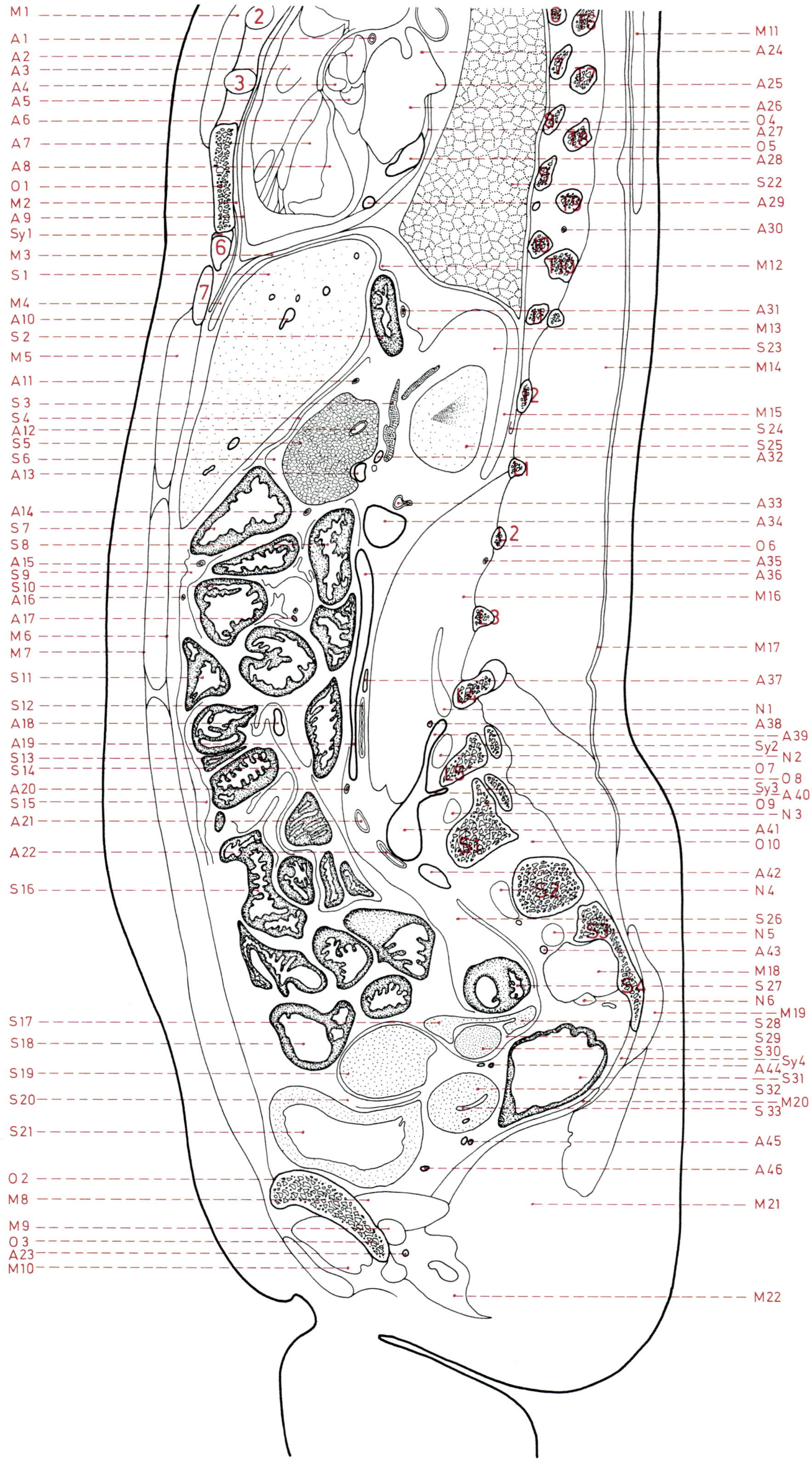
M1
A1
A2
A3
A4
A5
A6
A7
A8
O1
M2
A9
Sy1
M3
S1
M4
A10
S2
M5
A11
S3
S4
A12
S5
S6
A13
A14
S7
S8
A15
S9
S10
A16
A17
M6
M7
S11
S12
A18
A19
S13
S14
A20
S15
A21
A22
S16
S17
S18
S19
S20
S21
O2
M8
M9
O3
A23
M10
M11
A24
A25
A26
O4
A27
O5
A28
S22
A29
A30
M12
A31
M13
S23
M14
M15
S24
S25
A32
A33
A34
O6
A35
A36
M16
M17
A37
N1
A38
A39
Sy2
N2
O7
O8
Sy3
A40
O9
N3
A41
O10
A42
N4
S26
N5
A43
M18
S27
N6
M19
S28
S29
S30
Sy4
A44
S31
S32
M20
S33
A45
A46
M21
M22

Fig. SA♀ 1Si

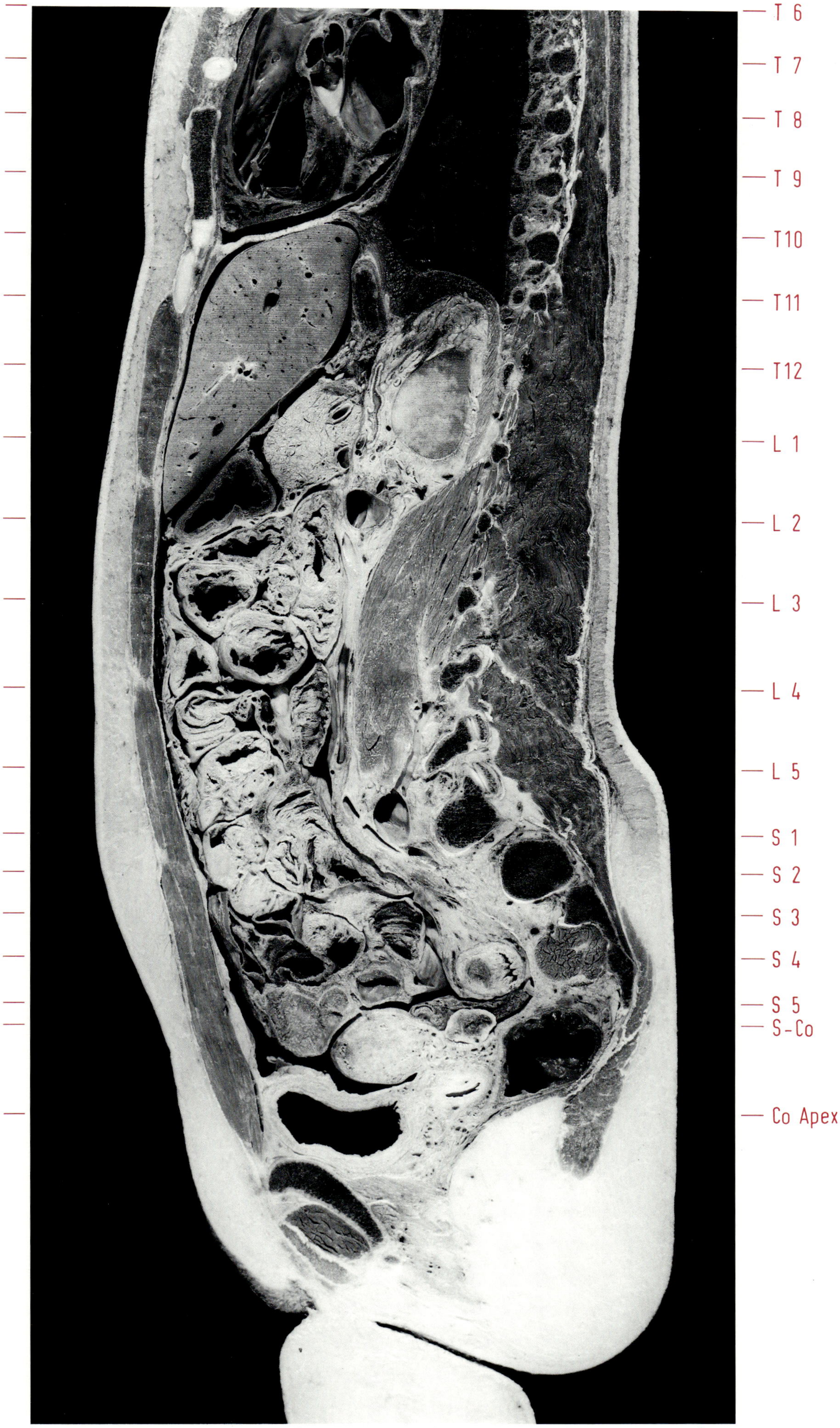
T 6
T 7
T 8
T 9
T10
T11
T12
L 1
L 2
L 3
L 4
L 5
S 1
S 2
S 3
S 4
S 5
S-Co
Co Apex
0
5
10
15
20
25

Fig. SA♀ 1 Si

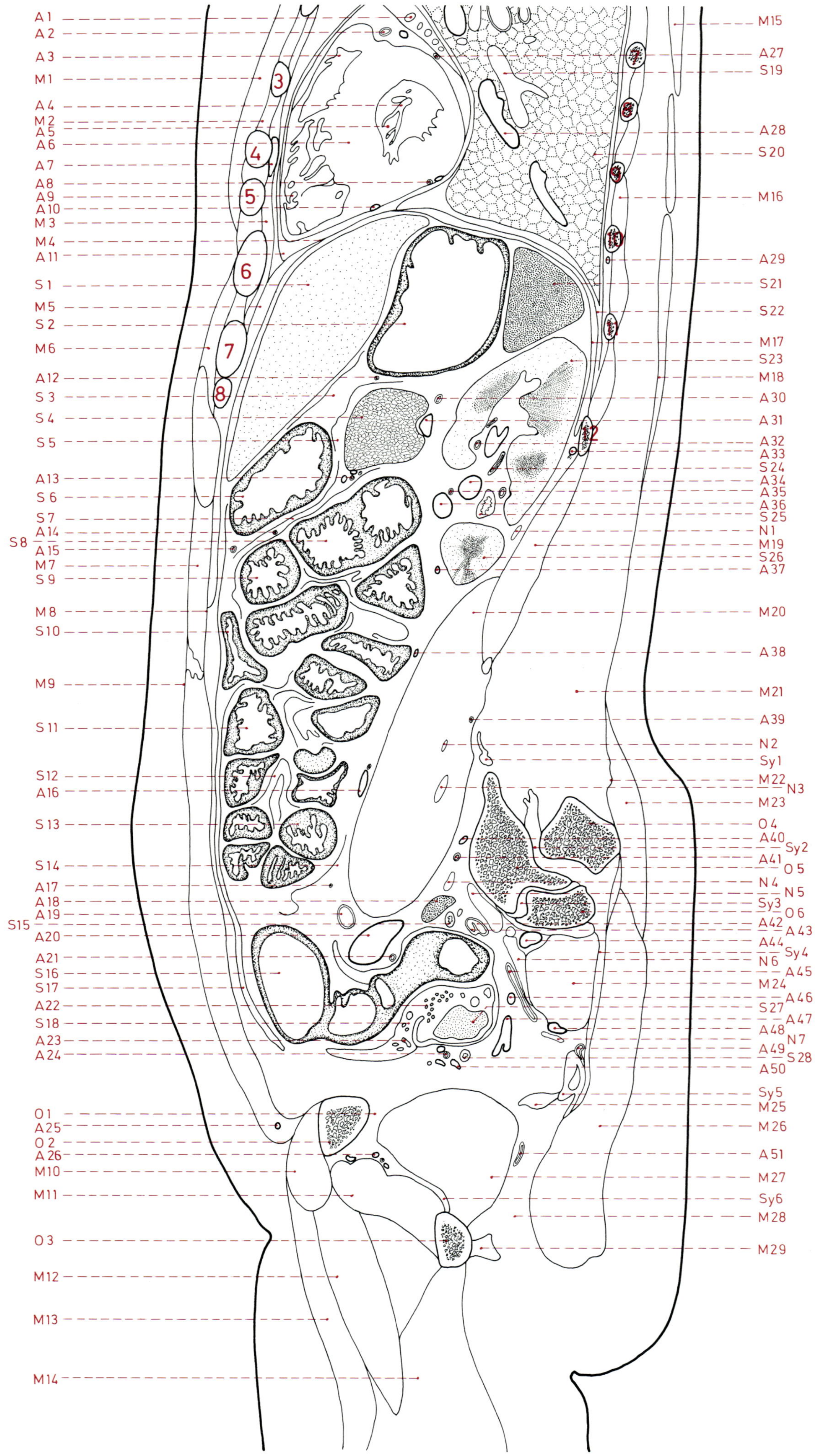
A1
A2
A3
M1
A4
M2
A5
A6
A7
A8
A9
A10
M3
M4
A11
S1
M5
S2
M6
A12
S3
S4
S5
A13
S6
S7
A14
S8
A15
M7
S9
M8
S10
M9
S11
S12
A16
S13
S14
A17
A18
A19
S15
A20
A21
S16
S17
A22
S18
A23
A24
O1
A25
O2
A26
M10
M11
O3
M12
M13
M14
M15
A27
S19
A28
S20
M16
A29
S21
S22
M17
S23
M18
A30
A31
A32
A33
S24
A34
A35
A36
S25
N1
M19
S26
A37
M20
A38
M21
A39
N2
Sy1
M22
N3
M23
O4
A40
Sy2
A41
O5
N4
N5
Sy3
O6
A42
A43
A44
Sy4
N6
A45
M24
A46
S27
A47
A48
N7
A49
S28
A50
Sy5
M25
M26
A51
M27
Sy6
M28
M29
3
4
5
6
7
8
7
8
9
10
11
12

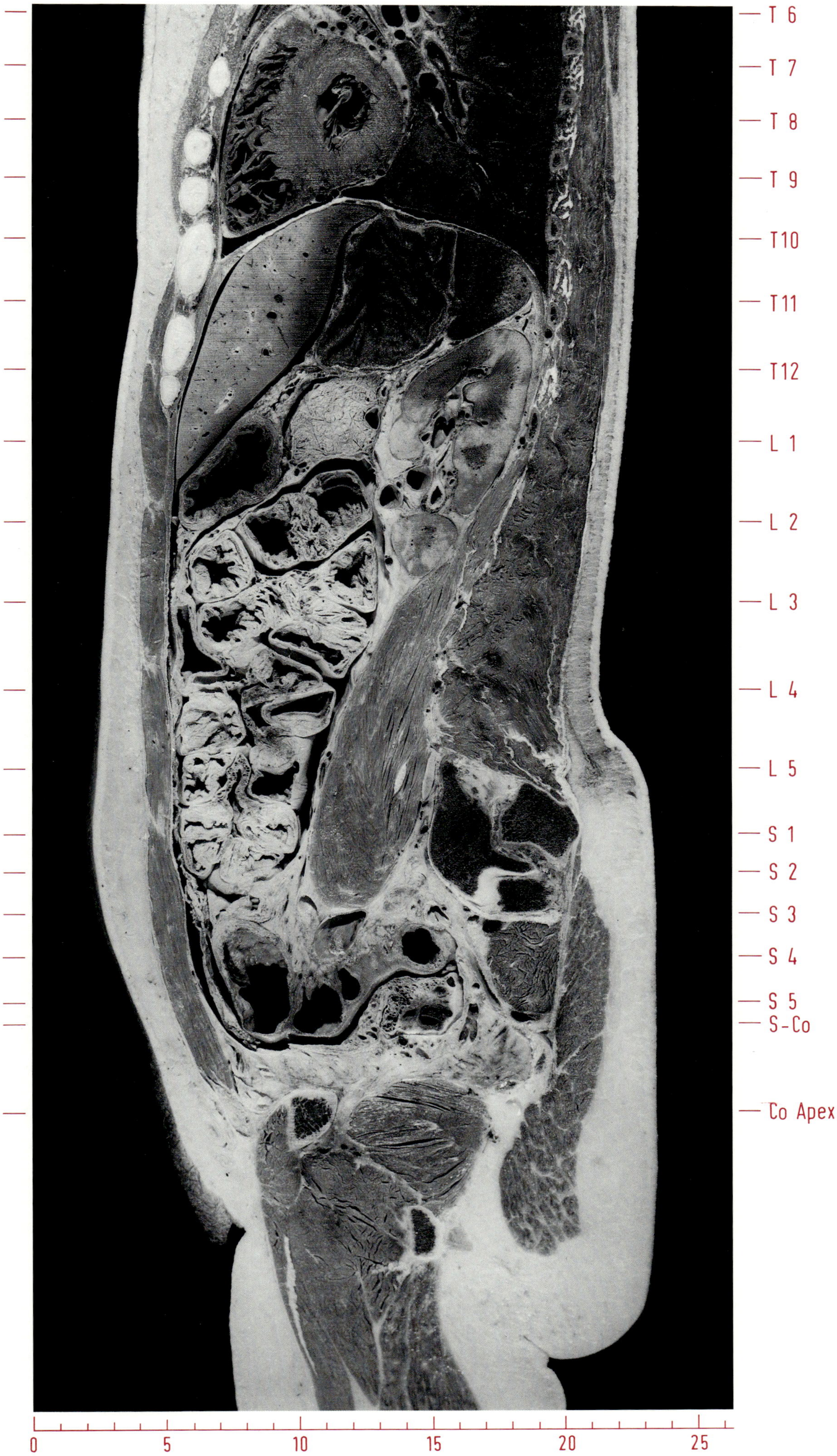
T 6
T 7
T 8
T 9
T10
T11
T12
L 1
L 2
L 3
L 4
L 5
S 1
S 2
S 3
S 4
S 5
S-Co
Co Apex
0
5
10
15
20
25

Fig. SA♀ 2 Si

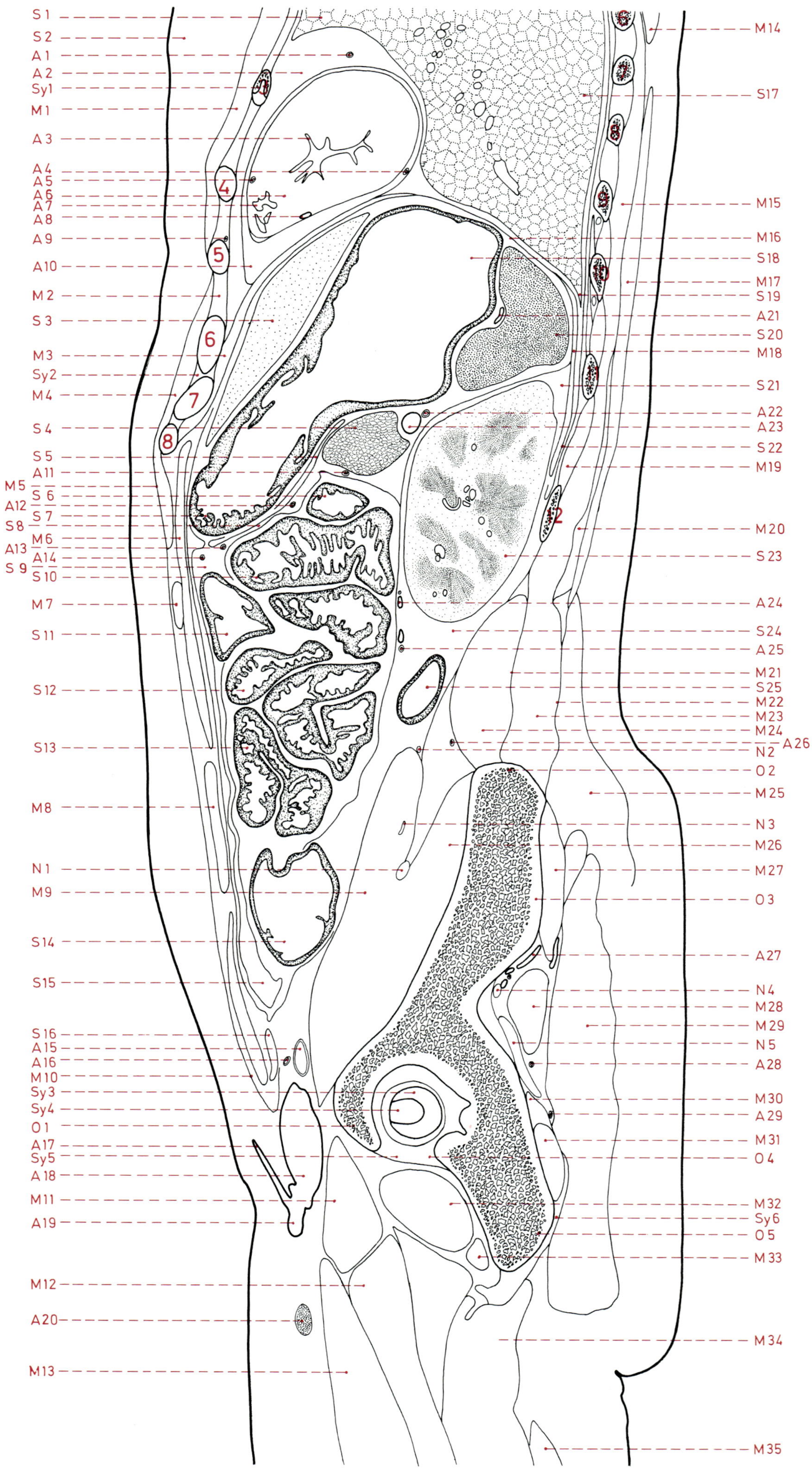

S1
S2
A1
A2
Sy1
M1
A3
A4
A5
A6
A7
A8
A9
A10
M2
S3
M3
Sy2
M4
S4
S5
A11
M5
S6
A12
S7
S8
M6
A13
A14
S9
S10
M7
S11
S12
S13
M8
N1
M9
S14
S15
S16
A15
A16
M10
Sy3
Sy4
O1
A17
Sy5
A18
M11
A19
M12
A20
M13
M14
S17
M15
M16
S18
M17
S19
A21
S20
M18
S21
A22
A23
S22
M19
M20
S23
A24
S24
A25
M21
S25
M22
M23
M24
A26
N2
O2
M25
N3
M26
M27
O3
A27
N4
M28
M29
N5
A28
M30
A29
M31
O4
M32
Sy6
O5
M33
M34
M35
4
5
6
7
8
12

Fig. SA♀ 3Si

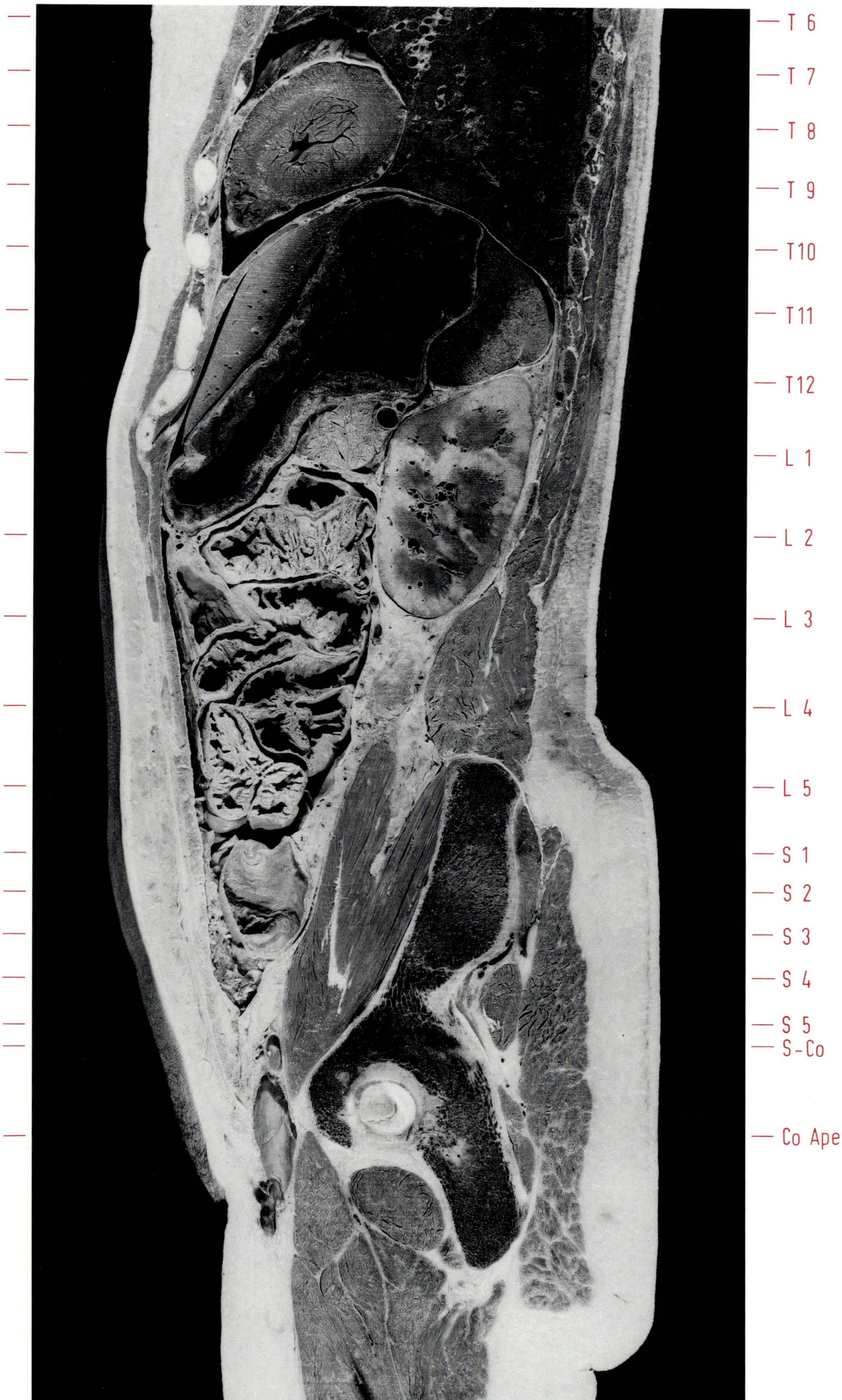
T 6
T 7
T 8
T 9
T10
T11
T12
L 1
L 2
L 3
L 4
L 5
S 1
S 2
S 3
S 4
S 5
S-Co
Co Apex
0
5
10
15
20
25

Fig. SA♀ 3 Si

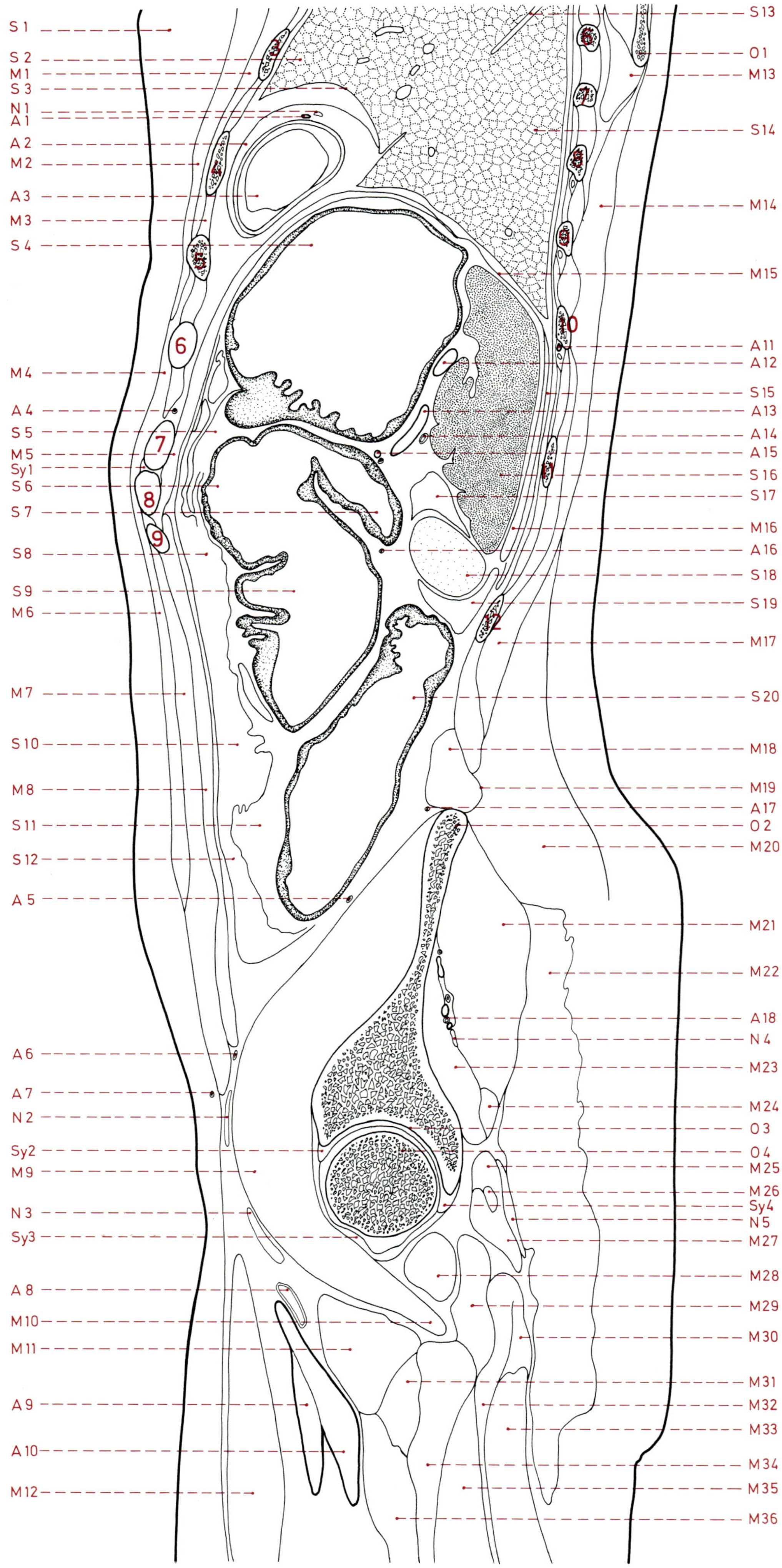
S1
S2
M1
S3
N1
A1
A2
M2
A3
M3
S4
M4
A4
S5
M5
Sy1
S6
S7
S8
S9
M6
M7
S10
M8
S11
S12
A5
A6
A7
N2
Sy2
M9
N3
Sy3
A8
M10
M11
A9
A10
M12
S13
O1
M13
S14
M14
M15
A11
A12
S15
A13
A14
A15
S16
S17
M16
A16
S18
S19
M17
S20
M18
M19
A17
O2
M20
M21
M22
A18
N4
M23
M24
O3
O4
M25
M26
Sy4
N5
M27
M28
M29
M30
M31
M32
M33
M34
M35
M36

Fig. SA♀ 4Si

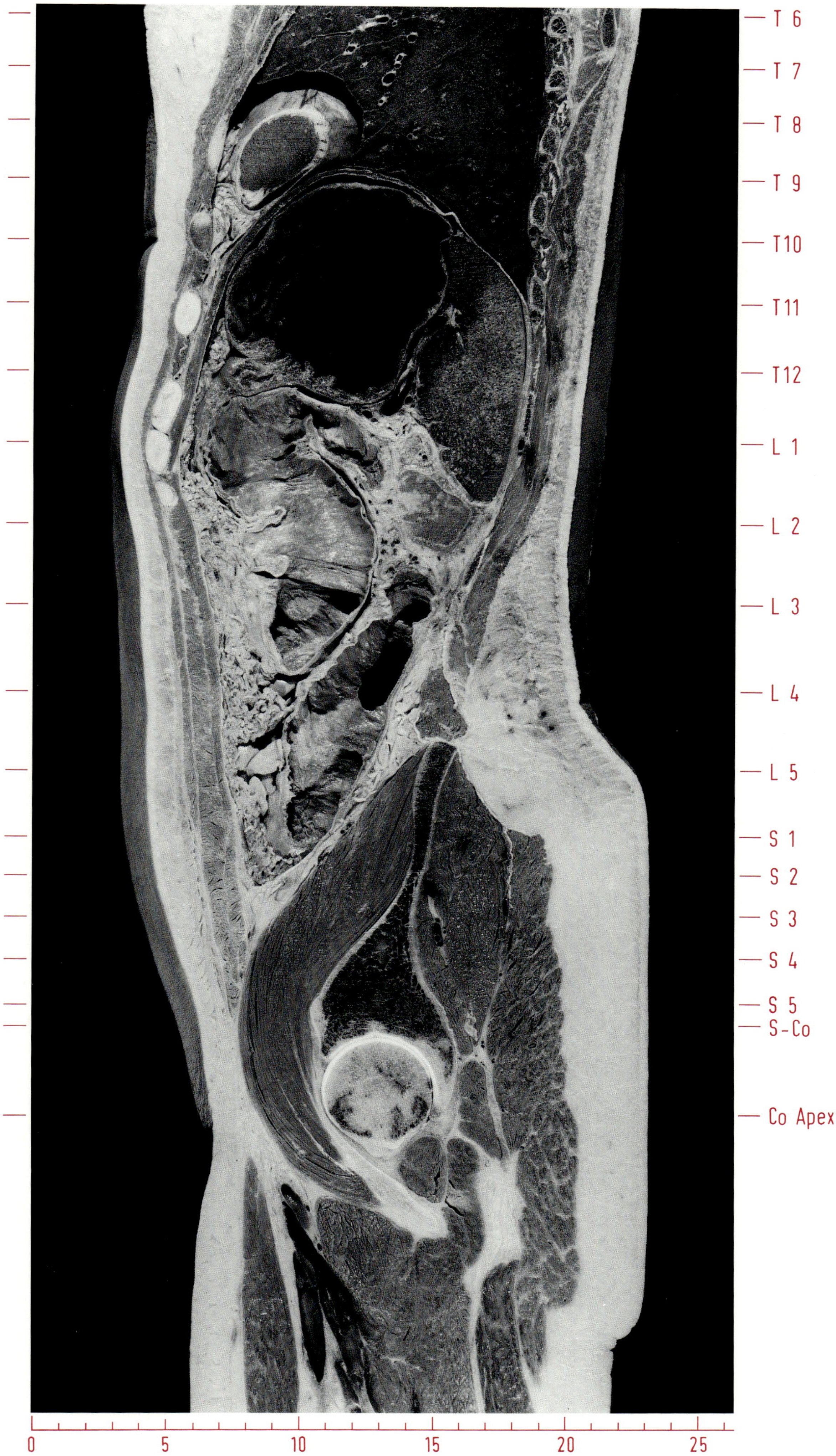
T 6
T 7
T 8
T 9
T10
T11
T12
L 1
L 2
L 3
L 4
L 5
S 1
S 2
S 3
S 4
S 5
S-Co
Co Apex
0
5
10
15
20
25

Fig. SA♀ 4 Si

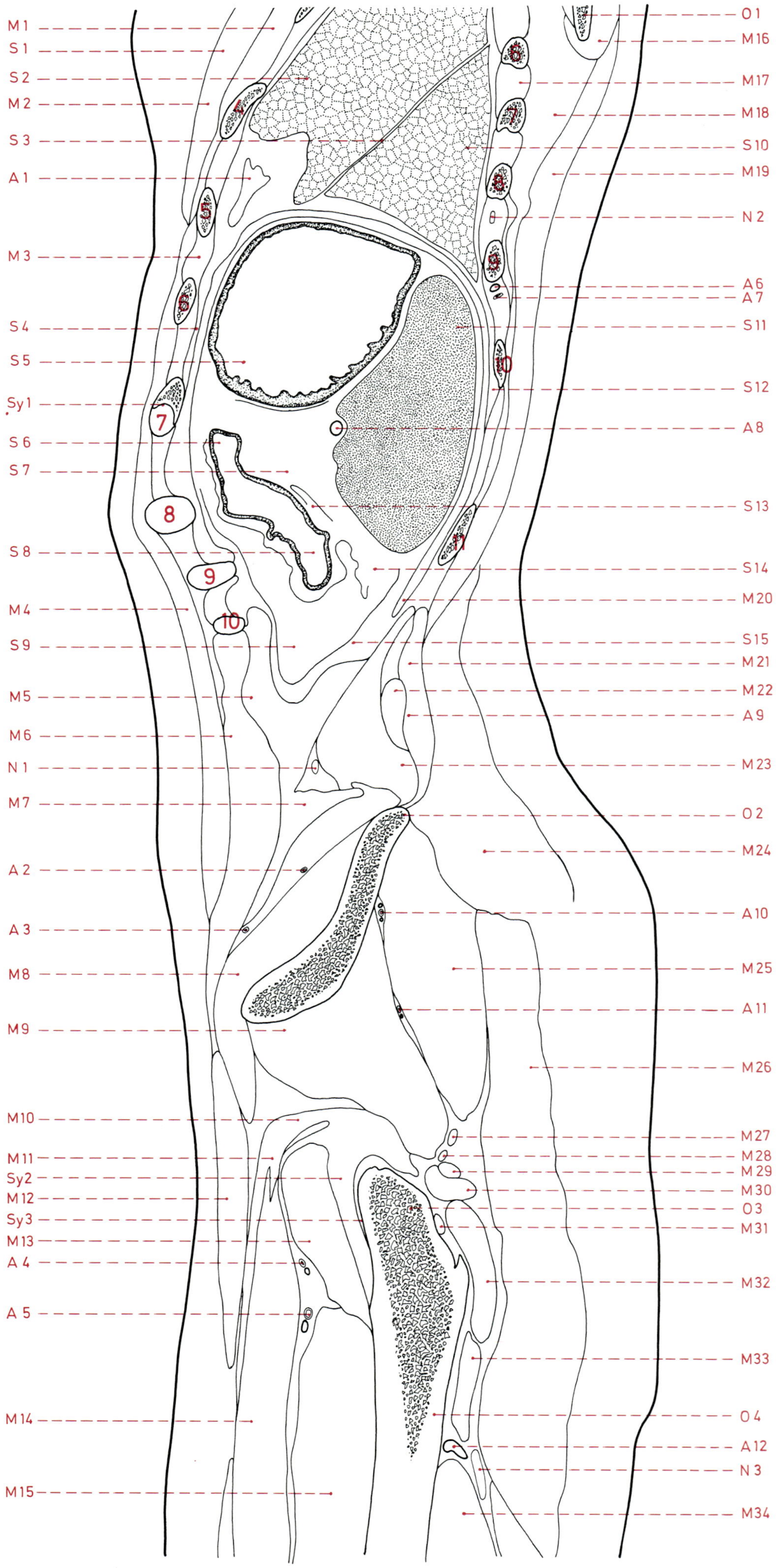
M1
S1
S2
M2
S3
A1
M3
S4
S5
Sy1
S6
S7
S8
M4
S9
M5
M6
N1
M7
A2
A3
M8
M9
M10
M11
Sy2
M12
Sy3
M13
A4
A5
M14
M15
O1
M16
M17
M18
S10
M19
N2
A6
A7
S11
S12
A8
S13
S14
M20
S15
M21
M22
A9
M23
O2
M24
A10
M25
A11
M26
M27
M28
M29
M30
O3
M31
M32
M33
O4
A12
N3
M34
4
5
6
7
8
9
10
11

Fig. SA♀ 5 Si

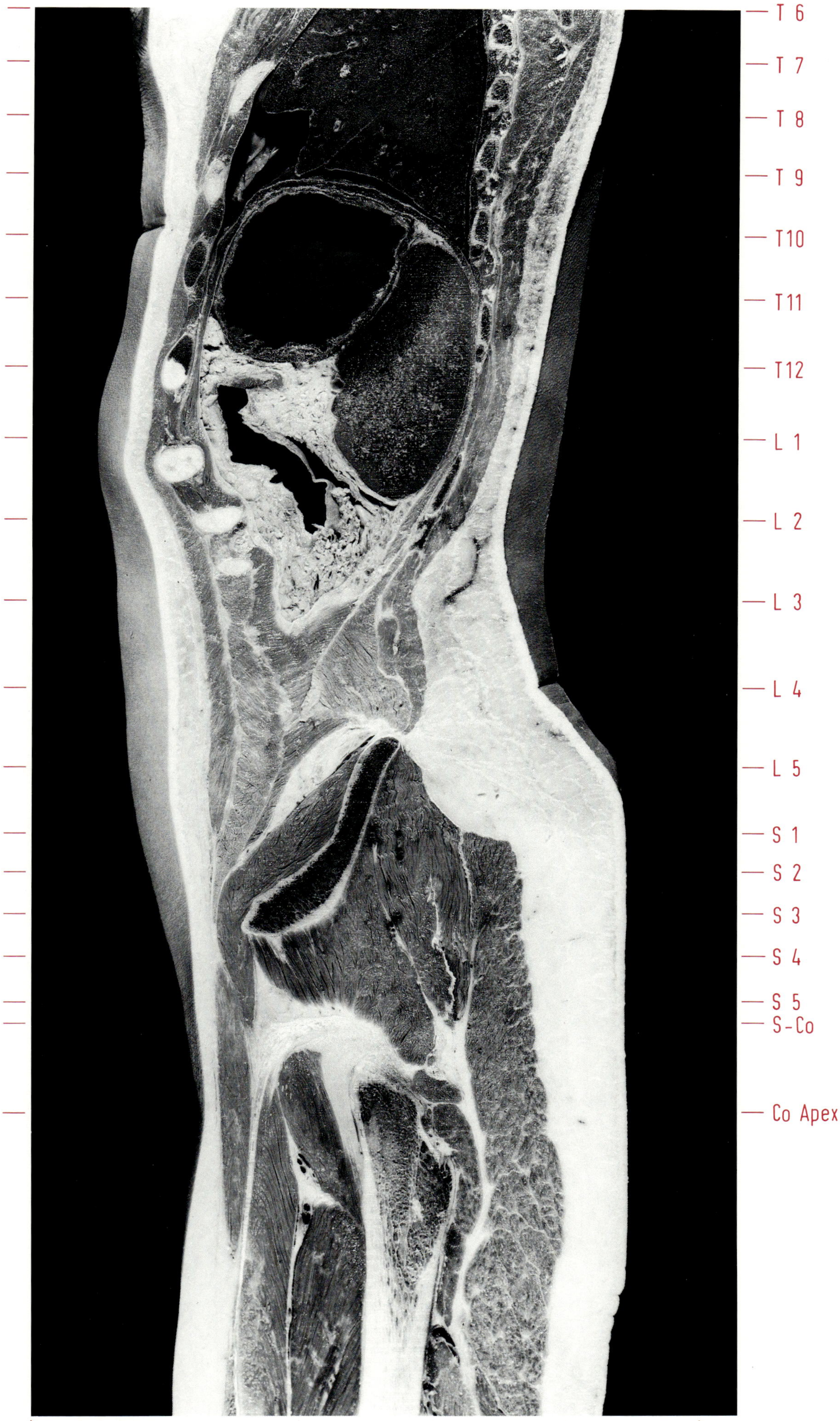
T 6
T 7
T 8
T 9
T10
T11
T12
L 1
L 2
L 3
L 4
L 5
S 1
S 2
S 3
S 4
S 5
S-Co
Co Apex
0
5
10
15
20
25

Fig. SA♀ 5 Si

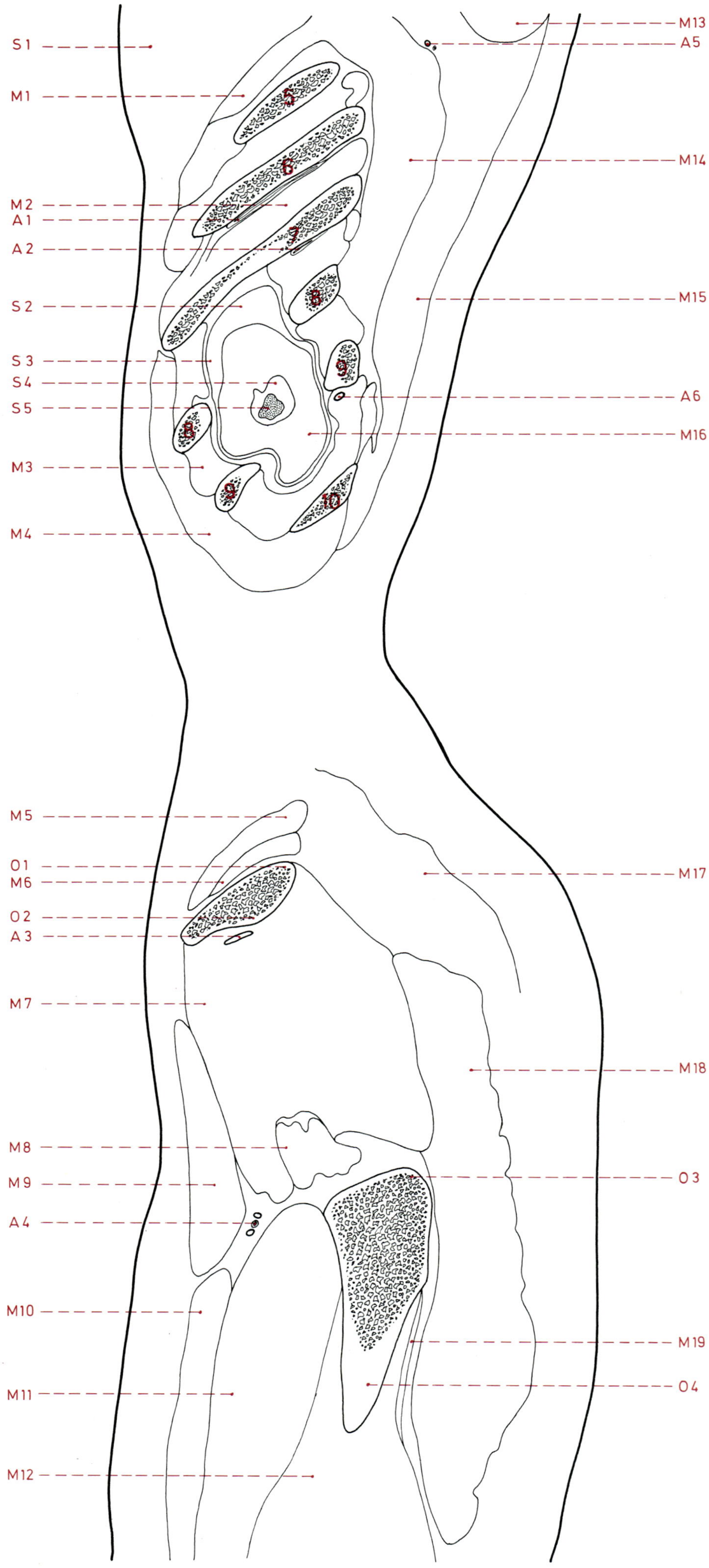
S1
M1
M2
A1
A2
S2
S3
S4
S5
M3
M4
M5
O1
M6
O2
A3
M7
M8
M9
A4
M10
M11
M12
M13
A5
M14
M15
A6
M16
M17
M18
O3
M19
O4
5
6
7
8
9
8
9
10

Fig. SA♀ 6 Si

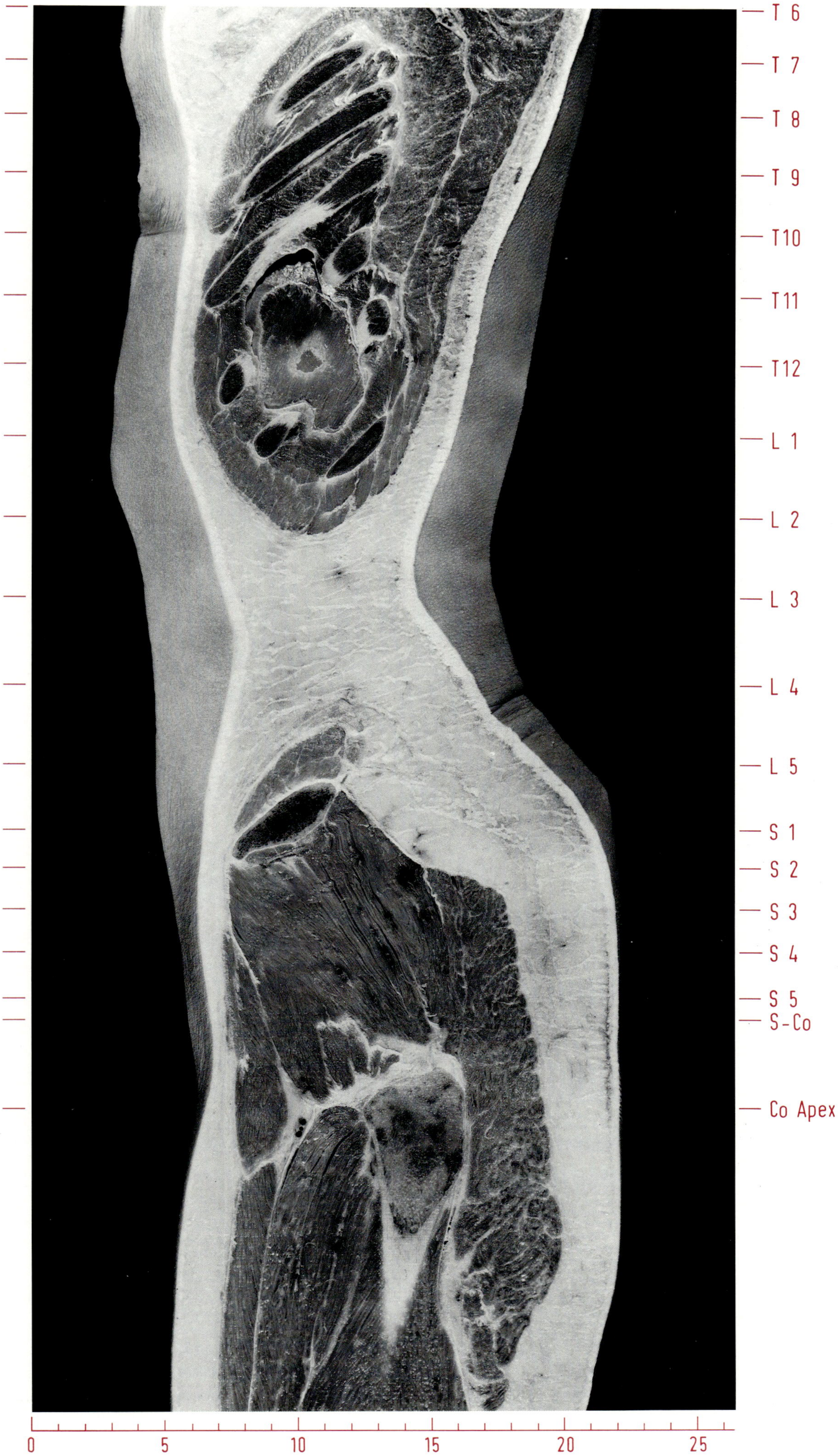
T 6
T 7
T 8
T 9
T10
T11
T12
L 1
L 2
L 3
L 4
L 5
S 1
S 2
S 3
S 4
S 5
S-Co
Co Apex
0
5
10
15
20
25

Fig. SA♀ 6 Si

HA

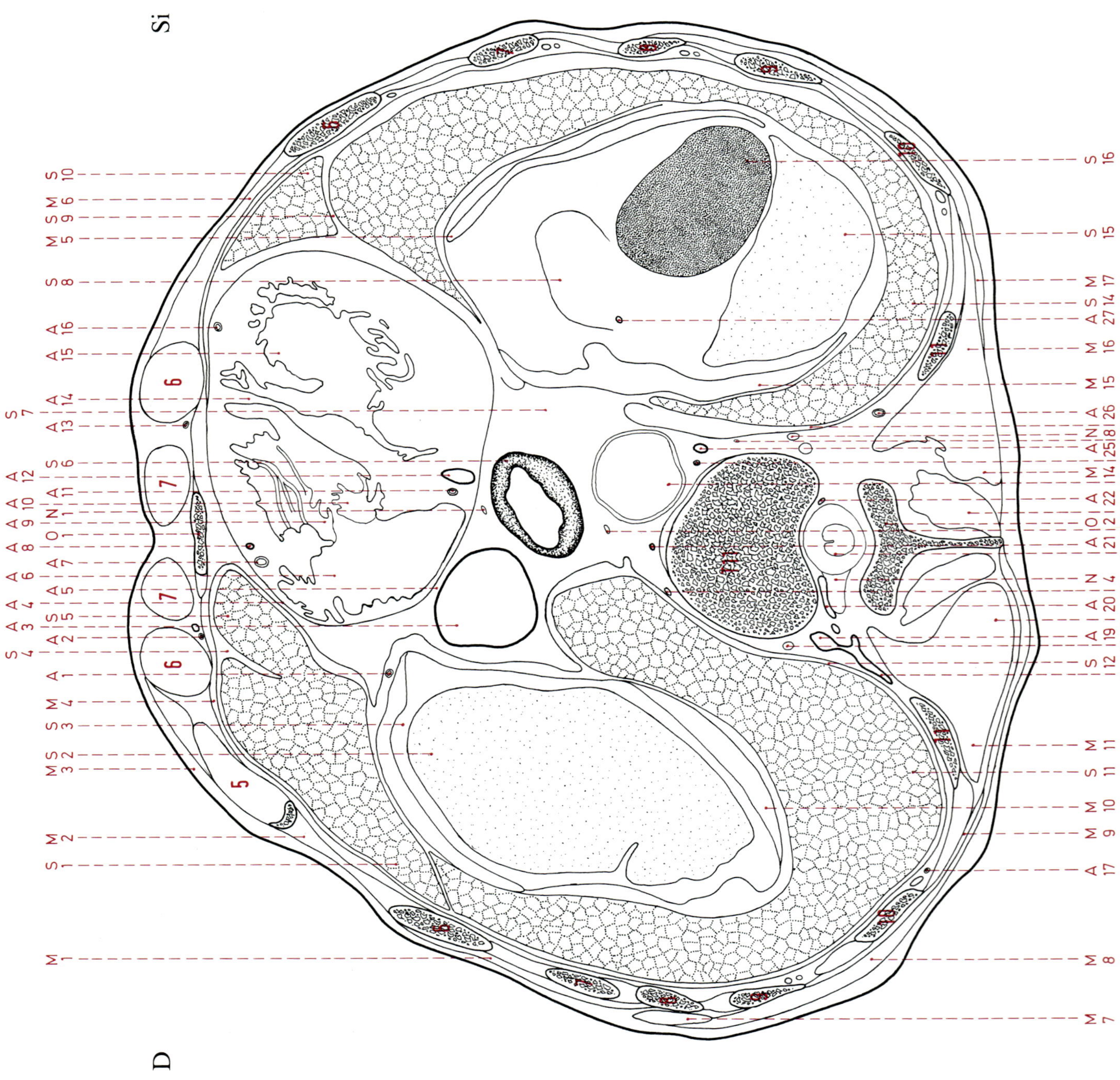

Fig. HA 1

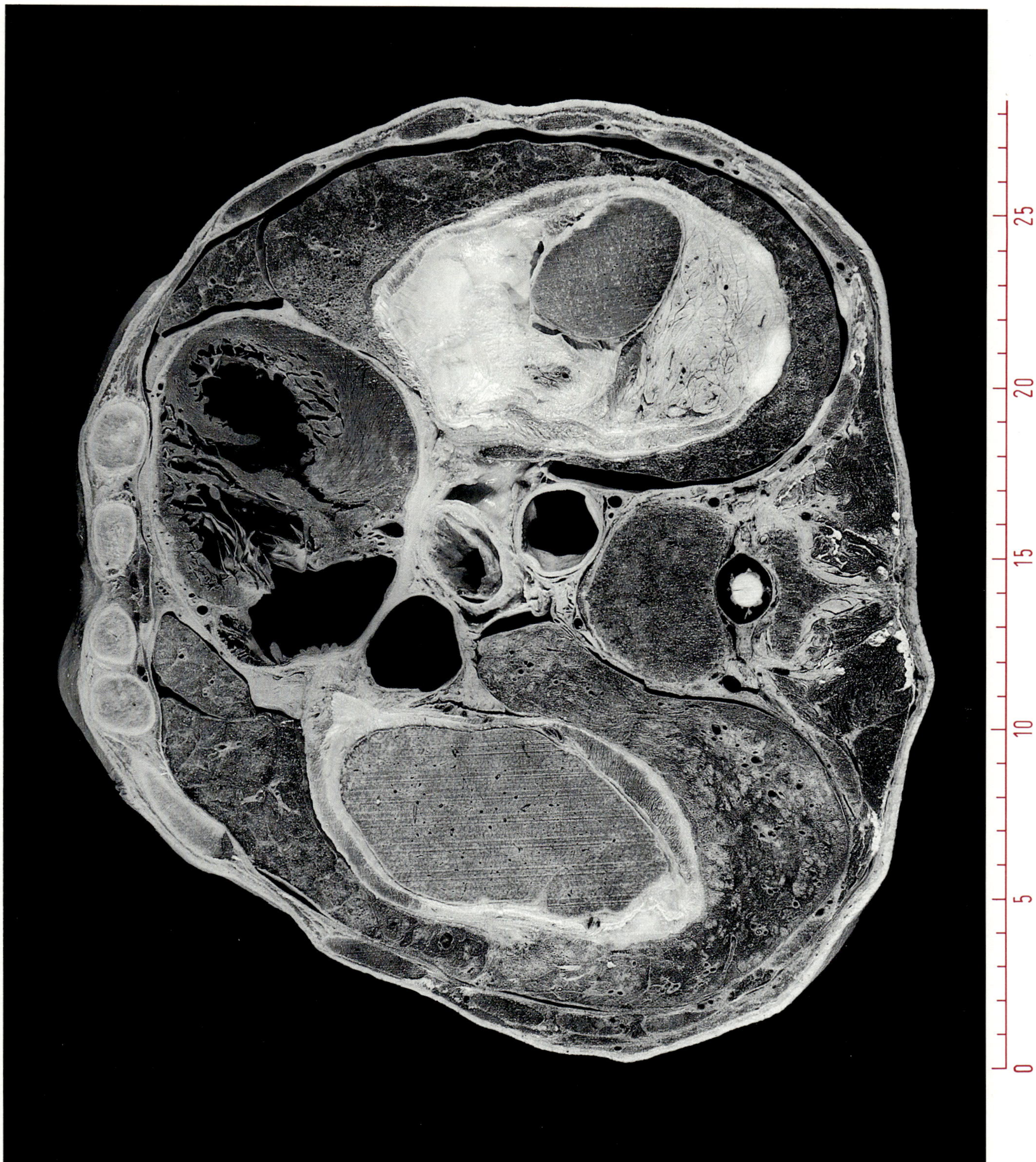

Fig. HA 1

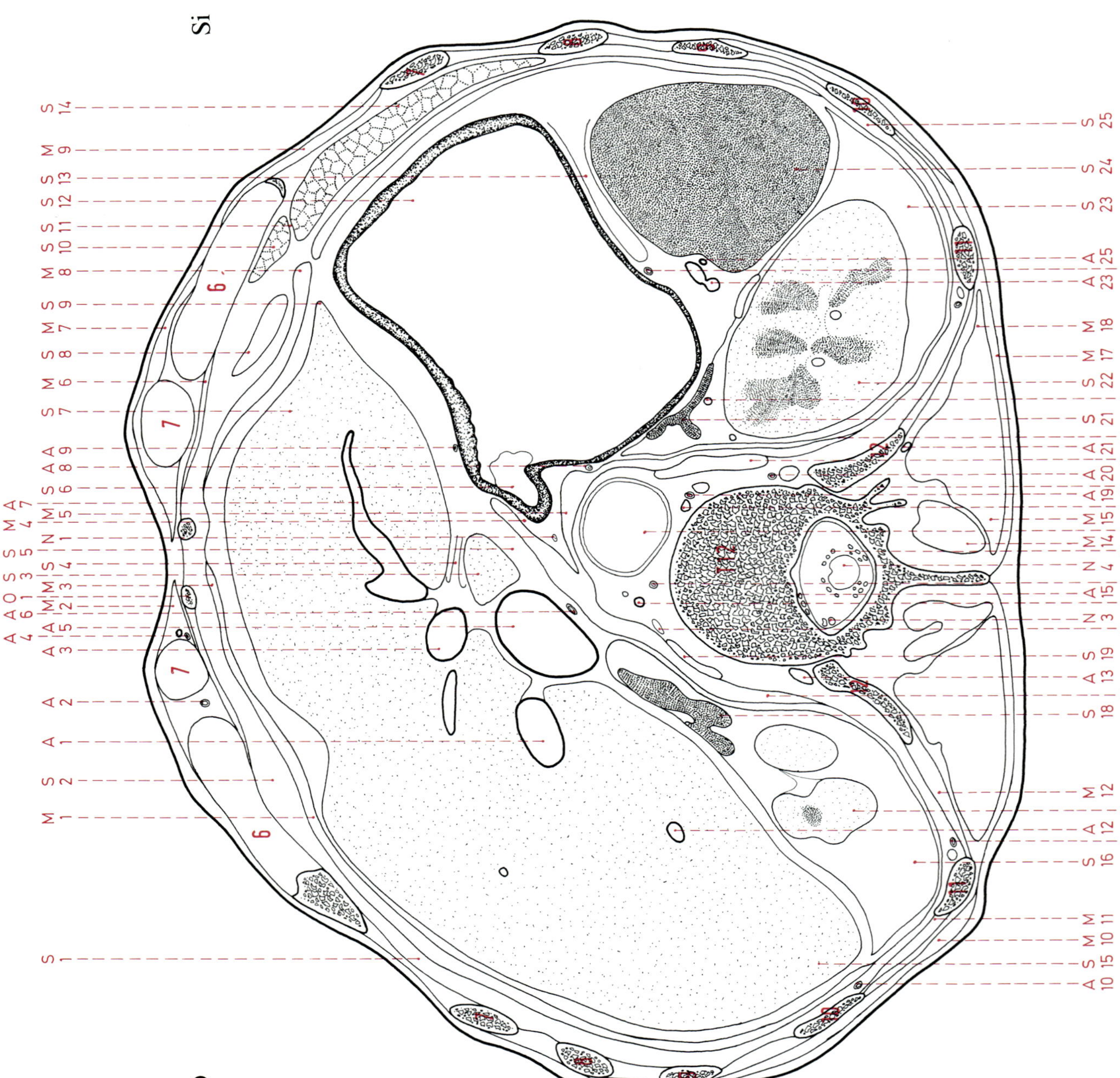

Fig. HA 2

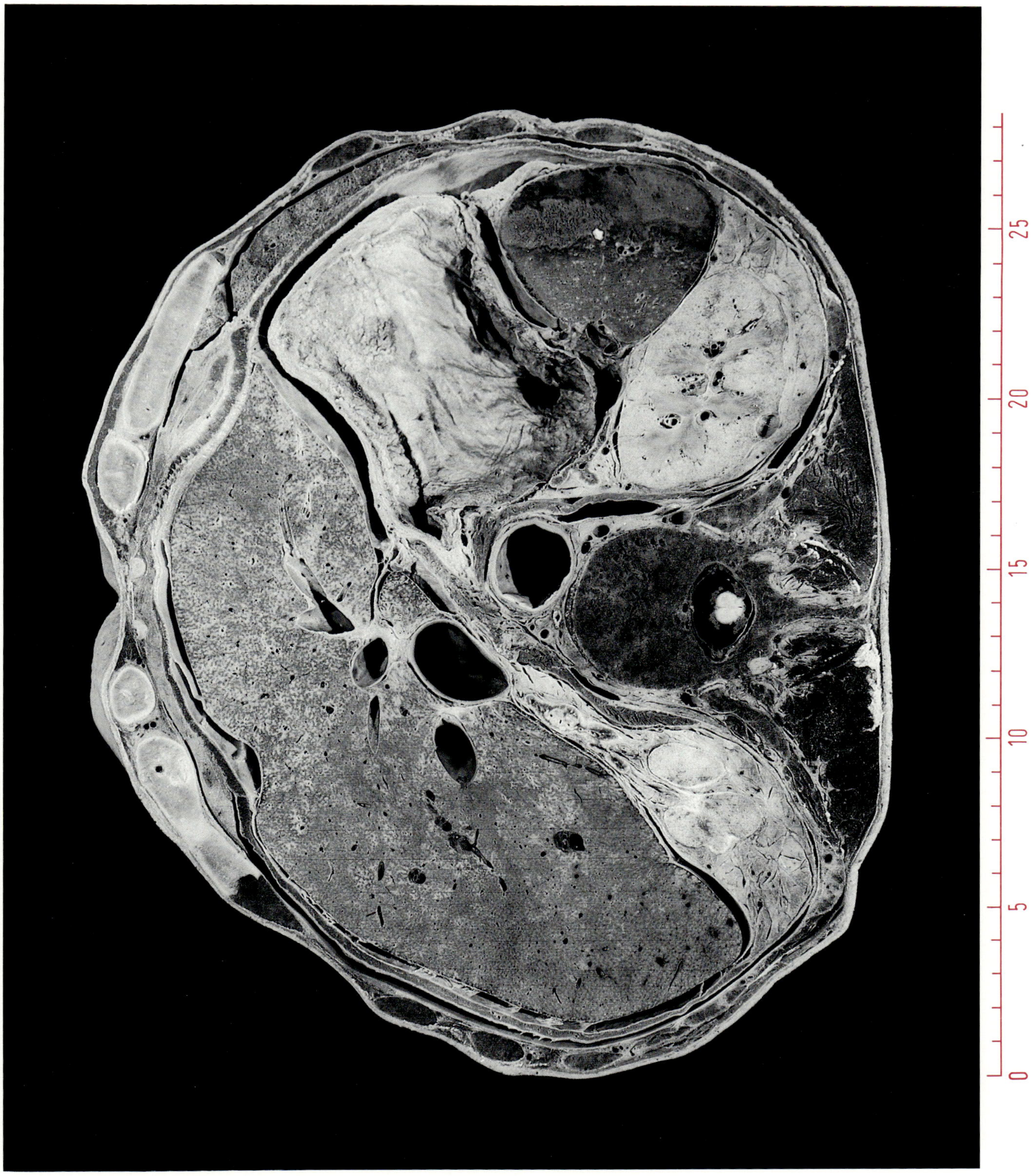

Fig. HA 2

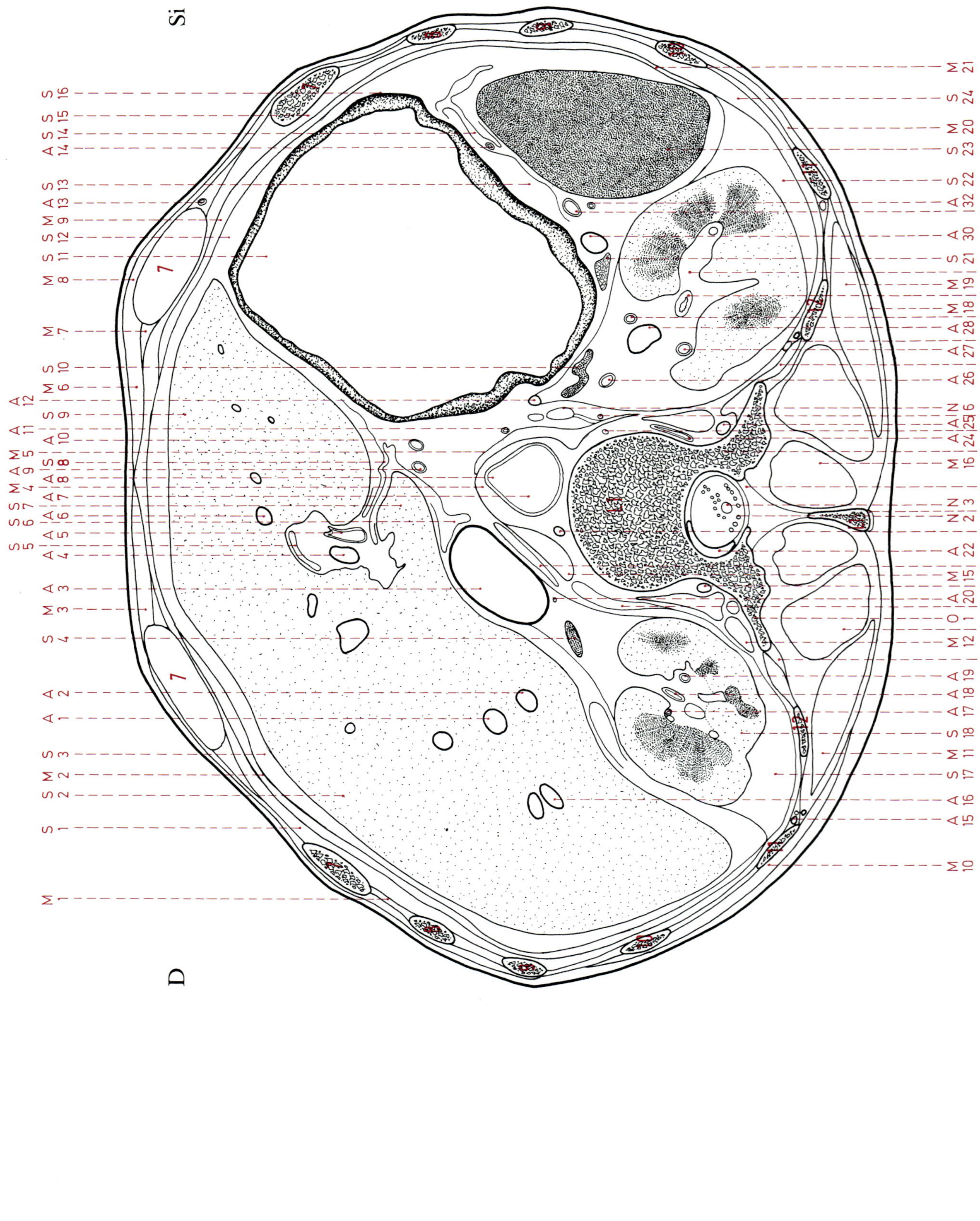

Fig. HA 3

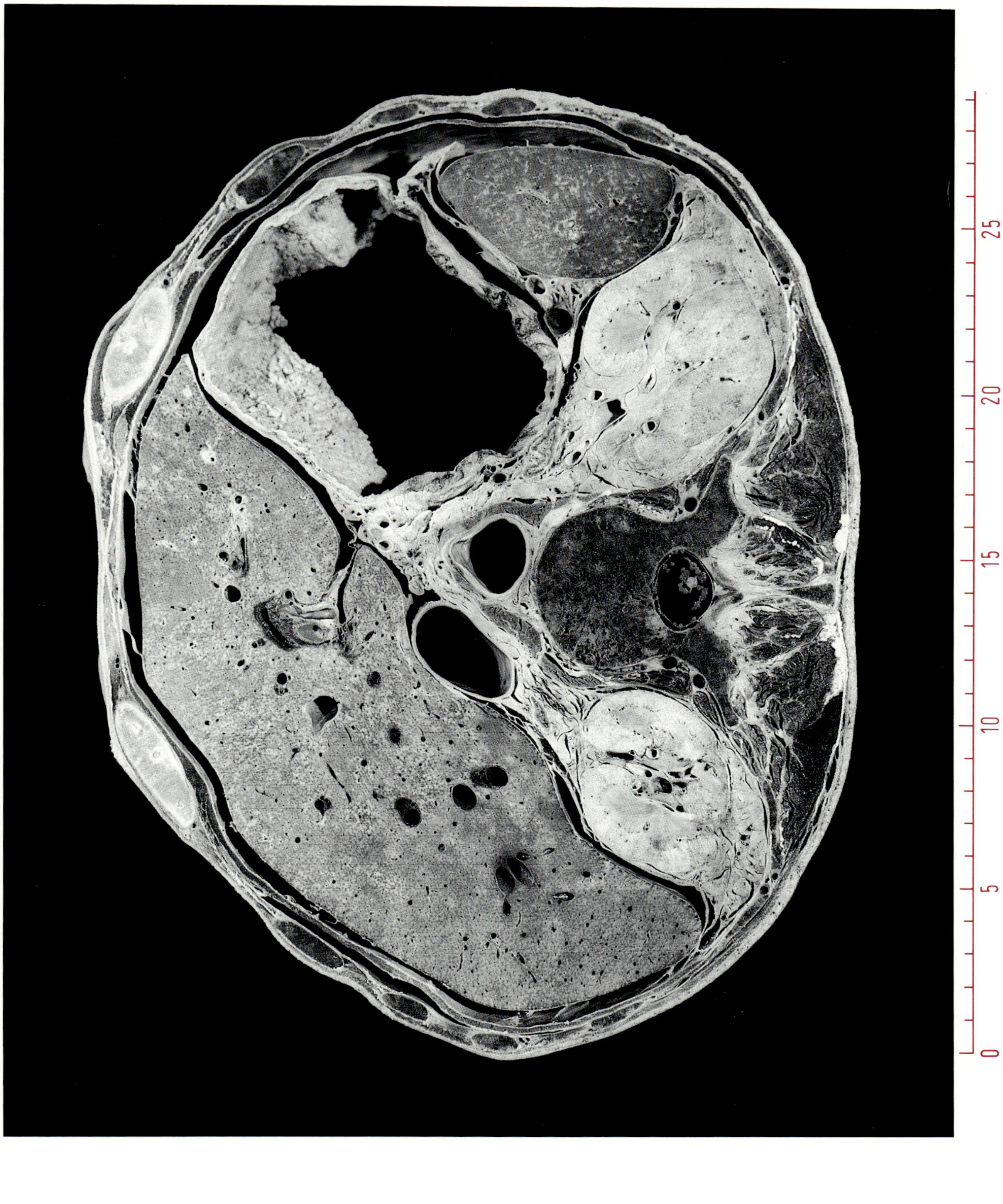

Fig. HA 3

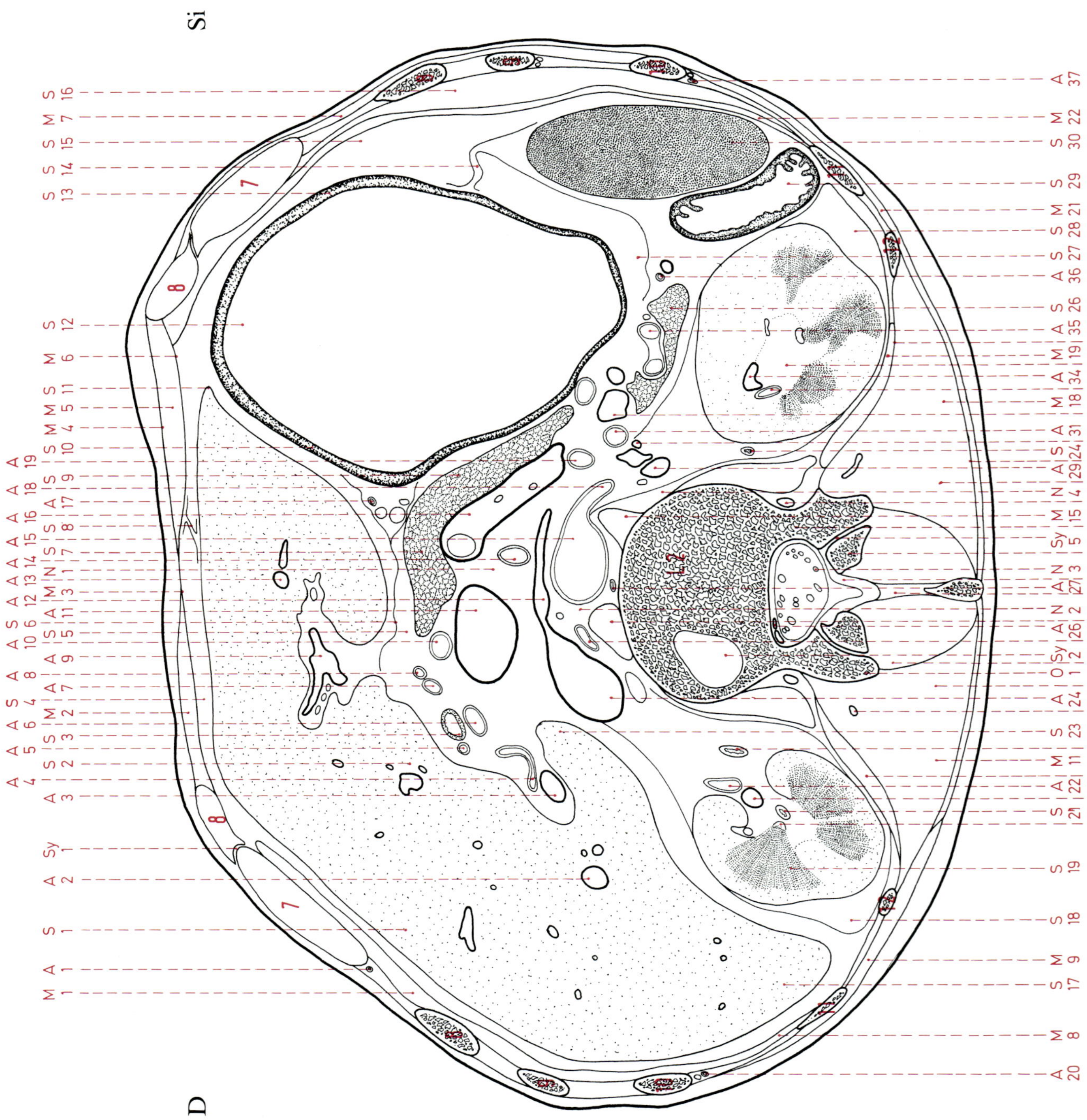

Fig. HA 4

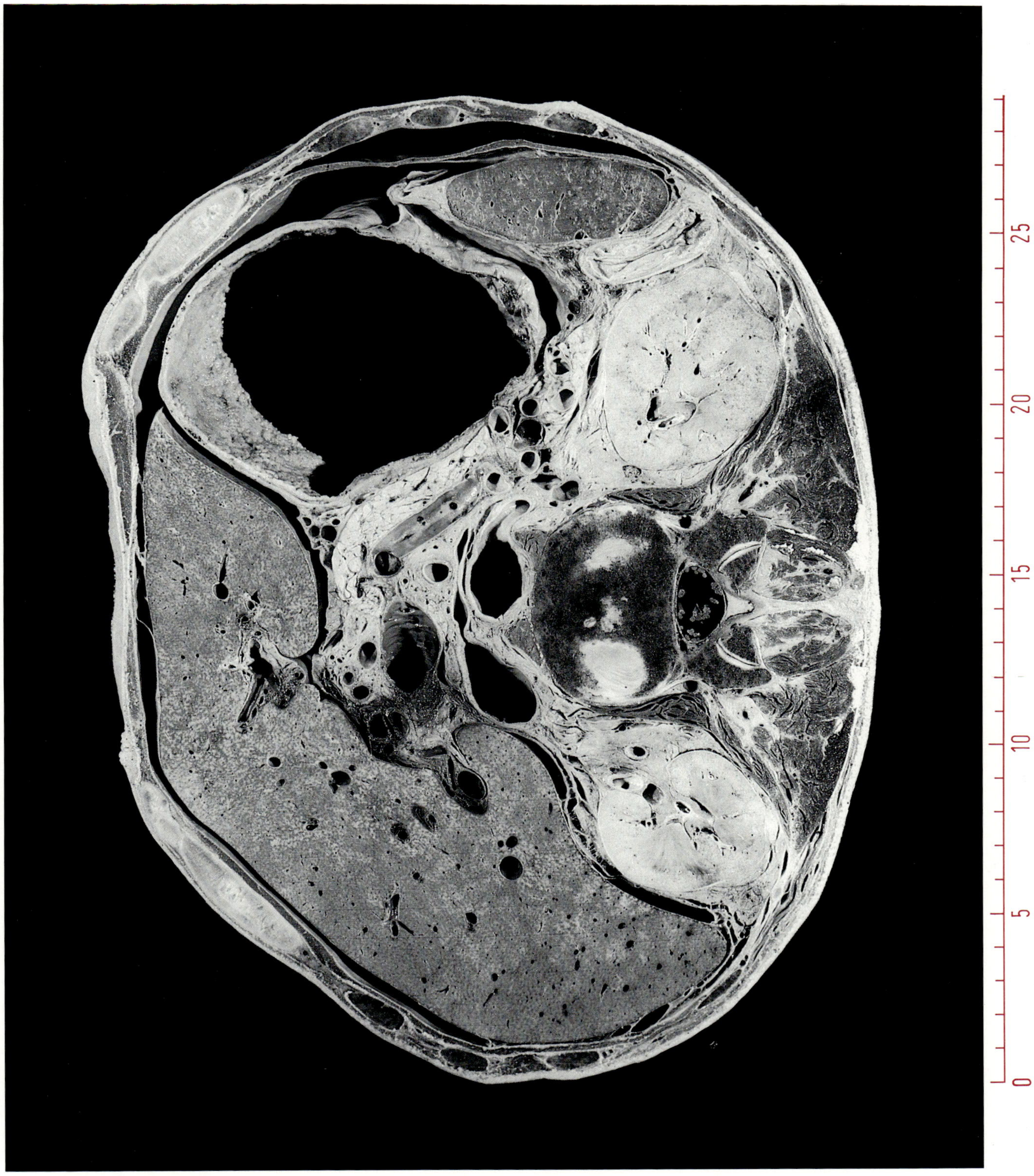

Fig. HA 4

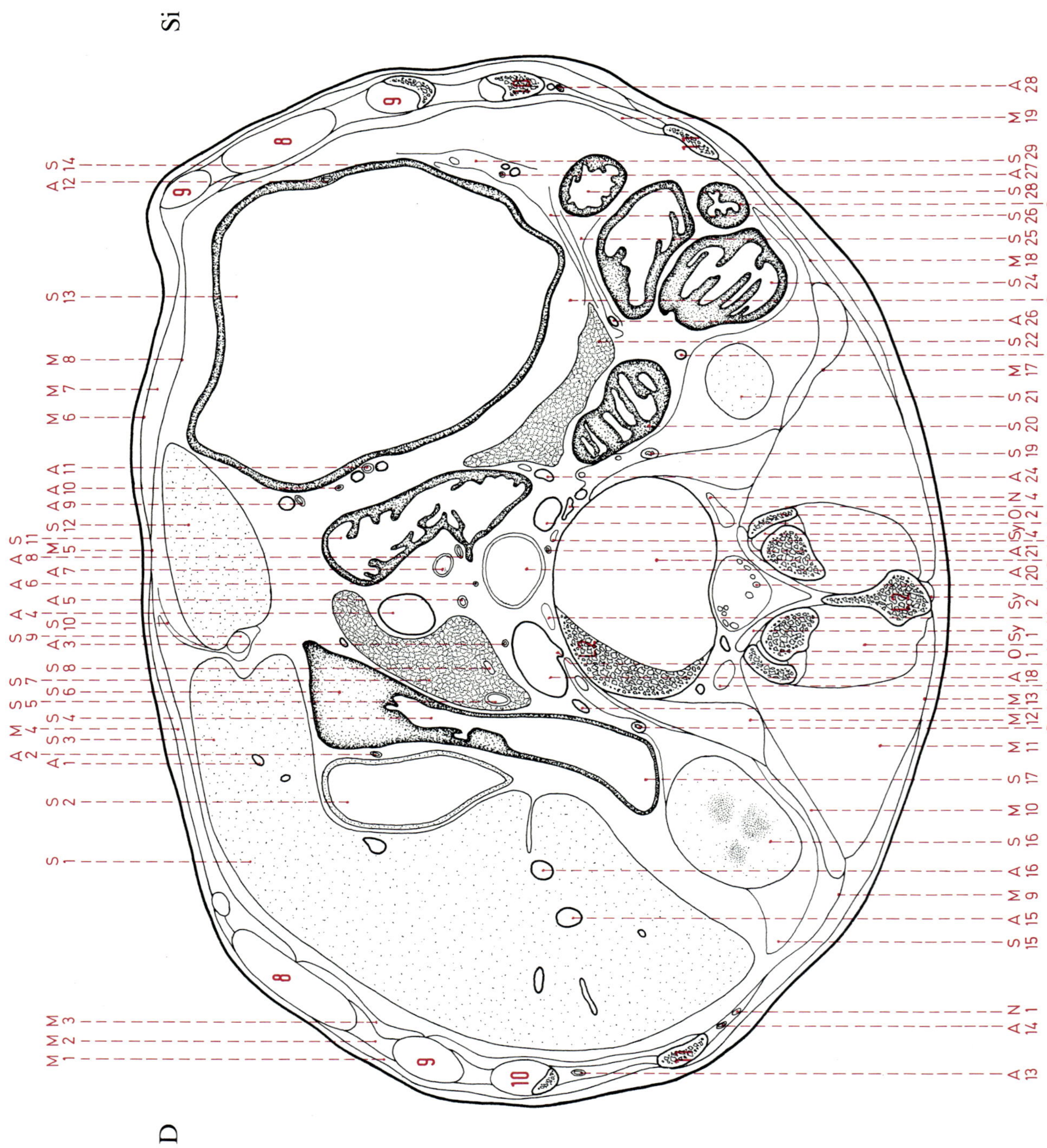

Fig. HA 5

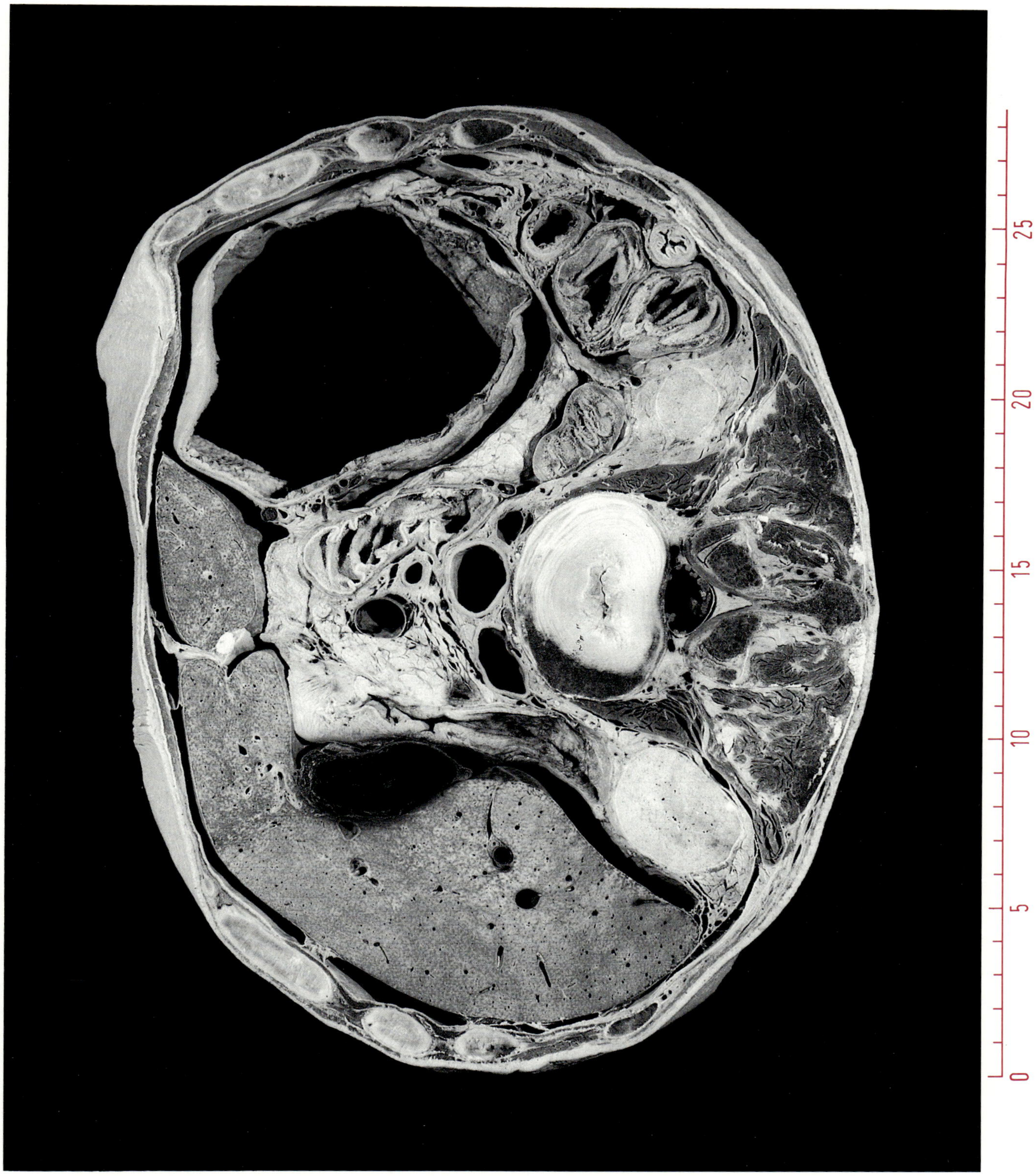

Fig. HA 5

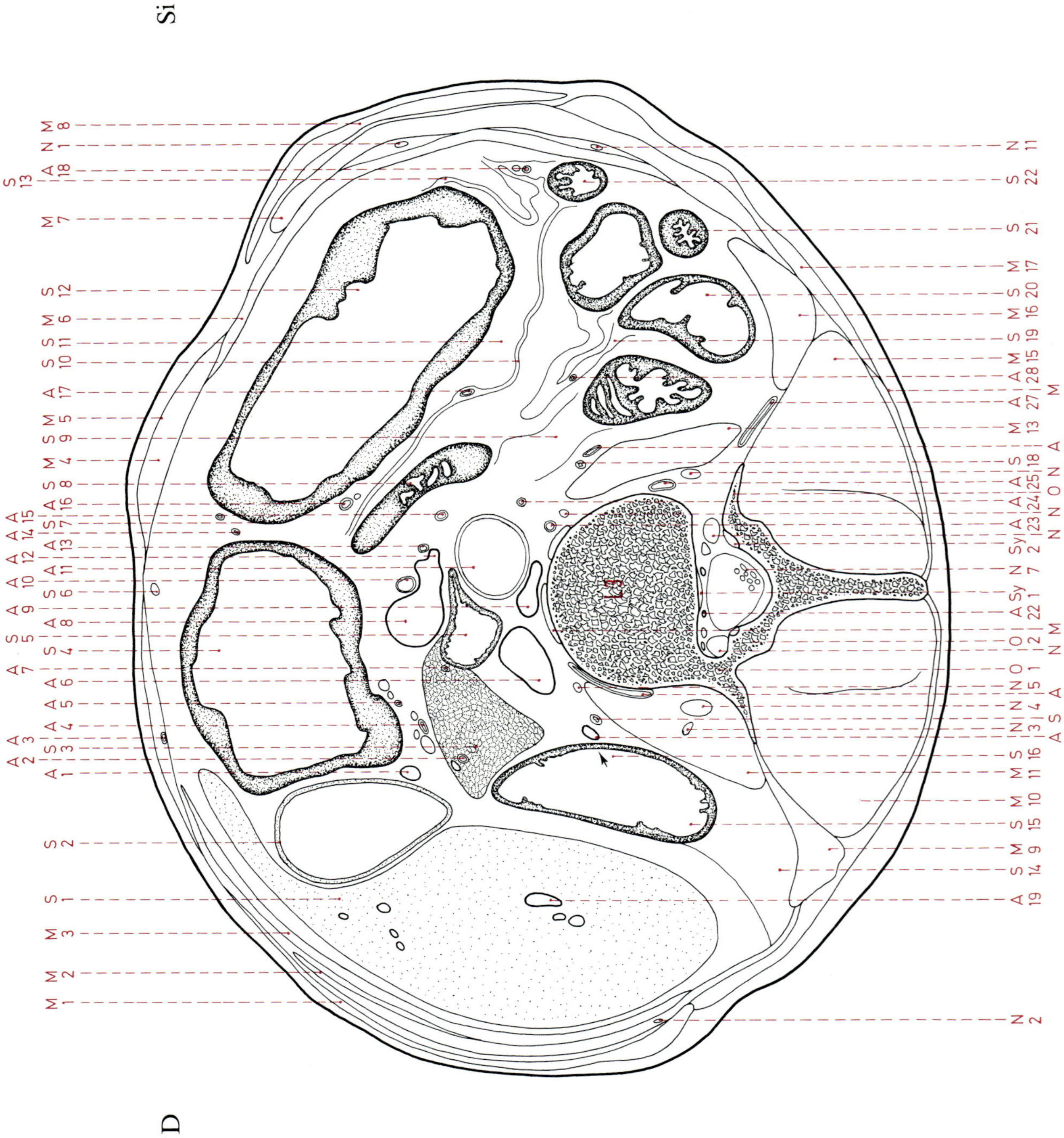

Fig. HA 6

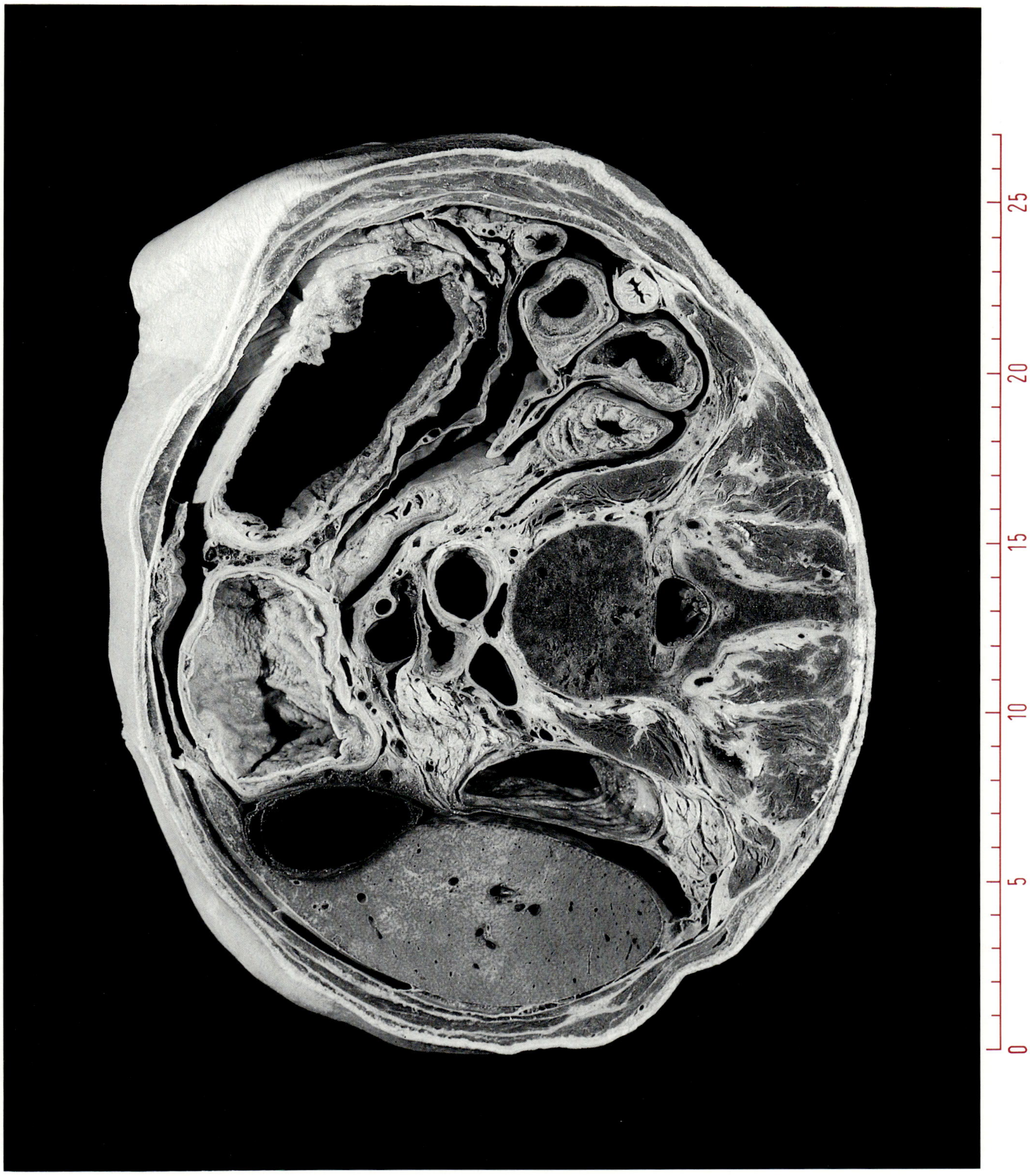

Fig. HA 6

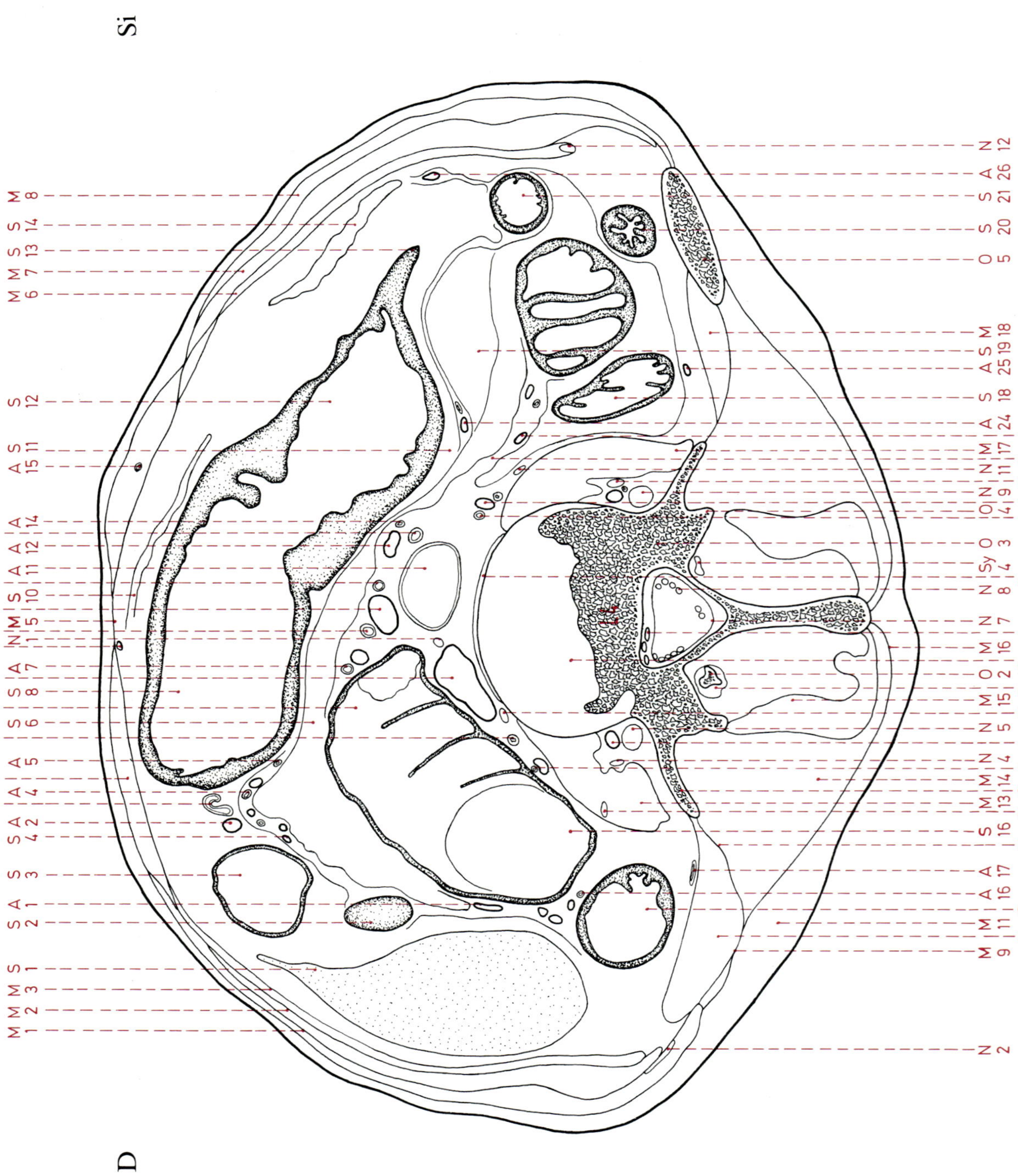

Fig. HA 7

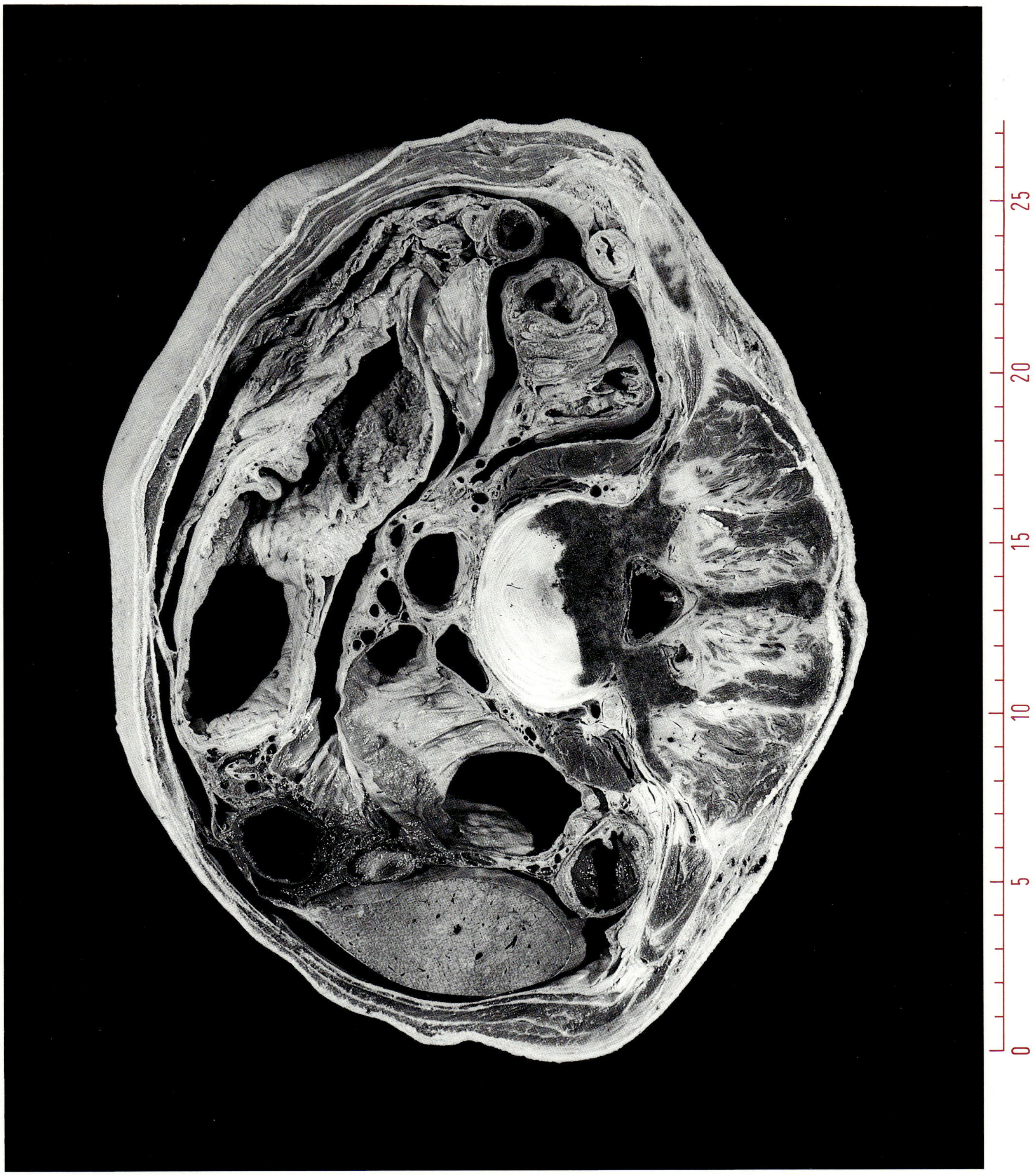

Fig. HA 7

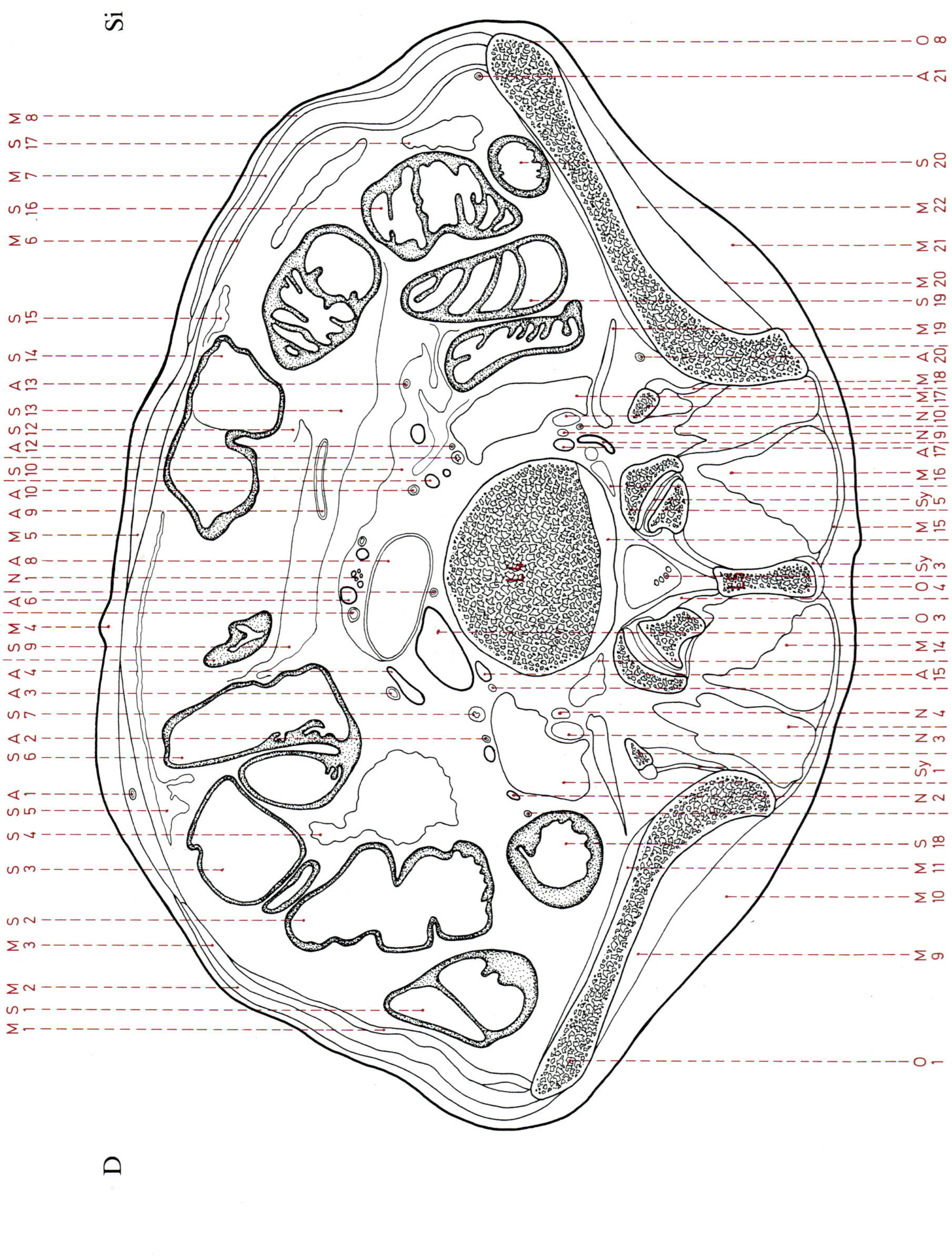

Fig. HA 8

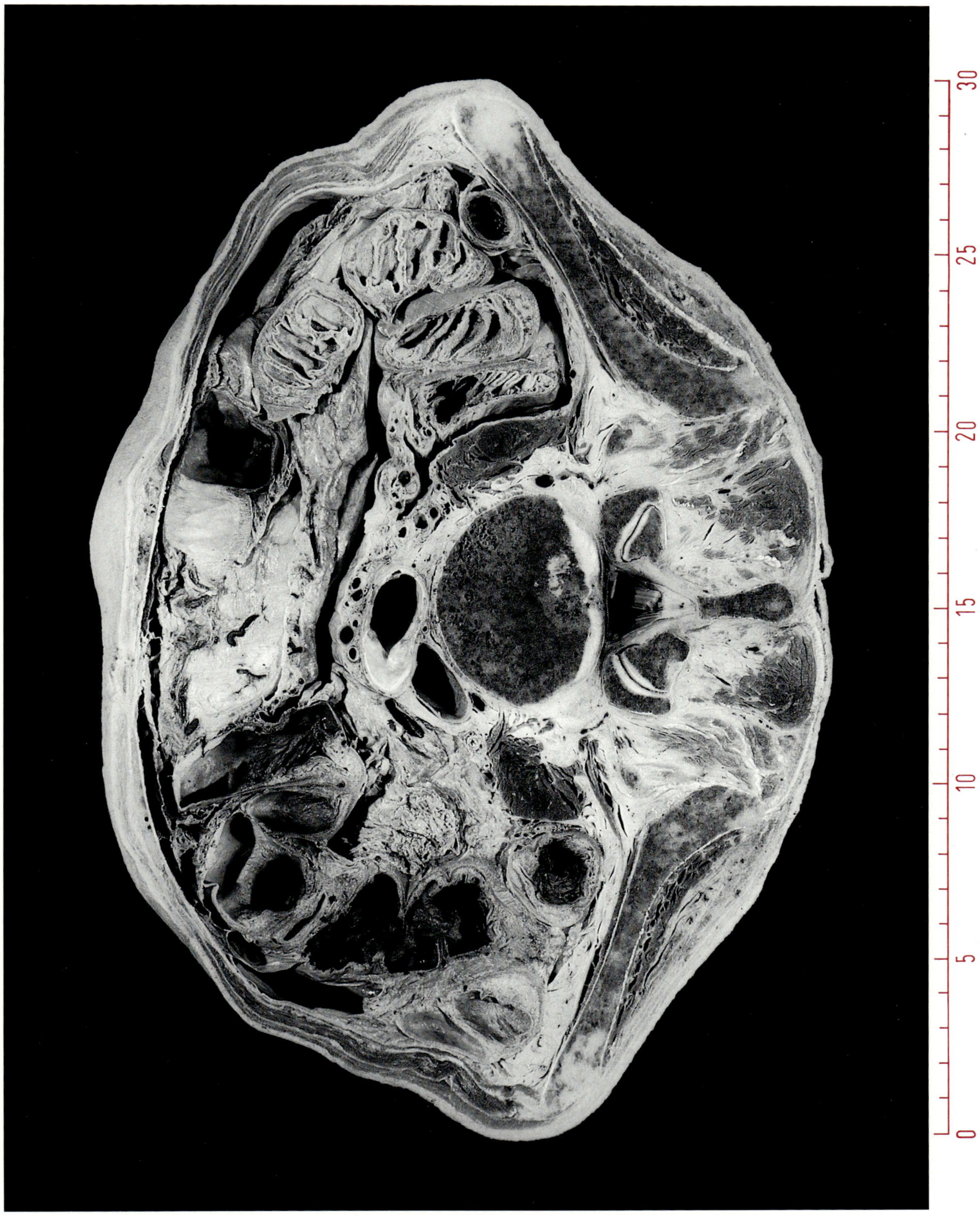

Fig. HA 8

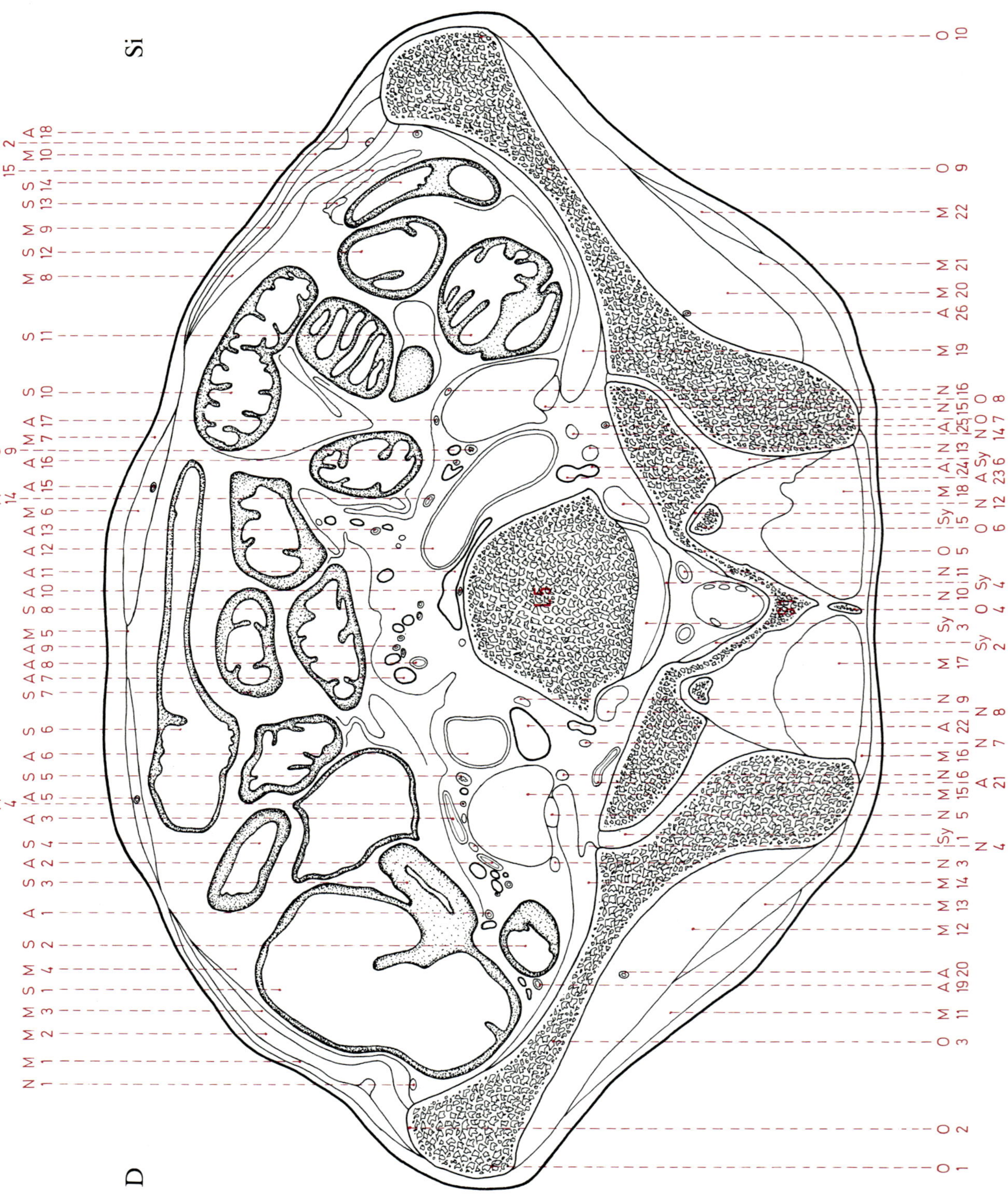

Fig. HA 9

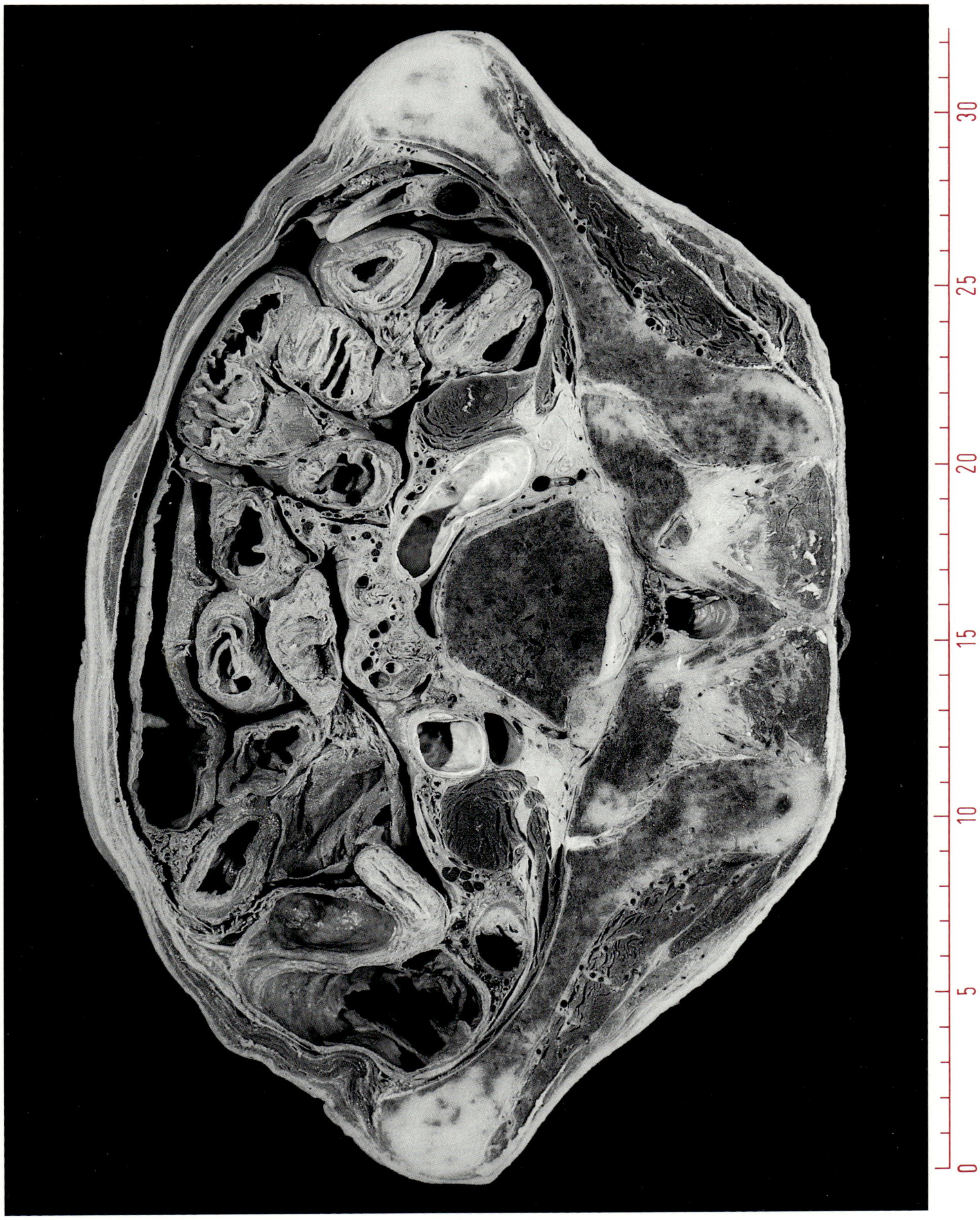

Fig. HA 9

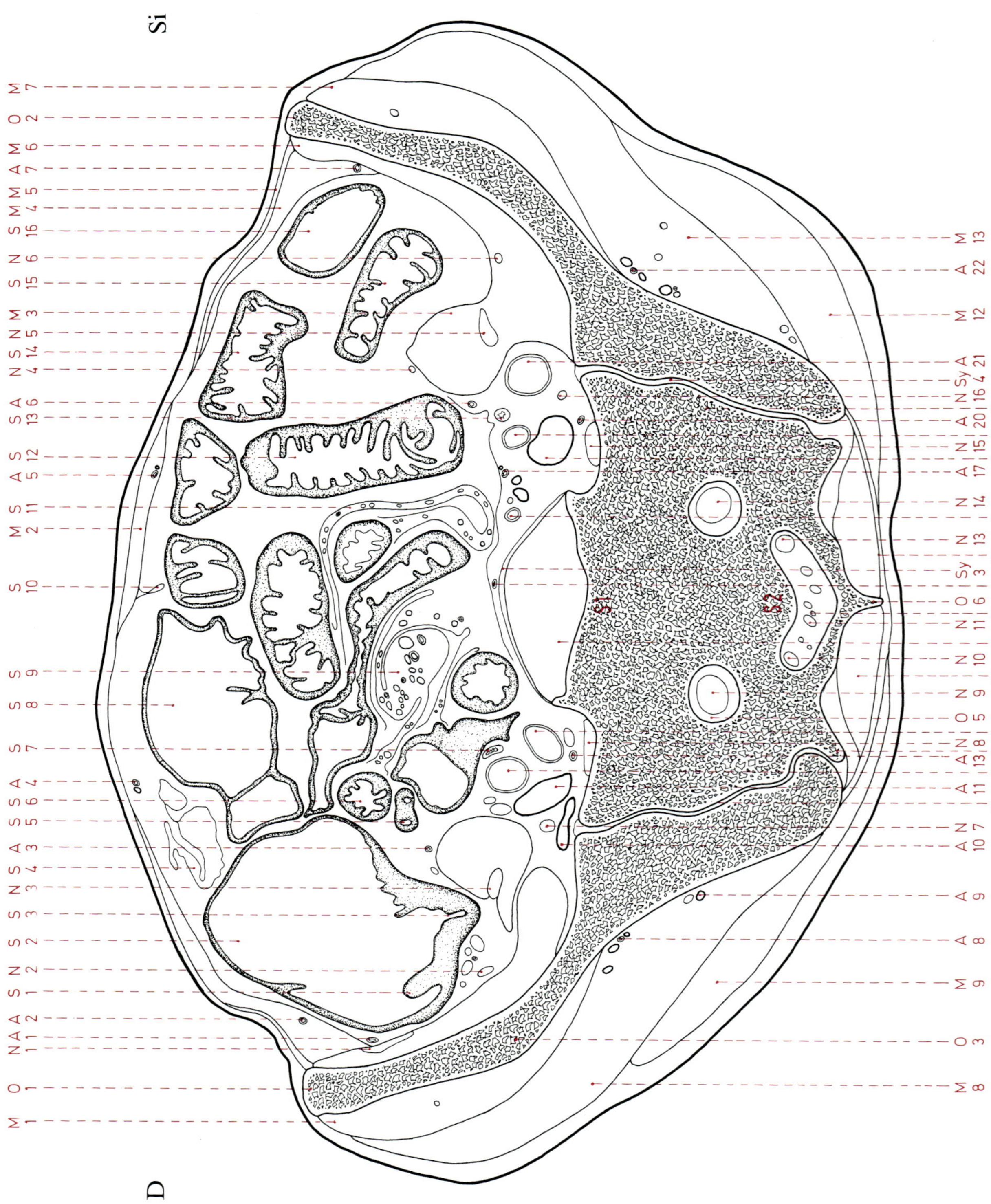

Fig. HA 10

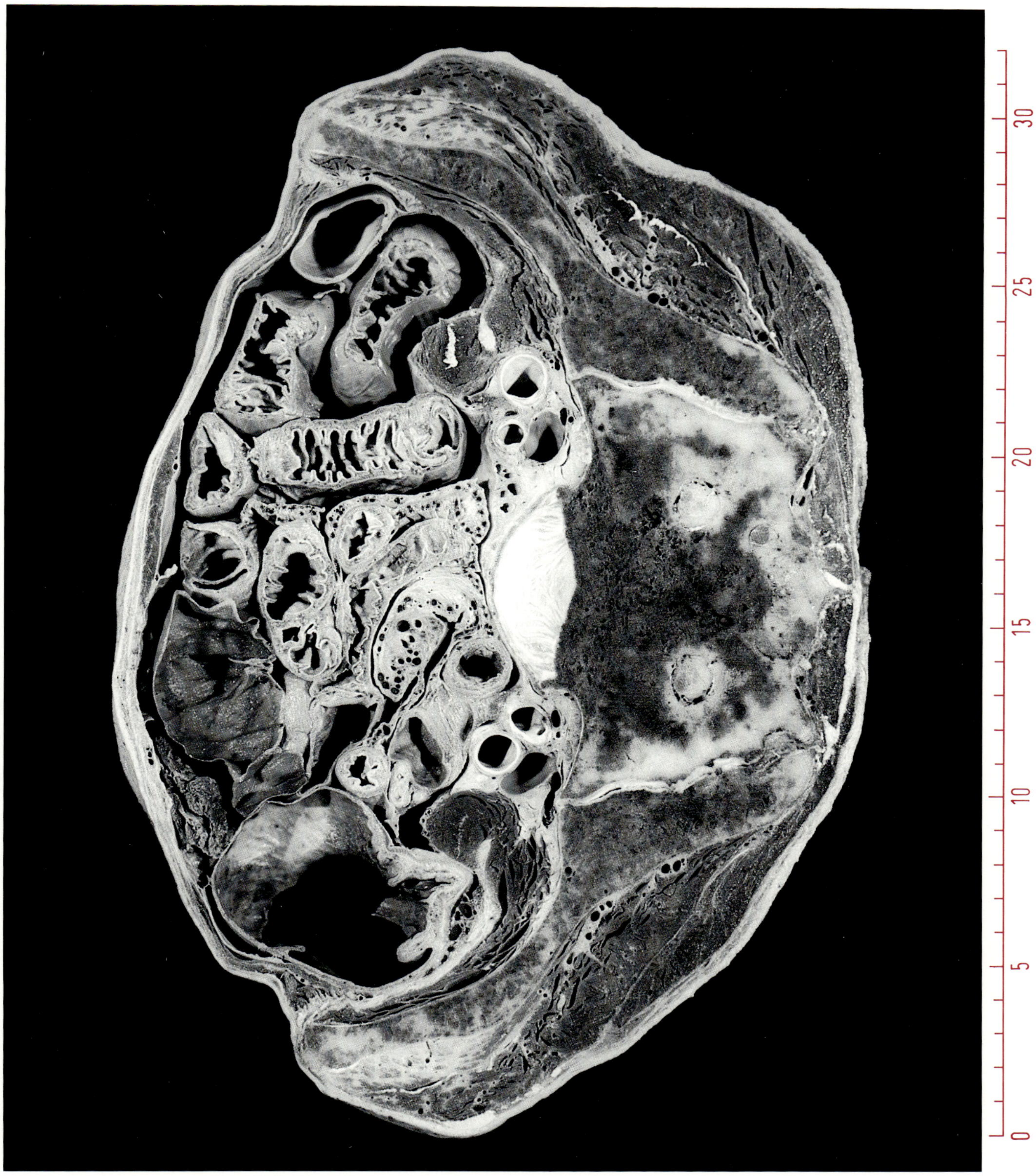

Fig. HA 10

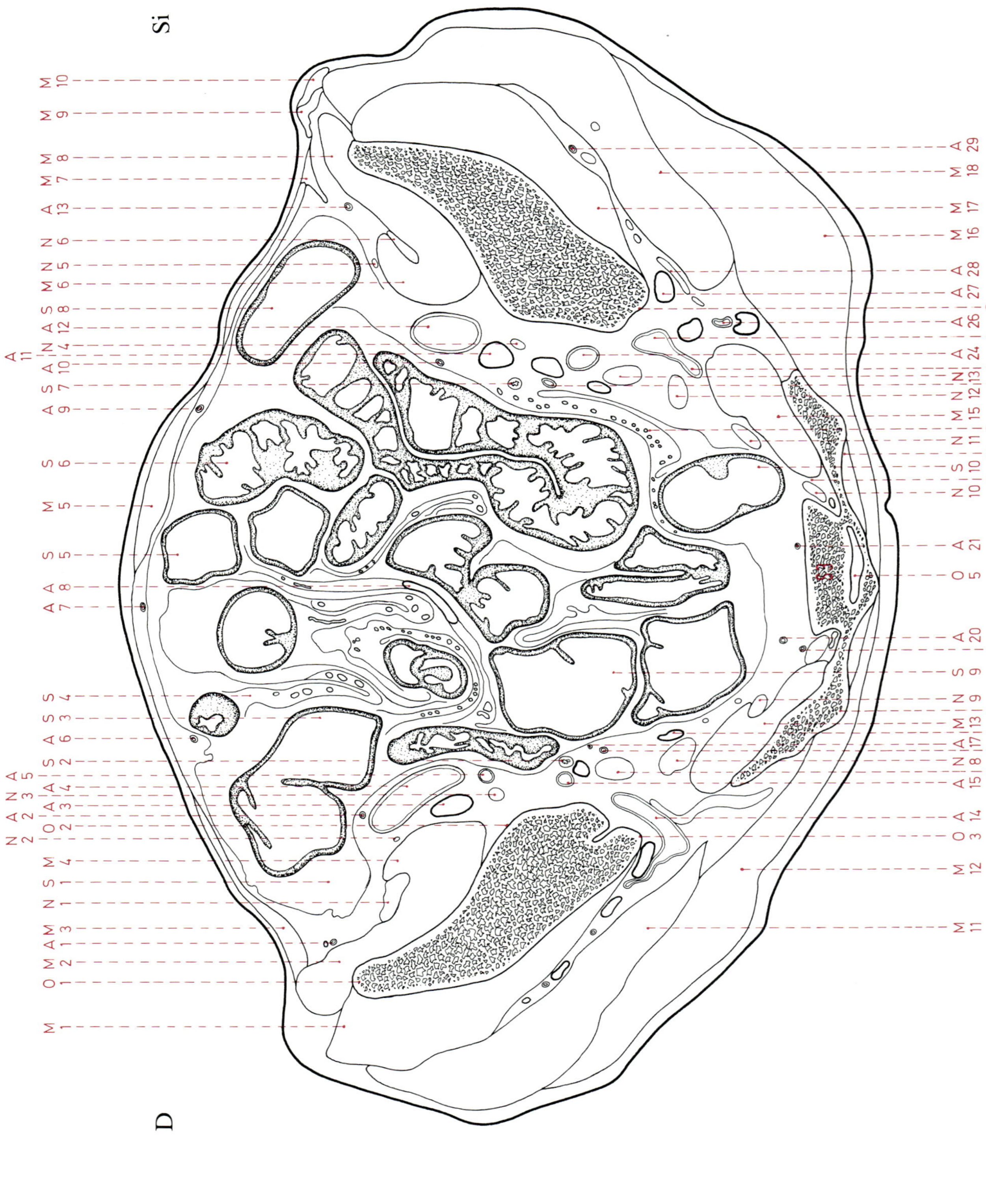

Fig. HA 11

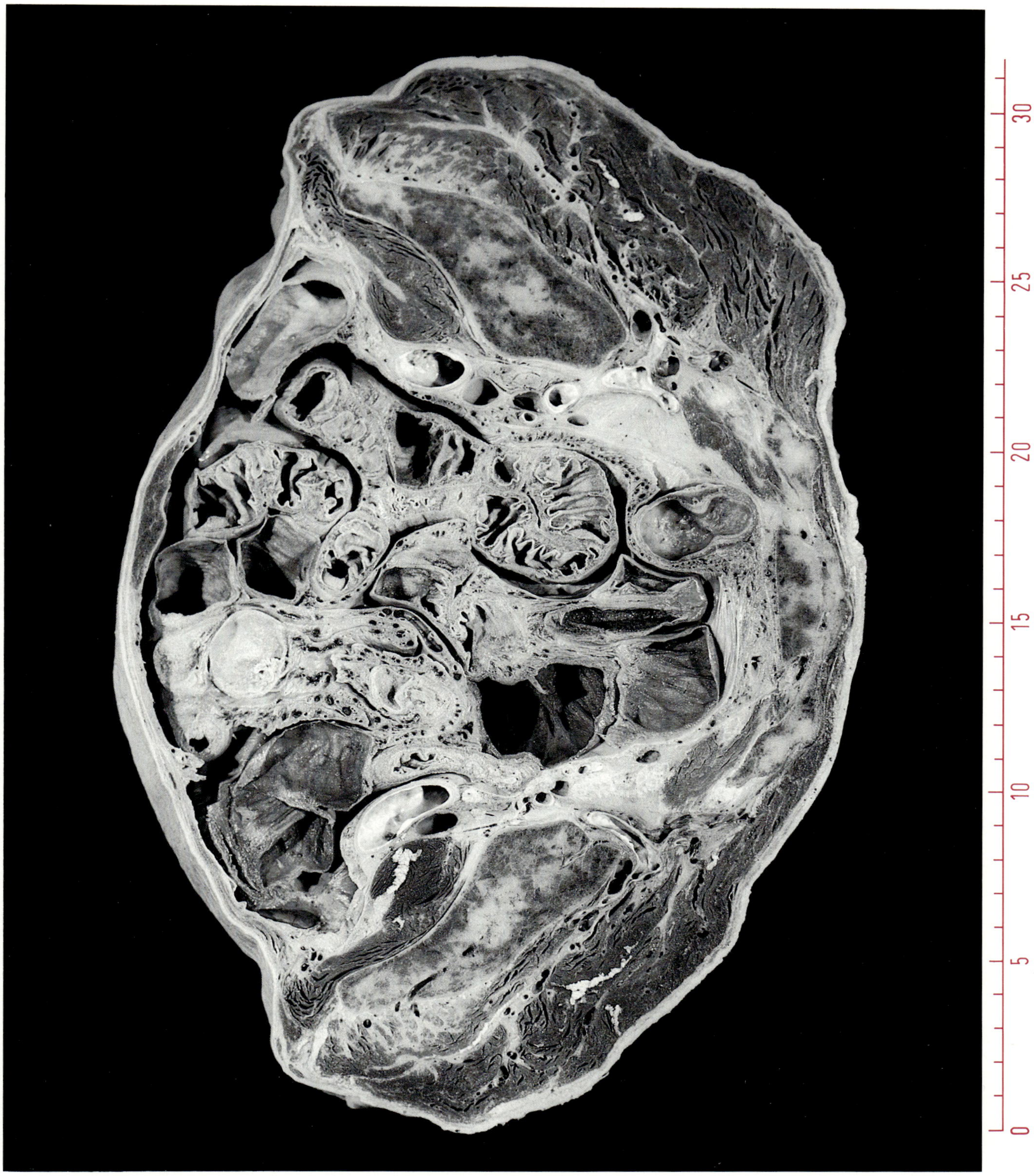

Fig. HA 11

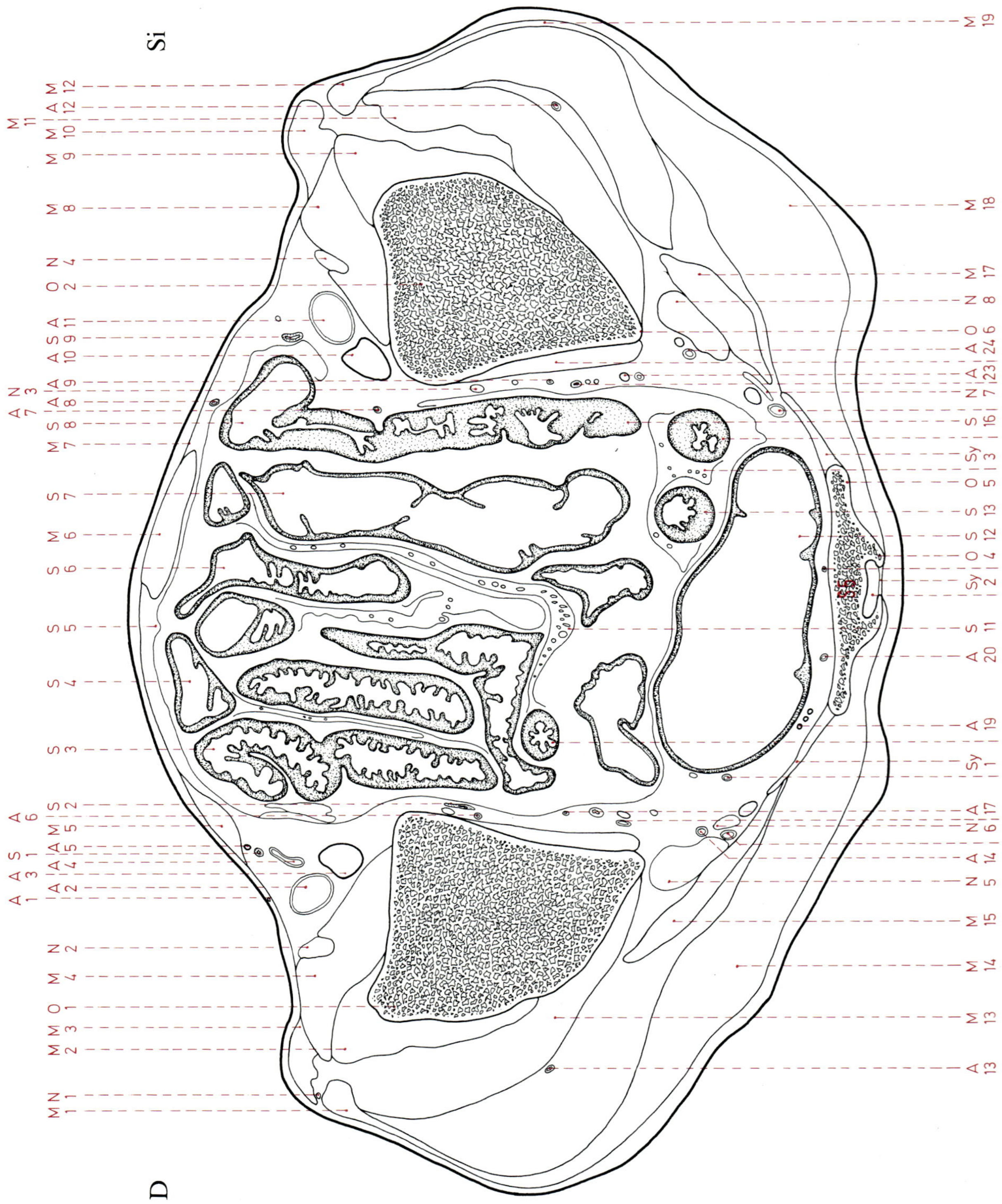

Fig. HA 12

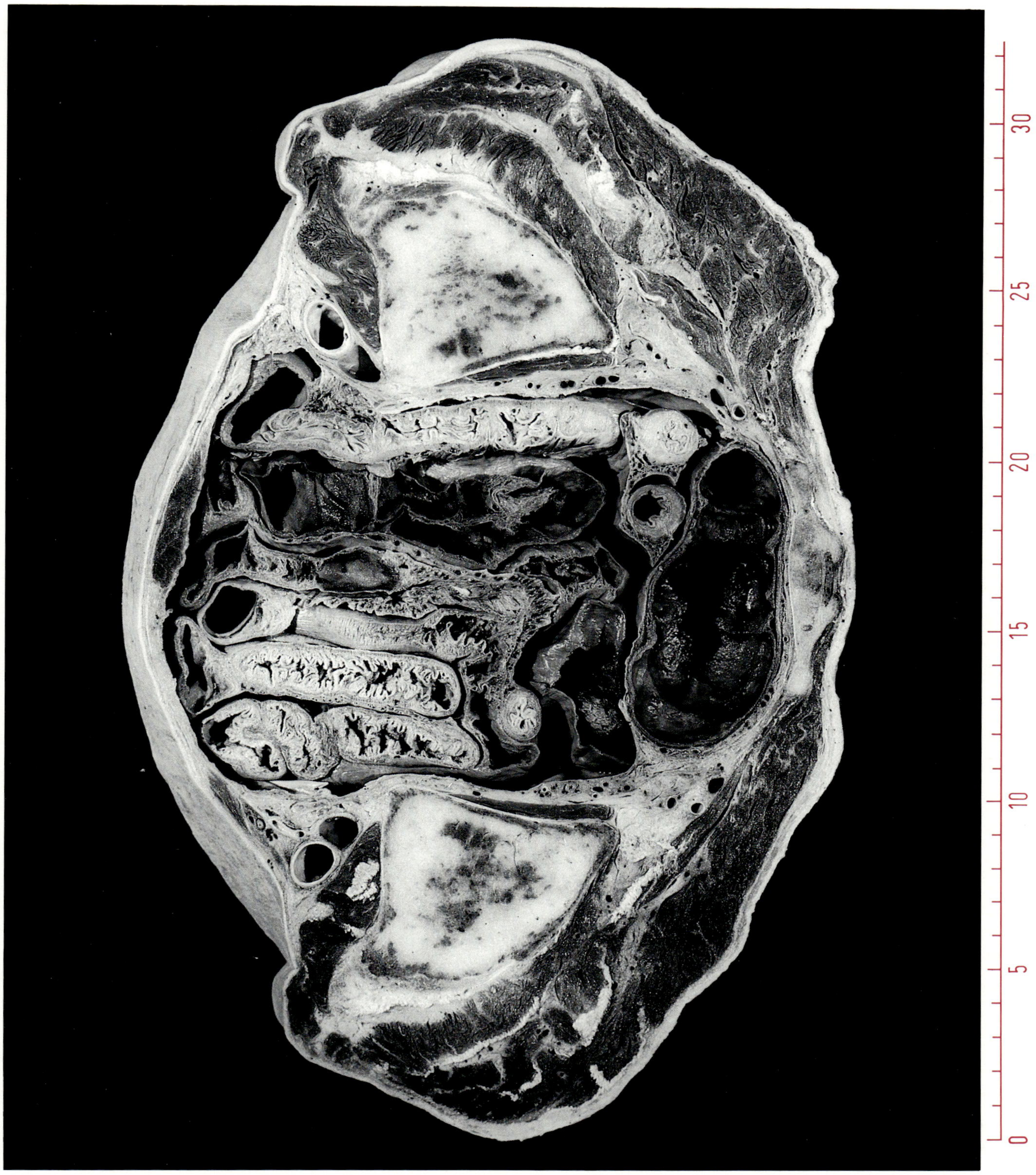

Fig. HA 12

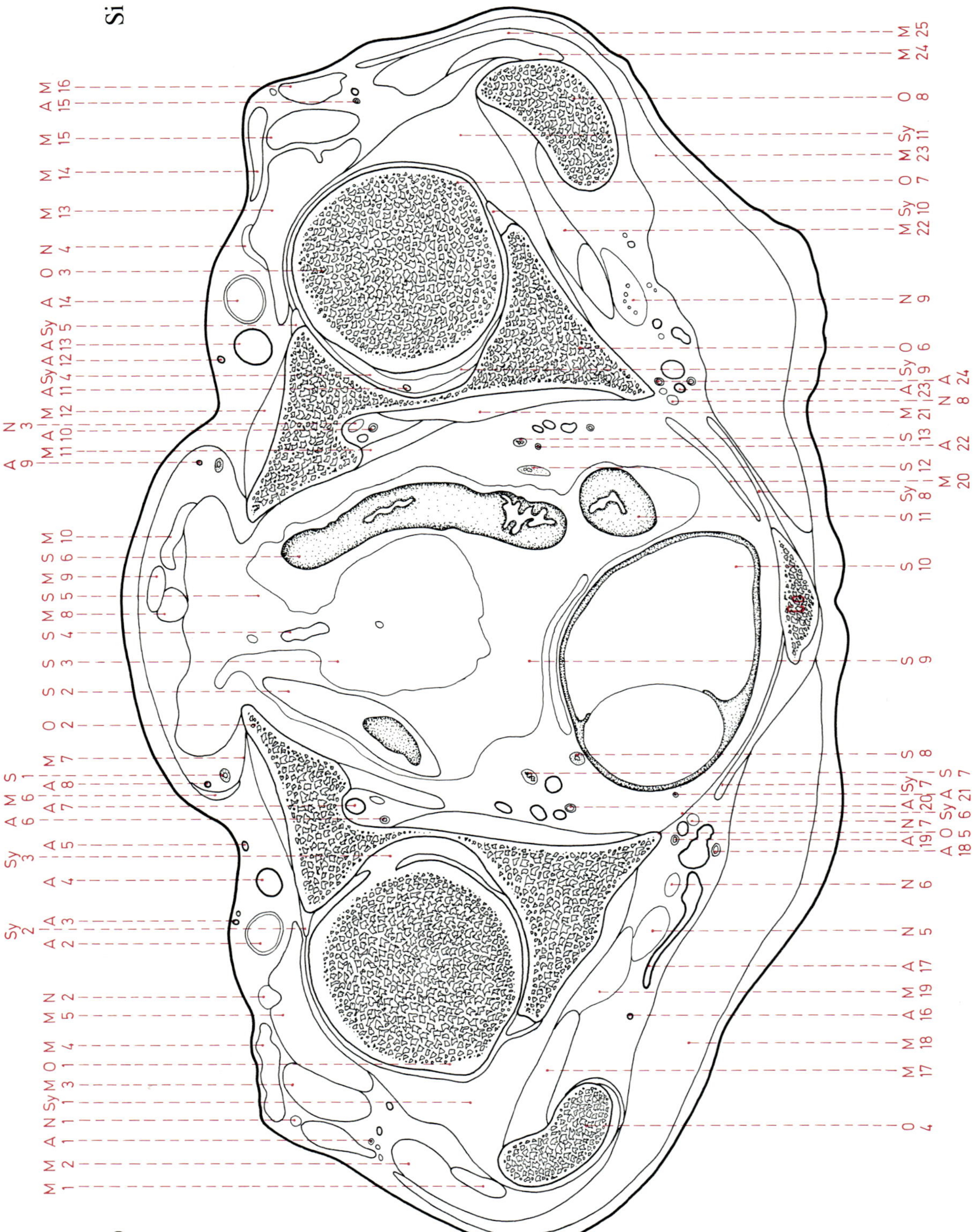

Fig. HA 13

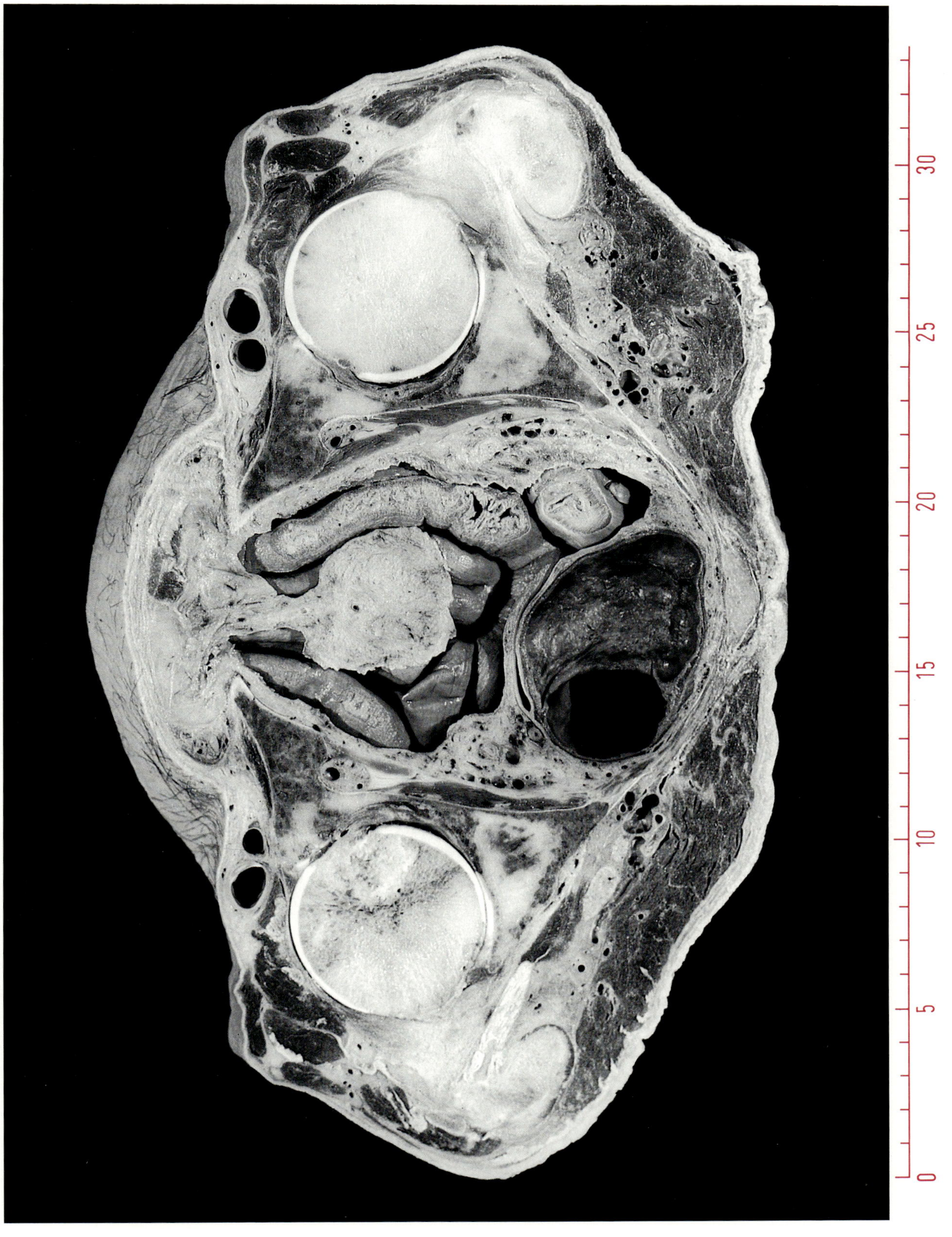

Fig. HA 13

HP

♂

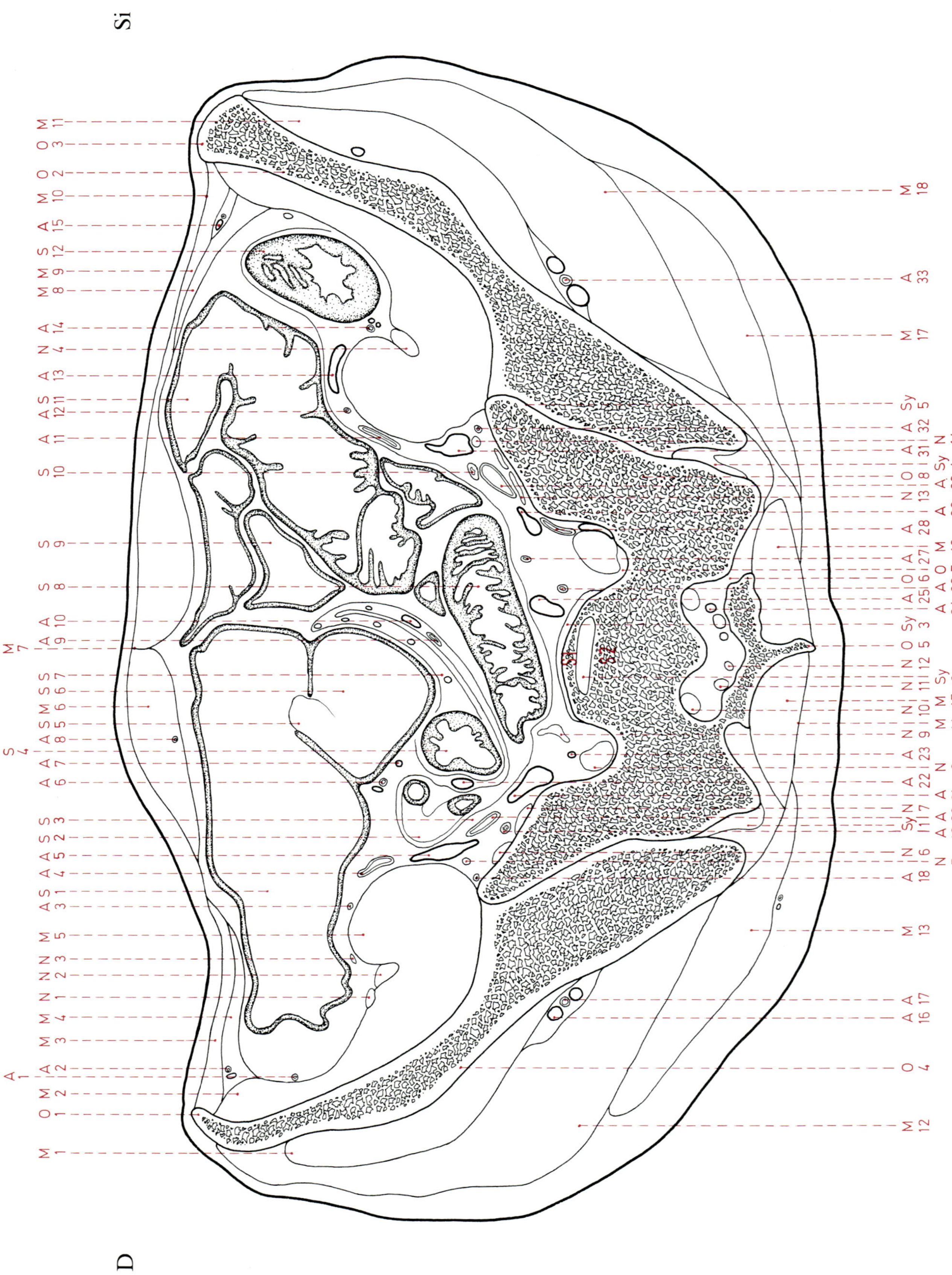

Fig. HP♂ 1

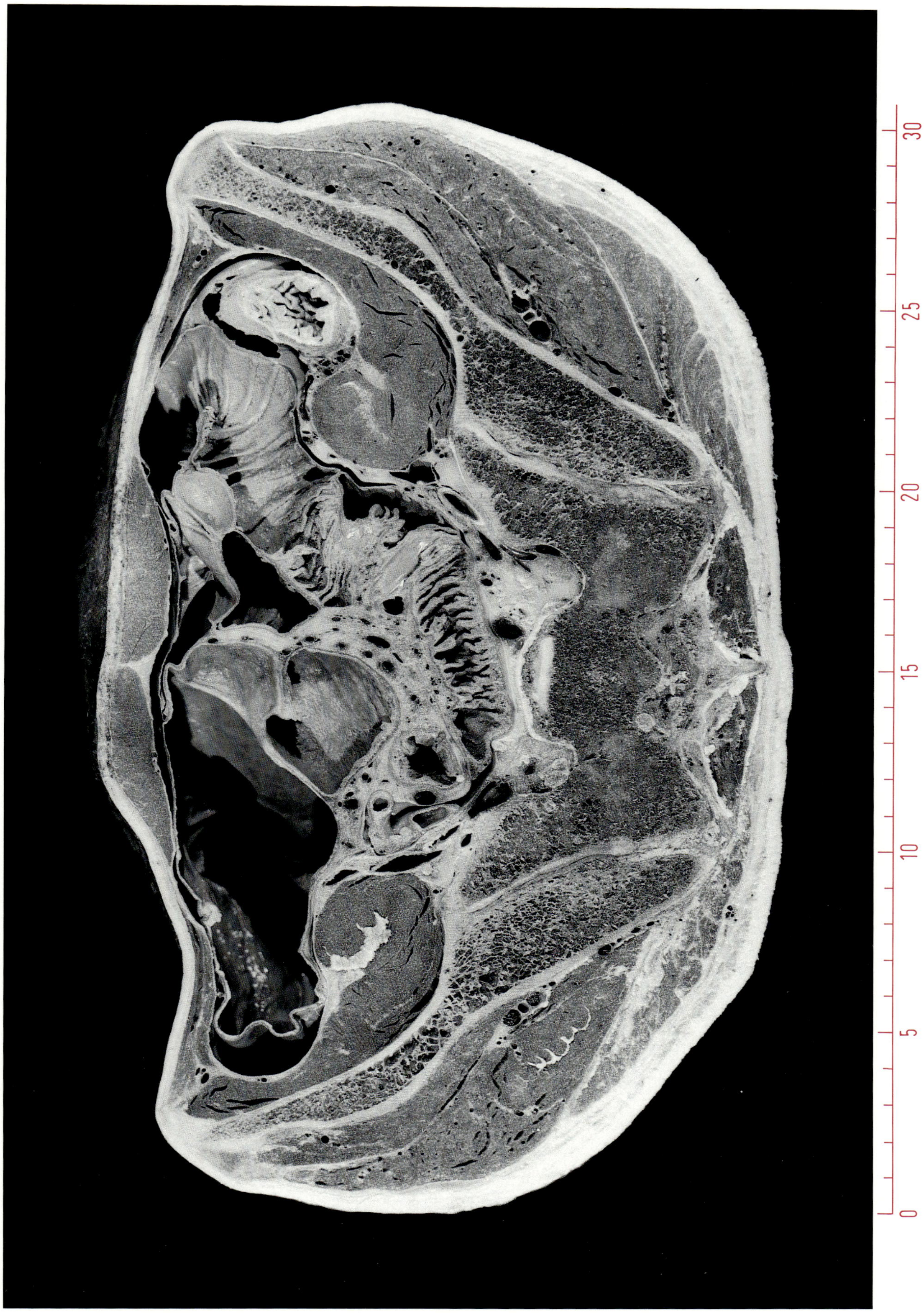

Fig. HP♂ 1

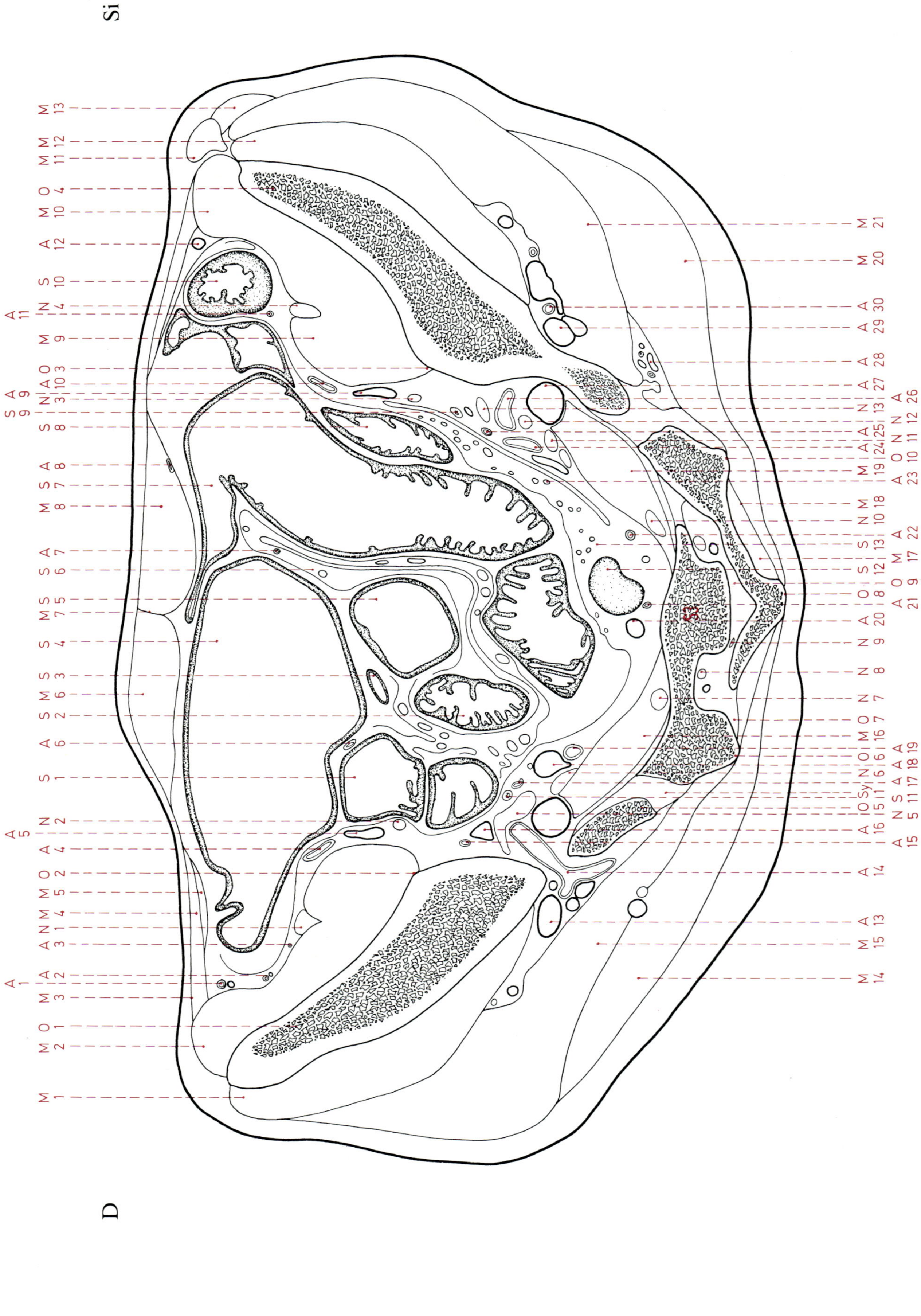

Fig. HP♂ 2

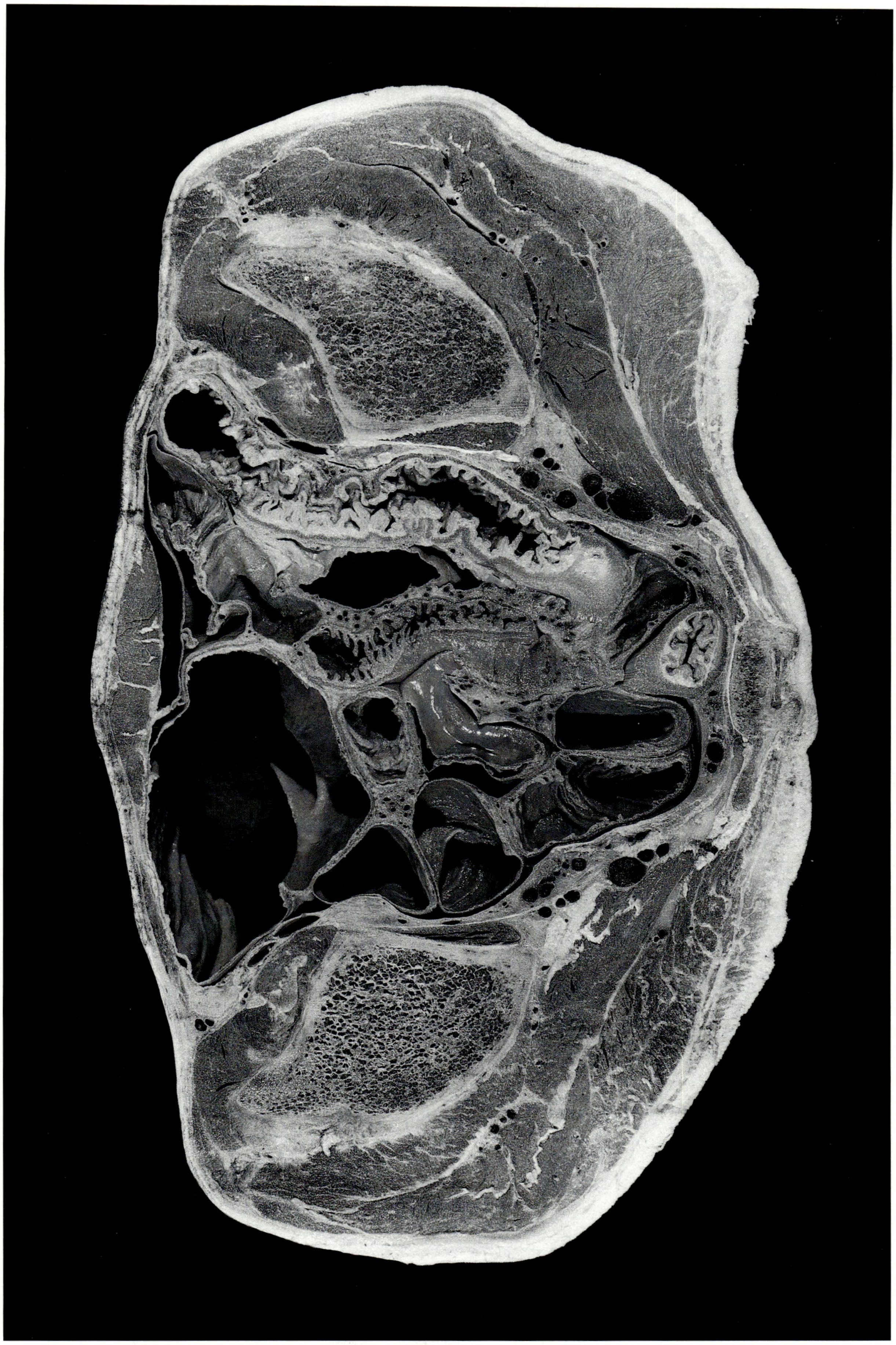

Fig. HP♂ 3

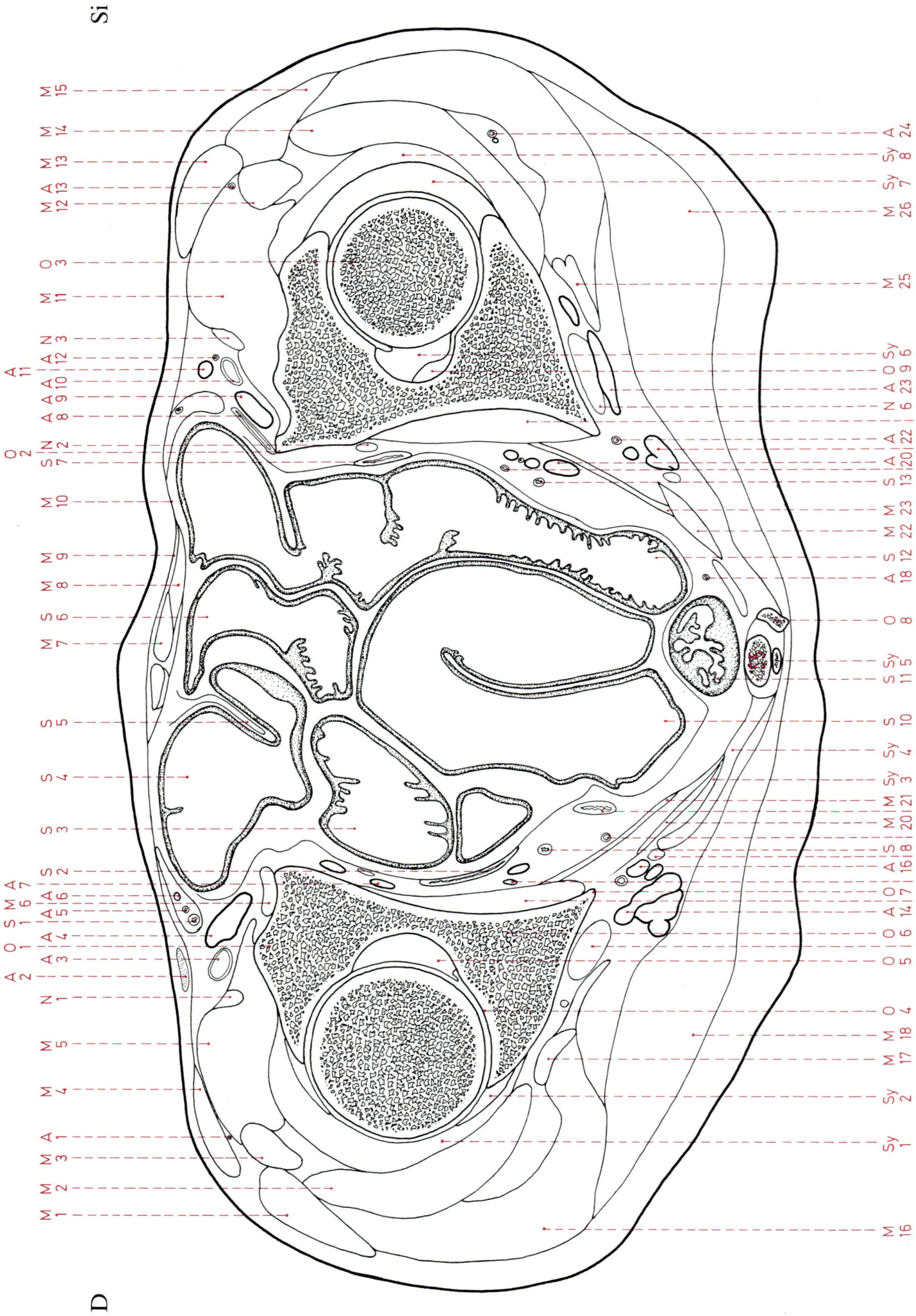

Fig. HP♂ 4

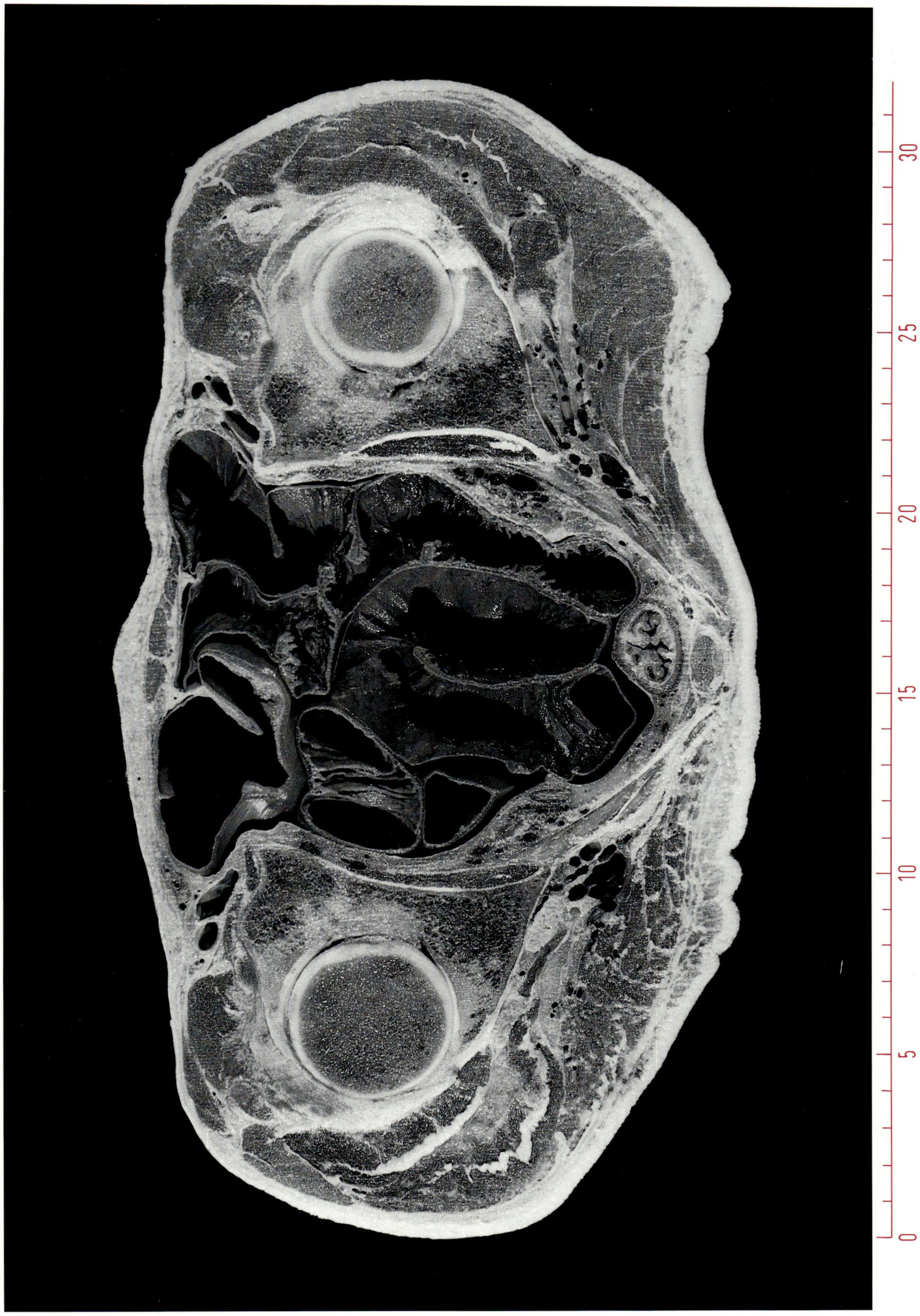

Fig. HP♂ 4

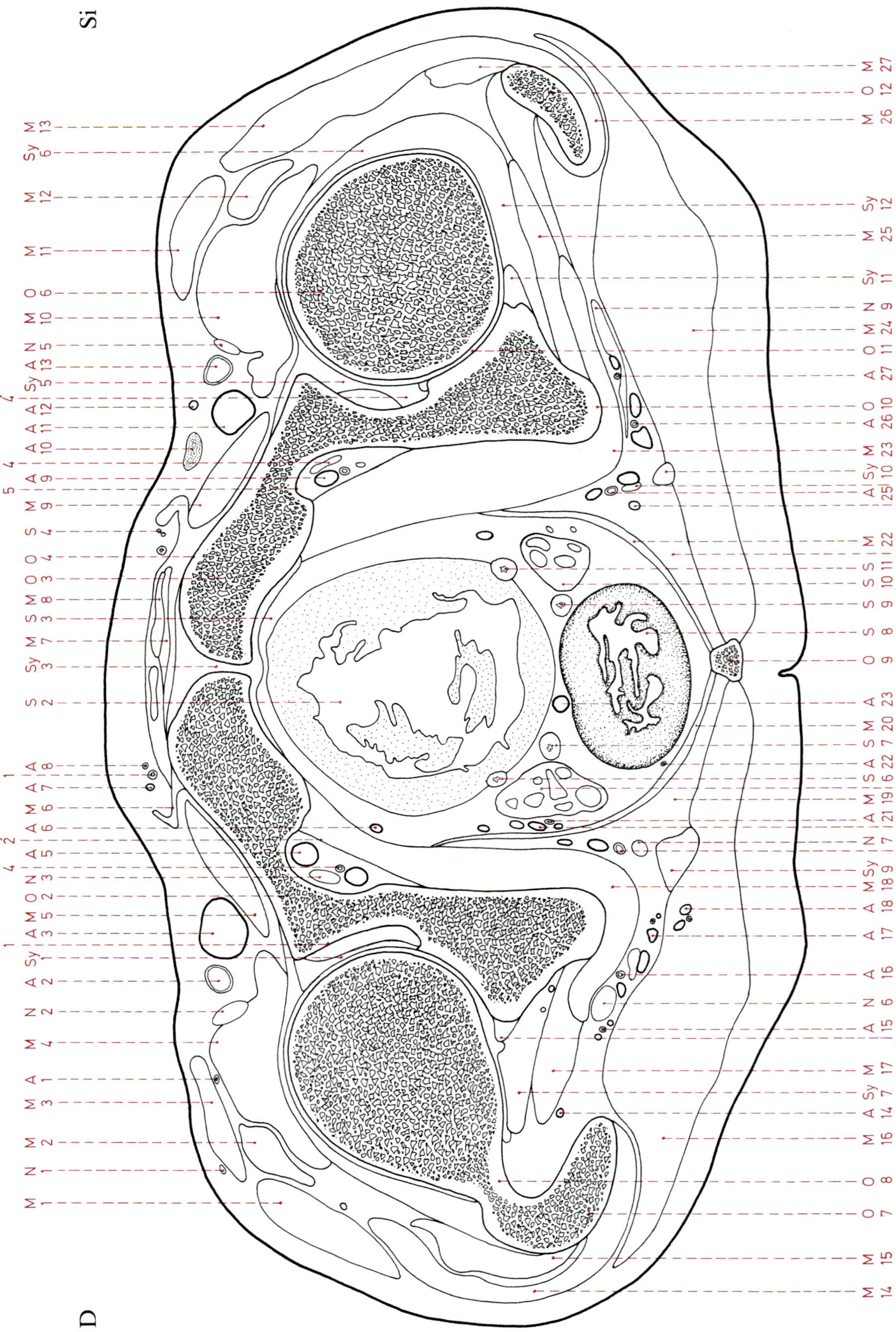

Fig. HP♂ 5

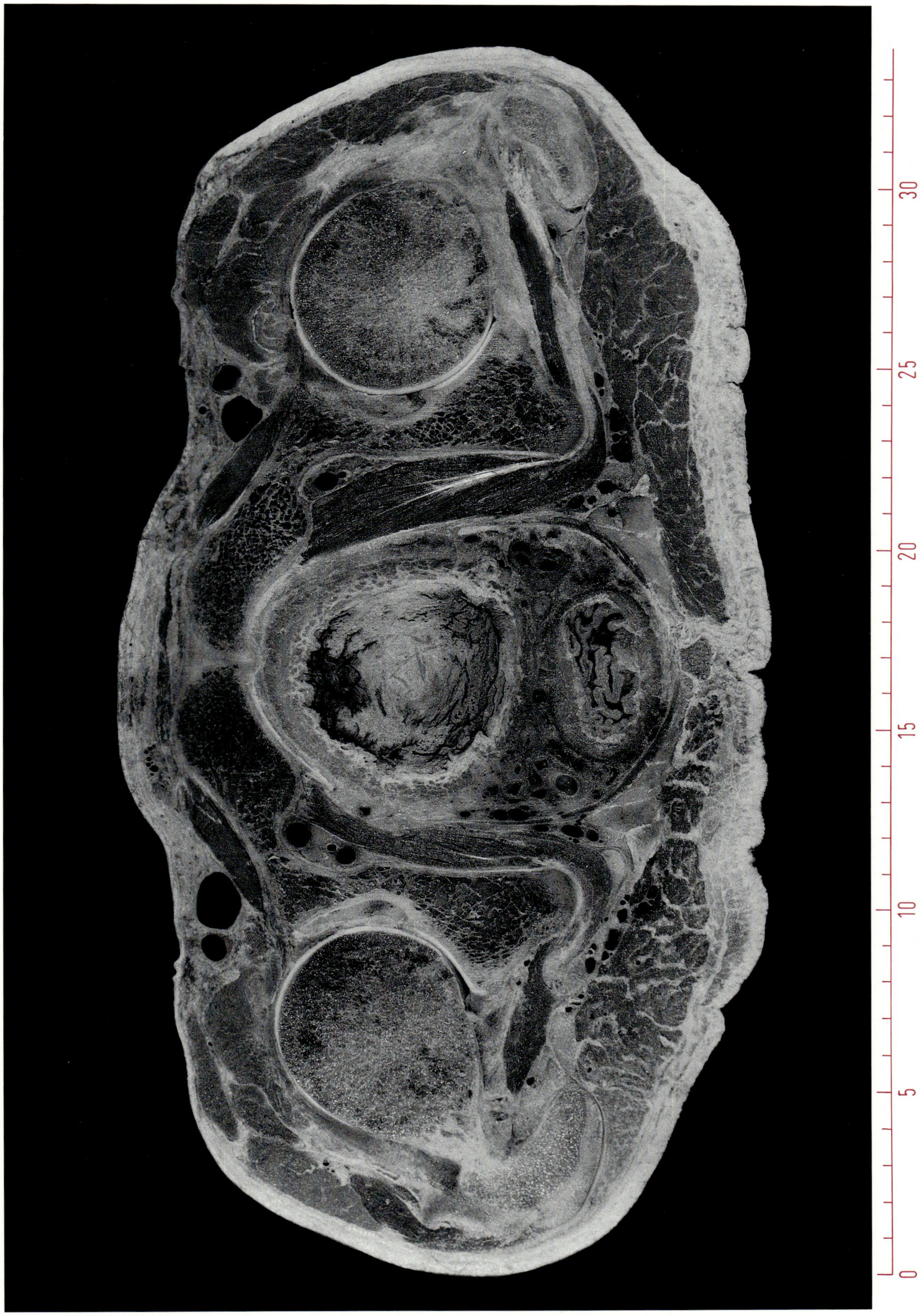

Fig. HP♂ 5

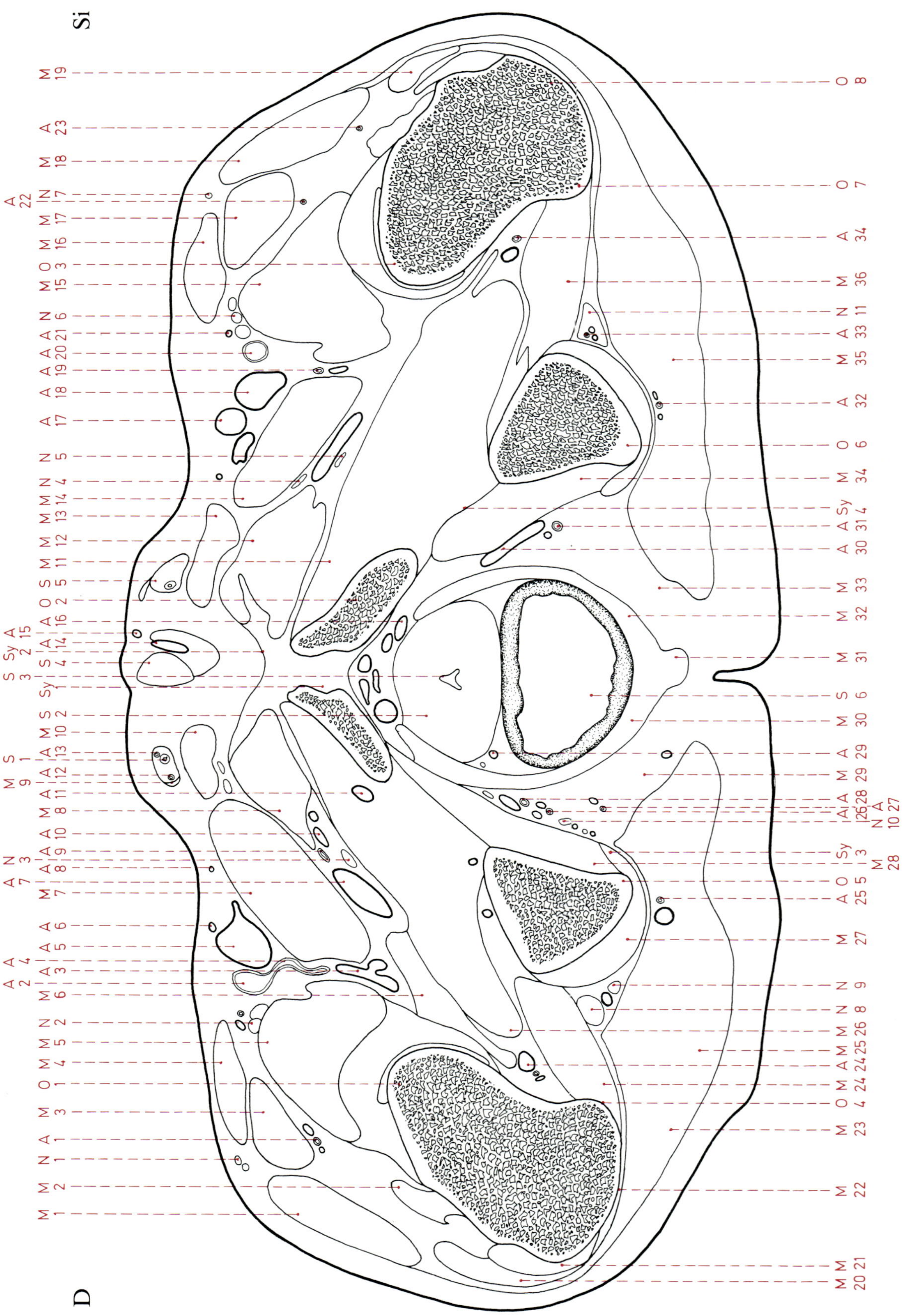

Fig. HP♂ 6

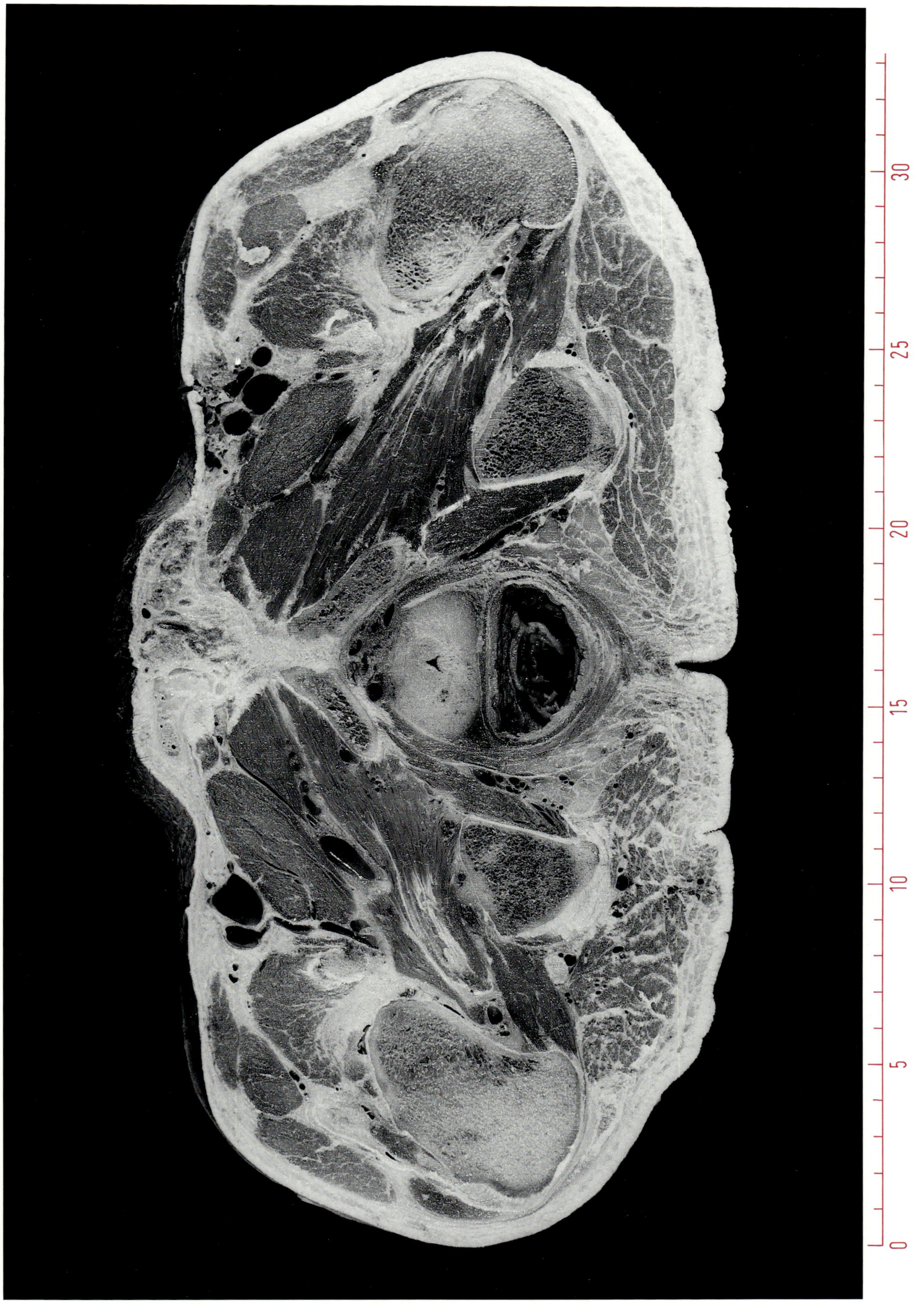

Fig. HP♂ 6

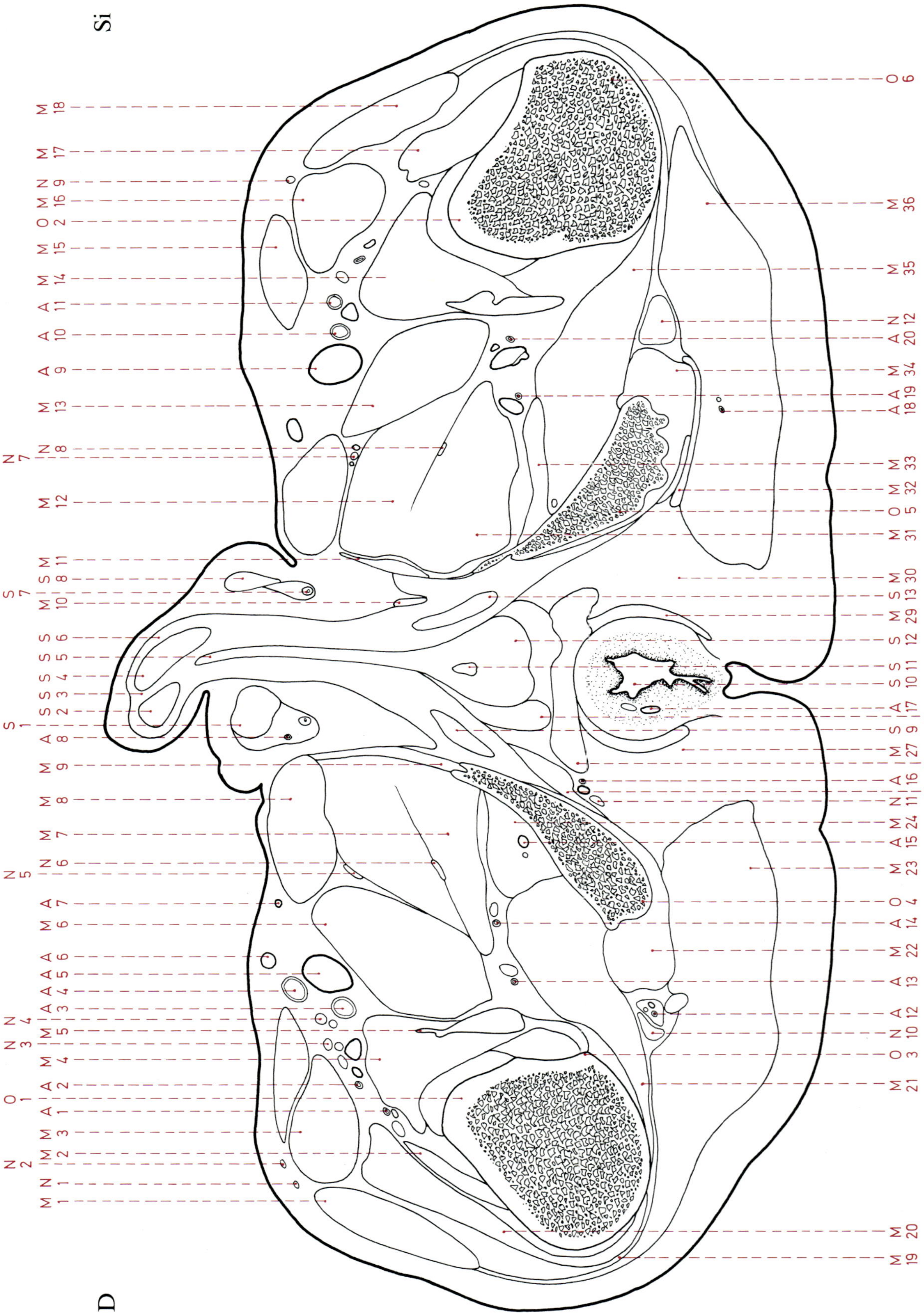
Si
D

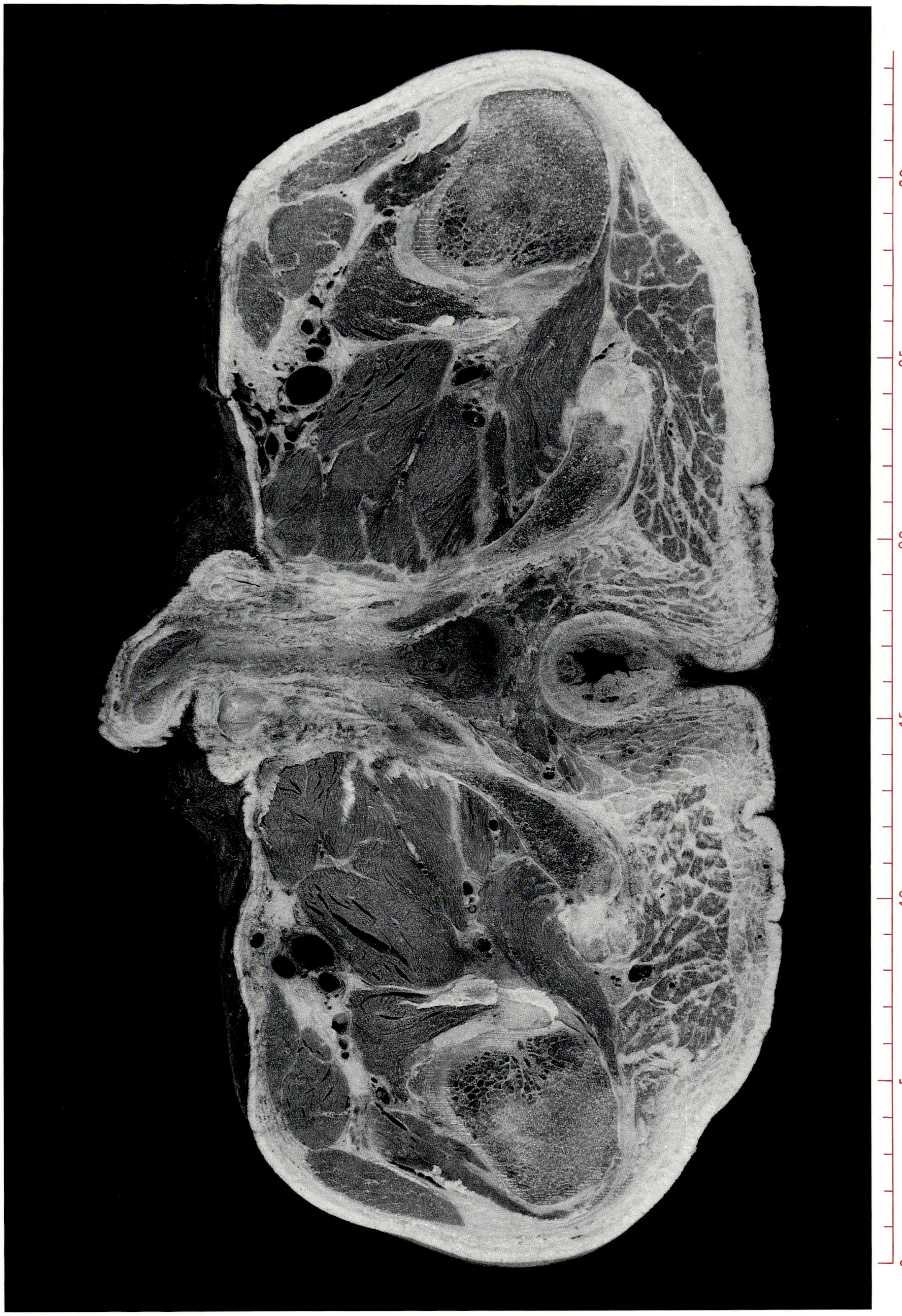

Fig. HP♂ 7

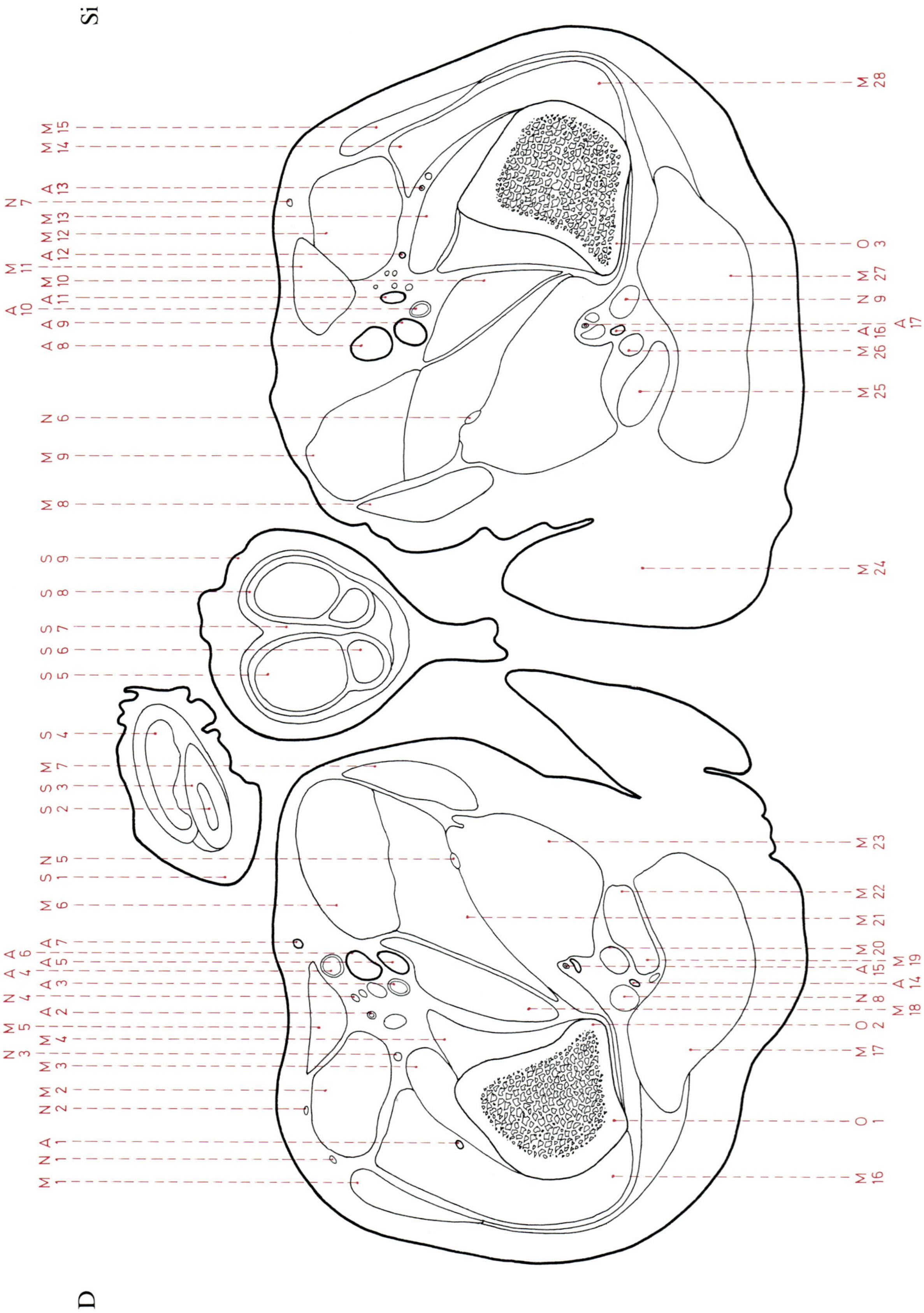

Fig. HP♂ 8

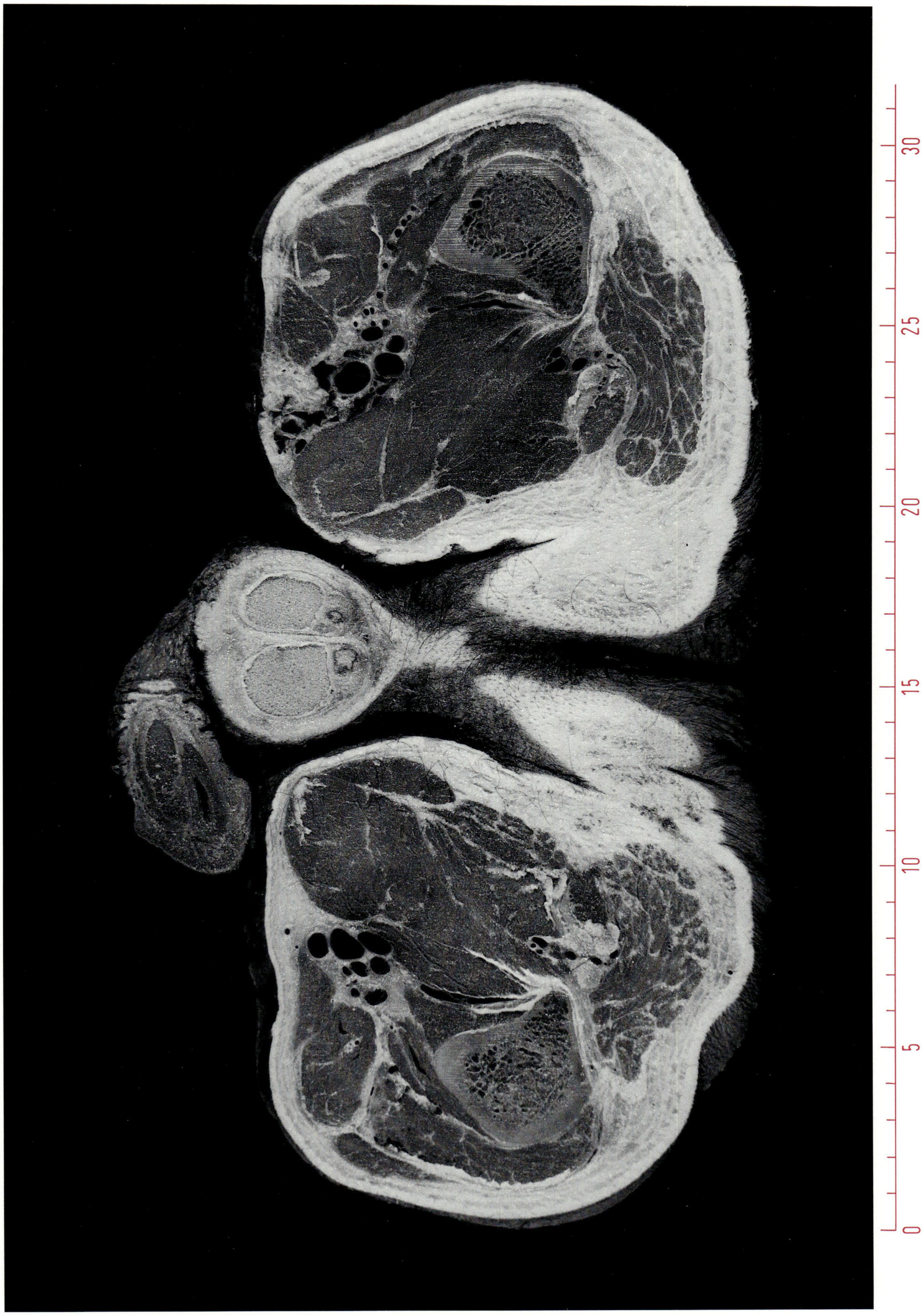

Fig. HP♂ 8

HP

♀

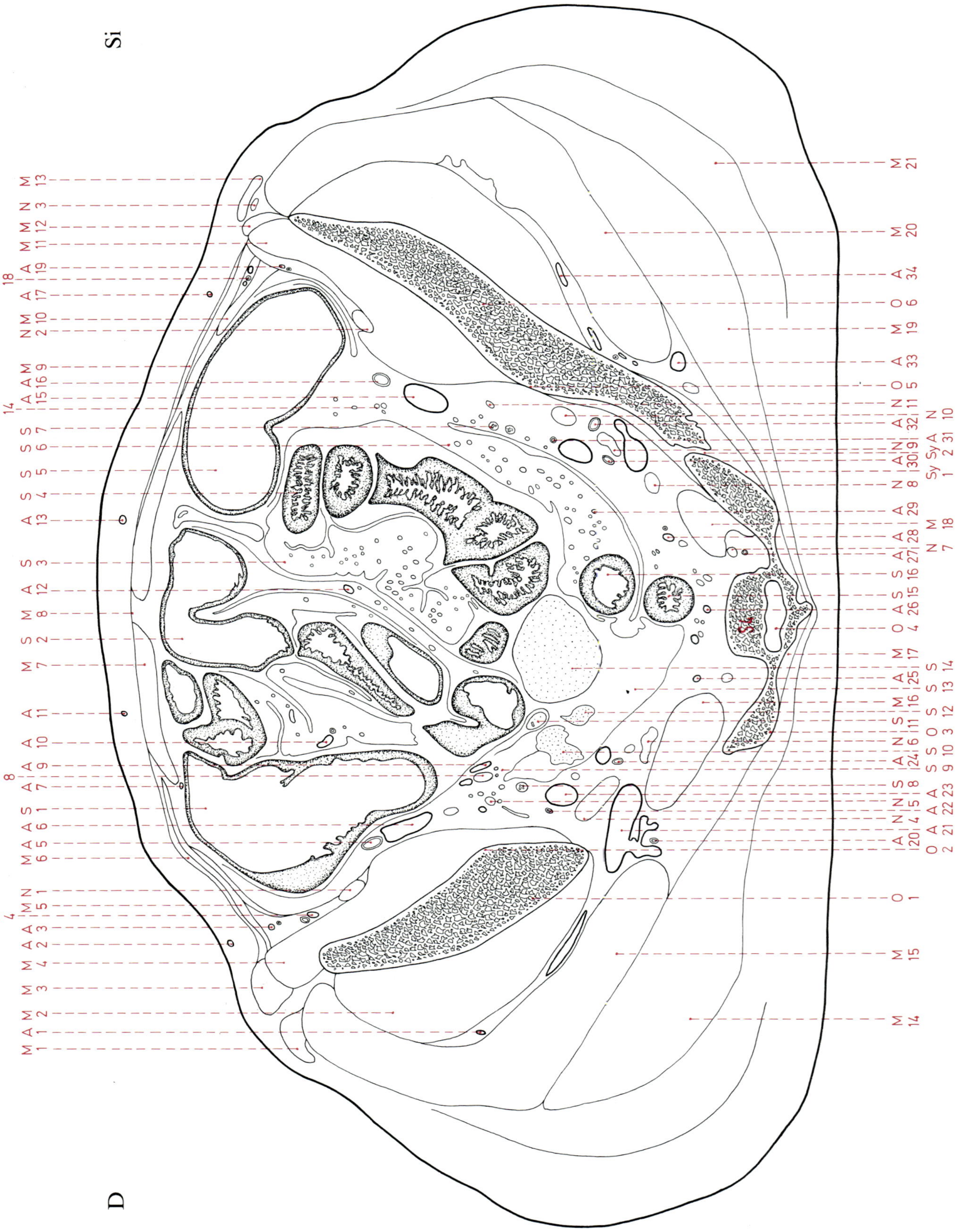

Fig. HP♀ 1

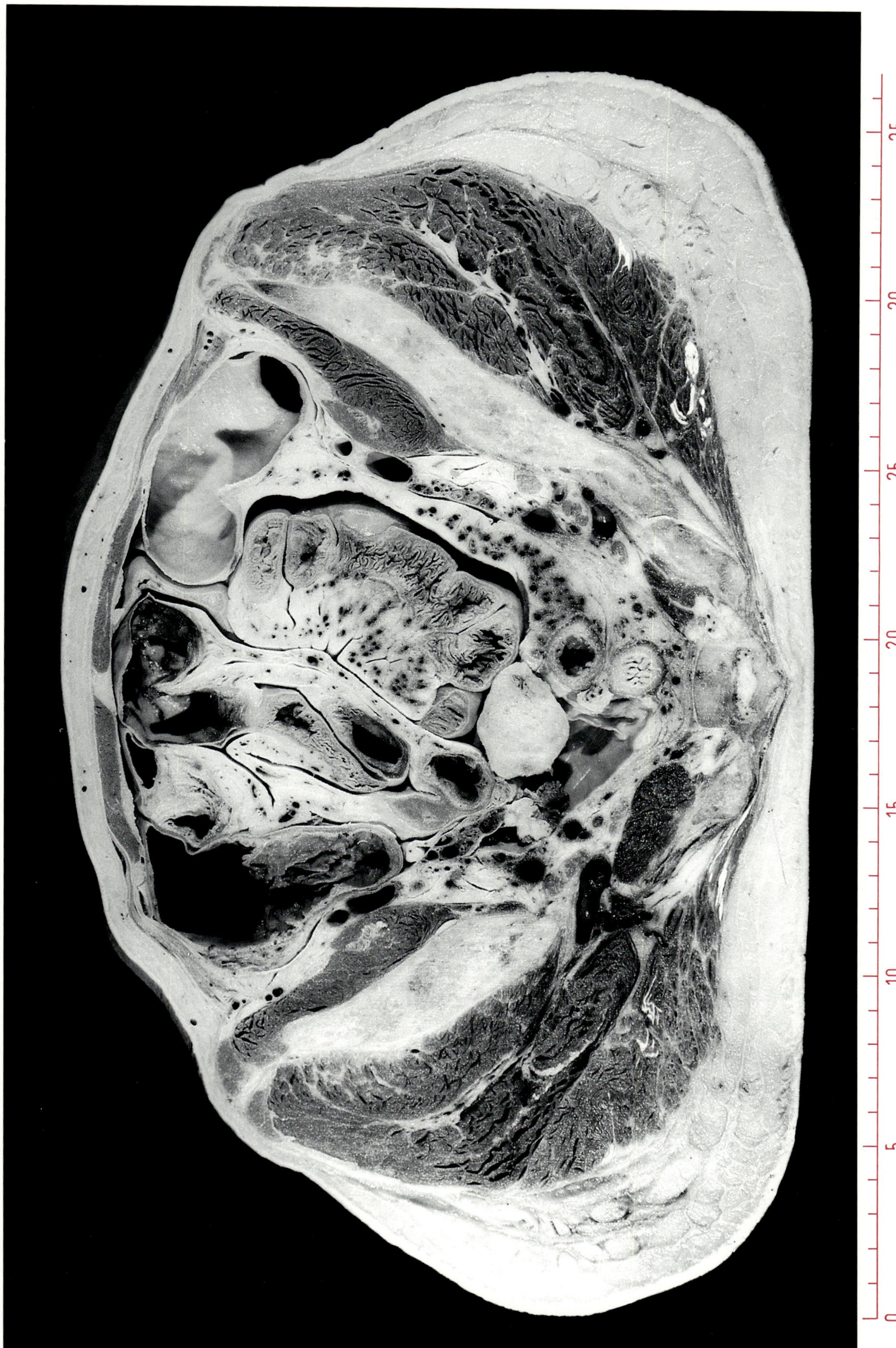

Fig. HP♀ 1

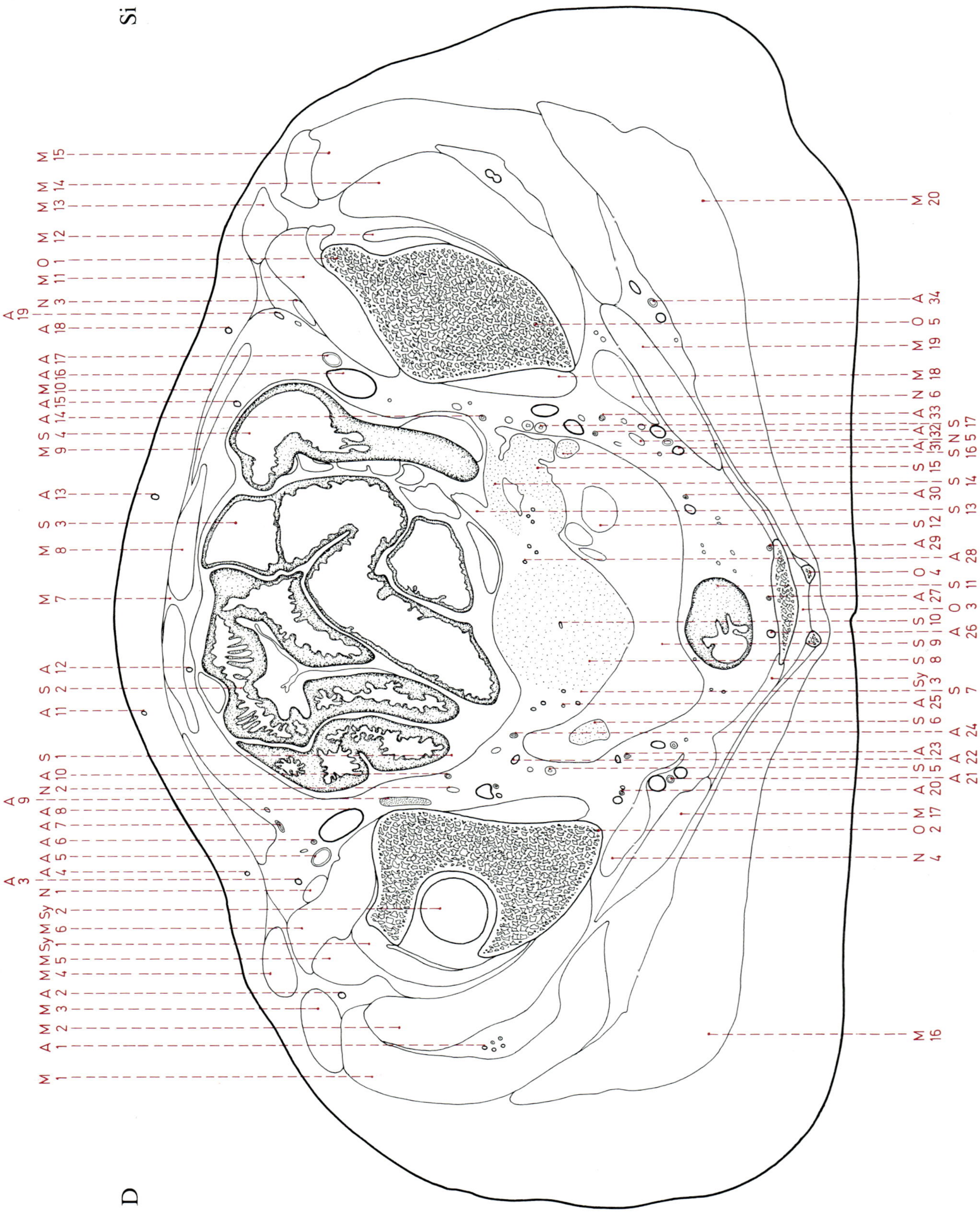

Fig. HP♀ 2

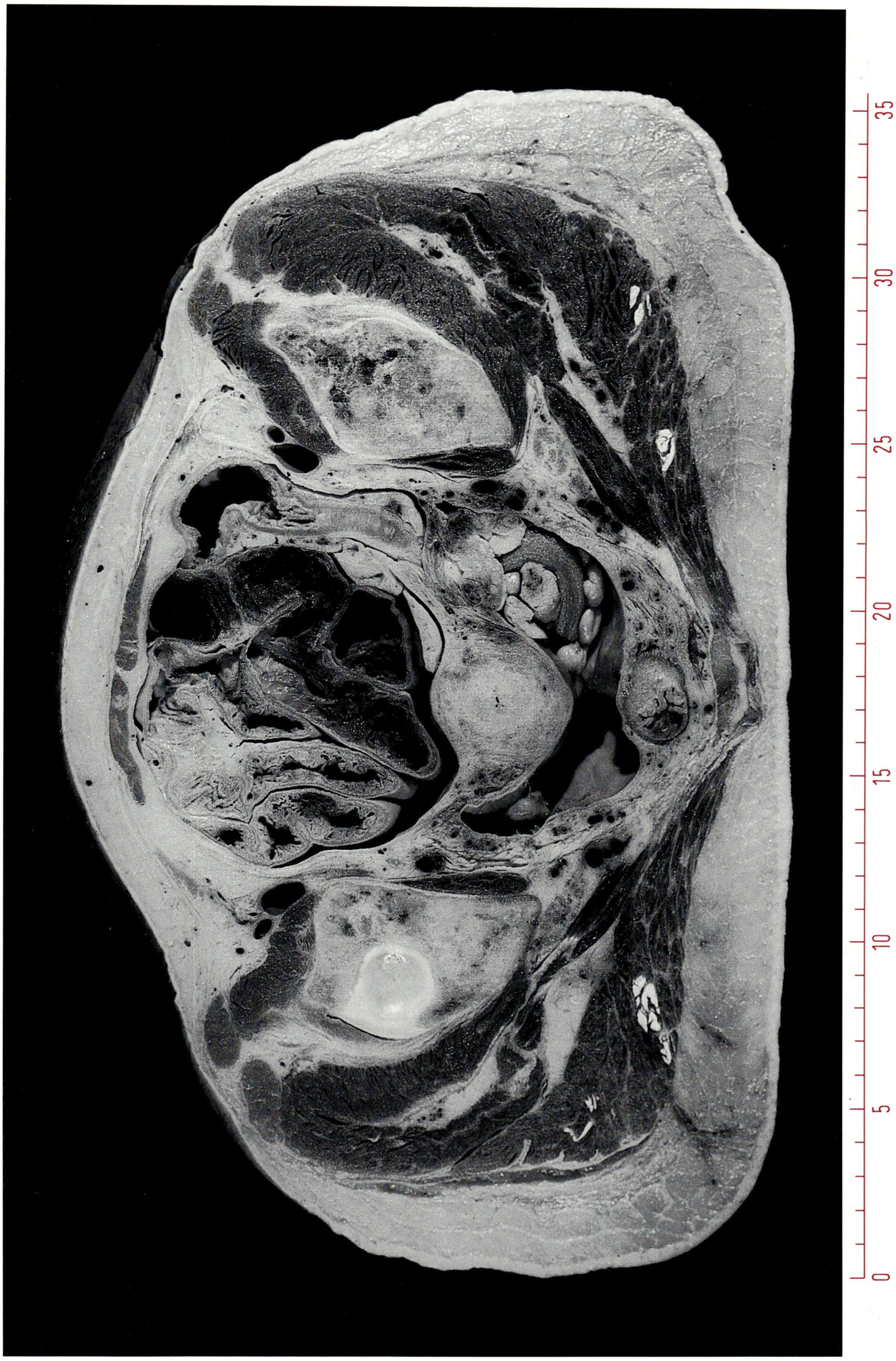

Fig. HP♀ 2

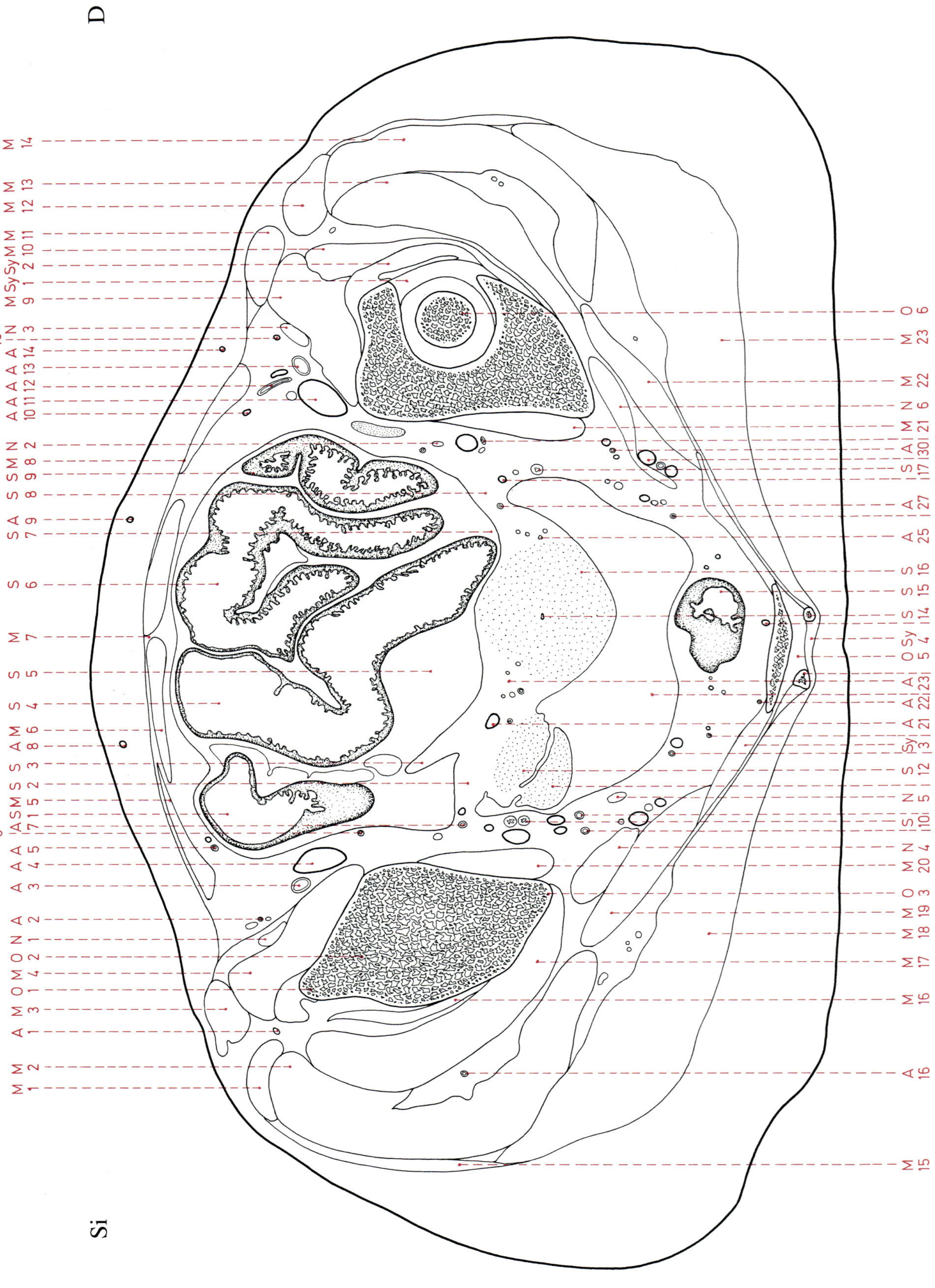

Fig. HP♀ 3

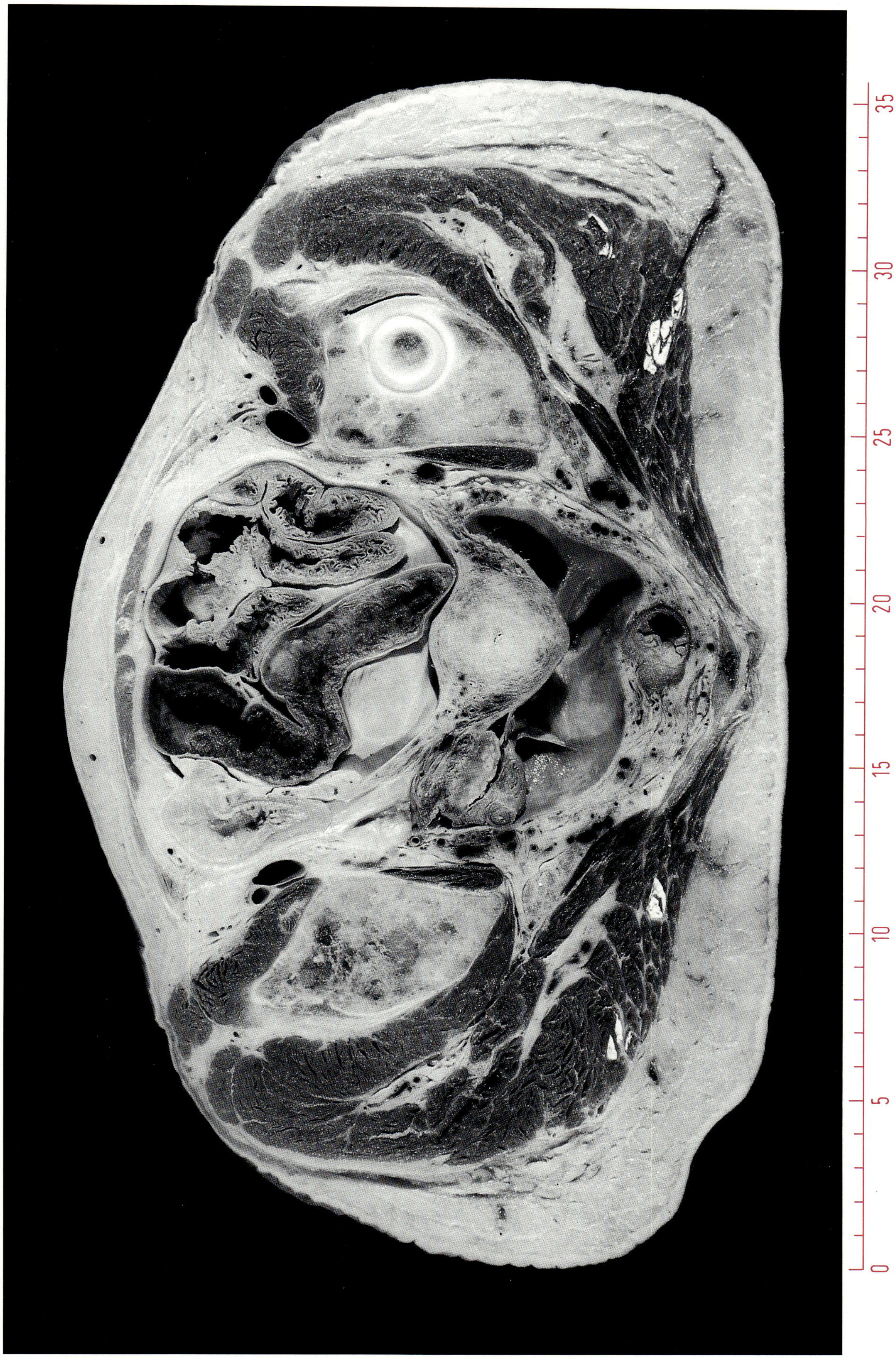

Fig. HP♀ 3

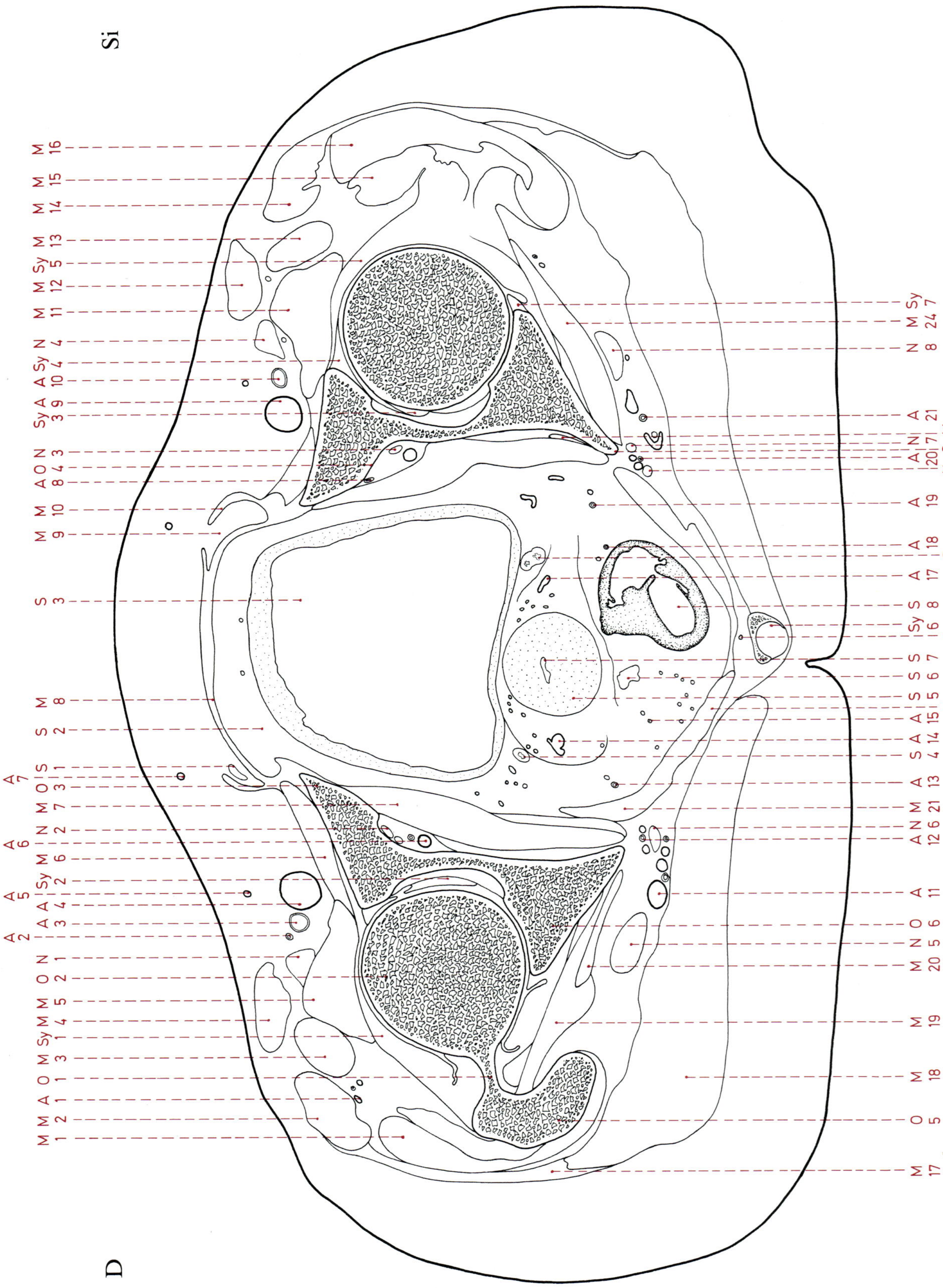

Fig. HP♀ 4

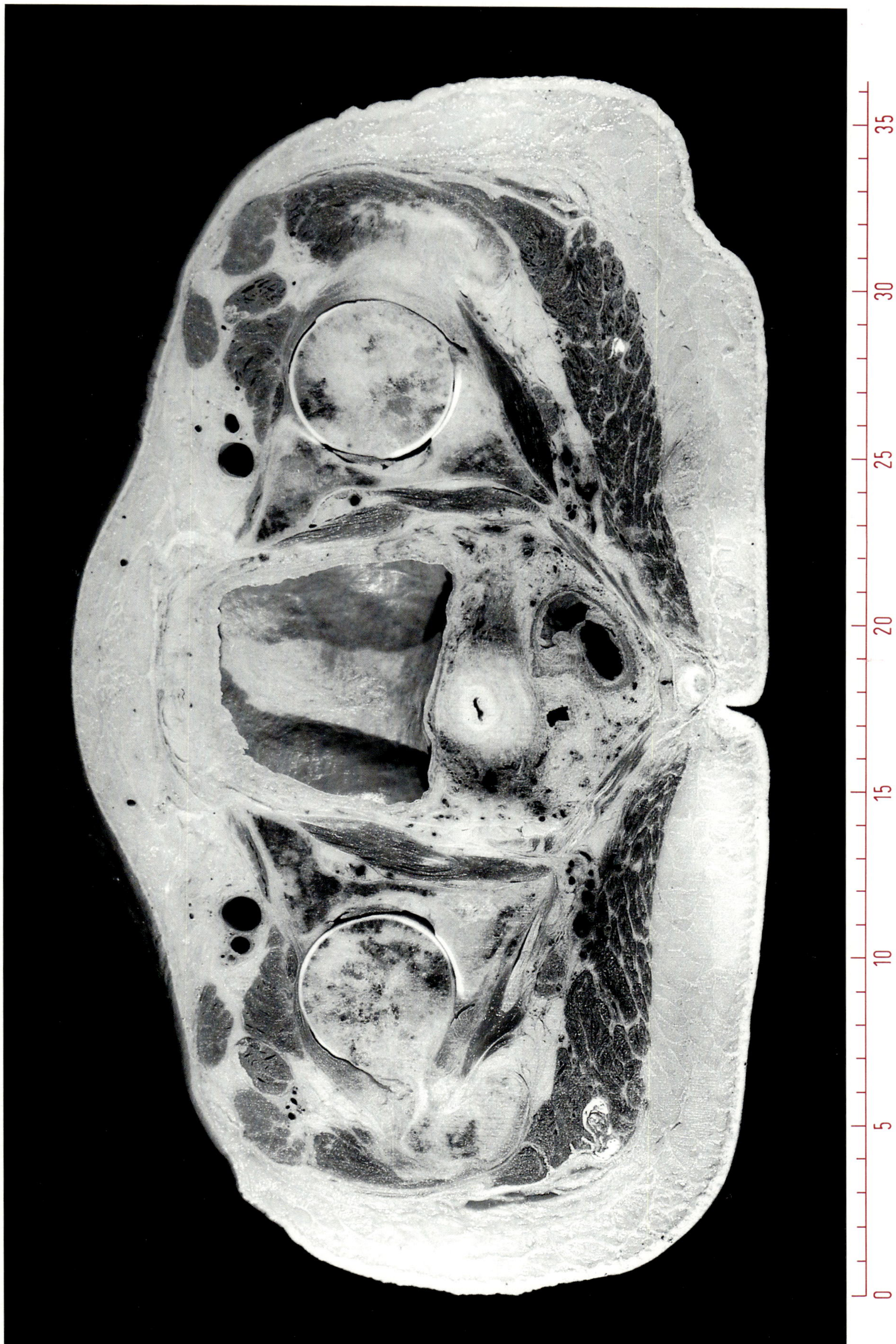

Fig. HP♀ 4

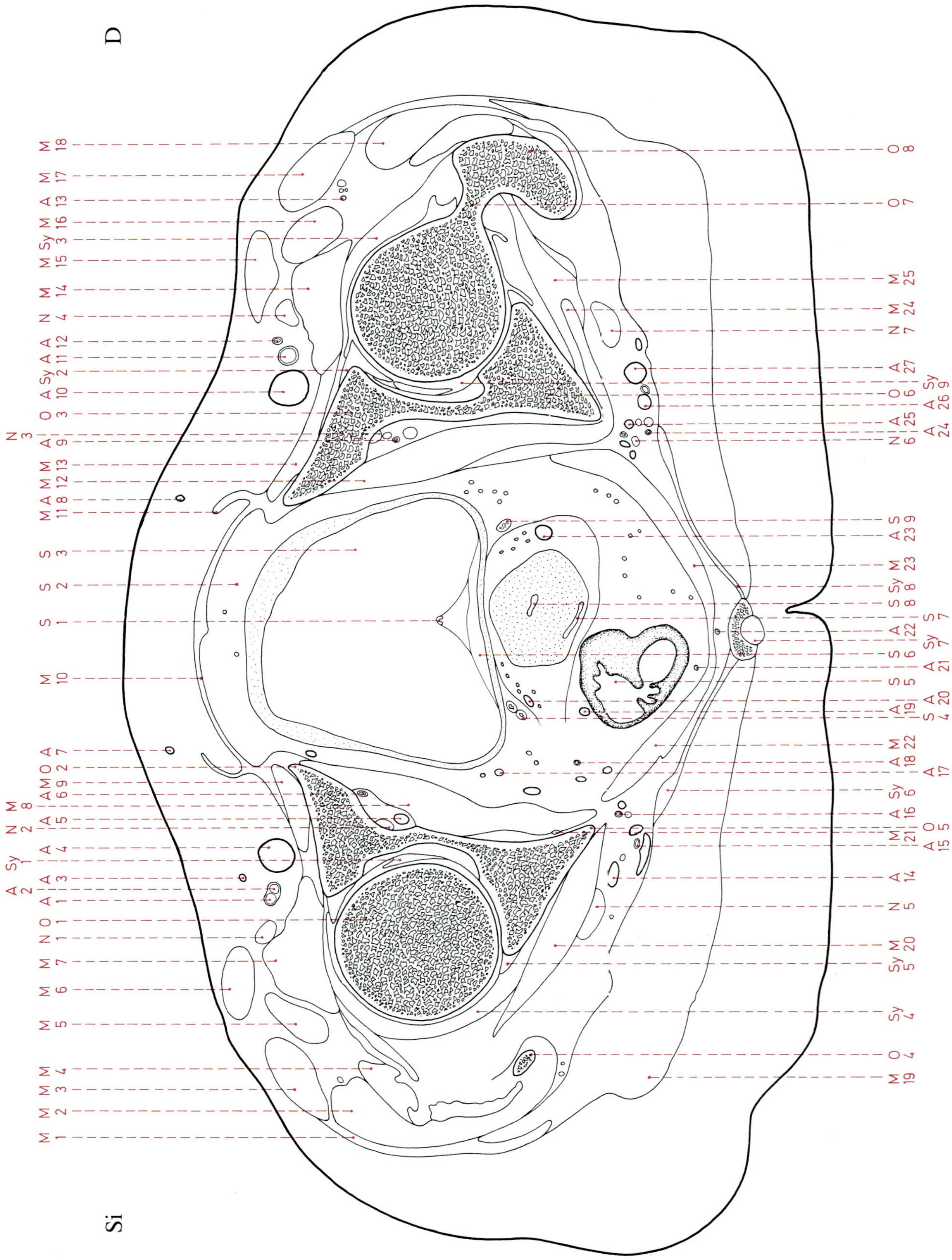

Fig. HP♀ 5

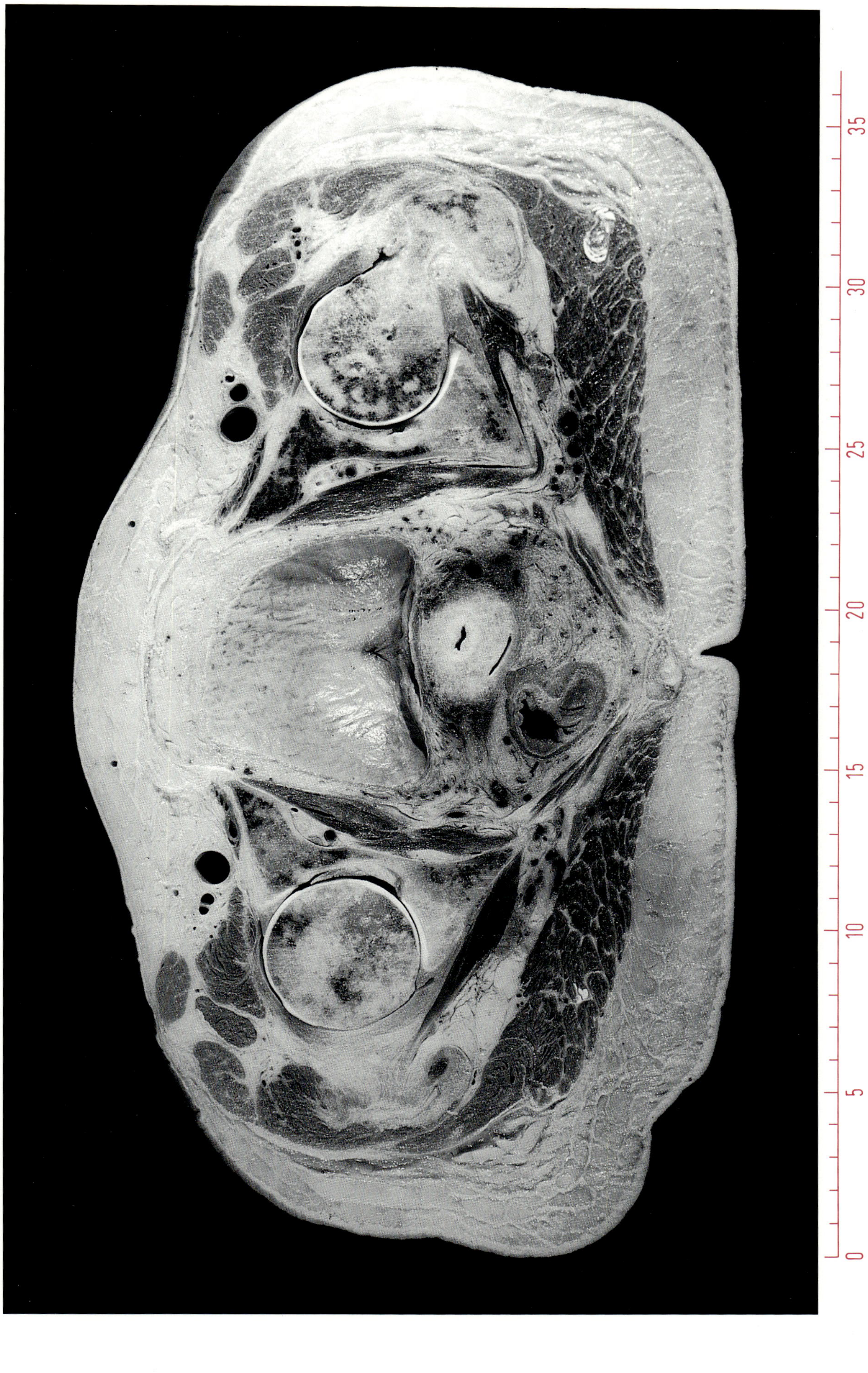

Fig. HP♀ 5

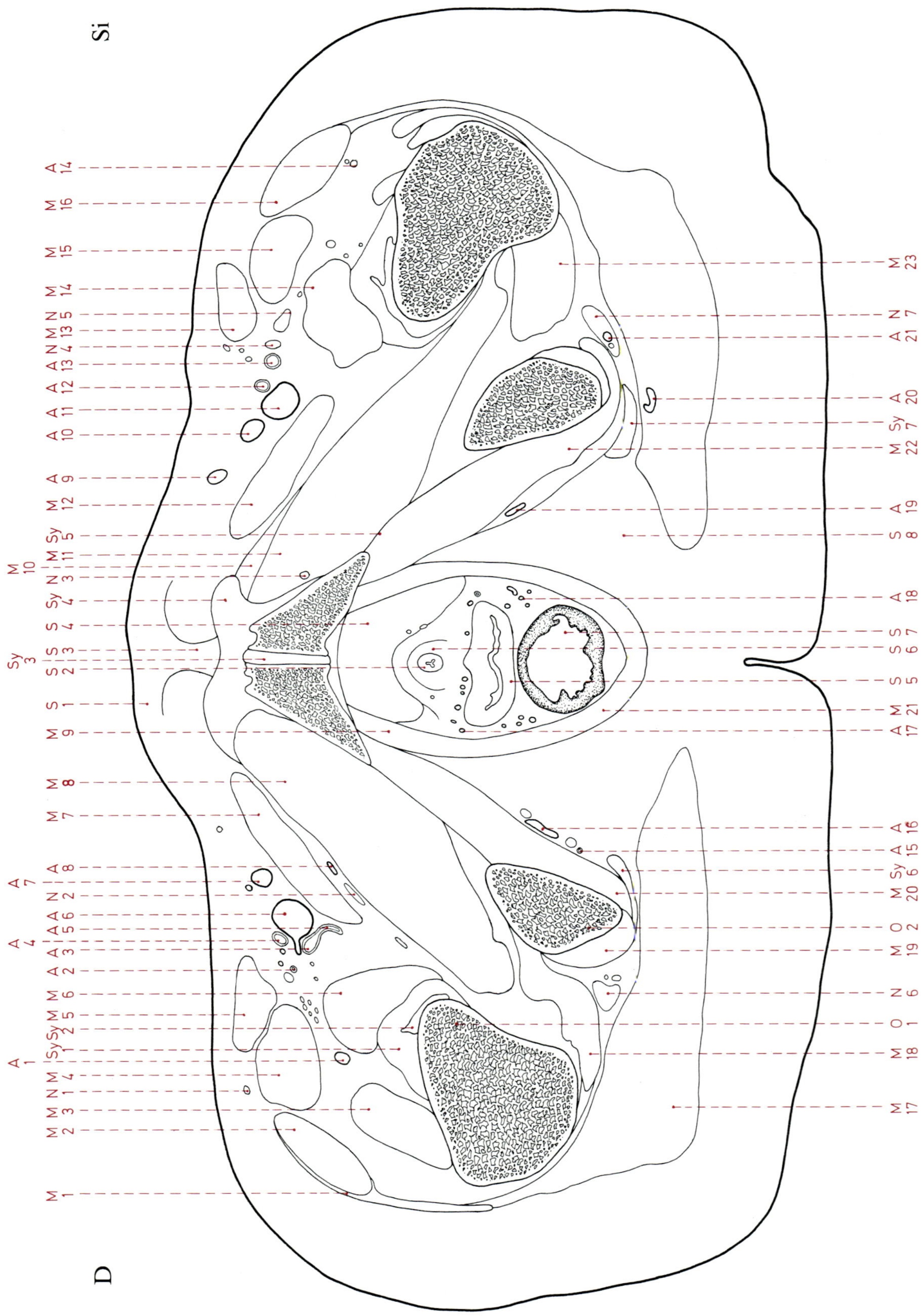

Fig. HP♀ 6

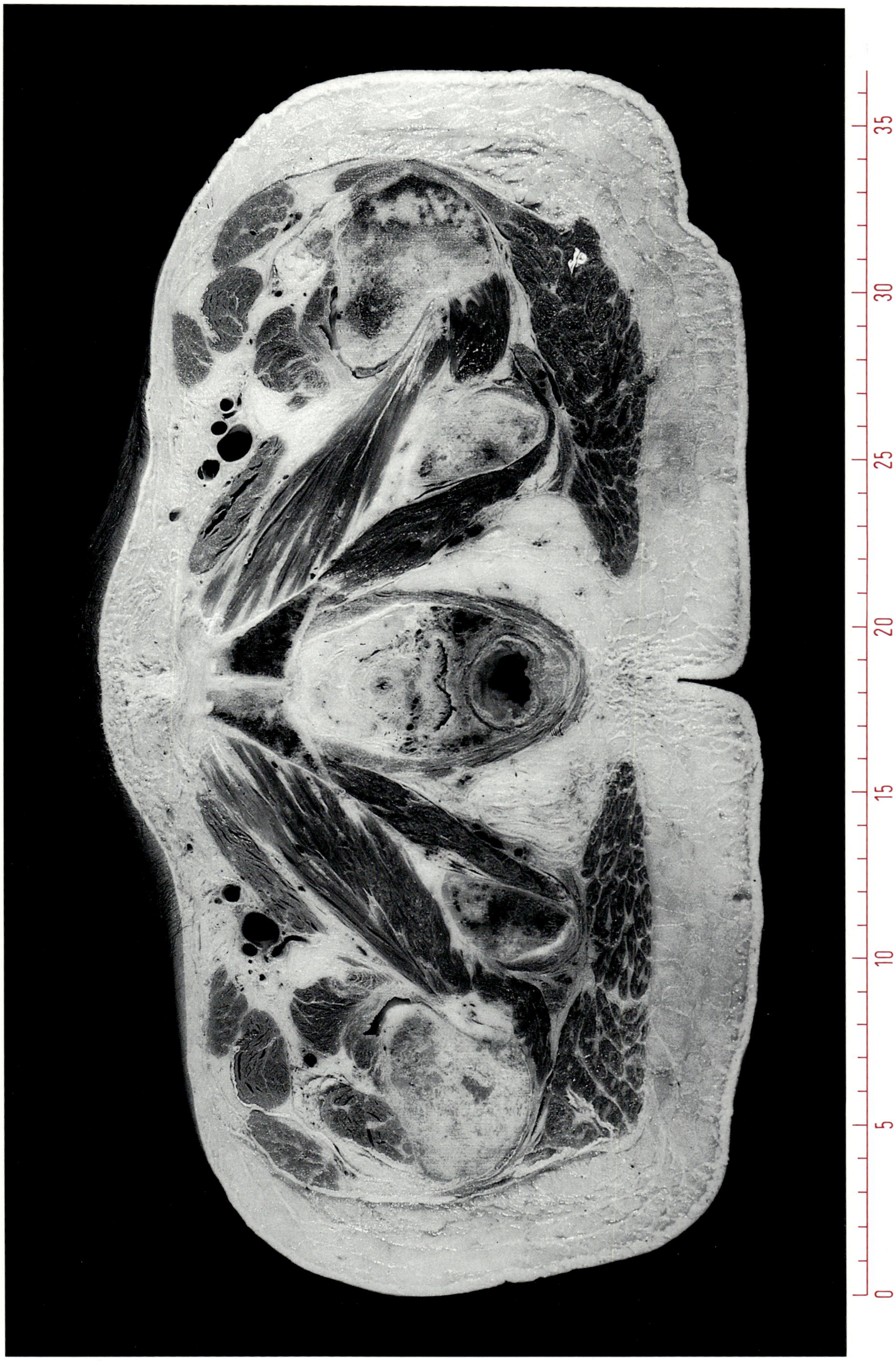

Fig. HP♀ 6

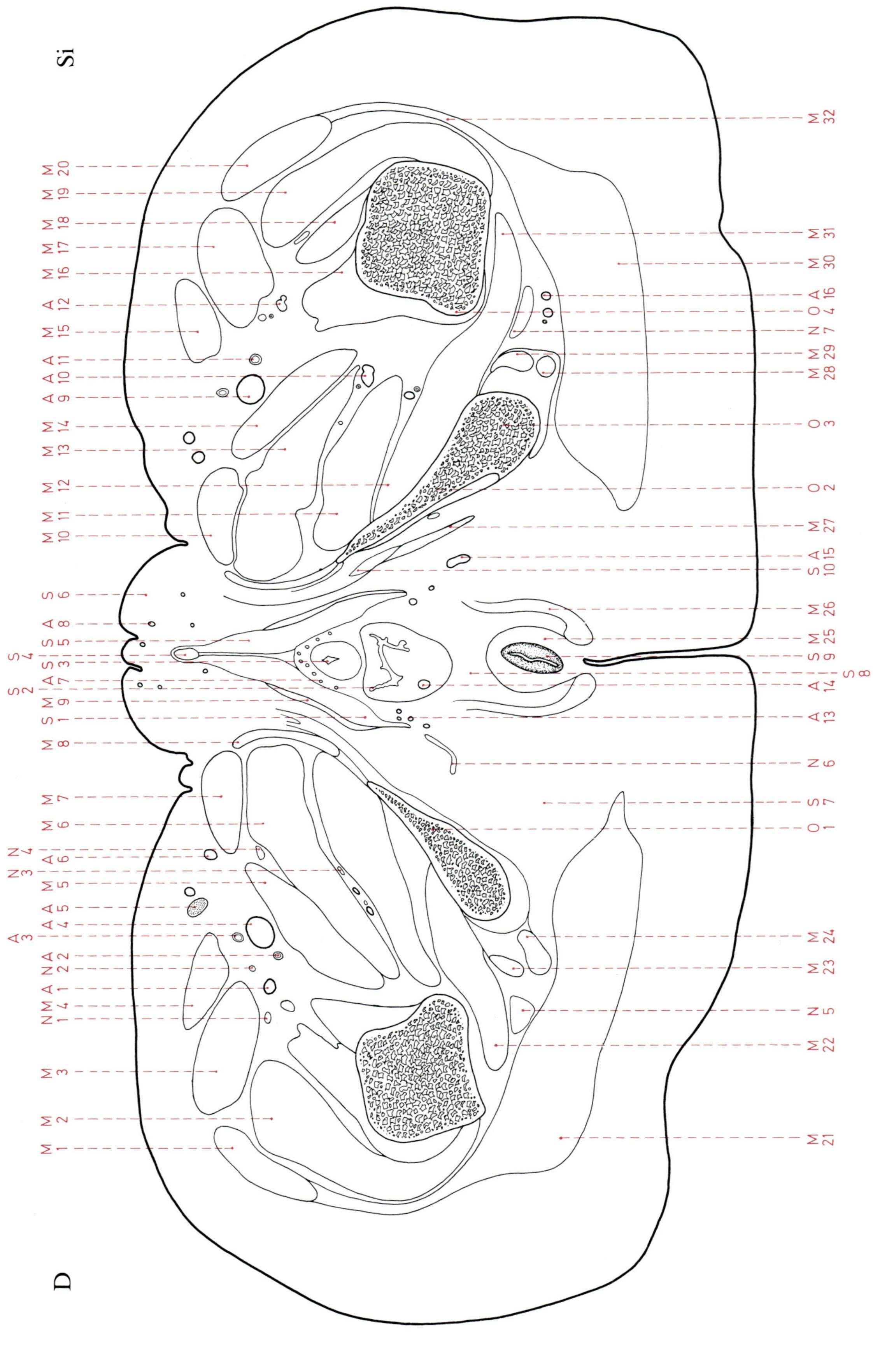

Fig. HP♀ 7

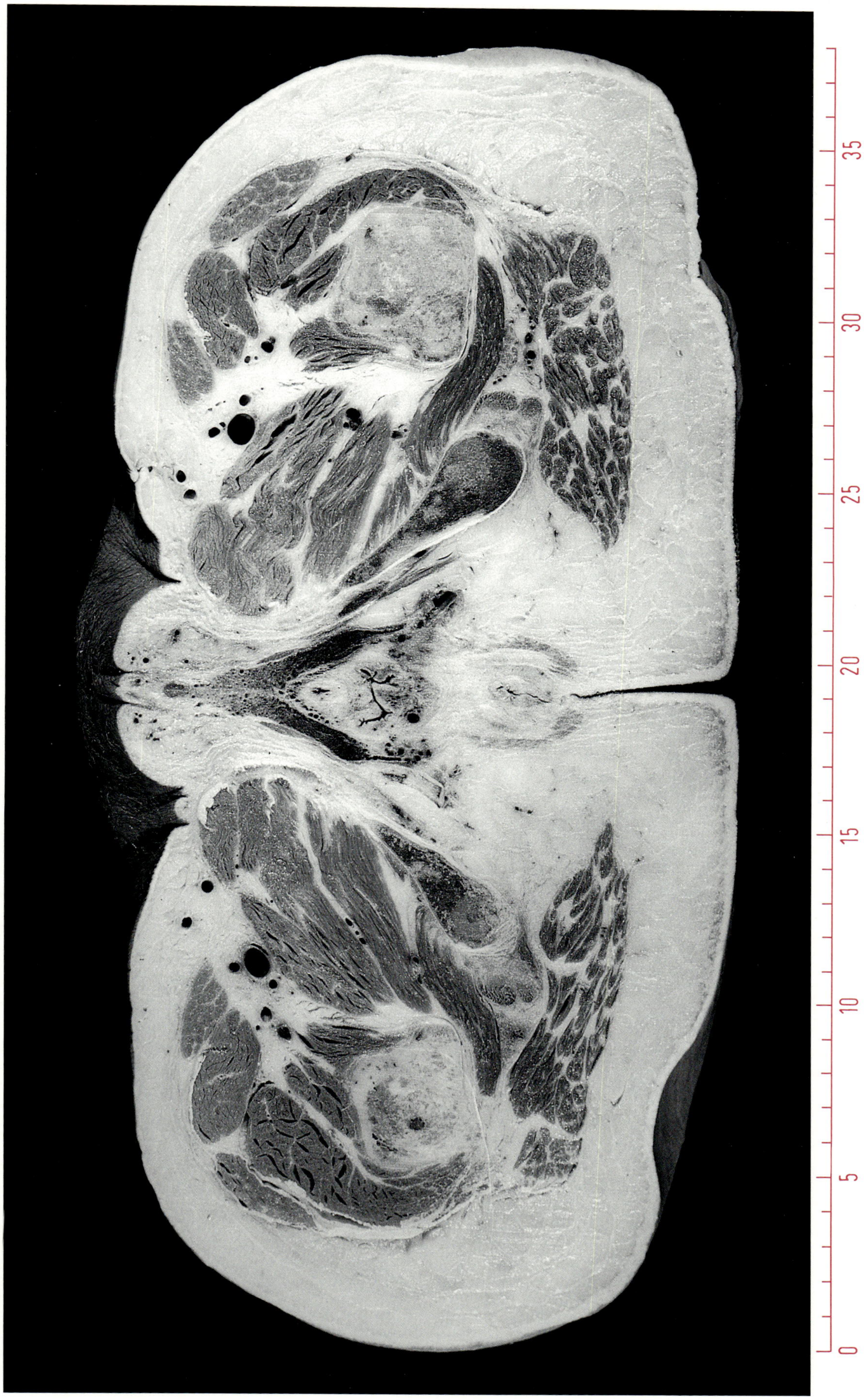

Fig. HP♀ 7

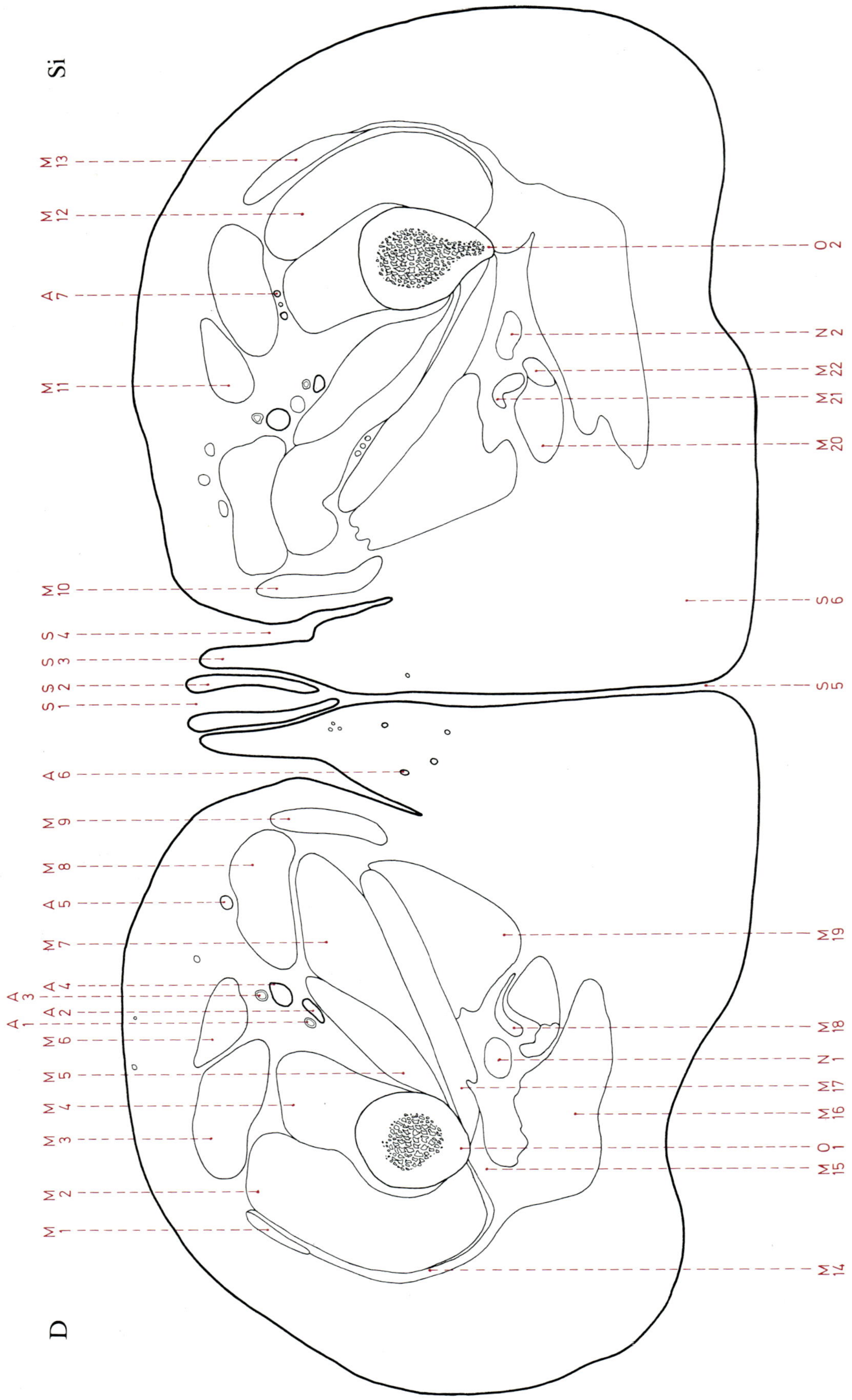
Si
D
M 13
M 12
A 7
M 11
M 10
S 4
S 3
S 2
S 1
A 6
M 9
M 8
A 5
M 7
A 4
A 3
A 2
A 1
M 6
M 5
M 4
M 3
M 2
M 1
O 2
N 2
M 22
M 21
M 20
S 6
S 5
M 19
M 18
N 1
M 17
M 16
O 1
M 15
M 14

Fig. HP♀ 8

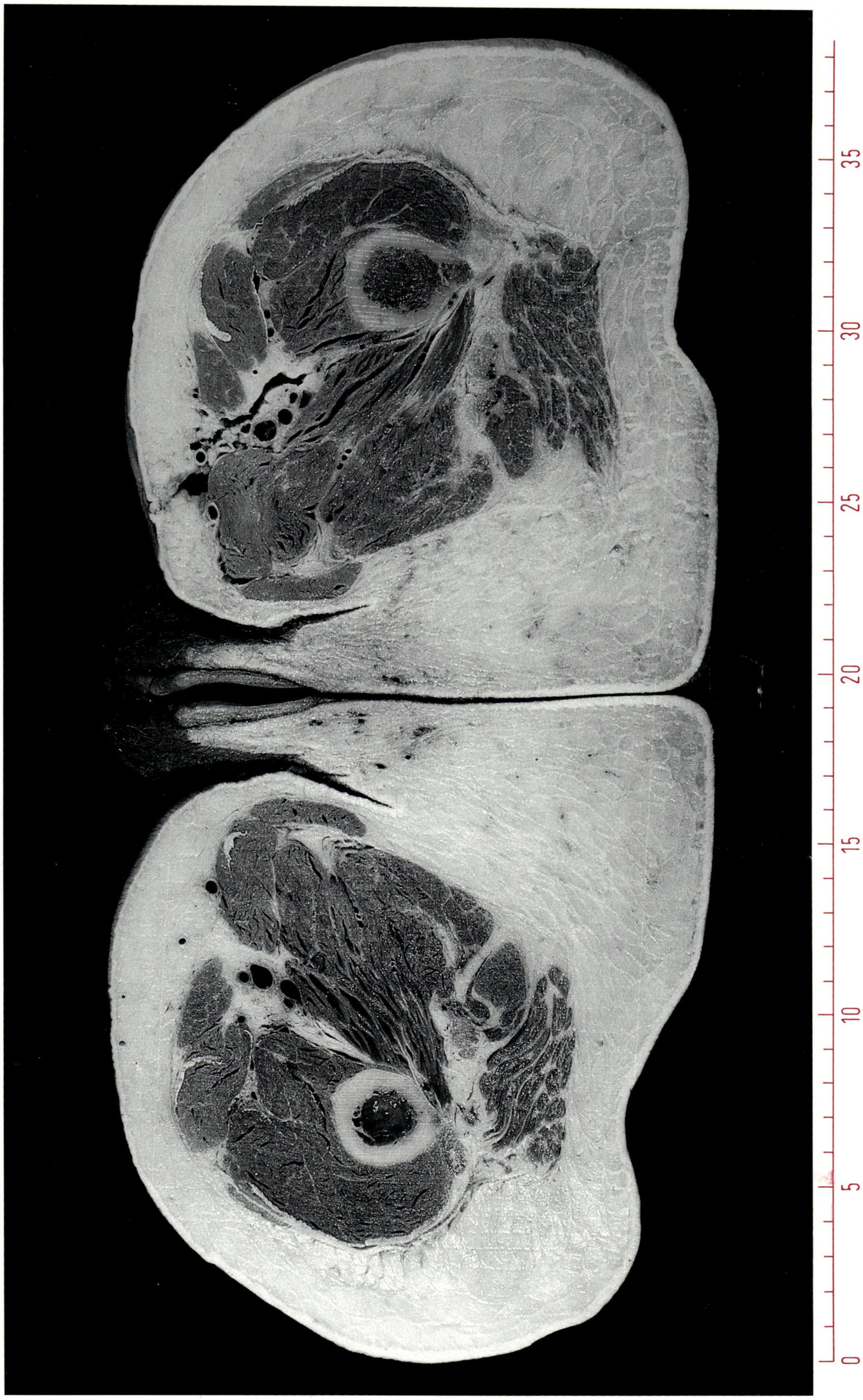

Fig. HP♀ 8

Koritké/Sick
Atlas of Sectional Human Anatomy
Index of Labels for Vol. 1

J. G. Koritké and H. Sick

Atlas of Sectional Human Anatomy

Index of Labels
Vol. 1: Head, Neck, Thorax

Urban & Schwarzenberg
Baltimore – Munich 1983

Authors' addresses:

Jean Georges Koritké
Chairman
Department of Anatomy
Faculty of Medicine
Louis Pasteur University
4, rue Kirschleger
F-67085 Strasbourg CEDEX
France

Henri Sick
Professor of Anatomy
Department of Anatomy
Faculty of Medicine
Louis Pasteur University
4, rue Kirschleger
F-67085 Strasbourg CEDEX
France

This atlas consists of two separate volumes:
Vol. 1: Head, Neck, Thorax (incl. Index)
Vol. 2: Abdomen, Pelvis (incl. Index)

 2–86

Printed in Germany by Kösel, Kempten

Vol. 1: ISBN 0-8067-1031-4 (Baltimore)
ISBN 3-541-71031-4 (Munich)
Complete set: ISBN 0-8067-1030-6 (Baltimore)
ISBN 3-541-71030-6 (Munich)

Table of Contents

Head and Neck

(cephalic and cervical regions)

Thorax

Directions for Use

The most important document is the photograph, a true reflection of the actual anatomy. From the very beginning, the observer is able to record a large amount of information, with the progression of the analysis indicated by the diagram and then the labels.

Photographs

The photographs reproduced on the right-hand pages convey different types of information: general appearance of the section, shape of its surfaces, positions, proportions, impressions of the density of the organs. The entire section or each of its elements can be situated within the body by referring to the following: the letters characterizing each section with respect to the planes of reference; the height of the vertebra shown on the osseous section as soon as it appears; the height of the center of the vertebral body indicated on both sides of all photographs of frontal and sagittal sections; the level of the ribs; and the scale shown at the bottom of each photograph.

Drawings

Each diagram, an interpretation of the actual anatomy, brings out the essential characteristic elements of the section. The thickness of the features and the character of the grid representing the organs indicate to which system the organs shown belong. References are also made using keys. Symmetric organs are not necessarily presented in a bilateral manner. This makes the drawing clearer. It was not always possible to superimpose extremely thin planes. A single feature can give an outline of several structures. Cuts of vertebrae and ribs are designated in any case.

Labels

The labels supply further information and help to check the identification of the photograph and the interpretation of the drawing.
Abbreviations have been used for the identification and orientation of the section, the classification of organs referred to by large systems, and finally in the labels. The first examination of the photograph, the interpretation of the drawing, and then reading the labels should logically lead the observer to re-examine the photograph in order to record the numerous elements that could not be referred to in the space allowed. It should then be possible to imagine and compare the image that will be supplied by CT, ultrasound, or any other new technique technology brings to us.

Abbreviations

1. Identification and Orientation of the Sections

a: anterior – anterior
p: posterior – posterior
D: right – dexter
Si: left – sinister
F: frontal – frontalis
H: horizontal – horizontalis
S: sagittal – sagittalis
C: head, neck – caput, collum
T: thorax – thorax
A: abdomen – abdomen
P: pelvis – pelvis
♀: female – femininum
♂: male – masculinum

The vertebrae are indicated by the letters C (cervical – cervicalis), T (thoracic – thoracicus), L (lumbar – lumbalis), S (sacral – sacralis) and Co (coccygeal – coccygea). The number that corresponds to their level is added to the letter. The ribs are marked by a number indicating their level ("3" refers to the third rib).

2. Association of the Organ to a Large System

A: Angiology – Angiologia
M: Myology – Myologia
N: Nervous System – Systema Nervosum
O: Osteology – Osteologia
OS: Sense Organs – Organa Sensuum
S: Splanchnology – Splanchnologia
Sy: Syndesmology, Arthrology – Syndesmologia

3. Abbreviations used in the Labels

a.: artery – arteria
art.: articulation – articulatio
b.: bursa – bursa
C.: cervical (vertebra) – cervicalis (vertebra)
Co.: coccyx – os coccygis
gl.: gland – glandula
L.: lumbar (vertebra) – lumbalis (vertebra)
lig.: ligament – ligamentum
m.: muscle – musculus
n.: nerve – nervus
no.: node (lymphatic) – nodus (lymphaticus)
r.: ramus – ramus
S.: sacral (vertebra) – sacralis (vertebra)
v.: vein – vena
( ): specifies the preceding term

FC
Frontal sections Head and Neck
(cephalo-cervical region)

49-year-old male ♂

5 sections spaced 2.5 cm apart and parallel to the posterior plane of reference upon which the subject is lying.
The inferior border of each section passes through the center of the 6th thoracic vertebral body [T 6].

FC 1 2.5 cm from the posterior plane of reference, postero-anterior view
FC 2 5.0 cm from the posterior plane of reference, postero-anterior view
FC 3 a 7.5 cm from the posterior plane of reference, antero-posterior view
FC 3 p 7.5 cm from the posterior plane of reference, postero-anterior view
FC 4 a 10 cm from the posterior plane of reference, antero-posterior view
FC 4 p 10 cm from the posterior plane of reference, postero-anterior view
FC 5 12.5 cm from the posterior plane of reference, antero-posterior view

FC 1

A 1 superior sagittal sinus
A 2 confluence of the sinuses
A 3 transverse sinus
A 4 occipital a.
A 5 transverse cervical a. (deep branch)
A 6 1st posterior intercostal a. (dorsal branch, medial branch)
A 7 5th posterior intercostal a.

M 1 semispinalis capitis m.
M 2 trapezius m.
M 3 trapezius m.
M 4 rhomboid minor
M 5 serratus posterior superior m.
M 6 levator scapulae m.
M 7 supraspinatus m.
M 8 erector spinae m.
M 9 subscapularis m.
M 10 deltoid m. (scapular portion)
M 11 infraspinatus m.
M 12 serratus anterior m.

N 1 arachnoid granulations
N 2 superior parietal lobe
N 3 inferior parietal lobe
N 4 precuneus
N 5 falx cerebri
N 6 angular gyrus
N 7 parietooccipital sulcus
N 8 cuneus
N 9 middle temporal gyrus
N 10 calcarine sulcus
N 11 posterior horn of lateral ventricle
N 12 gyrus occipitotemporalis medialis
N 13 inferior temporal gyrus
N 14 lateral occipitotemporal gyrus
N 15 tentorium cerebelli
N 16 falx cerebelli
N 17 cerebral hemisphere
N 18 greater occipital n.

O 1 parietal bone
O 2 squamous portion of occipital bone
O 3 spine of scapula
O 4 scapula (superior margin)
O 5 transverse process T 5
O 6 lamina T 6

S 1 right lung (superior lobe)
S 2 right lung (inferior lobe)

SY 1 sagittal suture
SY 2 lamboid suture
SY 3 ligamentum nuchae
SY 4 5th costotransverse articulation
SY 5 ligamentum flavum

FC 2

A 1 superior sagittal sinus
A 2 great cerebral v.
A 3 posterior cerebral a.
A 4 transverse sinus
A 5 occipital a.
A 6 transverse cervical a. (superficial branch)
A 7 suboccipital venous plexus
A 8 deep cervical v.
A 9 suprascapular a.
A 10 suprascapular a.
A 11 subscapular a.

M 1 obliquus capitis superior m.
M 2 rectus capitis posterior minor m.
M 3 rectus capitis posterior major m.
M 4 semispinalis capitis m.
M 5 obliquus capitis inferior m.
M 6 splenius capitis m.
M 7 interspinalis cervicis m.
M 8 transversospinalis m.
M 9 levator scapulae m.
M 10 trapezius m.
M 11 supraspinatus m.
M 12 deltoid m.
M 13 infraspinatus m.
M 14 subscapularis m.
M 15 serratus anterior m.
M 16 teres minor m.
M 17 triceps brachii (long head)
M 18 teres major m.

N 1 lateral sulcus
N 2 superior temporal gyrus
N 3 middle temporal gyrus
N 4 inferior temporal gyrus
N 5 lateral occipitotemporal gyrus
N 6 falx cerebri
N 7 precuneus
N 8 parietooccipital sulcus
N 9 cuneus
N 10 lateral ventricle (collateral triangle)
N 11 choroid plexus of lateral ventricle
N 12 medial occipitotemporal gyrus
N 13 vermis of cerebellum
N 14 tentorium cerebelli
N 15 cerebellar hemispheres
N 16 pyramis vermis
N 17 spinal dura mater
N 18 spinal cord

O 1 parietal bone
O 2 mastoid process
O 3 squamous portion of occipital bone
O 4 acromion
O 5 transverse process T 2

SY 1 ligamentum nuchae
SY 2 articular processes of vertebrae T 2–T 3
SY 3 articulation of head of 5th rib
SY 4 intervertebral disc T 5–T 6

FC 3 a

A 1 superior sagittal sinus
A 2 middle cerebral a.
A 3 posterior cerebral a.
A 4 superior petrosal sinus
A 5 sigmoid sinus
A 6 occipital a.

A 7 vertebral a.
A 8 vertebral a.
A 9 vertebral v.
A 10 subscapular a.

M 1 obliquus capitis inferior m.
M 2 sternocleidomastoid m.
M 3 splenius capitis m.
M 4 levator scapulae m.
M 5 splenius cervicis m.
M 6 transversospinalis m.
M 7 supraspinatus m.
M 8 scalenus posterior m.
M 9 infraspinatus m.
M 10 serratus anterior m.
M 11 teres minor m.
M 12 subscapularis m.
M 13 deltoid m.
M 14 teres major m.
M 15 latissimus dorsi m.

N 1 central portion of lateral ventricle
N 2 third ventricle
N 3 middle cerebellar peduncle
N 4 spinal cord (median fissure)
N 5 falx cerebri
N 6 superior frontal gyrus
N 7 precentral gyrus
N 8 central sulcus
N 9 gyrus cinguli
N 10 truncus corporis callosi
N 11 postcentral gyrus
N 12 body of fornix
N 13 insula
N 14 thalamus
N 15 substantia nigra
N 16 parahippocampal gyrus
N 17 tentorium cerebelli
N 18 flocculus
N 19 medulla oblongata
N 20 tonsil of the cerebellum
N 21 2nd cervical n.
N 22 spinal dura mater
N 23 cervical enlargement

O 1 parietal bone
O 2 temporal bone (squamous portion)
O 3 mastoid cells
O 4 occipital condyle
O 5 mastoid process
O 6 transverse process T 1
O 7 tubercle of 1st rib
O 8 pedicle T 1
O 9 coracoid process
O 10 head of humerus

SY 1 occipitomastoid suture
SY 2 articular processes of vertebrae C 4–C 5
SY 3 acromioclavicular articulation
SY 4 articulation of head of 2nd rib
SY 5 glenoid cavity
SY 6 intervertebral disc T 2–T 3

FC 3 p

A 1 superior sagittal sinus
A 2 middle cerebral a.
A 3 sigmoid sinus
A 4 vertebral a.
A 5 vertebral v.
A 6 superior petrosal sinus
A 7 vertebral a.
A 8 vertebral v.
A 9 1st intervertebral v.
A 10 vertebral v.
A 11 vertebral v.
A 12 suprascapular a.
A 13 scapular circumflex a.

M 1 temporalis m.
M 2 digastric m.
M 3 obliquus capitis inferior m.
M 4 sternocleidomastoid m.
M 5 levator scapulae m.
M 6 erector spinae m.
M 7 scalenus posterior m.
M 8 omohyoid m.
M 9 supraspinatus m.
M 10 infraspinatus m.
M 11 serratus anterior m.
M 12 teres minor m.
M 13 subscapularis m.
M 14 deltoid m. (acromial portion)
M 15 teres major m.
M 16 latissimus dorsi m.

N 1 falx cerebri
N 2 gyrus cinguli
N 3 corpus callosum
N 4 septum pellucidum
N 5 fornix
N 6 thalamus
N 7 substantia nigra
N 8 parahippocampal gyrus
N 9 tonsil of cerebellum
N 10 superior frontal gyrus
N 11 precentral gyrus
N 12 central sulcus
N 13 postcentral gyrus
N 14 caudate nucleus
N 15 lateral ventricle (central portion)
N 16 lateral sulcus
N 17 lentiform nucleus
N 18 third ventricle
N 19 lateral ventricle (inferior horn)
N 20 tentorium cerebelli
N 21 pons
N 22 middle cerebellar peduncle
N 23 cerebellar hemispheres
N 24 medulla oblongata
N 25 spinal root of accessory n.
N 26 dentate ligament
N 27 2nd cervical n.
N 28 5th cervical n. (dorsal root)
N 29 spinal cord

O 1 parietal bone
O 2 anterior semicircular duct
O 3 mastoid antrum
O 4 mastoid process
O 5 acromial end of clavicle
O 6 acromion
O 7 clavicle
O 8 coracoid process
O 9 head of humerus

SY 1 squamous suture
SY 2 articular process of vertebrae C 4–C 5
SY 3 articulation of head of 2nd rib
SY 4 subscapular bursa
SY 5 articulation of humerus (articular capsule)

FC 4 a

A 1 superior sagittal sinus
A 2 middle cerebral a.
A 3 anterior cerebral a.
A 4 internal carotid a.
A 5 cavernous sinus
A 6 internal carotid a.
A 7 pterygoid plexus
A 8 maxillary a.
A 9 maxillary a.
A 10 external carotid artery
A 11 deep cervical lymph node
A 12 internal jugular v.
A 13 internal carotid a.
A 14 external jugular v.
A 15 inferior thyroid a.
A 16 vertebral v.
A 17 external jugular v.
A 18 subclavian a.
A 19 posterior jugular v.
A 20 axillary v.
A 21 azygos v.
A 22 internal carotid a.
A 23 transverse facial a.
A 24 maxillary a.
A 25 internal jugular v.
A 26 common carotid a.
A 27 external jugular v.
A 28 vertebral v.
A 29 axillary v.
A 30 arch of aorta
A 31 left pulmonary a.

M 1 temporalis m.
M 2 longus capitis m.

M 3 styloglossus m.
M 4 stylopharyngeus m.
M 5 stylohyoid m., digastric m. (posterior belly)
M 6 sternocleidomastoid m.
M 7 longus colli m.
M 8 scalenus anterior m.
M 9 deltoid m. (clavicular portion)
M 10 subclavius m.
M 11 pectoralis minor m.
M 12 coracobrachialis m.
M 13 deltoid m.
M 14 lateral pterygoid m.
M 15 levator veli palatini m.
M 16 styloglossus m.
M 17 stylopharyngeus m.
M 18 stylohyoid m.
M 19 digastric m. (tendon)
M 20 inferior pharyngeal constrictor m.

N 1 falx cerebri
N 2 corpus callosum
N 3 septum pellucidum
N 4 thalamus
N 5 superior frontal gyrus
N 6 middle frontal gyrus
N 7 gyrus cinguli
N 8 inferior frontal gyrus
N 9 anterior horn of lateral ventricle
N 10 optic chiasm
N 11 hypophysis
N 12 temporal pole

O 1 mandible (condylar process)
O 2 styloid process
O 3 frontal bone
O 4 parietal bone
O 5 posterior clinoid process
O 6 temporal bone (squamous)
O 7 sphenoid bone (body)
O 8 zygomatic process
O 9 ramus of mandible
O 10 atlas
O 11 scapula (coracoid process)

S 1 parotid gland
S 2 thyroid gland (right lobe)
S 3 parotid gland (superficial portion)
S 4 parotid gland (deep portion)
S 5 pharynx
S 6 thyroid gland (left lobe)
S 7 esophagus
S 8 trachea

SY 1 squamous suture
SY 2 temporomandibular articulation (articular disc)
SY 3 anterior longitudinal ligament

FC 4p

A 1 superior sagittal sinus
A 2 anterior cerebral a.
A 3 anterior cerebral a.
A 4 internal carotid a.
A 5 cavernous sinus
A 6 internal carotid a.
A 7 pterygoid venous plexus
A 8 maxillary a.
A 9 facial a.
A 10 lingual a.
A 11 superior thyroid a.
A 12 internal jugular v.
A 13 common carotid a.
A 14 external jugular v.
A 15 internal thoracic a.
A 16 axillary v., cephalic v.
A 17 arch of aorta
A 18 left pulmonary a.
A 19 middle cerebral a.
A 20 cavernous sinus
A 21 deep cervical lymph node
A 22 external carotid a.
A 23 common carotid a.
A 24 internal jugular v.
A 25 vertebral a.
A 26 inferior thyroid a.
A 27 subclavian a.
A 28 posterior jugular v.
A 29 thoracoacromial a.
A 30 axillary v.

M 1 temporal m.
M 2 lateral pterygoid m.
M 3 levator veli palatini m.
M 4 longus capitis m.
M 5 medial pterygoid m.
M 6 styloglossus m.
M 7 superior pharyngeal constrictor m.
M 8 digastric m., stylohyoid m.
M 9 middle pharyngeal constrictor m.
M 10 sternocleidomastoid m.
M 11 inferior pharyngeal constrictor m.
M 12 omohyoid m.
M 13 pectoralis minor m.
M 14 pectoralis major m.
M 15 deltoid m.
M 16 styloglossus m.
M 17 stylopharyngeus m.
M 18 stylohyoideus m.
M 19 digastric m. (posterior belly)
M 20 scalenus anterior m.

N 1 falx cerebri
N 2 corpus callosum
N 3 septum pellucidum
N 4 thalamus
N 5 superior frontal gyrus
N 6 inferior frontal gyrus
N 7 lateral ventricle (anterior horn)
N 8 insula
N 9 optic chiasm
N 10 temporal pole
N 11 vagus n.

O 1 zygomatic process
O 2 ramus of mandible
O 3 hyoid (greater horn)
O 4 clavicle
O 5 humerus (surgical neck)
O 6 frontal bone
O 7 temporal bone (squamous portion)
O 8 sphenoid sinus
O 9 mandible (condylar process)

S 1 auditory tube
S 2 lingual tonsil
S 3 corniculate tubercle
S 4 parotid gland
S 5 posterior portion of soft palate
S 6 palatine uvula
S 7 epiglottis
S 8 piriformis recess
S 9 thyroid gland
S 10 trachea

FC 5

A 1 superior sagittal sinus
A 2 superior ophthalmic v.
A 3 maxillary a.
A 4 facial a.
A 5 facial v.
A 6 lingual a.
A 7 superior thyroid a.
A 8 impar thyroid v.
A 9 internal thoracic a.
A 10 right brachiocephalic v.
A 11 aortic arch

M 1 temporal m.
M 2 medial pterygoid m.
M 3 masseter m. (deep portion)
M 4 masseter m. (superficial portion)
M 5 mylohyoid m.
M 6 hyoglossus m.
M 7 digastric m.
M 8 omohyoid m.
M 9 thyrohyoid m.
M 10 sternocleidomastoid m.
M 11 subclavius m.
M 12 sternothyroid m.
M 13 pectoralis major m.
M 14 pectoralis minor m.
M 15 superior rectus m., levator palpebrae superioris m.
M 16 medial rectus m.
M 17 lateral rectus m.
M 18 inferior rectus m.

M 19 sternohyoid m.
M 20 sternocleidomastoid m.
M 21 sternothyroid m.

N 1 lingual n.
N 2 falx cerebri
N 3 frontal lobe
N 4 olfactory tract
N 5 optic n.
N 6 inferior alveolar n.

O 1 frontal lobe (orbital portion)
O 2 ethmoid bone (cribriform plate)
O 3 ethmoid cells
O 4 ethmoid bone (perpendicular plate)
O 5 maxillary sinus
O 6 zygomatic process
O 7 coronoid process
O 8 squamous portion of frontal bone
O 9 middle nasal choncha
O 10 inferior nasal choncha
O 11 pterygoid process
O 12 hyoid (lesser horn)
O 13 hyoid (greater horn)
O 14 clavicle

S 1 superior nasal meatus
S 2 middle nasal meatus
S 3 inferior nasal meatus
S 4 soft palate
S 5 retromolar space
S 6 median sulcus of tongue
S 7 lateral sublingual recess
S 8 septum of tongue
S 9 submandibular gland
S 10 thyrohyoid membrane
S 11 thyroid cartilage
S 12 vestibular fold
S 13 ventricle of larynx
S 14 vocal folds
S 15 infraglottic space
S 16 cricoid cartilage
S 17 1st tracheal cartilage
S 18 thyroid gland

SC
Sagittal sections Head and Neck
(cephalo-cervical region)

41-year-old male ♂

The plane of reference is the median sagittal section. 3 sections have been made on both sides of this plane.
The inferior border of each section passes through the center of the 7th thoracic vertebral body [T 7].
The outermost parts of the base of the neck are shown on the sagittal sections of the thorax.

SC 3D 5.5 cm to the right of the median plane, exterior view
SC 2D 3.5 cm to the right of the median plane, exterior view
SC 1D 1.5 cm to the right of the median plane, exterior view
SC MD median section, right segment, left-to-right view
SC MSi median section, left segment, right-to-left view
SC 1Si 2 cm to the left of the median plane, exterior view
SC 2Si 4 cm to the left of the median plane, exterior view
SC 3Si 6 cm to the left of the median plane, exterior view

SC 3D

A 1 occipital v.
A 2 suprascapular a.
A 3 right superior pulmonary v. (apical and posterior branches)
A 4 right pulmonary a. (apical branch of inferior lobe)
A 5 superficial temporal v.
A 6 external jugular v.
A 7 external jugular v.
A 8 suprascapular a.
A 9 external jugular v.
A 10 subclavian a.
A 11 subclavian v.
A 12 right pulmonary a. (anterior branch)
A 13 right pulmonary a. (middle lobe)

M 1 occipitofrontalis m. (occipital belly)
M 2 splenius capitis m.
M 3 sternocleidomastoid m.
M 4 sternocleidomastoid m.
M 5 sternocleidomastoid m.
M 6 trapezius m.
M 7 levator scapulae m.
M 8 scalenus posterior m.
M 9 scalenus medius m.
M 10 serratus anterior m.
M 11 supraspinatus m.
M 12 subscapularis m.
M 13 serratus posterior superior m.
M 14 trapezius m.
M 15 rhomboid m.
M 16 occipitofrontalis m. (frontal belly)
M 17 temporalis m.
M 18 orbicularis oculi m.
M 19 zygomaticus major m.
M 20 masseter m.
M 20 platysma m.
M 22 sternocleidomastoid m.
M 23 scalenus anterior m.
M 24 pectoralis major m.

N 1 postcentral gyrus
N 2 lateral sulcus
N 3 superior temporal gyrus
N 4 middle temporal gyrus
N 5 brachial plexus (superior trunk)
N 6 brachial plexus (inferior trunk)

O 1 parietal bone
O 2 external auditory meatus
O 3 mastoid cells
O 4 scapula (medial margin)
O 5 zygomatic process
O 6 mandible (condylar process)
O 7 clavicle

S 1 right lung (oblique fissure)
S 2 parotid gland (superficial portion)
S 3 parotid gland
S 4 anterior segmental bronchus
S 5 lateral segmental bronchus

SY 1 parietomastoid suture
SY 2 temporomandibular articulation (articular disc)

SC 2 D

A 1 transverse sinus
A 2 sigmoid sinus
A 3 maxillary a.
A 4 facial a.
A 5 internal jugular v.
A 6 external carotid a.
A 7 occipital a.
A 8 facial a.
A 9 facial a.
A 10 internal carotid a.
A 11 common carotid a.
A 12 anterior jugular v.
A 13 internal jugular v.
A 14 inferior thyroid v.
A 15 subclavian v.
A 16 superior vena cava
A 17 right pulmonary a. (apical branch)
A 18 right pulmonary a.
A 19 right superior pulmonary v.
A 20 right inferior pulmonary v.
A 21 right atrium

M 1 occipitofrontal m. (occipital belly)
M 2 obliquus capitis superior m.
M 3 obliquus capitis inferior m.
M 4 semispinalis capitis m.
M 5 splenius m.
M 6 trapezius m.
M 7 transversospinalis m.
M 8 levator scapulae m.
M 9 serratus posterior superior m.
M 10 rhomboid m.
M 11 trapezius m.
M 12 occipitofrontalis m. (frontal belly)
M 13 superior rectus m.
M 14 lateral rectus m.
M 15 inferior oblique m.
M 16 orbicularis oculi m.
M 17 temporal m.
M 18 lateral pterygoid m.
M 19 medial pterygoid m.
M 20 buccinator m.
M 21 stylopharyngeus m.
M 22 styloglossus m.
M 23 mylohyoid m.
M 24 digastric m. (tendon) and stylohyoid m.
M 25 longus capitis m.
M 26 sternocleidomastoid m.
M 27 omohyoid m.
M 28 scalenus anterior m.
M 29 sternothyroid m.
M 30 sternohyoid m.
M 31 pectoralis major m.

N 1 central sulcus
N 2 occipital pole
N 3 tentorium cerebelli
N 4 cerebellum
N 5 precentral sulcus
N 6 middle frontal gyrus
N 7 orbital gyrus
N 8 lateral sulcus
N 9 middle temporal gyrus
N 10 hypoglossal n.
N 11 vagus n.
N 12 6th cervical n. (ventral ramus)
N 13 8th cervical n. (ventral ramus)

O 1 parietal bone
O 2 inner ear
O 3 tympanic cavity
O 4 squama of occipital bone
O 5 atlas (transverse process)
O 6 frontal bone
O 7 sphenoid bone (greater wing)
O 8 maxilla (zygomatic process)
O 9 temporal bone (styloid process)
O 10 body of mandible
O 11 clavicle (sternal end)
O 12 manubrium of sternum

S 1 parotid gland
S 2 submandibular gland
S 3 right lung (apex)
S 4 right superior lobar bronchus
S 5 right lung (superior lobe)

SY 1 sternoclavicular articulation

SC 1 D

A 1 sigmoid sinus
A 2 transverse sinus
A 3 vertebral a.
A 4 deep cervical v.
A 5 angular v.
A 6 internal carotid a.
A 7 palatine tonsil
A 8 inferior thyroid a.
A 9 common carotid a.
A 10 subclavian a.
A 11 inferior thyroid v.
A 12 left brachiocephalic v.
A 13 azygos v.
A 14 pericardial cavity
A 15 right pulmonary a.
A 16 right auricle
A 17 superior vena cava
A 18 inferior tracheobronchial lymph node
A 19 left atrium
A 20 right atrium

M 1 rectus capitis posterior major m.
M 2 obliquus capitis inferior m.
M 3 semispinalis capitis m.
M 4 splenius m.
M 5 rhomboid m.
M 6 transversospinalis m.
M 7 trapezius m.
M 8 orbicularis oculi m.
M 9 levator palpebrae superioris m.
M 10 inferior rectus m.
M 11 medial pterygoid m.
M 12 tensor veli palatini m.
M 13 longus capitis m.
M 14 mylohyoid m.
M 15 geniohyoid m.
M 16 digastric m. (anterior belly)
M 17 hyoglossus m.
M 18 thyrohyoid m.
M 19 sternohyoid m.
M 10 cricothyroid m.
M 21 inferior pharyngeal constrictor m.
M 22 longus colli m.
M 23 sternothyroid m.
M 24 sternocleidomastoid m.
M 25 pectoralis major m.

N 1 precentral gyrus
N 2 central sulcus
N 3 glomus of choroid plexus
N 4 uncus parahippocampal gyrus
N 5 occipital pole
N 6 tentorium cerebelli
N 7 cerebellum
N 8 2nd cervical n. (dorsal ramus)
N 9 frontal pole
N 10 optic n.
N 11 temporal pole
N 12 trigeminal ganglion
N 13 3rd cervical n.

O 1 parietal bone
O 2 temporal bone (petrosal portion)
O 3 frontal sinus
O 4 sphenoid bone (lesser wing)
O 5 body of mandible
O 6 hyoid bone (greater horn)

S 1 right principal bronchus
S 2 right lung (superior lobe)
S 3 nasolacrimal duct
S 4 inferior nasal concha
S 5 auditory tube
S 6 pharyngeal recess
S 7 vestibule of oral cavity
S 8 sublingual gland
S 9 piriformis recess
S 10 thyroid cartilage
S 11 cricoid cartilage
S 12 thyroid gland
S 13 trachea
S 14 thymus (adipose body)

SY 1 lambdoidal suture
SY 2 atlantooccipital articulation
SY 3 lateral atlantoaxial articulation

SY 4 intervertebral disc C 6–C 7
SY 5 articular processes of vertebrae C 7–T 1
SY 6 manubriosternal synchondrosis
SY 7 3rd sternocostal articulation

SC MD

A 1 internal carotid a.
A 2 vertebral a.
A 3 lingual tonsil
A 4 inferior thyroid v.
A 5 brachiocephalic trunk
A 6 left brachiocephalic v.
A 7 right pulmonary a.
A 8 inferior tracheobronchial lymph node
A 9 ascending aorta
A 10 right auricle
A 11 left atrium
A 12 right coronary a.
A 13 transverse sinus

M 1 longus capitis m.
M 2 genioglossus m.
M 3 geniohyoid m.
M 4 digastric m. (anterior belly)
M 5 mylohyoid m.
M 6 sternohyoid m.
M 7 sternothyroid m.
M 8 rectus capitis posterior minor m.
M 9 semispinalis capitis m.
M 10 transversospinalis m.
M 11 transverse arythenoid m.
M 12 trapezius m.

N 1 frontal pole
N 2 optic tract
N 3 precentral gyrus
N 4 thalamus
N 5 mesencephalon
N 6 oculomotor n.
N 7 pons
N 8 middle cerebellar peduncle
N 9 occipital pole
N 10 tentorium cerebelli
N 11 olive
N 12 spinal medulla

O 1 frontal sinus
O 2 ethmoid cells
O 3 sphenoid sinus
O 4 manubrium of sternum
O 5 parietal bone
O 6 inion

S 1 nasal boundary
S 2 middle nasal concha
S 3 inferior nasal concha
S 4 hard palate
S 5 pharyngeal opening of auditory tube
S 6 palatine velum
S 7 palatine uvula
S 8 epiglottic vallecula
S 9 epiglottis
S 10 hyoid bone
S 11 tubercle of corniculate cartilage
S 12 vestibular fold
S 13 laryngeal ventricle
S 14 vocal fold
S 15 thyroid cartilage
S 16 cricoid cartilage
S 17 cricoid cartilage
S 18 thyroid gland
S 19 esophagus
S 20 trachea
S 21 thymus (adipose body)
S 22 carina of trachea

SY 1 median atlantoaxial joint
SY 2 manubriosternal synchondrosis
SY 3 intervertebral disc C 6–C 7
SY 4 ligamentum flavum

SC MSi

A 1 transverse sinus
A 2 anterior cerebral a.
A 3 internal carotid a.
A 4 posterior cerebral a.
A 5 pharyngeal tonsil
A 6 brachiocephalic trunk
A 7 left brachiocephalic v.
A 8 pericardial cavity
A 9 arch of aorta
A 10 right pulmonary a.
A 11 inferior tracheobronchial lymph node
A 12 right auricle
A 13 left atrium
A 14 right coronary a.

M 1 rectus capitis posterior minor m.
M 2 semispinalis capitis m.
M 3 transversospinalis m.
M 4 trapezius m.
M 5 orbicularis oris m.
M 6 superior longitudinal m.
M 7 genioglossus m.
M 8 geniohyoid m.
M 9 mylohyoid m.
M 10 transverse arythenoid m.
M 11 vocalis m.
M 12 sternohyoid m.
M 13 sternothyroid m.

N 1 central sulcus
N 2 right lateral ventricle (central portion)
N 3 thalamus
N 4 tectal lamina
N 5 pons
N 6 middle cerebellar peduncle
N 7 tentorium cerebelli
N 8 medulla oblongata (olive)
N 9 cerebellomedullary cisterna
N 10 spinal cord
N 11 falx cerebri
N 12 frontal pole
N 13 olfactory bulb
N 14 optic chiasm
N 15 hypophysis

O 1 parietal bone
O 2 occipital bone (basalar portion)
O 3 frontal sinus
O 4 nasal bone
O 5 sphenoidal sinus

S 1 nasal septum
S 2 hard palate
S 3 pharyngeal cavity (nasal portion)
S 4 palatine velum
S 5 palatine uvula
S 6 sublingual gland
S 7 epiglottis
S 8 epiglottic vallecula
S 9 tubercle of corniculate cartilage
S 10 vestibular fold
S 11 laryngeal ventricle
S 12 esophagus
S 13 trachea
S 14 isthmus of thyroid gland
S 15 left principal bronchus
S 16 carina of trachea

SY 1 lambdoidal suture
SY 2 transverse ligament of atlas
SY 3 ligamentum flavum
SY 4 intervertebral disc T 5–T 6
SY 5 manubriosternal synchondrosis

SC 1 Si

A 1 middle cerebral a.
A 2 internal carotid a.
A 3 maxillary a.
A 4 internal carotid a.
A 5 superior labial a.
A 6 palatine tonsil
A 7 lingual tonsil
A 8 vertebral a.
A 9 lingual a.
A 10 inferior thyroid a.
A 11 vertebral a.
A 12 inferior thyroid v.
A 13 common carotid a.
A 14 vertebral v.
A 15 anterior jugular v.

A 16 vertebral a. (opening)
A 17 subclavian a.
A 18 left brachiocephalic v.
A 19 arch of aorta
A 20 pericardial cavity
A 21 internal thoracic v.
A 22 pulmonary trunk
A 23 transverse pericardial sinus
A 24 right semilunar valve
A 25 aortic bulb
A 26 left atrium
A 27 right ventricle
A 28 superior sagittal sinus
A 29 cavernous sinus
A 30 confluence of sinuses
A 31 vertebral a.
A 32 suboccipital venous plexus
A 33 deep cervical v.
A 34 intervertebral v.
A 35 accessory hemiazygos v.
A 36 5th intercostal a. (opening)
A 37 descending aorta

M 1 superior oblique m. (tendon)
M 2 medial rectus m.
M 3 medial rectus m.
M 4 inferior rectus m.
M 5 levator labii superioris m.
M 6 medial pterygoid m.
M 7 tensor veli palatini m.
M 8 levator veli palatini m.
M 9 longus capitis m.
M 10 superior pharyngeal constrictor m.
M 11 hyoglossus m.
M 12 mylohyoid m.
M 13 digastric m. (anterior belly)
M 14 digastric m. (tendon)
M 15 platysma m.
M 16 thyrohyoid m.
M 17 inferior constrictor m.
M 18 cricothyroid m.
M 19 sternohyoid m.
M 20 sternocleidomastoid m.
M 21 sternothyroid m.
M 22 pectoralis major m.
M 23 rectus capitis posterior minor m.
M 24 rectus capitis posterior major m.
M 25 obliquus capitis inferior m.
M 26 semispinalis capitis m.
M 27 transversospinalis m.
M 28 splenius capitis m.
M 29 rhomboid m.
M 30 trapezius m.

N 1 frontal pole
N 2 caudate nucleus
N 3 optic n.
N 4 arachnoid granulations
N 5 central sulcus
N 6 lateral ventricle (central portion)
N 7 internal capsule
N 8 lentiform nucleus
N 9 splenium of corpus callosum
N 10 hippocampus
N 11 superior cerebellar peduncles
N 12 calcarine sulcus
N 13 fourth ventricle (lateral aperture)
N 14 6th cervical n.

O 1 parietal bone
O 2 frontal sinus
O 3 body of mandible
O 4 clavicle (sternal end)
O 5 manubrium of sternum
O 6 transverse process T 2
O 7 head of 3rd rib

S 1 cartilage of auditory tube
S 2 pharyngeal cavity
S 3 sublingual gland
S 4 hyoid bone
S 5 piriformis recess
S 6 thyroid cartilage
S 7 thyroid gland
S 8 left lung
S 9 left principal bronchus

SY 1 sternoclavicular articulation (articular disc)
SY 2 2nd costal cartilage
SY 3 atlantooccipital articulation
SY 4 articular processes of vertebrae C 4–C 5

SC 2 Si

A 1 middle cerebral a.
A 2 maxillary a.
A 3 middle meningeal a.
A 4 internal jugular v.
A 5 external carotid a.
A 6 retromandibular v.
A 7 facial a.
A 8 occipital a.
A 9 facial v.
A 10 internal jugular v.
A 11 transverse cervical a.
A 12 subclavian a.
A 13 anterior jugular v.
A 14 subclavian v.
A 15 bronchopulmonary lymph node
A 16 left pulmonary a.
A 17 left auricle
A 18 left superior pulmonary v.
A 19 transverse sinus
A 20 sigmoid sinus
A 21 suboccipital venous plexus

M 1 occipitofrontalis m. (anterior belly)
M 2 temporal m.
M 3 orbicularis oculi m.
M 4 lateral pterygoid m.
M 5 zygomaticus major m.
M 6 masseter m.
M 7 medial pterygoid m.
M 8 stylohyoid m.
M 9 digastric m. (posterior belly)
M 10 platysma m.
M 11 sternocleidomastoid m.
M 12 scalenus medius m.
M 13 scalenus anterior m.
M 14 pectoralis major m.
M 15 rectus capitis lateralis m.
M 16 rectus capitis posterior major m.
M 17 obliquus capitis superior m.
M 18 semispinalis capitis m.
M 19 semispinalis capitis m.
M 20 splenius m.
M 21 levator scapulae m.
M 22 trapezius m.
M 23 serratus posterior superior m.
M 24 rhomboid m.
M 25 erector spinae m.

N 1 frontal pole
N 2 temporal pole
N 3 cervical plexus
N 4 brachial plexus (superior trunk)
N 5 precentral gyrus
N 6 central sulcus
N 7 occipital pole
N 8 tentorium cerebelli
N 9 cerebellum (left hemisphere)
N 10 7th cervical n. (ventral ramus)
N 11 8th cervical n. (ventral ramus)
N 12 1st intercostal n.

O 1 frontal bone
O 2 sphenoid bone (lesser wing)
O 3 sphenoid bone (greater wing)
O 4 zygomatic bone
O 5 temporal bone (petrosal portion)
O 6 temporal bone (styloid process)
O 7 clavicle
O 8 parietal bone

S 1 lacrimal gland
S 2 parotid gland
S 3 submandibular gland
S 4 left lung (apex)
S 5 left lung (inferior lobe, apical segment)
S 6 oblique fissure
S 7 superior lobar bronchus
S 8 inferior lobar bronchus

SY 1 1st costochondral articulation
SY 2 lambdoidal suture

A 1 superficial temporal a.
A 2 superficial temporal v.
A 3 external jugular v.
A 4 transverse cervical a.
A 5 internal jugular v.
A 6 subclavian a.
A 7 subclavian v.
A 8 pericardial cavity
A 9 transverse sinus
A 10 descending scapular a.

M 1 orbicularis oculi m.
M 2 temporal m.
M 3 masseter m.
M 4 masseter m.
M 5 omohyoid m.
M 6 scalenus medius m.
M 7 subclavius m.
M 8 pectoralis major m.
M 9 splenius capitis m.
M 10 sternocleidomastoid m.
M 11 trapezius m.
M 12 scalenus posterior m.
M 13 supraspinatus m.
M 14 serratus anterior m.
M 15 serratus posterior superior m.
M 16 subscapularis m.
M 17 rhomboid m.

N 1 central sulcus
N 2 lateral sulcus
N 3 superior temporal sulcus
N 4 cervical plexus
N 5 brachial plexus
N 6 parietal lobe
N 7 temporal lobe
N 8 tentorium cerebelli

O 1 frontal bone
O 2 zygomatic process
O 3 clavicle
O 4 parietal bone
O 5 occipital bone
O 6 mastoid process and mastoid cells
O 7 scapula

S 1 external auditory meatus
S 2 parotid gland
S 3 left lung (apex)
S 4 oblique fissure
S 5 left lung (inferior lobe, apical segment)

SY 1 temporomandibular articulation (articular capsule)

HC
Horizontal sections Head and Neck
(cephalo-cervical region)

75-year-old male ♂

The plane of reference is the Frankfurt horizontal (eye-ear plane). The various horizontal sections are spaced 1.5 cm apart starting from this plane of reference.

HC 1 horizontal section of the head, 9 cm above the orbito-meatal line, inferior-superior view
HC 2 horizontal section of the head, 9 cm above the orbito-meatal line, superior-inferior view
HC 3 horizontal section of the head, 7.5 cm above the orbito-meatal line, inferior-superior view
HC 4 horizontal section of the head, 6 cm above the orbito-meatal line, inferior-superior view
HC 5 horizontal section of the head, 6 cm above the orbito-meatal line, superior-inferior view
HC 6 horizontal section of the head, 4.5 cm above the orbito-meatal line, inferior-superior view
HC 7 horizontal section of the head, 4.5 cm above the orbito-meatal line, superior-inferior view
HC 8 horizontal section of the head, 3 cm above the orbito-meatal line, inferior-superior view
HC 9 horizontal section of the head, 3 cm above the orbito-meatal line, superior-inferior view
HC 10 horizontal section of the head, 1.5 cm above the orbito-meatal line, inferior-superior view
HC 11 horizontal section of the head at the orbito-meatal line, inferior-superior view
HC 12 horizontal section of the head, 1.5 cm below the orbito-meatal line, inferior-superior view
HC 13 horizontal section of the head, 3 cm below the orbito-meatal line, inferior-superior view. The section passes through the lower part of the atlas [C 1].
HC 14 horizontal section of the head, 4.5 cm below the orbito-meatal line, inferior-superior view. The section passes through the lower third of the 2nd cervical vertebra [C 2].
HC 15 horizontal section of the head, 6 cm below the orbito-meatal line, inferior-superior view. The section passes through the upper third of the body of the 3rd cervical vertebra [C 3].
HC 16 horizontal section of the neck, 7.5 cm below the orbito-meatal line, inferior-superior view. The section passes through the lower third of the body of the 4th cervical vertebra [C 4].
HC 17 horizontal section of the neck, 9 cm below the orbito-meatal line, inferior-superior view. The section passes through the lower third of the body of the 5th cervical vertebra [C 5].
HC 18 horizontal section of the neck, 10.5 cm below the orbito-meatal line, inferior-superior view. The section passes through the 7th cervical vertebra [C 7].
HC 19 horizontal section of the

cervico-thoracic region, 12 cm below the orbito-meatal line, inferior-superior view. The section passes through the 1st thoracic vertebra [T 1].

HC 20 horizontal section of the cervico-thoracic region, 13.5 cm below the orbito-meatal line, inferior-superior view. The section passes through the 2nd thoracic vertebra [T 2].

HC 1

A 1 superior sagittal sinus
A 2 superficial temporal a. (parietal branch)
A 3 superior cerebral vv.
A 4 superior sagittal sinus

N 1 superior frontal gyrus
N 2 superior frontal sulcus
N 3 middle frontal gyrus
N 4 precentral gyrus
N 5 central sulcus
N 6 postcentral gyrus
N 7 falx cerebri
N 8 cranial dura mater
N 9 superior frontal gyrus
N 10 precentral gyrus
N 11 central sulcus
N 12 postcentral gyrus
N 13 superior parietal lobule

O 1 squama of frontal bone
O 2 parietal bone

HC 2

A 1 superior sagittal sinus
A 2 anterior cerebral a. (frontal branch)
A 3 superficial temporal a. (parietal branch)
A 4 superior sagittal sinus
A 5 supratrochlear v.
A 6 superficial temporal a. (frontal branch)
A 7 anterior cerebral a. (frontal branch)
A 8 superficial temporal a. (parietal branch)
A 9 anterior cerebral a. (parietal branch)

N 1 superior frontal gyrus
N 2 precentral gyrus
N 3 central sulcus
N 4 postcentral gyrus
N 5 superior parietal lobule
N 6 falx cerebri
N 7 middle frontal gyrus
N 8 superior frontal gyrus
N 9 precentral gyrus
N 10 central sulcus
N 11 postcentral gyrus
N 12 superior parietal lobe

O 1 squama of frontal bone
O 2 parietal bone

HC 3

A 1 superior cerebral v.
A 2 anterior cerebral a. (frontal branch)
A 3 callosomarginal a.
A 4 middle meningeal a. (parietal branch)
A 5 superficial temporal a. (parietal branch)
A 6 superficial temporal a. (parietal branch)
A 7 middle cerebral a. (parietal branch)
A 8 superior sagittal sinus
A 9 superior sagittal sinus
A 10 superficial temporal a. (frontal branch)
A 11 anterior cerebral a. (frontal branch)
A 12 middle cerebral a. (frontal branch)
A 13 anterior cerebral a. (parietal branch)
A 14 middle meningeal a. (parietal branch)
A 15 middle cerebral a. (parietal branch)

N 1 superior frontal gyrus
N 2 middle frontal gyrus
N 3 inferior frontal gyrus
N 4 precentral gyrus
N 5 centrum semiovale
N 6 gyrus cinguli
N 7 central sulcus
N 8 postcentral gyrus
N 9 postcentral sulcus
N 10 precuneus
N 11 superior parietal lobe
N 12 cuneus
N 13 arachnoid granulations
N 14 superior frontal gyrus
N 15 falx cerebri
N 16 cranial dura mater
N 17 precentral gyrus
N 18 central sulcus
N 19 postcentral gyrus

O 1 squama of frontal bone
O 2 parietal bone

HC 4

A 1 superior sagittal sinus
A 2 anterior cerebral a.
A 3 middle cerebral a. (frontal branch)
A 4 middle cerebral a. (parietal branch)
A 5 inferior sagittal sinus
A 6 posterior cerebral a. (parietooccipital branch)
A 7 superior sagittal sinus
A 8 superficial temporal a. (frontal branch)
A 9 anterior cerebral a. (marginal a. of corpus callosum)
A 10 middle cerebral a. (frontal branch)
A 11 superficial temporal a. (parietal branch)
A 12 middle cerebral a. (parietal branch)

N 1 superior frontal gyrus
N 2 sulcus cinguli
N 3 middle frontal gyrus
N 4 inferior frontal gyrus
N 5 precentral sulcus
N 6 precentral gyrus
N 7 caudate nucleus
N 8 central sulcus
N 9 postcentral gyrus
N 10 insula (cortex)
N 11 medullary center of corpus callosum
N 12 supramarginal gyrus
N 13 inferior parietal lobule
N 14 precuneus
N 15 cuneus
N 16 superior parietal lobule
N 17 frontal pole
N 18 superior frontal gyrus
N 19 middle frontal gyrus
N 20 inferior frontal gyrus
N 21 lateral ventricle (anterior horn)
N 22 genu of corpus callosum

N 23 precentral gyrus
N 24 septum pellucidum
N 25 central sulcus
N 26 medullary center of corpus callosum
N 27 lateral ventricle (central portion)
N 28 splenium of corpus callosum
N 29 collateral triangle
N 30 gyrus cinguli
N 31 falx cerebri

O 1 squama of frontal bone
O 2 parietal bone

HC 5

A 1 superior sagittal sinus
A 2 superficial temporal a. (frontal branch)
A 3 anterior cerebral a. (callosomarginal a.)
A 4 anterior cerebral a.
A 5 middle cerebral a. (frontal branch)
A 6 superficial temporal a. (parietal branch)
A 7 middle cerebral a. (parietal branch)
A 8 inferior sagittal sinus
A 9 posterior cerebral a. (parietooccipital branch)
A 10 superior sagittal sinus
A 11 middle cerebral a. (frontal branch)
A 12 thalamostriata v.
A 13 middle cerebral a. (frontal branch)
A 14 superficial temporal a. (parietal branch)
A 15 middle cerebral a. (parietal branch)
A 16 middle cerebral a. (parietal branch)

N 1 superior frontal gyrus
N 2 middle frontal gyrus
N 3 gyrus cinguli
N 4 genu of corpus callosum
N 5 inferior frontal gyrus
N 6 precentral gyrus
N 7 caudate nucleus
N 8 central sulcus
N 9 medullary center of corpus callosum
N 10 postcentral gyrus
N 11 insula (cortex)
N 12 thalamus
N 13 choroid glomus
N 14 lateral sulcus
N 15 collateral triangle
N 16 supramarginal gyrus
N 17 sulcus of corpus callosum
N 18 angular gyrus
N 19 frontal pole
N 20 superior frontal gyrus
N 21 sulcus cinguli
N 22 middle frontal gyrus
N 23 falx cerebri
N 24 sulcus of corpus callosum
N 25 lateral ventricle (anterior horn)
N 26 inferior frontal gyrus
N 27 head of caudate nucleus
N 28 septum pellucidum
N 29 precentral gyrus
N 30 medullary center of corpus callosum
N 31 central sulcus
N 32 body of fornix
N 33 insula
N 34 choroid plexus of lateral ventricle
N 35 lateral sulcus
N 36 splenium of corpus callosum
N 37 supramarginal gyrus
N 38 gyrus cinguli
N 39 falx cerebri
N 40 cuneus

O 1 squama of frontal bone
O 2 parietal bone

HC 6

A 1 right anterior cerebral a.
A 2 left anterior cerebral a.
A 3 middle cerebral a. (frontal branch)
A 4 superficial temporal a. (parietal branch)
A 5 middle cerebral a. (temporal branch)
A 6 internal cerebral v.
A 7 basal v.
A 8 posterior cerebral a. (occipital branch)
A 9 straight sinus
A 10 anterior cerebral a. (frontal branch)
A 11 middle meningeal a.
A 12 middle cerebral a.
A 13 middle cerebral a.
A 14 superficial temporal a.
A 15 middle cerebral a. (temporal branch)
A 16 posterior cerebral a. (occipital branch)
A 17 superior sagittal sinus

N 1 superior frontal gyrus
N 2 middle frontal gyrus
N 3 inferior frontal gyrus
N 4 head of caudate nucleus
N 5 anterior crus of internal capsule
N 6 precentral gyrus
N 7 rostrum of corpus callosum
N 8 superior temporal gyrus
N 9 claustrum
N 10 lentiform nucleus (putamen)
N 11 lateral sulcus
N 12 lentiform nucleus (globus pallidus)
N 13 third ventricle
N 14 habenula
N 15 pulvinar
N 16 fimbria of hippocampus
N 17 choroid glomus
N 18 middle temporal gyrus
N 19 occipital gyrus
N 20 cuneus
N 21 frontal pole
N 22 falx cerebri
N 23 middle frontal gyrus
N 24 gyrus cinguli
N 25 lateral ventricle (anterior horn)
N 26 lateral sulcus
N 27 inferior frontal gyrus
N 28 insula
N 29 column of fornix
N 30 genu of internal capsule
N 31 precentral gyrus
N 32 posterior crus of internal capsule
N 33 superior temporal gyrus
N 34 thalamus
N 35 retrolenticular portion of internal capsule
N 36 body of pineal
N 37 caudal portion of caudate nucleus
N 38 splenium of corpus callosum
N 39 middle temporal gyrus
N 40 collateral trigone
N 41 tentorium cerebelli
N 42 calcarine sulcus
N 43 lateral ventricle (posterior horn)
N 44 falx cerebri

O 1 frontal sinus
O 2 parietal bone
O 3 squama of occipital bone

HC 7

A 1 anterior cerebral a. (frontal branch)
A 2 middle cerebral a. (frontal branch)
A 3 anterior cerebral a. (frontal branch)
A 4 middle meningeal a.
A 5 anterior cerebral a.
A 6 superficial temporal a. (parietal branch)

A 7 middle cerebral a. (parietal branch)
A 8 posterior cerebral a.
A 9 posterior cerebral a. (occipital branch)
A 10 straight sinus
A 11 superior sagittal sinus
A 12 middle meningeal a.
A 13 middle cerebral a. (frontal branch)
A 14 middle cerebral a.
A 15 middle cerebral a.
A 16 posterior cerebral a.
A 17 basal v.

N 1 frontal pole
N 2 middle frontal gyrus
N 3 inferior frontal gyrus
N 4 anterior crus of internal capsule
N 5 insula
N 6 precentral gyrus
N 7 genu of internal capsule
N 8 claustrum
N 9 superior temporal gyrus
N 10 posterior crus of internal capsule
N 11 lateral sulcus (posterior ramus)
N 12 habenula
N 13 retrolentiform portion of internal capsule
N 14 caudal portion of caudate nucleus
N 15 middle temporal gyrus
N 16 body of pineal
N 17 vermis of cerebellum
N 18 calcarine sulcus
N 19 cuneus
N 20 occipital gyrus
N 21 longitudinal cerebral fissure
N 22 gyrus cinguli
N 23 lateral ventricle (anterior horn)
N 24 head of caudate nucleus
N 25 putamen
N 26 column of fornix
N 27 globus pallidus
N 28 third ventricle
N 29 interthalamic mass
N 30 third ventricle
N 31 posterior commissure
N 32 pulvinar
N 33 fimbria of hippocampus
N 34 subarachnoid cistern
N 35 glomus of choroid plexus
N 36 tentorium cerebelli
N 37 lateral ventricle (posterior horn)
N 38 calcarine sulcus
N 39 falx cerebri

O 1 frontal sinus
O 2 squama of frontal bone
O 3 parietal bone
O 4 squama of occipital bone

HC 8

A 1 supratrochlear v.
A 2 nasofrontal v.
A 3 nasofrontal v.
A 4 anterior cerebral a. (frontal branch)
A 5 middle cerebral a. (frontal branch)
A 6 middle cerebral a.
A 7 anterior cerebral a.
A 8 middle cerebral a.
A 9 anterior cerebral a.
A 10 internal carotid a.
A 11 superficial temporal a. (parietal branch)
A 12 posterior cerebral a.
A 13 posterior cerebral a.
A 14 superior cerebellar a.
A 15 straight sinus
A 16 occipital a. (occipital branch)
A 17 superficial temporal a. (frontal branch)
A 18 anterior cerebral a.
A 19 anterior cerebral a.
A 20 posterior cerebral a.
A 21 superficial temporal a. (parietal branch)
A 22 superior sagittal sinus

M 1 orbicularis oculi m.
M 2 levator palpebrae superioris m.
M 3 superior rectus m.
M 4 temporal m.
M 5 occipitofrontal m. (occipital belly)

N 1 falx cerebri
N 2 orbital gyrus
N 3 rectus gyrus
N 4 lamina terminalis
N 5 mammillary body
N 6 hippocampus
N 7 crus cerebri
N 8 parahippocampal gyrus
N 9 tentorium cerebelli
N 10 cerebellar hemisphere
N 11 lateral occipitotemporal gyrus
N 12 cerebral dura mater
N 13 vermis
N 14 medial occipitotemporal gyrus
N 15 longitudinal cerebral fissure
N 16 orbital gyrus
N 17 lateral sulcus
N 18 superior temporal gyrus
N 19 third ventricle
N 20 hypothalamus
N 21 optic tract
N 22 interpeduncular cistern
N 23 substantia nigra
N 24 transverse cerebral fissure
N 25 choroid plexus of lateral ventricle
N 26 inferior colliculus
N 27 cerebral aqueduct
N 28 lateral ventricle (inferior horn)
N 29 lateral occipitotemporal gyrus
N 30 falx cerebri
N 31 occipital gyrus

O 1 frontal bone (supraorbital margin)
O 2 frontal sinus
O 3 frontal bone (orbital portion)
O 4 parietal bone
O 5 temporal bone (squamous portion)
O 6 parietal bone
O 7 squama of occipital bone

HC 9

A 1 nasofrontal v.
A 2 superficial temporal a. (frontal branch)
A 3 middle cerebral a.
A 4 internal carotid a.
A 5 middle cerebral a.
A 6 anterior cerebral a. (opening)
A 7 posterior cerebral a.
A 8 superficial temporal a. (parietal branch)
A 9 posterior cerebral a.
A 10 superior cerebellar a.
A 11 occipital a. (occipital branch)
A 12 supratrochlear v.
A 13 nasofrontal v.
A 14 anterior cerebral a.
A 15 middle cerebral a.
A 16 anterior cerebral a.
A 17 internal carotid a.
A 18 posterior communicating a.
A 19 basilar a.
A 20 posterior cerebral a.
A 21 superior cerebellar a.
A 22 posterior cerebral a. (temporal branch)
A 23 straight sinus
A 24 confluence of sinuses

M 1 levator palpebrae superioris m.
M 2 temporal m.
M 3 occipitofrontal m. (occipital belly)
M 4 orbicularis oculi m.
M 5 superior rectus m.
M 6 levator palpebrae superioris m.

N 1 optic n.
N 2 optic tract
N 3 oculomotor n.
N 4 lateral ventricle (inferior horn)
N 5 uncus

N 6 cerebral aqueduct
N 7 tectum of midbrain
N 8 tentorium cerebelli
N 9 lateral occipitotemporal gyrus
N 10 medial occipitotemporal gyrus
N 11 occipital gyrus
N 12 falx cerebri
N 13 oribital gyrus
N 14 rectus gyrus
N 15 lateral sulcus
N 16 olfactory trigone
N 17 superior temporal gyrus
N 18 third ventricle (optic recess)
N 19 infundibular recess
N 20 hypothalamus
N 21 hippocampus
N 22 crus cerebri
N 23 substantia nigra
N 24 trochlear n.
N 25 inferior colliculus
N 26 cerebellar hemisphere
N 27 vermis
N 28 falx cerebri

O 1 frontal bone (supraorbital margin)
O 2 frontal bone (orbital portion)
O 3 squama of occipital bone
O 4 frontal sinus
O 5 crista galli
O 6 temporal bone (squamous portion)

S 1 lacrimal gland

HC 10

A 1 superior ophthalmic v.
A 2 middle meningeal a.
A 3 cavernous sinus
A 4 internal carotid a.
A 5 cavernous sinus
A 6 internal carotid a.
A 7 superior petrosal v.
A 8 transverse sinus
A 9 occipital a. (occipital branch)
A 10 angular a.
A 11 superior ophthalmic v.
A 12 middle cerebral a. (temporal branch)
A 13 internal carotid a.
A 14 superficial temporal a. (frontal branch)
A 15 basilar plexus
A 16 basilar a.
A 17 superficial temporal a. (parietal branch)
A 18 superior petrosal v.
A 19 superior petrosal sinus
A 20 transverse sinus

M 1 medial rectus m.
M 2 lateral rectus m.
M 3 temporal m.
M 4 inferior rectus m.
M 5 occipitofrontalis (occipital belly)

N 1 superior temporal gyrus
N 2 dura mater (hypophysial fossa)
N 3 trigeminal ganglion
N 4 parahippocampal gyrus
N 5 middle temporal gyrus
N 6 trigeminal n.
N 7 middle cerebellar peduncle
N 8 cerebellar tonsil
N 9 cerebellar hemisphere
N 10 falx cerebelli
N 11 oculomotor n.
N 12 lateral ventricle (inferior horn)
N 13 trigeminal cave
N 14 trigeminal n.
N 15 pons
N 16 medial eminence
N 17 fourth ventricle
N 18 roof (choroid plexus) of fourth ventricle
N 19 nodule
N 20 inferior temporal gyrus
N 21 tentorium cerebelli
N 22 dentate nucleus
N 23 cerebellar hemisphere
N 24 vermis

O 1 nasal bone
O 2 zygomatic bone
O 3 ethmoid labyrinth
O 4 sphenoid bone (greater wing)
O 5 sphenoid sinus
O 6 temporal bone (squamous portion)
O 7 sphenoid bone (body)
O 8 temporal bone (superior petrosal margin)
O 9 anterior semicircular canal
O 10 squama of occipital bone

OS 1 superior tarsus
OS 2 cornea
OS 3 lens
OS 4 superior palpebra
OS 5 lacrimal gland
OS 6 fat body of orbit
OS 7 triangular fossa
OS 8 cartilage of auricula

S 1 nasal septum
S 2 superior nasal concha
S 3 middle nasal concha
S 4 nasal cavity

HC 11

A 1 angular a.
A 2 infraorbital a.
A 3 maxillary a.
A 4 maxillary a.
A 5 deep temporal a.
A 6 emissary v. of foramen ovale
A 7 middle meningeal a.
A 8 superficial temporal a.
A 9 internal carotid a.
A 10 inferior petrosal sinus
A 11 anterior inferior cerebellar a.
A 12 anterior inferior cerebellar a.
A 13 bulb of superior jugular v.
A 14 sigmoid sinus
A 15 anterior inferior cerebellar a.
A 16 sigmoid sinus
A 17 occipital a.
A 18 infraorbital a.
A 19 infraorbital a.
A 20 sphenopalatine a.
A 21 superficial temporal a.
A 22 basilar plexus
A 23 internal carotid a.
A 24 right vertebral a.
A 25 left vertebral a.
A 26 bulb of superior jugular v.
A 27 sigmoid sinus

M 1 levator labii superioris m.
M 2 temporal m.
M 3 lateral pterygoid m.
M 4 masseter m.
M 5 sternocleidomastoid m.
M 6 semispinalis capitis m.
M 7 temporal m.
M 8 lateral pterygoid m.

N 1 mandibular n.
N 2 vagus n.
N 3 facial n.
N 4 maxillary n.
N 5 mandibular n.
N 6 pyramid of medulla oblongata
N 7 olive
N 8 inferior cerebellar peduncle
N 9 fourth ventricle
N 10 cerebellar tonsil
N 11 cerebellar hemisphere
N 12 falx cerebelli

O 1 ethmoid bone (perpendicular plate)
O 2 zygomatic bone
O 3 pterygoid process
O 4 foramen lacerum
O 5 mastoid cells
O 6 vomer
O 7 temporal bone (zygomatic process)
O 8 sphenoid bone (greater wing)

O 9 occipital bone (basilar portion)
O 10 temporal bone (mandibular fossa)
O 11 external occipital crest

OS 1 orbit
OS 2 auditory tube
OS 3 external auditory meatus
OS 4 nasolacrimal duct
OS 5 orbit
OS 6 layer of external auditory meatus
OS 7 tympanic cavity
OS 8 tympanic membrane

S 1 lateral nasal cartilage
S 2 nasal septal cartilage
S 3 middle nasal concha
S 4 superior nasal concha
S 5 maxillary sinus
S 6 superior nasal meatus

SY 1 temporomandibular articulation (articular disc)

HC 12

A 1 buccal a.
A 2 maxillary v.
A 3 maxillary a.
A 4 superficial temporal a.
A 5 internal carotid a.
A 6 maxillary v.
A 7 internal jugular v.
A 8 vertebral v.
A 9 vertebral a.
A 10 occipital a.
A 11 deep cervical v.
A 12 mastoid emissary v.
A 13 suboccipital plexus
A 14 facial v.
A 15 descending palatine a.
A 16 maxillary a.
A 17 pterygoid plexus
A 18 middle meningeal a.
A 19 internal carotid a.
A 20 external jugular v.
A 21 internal jugular v.
A 22 vertebral v.
A 23 vertebral a.

M 1 nasal m.
M 2 levator labii superioris nasi m.
M 3 zygomaticus major m.
M 4 temporal m.
M 5 masseter m.
M 6 temporal m.
M 7 medial pterygoid m.
M 8 tensor veli palatini m.
M 9 lateral pterygoid m.
M 10 levator veli palatini m.
M 11 longus capitis m.
M 12 digastric m.
M 13 tensor veli palatini m.
M 14 levator veli palatini m.
M 15 rectus capitis anterior m.
M 16 stylopharyngeus m.
M 17 rectus capitis lateralis m.
M 18 digastric m.
M 19 sternocleidomastoid m.
M 20 obliquus capitis superior m.
M 21 rectus capitis posterior major m.
M 22 rectus capitis posterior minor m.
M 23 semispinalis capitis m.
M 24 trapezius m.

N 1 mandibular n.
N 2 mandibular n.
N 3 facial n.
N 4 accessory n. (spinal roots)
N 5 mandibular n.
N 6 auriculotemporal n.
N 7 accessory n.
N 8 medulla oblongata
N 9 cerebellar tonsil
N 10 greater occipital n.

O 1 atlas (anterior arch)
O 2 mastoid process
O 3 vomer
O 4 mandible (coronoid process)
O 5 lateral lamina of pterygoid process
O 6 neck of mandible
O 7 styloid process
O 8 occipital condyle
O 9 mastoid process

S 1 major alar cartilage
S 2 nasal septal cartilage
S 3 nasal vestibule
S 4 maxillary sinus
S 5 inferior nasal concha
S 6 inferior nasal meatus
S 7 fornix of pharynx
S 8 pharyngeal opening of auditory tube
S 9 cartilage of auditory tube
S 10 pharyngeal recess
S 11 parotid

SY 1 atlantooccipital articulation
SY 2 ligamentum nuchae

HC 13

A 1 buccal a.
A 2 deep temporal a.
A 3 inferior alveolar a.
A 4 pharyngeal v.
A 5 ascending pharyngeal a.
A 6 external carotid a.
A 7 retromandibular v.
A 8 external jugular v.
A 9 anterior external vertebral venous plexus
A 10 internal jugular v.
A 11 vertebral v.
A 12 vertebral a.
A 13 facial a.
A 14 external palatine v.
A 15 ascending pharyngeal a.
A 16 internal carotid a.
A 17 external jugular v.
A 18 occipital a.
A 19 vertebral a.
A 20 intervertebral v.
A 21 anterior internal vertebral venous plexus

M 1 orbicularis oris m.
M 2 zygomaticus major m.
M 3 buccinator m.
M 4 masseter m.
M 5 buccinator m.
M 6 medial pterygoid m.
M 7 superior pharyngeal constrictor m.
M 8 styloglossus m.
M 9 stylopharyngeus m.
M 10 sternocleidomastoid m.
M 11 obliquus capitis inferior m.
M 12 splenius capitis m.
M 13 rectus capitis posterior major m.
M 14 semispinalis capitis m.
M 15 trapezius m.
M 16 superior pharyngeal constrictor m.
M 17 styloglossus m.
M 18 stylopharyngeus m.
M 19 longus capitis m.
M 20 stylohyoid m.
M 21 digastric m. (posterior belly)

N 1 spinal ganglion
N 2 lingual n.
N 3 inferior alveolar n.
N 4 superior cervical ganglion
N 5 hypoglossal n.
N 6 2nd cervical n. (ventral root)
N 7 2nd cervical n. (dorsal root)
N 8 spinal cord

O 1 pterygoid hamulus
O 2 ramus of mandible
O 3 axis (dens)
O 4 atlas (transverse process)
O 5 atlas (lateral mass)
O 6 atlas (posterior arch)

S 1 superior fold of lips
S 2 hard palate
S 3 oral vestibule

S 4 parotid duct
S 5 palatine velum
S 6 pharyngeal cavity
S 7 parotid gland

SY 1 medial atlantoaxial articulation
SY 2 lateral atlantoaxial articulation
SY 3 transverse ligament of the atlas

HC 14

A 1 superior labial a.
A 2 facial a.
A 3 inferior alveolar a.
A 4 retromandibular v.
A 5 ascending pharyngeal a.
A 6 external carotid a.
A 7 occipital a.
A 8 external jugular v.
A 9 internal carotid a.
A 10 internal jugular v.
A 11 vertebral v.
A 12 vertebral a.
A 13 anterior internal vertebral venous plexus
A 14 intervertebral v.
A 15 posterior internal vertebral venous plexus
A 16 deep cervical a.
A 17 lingual a.
A 18 superior labial a.
A 19 facial a.
A 20 facial a.
A 21 retromandibular v.
A 22 external carotid a.
A 23 external jugular v.
A 24 internal jugular v.
A 25 internal carotid a.
A 26 vertebral a.

M 1 orbicularis oris m.
M 2 depressor anguli oris m.
M 3 buccinator m.
M 4 mylohyoid m.
M 5 superior constrictor m. (glossopharyngeal portion)
M 6 masseter m.
M 7 medial pterygoid m.
M 8 stylohyoid m.
M 9 digastric m. (posterior belly)
M 10 semispinalis capitis m.
M 11 obliquus capitis inferior m.
M 12 splenius capitis m.
M 13 depressor anguli oris m.
M 14 genioglossus m.
M 15 hyoglossus m.
M 16 superior longitudinal m.
M 17 superior paryngeal constrictor m. (glossopharyngeal portion)
M 18 digastric m. (posterior belly)
M 19 longus colli m.
M 20 longus capitis m.
M 21 scalenus medius m.
M 22 levator scapulae m.
M 23 sternocleidomastoid m.
M 24 splenius capitis m.
M 25 semispinalis capitis m.
M 26 trapezius m.

N 1 lingual n.
N 2 superior cervical ganglion
N 3 3rd cervical n.
N 4 spinal cord
N 5 spinal dura mater
N 6 inferior alveolar n.
N 7 lingual n.
N 8 hypoglossal n.
N 9 vagus n.
N 10 accessory n.

O 1 angle of mandible
O 2 axis (transverse process)

S 1 oral cavity
S 2 submandibular gland
S 3 oral cavity
S 4 palatine tonsil
S 5 uvula
S 6 pharynx (oral portion)
S 7 parotid gland
S 8 inferior fold of lips
S 9 septum of tongue
S 10 submandibular gland

HC 15

A 1 deep lingual a.
A 2 inferior alveolar a.
A 3 sublingual v.
A 4 submandibular lymph node
A 5 lingual a.
A 6 lingual a.
A 7 sublingual v.
A 8 common trunk facial and lingual a.
A 9 external carotid a.
A 10 internal jugular v.
A 11 external jugular v.
A 12 internal carotid a.
A 13 vertebral a.
A 14 deep cervical v.
A 15 submental a.
A 16 submental a.
A 17 vena comitans of hypoglossal n.
A 18 submental a.
A 19 lingual a.
A 20 facial a.
A 21 lingual a.
A 22 common trunk facial and lingual vv.
A 23 superior superficial cervical lymph node
A 24 external carotid a.
A 25 internal jugular v.
A 26 carotid sinus
A 27 intervertebral v.
A 28 deep cervical a.

M 1 mentalis m.
M 2 depressor labii inferioris m.
M 3 platysma m.
M 4 genioglossus m.
M 5 mylohyoid m.
M 6 hyoglossus m.
M 7 superior pharyngeal constrictor m. (glossopharyngeal portion)
M 8 digastric m. (tendon)
M 9 stylohyoid m.
M 10 middle pharyngeal constrictor m.
M 11 sternocleidomastoid m.
M 12 longus colli m.
M 13 longus capitis m.
M 14 splenius capitis m.
M 15 semispinalis capitis m.
M 16 trapezius m.
M 17 digastric m. (intermediate tendon)
M 18 stylopharyngeus m.
M 19 sternocleidomastoid m.
M 20 scalenus medius m.
M 21 levator scapulae m.
M 22 transversopinalis m.
M 23 trapezius m.
M 24 semispinalis capitis m.
M 25 splenius capitis m.

N 1 hypoglossal n.
N 2 vagus n.
N 3 3rd cervical n. (ventral ramus)
N 4 accessory n.
N 5 spinal cord
N 6 inferior alveolar n.
N 7 3rd cervical n. (ventral ramus)
N 8 vagus n.

O 1 hyoid bone (greater horn)
O 2 transverse process C 3
O 3 transverse process C 3
O 4 inferior articulating process C 3

S 1 sublingual gland
S 2 submandibular gland
S 3 submandibular gland
S 4 dorsum of tongue
S 5 septum of tongue
S 6 epiglottis
S 7 pharyngeal cavity

SY 1 articular processes of vertebrae C 3–C 4
SY 2 ligamentum flavum
SY 3 ligamentum nuchae

HC 16

A 1 superior laryngeal v. (anterior branch)
A 2 superior laryngeal a.
A 3 anterior jugular v.
A 4 pharyngeal v.
A 5 superior thyroid v.
A 6 common carotid a.
A 7 pharyngeal v.
A 8 internal jugular v.
A 9 external jugular v.
A 10 vertebral a.
A 11 intervertebral v.
A 12 deep cervical v.
A 13 anterior jugular v.
A 14 superior thyroid a.
A 15 superior thyroid v.
A 16 internal jugular v.
A 17 common carotid a.
A 18 external jugular v.
A 19 deep cervical v.
A 20 deep cervical a.

M 1 omohyoid m.
M 2 longus capitis m.
M 3 sternocleidomastoid m.
M 4 levator scapulae m.
M 5 splenius m.
M 6 semispinalis capitis m.
M 7 transversospinalis m.
M 8 trapezius m.
M 9 sternohyoid m.
M 10 platysma m.
M 11 thyrohyoid m.
M 12 inferior pharyngeal constrictor m.
M 13 longus colli m.
M 14 longus capitis m.
M 15 sternocleidomastoid m.
M 16 scalenus medius m.
M 17 levator scapulae m.

N 1 vagus n.
N 2 dura mater of spinal cord
N 3 facial n. (cervical ramus)
N 4 sympathetic trunk
N 5 vagus n.
N 6 4th cervical n.
N 7 spinal cord
N 8 accessory n.

O 1 transverse process C 4 (anterior tubercle)
O 2 superior articulating process C 5
O 3 inferior articulating process C 4
O 4 spinous process C 5

S 1 thyrohyoid membrane
S 2 epiglottis
S 3 vestibule of larynx
S 4 piriformis recess
S 5 pharynx (laryngeal portion)
S 6 thyroid cartilage (superior horn)

SY 1 uncovertebral articulation C 4–C 5
SY 2 articular processes of vertebrae C 4–C 5
SY 3 ligamentum nuchae

HC 17

A 1 anterior jugular v.
A 2 superior thyroid v.
A 3 superior thyroid v.
A 4 superior thyroid v.
A 5 common carotid a.
A 6 internal jugular v.
A 7 external jugular v.
A 8 vertebral a.
A 9 deep cervical v.
A 10 anterior jugular v.
A 11 superior laryngeal v. (internal branch)
A 12 superior thyroid a.
A 13 pharyngeal plexus
A 14 common carotid a.
A 15 pharyngeal v.
A 16 internal jugular v.
A 17 vertebral a.
A 18 deep cervical v.
A 19 deep cervical a.
A 20 superficial cervical v.

M 1 sternohyoid m.
M 2 omohyoid m.
M 3 thyrohyoid m.
M 4 platysma m.
M 5 scalenus medius m.
M 6 levator scapulae m.
M 7 transversospinalis m.
M 8 semispinalis capitis m.
M 9 trapezius m.
M 10 splenius m.
M 11 sternohyoid m.
M 12 thyroarytenoid m.
M 13 sternocleidomastoid m.
M 14 transverse arytenoid and oblique mm.
M 15 inferior pharyngeal constrictor m.
M 16 longus colli m.
M 17 longus capitis m.
M 18 scalenus anterior m.
M 19 scalenus medius m.
M 20 scalenus posterior m.
M 21 levator scapulae m.
M 22 trapezius m.
M 23 splenius m.
M 24 semispinalis capitis m.

N 1 superior laryngeal n.
N 2 vagus n.
N 3 sympathetic trunk
N 4 cervical plexus
N 5 cervical plexus
N 6 5th cervical n.
N 7 lesser occipital n.
N 8 6th cervical n.
N 9 cervical plexus
N 10 dentate ligament
N 11 spinal cord
N 12 accessory n.

S 1 vestibule of larynx
S 2 thyroid cartilage
S 3 arytenoid cartilage
S 4 piriformis recess
S 5 ventricle of larynx
S 6 vestibular folds
S 7 piriformis recess
S 8 pharyngeal cavity

SY 1 uncovertebral articulation C 5–C 6
SY 2 intervertebral disc C 5–C 6
SY 3 articular processes of vertebrae C 6–C 7
SY 4 ligamentum flavum

HC 18

A 1 external jugular v.
A 2 internal jugular v.
A 3 anterior jugular v.
A 4 superior thyroid a.
A 5 common carotid a.
A 6 external jugular v.
A 7 deep cervical v.
A 8 transverse cervical a.
A 9 vertebral a.
A 10 deep cervical v.
A 11 superficial cervical v.

M 1 scalenus medius m.
M 2 scalenus anterior m.
M 3 sternocleidomastoid m.
M 4 longus colli m.
M 5 cricothyroid m.
M 6 posterior cricoarytenoid m.
M 7 sternohyoid m.
M 8 inferior pharyngeal constrictor m.
M 9 sternothyroid m.
M 10 scalenus anterior m.
M 11 omohyoid m.
M 12 trapezius m.
M 13 levator scapulae m.
M 14 scalenus posterior m.
M 15 transversospinalis m.
M 16 rhomboid m.
M 17 splenius m.
M 18 serratus posterior superior m.
M 19 levator scapulae m.
M 20 scalenus posterior m.
M 21 trapezius m.

N 1 5th cervical n. (ventral ramus)
N 2 vagus n.
N 3 sympathetic trunk
N 4 6th cervical n. (ventral ramus)
N 5 5th cervical n. (ventral ramus)
N 6 accessory n.
N 7 6th cervical n. (ventral ramus)
N 8 7th cervical n.
N 9 spinal cord
N 10 cervical plexus (supraclavicular n.)

O 1 acromion
O 2 clavicle (acromial end)
O 3 transverse process C 7 (posterior tubercle)
O 4 lamina
O 5 superior articulating process T 1
O 6 transverse process C 7 (anterior tubercle)
O 7 transverse process T 1
O 8 clavicle (acromial end)
O 9 acromion

S 1 thyroid gland (right lobe)
S 2 cricoid cartilage
S 3 pharyngeal cavity (laryngeal portion)
S 4 vocal fold
S 5 infraglottic cavity
S 6 cricothyroid ligament
S 7 thyroid cartilage (inferior horn)

SY 1 acromioclavicular articulation
SY 2 intervertebral disc C 6–C 7
SY 3 anterior longitudinal ligament
SY 4 articular processes of vertebrae C 7–T 1
SY 5 1st costotransverse articulation
SY 6 acromioclavicular articulation

HC 19

A 1 external jugular v.
A 2 inferior thyroid a.
A 3 vertebral v.
A 4 deep cervical v.
A 5 common carotid a.
A 6 inferior thyroid a.
A 7 anterior jugular v.
A 8 thyroid venous plexus
A 9 inferior thyroid a.
A 10 common carotid a.
A 11 deep cervical v.
A 12 thyrocervical trunk
A 13 internal jugular v.
A 14 external jugular v.
A 15 transverse cervical v.
A 16 transverse cervical a.
A 17 2nd intercostal v. (dorsal branch)

M 1 deltoid m. (clavicular portion)
M 2 scalenus anterior m.
M 3 sternocleidomastoid m.
M 4 longus colli m.
M 5 sternohyoid m.
M 6 sternothyroid m.
M 7 scalenus anterior m.
M 8 omohyoid m.
M 9 scalenus medius m.
M 10 omohyoid m.
M 11 subclavius m.
M 12 supraspinatus m.
M 13 deltoid m. (acromial portion)
M 14 infraspinatus m.
M 15 supraspinatus m.
M 16 trapezius m.
M 17 levator scapulae m.
M 18 serratus posterior superior m.
M 19 transversospinalis m.
M 20 splenius cervicis m.
M 21 rhomboid m.
M 22 levator scapulae m.
M 23 intercostal mm.
M 24 scalenus posterior m.
M 25 serratus anterior m.
M 26 trapezius m.
M 2o supraspinatus m.
M 28 infraspinatus m.
M 29 deltoid m. (acromial portion)

N 1 cervical plexus (supraclavicular n.)
N 2 brachial plexus
N 3 phrenic n.
N 4 8th cervical n. (ventral ramus)
N 5 vagus n.
N 6 right recurrent laryngeal n.
N 7 left recurrent laryngeal n.
N 8 cervicothoracic ganglion
N 9 vagus n.
N 10 phrenic n.
N 11 8th cervical n. (ventral ramus)
N 12 7th cervical n. (ventral ramus)
N 13 brachial plexus (superior trunk)
N 14 long thoracic n.
N 15 spinal cord
N 16 1st thoracic n.

O 1 scapula (coracoid process)
O 2 clavicle
O 3 clavicle
O 4 scapula (coracoid process)
O 5 acromion
O 6 superior articulating process T 2
O 7 transverse process T 2
O 8 scapula
O 9 acromion

S 1 thyroid gland (isthmus)
S 2 esophagus
S 3 trachea

SY 1 articulation of humerus (articular capsule)
SY 2 head of humerus
SY 3 coracoacromial ligament
SY 4 trapezoid ligament
SY 5 conoid ligament
SY 6 articulation of head of 1st rib
SY 7 articular processes of vertebrae T 1–T 2
SY 8 2nd costotransverse articulation

HC 20

A 1 cephalic v.
A 2 transverse cervical a.
A 3 transverse cervical a.
A 4 transverse cervical v.
A 5 transverse cervical a.
A 6 external jugular v.
A 7 subclavian a.
A 8 internal jugular v. (inferior bulb)
A 9 jugular venous arch
A 10 vertebral a.
A 11 common carotid a.
A 12 inferior thyroid a.
A 13 thyroid venous plexus
A 14 thyroid venous plexus
A 15 external jugular v.
A 16 subclavian a.
A 17 common carotid a.
A 18 anterior jugular v.
A 19 vertebral v.
A 20 internal jugular v. (inferior bulb)
A 21 internal thoracic a.
A 22 external jugular v.
A 23 transverse cervical v.
A 24 subclavian a.
A 25 cephalic v.
A 26 thoracoacromial a.
A 27 transverse cervical a.
A 28 superior intercostal a.
A 29 2nd intercostal v.
A 30 transverse cervical a. (descending scapular)
A 31 suprascapular a.

M 1 biceps brachii (long head)
M 2 subscapularis m.
M 3 pectoralis minor m.
M 4 scalenus medius m.
M 5 scalenus anterior m.
M 6 sternocleidomastoid m.
M 7 sternothyroid m.
M 8 subclavius m.
M 9 pectoralis major m.
M 10 serratus anterior m.
M 11 deltoid m. (clavicular portion)
M 12 biceps brachii (long head)
M 13 infraspinatus m.
M 14 subscapularis m.

M 15 scalenus posterior m.
M 16 serratus anterior m.
M 17 intercostal m.
M 18 transversospinalis m.
M 19 trapezius m.
M 20 rhomboid m.
M 21 serratus posterior superior m.
M 22 supraspinatus m.
M 23 subscapularis m.
M 24 infraspinatus m.
M 25 deltoid m. (scapular portion)

N 1 brachial plexus (superior trunk)
N 2 brachial plexus (middle trunk)
N 3 brachial plexus (inferior trunk)
N 4 phrenic n.
N 5 vagus n.
N 6 recurrent laryngeal n.
N 7 recurrent laryngeal n.
N 8 vagus n.
N 9 brachial plexus (inferior trunk)
N 10 brachial plexus (middle trunk)
N 11 brachial plexus (superior trunk)
N 12 suprascapular n.
N 13 sympathetic trunk
N 14 spinal cord
N 15 spinal dura mater
N 16 2nd thoracic n. (dorsal ramus)
N 17 2nd intercostal n.

O 1 humerus (greater tubercle)
O 2 head of humerus
O 3 scapula (coracoid process)
O 4 clavicle
O 5 clavicle
O 6 scapula (coracoid process)
O 7 neck of scapula
O 8 spine of scapula
O 9 transverse process T 3
O 10 scapula (medial or vertebral margin)
O 11 spine of scapula
O 12 spine of scapula

S 1 cupula of pleura
S 2 esophagus
S 3 trachea
S 4 apex of lung

SY 1 glenoid labrum
SY 2 glenoid cavity
SY 3 articulation of head of 2nd rib
SY 4 intervertebral disc T 1–T 2
SY 5 subscapularis bursa
SY 6 glenoid labrum

FT
Frontal sections
Thorax

35-year-old female ♀
6 sections, spaced 2.5 cm apart, parallel to the posterior plane of reference upon which the subject is lying.
The superior border of most sections passes through the center of the 5th cervical vertebral body [C 5]; the inferior border of each section passes through the center of the 2nd lumbar vertebral body [L 2].

FT 1 2.5 cm from the posterior plane of reference, postero-anterior view
FT 2 5 cm from the posterior plane of reference, postero-anterior view
FT 3 a 7.5 cm from the posterior plane of reference, antero-posterior view
FT 3 p 7.5 cm from the posterior plane of reference, postero-anterior view
FT 4 10 cm from the posterior plane of reference, antero-posterior view
FT 5 12.5 cm from the posterior plane of reference, antero-posterior view
FT 6 15 cm from the posterior plane of reference, antero-posterior view

FT 1

A 1 superficial cervical a.
A 2 descending scapular a.
A 3 5th dorsal intercostal a.
A 4 lateral thoracic a.
A 5 posterior spinal v.

M 1 trapezius m.
M 2 levator scapulae m.
M 3 splenius m.
M 4 supraspinatus m.
M 5 erector spinae m.
M 6 serratus posterior superior m.
M 7 subscapularis m.
M 8 infraspinatus m.
M 9 teres minor m.
M 10 deltoid m. (scapular portion)
M 11 teres major m.
M 12 internal intercostal m.
M 13 external intercostal m.
M 14 triceps brachii m. (long head)
M 15 serratus anterior m.
M 16 latissimus dorsi m.
M 17 serratus posterior inferior m.
M 18 longissimus m.
M 19 iliocostalis m.
M 20 transversospinalis m.

O 1 spine of scapula
O 2 lateral margin of scapula
O 3 superior articular process T 10
O 4 transverse process T 10
O 5 inferior articular process T 10

S 1 right lung (superior lobe)
S 2 oblique fissure
S 3 right lung (inferior lobe)
S 4 diaphragmatic pleura
S 5 liver
S 6 pleural recess

SY 1 articulating processes T 8–T 9
SY 2 10th costotransverse articulation
SY 3 ligamentum flavum

FT 2

A 1 superficial cervical a.
A 2 descending scapular a.
A 3 anterior humeral circumflex a.
A 4 posterior humeral circumflex a.
A 5 subscapular a.
A 6 subscapular v.
A 7 deep brachial a.
A 8 9th dorsal intercostal a.

M 1 supraspinatus m.
M 2 serratus anterior m.
M 3 infraspinatus m.
M 4 subscapularis m.
M 5 teres minor m.
M 6 triceps brachii m. (long head)
M 7 teres major m.
M 8 latissimus dorsi m.
M 9 diaphragm (lumbar portion)
M 10 sternocleidomastoid m.
M 11 splenius cervicis m.
M 12 erector spinae m.
M 13 scalenus posterior m.
M 14 omohyoid m.
M 15 supraspinalis m.

M 16 subscapularis m.
M 17 teres major m.
M 18 latissimus dorsi m.
M 19 deltoid m. (scapular portion)
M 20 triceps brachii m. (long head)
M 21 triceps brachii m. (medial head)
M 22 triceps brachii m. (lateral head)
M 23 intercostal mm.
M 24 serratus anterior m.
M 25 latissimus dorsi m.
M 26 diaphragm
M 27 psoas major m.

N 1 1st thoracic spinal n. (roots)
N 2 spinal dura mater
N 3 radial n.
N 4 ulnar n.

O 1 acromion
O 2 neck of scapula
O 3 pedicle of vertebra L 1

S 1 left lung (superior lobe)
S 2 oblique fissure
S 3 left lung (inferior lobe)
S 4 spleen
S 5 perirenal fat
S 6 left kidney
S 7 perirenal fat
S 8 liver
S 9 mediastinal-diaphragmatic recess
S 10 right suprarenal gland
S 11 costodiaphragmatic recess
S 12 peritoneal cavity

SY 1 acromioclavicular articulation

FT 3a

A 1 external jugular v.
A 2 transverse cervical a.
A 3 suprascapular v.
A 4 superior intercostal a.
A 5 axillary a.
A 6 axillary v.
A 7 azygos v.
A 8 pulmonary a. (inferior branch of apical lobe)
A 9 pulmonary a. (basal portion)
A 10 inferior right pulmonary v.
A 11 inferior right pulmonary v.
A 12 abdominal aorta
A 13 inferior vena cava
A 14 vertebral a.
A 15 subclavian a.
A 16 subclavian a.
A 17 axillary a.
A 18 arch of aorta
A 19 axillary v.
A 20 left pulmonary a.
A 21 left atrium
A 22 pericardial cavity
A 23 splenic a.
A 24 splenic v.

M 1 sternocleidomastoid m.
M 2 scalenus posterior m.
M 3 longus colli m.
M 4 scalenus medius m.
M 5 deltoid m. (clavicular portion)
M 6 diaphragm (costal portion)
M 7 serratus anterior m.
M 8 diaphragm (right crus)
M 9 deltoid m.
M 10 biceps brachii m. (long head)
M 11 subscapularis m.
M 12 serratus anterior m.
M 13 coracobrachialis m.
M 14 serratus anterior m.
M 15 psoas m.

N 1 5th cervical n. (ventral ramus)
N 2 cervicothoracic ganglion
N 3 brachial plexus
N 4 axillary n.
N 5 median n.
N 6 brachial plexus (posterior cord)
N 7 ulnar n.
N 8 median n.

O 1 carotid tubercle
O 2 clavicle
O 3 coracoid process
O 4 head of humerus

S 1 apex of right lung
S 2 esophagus
S 3 trachea
S 4 right main bronchus
S 5 carina of trachea
S 6 right superior lobar bronchus
S 7 intermediate bronchus
S 8 oblique fissure
S 9 esophagus
S 10 liver
S 11 right suprarenal gland
S 12 costodiaphragmatic recess
S 13 right kidney
S 14 apical-posterior bronchial segment
S 15 left main bronchus
S 16 left superior lobe bronchus
S 17 oblique fissure
S 18 fundus of stomach
S 19 spleen
S 20 left suprarenal gland
S 21 minor renal calyx

FT 3p

A 1 vertebral v.
A 2 subclavian a.
A 3 vertebral a.
A 4 subclavian a.
A 5 axillary a.
A 6 arch of aorta
A 7 left pulmonary a.
A 8 left pulmonary a. (lingular branch)
A 9 left superior pulmonary v.
A 10 left atrium
A 11 coronary sinus
A 12 abdominal aorta
A 13 splenic a.
A 14 splenic v.
A 15 left renal v.
A 16 suprascapular v.
A 17 axillary a.
A 18 axillary v.
A 19 azygos v.
A 20 pulmonary a. (inferior branch to apical lobe)
A 21 inferior tracheobronchial lymph node
A 22 inferior lobar a.
A 23 right inferior pulmonary v.
A 24 inferior vena cava
A 25 thoracic duct

M 1 sternocleidomastoid m.
M 2 scalenus anterior m.
M 3 longus colli m.
M 4 omohyoid m.
M 5 deltoid m. (clavicular portion)
M 6 subclavius m.
M 7 subscapularis m. (tendon)
M 8 serratus anterior m.
M 9 deltoid m.
M 10 coracobrachialis m.
M 11 biceps brachii m. (short head)
M 12 biceps brachii m. (long head)
M 13 diaphragm (esophageal hiatus)
M 14 diaphragm (left crus)
M 15 psoas m.

N 1 5th cervical n. (ventral ramus)
N 2 trunk of brachial plexus
N 3 brachial plexus (posterior cord)
N 4 median n.

S 1 carina of trachea
S 2 left main bronchus
S 3 apicoposterior bronchial segment
S 4 left superior lobar bronchus
S 5 oblique fissure
S 6 left lung (inferior lobe)
S 7 esophagus
S 8 fundus of stomach
S 9 spleen
S 10 left suprarenal gland
S 11 left kidney (superior pole)
S 12 renal pelvis
S 13 esophagus
S 14 trachea
S 15 apical bronchial segment

S 16 posterior bronchial segment
S 17 right superior lobar bronchus
S 18 right medial lobar bronchus
S 19 oblique fissure
S 20 right inferior lobar bronchus
S 21 liver (caudate lobe)
S 22 right lobe of liver
S 23 right suprarenal gland
S 24 right kidney

FT 4

A 1 internal jugular v.
A 2 right subclavian v.
A 3 right brachiocephalic v.
A 4 brachiocephalic trunk (opening)
A 5 superior vena cava
A 6 arch of aorta
A 7 azygos v. (opening)
A 8 right superior pulmonary v. (apical branch)
A 9 right pulmonary a. (middle lobe branches)
A 10 right superior pulmonary v. (middle lobe branch)
A 11 fossa ovalis
A 12 right atrium
A 13 hepatic v.
A 14 celiac trunk
A 15 superior mesenteric a.
A 16 inferior vena cava
A 17 internal jugular v.
A 18 thoracoacromial v.
A 19 left brachiocephalic v.
A 20 internal thoracic a.
A 21 pericardial cavity
A 22 pulmonary trunk
A 23 left coronary a.
A 24 aortic bulb
A 25 left atrium
A 26 left atrioventricular valve (posterior cusp)
A 27 left ventricle
A 28 coronary sinus
A 29 right coronary a.
A 30 pericardial cavity
A 31 splenic a.
A 32 splenic v.
A 33 left renal v.
A 34 inferior mesenteric v.
A 35 left colic v.
A 36 aorta

M 1 sternocleidomastoid m.
M 2 sternohyoid m.
M 3 subclavius m.
M 4 pectoralis major m. (clavicular portion)
M 5 pectoralis minor m.
M 6 serratus anterior m.
M 7 diaphragm
M 8 external oblique m.
M 9 pectoralis minor m.
M 10 pectoralis major m.

O 1 clavicle

S 1 thyroid gland (right lobe)
S 2 trachea
S 3 right lung (superior lobe)
S 4 horizontal fissure
S 5 right lung (middle lobe)
S 6 oblique fissure
S 7 right lung (inferior lobe)
S 8 mediastinodiaphragmatic recess
S 9 caudate lobe
S 10 costodiaphragmatic recess
S 11 thyroid cartilage
S 12 cricoid cartilage
S 13 thyroid gland (left lobe)
S 14 left lung (apex)
S 15 left lung (superior lobe)
S 16 oblique fissure
S 17 left lung (inferior lobe)
S 18 left hepatic lobe
S 19 fundus of stomach
S 20 vestibule of omental bursa
S 21 gastrosplenic ligament
S 22 spleen
S 23 inferior recess of omental bursa
S 24 body of pancreas
S 25 duodenojejunal flexure

FT 5

A 1 internal thoracic a.
A 2 pericardiosternal ligament
A 3 right auricle
A 4 right coronary a.
A 5 right atrium
A 6 pericardium
A 7 right coronary a.
A 8 hepatic v.
A 9 hepatic v.
A 10 portal v. (left branch)
A 11 proper hepatic a.
A 12 pulmonary valve
A 13 left coronary a. (circumflex branch)
A 14 conus arteriosus pulmonary trunk
A 15 right atrioventricular valve (septal cusp)
A 16 right ventricle
A 17 left ventricle
A 18 pericardial cavity
A 19 hepatic v.
A 20 left gastric a.
A 21 gastroduodenal a.
A 22 portal v.

M 1 pectoralis major m.
M 2 pectoralis minor m.
M 3 serratus anterior m.
M 4 diaphragm

O 1 manubrium

S 1 right lung (superior lobe)
S 2 right lung (middle lobe)
S 3 mammary gland
S 4 right lung (inferior lobe)
S 5 right hepatic lobe
S 6 common hepatic duct
S 7 gall bladder
S 8 duodenum (superior portion)
S 9 head of pancreas
S 10 left lung (superior lobe)
S 11 costodiaphragmatic recess
S 12 left hepatic lobe
S 13 body of stomach
S 14 caudate lobe
S 15 lesser omentum
S 16 head of pancreas
S 17 jejunum
S 18 descending colon

FT 6

A 1 internal thoracic a.
A 2 pericardiosternal ligament
A 3 right coronary a.
A 4 pericardium
A 5 right ventricle
A 6 left ventricle
A 7 portal v. (left branch)
A 8 hepatic v.
A 9 left colic a.

M 1 internal intercostal m.
M 2 pectoralis major m.
M 3 diaphragm
M 4 external oblique m.

O 1 body of sternum
O 2 3rd costal cartilage

S 1 right lung (middle lobe)
S 2 costodiaphragmatic recess
S 3 right lobe of liver
S 4 mammary gland
S 5 left lung (superior lobe)
S 6 left hepatic lobe
S 7 body of stomach
S 8 inferior recess of omental bursa
S 9 transverse mesocolon
S 10 jejunum
S 11 pyloric antrum
S 12 descending colon

SY 1 3rd sternocostal articulation
SY 2 3rd costochondral articulation

ST♂ Sagittal sections Thorax

56-year-old male ♂

The plane of reference is the median sagittal section. The following sections have been made on both sides of this plane of reference:
7 sections, spaced 2 cm apart, to the right of the median plane;
6 sections, spaced 2 cm apart, to the left of the median plane.
The superior border of most sections passes through the center of the 5th cervical vertebral body [C 5]; the inferior border of each section passes through the center of the 4th lumbar vertebral body [L 4].

ST♂7D	sagittal section 14 cm to the right of the median plane, exterior view
ST♂6D	sagittal section 12 cm to the right of the median plane, exterior view
ST♂5D	sagittal section 10 cm to the right of the median plane, exterior view
ST♂4D	sagittal section 8 cm to the right of the median plane, exterior view
ST♂3D	sagittal section 6 cm to the right of the median plane, exterior view
ST♂2D	sagittal section 4 cm to the right of the median plane, exterior view
ST♂1D	sagittal section 2 cm to the right of the median plane, exterior view
ST♂MD	median sagittal section, right segment of the body, left-to-right view
ST♂MSi	median sagittal section, left segment of the body, right-to-left view
ST♂1Si	2 cm to the left of the median plane, exterior view
ST♂2Si	4 cm to the left of the median plane, exterior view
ST♂3Si	6 cm to the left of the median plane, exterior view
ST♂4Si	8 cm to the left of the median plane, exterior view
ST♂5Si	10 cm to the left of the median plane, exterior view
ST♂6Si	12 cm to the left of the median plane, exterior view

ST♂ 7D

A 1 suprascapular a.
A 2 suprascapular v.
A 3 posterior humeral circumflex a.
A 4 axillary a.
A 5 axillary v.

M 1 trapezius m.
M 2 supraspinatus m.
M 3 deltoid m.
M 4 infraspinatus m.
M 5 teres major m.
M 6 latissimus dorsi m.
M 7 serratus anterior m.
M 8 external oblique m.
M 9 external oblique m.
M 10 pectoralis major m.
M 11 coracobrachialis m.
M 12 subscapularis m.
M 13 axillary fossa
M 14 latissimus dorsi m. (tendon)
M 15 teres major m.

N 1 radial n.
N 2 ulnar n.
N 3 median n.

O 1 spine of scapula
O 2 scapula (lateral margin)
O 3 clavicle
O 4 coracoid process
O 5 glenoid cavity

ST♂ 6D

A 1 suprascapular v.
A 2 lateral thoracic a. (lateral branch)
A 3 acromial arterial network
A 4 suprascapular a.
A 5 axillary a.
A 6 lateral thoracic v.
A 7 axillary v.

M 1 trapezius m.
M 2 supraspinatus m.
M 3 infraspinatus m.
M 4 subscapularis m.
M 5 serratus anterior m.
M 6 latissimus dorsi m.
M 7 diaphragm (costal portion)
M 8 external oblique m.
M 9 internal oblique m.
M 10 transverse abdominis m.
M 11 pectoralis major m.
M 12 pectoralis minor m.
M 13 pectoralis major m.
M 14 axillary fossa
M 15 intercostal mm.
M 16 external oblique m.
M 17 diaphragm (costal portion)
M 18 intercostal mm.
M 19 external oblique m.
M 20 transverse abdominis m.
M 21 internal oblique m.

N 1 brachial plexus (lateral cord)
N 2 radial n.
N 3 median nerve (medial root)
N 4 ulnar n.

O 1 spine of scapula
O 2 inferior angle of scapula
O 3 clavicle
O 4 coracoid process

S 1 costodiaphragmatic pleural recess
S 2 liver (right lobe)

SY 1 coracoclavicular ligament

ST♂ 5D

A 1 suprascapular v.
A 2 suprascapular a.
A 3 thoracoacromial a. (acromial branch)
A 4 axillary a.
A 5 axillary v.
A 6 lateral thoracic a.

M 1 trapezius m.
M 2 supraspinatus m.

M 3 subscapularis m.
M 4 infraspinatus m.
M 5 serratus anterior m.
M 6 latissimus dorsi m.
M 7 serratus posterior inferior m.
M 8 internal oblique m.
M 9 fat body of lumbar triangle
M 10 subclavius m.
M 11 axillary fossa
M 12 pectoralis major m.
M 13 pectoralis minor m.
M 14 diaphragm (costal portion)
M 15 transverse abdominis m.
M 16 internal oblique m.

N 1 suprascapular n.
N 2 brachial plexus
N 3 brachial plexus
N 4 brachial plexus

O 1 spine of scapula
O 2 clavicle

S 1 right lung (superior lobe)
S 2 horizontal fissure
S 3 right lung (inferior lobe)
S 4 costodiaphragmatic recess
S 5 perirenal fat
S 6 right kidney
S 7 transverse colon
S 8 right lung (middle lobe)
S 9 oblique fissure
S 10 right hepatic lobe
S 11 gall bladder
S 12 transverse colon

ST♂ 4D

A 1 transverse cervical v.
A 2 5th posterior intercostal v.
A 3 10th posterior intercostal v.
A 4 portal v. (right branch)
A 5 hepatic v.
A 6 subscapular v.
A 7 suprascapular a. (opening)
A 8 subclavian a.
A 9 subclavian v.
A 10 pericardiophrenic v.

M 1 trapezius m.
M 2 supraspinatus m.
M 3 subscapularis m.
M 4 serratus anterior m.
M 5 serratus anterior m.
M 6 latissimus dorsi m.
M 7 diaphragm (costal portion)
M 8 thoracolumbar fascia
M 9 psoas m.
M 10 fat body of lumbar triangle
M 11 platysma m.
M 12 omohyoid m.
M 13 serratus anterior m.
M 14 subclavius m.
M 15 pectoralis major m.
M 16 pectoralis minor m.
M 17 diaphragm (costal portion)
M 18 sheath of rectus abdominis m.
M 19 transverse abdominis m.

N 1 trunk of brachial plexus

O 1 spine of scapula
O 2 clavicle

S 1 right lung (superior lobe)
S 2 right lung (inferior lobe)
S 3 oblique fissure
S 4 costodiaphragmatic recess
S 5 perirenal fat
S 6 right kidney
S 7 right lobe of liver
S 8 gall bladder
S 9 greater omentum
S 10 transverse colon

ST♂ 3D

A 1 right pulmonary a. (lateral basal branch)
A 2 right inferior pulmonary v. (superior basal branch)
A 3 right pulmonary a. (posterior basal branch)
A 4 right hepatic v.
A 5 portal v. (right branch)
A 6 renal v.
A 7 renal v.
A 8 external jugular v.
A 9 external jugular v.
A 10 deep cervical supraclavicular lymph node
A 11 subclavian a.
A 12 suprascapular v.
A 13 subclavian v.
A 14 right superior pulmonary v.
A 15 right superior pulmonary v. (apical branch)
A 16 right pulmonary a. (anterior branch)
A 17 right superior pulmonary v. (anterior branch)
A 18 right pulmonary a. (middle lobe branch)
A 19 right superior pulmonary v. (middle lobe branch)
A 20 pericardiophrenic a.
A 21 hepatic v.

M 1 trapezius m.
M 2 scalenus posterior m.
M 3 levator scapulae m.
M 4 rhomboid m.
M 5 intercostal mm.
M 6 latissimus dorsi m.
M 7 diaphragm (lumbar portion)
M 8 erector spinae m.
M 9 quadratus lumborum m.
M 10 sternocleidomastoid m.
M 11 scalenus medius m.
M 12 omohyoid m.
M 13 subclavius m.
M 14 pectoralis major m.
M 15 intercostal mm.
M 16 diaphragm (costal portion)
M 17 rectus abdominis m.
M 18 transverse abdominis m.

N 1 brachial plexus
N 2 brachial plexus
N 3 brachial plexus

O 1 clavicle

S 1 right lung (superior lobe)
S 2 right lung (inferior lobe)
S 3 costodiaphragmatic recess
S 4 right lobe of liver
S 5 right kidney (superior pole)
S 6 minor calyx
S 7 renal sinus
S 8 renal pelvis
S 9 duodenum (descending portion)
S 10 perirenal fat
S 11 anterior bronchial segment
S 12 anterior bronchial segment
S 13 oblique fissure
S 14 right lung (middle lobe)
S 15 right lobe of liver
S 16 cystic duct
S 17 gall bladder
S 18 duodenum (superior portion)
S 19 pylorus
S 20 greater omentum
S 21 transverse colon
S 22 transverse mesocolon
S 23 small intestine

ST♂ 2D

A 1 bronchopulmonary lymph node
A 2 right pulmonary a.
A 3 right inferior pulmonary v. (apical branch)
A 4 right inferior pulmonary v. (common basal branch)
A 5 10th posterior intercostal v.
A 6 inferior vena cava
A 7 renal a. (branch)
A 8 renal v.
A 9 renal a. (branch)

A 10 right testicular v.
A 11 internal jugular v.
A 12 subclavian a.
A 13 internal thoracic a.
A 14 jugular venous arch
A 15 internal thoracic a.
A 16 right pulmonary a. (apical branch)
A 17 right superior pulmonary v. (anterior branch)
A 18 right pulmonary a. (anterior branch)
A 19 right superior pulmonary v. (posterior branch)
A 20 right superior pulmonary v. (middle lobe branch)
A 21 right atrium
A 22 pericardial cavity
A 23 hepatic v.
A 24 portal v. (left branch)
A 25 portal v.
A 26 proper hepatic a.

M 1 trapezius m.
M 2 serratus posterior superior m.
M 3 trapezius m.
M 4 rhomboid m.
M 5 erector spinae m.
M 6 intercostal mm.
M 7 diaphragm (lateral crus)
M 8 erector spinae m.
M 9 quadratus lumborum m.
M 10 psoas m.
M 11 sternocleidomastoid m.
M 12 scalenus posterior m.
M 13 scalenus medius m.
M 14 omohyoid m.
M 15 scalenus anterior m.
M 16 pectoralis major m.
M 17 diaphragm (costal portion)
M 18 rectus abdominis m.

N 1 brachial plexus (superior trunk)
N 2 brachial plexus (inferior trunk)

O 1 transverse process T 1
O 2 costal process L 1
O 3 clavicle
O 4 1st costal cartilage

S 1 right lung (apex)
S 2 apical bronchial segment
S 3 dorsal bronchial segment
S 4 ventral bronchial segment
S 5 right middle lobe bronchus
S 6 right inferior lobe bronchus
S 7 right lung (inferior lobe)
S 8 costodiaphragmatic pleural recess
S 9 right lobe of liver
S 10 suprarenal gland
S 11 perirenal fat
S 12 right kidney
S 13 right lung (superior lobe)
S 14 quadrate lobe of liver
S 15 common hepatic duct
S 16 cystic duct
S 17 greater omentum
S 18 duodenum (superior portion)
S 19 pyloric antrum
S 20 duodenum (descending portion)
S 21 transverse colon
S 22 small intestine

ST♂ 1D

A 1 right superior intercostal v.
A 2 bronchopulmonary lymph node
A 3 left atrium
A 4 right inferior pulmonary v.
A 5 pericardial cavity
A 6 10th posterior superior intercostal v. (spinal branch)
A 7 inferior vena cava
A 8 inferior phrenic a.
A 9 right renal a.
A 10 common trunk 1st and 2nd lumbar vv.
A 11 3rd lumbar a.
A 12 middle thyroid v.
A 13 posterior mediastinal lymph node
A 14 jugular venous arch
A 15 common carotid a.
A 16 subclavian a.
A 17 tracheal lymph node
A 18 superior vena cava
A 19 azygos v.
A 20 azygos v.
A 21 ascending aorta
A 22 right pulmonary a.
A 23 superior vena cava
A 24 right auricle
A 25 right atrium
A 26 right coronary a.
A 27 tricuspid valve (anterior cusp)
A 28 right ventricle
A 29 right coronary a.
A 30 proper hepatic a.
A 31 gastroduodenal a.
A 32 portal v.
A 33 inferior vena cava

M 1 splenius m.
M 2 semispinalis capitis m.
M 3 trapezius m.
M 4 diaphragm (lumbar portion)
M 5 erector spinae m.
M 6 erector spinae m.
M 7 thyrohyoid m.
M 8 inferior pharyngeal constrictor m.
M 9 sternohyoid m.
M 10 longus colli m.
M 11 sternothyroid m.
M 12 sternocleidomastoid m.
M 13 pectoralis major m.
M 14 transversus thoracis m.
M 15 diaphragm (costal portion)
M 16 rectus abdominis m.
M 17 rectus abdominis m.

N 1 8th cervical n. (ventral ramus)
N 2 11th thoracic n.
N 3 2nd lumbar n.

O 1 transverse process T 5
O 2 neck of rib
O 3 pedicle T 12
O 4 lamina L 1
O 5 clavicle (sternal end)
O 6 manubrium
O 7 7th costal cartilage

S 1 right lung (medial side, vertebral portion)
S 2 right main bronchus
S 3 right lung (medial side)
S 4 thyroid cartilage
S 5 retrovisceral space
S 6 thyroid gland (right lobe)
S 7 anterior mediastinum
S 8 lesser omentum
S 9 left lobe of liver
S 10 caudate lobe of liver
S 11 lesser omentum
S 12 greater omentum
S 13 stomach (pyloric portion)
S 14 head of pancreas
S 15 transverse colon
S 16 small intestine
S 17 duodenum (horizontal portion)
S 18 mesentery

SY 1 articulation of head of 9th rib
SY 2 intervertebral disc L 1–L 2
SY 3 sternoclavicular articulation (articular disc)
SY 4 4th sternocostal articulation

ST♂ MD

A 1 anterior mediastinal lymph node
A 2 brachiocephalic trunk
A 3 left brachiocephalic v.
A 4 pericardial cavity
A 5 pericardial cavity
A 6 ascending aorta
A 7 right pulmonary a.
A 8 right auricle
A 9 transverse pericardial sinus
A 10 right coronary a. (branch)
A 11 right coronary a.
A 12 tricuspid valve (anterior cusp)

A 13 right coronary a. (posterior interventricular branch)
A 14 right ventricle
A 15 right coronary a. (terminal branch)
A 16 hepatic v.
A 17 common hepatic a.
A 18 right renal a.
A 19 origin of portal v.
A 20 left renal v.
A 21 abdominal aorta
A 22 superior mesenteric a.
A 23 azygos v.
A 24 inferior tracheobronchial lymph node
A 25 oblique pericardial sinus
A 26 left atrium
A 27 interatrial septum
A 28 right atrium

M 1 sternohyoid m.
M 2 diaphragm (lumbar portion)
M 3 trapezius m.
M 4 erector spinae m.
M 5 trapezius m.

N 1 spinal cord
N 2 spinal n. T 1 (roots)
N 3 spinal dura mater
N 4 spinal cord
N 5 conus medullaris
N 6 cauda equina

O 1 manubrium
O 2 sternal foramen
O 3 xiphoid process
O 4 spinous process C 7
O 5 spinous process T 12

S 1 ventricle of larynx
S 2 pharynx (laryngeal portion)
S 3 larynx (infraglottic space)
S 4 thyroid gland (isthmus)
S 5 trachea
S 6 esophagus
S 7 carina of trachea
S 8 liver (left lobe)
S 9 liver (caudate lobe)
S 10 lesser omentum
S 11 body of pancreas
S 12 stomach (pyloric portion)
S 13 inferior omental recess
S 14 transverse mesocolon
S 15 transverse colon
S 16 small intestine
S 17 mesentery
S 18 duodenum (horizontal portion)
S 19 esophagus

SY 1 ligamentum flavum
SY 2 articulating processes T 3–T 4
SY 3 ligamentum flavum
SY 4 nucleus pulposus
SY 5 interspinal ligament
SY 6 supraspinal ligament

ST♂ MSi

A 1 azygos v.
A 2 tracheobronchial lymph node
A 3 oblique pericardial sinus
A 4 left atrium
A 5 interatrial septum
A 6 right renal a. (opening)
A 7 abdominal aorta
A 8 left renal v.
A 9 3rd lumbar a. (opening)
A 10 anterior mediastinal lymph node
A 11 brachiocephalic trunk
A 12 left brachiocephalic v.
A 13 left common carotid a.
A 14 arch of aorta
A 15 pericardial cavity
A 16 ascending aorta
A 17 right pulmonary a.
A 18 aortic bulb
A 19 right coronary a. (ventricular branch)
A 20 right coronary a. (ventricular branch)
A 21 right coronary a.
A 22 aortic valve
A 23 tricuspid valve (anterior cusp)
A 24 anterior papillary m.
A 25 tricuspid valve (posterior cusp)
A 26 right coronary a. (posterior interventricular)
A 27 right coronary a. (terminal branch)
A 28 hepatic v.
A 29 left gastric v.
A 30 common hepatic a.
A 31 splenic v.
A 32 superior mesenteric v.
A 33 inferior mesenteric a. (opening)
A 34 superior mesenteric a. (branch)

M 1 trapezius m.
M 2 transversospinalis m.
M 3 vocalis m.
M 4 diaphragm (lumbar portion)
M 5 linea alba

N 1 spinal cord
N 2 spinal dura mater
N 3 conus medullaris
N 4 cauda equina
N 5 spinal dura mater

O 1 spinous process C 7
O 2 manubrium
O 3 sternal foramen
O 4 xiphoid process

S 1 pharynx (laryngeal portion)
S 2 thyroid cartilage
S 3 lamina of cricoid cartilage
S 4 arch of cricoid cartilage
S 5 esophagus
S 6 isthmus of thyroid gland
S 7 trachea
S 8 carina of trachea
S 9 esophagus
S 10 greater omentum
S 11 left lobe of liver
S 12 lesser omentum
S 13 body of pancreas
S 14 stomach (pyloric portion)
S 15 transverse colon
S 16 duodenum (horizontal portion)
S 17 mesentery

SY 1 supraspinal ligament
SY 2 ligamentum flavum
SY 3 interspinal ligament
SY 4 intervertebral disc

ST♂ 1 Si

A 1 inferior thyroid a.
A 2 jugular venous arch
A 3 left common carotid a.
A 4 left brachiocephalic v.
A 5 left subclavian a.
A 6 arch of aorta
A 7 pericardial cavity
A 8 pericardial cavity
A 9 pulmonary trunk
A 10 left coronary a.
A 11 right semilunar valve
A 12 aortic bulb
A 13 aortic valve
A 14 right ventricle (conus arteriosus)
A 15 left ventricle
A 16 tricuspid valve (anterior cusp)
A 17 anterior papillary m.
A 18 middle cardiac v.
A 19 splenic a.
A 20 splenic v.
A 21 superior mesenteric a.
A 22 jejunal v.
A 23 jejunal v.
A 24 vertebral a.
A 25 superior tracheobronchial lymph node
A 26 transverse pericardial sinus
A 27 descending aorta
A 28 oblique pericardial sinus
A 29 left atrium
A 30 coronary sinus
A 31 11th intercostal a. (opening)

A 32 12th intercostal v.
A 33 inferior phrenic a.
A 34 left renal a.
A 35 inferior phrenic v.
A 36 lumbar azygos v.
A 37 2nd lumbar v.
A 38 left renal v.

M 1 sternohyoid m.
M 2 inferior pharyngeal constrictor m.
M 3 longus colli m.
M 4 sternothyroid m.
M 5 sternocleidomastoid m.
M 6 pectoralis major m.
M 7 diaphragm (esophageal hiatus)
M 8 rectus abdominis m.
M 9 semispinalis capitis m.
M 10 splenius m.
M 11 trapezius m.
M 12 diaphragm (lumbar portion)
M 13 diaphragm (medial crus)
M 14 erector spinae m.

N 1 spinal n. T 3
N 2 spinal n. L 3

O 1 head of 1st rib
O 2 manubrium
O 3 3rd costal cartilage
O 4 xiphoid process
O 5 superior articulating process T 2
O 6 pedicle T 2
O 7 transverse process T 8
O 8 transverse process T 11
O 9 costal process L 1

S 1 thyroid cartilage
S 2 thyroid gland (left lobe)
S 3 greater omentum
S 4 left lobe of liver
S 5 stomach (pyloric portion)
S 6 inferior omental recess
S 7 head of pancreas
S 8 small intestine
S 9 transverse colon
S 10 mesentery
S 11 duodenum (horizontal portion)
S 12 left main bronchus
S 13 esophagus (thoracic portion)
S 14 esophagus (thoracic portion)
S 15 esophagus (thoracic portion)
S 16 stomach (cardiac portion)

SY 1 sternoclavicular articulation (articular disc)
SY 2 articulating processes T 3–T 4
SY 3 articulation of head of 7th rib
SY 4 intervertebral disc L 2–L 3

ST♂ 2 Si

A 1 internal jugular v.
A 2 deep cervical lymph node
A 3 subclavian a.
A 4 jugular venous arch
A 5 subclavian v.
A 6 internal thoracic a.
A 7 left auricle
A 8 left coronary a. (anterior interventricular branch)
A 9 left coronary a. (circumflex branch)
A 10 right ventricle (conus arteriosus)
A 11 mitral valve (anterior cusp)
A 12 interventricular septum
A 13 papillary muscle
A 14 left ventricle
A 15 aortic arch
A 16 left pulmonary a.
A 17 transverse pericardial sinus
A 18 left pulmonary v.
A 19 thoracic aorta
A 20 left inferior pulmonary v.
A 21 left atrium
A 22 coronary sinus
A 23 mitral valve (posterior cusp)
A 24 10th posterior intercostal v.
A 25 posterior papillary mm.
A 26 inferior phrenic a.
A 27 splenic a.
A 28 left renal v.
A 29 left renal a. (anterior branch)
A 30 left renal a. (posterior branch)
A 31 left renal v.
A 32 left colic a.
A 33 inferior mesenteric v.

M 1 sternocleidomastoid m.
M 2 omohyoid m.
M 3 scalenus anterior m.
M 4 pectoralis major m.
M 5 transversus thoracis m.
M 6 diaphragm (costal portion)
M 7 rectus abdominis m.
M 8 transversospinalis m.
M 9 splenius m.
M 10 scalenus medius m.
M 11 trapezius m.
M 12 rhomboid m.
M 13 erector spinae m.
M 14 quadratus lumborum m.
M 15 quadratus lumborum m.
M 16 psoas major m.

N 1 brachial plexus (superior trunk)
N 2 8th cervical n. (ventral ramus)
N 3 1st thoracic n. (ventral ramus)

O 1 clavicle
O 2 1st costal cartilage
O 3 transverse process T 3
O 4 3rd costal tubercle
O 5 costal process L 3

S 1 left lung (superior lobe)
S 2 left lobe of liver
S 3 greater omentum
S 4 body of stomach
S 5 stomach (pyloric portion)
S 6 greater curvature of stomach
S 7 transverse mesocolon
S 8 body of pancreas
S 9 transverse colon
S 10 duodenojejunal flexure
S 11 duodenum (ascending portion)
S 12 jejunum
S 13 mesentery
S 14 left lung (apex)
S 15 oblique fissure
S 16 left lung (inferior lobe)
S 17 fundus of stomach
S 18 costodiaphragmatic pleural recess
S 19 perirenal fat
S 20 left suprarenal gland
S 21 left kidney

SY 1 1st costotransverse articulation
SY 2 4th costotransverse articulation

ST♂ 3 Si

A 1 subclavian a.
A 2 subclavian v.
A 3 left superior pulmonary v. (anterior branch)
A 4 left superior pulmonary v. (anterior branch)
A 5 left pulmonary a. (anterior ascending)
A 6 left superior pulmonary v. (apical posterior branch)
A 7 left pulmonary a. (lingular branch)
A 8 left superior pulmonary v. (lingular branch)
A 9 left superior pulmonary v. (lingular branch)
A 10 left coronary a. (anterior interventricular branch)
A 11 anterior papillary m.
A 12 left coronary a. (circumflex branch)
A 13 left ventricle
A 14 interventricular septum
A 15 right ventricle
A 16 bronchopulmonary lymph node
A 17 left pulmonary a. (apical branch of inferior lobe)
A 18 left pulmonary a. (medial basal branch)

A 19 left inferior pulmonary v. (inferior basal v.)
A 20 pulmonary lymph node
A 21 left inferior pulmonary v. (inferior basal v.)
A 22 splenic v.
A 23 left renal a.
A 24 left renal a.
A 25 left renal v.
A 26 left renal a.
A 27 left renal v.

M 1 platysma m.
M 2 sternocleidomastoid m.
M 3 scalenus medius m.
M 4 omohyoid m.
M 5 scalenus anterior m.
M 6 subclavius m.
M 7 pectoralis major m.
M 8 diaphragm (costal portion)
M 9 rectus abdominis m.
M 10 levator scapulae m.
M 11 scalenus posterior m.
M 12 serratus anterior m.
M 13 intercostal mm.
M 14 serratus posterior superior m.
M 15 trapezius m.
M 16 rhomboid major m.
M 17 latissimus dorsi m.
M 18 diaphragm
M 19 quadratus lumborum m.
M 20 erector spinae m.

N 1 trunk of brachial plexus
N 2 trunk of brachial plexus

O 1 clavicle
O 2 scapula (medial margin)

S 1 superior lingular bronchus
S 2 inferior lingular bronchus
S 3 left lobe of liver
S 4 greater omentum
S 5 transverse colon
S 6 body of stomach
S 7 transverse colon
S 8 transverse mesocolon
S 9 body of pancreas
S 10 small intestine
S 11 mesentery
S 12 jejunum
S 13 mesentery
S 14 left lung (apex)
S 15 oblique fissure
S 16 left lung (inferior lobe, apical segment)
S 17 left superior lobar bronchus
S 18 anterior basal segmental bronchus
S 19 lateral basal segmental bronchus
S 20 posterior basal segment
S 21 costodiaphragmatic pleural recess
S 22 spleen
S 23 inferior omental recess
S 24 left kidney
S 25 renal pelvis
S 26 perirenal fat
S 27 descending colon

SY 1 5th & 6th interchondral articulation
SY 2 6th & 7th interchondral articulation

ST♂ 4 Si

A 1 superficial cervical v.
A 2 transverse cervical v.
A 3 subclavian a.
A 4 subclavian v.
A 5 left superior pulmonary v. (apical posterior branch)
A 6 left superior pulmonary v. (lingular branch)
A 7 left superior pulmonary v. (lingular branch)
A 8 left pulmonary a. (inferior lingular branch)
A 9 left superior pulmonary v. (lingular branch)
A 10 left coronary a. (ventricular branch)
A 11 left ventricle
A 12 fibrous pericardium
A 13 8th posterior intercostal v.
A 14 left pulmonary a. (lateral basal branch)
A 15 left inferior pulmonary v. (inferior basal v.)
A 16 left inferior pulmonary v. (inferior basal v.)
A 17 splenic v.
A 18 left colic a.

M 1 omohyoid m.
M 2 subclavius m.
M 3 external intercostal m.
M 4 pectoralis minor m.
M 5 pectoralis major m.
M 6 diaphragm (costal portion)
M 7 sheath of rectus abdominis
M 8 transverse abdominis m.
M 9 rectus abdominis m.
M 10 trapezius m.
M 11 serratus anterior m.
M 12 supraspinatus m.
M 13 subscapularis m.
M 14 serratus anterior m.
M 15 infraspinatus m.
M 16 serratus anterior m.
M 17 serratus anterior m.
M 18 latissimus dorsi m.
M 19 diaphragm
M 20 thoracolumbar fascia
M 21 quadratus lumborum m.
M 22 fat body of lumbar triangle

N 1 trunk of brachial plexus
N 2 trunk of brachial plexus

O 1 clavicle
O 2 spine of scapula
O 3 scapula (medial margin)

S 1 left lung (superior lobe)
S 2 left lung (lingula)
S 3 anterior mediastinum
S 4 transverse colon
S 5 greater omentum
S 6 body of stomach
S 7 inferior omental recess
S 8 tail of pancreas
S 9 transverse mesocolon
S 10 jejunum
S 11 mesentery
S 12 small intestine
S 13 small intestine
S 14 jejunum
S 15 left lung (inferior lobe, apical segment)
S 16 oblique fissure
S 17 peritoneal cavity
S 18 costodiaphragmatic pleural recess
S 19 spleen
S 20 perirenal fat
S 21 left kidney
S 22 descending colon

ST♂ 5 Si

A 1 suprascapular a.
A 2 thoracoacromial a.
A 3 axillary a.
A 4 axillary v.
A 5 musculophrenic a.

M 1 subclavius m.
M 2 intercostal mm.
M 3 pectoralis major m.
M 4 serratus anterior m.
M 5 pectoralis minor m.
M 6 pectoralis major m.
M 7 pectoralis major m. (abdominal portion)
M 8 diaphragm (costal portion)
M 9 transverse abdominis m.
M 10 internal oblique m.
M 11 external oblique m.
M 12 trapezius m.
M 13 supraspinatus m.
M 14 subscapularis m.

M 15 serratus anterior m.
M 16 subscapularis m.
M 17 infraspinatus m.
M 18 subscapularis m.
M 19 serratus anterior m.
M 20 latissimus dorsi m.
M 21 diaphragm (costal portion)
M 22 external oblique m.
M 23 internal oblique m.
M 24 transverse abdominis m.
M 25 fat body of lumbar triangle

N 1 brachial plexus (lateral root)
N 2 brachial plexus (medial root)

O 1 clavicle
O 2 scapula (superior margin)
O 3 spine of scapula
O 4 scapula (inferior angle)

S 1 greater omentum
S 2 jejunum
S 3 left lung (superior lobe)
S 4 oblique fissure
S 5 left lung (inferior lobe)
S 6 costodiaphragmatic recess
S 7 spleen
S 8 left colic flexure
S 9 left kidney
S 10 perirenal fat
S 11 descending colon
S 12 small intestine

SY 1 subdeltoid bursa
SY 2 7th & 8th interchondral articulation

ST♂ 6 Si

A 1 thoracoacromial a. (pectoral branch)
A 2 axillary a.
A 3 thoracoacromial v.
A 4 axillary v.
A 5 subscapular a.
A 6 axillary lymph node
A 7 suprascapular a.

M 1 deltoid m. (clavicular portion)
M 2 pectoralis minor m.
M 3 pectoralis major m.
M 4 axillary fossa
M 5 serratus anterior m.
M 6 external intercostal m.
M 7 serratus anterior m.
M 8 internal intercostal m.
M 9 endothoracic fascia
M 10 transverse abdominis m.
M 11 external oblique m.
M 12 internal oblique m.
M 13 transverse abdominis m.
M 14 trapezius m.
M 15 supraspinatus m.
M 16 deltoid m. (scapular portion)
M 17 infraspinatus m.
M 18 subscapularis m.
M 19 teres major m.
M 20 latissimus dorsi m.
M 21 serratus anterior m.
M 22 diaphragm (costal portion)
M 23 latissimus dorsi m.
M 24 fat body of lumbar triangle
M 25 external oblique m.
M 26 internal oblique m.
M 27 transverse abdominis m.

N 1 brachial plexus (lateral root)
N 2 brachial plexus (posterior root)
N 3 brachial plexus (medial root)

O 1 clavicle
O 2 coracoid process
O 3 spine of scapula
O 4 scapula (lateral margin)

S 1 spleen
S 2 greater omentum
S 3 transverse colon
S 4 small intestine

SY 1 coracoclavicular ligament
SY 2 subscapular bursa

ST♀ Sagittal sections Thorax

25-year-old female ♀

The plane of reference is the median plane. The section 9 cm to the left and the right of the median plane deals with the nipple. On the right-hand side, a section has been made on both sides 2 cm away from the nipple. The inferior border of each section passes through the intervertebral disc L 3–L 4.

ST♀ 1 D sagittal section of the female thorax, 7 cm to the right of the median plane, 2 cm to the left of the nipple, exterior view

ST♀ 2 D sagittal section of the female thorax, 9 cm to the right of the median plane, passing through the nipple, exterior view

ST♀ 3 D sagittal section of the female thorax, 11 cm to the right of the median plane, exterior view

ST♀ Si sagittal section of the female thorax, 9 cm to the left of the median plane, passing through the left nipple, exterior view

ST♀ 1 D

A 1 transverse cervical v. (superficial branch)
A 2 4th posterior intercostal a.
A 3 7th posterior intercostal a.
A 4 right hepatic v.
A 5 1st lumbar v.
A 6 transverse cervical a. (superficial branch)
A 7 transverse cervical a. (deep branch)
A 8 transverse cervical a. (deep branch)
A 9 transverse cervical a. (deep branch)
A 10 transverse cervical a.

A 11 cephalic v.
A 12 axillary a.
A 13 axillary v.
A 14 portal v. (right branch)
A 15 right hepatic v.
A 16 right colic a.

M 1 trapezius m.
M 2 supraspinatus m.
M 3 serratus anterior m.
M 4 serratus posterior superior m.
M 5 rhomboid m.
M 6 diaphragm (central tendon)
M 7 latissimus dorsi m.
M 8 diaphragm (costal portion)
M 9 serratus posterior inferior m.
M 10 erector spinae m.
M 11 thoracolumbar fascia (deep layer)
M 12 quadratus lumborum m.
M 13 supraclavicular fossa
M 14 platysma m.
M 15 omohyoid m.
M 16 subclavius m.
M 17 pectoralis major m.
M 18 pectoralis minor m.
M 19 intercostal mm.
M 20 diaphragm (costal portion)
M 21 rectus abdominis m.
M 22 transverse abdominis m.
M 23 external oblique m.
M 24 internal oblique m.

N 1 brachial plexus

O 1 scapula (medial margin)

S 1 right lung (superior lobe)
S 2 oblique fissure
S 3 right lung (inferior lobe)
S 4 costodiaphragmatic pleural recess
S 5 right kidney
S 6 renal calyx
S 7 duodenum (descending portion)
S 8 perirenal fat
S 9 ascending colon
S 10 horizontal fissure
S 11 mammary gland
S 12 oblique fissure
S 13 right lung (middle lobe)
S 14 liver (right lobe)
S 15 costodiaphragmatic pleural recess
S 16 gall bladder
S 17 stomach (pyloric portion)
S 18 transverse colon
S 19 small intestine

SY 1 5th costochondral articulation
SY 2 7th & 8th interchondral articulation

ST♀2D

A 1 transverse cervical a.
A 2 8th posterior intercostal a.
A 3 1st lumbar a.
A 4 transverse cervical a.
A 5 cephalic v.
A 6 thoracoacromial a.
A 7 thoracoacromial a.
A 8 axillary a.
A 9 axillary v.
A 10 lateral thoracic a.
A 11 portal v.
A 12 hepatic v.

M 1 trapezius m.
M 2 supraspinatus m.
M 3 subscapularis m.
M 4 infraspinatus m.
M 5 serratus anterior m.
M 6 intercostal mm.
M 7 serratus anterior m.
M 8 latissimus dorsi m.
M 9 serratus posterior inferior m.
M 10 diaphragm (costal portion)
M 11 erector spinae m.
M 12 serratus posterior inferior m.
M 13 latissimus dorsi m.
M 14 fat body of lumbar triangle
M 15 internal oblique m.
M 16 omohyoid m.
M 17 subclavius m.
M 18 serratus anterior m.
M 19 intercostal mm.
M 20 pectoralis major m.
M 21 pectoralis minor m.
M 22 intercostal mm.
M 23 external oblique m.
M 24 diaphragm (costal portion)
M 25 transverse abdominis m.
M 26 internal oblique m.
M 27 external oblique m.

N 1 brachial plexus

O 1 scapula (superior margin)
O 2 spine of scapula
O 3 scapula (inferior angle)
O 4 clavicle

S 1 right lung (superior lobe)
S 2 oblique fissure
S 3 right lung (inferior lobe)
S 4 costodiaphragmatic pleural recess
S 5 right kidney
S 6 perirenal fat
S 7 right colic flexure
S 8 ascending colon
S 9 horizontal fissure
S 10 oblique fissure
S 11 right lung (middle lobe)
S 12 areola of mammary gland
S 13 mammary gland
S 14 papilla of mammary gland
S 15 liver (right lobe)
S 16 stomach (pyloric portion)
S 17 transverse colon
S 18 small intestine

SY 1 7th costochondral articulation

ST♀3D

A 1 suprascapular v.
A 2 8th posterior intercostal a.
A 3 inferior phrenic v.
A 4 11th posterior intercostal a.
A 5 thoracoacromial v.
A 6 cephalic v.
A 7 thoracoacromial a. (pectoralis branch)
A 8 axillary a.
A 9 subscapular a. (opening)
A 10 axillary v.
A 11 lateral thoracic a.
A 12 superficial epigastric v.

M 1 trapezius m.
M 2 supraspinatus m.
M 3 infraspinatus m.
M 4 subscapularis m.
M 5 serratus anterior m.
M 6 teres major m.
M 7 serratus anterior m.
M 8 latissimus dorsi m.
M 9 serratus posterior inferior m.
M 10 intercostal mm.
M 11 internal oblique m.
M 12 fat body of lumbar triangle
M 13 external oblique m.
M 14 deltoid m. (clavicular portion)
M 15 axillary fossa
M 16 pectoralis major m.
M 17 pectoralis minor m.
M 18 serratus anterior m.
M 19 diaphragm
M 20 intercostal mm.
M 21 external oblique m.
M 22 intercostal mm.
M 23 transverse abdominis m.
M 24 internal oblique m.

N 1 brachial plexus

O 1 spine of scapula
O 2 scapula (lateral margin)
O 3 clavicle
O 4 coracoid process

S 1 right lung (inferior lobe)
S 2 costodiaphragmatic pleural recess

S 3 mammary gland
S 4 papilla of mammary gland
S 5 costodiaphragmatic recess
S 6 liver (right lobe)
S 7 peritoneal cavity

SY 1 coracoclavicular ligament
SY 2 9th costochondral articulation

ST♀Si

A 1 cephalic v.
A 2 thoracoacromial a.
A 3 axillary a.
A 4 axillary v.
A 5 pericardiophrenic a.
A 6 transverse cervical v.
A 7 transverse cervical a.
A 8 8th posterior intercostal a.

M 1 omohyoid m.
M 2 subclavius m.
M 3 platysma m.
M 4 pectoralis major m.
M 5 pectoralis minor m.
M 6 intercostal mm.
M 7 diaphragm (costal portion)
M 8 external oblique m.
M 9 intercostal mm.
M 10 internal oblique m.
M 11 transverse abdominis m.
M 12 trapezius m.
M 13 supraspinatus m.
M 14 infraspinatus m.
M 15 subscapularis m.
M 16 serratus anterior m.
M 17 latissimus dorsi m.
M 18 intercostal mm.
M 19 diaphragm (costal portion)
M 20 serratus posterior inferior m.
M 21 latissimus dorsi m.
M 22 internal oblique m.
M 23 transverse abdominis m.
M 24 external oblique m.
M 25 fat body of lumbar triangle

N 1 brachial plexus

O 1 clavicle
O 2 spine of scapula
O 3 scapula (inferior angle)

S 1 left lung (superior lobe)
S 2 mammary gland
S 3 areola of mammary gland
S 4 papilla of mammary gland
S 5 costodiaphragmatic pleural recess
S 6 left colic flexure
S 7 greater omentum
S 8 oblique fissure
S 9 left lung (inferior lobe)
S 10 body of stomach
S 11 costodiaphragmatic recess
S 12 spleen

SY 1 8th costochondral articulation

HT
Horizontal sections Thorax

40-year-old male ♂

Transverse sections of the trunk, perpendicular to the horizontal plane upon which the subject is lying. The plane of reference passes through the antero-superior iliac spines. The sections are spaced 2.5 cm apart. Only sections of interest for the thoracic region have been retained. All of the sections are viewed from the inferior side.

HT 1 horizontal section of the thorax passing through the upper part of the 1st thoracic vertebra [T 1] and the disc C 7–T 1, inferior-superior view
HT 2 horizontal section of the thorax passing through the upper part of the body of the 2nd thoracic vertebra [T 2], inferior-superior view
HT 3 horizontal section of the thorax passing through the upper part of the body of the 3rd thoracic vertebra [T 3] and the disc T 2–T 3, inferior-superior view
HT 4 horizontal section of the thorax passing through the lower part of the body of the 4th thoracic vertebra [T 4], inferior-superior view
HT 5 horizontal section of the thorax passing through the lower part of the body of the 5th thoracic vertebra [T 5], inferior-superior view
HT 6 horizontal section of the thorax passing through the lower part of the body of the 6th thoracic vertebra [T 6], inferior-superior view
HT 7 horizontal section of the thorax passing through the lower part of the body of the 7th thoracic vertebra [T 7], inferior-superior view
HT 8 horizontal section of the thorax passing through the lower part of the body of the 8th thoracic vertebra [T 8], inferior-superior view
HT 9 horizontal section of the thorax passing through the center of the body of the 9th thoracic vertebra [T 9], inferior-superior view
HT 10 horizontal section of the thoraco-abdominal region passing through the upper part of the body of the 10th thoracic vertebra [T 10], inferior-superior view
HT 11 horizontal section of the thoraco-abdominal region passing through the upper part of the body of the 11th thoracic vertebra [T 11], inferior-superior view
HT 12 horizontal section of the thoraco-abdominal region passing through the upper part of the body of the 12th thoracic vertebra [T 12 sup.], inferior-superior view
HT 13 horizontal section of the thoraco-abdominal region passing through the lower part of the

body of the 12th thoracic vertebra [T 12 inf.], inferior-superior view

HT 14 horizontal section of the thoraco-abdominal region passing through the central part of the body of the 1st lumbar vertebra [L 1], inferior-superior view

HT 15 horizontal section of the thoraco-abdominal region passing through the upper part of the body of the 2nd lumbar vertebra [L 2], inferior-superior view

HT 1

A 1 external jugular v.
A 2 internal jugular v.
A 3 vertebral v.
A 4 inferior thyroid a.
A 5 common carotid a.
A 6 inferior thyroid a. (glandular branch)
A 7 inferior thyroid a.
A 8 common carotid a.
A 9 vertebral a.
A 10 ascending cervical a.
A 11 external jugular v.
A 12 transverse cervical v.
A 13 deep cervical v.
A 14 deep cervical v.
A 15 internal vertebral venous plexus
A 16 transverse cervical v.

M 1 platysma m.
M 2 scalenus medius m.
M 3 scalenus anterior m.
M 4 sternocleidomastoid m.
M 5 subclavius m.
M 6 deltoid m.
M 7 trapezius m.
M 8 scalenus posterior m.
M 9 transversospinal m.
M 10 semispinalis capitis m.
M 11 splenius m.
M 12 rhomboid minor m.
M 13 serratus posterior superior m.
M 14 levator scapulae m.
M 15 trapezius m.
M 16 supraspinatus m.
M 17 infraspinatus m.
M 18 deltoid m.

N 1 6th cervical n. (ventral ramus)
N 2 7th cervical n. (ventral ramus)
N 3 8th cervical n.
N 4 phrenic n.

O 1 clavicle
O 2 clavicle
O 3 head of femur
O 4 acromion
O 5 transverse process T 1
O 6 acromion

S 1 thyroid gland (right lobe)
S 2 isthmus of thyroid gland
S 3 esophagus
S 4 trachea

SY 1 acromioclavicular articulation
SY 2 intervertebral disc C 7–T 1
SY 3 coracoclavicular ligament
SY 4 coracoacromial ligament
SY 5 articular capsule of humerus
SY 6 1st costotransverse articulation

HT 2

A 1 suprascapular v.
A 2 external jugular v.
A 3 external jugular v. (branch)
A 4 superficial cervical v.
A 5 subclavian a.
A 6 internal jugular v.
A 7 vertebral v.
A 8 inferior thyroid v.
A 9 common carotid a.
A 10 anterior jugular v.
A 11 inferior thyroid v.
A 12 common carotid a.
A 13 vertebral a.
A 14 vertebral v.
A 15 costo-cervical trunk
A 16 left superior intercostal v.
A 17 suprascapular a.
A 18 subclavian a.
A 19 external jugular v.
A 20 external jugular v.
A 21 thoracoacromial v. (branch)
A 22 transverse cervical a.
A 23 left superior intercostal v.
A 24 intervertebral v.
A 25 posterior intercostal v. (spinal branch)
A 26 dorsal scapular v.
A 27 suprascapular v.

M 1 biceps brachii m. (long head)
M 2 deltoid m.
M 3 subclavius m.
M 4 longus colli m.
M 5 sternothyroid m.
M 6 sternohyoid m.
M 7 sternocleidomastoid m.
M 8 scalenus anterior m.
M 9 subclavius m.
M 10 pectoralis major m.
M 11 pectoralis minor m.
M 12 subscapularis m.
M 13 coracobrachialis m.
M 14 biceps brachialis m. (long head)
M 15 deltoid m.
M 16 deltoid m.
M 17 infraspinatus m.
M 18 supraspinatus m.
M 19 omohyoid m.
M 20 serratus anterior m.
M 21 serratus posterior superior m.
M 22 trapezius m.
M 23 intercostal mm.
M 24 levator scapulae m.
M 25 rhomboid m.
M 26 semispinalis capitis m.
M 27 transversospinalis m.
M 28 splenius m.
M 29 trapezius m.
M 30 supraspinatus m.
M 31 serratus anterior m.
M 32 axillary fossa
M 33 infraspinatus m.

N 1 brachial plexus (superior trunk)
N 2 brachial plexus (inferior trunk)
N 3 vagus n.
N 4 reccurrent laryngeal n.
N 5 vagus n.
N 6 brachial plexus
N 7 suprascapular n.
N 8 spinal cord
N 9 spinal dura mater

O 1 head of humerus
O 2 coracoid process
O 3 clavicle
O 4 clavicle
O 5 humerus (lesser tubercle)
O 6 humerus (greater tubercle)
O 7 glenoid cavity
O 8 acromion
O 9 scapula (superior angle)
O 10 spine of scapula
O 11 glenoid cavity

S 1 right lung (apex)
S 2 trachea
S 3 esophagus

SY 1 articular capsule
SY 2 glenoid labrum
SY 3 intervertebral disc T 1–T 2
SY 4 articular processes of vertebrae T 2–T 3
SY 5 articulation of head of 2nd rib
SY 6 articular capsule

HT 3

A 1 thoracoacromial a.
A 2 lateral thoracic v.
A 3 axillary a.
A 4 axillary v.
A 5 internal thoracic a.
A 6 right angular v.
A 7 jugular venous arch
A 8 brachiocephalic trunk
A 9 inferior thyroid v. (opening)
A 10 common carotid a.
A 11 subclavian a.
A 12 left brachiocephalic v.
A 13 internal thoracic a.
A 14 subclavian v.
A 15 venous valve
A 16 thoracoacromial a. (thoracic branch)
A 17 axillary a.
A 18 thoracoacromial a.
A 19 posterior humeral circumflex a.
A 20 suprascapular a.
A 21 right superior intercostal v.
A 22 dorsal scapular v.

M 1 teres minor m.
M 2 biceps brachii m. (long head)
M 3 deltoid m.
M 4 biceps brachii m. (short head) coracobrachialis m.
M 5 pectoralis minor m.
M 6 axillary fossa
M 7 subclavius m.
M 8 pectoralis major m.
M 9 sternohyoid m.
M 10 sternothyroid m.
M 11 sternocleidomastoid m.
M 12 subclavius m.
M 13 pectoralis minor m.
M 14 pectoralis major m.
M 15 biceps brachii m. (short head) coracobrachialis m.
M 16 subscapularis m.
M 17 biceps brachii m. (long head)
M 18 deltoid m.
M 19 infraspinatus m.
M 20 serratus anterior m.
M 21 subscapularis m.
M 22 supraspinatus m.
M 23 levator scapulae m.
M 24 trapezius m.
M 25 erector spinae m.
M 26 trapezius m.
M 27 rhomboid m.
M 28 intercostal mm.
M 29 serratus anterior m.
M 30 subscapularis m.
M 31 infraspinatus m.
M 32 teres minor m.

N 1 brachial plexus (lateral cord)
N 2 brachial plexus (posterior and medial cords)
N 3 vagus n.
N 4 vagus n.
N 5 brachial plexus (medial cord, posterior cord)
N 6 brachial plexus (lateral cord)
N 7 spinal cord
N 8 spinal dura mater

O 1 humerus (greater trochanter)
O 2 humerus (lesser trochanter)
O 3 clavicle
O 4 humerus (surgical neck)
O 5 spine of scapula
O 6 scapula (medial margin)
O 7 pedicle T 3
O 8 scapula (medial margin)
O 9 infraglenoid tubercle

S 1 trachea
S 2 esophagus
S 3 superior lobe of left lung

SY 1 glenoid labrum
SY 2 sternoclavicular articulation (articular disc)
SY 3 glenoid labrum
SY 4 humerus articulation
SY 5 intervertebral disc T 2–T 3
SY 6 articulation of head of 3rd rib
SY 7 glenoid labrum
SY 8 humerus articulation

HT 4

A 1 humeral circumflex v.
A 2 axillary a.
A 3 axillary v.
A 4 internal thoracic a.
A 5 internal thoracic v.
A 6 superior vena cava
A 7 anterior mediastinal lymph node
A 8 left brachiocephalic v.
A 9 arch of aorta
A 10 internal thoracic a.
A 11 lateral thoracic v.
A 12 posterior humeral circumflex a.
A 13 3rd posterior intercostal v.
A 14 azygos v.
A 15 right superior intercostal v.
A 16 internal vertebral plexus of veins
A 17 axillary v.
A 18 scapular circumflex a.
A 19 scapular circumflex a.
A 20 axillary a.

M 1 deltoid m.
M 2 pectoralis major m. (tendon)
M 3 biceps brachii m. (long head)
M 4 latissimus dorsi and teres major tendons
M 5 coracobrachialis m. and biceps brachii m. (short head)
M 6 pectoralis major m.
M 7 pectoralis minor m.
M 8 coracobrachialis m. and biceps brachii m. (short head)
M 9 biceps brachii m. (long head)
M 10 deltoid m.
M 11 subscapularis m.
M 12 triceps brachii m. (long head)
M 13 teres minor m.
M 14 infraspinatus m.
M 15 subscapularis m.
M 16 serratus anterior m.
M 17 intercostal m.
M 18 erector spinae m.
M 19 trapezius m.
M 20 rhomboid m.
M 21 serratus anterior m.
M 22 subscapularis m.
M 23 infraspinatus m.
M 24 teres minor m.
M 25 latissimus dorsi m. (tendon)
M 26 teres major m.
M 27 triceps brachii m. (long head)
M 28 triceps brachii m. (lateral head)
M 29 deltoid m.

N 1 brachial plexus
N 2 brachial plexus
N 3 brachial plexus
N 4 right phrenic n.
N 5 right vagus n.
N 6 left recurrent laryngeal n.
N 7 left phrenic n.
N 8 left vagus n.
N 9 ulnar n.
N 10 radial n.
N 11 median n.

O 1 body of humerus
O 2 1st costal cartilage
O 3 manubrium
O 4 scapula (lateral margin)
O 5 scapula (medial margin)
O 6 superior articulating process T 5
O 7 transverse process T 5
O 8 costal tubercle

S 1 superior lobe of left lung
S 2 costal pleura
S 3 right lung (anterior margin)
S 4 trachea
S 5 esophagus
S 6 superior lobe of left lung

SY 1 supraspinal ligament
SY 2 articular processes of vertebrae T 4–T 5
SY 3 articulation of head of 5th rib

HT 5

A 1 pectoral lymph node
A 2 lateral thoraci v.
A 3 thoracoacromial a. (branch)
A 4 right superior pulmonary v. (branch)
A 5 right pulmonary a. (branch)
A 6 internal thoracic a.
A 7 superior vena cava
A 8 ascending aorta
A 9 right tracheobronchial lymph node
A 10 mediastinal v.
A 11 pericardial cavity
A 12 left tracheobronchial lymph node
A 13 pulmonary trunk
A 14 bronchopulmonary lymph node
A 15 left pulmonary a. (apical branch)
A 16 lateral thoracic v.
A 17 deep brachial v.
A 18 axillary v.
A 19 axillary a.
A 20 axillary v.
A 21 scapular circumflex v.
A 22 thoracodorsal v.
A 23 5th posterior intercostal v.
A 24 right pulmonary a. (apical branch)
A 25 azygos v.
A 26 descending aorta
A 27 thoracodorsal a.
A 28 scapular circumflex
A 29 brachial a.

M 1 biceps brachii m. (long head)
M 2 coracobrachialis m.
M 3 biceps brachii m. (short head)
M 4 pectoralis major m.
M 5 intercostal mm.
M 6 pectoralis minor m.
M 7 coracobrachialis m.
M 8 biceps brachii m. (short head)
M 9 biceps brachii m. (long head)
M 10 pectoralis major m.
M 11 deltoid m.
M 12 triceps brachii (lateral head)
M 13 triceps brachii (long head)
M 14 teres major m.
M 15 latissimus dorsi m.
M 16 teres minor m.
M 17 subscapularis m.
M 18 infraspinatus m.
M 19 serratus anterior m.
M 20 rhomboid m.
M 21 trapezius m.
M 22 erector spinae m.
M 23 trapezius m.
M 24 rhomboid m.
M 25 intercostal mm.
M 26 serratus anterior m.
M 27 infraspinatus m.
M 28 subscapularis m.
M 29 teres major m.
M 30 latissimus dorsi m.
M 31 triceps brachii (long head)
M 32 triceps brachii (lateral head)
M 33 deltoid m.

N 1 median n.
N 2 right phrenic n.
N 3 left phrenic n.
N 4 brachial plexus (posterior cord)
N 5 brachial plexus (medial cord)
N 6 right vagus n.
N 7 left vagus n.

O 1 manubrium
O 2 2nd costal cartilage
O 3 body of humerus
O 4 scapula (lateral margin)
O 5 scapula (medial margin)
O 6 superior articular process T 6
O 7 lamina T 5
O 8 spinous process T 4

S 1 right superior lobar bronchus
S 2 right main bronchus
S 3 bifurcation of trachea
S 4 pleural recess
S 5 right lung (right lobe)
S 6 esophagus
S 7 left main bronchus
S 8 parietal pleura
S 9 oblique fissure
S 10 inferior lobe (apical segment)
S 11 left lung (superior lobe)

SY 1 articular process of vertebrae T 5–T 6
SY 2 articulation of head of 6th rib

HT 6

A 1 lateral thoracic v. (branch)
A 2 right superior pulmonary v.
A 3 internal thoracic v.
A 4 right pulmonary a.
A 5 superior vena cava
A 6 pericardial cavity
A 7 transverse pericardial sinus
A 8 ascending aorta
A 9 left coronary a. (orifice)
A 10 valve of pulmonary trunk (right semilunar valve)
A 11 pulmonary trunk
A 12 left superior pulmonary v.
A 13 transverse pericardial sinus
A 14 left superior pulmonary v. (apical posterior branch)
A 15 left pulmonary a. (lingular branch)
A 16 deep brachial a.
A 17 brachial a.
A 18 thoracodorsal a.
A 19 right superior pulmonary v. (middle lobe)
A 20 right pulmonary a. (posterior ascending)
A 21 bronchopulmonary lymph node
A 22 6th intercostal v.
A 23 azygos v.
A 24 descending aorta
A 25 accessory hemiazygos v.
A 26 left pulmonary a.
A 27 thoracodorsal v.

M 1 deltoid m.
M 2 biceps brachii m. (long head)
M 3 biceps brachii m. (short head)
M 4 pectoralis minor m.
M 5 intercostal mm.
M 6 pectoralis major m.
M 7 triceps brachii m. (lateral head)
M 8 triceps brachii m. (long head)
M 9 coracobrachialis m.
M 10 latissimus dorsi m.
M 11 teres major m.
M 12 teres minor m.
M 13 subscapularis m.
M 14 infraspinatus m.
M 15 serratus anterior m.
M 16 rhomboid m.
M 17 trapezius m.
M 18 transversospinal m.
M 19 trapezius m.
M 20 longissimus thoracis m.
M 21 rhomboid m.
M 22 intercostal mm.
M 23 serratus anterior m.
M 24 subscapularis m.
M 25 infraspinatus m.
M 26 teres major m.
M 27 latissimus dorsi m.

N 1 right phrenic n.
N 2 left phrenic n.
N 3 radial n.
N 4 ulnar n.
N 5 medial antibrachial cutaneous n.
N 6 sympathetic trunk
N 7 right vagus n.
N 8 spinal cord
N 9 left vagus n.
N 10 median n.
N 11 musculocutanenous n.
N 12 radial n.

O 1 body of humerus
O 2 scapula (medial margin)
O 3 lamina T 6
O 4 spinous process T 5
O 5 superior articulating process T 7
O 6 inferior articulating process T 6

S 1 superior lobe of right lung
S 2 right inferior lobar bronchus
S 3 mediastinal pleura
S 4 costomediastinal recess
S 5 esophagus
S 6 left superior lobar bronchus
S 7 oblique fissure
S 8 inferior lobe of right lung (apical segment)
S 9 inferior lobe of left lung (apical segment)
S 10 oblique fissure
S 11 superior lobe of left lung

SY 1 articulation of head of 7th rib
SY 2 articular processes of vertebrae (T 6–T 7)

HT 7

A 1 right superior pulmonary v. (anterior branch)
A 2 internal thoracic a.
A 3 right superior pulmonary v. (middle lobe)
A 4 right atrium
A 5 superior vena cava
A 6 right auricle
A 7 right coronary a.
A 8 aortic valve (posterior semilunar valve)
A 9 aortic valve (right semilunar valve)
A 10 right ventricle (conus arteriosus)
A 11 aortic valve (left semilunar valve)
A 12 left auricle
A 13 left coronary a. (circumflex branch)
A 14 coronary sinus
A 15 left coronary a. (anterior interventricular branch)
A 16 left coronary a. (branch)
A 17 left pulmonary a. (posterior descending branch)
A 18 thoracodorsal a.
A 19 thoracodorsal a. (branch)
A 20 right pulmonary a. (anterior basal)
A 21 right pulmonary a. (lateral basal and posterior basal)
A 22 right pulmonary a. (medial basal)
A 23 7th intercostal a.
A 24 7th intercostal v.
A 25 right inferior pulmonary v. (apical branch)
A 26 7th intercostal v. (opening)
A 27 azygos v.
A 28 left atrium
A 29 descending aorta
A 30 accessory hemiazygos v.
A 31 left inferior pulmonary v.
A 32 left pulmonary a. (basal portion)

M 1 pectoralis minor m.
M 2 pectoralis major m.
M 3 latissimus dorsi m.
M 4 teres major m.
M 5 subscapularis m.
M 6 infraspinatus m.
M 7 serratus anterior m.
M 8 rhomboid m.
M 9 external intercostal m.
M 10 internal intercostal m.
M 11 transversospinal m.
M 12 trapezius m.
M 13 longissimus thoracis m.
M 14 rhomboid m.
M 15 serratus anterior m.
M 16 teres major m.
M 17 latissimus dorsi m.

N 1 right phrenic n.
N 2 left pherenic n.
N 3 sympathetic trunk
N 4 right vagus n.
N 5 left vagus n.

O 1 scapula (medial margin)
O 2 lamina T 7
O 3 spinous process T 6
O 4 superior articulating process T 8
O 5 scapula (lateral margin)

S 1 oblique fissure
S 2 right lung (middle lobe)
S 3 horizontal fissure
S 4 right lung (superior lobe)
S 5 lateral segmental bronchus
S 6 medial segmental bronchus
S 7 lingula of left lung
S 8 inferior bronchus of lingula
S 9 left lung (superior lobe)
S 10 oblique fissure
S 11 right inferior lobar bronchus
S 12 medial basal segmental bronchus
S 13 esophagus
S 14 anterior basal segmental bronchus and medial basal segmental bronchus
S 15 posterior basal segmental bronchus and lateral basal segmental bronchus
S 16 left lung (inferior lobe)

HT 8

A 1 right superior pulmonary v. (middle lobe)
A 2 pericardium
A 3 right coronary a.
A 4 right atrium
A 5 right atrioventricular valve (anterior cusp)
A 6 fossa ovalis
A 7 limbus fossae ovalis
A 8 right ventricle
A 9 coronary sinus
A 10 left atrium
A 11 left atrioventricular valve (anterior cusp)
A 12 interventricular septum
A 13 left atrioventricular valve (posterior cusp)
A 14 internal thoracic v.
A 15 left coronary a. (anterior interventricular branch)
A 16 anterior papillary m.
A 17 left superior pulmonary v. (anterior branch)
A 18 6th posterior intercostal a.
A 19 right inferior pulmonary v. (anterior basal)
A 20 right pulmonary a. (posterior basal)
A 21 8th posterior intercostal a.
A 22 right inferior pulmonary v. (common vessel)
A 23 azygos v.
A 24 internal vertebral venous plexus
A 25 hemiazygos v.
A 26 descending aorta
A 27 left coronary a. (circumflex branch)
A 28 left inferior pulmonary v. (inferior basal branch)
A 29 left inferior pulmonary v. (inferior basal branch)
A 30 left inferior pulmonary v. (anterior basal branch)
A 31 thoracodorsal a.

M 1 pectoralis major m.
M 2 pectoralis major m.
M 3 latissimus dorsi m.
M 4 serratus anterior m.
M 5 internal intercostal m.
M 6 external intercostal m.
M 7 intercostal mm.
M 8 longissimus thoracis m.
M 9 trapezius m.
M 10 transversospinal m.
M 11 serratus anterior m.
M 12 latissimus dorsi m.

N 1 right phrenic n.
N 2 left phrenic n.
N 3 sympathetic trunk
N 4 right vagus n.
N 5 dentate ligament
N 6 left vagus n.

O 1 scapula (inferior angle)

O 2 spinous process T 7
O 3 lamina T 8
O 4 superior articulating process T 9

S 1 oblique fissure
S 2 right lung (middle lobe)
S 3 left lung (lingula)
S 4 costal pleura
S 5 oblique fissure
S 6 papilla of mammary gland
S 7 right lung (inferior lobe)
S 8 lateral basal segmental bronchus
S 9 posterior basal segmental bronchus
S 10 pulmonary ligament
S 11 esophagus
S 12 pulmonary ligament
S 13 dorsal basal segmental bronchus
S 14 lateral basal segmental bronchus
S 15 left lung (inferior lobe)

HT 9

A 1 inferior phrenic a.
A 2 internal thoracic v.
A 3 right coronary a.
A 4 right atrium
A 5 right coronary a. (ventricular branch)
A 6 inferior vena cava
A 7 right coronary a. (posterior interventricular branch)
A 8 right atrioventricular valve (posterior cusp)
A 9 middle cardiac v.
A 10 right ventricle
A 11 anterior papillary m.
A 12 interventricular septum
A 13 left ventricle
A 14 left coronary a. (anterior interventricular branch)
A 15 anterior papillary m.
A 16 right inferior pulmonary v. (inferior basal v.)
A 17 9th posterior intercostal v.
A 18 posterior intercostal v. (dorsal branch)
A 19 intervertebral v.
A 20 azygos v.
A 21 internal vertebral venous plexus
A 22 hemiazygos v.
A 23 descending aorta
A 24 9th posterior intercostal v.
A 25 left inferior pulmonary v. (inferior basal v.)

M 1 diaphragm
M 2 pectoralis major m.
M 3 transversus thoracis m.
M 4 latissimus dorsi m.
M 5 serratus anterior m.
M 6 external intercostal m. and internal intercostal m.
M 7 internal intercostal m.
M 8 longissimus thoracis m.
M 9 trapezius m.
M 10 transversospinal m.

N 1 right phrenic n.
N 2 vagus n.
N 3 left phrenic n.
N 4 spinal ganglion
N 5 spinal n. (dorsal root)
N 6 vagus n.
N 7 spinal dura mater

O 1 spinous process T 8
O 2 transverse process T 9

S 1 parietal peritoneum
S 2 oblique fissure
S 3 diaphragmatic pleura
S 4 right lung (middle lobe)
S 5 esophagus
S 6 pulmonary ligament
S 7 lingula of left lobe
S 8 oblique fissure
S 9 right lung (inferior lobe)
S 10 pleural recess
S 11 left lung (inferior lobe)

SY 1 6th sternocostal articulation
SY 2 4th costochondral articulation

HT 10

A 1 right hepatic v.
A 2 left hepatic v.
A 3 left gastric a. (esophageal branch)
A 4 right ventricle
A 5 internal thoracic a.
A 6 left gastric a. (anterior gastric branch)
A 7 interventricular septum
A 8 left coronary a. (anterior interventricular branch)
A 9 6th anterior intercostal a.
A 10 left ventricle
A 11 fibrous pericardium
A 12 right hepatic v.
A 13 10th intercostal v.
A 14 intervertebral v.
A 15 inferior vena cava
A 16 inferior phrenic v.
A 17 azygos v.
A 18 descending aorta
A 19 esophageal v.
A 20 hemiazygos v.
A 21 splenic v. (short gastric v.)
A 22 9th posterior intercostal v.
A 23 splenic v. (short gastric v.)

M 1 pectoralis major m.
M 2 diaphragm (costal portion)
M 3 rectus abdominis m.
M 4 transversus thoracis m.
M 5 diaphragm (crus)
M 6 diaphragm (central tendon)
M 7 diaphragm (costal portion)
M 8 latissimus dorsi m.
M 9 serratus anterior m.
M 10 diaphragm (central tendon)
M 11 trapezius m.
M 12 erector spinae m.
M 13 diaphragm (crus)

O 1 xiphoid process
O 2 spinous process T 9
O 3 lamina T 10
O 4 pedicle T 10
O 5 transverse process T 10

S 1 costodiaphragmatic recess
S 2 peritoneal cavity
S 3 liver (right lobe)
S 4 right lung (middle lobe)
S 5 liver (caudate lobe)
S 6 costomediastinal recess
S 7 esophagus (abdominal portion)
S 8 costomediastinal recess
S 9 liver (left lobe)
S 10 body of stomach
S 11 left lung (superior lobe)
S 12 oblique fissure
S 13 right lung (inferior lobe)
S 14 pleural recess
S 15 cardiac incisura
S 16 peritoneal cavity
S 17 pleural cavity

HT 11

A 1 right hepatic v.
A 2 portal v. (branch)
A 3 left gastric a. (esophageal branch)
A 4 superior epigastric a.
A 5 musculophrenic a.
A 6 pericardium
A 7 left gastric a. (anterior gastric branch)
A 8 10th posterior intercostal a.
A 9 portal v. (dorsal branch)
A 10 11th posterior intercostal v.
A 11 inferior vena cava
A 12 thoracic aorta
A 13 hemiazygos v.
A 14 splenic v.
A 15 spleen
A 16 splenic v.

A 17 splenic a.
A 18 splenic a. (short gastric a.)

M 1 transversus thoracis m.
M 2 rectus abdominis m.
M 3 diaphragm (sternal portion)
M 4 diaphragm (esophageal hiatus)
M 5 diaphragm (costal portion)
M 6 diaphragm (costal portion)
M 7 erector spinae m.
M 8 transversospinal m.
M 9 diaphragm (lumbar portion)
M 10 intercostal mm.
M 11 serratus posterior inferior m.

O 1 xiphoid process
O 2 xiphoid process
O 3 transverse process T 11
O 4 pedicle T 11

S 1 liver (right lobe)
S 2 costodiaphragmatic recess
S 3 falciform ligament
S 4 stomach (cardiac portion)
S 5 liver (superior lobe)
S 6 costodiaphragmatic recess
S 7 left lung (lingula)
S 8 fundus of stomach
S 9 coronary ligament
S 10 liver (caudate lobe)
S 11 vestibule of omental bursa
S 12 perirenal fat
S 13 inferior omental recess
S 14 greater omentum

SY 1 articulation of head of 11th rib
SY 2 posterior longitudinal ligament
SY 3 interspinous ligament
SY 4 anterior longitudinal ligament

HT 12

A 1 portal v. (right branch, right ventral)
A 2 musculophrenic a.
A 3 right hepatic v.
A 4 portal v. (right branch, ventral branch)
A 5 portal v. (left branch)
A 6 superior epigastric a.
A 7 hepatic a. (left branch)
A 8 left gastric a.
A 9 left gastric v.
A 10 left gastric a.
A 11 left gastric a. (branch)
A 12 9th intercostal v.
A 13 portal v. (right branch, dorsal branch)
A 14 11th posterior intercostal v.
A 15 11th posterior intercostal a.
A 16 inferior vena cava
A 17 inferior phrenic a.
A 18 thoracic duct
A 19 internal vertebral venous plexus
A 20 thoracic aorta
A 21 hemiazygos v.
A 22 inferior phrenic a.
A 23 left gastric a. (posterior gastric branch)
A 24 left gastric v. (posterior gastric branch)
A 25 splenic a.
A 26 splenic a. (branch)
A 27 splenic v.
A 28 splenic a.
A 29 splenic v.

M 1 intercostal mm.
M 2 external oblique m.
M 3 intercostal mm.
M 4 rectus abdominis m.
M 5 diaphragm (costal part)
M 6 internal oblique m. and transverse abdominis m.
M 7 diaphragm (crus)
M 8 rectus abdominis m.
M 9 latissimus dorsi m.
M 10 serratus posterior inferior m.
M 11 erector spinae m.

N 1 celiac plexus
N 2 sympathetic trunk

O 1 transverse process T 12
O 2 inferior articulating process T 11
O 3 spinous process T 11

S 1 liver (right lobe)
S 2 costodiaphragmatic recess
S 3 falciform ligament
S 4 right hepatic duct
S 5 ligamentum teres
S 6 liver (caudate lobe)
S 7 fundus of stomach
S 8 inferior omental recess
S 9 greater omentum
S 10 liver (right lobe)
S 11 perirenal fat
S 12 right suprarenal gland
S 13 costomediastinal recess
S 14 left suprarenal gland
S 15 pleural recess
S 16 inferior omental recess
S 17 left kidney
S 18 perirenal fat
S 19 spleen
S 20 pleural cavity

HT 13

A 1 portal v. (left branch)
A 2 portal v.
A 3 portal v. (left branch)
A 4 proper hepatic a.
A 5 left gastric v.
A 6 left gastric a.
A 7 left gastric a.
A 8 left gastric v.
A 9 superior epigastric a.
A 10 left colic v.
A 11 11th intercostal v.
A 12 portal v. (right branch)
A 13 hepatic a. (superior mesenteric a. variation)
A 14 inferior vena cava
A 15 intervertebral v.
A 16 inferior phrenic a.
A 17 abdominal aorta
A 18 celiac trunk (opening)
A 19 anterior internal vertebral venous plexus
A 20 splenic a.
A 21 left suprarenal v.
A 22 splenic a.
A 23 splenic v.
A 24 left colic v.
A 25 renal v.

M 1 external oblique m.
M 2 diaphragm (costal portion)
M 3 transverse abdominis m.
M 4 rectus abdominis m.
M 5 erector spinae m.
M 6 transversospinal m.
M 7 diaphragm (aortic hiatus)
M 8 serratus posterior inferior m.
M 9 latissimus dorsi m.

O 1 lamina T 12
O 2 spinous process T 12

S 1 right lobe of liver
S 2 liver (quadrate lobe)
S 3 hepatic duct
S 4 ligamentum teres
S 5 left lobe of liver
S 6 body of stomach
S 7 peritoneal cavity
S 8 transverse mesocolon
S 9 jejunum
S 10 left colic flexure
S 11 gastrocolic ligament
S 12 greater omentum
S 13 costodiaphragmatic recess
S 14 peritoneal cavity
S 15 right kidney (superior pole)
S 16 perirenal fat
S 17 suprarenal gland
S 18 pleural recess
S 19 left suprarenal gland

S 20 tail of pancreas
S 21 left kidney
S 22 perirenal fat
S 23 spleen

SY 1 7th costochondral articulation

HT 14

A 1 portal v. (branch)
A 2 cystic a.
A 3 gastroduodenal a.
A 4 portal v.
A 5 left renal v. (opening)
A 6 hepatic a. (right branch, superior mesenteric a., variation)
A 7 superior mesenteric a.
A 8 right gastric a.
A 9 splenic v.
A 10 left colic a.
A 11 11th intercostal a.
A 12 renal a.
A 13 renal v.
A 14 inferior vena cava
A 15 ascending lumbar v.
A 16 pancreaticoduodenal v.
A 17 anterior internal vertebral venous plexus
A 18 posterior internal vertebral venous plexus
A 19 abdominal aorta
A 20 left suprarenal v.
A 21 renal a.
A 22 renal v.
A 23 left colic a.

M 1 external oblique m.
M 2 intercostal mm.
M 3 transverse abdominis m.
M 4 rectus abdominis m.
M 5 linea alba
M 6 latissimus dorsi m.
M 7 diaphragm (costal portion)
M 8 quadratus lumborum m.
M 9 erector spinae m.
M 10 psoas m.
M 11 diaphragm (lumbar portion)
M 12 diaphragm (crus)
M 13 serratus posterior inferior m.

N 1 cauda equina
N 2 conus medullaris

O 1 spinous process L 1
O 2 lamina L 1
O 3 accessory process L 1
O 4 costal process L 1

S 1 liver (quadrate lobe)
S 2 ligamentum teres
S 3 duodenum (superior portion)
S 4 common bile duct
S 5 head of pancreas
S 6 left lobe of liver
S 7 duodenal jejunal flexure
S 8 body of stomach
S 9 jejunum
S 10 transverse mesocolon
S 11 jejunum
S 12 transverse colon
S 13 greater omentum
S 14 right lobe of liver
S 15 costodiaphragmatic recess
S 16 peritoneal cavity
S 17 right kidney
S 18 gall bladder
S 19 ureter
S 20 small intestine
S 21 renal sinus
S 22 small intestine
S 23 small intestine
S 24 descending colon

SY 1 7th & 8th interchondral articulation

HT 15

A 1 pancreaticoduodenal v.
A 2 right gastroepiploic a.
A 3 inferior vena cava
A 4 superior mesenteric v.
A 5 superior mesenteric a.
A 6 abdominal aorta
A 7 left colic a.
A 8 jejunal a.
A 9 11th intercostal a.
A 10 renal v.
A 11 renal a. (anterior inferior segmental a.)
A 12 2nd lumbar v.
A 13 ascending lumbar v.
A 14 lumbar v. (dorsal branch)
A 15 inferior mesenteric v.
A 16 left colic a.
A 17 subcostal a.

M 1 intercostal mm.
M 2 external oblique m.
M 3 transverse abdominis m.
M 4 rectus abdominis m.
M 5 linea alba
M 6 rectus abdominis m.
M 7 latissimus dorsi m.
M 8 diaphragm (costal portion)
M 9 serratus posterior inferior m.
M 10 quadratus lumborum m.
M 11 thoracolumbar fascia (deep layer)
M 12 erector spinae m.
M 13 psoas m.
M 14 diaphragm (crus)
M 15 transversospinal m.

N 1 lumbar plexus
N 2 cauda equina
N 3 sympathetic trunk
N 4 lumbar plexus

O 1 accessory process L 2
O 2 inferior articular process L 1
O 3 costal process L 2

S 1 right colic flexure
S 2 gall bladder
S 3 duodenum (descending portion)
S 4 major duodenal papilla
S 5 duodenum (superior portion)
S 6 head of pancreas
S 7 pylorus
S 8 left lobe of liver
S 9 stomach (pyloric portion)
S 10 duodenum (ascending portion)
S 11 jejunum
S 12 jejunum
S 13 mesentery
S 14 transverse colon
S 15 small intestine
S 16 greater omentum
S 17 right lobe of liver
S 18 peritoneal cavity
S 19 right kidney
S 20 renal pelvis
S 21 perirenal fat
S 22 left kidney
S 23 jejunum
S 24 small intestine
S 25 descending colon
S 26 small intestine

SY 1 ligamentum flavum
SY 2 interspinal ligament

Koritké/Sick

Atlas of Sectional Human Anatomy

Index of Labels for Vol. 2

J. G. Koritké and H. Sick

Atlas of Sectional Human Anatomy

Index of Labels
Vol. 2: Abdomen, Pelvis

Urban & Schwarzenberg
Baltimore – Munich 1983

Authors' adresses:

Jean Georges Koritké
Chairman
Department of Anatomy
Faculty of Medicine
Louis Pasteur University
4, rue Kirschleger
F-67085 Strasbourg CEDEX
France

Henri Sick
Professor of Anatomy
Department of Anatomy
Faculty of Medicine
Louis Pasteur University
4, rue Kirschleger
F-67085 Strasbourg CEDEX
France

This atlas consists of two separate volumes:
Vol. 1: Head, Neck, Thorax (incl. Index)
Vol. 2: Abdomen, Pelvis (incl. Index)

2–86

Printed in Germany by Kösel, Kempten

Vol. 2: ISBN 0-8067-1032-2 (Baltimore)
ISBN 3-541-71032-2 (Munich)
Complete set: ISBN 0-8067-1030-6 (Baltimore)
ISBN 3-541-71030-6 (Munich)

Table of Contents

Abdomen (Abdomen and Pelvis)

Pelvis

Directions for Use

The most important document is the photograph, a true reflection of the actual anatomy. From the very beginning, the observer is able to record a large amount of information, with the progression of the analysis indicated by the diagram and then the labels.

Photographs

The photographs reproduced on the right-hand pages convey different types of information: general appearance of the section, shape of its surfaces, positions, proportions, impressions of the density of the organs. The entire section or each of its elements can be situated within the body by referring to the following: the letters characterizing each section with respect to the planes of reference; the height of the vertebra shown on the osseous section as soon as it appears; the height of the center of the vertebral body indicated on both sides of all photographs of frontal and sagittal sections; the level of the ribs; and the scale shown at the bottom of each photograph.

Drawings

Each diagram, an interpretation of the actual anatomy, brings out the essential characteristic elements of the section. The thickness of the features and the character of the grid representing the organs indicate to which system the organs shown belong. References are also made using keys. Symmetric organs are not necessarily presented in a bilateral manner. This makes the drawing clearer. It was not always possible to superimpose extremely thin planes. A single feature can give an outline of several structures. Cuts of vertebrae and ribs are designated in any case.

Labels

The labels supply further information and help to check the identification of the photograph and the interpretation of the drawing.
Abbreviations have been used for the identification and orientation of the section, the classification of organs referred to by large systems, and finally in the labels. The first examination of the photograph, the interpretation of the drawing, and then reading the labels should logically lead the observer to re-examine the photograph in order to record the numerous elements that could not be referred to in the space allowed. It should then be possible to imagine and compare the image that will be supplied by CT, ultrasound, or any other new technique technology brings to us.

Abbreviations

1. Identification and Orientation of the Sections

a: anterior – anterior
p: posterior – posterior
D: right – dexter
Si: left – sinister
F: frontal – frontalis
H: horizontal – horizontalis
S: sagittal – sagittalis
C: head, neck – caput, collum
T: thorax – thorax
A: abdomen – abdomen
P: pelvis – pelvis
♀: female – femininum
♂: male – masculinum

The vertebrae are indicated by the letters C (cervical – cervicalis), T (thoracic – thoracicus), L (lumbar – lumbalis), S (sacral – sacralis) and Co (coccygeal – coccygea). The number that corresponds to their level is added to the letter. The ribs are marked by a number indicating their level ("3" refers to the third rib).

2. Association of the Organ to a Large System

A: Angiology – Angiologia
M: Myology – Myologia
N: Nervous System – Systema Nervosum
O: Osteology – Osteologia
OS: Sense Organs – Organa Sensuum
S: Splanchnology – Splanchnologia
Sy: Syndesmology, Arthrology – Syndesmologia

3. Abbreviations used in the Labels

a.: artery – arteria
art.: articulation – articulatio
b.: bursa – bursa
C.: cervical (vertebra) – cervicalis (vertebra)
Co.: coccyx – os coccygis
gl.: gland – glandula
L.: lumbar (vertebra) – lumbalis (vertebra)
lig.: ligament – ligamentum
m.: muscle – musculus
n.: nerve – nervus
no.: node (lymphatic) – nodus (lymphaticus)
r.: ramus – ramus
S.: sacral (vertebra) – sacralis (vertebra)
v.: vein – vena
( ): specifies the preceding term

FA♂
Frontal sections Abdomen

Abdomen and Pelvis

42-year-old male

plane of reference: subject lying on a horizontal plane; 7 sections, spaced 2.5 cm apart, parallel to the posterior plane of reference. The superior border of each section passes through the center of the body of the 8th thoracic vertebra [T 8].

FA♂ 1 2.5 cm from the posterior plane of reference, antero-posterior view
FA♂ 2 5 cm from the posterior plane of reference, antero-posterior view
FA♂ 3 7.5 cm from the posterior plane of reference, antero-posterior view
FA♂ 4 10 cm from the posterior plane of reference, antero-posterior view
FA♂ 5 12.5 cm from the posterior plane of reference, antero-posterior view
FA♂ 6 15 cm from the posterior plane of reference, antero-posterior view
FA♂ 7 17.5 cm from the posterior plane of reference, antero-posterior view

FA♂ 1

A 1 lateral sacral a.
A 2 11th posterior intercostal a.

M 1 subcostalis m.
M 2 transversospinal m.
M 3 serratus posterior inferior m.
M 4 latissimus dorsi m.
M 5 longissimus m.
M 6 iliocostalis m.
M 7 thoracolumbar fascia
M 8 erector spinae m.
M 9 piriformis m.
M 10 gluteus maximus m.
M 11 teres major m.
M 12 diaphragm (costal portion)

N 1 10th intercostal n.
N 2 4th sacral n.
N 3 5th sacral n. (ventral ramus)
N 4 11th intercostal n.

O 1 transverse process T 9
O 2 transverse process T 10
O 3 iliac tuberosity
O 4 middle sacral crest (S 2)
O 5 sacrum (lateral portion)
O 6 inferior angle of scapula
O 7 sacral tuberosity

S 1 left lung
S 2 retrorectal space

SY 1 interspinous ligament
SY 2 supraspinous ligament
SY 3 dorsal sacroiliac ligament
SY 4 sacroiliac articulation

FA♂ 2

A 1 9th posterior intercostal a.
A 2 inferior phrenic a.
A 3 lateral sacral v.
A 4 inferior gluteal a.
A 5 superior gluteal a.
A 6 inferior gluteal a.
A 7 inferior gluteal a.

M 1 diaphragm (lumbar portion)
M 2 diaphragm (costal portion)
M 3 quadratus lumborum m.
M 4 serratus posterior inferior m.
M 5 quadratus lumborum m.
M 6 latissimus dorsi m.
M 7 iliocostalis thoracis m.
M 8 longissimus thoracis m.
M 9 transversospinal m.
M 10 fat body of lumbar triangle
M 11 gluteus medius m.
M 12 ischiococcygeus m.
M 13 gluteus maximus m.
M 14 sphincter ani externus m.
M 15 latissimus dorsi m.
M 16 serratus anterior m.
M 17 gluteus minimus m.

N 1 spinal cord
N 2 spinal dura mater
N 3 12th intercostal n.
N 4 cauda equina
N 5 2nd sacral n.
N 6 2nd sacral n.
N 7 11th intercostal n.

O 1 11th rib
O 2 iliac crest
O 3 intermediate sacral crest
O 4 lateral sacral crest
O 5 iliac tuberosity

S 1 right lung (inferior lobe)
S 2 liver (right lobe)
S 3 perirenal fat
S 4 spleen
S 5 left kidney
S 6 perirenal fat
S 7 rectal ampulla
S 8 transverse rectal folds
S 9 ischiorectal fossa

SY 1 11th costotransverse articulation
SY 2 superior articulating process T 12
SY 3 ligamentum flavum
SY 4 articulating processes of vertebrae T 12–L 1
SY 5 interspinous ligament
SY 6 sacroiliac articulation

FA♂ 3

A 1 lateral sacral a.
A 2 inferior gluteal a.
A 3 11th posterior intercostal a.
A 4 splenic v.
A 5 internal vertebral venous plexus
A 6 superior gluteal v.
A 7 inferior gluteal a.
A 8 superior gluteal a.
A 9 superior gluteal v.
A 10 inferior gluteal v.
A 11 internal pudendal a.
A 12 inferior gluteal a.
A 13 internal pudendal a.

M 1 latissimus dorsi m.
M 2 serratus anterior m.
M 3 diaphragm (lumbar portion)
M 4 diaphragm (costal portion)
M 5 psoas major m.
M 6 transverse abdominis m.
M 7 internal oblique m.
M 8 external oblique m.
M 9 quadratus lumborum m.
M 10 transversospinalis m.
M 11 iliacus m.
M 12 gluteus medius m.
M 13 gluteus medius m.
M 14 gluteus maximus m.
M 15 gluteus minimus m.
M 16 piriformis m.
M 17 obturator internus m.

M 18 superior gemellus m.
M 19 obturator internus m.
M 20 inferior gemellus m.
M 21 levator ani m.
M 22 quadratus femoris m.
M 23 ischiorectal fossa
M 24 ischiocrurales mm.
M 25 gluteus maximus m.
M 26 piriformis m.
M 27 obturator internus m.
M 28 gluteus maximus m.

N 1 superior gluteal n.
N 2 sacral plexus
N 3 sciatic n.
N 4 cauda equina
N 5 iliohypogastric n.
N 6 cauda equina
N 7 sacral plexus
N 8 sciatic n.

O 1 intervertebral foramen
O 2 pedicle L 2
O 3 inferior articulating process L 2
O 4 costal process L 3
O 5 iliac crest
O 6 costal process L 5
O 7 sacrum (superior articulating process)
O 8 ala of ilium
O 9 lateral sacral portion
O 10 greater sciatic foramen
O 11 greater trochanter
O 12 tuberosity of ischium

S 1 right lung (inferior lobe)
S 2 right lobe of liver
S 3 right suprarenal gland
S 4 right kidney
S 5 left lung (inferior lobe)
S 6 fundus of stomach
S 7 spleen
S 8 left suprarenal gland
S 9 descending colon
S 10 left kidney
S 11 descending colon
S 12 rectum
S 13 transverse rectal folds

SY 1 ligamentum flavum
SY 2 processes of vertebrae L 2–L 3
SY 3 dorsal sacroiliac ligament
SY 4 processes of vertebrae L 5–S 1
SY 5 sacroiliac articulation
SY 6 intervertebral disc S 1–S 2
SY 7 sacrotuberous ligament

FA ♂ 4

A 1 inferior vena cava
A 2 inferior vena cava
A 3 right hepatic v.
A 4 thoracic aorta
A 5 11th posterior intercostal a. (opening)
A 6 renal a. (anterior superior segmental a.)
A 7 renal a. (superior segmental a.)
A 8 renal v.
A 9 renal a. (posterior segmental a.)
A 10 renal a. (anterior inferior segmental a.)
A 11 iliolumbar a. (iliac branch)
A 12 iliolumbar v.
A 13 iliolumbar a.
A 14 internal iliac v.
A 15 superior gluteal a.
A 16 inferior gluteal a.
A 17 internal iliac v.
A 18 superior vesical a.
A 19 vesical venous plexus
A 20 internal pudendal a. (bifurcation)
A 21 pericardium
A 22 short gastric a.
A 23 splenic v.
A 24 left gastroepiploic a.
A 25 splenic v.
A 26 left colic v.
A 27 3rd lumbar a.
A 28 superior gluteal v.
A 29 ascending lumbar v.
A 30 inferior gluteal a.
A 31 superior gluteal a.
A 32 internal iliac v.

M 1 diaphragm (central tendon)
M 2 diaphragm (costal portion)
M 3 serratus anterior m.
M 4 diaphragm (lumbar portion)
M 5 external oblique m.
M 6 internal oblique m.
M 7 transverse abdominis m.
M 8 psoas major m.
M 9 iliacus m.
M 10 gluteus medius m.
M 11 gluteus minimus m.
M 12 obturator internus m.
M 13 obturator externus m.
M 14 levator ani m.
M 15 ischiocavernosus m.
M 16 quadratus femoris m.
M 17 superficial perineal space
M 18 sphincter ani externus m.
M 19 adductor magnus m.
M 20 vastus lateralis m.
M 21 diaphragm (crus)
M 22 diaphragm (costal portion)
M 23 quadratus femoris m.
M 24 ischiorectal fossa
M 25 ischiocrurales mm.

N 1 femoral n.
N 2 obturator n.
N 3 femoral n.
N 4 sciatic n.

O 1 iliac crest
O 2 arcuate line
O 3 head of femur
O 4 neck of femur
O 5 greater trochanter
O 6 lesser trochanter
O 7 greater trochanter

S 1 right lung (inferior lobe)
S 2 right lobe of liver
S 3 right suprarenal gland
S 4 right ureter
S 5 right kidney
S 6 perirenal fat
S 7 ascending colon
S 8 sigmoid colon
S 9 right ureter
S 10 ductus deferens
S 11 urinary bladder
S 12 right ureter
S 13 esophagus
S 14 left lung (inferior lobe)
S 15 body of stomach
S 16 spleen
S 17 left suprarenal gland
S 18 head of pancreas
S 19 left colic flexure
S 20 left kidney (superior pole)
S 21 small intestine
S 22 small intestine
S 23 descending colon
S 24 sigmoid colon
S 25 rectal ampulla
S 26 seminal vesicles
S 27 ampulla of ductus deferens
S 28 rectal ampulla
S 29 perineal rectal flexure

SY 1 hip joint
SY 2 lumbosacral articulation
SY 3 sacroiliac articulation

FA ♂ 5

A 1 right atrium
A 2 hepatic v.
A 3 portal v. (right branch)
A 4 inferior vena cava
A 5 renal a. (anterior superior segmental a.)
A 6 renal a. (anterior inferior segmental a.)

A 7 renal v.
A 8 renal v.
A 9 common iliac v.
A 10 internal iliac a.
A 11 external iliac v.
A 12 umbilical a.
A 13 obturator a.
A 14 inferior vesical v.
A 15 prostatic venous plexus
A 16 internal pudendal a.
A 17 left ventricle
A 18 splenic a.
A 19 splenic v.
A 20 celiac trunk
A 21 abdominal aorta
A 22 suparenal v.
A 23 renal a.
A 24 renal v.
A 25 left colic a.
A 26 left colic v.
A 27 renal a. (anterior inferior segmental a.)
A 28 left colic a.
A 29 3rd lumbar a.
A 30 left common iliac v.
A 31 superior rectal a.
A 32 external iliac v.
A 33 internal iliac a.
A 34 obturator a.
A 35 internal pudendal v.

M 1 serratus anterior m.
M 2 diaphragm (costal portion)
M 3 diaphragm (crus)
M 4 external oblique m.
M 5 internal oblique m.
M 6 transverse abdominis m.
M 7 psoas major m.
M 8 iliacus m.
M 9 gluteus medius m.
M 10 gluteus minimus m.
M 11 obturator internus m.
M 12 obturator externus m.
M 13 levator ani m.
M 14 iliopsoas m.
M 15 vastus lateralis m.
M 16 ischiocavernosus m.
M 17 bulbospongiosus m.
M 18 adductor magnus m.

N 1 obturator n.
N 2 femoral n.

O 1 iliac crest
O 2 arcuate line
O 3 head of femur
O 4 ramus of ischium
O 5 head of femur
O 6 acetabular fossa
O 7 neck of femur
O 8 greater trochanter
O 9 lesser trochanter

S 1 right lung (inferior lobe)
S 2 liver (caudate lobe)
S 3 superior duodenal flexure
S 4 duodenum (descending portion)
S 5 ascending colon
S 6 small intestine
S 7 small intestine
S 8 mesoappendix
S 9 appendix
S 10 ductus deferens
S 11 left lung (superior lobe)
S 12 left lung (inferior lobe)
S 13 left lobe of liver
S 14 body of stomach
S 15 body of pancreas
S 16 descending colon
S 17 left ureter
S 18 jejunum
S 19 greater omentum
S 20 descending colon
S 21 descending colon
S 22 sigmoid colon
S 23 sigmoid colon
S 24 urinary bladder
S 25 right ureter
S 26 fundus of bladder
S 27 prostate
S 28 urethra

SY 1 acetabular labrum
SY 2 articular capsule
SY 3 obturator membrane

FA♂ 6

A 1 right atrium
A 2 hepatic v.
A 3 portal v. (left branch)
A 4 common hepatic a.
A 5 gastroduodenal a.
A 6 portal v.
A 7 right inferior pancreaticoduodenal v.
A 8 inferior vena cava
A 9 ileocolic v.
A 10 right external iliac a.
A 11 right external iliac v.
A 12 obturator v.
A 13 lateral femoral circumflex a.
A 14 deep dorsal vein of penis
A 15 left ventricle
A 16 right ventricle
A 17 right coronary a.
A 18 left gastric a.
A 19 splenic a.
A 20 splenic a.
A 21 right gastric v.
A 22 splenic v.
A 23 inferior mesenteric v.
A 24 superior mesenteric a.
A 25 lumbar a. (opening)
A 26 left colic a.
A 27 abdominal aorta
A 28 sigmoid a.
A 29 left colic a.
A 30 left common iliac a.
A 31 left external iliac a.
A 32 left external iliac v.
A 33 obturator v.
A 34 obturator a.

M 1 diaphragm (costal portion)
M 2 serratus anterior m.
M 3 external oblique m.
M 4 internal oblique m.
M 5 transverse abdominis m.
M 6 psoas major m.
M 7 iliacus m.
M 8 gluteus medius m.
M 9 gluteus minimus m.
M 10 obturator internus m.
M 11 obturator externus m.
M 12 iliopsoas m.
M 13 rectus femoris m.
M 14 tensor fasciae latae m.
M 15 ischiocavernosus m.
M 16 quadriceps femoris m. (vastus intermedius m.)
M 17 bulbospongiosus m.
M 18 adductor magnus m.
M 19 gracilis m.

N 1 femoral n.

O 1 body of penis
O 2 head of femur
O 3 iliac crest
O 4 arcuate line
O 5 acetabular fossa
O 6 greater trochanter
O 7 inferior ramus of pubis
O 8 neck of femur

S 1 right lung (inferior lobe)
S 2 right lobe of liver
S 3 gall bladder
S 4 superior duodenal portion
S 5 head of pancreas
S 6 right colic flexure
S 7 horizontal portion of duodenum
S 8 ascending colon
S 9 cecum
S 10 appendix
S 11 small intestine
S 12 left lung (superior lobe)
S 13 left lobe of liver
S 14 body of stomach
S 15 body of pancreas
S 16 transverse colon
S 17 ascending portion of duodenum
S 18 small intestine
S 19 descending colon

S 20 small intestine
S 21 sigmoid colon
S 22 urinary bladder
S 23 corpus cavernosum penis
S 24 portion of corpus spongiosum penis
S 25 corpus spongiosum penis

SY 1 8th costochondral articulation
SY 2 acetabular labrum
SY 3 articular capsule
SY 4 obturator membrane

FA♂7

A 1 right coronary a.
A 2 portal v. (left branch)
A 3 gastroduodenal a.
A 4 gastroduodenal a.
A 5 right gastroepiploic v.
A 6 right inferior pancreaticoduodenal a.
A 7 right colic a.
A 8 jejunal v.
A 9 jejunal a.
A 10 ileal v.
A 11 deep circumflex iliac a.
A 12 testicular v. and a.
A 13 external iliac a.
A 14 external iliac v.
A 15 deep inguinal lymph node
A 16 deep femoral a.
A 17 lateral femoral circumflex a.
A 18 femoral v.
A 19 femoral a.
A 20 left ventricle
A 21 right ventricle
A 22 left gastric lymph node
A 23 left gastric a.
A 24 left colic a.
A 25 left colic a.
A 26 superior mesenteric lymph node
A 27 external iliac a.
A 28 external iliac v.

M 1 serratus anterior m.
M 2 external intercostal m.
M 3 external oblique m.
M 4 internal intercostal m.
M 5 transverse abdominis m.
M 6 internal oblique m.
M 7 external oblique m.
M 8 tensor fasciae latae m.
M 9 sartorius m.
M 10 iliacus m.
M 11 obturator externus m.
M 12 pectineus m.
M 13 adductor minimus m.
M 14 rectus femoris m.
M 15 sartorius m.
M 16 adductor longus m.
M 17 gracilis m.
M 18 psoas m.
M 19 rectus femoris m. (tendon)
M 20 gluteus medius m.
M 21 gluteus minimus m.
M 22 tensor fasciae latae m.
M 23 iliacus m.
M 24 vastus lateralis m.
M 25 bulbospongiosus m.
M 26 vastus intermedius m.

N 1 femoral n.
N 2 femoral n.

O 1 anterior superior iliac spine
O 2 iliac crest
O 3 anterior inferior iliac spine
O 4 head of femur
O 5 obturator crest

S 1 right lung (middle lobe)
S 2 falciform ligament
S 3 right lobe of liver
S 4 gall bladder
S 5 pylorus
S 6 right colic flexure
S 7 mesentery
S 8 ileum
S 9 ileocecal valve
S 10 cecum
S 11 small intestine
S 12 ductus deferens
S 13 urinary bladder
S 14 suspensory ligament of penis
S 15 corpus cavernosum penis
S 16 corpus spongiosum penis
S 17 left lung (superior lobe)
S 18 body of stomach
S 19 inferior omental recess
S 20 lesser omentum
S 21 body of pancreas
S 22 greater omentum
S 23 duodenal jejunal flexure
S 24 transverse mesocolon
S 25 1st loop of jejunum
S 26 duodenum (ascending portion)
S 27 transverse colon
S 28 jejunum
S 29 descending colon
S 30 sigmoid colon
S 31 urethra

FA♀
Frontal sections Abdomen

Abdomen and Pelvis

39-year-old female

plane of reference: subject lying on a horizontal plane; 6 sections, spaced 2.0 or 2.5 cm apart, parallel to the posterior plane of reference.

The superior border of each section passes through the center of the body of the 9th thoracic vertebra [T 9].

FA♀1 4 cm from the posterior plane of reference, postero-anterior view
FA♀2 6 cm from the posterior plane of reference, antero-posterior view
FA♀3 8.5 cm from the posterior plane of reference, antero-posterior view
FA♀4 10.5 cm from the posterior plane of reference, antero-posterior view
FA♀5a 13 cm from the posterior plane of reference, antero-posterior view
FA♀5p 13 cm from the posterior plane of reference, postero-anterior view
FA♀6 15 cm from the posterior plane of reference, antero-posterior view

FA♀1

A 1 middle sacral a.

M 1 diaphragm (costal portion)
M 2 intercostal mm.
M 3 transversospinal m.
M 4 latissimus dorsi m.
M 5 serratus posterior inferior m.

M 6 iliocostalis m.
M 7 longissimus m.
M 8 piriformis m.
M 9 levator ani m.
M 10 gluteus maximus m.
M 11 sphincter ani externus m.

O 1 transverse process T 10
O 2 iliac crest
O 3 lateral sacral crest
O 4 ala of sacrum
O 5 body of vertebra S 4
O 6 posterior superior iliac spine
O 7 3rd ventral sacral foramen

S 1 left lung (inferior lobe)
S 2 spleen
S 3 small intestine
S 4 rectum

SY 1 ligamentum flavum
SY 2 10th costotransverse articulation
SY 3 interspinal ligament

FA♀2

A 1 10th posterior intercostal a.
A 2 inferior gluteal a.

M 1 serratus anterior m.
M 2 diaphragm (costal portion)
M 3 quadratus lumborum m.
M 4 serratus posterior inferior m.
M 5 latissimus dorsi m.
M 6 iliocostalis m.
M 7 longissimus m.
M 8 transversospinal m.
M 9 fat body of lumbar triangle
M 10 gluteus maximus m.
M 11 piriformis m.
M 12 levator ani m.
M 13 quadratus lumborum m.
M 14 transverse abdominis m.
M 15 obturator internus m.
M 16 semitendinosus m.
M 17 biceps femoris m.
M 18 adductor magnus m.

N 1 spinal dura mater
N 2 spinal cord
N 3 dorsal spinal nerve root
N 4 spinal dura mater
N 5 2nd sacral n. (ventral ramus)
N 6 2nd sacral n. (ventral ramus)

O 1 costal process L 1
O 2 superior articulating process L 2
O 3 costal process L 2
O 4 ala of ilium
O 5 intermediate sacral crest
O 6 lateral sacral crest
O 7 ala of sacrum
O 8 body of vertebra S 2
O 9 tuberosity of ischium

S 1 right lobe of liver
S 2 perirenal fat
S 3 left lung (inferior lobe)
S 4 spleen
S 5 left kidney
S 6 perirenal fat
S 7 sigmoid colon
S 8 small intestine
S 9 rectum
S 10 ischiorectal fossa

SY 1 sacrotuberous ligament
SY 2 articulation of head of 11th rib
SY 3 processes of vertebrae L 1–L 2
SY 4 interspinal ligament
SY 5 sacroiliac articulation
SY 6 sacrotuberous ligament

FA♀3

A 1 thoracic aorta
A 2 12th posterior intercostal a.
A 3 superior gluteal v.
A 4 internal iliac v.
A 5 uterine v.
A 6 uterine venous plexus
A 7 vaginal venous plexus
A 8 vaginal v.
A 9 splenic v.
A 10 splenic a.
A 11 sigmoidal a. and v.
A 12 internal iliac v.
A 13 uterine v.

M 1 serratus anterior m.
M 2 latissimus dorsi m.
M 3 diaphragm
M 4 psoas major m.
M 5 external oblique m.
M 6 internal oblique m.
M 7 transverse abdominis m.
M 8 quadratus lumborum m.
M 9 transversospinal m.
M 10 gluteus medius m.
M 11 gluteus maximus m.
M 12 piriformis m.
M 13 gemellus superior m.
M 14 obturator internus m.
M 15 gemellus inferior m.
M 16 obturator internus m.
M 17 quadratus femoris m.
M 18 levator ani m.
M 19 inferior fascia of urogenital diaphragm
M 20 superficial transverse perineal m.
M 21 pubovaginalis m.
M 22 adductor magnus m.
M 23 gluteus maximus m.

N 1 spinal dura mater
N 2 cauda equina
N 3 spinal dura mater
N 4 sacral n. (ventral ramus)
N 5 sciatic n.
N 6 sciatic n.

O 1 costal process L 3
O 2 sacrum (superior articulating process)
O 3 sacrum (lateral portion)
O 4 sacral tuberosity
O 5 ala of ilium
O 6 ischium
O 7 greater trochanter

S 1 right lobe of liver
S 2 mediastinodiaphragmatic recess
S 3 right suprarenal gland
S 4 right kidney
S 5 perirenal fat
S 6 sigmoid colon
S 7 mesentery
S 8 small intestine
S 9 fornix of vagina
S 10 left lung (inferior lobe)
S 11 left lobe of liver
S 12 stomach
S 13 spleen
S 14 left suprarenal gland
S 15 tail of pancreas
S 16 left kidney
S 17 ileum
S 18 cervical canal
S 19 fornix of vagina
S 20 ischiorectal fossa
S 21 dorsal vaginal wall

SY 1 intervertebral disc T 12–L 1
SY 2 processes of vertebrae L 3–L 4
SY 3 ligamentum flavum
SY 4 processes of vertebrae L 5–S1
SY 5 interosseous sacroiliac ligament
SY 6 sacroiliac articulation

FA♀4

A 1 right atrium
A 2 inferior vena cava
A 3 hepatic v.
A 4 right internal iliac v.
A 5 superior gluteal v. (deep branch)
A 6 internal iliac a.
A 7 obturator v.
A 8 uterine v.

A 9 inferior vesical v.
A 10 internal pudendal v.
A 11 labia majora v.
A 12 left ventricle
A 13 coronary sinus
A 14 abdominal aorta
A 15 splenic v.
A 16 splenic a.
A 17 left renal a.
A 18 left renal v.
A 19 3rd lumbar a.
A 20 ascending lumbar v.
A 21 iliolumbar v.
A 22 superior rectal a.
A 23 left internal iliac v.
A 24 iliolumbar v.
A 25 left internal iliac a.
A 26 uterine a.
A 27 uterine v.
A 28 vaginal plexus of veins

M 1 diaphragm (central tendon)
M 2 diaphragm (esophageal hiatus)
M 3 serratus anterior m.
M 4 diaphragm (lumbar portion)
M 5 external oblique m.
M 6 internal oblique m.
M 7 transverse abdominis m.
M 8 psoas major m.
M 9 iliacus m.
M 10 gluteus medius m.
M 11 gluteus minimus m.
M 12 gluteus maximus m.
M 13 obturator internus m.
M 14 levator ani m.
M 15 obturator externus m.
M 16 iliopsoas m.
M 17 adductor magnus m.
M 18 vastus lateralis m.
M 19 gracilis m.

N 1 lumbar plexus
N 2 obturator n.

O 1 iliac crest
O 2 ala of ilium
O 3 head of femur
O 4 acetabular fossa
O 5 neck of femur
O 6 greater trochanter
O 7 ramus of ischium

S 1 right lung (inferior lobe)
S 2 liver (caudate lobe)
S 3 right lobe of liver
S 4 costodiaphragmatic pleural recess
S 5 right suprarenal gland
S 6 right kidney
S 7 renal pelvis
S 8 perirenal fat
S 9 ascending colon
S 10 sigmoid colon
S 11 ureter
S 12 appendix
S 13 infundibulum of uterine tube
S 14 ileum
S 15 body of uterus
S 16 urinary bladder
S 17 left lung (inferior lobe)
S 18 left lobe of liver
S 19 stomach
S 20 spleen
S 21 left suprarenal gland
S 22 body of pancreas
S 23 descending colon
S 24 renal pelvis
S 25 left kidney
S 26 ureter
S 27 ileum
S 28 fornix of vagina
S 29 vagina
S 30 labia minora
S 31 pudendal fissure
S 32 labia majora

SY 1 sacroiliac articulation (anterior portion)
SY 2 lumbosacral articulation
SY 3 lunate surface
SY 4 acetabular labrum
SY 5 articular capsule

FA♀5 a

A 1 right atrium
A 2 portal v. (right branch)
A 3 portal v.
A 4 inferior vena cava
A 5 right common iliac v.
A 6 right common iliac a.
A 7 right external iliac v.
A 8 obturator a.
A 9 dorsal vein of the clitoris
A 10 deep femoral a.
A 11 left ventricle
A 12 coronary sinus
A 13 hepatic v.
A 14 right gastric a.
A 15 splenic a.
A 16 common hepatic a.
A 17 splenic v.
A 18 superior mesenteric a.
A 19 inferior mesenteric v.
A 20 abdominal a.
A 21 lumbar a. (opening)
A 22 left ovarian v.
A 23 left common iliac v.
A 24 superior rectal a.
A 25 left common iliac a.
A 26 left external iliac v.
A 27 obturator a.
A 28 uterine a.
A 29 deep femoral a.

M 1 diaphragm (central tendon)
M 2 diaphragm (costal portion)
M 3 external oblique m.
M 4 internal oblique m.
M 5 transverse abdominis m.
M 6 psoas major m.
M 7 iliacus m.
M 8 gluteus minimus m.
M 9 gluteus medius m.
M 10 fascia lata, iliotibial tract
M 11 obturator internus m.
M 12 levator ani m.
M 13 obturator externus m.
M 14 transverse perineal ligament
M 15 adductor brevis m.
M 16 iliopsoas m.
M 17 pectineus m.
M 18 adductor brevis m.
M 19 vastus lateralis m.
M 20 gracilis m.
M 21 vastus intermedius m.
M 22 adductor longus m.

N 1 femoral n.
N 2 obturator n.

O 1 iliac crest
O 2 arcuate line
O 3 acetabular fossa
O 4 head of femur
O 5 inferior ramus of pelvis

S 1 right lung (inferior lobe)
S 2 right hepatic lobe
S 3 gall bladder
S 4 superior duodenal flexure
S 5 duodenum (descending portion)
S 6 right colic flexure
S 7 right ureter
S 8 ascending colon
S 9 right ureter
S 10 small intestine
S 11 mesentery
S 12 appendix
S 13 ampulla of uterine tube
S 14 body of uterus
S 15 ligamentum teres
S 16 uterovesical space
S 17 urinary bladder
S 18 left lobe of liver
S 19 stomach
S 20 lesser omentum
S 21 liver (caudate lobe)
S 22 left colic flexure
S 23 body of pancreas
S 24 descending colon
S 25 duodenum (ascending portion)
S 26 jejunum
S 27 left ureter

S 28 sigmoid colon
S 29 left ureter
S 30 infundibulum of uterine tube
S 31 ovary
S 32 broad ligament
S 33 crus of clitoris
S 34 vestibular bulb
S 35 labia minora
S 36 labia majora

SY 1 anterior longitudinal ligament
SY 2 acetabular labrum
SY 3 ligamentum teres
SY 4 transverse acetabular ligament
SY 5 articular capsule

FA♀5p

A 1 left ventricle
A 2 pericardium
A 3 left hepatic v.
A 4 right gastric a.
A 5 common hepatic a.
A 6 left gastric v.
A 7 splenic v.
A 8 superior mesenteric a.
A 9 inferior mesenteric v.
A 10 inferior mesenteric v.
A 11 inferior mesenteric a. (opening)
A 12 abdominal aorta
A 13 left common iliac a.
A 14 superior rectal v.
A 15 left external iliac v.
A 16 obturator v.
A 17 dorsal artery of clitoris
A 18 deep femoral v. (branch)
A 19 portal v. (left branch)
A 20 portal v. (right branch)
A 21 portal v.
A 22 inferior vena cava
A 23 right colic venous arch
A 24 right common iliac a.
A 25 right common iliac v.
A 26 right external iliac v.
A 27 uterine a.

M 1 diaphragm (costal portion)
M 2 serratus anterior m.
M 3 external oblique m.
M 4 internal oblique m.
M 5 transverse abdominis m.
M 6 psoas major m.
M 7 iliacus m.
M 8 gluteus medius m.
M 9 gluteus minimus m.
M 10 obturator internus m.
M 11 levator ani m.
M 12 obturator externus m.
M 13 iliopsoas m.
M 14 pectineus m.
M 15 transverse perineal ligament
M 16 tensor fasciae latae m.
M 17 adductor brevis m.
M 18 gracilis m.
M 19 vastus lateralis m.
M 20 rectus femoris m.
M 21 adductor longus m.

N 1 femoral n.
N 2 obturator n.

O 1 iliac crest
O 2 inferior ramus of pubis

S 1 left lung (inferior lobe)
S 2 left lobe of liver
S 3 stomach
S 4 inferior omental recess
S 5 left colic flexure
S 6 lesser omentum
S 7 transverse mesocolon
S 8 duodenal jejunal flexure
S 9 body of pancreas
S 10 duodenum (ascending portion)
S 11 jejunum
S 12 greater omentum
S 13 left ureter
S 14 left ureter
S 15 sigmoid colon
S 16 mesentery
S 17 infundibulum of uterine tube
S 18 left ovary
S 19 ileum
S 20 ovarian ligament
S 21 uterine tube
S 22 body of uterus
S 23 uterovesical space
S 24 urinary bladder
S 25 right lung
S 26 neck of gall bladder
S 27 gall bladder
S 28 superior duodenal flexure
S 29 right colic flexure
S 30 duodenum (descending portion)
S 31 right ureter
S 32 ascending colon
S 33 right ureter
S 34 ileum
S 35 appendix
S 36 infundibulum of uterine tube
S 37 round ligament
S 38 crus of clitoris
S 39 vestibular bulb
S 40 labia minora
S 41 labia majora
S 42 pudendal fissure

SY 1 anterior longitudinal ligament
SY 2 intervertebral disc L 4–L 5
SY 3 acetabular labrum
SY 4 ligamentum teres
SY 5 articular capsule
SY 6 transverse acetabular ligament

FA♀6

A 1 right coronary a.
A 2 pericardium
A 3 right gastric v.
A 4 gastroduodenal a.
A 5 ileocolic a.
A 6 external iliac a.
A 7 external iliac v.
A 8 femoral v.
A 9 femoral a.
A 10 greater saphenous v.
A 11 left ventricle
A 12 right ventricle
A 13 portal vein (left branch)
A 14 left gastric v.
A 15 middle colic a. (branch)
A 16 jejunal v.
A 17 right gastroepiploic v.
A 18 superior mesenteric v.
A 19 ileal v.
A 20 external iliac a.
A 21 external iliac v.

M 1 serratus anterior m.
M 2 external oblique m.
M 3 internal oblique m.
M 4 transverse abdominis m.
M 5 psoas major m.
M 6 iliacus m.
M 7 gluteus minimus m.
M 8 gluteus medius m.
M 9 iliopsoas m.
M 10 obturator externus m.
M 11 pectineus m.
M 12 adductor brevis m.
M 13 rectus femoris m.
M 14 tensor fasciae latae m.
M 15 vastus lateralis m.
M 16 adductor longus m.
M 17 sartorius m.

N 1 femoral n.

O 1 iliac crest
O 2 head of femur
O 3 superior ramus of pubis
O 4 iliopubic eminence

S 1 mammary gland
S 2 right lung (inferior lobe)
S 3 right lobe of liver
S 4 liver (quadrate lobe)
S 5 lesser omentum (pyloric portion of stomach)
S 6 gall bladder
S 7 pylorus

S 8 body of pancreas
S 9 head of pancreas
S 10 major duodenal papilla
S 11 duodenum (descending portion)
S 12 duodenum (horizontal portion)
S 13 transverse colon
S 14 ileum
S 15 ileum (terminal portion)
S 16 cecum
S 17 left lung (inferior lobe)
S 18 left hepatic lobe
S 19 inferior omental recess
S 20 stomach
S 21 transverse mesocolon
S 22 jejunum (superior portion)
S 23 jejunum
S 24 duodenum (ascending part)
S 25 descending colon
S 26 mesentery
S 27 descending colon
S 28 sigmoid colon
S 29 urinary bladder
S 30 retropubic space
S 31 body of clitoris
S 32 foreskin of clitoris
S 33 labia majora

SY 1 acetabular labrum
SY 2 hip joint (articular capsule)
SY 3 pubic symphysis

SA♂
Sagittal sections Abdomen

Abdomen and Pelvis

50-year-old male

The median sagittal section is the plane of reference. It is not shown here. The first section on each side of the median plane is 1 cm from this plane. The remaining sections are spaced 2 cm apart.

The superior border of each section passes through the center of the body of the 7th thoracic vertebra [T 7].

The subject has a long sigmoid colon.

SA♂ 6D sagittal section, 11 cm to the right of the median plane, exterior view
SA♂ 5D sagittal section, 9 cm to the right of the median plane, exterior view
SA♂ 4D sagittal section, 7 cm to the right of the median plane, exterior view
SA♂ 3D sagittal section, 5 cm to the right of the median plane, exterior view
SA♂ 2D sagittal section, 3 cm to the right of the median plane, exterior view
SA♂ 1D sagittal section, 1 cm to the right of the median plane, exterior view
SA♂ 1Si sagittal section, 1 cm to the left of the median plane, exterior view
SA♂ 2Si sagittal section, 3 cm to the left of the median plane, exterior view
SA♂ 3Si sagittal section, 5 cm to the left of the median plane, exterior view
SA♂ 4Si sagittal section, 7 cm to the left of the median plane, exterior view
SA♂ 5Si sagittal section, 9 cm to the left of the median plane, exterior view
SA♂ 6Si sagittal section, 11 cm to the left of the median plane, exterior view

SA♂ 6D

A 1 superior gluteal a. (deep branch)
A 2 iliolumbar a. (iliacus branch)
A 3 lateral femoral circumflex v.

M 1 serratus anterior m.
M 2 latissimus dorsi m.
M 3 diaphragm (costal portion)
M 4 external oblique m.
M 5 internal oblique m.
M 6 thoracolumbar fascia
M 7 fat body of lumbar triangle
M 8 gluteus medius m.
M 9 gluteus maximus m.
M 10 pectoralis major m.
M 11 pectoralis minor m.
M 12 intercostal mm.
M 13 transverse abdominis m.
M 14 internal oblique m.
M 15 external oblique m.
M 16 iliacus m.
M 17 gluteus minimus m.
M 18 sartorius m.
M 19 rectus femoris m.
M 20 vastus medialis m.

O 1 inferior angle of scapula
O 2 iliac crest
O 3 greater trochanter
O 4 anterior superior iliac spine

S 1 right lung (inferior lobe)
S 2 right lobe of liver
S 3 perirenal fat
S 4 ascending colon
S 5 right lung (middle lobe)
S 6 transverse colon
S 7 greater omentum
S 8 cecum
S 9 appendix
S 10 small intestine

SA♂ 5D

A 1 right colic a.
A 2 iliolumbar a. (iliac branch)
A 3 superior gluteal a. (deep branch)
A 4 superior gluteal a. (superficial branch)
A 5 portal v. (right branch)
A 6 superficial circumflex iliac v.
A 7 superficial inguinal lymph node
A 8 lateral femoral circumflex v.
A 9 lateral femoral circumflex a.
A 10 lateral femoral circumflex v.
A 11 deep femoral v.
A 12 deep femoral a.
A 13 femoral v.

M 1 latissimus dorsi m.
M 2 serratus posterior inferior m.
M 3 diaphragm (costal portion)
M 4 thoracolumbar fascia (deep layer)
M 5 latissimus dorsi m.
M 6 quadratus lumborum m.
M 7 fat body of lumbar triangle
M 8 gluteus medius m.
M 9 gluteus minimus m.
M 10 gluteus maximus m.
M 11 piriformis m.
M 12 obturator internus m.
M 13 obturator externus m.
M 14 quadratus femoris m.

M 15 adductor magnus m.
M 16 biceps femoris m.
M 17 pectoralis major m.
M 18 transverse abdominis m.
M 19 rectus abdominis m.
M 20 external oblique m.
M 21 internal oblique m.
M 22 iliacus m.
M 23 sartorius m.
M 24 adductor brevis m.
M 25 pectineus m.
M 26 adductor longus m.

N 1 sciatic n.

O 1 iliac crest
O 2 body of ischium
O 3 greater trochanter
O 4 body of pubis
O 5 head of femur
O 6 neck of femur

S 1 right lung (inferior lobe)
S 2 right lobe of liver
S 3 perirenal fat
S 4 right colic flexure
S 5 right kidney
S 6 ascending colon
S 7 right lung (middle lobe)
S 8 gall bladder
S 9 transverse colon
S 10 greater omentum
S 11 cecum
S 12 small intestine
S 13 appendix
S 14 mesentery

SY 1 acetabular labrum
SY 2 articular capsule

SA♂ 4D

A 1 portal v. (right anterior branch)
A 2 portal v. (right posterior branch)
A 3 right colic a. (superior branch)
A 4 right colic a.
A 5 iliolumbar a. (iliac branch)
A 6 superior gluteal a. (deep branch)
A 7 superior gluteal a. (superficial branch)
A 8 inferior gluteal a.
A 9 superficial epigastric v.
A 10 inferior epigastric a.
A 11 femoral a.
A 12 femoral v.
A 13 external pudendal v.
A 14 medial femoral circumflex v.
A 15 medial femoral circumflex a.
A 16 superficial inguinal lymph node

M 1 latissimus dorsi m.
M 2 diaphragm (costal portion)
M 3 iliocostalis m.
M 4 quadratus lumborum m.
M 5 fat body of lumbar triangle
M 6 psoas major m.
M 7 iliacus m.
M 8 gluteus medius m.
M 9 gluteus maximus m.
M 10 gluteus minimus m.
M 11 piriformis m.
M 12 gemellus superior m.
M 13 obturator internus m.
M 14 gemellus inferior m.
M 15 obturator externus m.
M 16 quadratus femoris m.
M 17 semimembranosus m.
M 18 semitendinosus m.
M 19 pectoralis major m.
M 20 transverse thoracic m.
M 21 transverse abdominis m.
M 22 rectus abdominis m.
M 23 psoas major m.
M 24 internal oblique m.
M 25 inguinal canal
M 26 reflected inguinal ligament, external oblique m.
M 27 pectineus m.
M 28 adductor longus m.
M 29 adductor brevis m.
M 30 adductor magnus m.

N 1 femoral n.
N 2 sciatic n.

O 1 iliac crest
O 2 body of ilium
O 3 body of ischium

S 1 right lung (inferior lobe)
S 2 right lobe of liver
S 3 right kidney
S 4 right colic flexure
S 5 perirenal fat
S 6 right lung (middle lobe)
S 7 gall bladder
S 8 body of gall bladder
S 9 transverse colon
S 10 transverse colon
S 11 ascending colon
S 12 greater omentum
S 13 mesentery
S 14 small intestine

SA♂ 3D

A 1 hepatic v.
A 2 anterior superior segment of renal a.
A 3 renal v.
A 4 renal a. (posterior branch)
A 5 iliolumbar a.
A 6 superior gluteal a.
A 7 superior gluteal a.
A 8 inferior gluteal v.
A 9 right ventricle
A 10 portal v. (left branch)
A 11 portal v. (right branch)
A 12 hepatic a. (right branch)
A 13 right gastroepiploic a.
A 14 right colic a.
A 15 right colic a.
A 16 inferior mesenteric a.
A 17 superior mesenteric a. (ilial branch)
A 18 external iliac a.
A 19 external iliac v.
A 20 pampiniform plexus
A 21 obturator a. (acetabular branch)
A 22 superficial inguinal lymph node
A 23 external pudendal v.

M 1 trapezius m.
M 2 latissimus dorsi m.
M 3 diaphragm (costal portion)
M 4 quadratus lumborum m.
M 5 iliocostalis lumborum m.
M 6 psoas m.
M 7 gluteus medius m.
M 8 iliacus m.
M 9 obturator internus m.
M 10 piriformis m.
M 11 gemellus superior m.
M 12 obturator internus m.
M 13 gemellus inferior m.
M 14 gluteus maximus m.
M 15 quadratus femoris m.
M 16 semitendinosus m.
M 17 adductor magnus m.
M 18 semimembranosus m.
M 19 pectoralis major m.
M 20 diaphragm (costal portion)
M 21 rectus abdominis m.
M 22 reflected inguinal ligament
M 23 pectineus m.
M 24 obturator externus m.
M 25 adductor brevis m.
M 26 adductor magnus m.
M 27 adductor longus m.

N 1 femoral n.
N 2 sciatic n.

O 1 iliac crest
O 2 sacrum
O 3 iliac tuberosity
O 4 body of ilium
O 5 tuberosity of ischium
O 6 body of pubis

S 1 right lung (inferior lobe)
S 2 perirenal fat

S 3 right kidney
S 4 duodenum (descending portion)
S 5 right lung (inferior lobe)
S 6 left hepatic duct
S 7 common hepatic duct
S 8 liver (quadrate lobe)
S 9 gall bladder
S 10 left lobe of liver
S 11 falciform ligament
S 12 ligamentum teres
S 13 transverse colon
S 14 jejunum
S 15 transverse colon
S 16 jejunum
S 17 small intestine
S 18 mesentery

SY 1 sacroiliac articulation
SY 2 acetabular fossa

SA♂ 2D

A 1 left atrium
A 2 right atrium
A 3 inferior vena cava
A 4 right renal a.
A 5 right inferior colic a.
A 6 right common iliac v.
A 7 right common iliac a.
A 8 superior gluteal v.
A 9 external iliac a.
A 10 internal iliac a.
A 11 internal iliac v.
A 12 inferior vesical v.
A 13 internal pudendal a.
A 14 tricuspid valve
A 15 right ventricle
A 16 right coronary a.
A 17 hepatic v.
A 18 v. of lesser omentum
A 19 hepatic a.
A 20 portal v.
A 21 right gastric v.
A 22 right gastroepiploic a.
A 23 right colic a.
A 24 right colic a.
A 25 superior mesenteric a. (branch)
A 26 superior mesenteric a. (branch)
A 27 pampiniform plexus

M 1 trapezius m.
M 2 diaphragm (costal portion)
M 3 longissimus m.
M 4 intertransverse m.
M 5 psoas m.
M 6 piriformis m.
M 7 coccygeus m.
M 8 obturator internus m.
M 9 gluteus maximus m.
M 10 ischiorectal fossa
M 11 adductor magnus m.
M 12 diaphragm (sternal portion)
M 13 rectus abdominis m.
M 14 rectus abdominis m.
M 15 inguinal canal
M 16 pectineus m.
M 17 obturator externus m.
M 18 adductor longus m.
M 19 adductor brevis m.
M 20 gracilis m.

N 1 3rd lumbar n. (ventral ramus)
N 2 4th lumbar n. (ventral ramus)
N 3 5th lumbar n. (ventral ramus)
N 4 1st sacral n. (ventral ramus)
N 5 2nd sacral n. (ventral ramus)

O 1 10th rib
O 2 costal process L 5
O 3 posterior superior iliac spine
O 4 lateral sacral crest
O 5 ramus of ischium
O 6 body of sternum
O 7 7th costal cartilage
O 8 superior ramus of pubis

S 1 right lung (inferior lobe)
S 2 liver (caudate lobe)
S 3 right suprarenal gland
S 4 superior duodenal flexure
S 5 perirenal fat
S 6 duodenum (descending portion)
S 7 small intestine
S 8 urinary bladder
S 9 ureter
S 10 left lobe of liver
S 11 liver (caudate lobe)
S 12 lesser omentum
S 13 common bile duct
S 14 duodenum (superior portion)
S 15 pylorus
S 16 pyloric antrum
S 17 head of pancreas
S 18 jejunum
S 19 transverse colon
S 20 greater omentum
S 21 small intestine
S 22 spermatic cord

SY 1 dorsal sacroiliac ligament
SY 2 sacrotuberous ligament
SY 3 obturator membrane
SY 4 7th sternocostal articulation

SA♂ 1D

A 1 left atrium
A 2 coronary sinus
A 3 fibrous pericardium
A 4 inferior phrenic a.
A 5 1st lumbar a.
A 6 3rd lumbar a.
A 7 inferior vena cava
A 8 right common iliac a.
A 9 inferior vena cava (origin)
A 10 right ventricle
A 11 pericardial cavity
A 12 left gastric v.
A 13 left gastric a.
A 14 common hepatic a.
A 15 right gastroepiploic a.
A 16 portal v.
A 17 left renal v.
A 18 marginal artery of Drummond
A 19 left renal a.
A 20 common trunk, right gastroepiploic v. and right inferior pancreaticoduodenal a.
A 21 right colic a. (branch)
A 22 right colic a.
A 23 superior mesenteric v.
A 24 superior mesenteric a.
A 25 external pudendal v.
A 26 dorsal vein of the penis
A 27 prostatic venous plexus

M 1 trapezius m.
M 2 longissimus m.
M 3 transversospinal m.
M 4 longissimus m.
M 5 diaphragm (crus)
M 6 piriformis m.
M 7 gluteus maximus m.
M 8 coccygeus m.
M 9 gluteus maximus m.
M 10 levator ani m.
M 11 sphincter ani externus m.
M 12 diaphragm (sternal portion)
M 13 rectus abdominis m.
M 14 sheath of rectus abdominis m.
M 15 umbilicus
M 16 rectus abdominis m.
M 17 levator ani m.

N 1 spinal dura mater
N 2 3rd sacral n.

O 1 superior articulating process T 12
O 2 pedicle T 12
O 3 lamina T 12
O 4 intermediate sacral crest
O 5 3rd sacral canal
O 6 coccyx
O 7 body of sternum
O 8 xiphoid process
O 9 body of pubis

S 1 esophagus (thoracic portion)
S 2 liver (caudate lobe)
S 3 head of pancreas
S 4 duodenum (horizontal portion)
S 5 small intestine

S 6 rectovesical pouch
S 7 urinary bladder
S 8 ampulla of ductus deferens
S 9 seminal vesicles
S 10 prostate
S 11 left lobe of liver
S 12 stomach
S 13 transverse colon
S 14 jejunum
S 15 greater omentum
S 16 small intestine
S 17 mesentery
S 18 small intestine
S 19 corpus cavernosum penis
S 20 epididymis
S 21 corpus spongiosum penis
S 22 glans penis
S 23 right testis
S 24 prepuce of penis
S 25 left testis

SY 1 processes of vertebrae T 10–T 11
SY 2 intervertebral disc L 5–S 1
SY 3 xiphosternal synchondrosis
SY 4 arcuate ligament of pubis

SA♂ 1 Si

A 1 left ventricle
A 2 right ventricle
A 3 interventricular septum
A 4 pericardial cavity
A 5 left gastric a.
A 6 celiac trunk
A 7 splenic a.
A 8 splenic v.
A 9 superior mesenteric a.
A 10 inferior mesenteric v.
A 11 left renal v.
A 12 abdominal aorta
A 13 superior mesenteric a. (branch)
A 14 inferior mesenteric a.
A 15 superior mesenteric a. (branch)
A 16 4th lumbar a.
A 17 bifurcation of aorta
A 18 left common iliac v.
A 19 external pudendal v.
A 20 prostatic venous plexus
A 21 thoracic aorta
A 22 left atrium
A 23 coronary sinus
A 24 mitral valve
A 25 thoracic aorta
A 26 superior rectal a.

M 1 pectoralis major m.
M 2 diaphragm (costal portion)
M 3 rectus abdominis m.
M 4 trapezius m.
M 5 diaphragm (lumbar portion)
M 6 erector spinae m.
M 7 erector spinae m.
M 8 levator ani m.
M 9 bulbospongiosus m.
M 10 sphincter ani externus m.

N 1 spinal dura mater
N 2 cauda equina
N 3 1st sacral n.
N 4 2nd sacral n.

O 1 body of pubis
O 2 superior articulating process T 11
O 3 pedicle T 11
O 4 lamina T 11
O 5 intermediate sacral crest

S 1 esophagus (abdominal portion)
S 2 liver (left lobe)
S 3 body of stomach
S 4 transverse colon
S 5 body of pancreas
S 6 transverse colon
S 7 sigmoid colon (variation)
S 8 jejunum (proximal loop)
S 9 duodenum (ascending portion)
S 10 small intestine
S 11 mesentery
S 12 small intestine
S 13 small intestine
S 14 sigmoid colon
S 15 urinary bladder
S 16 corpus cavernosum penis
S 17 corpus spongiosum penis
S 18 head of epididymis
S 19 left testis
S 20 transverse rectal folds
S 21 seminal vesicle
S 22 prostate
S 23 rectal ampulla
S 24 rectum (perineal flexure)
S 25 anal canal

SY 1 processes of vertebrae T 9–T 10
SY 2 intervertebral disc T 12–L 1
SY 3 ligamentum flavum
SY 4 lumbosacral articulation

SA♂ 2 Si

A 1 left ventricle
A 2 fibrous pericardium
A 3 sigmoid v.
A 4 external iliac a.
A 5 external iliac v.
A 6 internal iliac a.
A 7 testicular v.
A 8 external pudendal v.
A 9 9th posterior intercostal a.
A 10 splenic a.
A 11 splenic v.
A 12 renal v.
A 13 renal a.
A 14 inferior mesenteric v.
A 15 left colic a.
A 16 inferior mesenteric v.
A 17 superior lateral sacral a.
A 18 superior rectal v.
A 19 internal pudendal a.

M 1 pectoralis major m.
M 2 transverse thoracic m.
M 3 diaphragm (costal portion)
M 4 rectus abdominis m.
M 5 inguinal canal
M 6 pectineus m.
M 7 obturator internus m.
M 8 obturator externus m.
M 9 adductor brevis m.
M 10 adductor magnus m.
M 11 adductor longus m.
M 12 gracilis m.
M 13 trapezius m.
M 14 longissimus thoracis m.
M 15 longissimus thoracis m.
M 16 transversospinalis m.
M 17 intertransverse m.
M 18 psoas m.
M 19 iliocostalis m.
M 20 piriformis m.
M 21 gluteus maximus m.
M 22 levator ani m.
M 23 ischiorectal fossa

N 1 2nd lumbar n. (ventral ramus)
N 2 3rd lumbar n. (ventral ramus)
N 3 4th lumbar n. (ventral ramus)
N 4 5th lumbar n. (ventral ramus)
N 5 1st sacral n. (ventral ramus)
N 6 2nd sacral n. (ventral ramus)
N 7 3rd sacral n. (ventral ramus)
N 8 4th sacral n. (ventral ramus)

O 1 pecten pubis
O 2 inferior ramus of pubis
O 3 transverse process T 9
O 4 costal process L 1
O 5 costal process L 5
O 6 costal process of 1st sacral vertebra
O 7 2nd pelvic sacral foramen
O 8 coccyx

S 1 body of stomach
S 2 transverse colon
S 3 body of pancreas
S 4 transverse colon
S 5 sigmoid colon (variation)
S 6 small intestine
S 7 greater omentum
S 8 mesentery
S 9 left ureter

S 10 small intestine
S 11 urinary bladder
S 12 ductus deferens
S 13 left lung (inferior lobe)
S 14 spleen
S 15 left suprarenal gland
S 16 perirenal fat
S 17 duodenal jejunal flexure
S 18 jejunum (proximal loop)
S 19 seminal vesicle
S 20 ureter (pelvic portion)

SA ♂ 3 Si

A 1 left ventricle
A 2 fibrous pericardium
A 3 left colic a. (branch)
A 4 left colic a. (branch)
A 5 superficial epigastric v.
A 6 external iliac a.
A 7 inferior epigastric a.
A 8 external iliac v.
A 9 obturator a.
A 10 superficial epigastric v.
A 11 superficial inguinal lymph node
A 12 external pudendal v.
A 13 greater saphenous v.
A 14 short gastric a.
A 15 splenic a.
A 16 short gastric a.
A 17 splenic v.
A 18 renal a. (anterior branch)
A 19 renal v.
A 20 renal a. (posterior branch)
A 21 superior gluteal a.
A 22 inferior gluteal a.
A 23 inferior gluteal v.
A 24 inferior gluteal a.

M 1 pectoralis major m.
M 2 diaphragm (costal portion)
M 3 rectus abdominis m.
M 4 transverse abdominis m.
M 5 rectus abdominis m.
M 6 psoas m.
M 7 transverse abdominis m.
M 8 obturator canal
M 9 pectineus m.
M 10 obturator externus m.
M 11 adductor brevis m.
M 12 adductor magnus m.
M 13 adductor magnus m.
M 14 adductor longus m.
M 15 trapezius m.
M 16 latissimus dorsi m.
M 17 iliocostalis m.
M 18 quadratus lumborum m.
M 19 iliocostalis m.
M 20 piriformis m.
M 21 obturator internus m.
M 22 gluteus maximus m.

N 1 3rd lumbar n. (ventral ramus)
N 2 4th lumbar n. (ventral ramus)
N 3 lumbosacral trunk
N 4 1st sacral n. (ventral ramus)
N 5 obturator n.

O 1 superior ramus of pubis
O 2 iliac tuberosity
O 3 sacral tuberosity
O 4 posterior inferior iliac spine
O 5 sciatic spine
O 6 ramus of ischium

S 1 transverse colon
S 2 transverse mesocolon
S 3 transverse colon
S 4 sigmoid colon (variation)
S 5 greater omentum
S 6 small intestine
S 7 sigmoid colon (variation)
S 8 small intestine
S 9 mesentery
S 10 left lung (inferior lobe)
S 11 body of stomach
S 12 spleen
S 13 body of pancreas
S 14 left kidney
S 15 left ureter

SY 1 sacroiliac articulation
SY 2 sacrotuberous ligament

SA ♂ 4 Si

A 1 left colic a. (branch)
A 2 splenic a.
A 3 splenic v.
A 4 left gastroepiploic v.
A 5 left colic a. (branch)
A 6 deep circumflex iliac a.
A 7 femoral a.
A 8 medial femoral circumflex a.
A 9 superficial circumflex iliac v.
A 10 lateral femoral circumflex a. (descending branch)
A 11 femoral v.
A 12 11th intercostal a.
A 13 superior gluteal a.

M 1 pectoralis major m.
M 2 diaphragm (costal portion)
M 3 external oblique m.
M 4 transversus abdominis m.
M 5 internal oblique m.
M 6 psoas m.
M 7 external oblique m.
M 8 pectineus m.
M 9 adductor brevis m.
M 10 adductor longus m.
M 11 rhomboideus m.
M 12 latissimus dorsi m.
M 13 iliocostalis m.
M 14 thoracolumbar fascia
M 15 quadratus lumborum m.
M 16 fat body of lumbar triangle
M 17 iliacus m.
M 18 gluteus medius m.
M 19 gluteus maximus m.
M 20 gluteus minimus m.
M 21 piriformis m.
M 22 gemellus superior m.
M 23 obturator internus m.
M 24 gemellus inferior m.
M 25 biceps femoris m.
M 26 quadratus femoris m.
M 27 adductor magnus m.
M 28 ischiocrurales mm.

N 1 femoral n.
N 2 sciatic n.

O 1 iliac crest
O 2 body of ischium
O 3 tuberosity of ischium

S 1 left lung (inferior lobe)
S 2 body of stomach
S 3 transverse colon
S 4 sigmoid colon (variation)
S 5 descending colon
S 6 greater omentum
S 7 peritoneal cavity
S 8 small intestine
S 9 spleen
S 10 left kidney
S 11 perirenal fat

SY 1 7th & 8th interchondral articulation
SY 2 acetabular labrum

SA ♂ 5 Si

A 1 splenic v. (branch)
A 2 splenic v.
A 3 iliolumbar a. (iliac branch)
A 4 deep circumflex iliac a.
A 5 superficial epigastric v.
A 6 lateral femoral circumflex a.
A 7 deep femoral a.

M 1 pectoralis major m.
M 2 diaphragm (costal portion)
M 3 transverse abdominis m.
M 4 internal oblique m.
M 5 external oblique m.
M 6 iliacus m.

M 7 rectus femoris m.
M 8 pectineus m.
M 9 sartorius m.
M 10 adductor brevis m.
M 11 adductor magnus m.
M 12 vastus medialis m.
M 13 serratus anterior m.
M 14 latissimus dorsi m.
M 15 serratus posterior inferior m.
M 16 serratus posterior inferior m.
M 17 diaphragm (costal portion)
M 18 thoracolumbar fascia
M 19 quadratus lumborum m.
M 20 fat body of lumbar triangle
M 21 gluteus medius m.
M 22 gluteus maximus m.
M 23 gluteus minimus m.
M 24 piriformis m.
M 25 gemellus superior m.
M 26 obturator internus m.
M 27 gemellus inferior m.
M 28 obturator externus m.
M 29 quadratus femoris m.
M 30 semitendinosus m.
M 31 ischiocrurales mm.

N 1 sciatic n.

O 1 inferior angle of scapula
O 2 iliac crest
O 3 anterior inferior iliac spine
O 4 lesser trochanter

S 1 left lung (superior lobe)
S 2 left lung (inferior lobe)
S 3 left colic flexure
S 4 sigmoid colon (variation)
S 5 descending colon
S 6 sigmoid colon
S 7 greater omentum
S 8 small intestine
S 9 spleen
S 10 left kidney
S 11 perirenal fat

SY 1 7th costochondral articulation
SY 2 interchondral articulation 8-9
SY 3 articulate capsule
SY 4 acetabular labrum

SA♂ 6Si

A 1 superior gluteal a. (deep branch)
A 2 lateral femoral circumflex a.
A 3 lateral femoral circumflex a. (descending branch)

M 1 pectoralis major m.
M 2 pectoralis minor m.
M 3 serratus anterior m.
M 4 external intercostal m.
M 5 serratus anterior m.
M 6 internal intercostal m.
M 7 diaphragm (costal portion)
M 8 transverse abdominis m.
M 9 internal oblique m.
M 10 external oblique m.
M 11 transverse abdominis m.
M 12 iliacus m.
M 13 gluteus minimus m.
M 14 sartorius m.
M 15 rectus femoris m.
M 16 vastus medialis m.
M 17 subscapularis m.
M 18 teres major m.
M 19 serratus anterior m.
M 20 latissimus dorsi m.
M 21 diaphragm (costal portion)
M 22 external oblique m.
M 23 fat body of lumbar triangle
M 24 gluteus medius m.
M 25 piriformis m.
M 26 obturator internus m.
M 27 gluteus maximus m.
M 28 quadratus femoris m.
M 29 adductor magnus m.
M 30 biceps femoris m. (long head)

N 1 sciatic n.

O 1 iliac crest
O 2 anterior superior iliac spine
O 3 axillary margin of scapula
O 4 greater trochanter
O 5 neck of femur
O 6 femoral shaft

S 1 spleen
S 2 costodiaphragmatic recess
S 3 left colic flexure
S 4 descending colon
S 5 left lung (inferior lobe)
S 6 accessory spleen
S 7 perirenal fat

SY 1 9th costochondral articulation
SY 2 articular capsule of hip joint

SA♀
Sagittal sections Abdomen
Abdomen and Pelvis

25-year-old female

The median sagittal section is the plane of reference. Sections have been made on both sides of this plane. They are spaced 2 cm apart.
The superior border of each section passes through the upper part of the body of the 6th thoracic vertebra [T 6].

SA♀6D sagittal section, 12 cm to the right of the median plane, exterior view
SA♀5D sagittal section, 10 cm to the right of the median plane, exterior view
SA♀4D sagittal section, 8 cm to the right of the median plane, exterior view
SA♀3D sagittal section, 6 cm to the right of the median plane, exterior view
SA♀2D sagittal section, 4 cm to the right of the median plane, exterior view
SA♀1D sagittal section, 2 cm to the right of the median plane, exterior view
SA♀MD median section, right segment, left-to-right view
SA♀MSi median section, left segment, right-to-left view
SA♀1Si sagittal section, 2 cm to the left of the median plane, exterior view
SA♀2Si sagittal section, 4 cm to the left of the median plane, exterior view
SA♀3Si sagittal section, 6 cm to the left of the median plane, exterior view
SA♀4Si sagittal section, 8 cm to the left of the median plane, exterior view

SA♀5Si sagittal section, 10 cm to the left of the median plane, exterior view

SA♀6Si sagittal section, 12 cm to the left of the median plane, exterior view

SA♀6D

A 1 8th posterior intercostal a.
A 2 1st perforating a.
A 3 pectoral lymph node
A 4 lateral thoracic v.
A 5 6th intercostal v.
A 6 4th lumbar v.
A 7 superior gluteal v. (deep branch)
A 8 superior gluteal a. (deep branch)
A 9 lateral femoral circumflex a. (ascending branch)
A 10 lateral femoral circumflex a. (ascending branch)
A 11 lateral femoral circumflex a. (descending branch)

M 1 teres major m.
M 2 serratus anterior m.
M 3 external intercostal m.
M 4 latissimus dorsi m.
M 5 external intercostal m.
M 6 serratus posterior inferior m.
M 7 diaphragm (costal portion)
M 8 fat body of triangle
M 9 latissimus dorsi m.
M 10 gluteus medius m.
M 11 gluteus maximus m.
M 12 piriformis m. (tendon)
M 13 obturator internus and gemelli mm. (tendon)
M 14 gluteus maximus m. (tendon)
M 15 vastus lateralis m.
M 16 serratus anterior m.
M 17 internal intercostal m.
M 18 diaphragm
M 19 external oblique m.
M 20 external oblique m.
M 21 external oblique m.
M 22 internal oblique m.
M 23 gluteus minimus m.
M 24 gluteus medius m.
M 25 gluteus minimus m.
M 26 tensor fasciae latae m.
M 27 rectus femoris m.
M 28 vastus lateralis m.
M 29 vastus intermedius m.

O 1 iliac crest
O 2 greater trochanter
O 3 shaft of femur
O 4 neck of femur

S 1 right lung (inferior lobe)
S 2 diaphragmatic pleura
S 3 costodiaphragmatic pleural recess
S 4 mammary gland
S 5 peritoneal cavity
S 6 right lobe of liver
S 7 costodiaphragmatic pleural recess

SY 1 hip joint (articular capsule)
SY 2 hip joint (articular cavity)

SA♀5D

A 1 7th posterior intercostal v.
A 2 right hepatic v.
A 3 11th posterior intercostal a.
A 4 subcostal a.
A 5 4th lumbar v.
A 6 superior gluteal a. (deep branch)
A 7 superior gluteal a. (deep branch)
A 8 1st perforating a.
A 9 deep circumflex iliac a.
A 10 lateral femoral circumflex a. (ascending branch)
A 11 lateral femoral circumflex a. (descending branch)

M 1 infraspinatus m.
M 2 subscapularis m.
M 3 serratus anterior m.
M 4 latissimus dorsi m.
M 5 serratus posterior inferior m.
M 6 diaphragm (costal portion)
M 7 latissimus dorsi m.
M 8 external oblique m.
M 9 internal oblique m.
M 10 transverse abdominis m.
M 11 fat body of lumbar triangle
M 12 gluteus medius m.
M 13 gluteus minimus m.
M 14 gluteus maximus m.
M 15 piriformis m.
M 16 gemellus superior m.
M 17 obturator internus m.
M 18 gemellus inferior m.
M 19 obturator externus m.
M 20 quadratus femoris m.
M 21 adductor magnus m.
M 22 adductor magnus m.
M 23 pectoralis minor m.
M 24 pectoralis major m.
M 25 serratus anterior m.
M 26 intercostal mm.
M 27 external oblique m.
M 28 diaphragm (costal portion)
M 29 internal oblique m.
M 30 transverse abdominis m.
M 31 external oblique m.
M 32 internal oblique m.
M 33 iliacus m.
M 34 inguinal ligament
M 35 gluteus minimus m.
M 36 rectus femoris m. (tendon)
M 37 sartorius m.
M 38 iliopsoas m.
M 39 rectus femoris m.
M 40 vastus intermedius m.

N 1 sciatic n.
N 2 lateral femoral cutaneous n.
N 3 femoral n. (muscular ramus)

O 1 scapula (inferior angle)
O 2 iliac crest
O 3 shaft of femur
O 4 neck of femur

S 1 right lung (inferior lobe)
S 2 costodiaphragmatic pleural recess
S 3 mammary gland
S 4 right lung (middle lobe)
S 5 oblique fissure
S 6 right lobe of liver
S 7 transverse colon
S 8 greater omentum
S 9 right colic flexure
S 10 ascending colon

SY 1 8th costochondral articulation
SY 2 hip joint (articular capsule)

SA♀4D

A 1 8th intercostal a.
A 2 portal v. (right branch)
A 3 right hepatic v.
A 4 portal v. (right branch)
A 5 subcostal v.
A 6 4th lumbar a.
A 7 iliolumbar a. (iliac branch)
A 8 superior gluteal a. (deep branch)
A 9 superior gluteal a. (deep branch)
A 10 superior gluteal a. (superficial branch)
A 11 hepatic a. (right branch)
A 12 portal v. (right anterior branch)
A 13 right colic a. (branch)
A 14 deep circumflex iliac a.
A 15 deep circumflex iliac a. (branch)
A 16 deep femoral a.
A 17 femoral v.
A 18 deep femoral v.

M 1 rhomboid m.
M 2 latissimus dorsi m.
M 3 diaphragm
M 4 serratus posterior inferior m.
M 5 iliocostalis m.
M 6 latissimus dorsi m.
M 7 quadratus lumborum m.
M 8 thoracolumbar fascia (deep layer)
M 9 fat body of lumbar triangle
M 10 gluteus medius m.
M 11 gluteus minimus m.
M 12 gluteus maximus m.
M 13 piriformis m.
M 14 gemellus superior m.
M 15 obturator internus m.
M 16 gemellus inferior m.
M 17 obturator externus m.
M 18 quadratus femoris m.
M 19 ischiocrural mm. (tendon)
M 20 adductor magnus m.
M 21 semimembranosus m.
M 22 semitendinosus m.
M 23 pectoralis major m.
M 24 pectoralis minor m.
M 25 diaphragm (costal portion)
M 26 intercostal mm.
M 27 external oblique m.
M 28 transverse abdominis m.
M 29 internal oblique m.
M 30 external oblique m.
M 31 iliacus m.
M 32 pectineus m.
M 33 sartorius m.
M 34 adductor longus m.
M 35 vastus medialis m.
M 36 rectus femoris m.

N 1 lateral femoral cutaneous n.
N 2 sciatic n.
N 3 femoral n.

O 1 iliac crest
O 2 body of ischium
O 3 body of pubis
O 4 head of femur

S 1 right lung (inferior lobe)
S 2 right hepatic lobe
S 3 costodiaphragmatic pleural recess
S 4 perirenal fat
S 5 right kidney
S 6 perirenal fat
S 7 right colic flexure
S 8 ascending colon
S 9 papilla of mammary gland
S 10 mammary gland
S 11 oblique fissure
S 12 right lung (middle lobe)
S 13 costodiaphragmatic pleural recess
S 14 right lobe of liver
S 15 transverse colon
S 16 small intestine
S 17 greater omentum
S 18 ileocecal opening
S 19 cecum

SY 1 acetabular labrum
SY 2 interchondral articulation 8-9
SY 3 acetabular labrum
SY 4 hip joint (articular capsule)

SA♀3D

A 1 right pulmonary a. (lateral basal branch)
A 2 right hepatic v.
A 3 11th posterior intercostal a.
A 4 renal v.
A 5 renal a. (superior segmental branch)
A 6 renal v.
A 7 renal v.
A 8 right colic v. (branch)
A 9 4th lumbar a.
A 10 iliolumbar a. (iliac branch)
A 11 superior gluteal v.
A 12 superior gluteal a.
A 13 superior gluteal a. (superficial branch)
A 14 inferior gluteal a. (branch)
A 15 inferior gluteal a. (branch)
A 16 inferior gluteal a. (branch)
A 17 right pulmonary a. (middle lobe, lateral branch)
A 18 middle hepatic v.
A 19 accessory hepatic a.
A 20 portal v. (right branch)
A 21 right colic a. (branch)
A 22 ileal a.
A 23 right colic a. (branch)
A 24 iliocolic a. (branch)
A 25 posterior cecal a.
A 26 external iliac a.
A 27 deep circumflex iliac v.
A 28 inferior epigastric a. (opening)
A 29 femoral v.
A 30 great saphenous v. (opening)
A 31 medial femoral circumflex a. (deep branch)
A 32 superficial inguinal lymph node
A 33 superficial inguinal lymph node

M 1 trapezius m.
M 2 erector spinae m.
M 3 latissimus dorsi m.
M 4 diaphragm (central tendon)
M 5 intercostal mm.
M 6 erector spinae m.
M 7 diaphragm (crus)
M 8 lateral arcuate ligament
M 9 erector spinae m.
M 10 quadratus lumborum m.
M 11 psoas major m.
M 12 thoracolumbar fascia
M 13 iliacus m.
M 14 fat body of lumbar triangle
M 15 piriformis m.
M 16 obturator internus m.
M 17 gluteus maximus m.
M 18 obturator internus m.
M 19 obturator externus m.
M 20 quadratus femoris m.
M 21 adductor magnus m.
M 22 adductor magnus m.
M 23 adductor magnus m.
M 24 pectoralis major m.
M 25 intercostal mm.
M 26 diaphragm (costal portion)
M 27 rectus abdominis m.
M 28 transverse abdominis m.
M 29 internal oblique m.
M 30 external oblique m.
M 31 iliopsoas m.
M 32 inguinal ligament
M 33 pectineus m.
M 34 adductor longus m.
M 35 adductor brevis m.

N 1 lateral femoral cutaneous n.
N 2 femoral n.
N 3 superior gluteal n.
N 4 sacral plexus
N 5 sciatic n.
N 6 genitofemoral n.

O 1 posterior superior iliac spine
O 2 sacrum (lateral portion)
O 3 acetabular fossa
O 4 lesser sciatic notch
O 5 tuberosity of ischium
O 6 iliopubic eminence
O 7 acetabular notch

S 1 right lung (inferior lobe)
S 2 right lobe of liver
S 3 costodiaphragmatic pleural recess
S 4 perirenal fat
S 5 right kidney (superior pole)
S 6 right lobe of liver
S 7 minor calyx
S 8 perirenal fat
S 9 right lung (middle lobe)
S 10 neck of gall bladder
S 11 body of gall bladder
S 12 superior duodenal flexure
S 13 duodenum (descending portion)
S 14 major duodenal papilla
S 15 transverse colon
S 16 small intestine

S 17 inferior duodenal flexure
S 18 greater omentum
S 19 mesentery
S 20 small intestine
S 21 ileum (terminal portion)
S 22 cecum
S 23 opening of appendix
S 24 small intestine
S 25 ligamentum teres

SY 1 iliolumbar ligament
SY 2 dorsal sacroiliac ligament
SY 3 sacroiliac articulation
SY 4 ligamentum teres
SY 5 sacrotuberous ligament
SY 6 3rd costochondral articulation
SY 7 interchondral articulation 7-8

SA♀2D

A 1 right pulmonary a.
A 2 right inferior v. (anterior basal branch)
A 3 right inferior pulmonary v. (posterior basal branch)
A 4 9th posterior intercostal a.
A 5 right hepatic v.
A 6 right renal v.
A 7 right renal a.
A 8 renal a. (inferior segmental a.)
A 9 4th lumbar a.
A 10 iliolumbar v. (lumbar branch)
A 11 iliolumbar v.
A 12 iliolumbar v.
A 13 iliolumbar a. (lumbar branch)
A 14 iliolumbar a. (opening)
A 15 internal iliac a.
A 16 internal iliac v.
A 17 superior gluteal v.
A 18 internal iliac v.
A 19 uterine v.
A 20 middle rectal a.
A 21 uterine venous plexus
A 22 vaginal venous plexus
A 23 vesical venous plexus
A 24 internal pudendal a.
A 25 right superior pulmonary v. (middle lobe)
A 26 3rd intercostal a.
A 27 right atrium
A 28 middle hepatic v.
A 29 portal v. (right branch)
A 30 accessory hepatic a. (branch of superior mesenteric a.)
A 31 cystic a.
A 32 portal v.
A 33 right colic v.
A 34 right colic v.
A 35 jejunal v.
A 36 right ovarian v. (branch to perirenal fat)
A 37 superficial epigastric v.
A 38 ileocolic v.
A 39 right ovarian v.
A 40 common iliac v.
A 41 external iliac a.
A 42 external iliac lymph node
A 43 external iliac v.
A 44 right ovarian v.
A 45 inferior vesical a.
A 46 vein to ligamentum teres
A 47 obturator v.
A 48 external pudenal v.

M 1 trapezius m.
M 2 diaphragm (central tendon)
M 3 diaphragm (crus)
M 4 intertransverse m.
M 5 erector spinae m.
M 6 psoas m.
M 7 thoracolumbar fascia
M 8 fat body of lumbar triangle
M 9 piriformis m.
M 10 coccygeus m.
M 11 levator ani m.
M 12 obturator internus m.
M 13 ischiorectal fossa
M 14 gluteus maximus m.
M 15 transverse perineal m.
M 16 intercostal mm.
M 17 pectoralis major m.
M 18 transverse thoracic m.
M 19 transverse abdominis m.
M 20 rectus abdominis m.
M 21 sheath of rectus abdominis m. (anterior layer)
M 22 sheath of rectus abdominis m. (posterior layer)
M 23 reflected inguinal ligament
M 24 pectineus m.
M 25 obturator externus m.
M 26 adductor magnus m.
M 27 adductor brevis m.
M 28 adductor magnus m.
M 29 adductor longus m.

N 1 iliohypogastric n.
N 2 lateral femoral cutaneous n.
N 3 femoral n.
N 4 lumbosacral trunk
N 5 obturator n.
N 6 1st sacral n. (ventral ramus)
N 7 2nd sacral n. (ventral ramus)
N 8 3rd sacral n. (ventral ramus)

O 1 body of 9th rib
O 2 costal process L 3
O 3 sacrum (lateral portion)
O 4 superior ramus of pubis
O 5 ramus of ischium

S 1 right lung (inferior lobe)
S 2 trunk of bronchopulmonary segment (lateral basal and posterior basal)
S 3 right lobe of liver
S 4 right suprarenal gland
S 5 costodiaphragmatic pleural recess
S 6 caudate process
S 7 right kidney (superior pole)
S 8 perirenal fat
S 9 ureter
S 10 ureter
S 11 infundibulum of uterine tube
S 12 ureter
S 13 right lung (superior lobe)
S 14 right middle lobar bronchus
S 15 right lung (middle lobe)
S 16 costodiaphragmatic recess
S 17 left hepatic duct
S 18 liver (quadrate lobe)
S 19 cystic duct
S 20 common hepatic duct
S 21 duodenum (superior portion)
S 22 cystic duct
S 23 head of pancreas
S 24 transverse colon
S 25 duodenum (horizontal portion)
S 26 small intestine
S 27 transverse colon
S 28 small intestine
S 29 mesentery
S 30 ileum (terminal portion)
S 31 appendix
S 32 mesoappendix
S 33 appendix
S 34 small intestine
S 35 broad ligament
S 36 ovary
S 37 broad ligament
S 38 ligamentum teres

SY 1 dorsal sacroiliac ligament
SY 2 sacrotuberous ligament

SA♀1D

A 1 right pulmonary a.
A 2 right superior pulmonary v.
A 3 right inferior pulmonary v.
A 4 8th posterior intercostal v.
A 5 intervertebral v.
A 6 inferior vena cava
A 7 posterior internal vertebral venous plexus
A 8 anterior internal vertebral venous plexus
A 9 3rd lumbar v. (opening)
A 10 right common iliac v. (opening)
A 11 right common iliac a.

A 12 lateral sacral v.
A 13 uterine venous plexus
A 14 uterine a.
A 15 vaginal venous plexus
A 16 v. of vestibular bulb
A 17 right auricle
A 18 internal thoracic v.
A 19 superior vena cava
A 20 right atrium
A 21 anterior cardiac v.
A 22 valve of inferior vena cava
A 23 pectinati mm.
A 24 pericardial cavity
A 25 left hepatic v.
A 26 portal v. (left branch)
A 27 proper hepatic a. (left branch)
A 28 common hepatic a.
A 29 portal v.
A 30 left renal v. (opening)
A 31 right gastroepiploic a.
A 32 right inferior pancreaticoduodenal v.
A 33 right gastroepiploic v.
A 34 right colic a. (branch)
A 35 right colic a. (branch)
A 36 right colic a. (branch)
A 37 jejunal v.
A 38 superior mesenteric v. (branch)
A 39 4th lumbar v. (opening)
A 40 mesenteric lymph node
A 41 uterine a. (tubular branch)
A 42 vesical v.
A 43 obturator v. (branch)

M 1 trapezius m.
M 2 transversospinal m.
M 3 erector spinae m.
M 4 diaphragm (crus)
M 5 piriformis m.
M 6 gluteus maximus m.
M 7 coccygeus m.
M 8 levator ani m.
M 9 ischiorectal fossa
M 10 levator ani m.
M 11 bulbospongiosus m.
M 12 transverse perineal mm.
M 13 pectoralis major m.
M 14 intercostal mm.
M 15 transverse thoracic m.
M 16 diaphragm (costal portion)
M 17 rectus abdominis m.
M 18 sheath of rectus abdominis m. (anterior layer)
M 19 sheath of rectus abdominis m. (posterior layer)
M 20 adductor longus m.
M 21 obturator externus m.
M 22 obturator internus m.
M 23 adductor brevis m.

N 1 8th thoracic n.
N 2 spinal dura mater
N 3 4th lumbar n.
N 4 1st sacral n.
N 5 2nd sacral n.

O 1 transverse process T 7
O 2 neck of 8th rib
O 3 head of 9th rib
O 4 pedicle T 11
O 5 lamina L 2
O 6 pedicle L 3
O 7 intermediate sacral crest
O 8 coccyx (transverse process)
O 9 pubic tubercle

S 1 right inferior lobar bronchus
S 2 right lung (inferior lobe)
S 3 costodiaphragmatic pleural recess
S 4 accessory suprarenal gland
S 5 small intestine
S 6 rectal ampulla
S 7 uterine tube
S 8 rectouterine pouch
S 9 rectal ampulla
S 10 vagina
S 11 vestibular bulb
S 12 left lobe of liver
S 13 liver (caudate lobe)
S 14 superior omental recess
S 15 left hepatic lobe
S 16 falciform ligament
S 17 pylorus
S 18 head of pancreas
S 19 ligamentum teres
S 20 jejunum
S 21 duodenum (horizontal portion)
S 22 transverse mesocolon
S 23 transverse colon
S 24 mesentery
S 25 mesentery
S 26 small intestine
S 27 uterine tube
S 28 vesicouterine pouch
S 29 ureter (opening)
S 30 urinary bladder
S 31 mons pubis
S 32 crus of clitoris

SY 1 joint of the head of the 9th rib
SY 2 vertebral articulating processes T 9–T 10
SY 3 intervertebral disc T 11–T 12

SA♀MD

A 1 ascending aorta
A 2 right auricle
A 3 right coronary a.
A 4 right atrium
A 5 valve of inferior vena cava
A 6 right ventricle
A 7 pericardial cavity
A 8 right coronary a.
A 9 left hepatic v.
A 10 portal v. (branch)
A 11 left gastric v.
A 12 common hepatic a.
A 13 superior mesenteric a.
A 14 portal v. (origin)
A 15 left renal v.
A 16 right gastroepiploic a.
A 17 superior mesenteric a.
A 18 jejunal v.
A 19 4th lumbar v.
A 20 superior mesenteric a.
A 21 abdominal a. (bifurcation)
A 22 mesenteric lymph node
A 23 left common iliac v.
A 24 ileal v.
A 25 vesical venous plexus
A 26 dorsal v. of clitoris
A 27 accessory hemiazygos v.
A 28 left atrium
A 29 valve of foramen ovale
A 30 interatrial septum
A 31 hemiazygos v.
A 32 basivertebral v.
A 33 anterior internal vertebral venous plexus
A 34 10th posterior intercostal a. (opening)
A 35 thoracic aorta
A 36 anterior internal vertebral venous plexus
A 37 superior rectal a. (branch)
A 38 superior rectal a. (branch)

M 1 diaphragm (cental tendon)
M 2 diaphragm (crus)
M 3 diaphragm (sternal portion)
M 4 rectus abdominis m.
M 5 diaphragm (aortic hiatus)
M 6 diaphragm (left crus)
M 7 linea alba
M 8 umbilicus
M 9 sheath of rectus abdominis (anterior layer)
M 10 rectus abdominis m.
M 11 transversospinal m.
M 12 erector spinae m.
M 13 levator ani m.
M 14 ischiorectal fossa
M 15 sphincter ani externus m.
M 16 central tendon of perineum (perineal body)

N 1 spinal cord
N 2 spinal dura mater
N 3 cauda equina

O 1 body of sternum
O 2 xiphoid process

O 3 pedicle T 12
O 4 spinous process L 2
O 5 middle sacral crest

S 1 liver (caudate lobe)
S 2 left lobe of liver
S 3 lesser omentum
S 4 body of pancreas
S 5 stomach (pyloric portion)
S 6 transverse mesocolon
S 7 duodenum (ascending portion)
S 8 jejunum
S 9 transverse colon
S 10 greater omentum
S 11 mesentery
S 12 small intestine
S 13 uterine cavity
S 14 vesicouterine space (pouch)
S 15 fornix of vagina
S 16 urinary bladder
S 17 internal urethral opening
S 18 vagina
S 19 body of clitoris
S 20 urethra
S 21 glans clitoris
S 22 prepuce of clitoris
S 23 carunculae hymenales
S 24 labia minora
S 25 labia majora
S 26 esophagus (thoracic portion)
S 27 sigmoid colon
S 28 rectal ampulla
S 29 rectouterine fold
S 30 rectouterine space
S 31 cervix
S 32 transverse rectal folds
S 33 fornix of vagina
S 34 cervix (vaginal portion)
S 35 anal canal
S 36 anus

SY 1 symphysis pubis
SY 2 ligamentum flavum
SY 3 intervertebral disc L 2–L 3 (nucleus pulposus)
SY 4 intervertebral disc L 5–S 1
SY 5 dorsal sacrococcygeal ligament

SA♀MSi

A 1 accessory hemiazygos v.
A 2 left atrium
A 3 valve of foramen ovale
A 4 oblique pericardial sinus
A 5 hemiazygos v.
A 6 right atrium
A 7 valve of inferior vena cava
A 8 right coronary a.
A 9 thoracic aorta
A 10 11th posterior intercostal a.
A 11 intervertebral v.
A 12 abdominal aorta
A 13 left renal a. (opening)
A 14 4th left lumbar v.
A 15 anterior internal vertebral venous plexus
A 16 left common iliac v.
A 17 superior rectal a.
A 18 vaginal venous plexus
A 19 ascending aorta
A 20 right auricle
A 21 right coronary a.
A 22 right atrioventricular valve (anterior cusp)
A 23 right ventricle
A 24 pericardial cavity
A 25 left hepatic v.
A 26 left gastric v.
A 27 left gastric a.
A 28 splenic a.
A 29 celiac trunk
A 30 trunk of splenic and inferior mesenteric vv.
A 31 left renal v.
A 32 middle colic a.
A 33 right gastroepiploic a.
A 34 jejunal v.
A 35 abdominal a.
A 36 superior mesenteric v.
A 37 jejunal v.
A 38 inferior mesenteric a. (opening)
A 39 superior mesenteric v.
A 40 superior mesenteric a.
A 41 mesenteric lymph node
A 42 vesical venous plexus

M 1 erector spinae m.
M 2 diaphragm (lumbar portion)
M 3 transversospinal m.
M 4 diaphragm (left crus)
M 5 thoracolumbar fascia (superficial layer)
M 6 erector spinae m.
M 7 ischiorectal fossa
M 8 levator ani m.
M 9 sphincter ani externus m.
M 10 central tendon of perineum (perineal body)
M 11 sphincter ani externus m.
M 12 diaphragm (central tendon)
M 13 rectus abdominis m.
M 14 sheath of rectus abdominis m.
M 15 linea alba
M 16 umbilicus
M 17 rectus abdominis m.

N 1 spinal cord
N 2 12th thoracic n.
N 3 root of 4th lumbar n.
N 4 1st sacral n.
N 5 3rd sacral n.

O 1 spinous process T 9
O 2 pedicle T 11
O 3 superior articulating process T 12
O 4 spinous process L 2
O 5 lamina L 3
O 6 middle sacral crest
O 7 body of sternum
O 8 xiphoid process

S 1 esophagus (thoracic portion)
S 2 liver (caudate lobe)
S 3 sigmoid colon
S 4 rectal ampulla
S 5 uterine tube
S 6 rectouterine fold
S 7 rectouterine space (pouch)
S 8 cervical canal
S 9 transverse rectal folds
S 10 fornix of vagina
S 11 fornix of vagina
S 12 uterine opening
S 13 vagina
S 14 rectum (perineal flexure)
S 15 anal canal
S 16 coronary ligament
S 17 left lobe of liver
S 18 lesser omentum
S 19 inferior omental recess
S 20 body of pancreas
S 21 inferior omental recess
S 22 stomach (pyloric portion)
S 23 small intestine
S 24 transverse mesocolon
S 25 duodenum (ascending portion)
S 26 transverse colon
S 27 greater omentum
S 28 mesentery
S 29 small intestine
S 30 small intestine
S 31 small intestine
S 32 uterine cavity
S 33 body of uterus
S 34 vesicouterine space
S 35 urinary bladder
S 36 fundus of urinary bladder
S 37 mons pubis
S 38 urethra
S 39 clitoris
S 40 external urethral opening
S 41 caronculae hymenales
S 42 labia minora
S 43 labia majora

SY 1 supraspinal ligament
SY 2 intervertebral disc
SY 3 articulating processes of vertebrae T 11–T 12
SY 4 ligamentum flavum
SY 5 sacrococcygeal articulation
SY 6 symphysis pubis
SY 7 arcuate pubic ligament

SA♀1 Si

A 1 left coronary a.
A 2 aortic bulb
A 3 right ventricle (conus arteriosus)
A 4 semilunar valve
A 5 left ventricle (aortic opening)
A 6 right atrioventricular valve (anterior cusp)
A 7 right atrioventricular opening
A 8 right atrioventricular valve (septal cusp)
A 9 pericardial cavity
A 10 left hepatic v.
A 11 left gastric a.
A 12 splenic a.
A 13 splenic v.
A 14 middle colic a.
A 15 gastroepiploic a.
A 16 left colic a. (branch)
A 17 superior mesenteric v. (branch)
A 18 superior mesenteric v. (branch)
A 19 inferior mesenteric v.
A 20 inferior mesenteric a. (branch)
A 21 external iliac a.
A 22 internal iliac a.
A 23 obturator v. (branch)
A 24 left superior pulmonary v.
A 25 left inferior pulmonary v.
A 26 left atrium
A 27 oblique pericardial sinus
A 28 coronary sinus
A 29 middle cardiac v.
A 30 posterior intercostal a. (dorsal branch)
A 31 inferior phrenic a.
A 32 left suprarenal v.
A 33 left renal a.
A 34 left renal v.
A 35 2nd lumbar a.
A 36 inferior mesenteric v.
A 37 left ovarian v.
A 38 4th lumbar v.
A 39 ascending lumbar v.
A 40 intervertebral v. L 5–S 1
A 41 external iliac v.
A 42 internal iliac v.
A 43 lateral sacral v. (branch)
A 44 uterine venous plexus
A 45 vaginal venous plexus
A 46 vesical v.

M 1 pectoralis major m.
M 2 transverse thoracis m.
M 3 diaphragm (central tendon)
M 4 diaphragm (costal portion)
M 5 rectus abdominis m.
M 6 sheath of rectus abdominis m. (posterior layer)
M 7 sheath of rectus abdominis m. (anterior layer)
M 8 levator ani m.
M 9 obturator internus m.
M 10 adductor brevis m.
M 11 trapezius m.
M 12 diaphragm (esophageal hiatus)
M 13 diaphragm (left crus)
M 14 erector spinae m.
M 15 diaphragm (lumbar portion)
M 16 psoas m.
M 17 thoracolumbar fascia
M 18 piriformis m.
M 19 gluteus maximus m.
M 20 levator ani m.
M 21 ischiorectal fossa
M 22 transverse perineal m.

N 1 lumbar plexus
N 2 4th lumbar n. (ventral ramus)
N 3 5th lumbar n. (ventral ramus)
N 4 1st sacral n. (ventral ramus)
N 5 2nd sacral n. (ventral ramus)
N 6 3rd sacral n. (ventral ramus)

O 1 body of sternum
O 2 pubic crest
O 3 inferior ramus of pubis
O 4 neck of 8th rib
O 5 transverse process T 8
O 6 costal process L 2
O 7 superior articulating process L 5
O 8 inferior articulating process L 5
O 9 sacrum (superior articulating process)
O 10 1st dorsal sacral foramen

S 1 left hepatic lobe
S 2 esophagus (abdominal portion)
S 3 suprarenal gland
S 4 lesser omentum
S 5 head of pancreas
S 6 inferior omental recess
S 7 stomach (pyloric portion)
S 8 duodenal jejunal flexure
S 9 gastrocolic ligament
S 10 small intestine
S 11 transverse colon
S 12 ureter
S 13 small intestine
S 14 mesentery
S 15 greater omentum
S 16 small intestine
S 17 uterine tube
S 18 sigmoid colon
S 19 body of uterus
S 20 vesicouterine space (pouch)
S 21 urinary bladder
S 22 left lung (inferior lobe)
S 23 perirenal fat
S 24 costodiaphragmatic recess
S 25 left kidney
S 26 sigmoid mesocolon
S 27 sigmoid colon
S 28 infundibulum of uterine tube
S 29 rectouterine space (pouch)
S 30 ovary
S 31 rectum
S 32 cervix uteri
S 33 fornix of vagina

SY 1 6th sternocostal articulation
SY 2 articulating processes of L 4–L 5
SY 3 lumbosacral articulation
SY 4 sacrotuberous ligament

SA♀2 Si

A 1 left auricle
A 2 left coronary a. (anterior interventricular branch)
A 3 right ventricle
A 4 left atrioventricular valve (anterior cusp)
A 5 left ventricle
A 6 interventricular septum
A 7 internal thoracic v.
A 8 right coronary a. (ventricular branch)
A 9 anterior papillary m.
A 10 middle cardiac v.
A 11 pericardial cavity
A 12 left gastric a. (branch)
A 13 middle colic a.
A 14 left colic a. (branch)
A 15 gastroepiploic a.
A 16 left colic v. (branch)
A 17 sigmoid a.
A 18 internal iliac lymph node
A 19 external iliac a.
A 20 external iliac v.
A 21 ovarian a.
A 22 ovarian v. (branch)
A 23 uterine venous plexus
A 24 uterine a.
A 25 inferior epigastric v. (branch)
A 26 obturator v.
A 27 left coronary a. (circumflex branch)
A 28 left inferior pulmonary v.
A 29 10th posterior intercostal v.
A 30 splenic a.
A 31 splenic v.
A 32 renal a. (branch)
A 33 subcostal v.
A 34 renal v. (branch)
A 35 renal a. (branch)
A 36 renal v. (branch)
A 37 left colic v. (branch)
A 38 left colic v. (branch)
A 39 4th lumbar a.
A 40 ascending lumbar v.
A 41 iliolumbar a. (lumbar branch)
A 42 inferior gluteal a.
A 43 uterine a.

A 44 superior gluteal v.
A 45 inferior gluteal a.
A 46 obturator v.
A 47 internal iliac v.
A 48 inferior gluteal v.
A 49 inferior gluteal a. (branch)
A 50 uterine venous plexus
A 51 internal pudendal a.

M 1 pectoralis major m.
M 2 intercostal mm.
M 3 transverse thoracis m.
M 4 diaphragm (central tendon)
M 5 diaphragm (costal portion)
M 6 rectus abdominis m.
M 7 rectus abdominis m.
M 8 sheath of rectus abdominis m. (posterior layer)
M 9 sheath of rectus abdominis m. (anterior layer)
M 10 pectineus m.
M 11 obturator externus m.
M 12 adductor brevis m.
M 13 adductor longus m.
M 14 adductor magnus m.
M 15 trapezius m.
M 16 intercostal mm.
M 17 diaphragm (lumbar portion)
M 18 latissimus dorsi m.
M 19 quadratus lumborum m.
M 20 psoas m.
M 21 erector spinae m.
M 22 thoracolumbar fascia
M 23 fat body of lumbar triangle
M 24 piriformis m.
M 25 levator ani m.
M 26 gluteus maximus m.
M 27 obturator internus m.
M 28 ischiorectal fossa
M 29 transverse perineal mm.

N 1 iliohypogastric n.
N 2 lateral femoral cutaneous n.
N 3 femoral n.
N 4 obturator n.
N 5 lumbosacral trunk
N 6 sacral plexus
N 7 pudendal plexus

O 1 obturator canal
O 2 superior ramus of pubis
O 3 ramus of ischium
O 4 iliac tuberosity
O 5 ala of sacrum
O 6 ilium (auricular surface)

S 1 left lobe of liver
S 2 fundus of stomach
S 3 lesser omentum
S 4 body of pancreas
S 5 inferior omental recess
S 6 stomach (pyloric portion)
S 7 transverse mesocolon
S 8 jejunum
S 9 small intestine
S 10 transverse colon
S 11 small intestine
S 12 mesentery
S 13 small intestine
S 14 sigmoid mesocolon
S 15 ureter
S 16 sigmoid colon
S 17 greater omentum
S 18 broad ligament
S 19 left inferior lobar bronchus
S 20 left lung (inferior lobe)
S 21 spleen
S 22 costodiaphragmatic recess
S 23 left kidney
S 24 major renal calyx
S 25 renal pelvis
S 26 left kidney
S 27 ovary
S 28 ureter

SY 1 iliolumbar ligament
SY 2 dorsal sacroiliac ligament
SY 3 sacroiliac articulation
SY 4 sacrotuberous ligament
SY 5 sacrospinous ligament
SY 6 obturator membrane

SA♀3 Si

A 1 pericardiophrenic a.
A 2 fibrous pericardium
A 3 left ventricle
A 4 right coronary a. (ventricular branch)
A 5 left coronary a. (anterior interventricular branch)
A 6 interventricular septum
A 7 right ventricle
A 8 middle cardiac v.
A 9 4th intercostal a.
A 10 pericardial cavity
A 11 left colic a.
A 12 left colic a.
A 13 left colic a.
A 14 left gastroepiploic a.
A 15 external iliac a.
A 16 inferior epigastric a.
A 17 superficial epigastric v.
A 18 femoral v.
A 19 great saphenous v.
A 20 superficial inguinal lymph node
A 21 splenic v. (branch)
A 22 splenic a.
A 23 splenic v.
A 24 left colic v. (branch)
A 25 left colic a.
A 26 4th lumbar a.
A 27 superior gluteal v. (branch)
A 28 inferior gluteal a. (branch)
A 29 inferior gluteal a. (branch)

M 1 pectoralis major m.
M 2 intercostal mm.
M 3 diaphragm (costal portion)
M 4 rectus abdominis m.
M 5 sheath of rectus abdominis m.
M 6 transverse abdominis m.
M 7 rectus abdominis m.
M 8 internal oblique m.
M 9 psoas m.
M 10 external oblique m.
M 11 pectineus m.
M 12 adductor brevis m.
M 13 adductor longus m.
M 14 rhomboid major m.
M 15 erector spinae m.
M 16 diaphragm (central tendon)
M 17 latissimus dorsi m.
M 18 diaphragm (lumbar portion)
M 19 intercostal mm.
M 20 serratus posterior inferior m.
M 21 thoracolumbar fascia (deep layer)
M 22 thoracolumbar fascia (superficial layer)
M 23 erector spinae m.
M 24 quadratus lumborum m.
M 25 fat body of lumbar triangle
M 26 iliacus m.
M 27 gluteus medius m.
M 28 piriformis m.
M 29 gluteus maximus m.
M 30 gemellus superior m.
M 31 obturator internus m.
M 32 obturator externus m.
M 33 quadratus femoris m.
M 34 adductor magnus m.
M 35 ischiocrural mm.

N 1 femoral n.
N 2 iliohypogastric n.
N 3 lateral femoral cutaneous n.
N 4 superior gluteal n.
N 5 sciatic n.

O 1 body of pubis
O 2 iliac crest
O 3 ala of ilium
O 4 acetabular notch
O 5 tuberosity of ischium

S 1 left lung (cardiac notch)
S 2 mammary gland
S 3 left lobe of liver
S 4 body of pancreas
S 5 inferior omental recess
S 6 jejunum
S 7 stomach (pyloric portion)
S 8 transverse mesocolon
S 9 gastrocolic ligament

S 10 small intestine
S 11 transverse colon
S 12 small intestine
S 13 small intestine
S 14 descending colon
S 15 greater omentum
S 16 ligamentum teres
S 17 left lung (inferior lobe)
S 18 fundus of stomach
S 19 left lung (inferior margin)
S 20 spleen
S 21 perirenal fat
S 22 costodiaphragmatic recess
S 23 left kidney
S 24 perirenal fat
S 25 descending colon

SY 1 3rd costochondral articulation
SY 2 interchondral articulation 6-7
SY 3 head of femur (articular cartilage)
SY 4 ligamentum teres
SY 5 transverse acetabular ligament
SY 6 sacrotuberous ligament

SA♀4 Si

A 1 pericardiophrenic v.
A 2 fibrous pericardium
A 3 apex of heart
A 4 6th intercostal a.
A 5 iliolumbar a.
A 6 deep iliac circumflex a.
A 7 superficial epigastric a. (branch)
A 8 deep femoral a.
A 9 femoral v.
A 10 deep femoral v.
A 11 10th posterior intercostal v.
A 12 splenic v.
A 13 splenic v.
A 14 splenic a.
A 15 right gastroepiploic v.
A 16 left colic a. (branch)
A 17 4th lumbar a.
A 18 superior gluteal a. (deep branch)

M 1 pectoralis minor m.
M 2 pectoralis major m.
M 3 intercostal mm.
M 4 external oblique m.
M 5 diaphragm (costal portion)
M 6 external oblique m.
M 7 internal oblique m.
M 8 transverse abdominis m.
M 9 iliopsoas m.
M 10 psoas m. (tendon)
M 11 pectineus m.
M 12 sartorius m.
M 13 serratus anterior m.
M 14 lattissimus dorsi m.
M 15 central tendon of diaphragm
M 16 diaphragm (lumbar portion)
M 17 latissimus dorsi m.
M 18 quadratus lumborum m.
M 19 thoracolumbar fascia
M 20 fat body of lumbar triangle
M 21 gluteus medius m.
M 22 gluteus maximus m.
M 23 gluteus minimus m.
M 24 piriformis m.
M 25 gemellus superior m.
M 26 obturator internus m.
M 27 inferior gemellus m.
M 28 obturator externus m.
M 29 quadratus femoris m.
M 30 ischiocrural mm.
M 31 adductor brevis m.
M 32 semimembranosus m.
M 33 ischiocrural mm.
M 34 adductor magnus m.
M 35 adductor magnus m.
M 36 adductor longus m.

N 1 phrenic n.
N 2 lateral femoral cutaneous n.
N 3 femoral n.
N 4 superior gluteal n.
N 5 sciatic n.

O 1 medial margin of scapula
O 2 iliac crest
O 3 acetabulum
O 4 head of femur

S 1 mammary gland
S 2 left lung (superior lobe)
S 3 left lung (cardiac impression)
S 4 stomach
S 5 greater curvature of stomach
S 6 left colic flexure
S 7 descending colon
S 8 gastrocolic ligament
S 9 transverse colon
S 10 greater omentum
S 11 small intestine
S 12 greater omentum
S 13 oblique fissure
S 14 left lung (inferior lobe)
S 15 costodiaphragmatic recess
S 16 spleen
S 17 perirenal fat
S 18 left kidney
S 19 perirenal fat
S 20 descending colon

SY 1 interchondral joint 7-8
SY 2 acetabular labrum
SY 3 articular capsule
SY 4 acetabular labrum

SA♀5 Si

A 1 fibrosus pericardium
A 2 iliolumbar a. (iliac branch)
A 3 deep iliac circumflex a.
A 4 lateral femoral circumflex a. (ascending branch)
A 5 lateral femoral circumflex a. (descending branch)
A 6 9th posterior intercostal v.
A 7 9th posterior intercostal a.
A 8 splenic v.
A 9 subcostal a. (branch)
A 10 superior gluteal a. (deep branch)
A 11 superior gluteal a. (deep branch)
A 12 perforating v.

M 1 pectoralis minor m.
M 2 pectoralis major m.
M 3 intercostal mm.
M 4 external oblique m.
M 5 transverse abdominis m.
M 6 internal oblique m.
M 7 transverse abdominis m.
M 8 iliacus m.
M 9 gluteus minimus m.
M 10 rectus femoris m. (reflected head)
M 11 rectus femoris m.
M 12 sartorius m.
M 13 iliopsoas m.
M 14 rectus femoris m.
M 15 vastus intermedius m.
M 16 teres major m.
M 17 intercostal mm.
M 18 serratus anterior m.
M 19 latissimus dorsi m.
M 20 diaphragm
M 21 latissimus dorsi m.
M 22 external oblique m.
M 23 internal oblique m.
M 24 fat body of lumbar triangle
M 25 gluteus medius m.
M 26 gluteus maximus m.
M 27 piriformis m.
M 28 superior gemellus m.
M 29 obturator internus m.
M 30 inferior gemellus m.
M 31 obturator externus m.
M 32 quadratus femoris m.
M 33 adductor magnus m.
M 34 adductor magnus m.

N 1 subcostal n.
N 2 8th intercostal n.
N 3 sciatic n.

O 1 scapula
O 2 iliac crest
O 3 neck of femur
O 4 body of femur

S 1 mammary gland

S 2 left lung (superior lobe)
S 3 oblique fissure
S 4 costodiaphragmatic recess
S 5 stomach
S 6 left colic flexure
S 7 transverse mesocolon
S 8 transverse colon
S 9 greater omentum
S 10 left lung (inferior lobe)
S 11 spleen
S 12 costodiaphragmatic recess
S 13 peritoneal cavity
S 14 descending colon
S 15 parietal peritoneum

SY 1 7th costochondral articulation
SY 2 articular capsule of hip joint
SY 3 articular cavity of hip joint

SA♀6 Si

A 1 6th posterior intercostal a.
A 2 7th posterior intercostal a.
A 3 superior gluteal a. (deep branch)
A 4 lateral femoral circumflex a. (ascending branch)
A 5 thoracodorsal v.
A 6 9th posterior intercostal v.

M 1 serratus anterior m.
M 2 intercostalis intimus m.
M 3 intercostal mm.
M 4 external oblique m.
M 5 external oblique m.
M 6 internal oblique m.
M 7 gluteus medius m.
M 8 gluteus minimus m.
M 9 tensor fasciae latae m.
M 10 rectus femoris m.
M 11 vastus lateralis m.
M 12 vastus intermedius m.
M 13 teres major m.
M 14 serratus anterior m.
M 15 latissimus dorsi m.
M 16 diaphragm (costal portion)
M 17 fat body of lumbar triangle
M 18 gluteus maximus m.
M 19 trochanteric bursa, gluteus maximus m.

O 1 iliac crest
O 2 iliac tubercle
O 3 greater trochanter
O 4 body of femur

S 1 mammary gland
S 2 pleural cavity
S 3 costodiaphramatic recess
S 4 parietal peritoneum
S 5 spleen

HA
Horizontal sections Abdomen

75-year-old male

Transverse sections of the trunk, perpendicular to the plane upon which the subject is lying. The plane of reference passes through the antero-superior iliac spines. The sections are spaced 2.5 cm apart. All of the sections are viewed from the inferior side.

HA 1 horizontal section, 22.5 cm above the antero-superior iliac spines, inferior-superior view. The section passes through the lower part of the 11th thoracic vertebra [T 11].

HA 2 horizontal section, 20 cm above the antero-superior iliac spines, inferior-superior view. The section passes through the upper third of the body of the 12th thoracic vertebra [T 12].

HA 3 horizontal section, 17.5 cm above the antero-superior iliac spines, inferior-superior view. The section passes through the central part of the body of the 1st lumbar vertebra [L 1].

HA 4 horizontal section, 15 cm above the antero-superior iliac spines, inferior-superior view. The section passes through the upper part of the 2nd lumbar vertebra [L 2].

HA 5 horizontal section, 12.5 cm above the antero-superior iliac spines, inferior-superior view. The section passes through the intervertebral disc L 2–L 3, 7.5 cm above the umbilicus.

HA 6 horizontal section, 10 cm above the antero-superior iliac spines, inferior-superior view. The section passes through the center of the 3rd lumbar vertebral body [L 3].

HA 7 horizontal section, 7.5 cm above the antero-superior iliac spines, inferior-superior view. The section passes through the upper part of the body of L 4, the spinal process L 4, and the top of the iliac crest.

HA 8 horizontal section, 5 cm above the antero-superior iliac spines, inferior-superior view. The section passes through the lower part of the body of L 4, the spinal process L 5, the umbilicus, and the aortic bifurcation.

HA 9 horizontal section, 2.5 cm above the antero-superior iliac spines, inferior-superior view. The section passes through the anterior part of the body of L 5, the posterior part of the lumbosacral articulation, and the 1st median sacral tubercle.

HA 10 horizontal section passing through the antero-superior iliac spines, inferior-superior view. The section passes through the promontory and the sacroiliac articulation.

HA 11 horizontal section, 2.5 cm below the antero-superior iliac spines, inferior-superior view. The section passes through the body of S 3 and the postero-inferior iliac spine.

HA 12 horizontal section, 5 cm below the antero-superior iliac spines, inferior-superior view. The section passes through the top of the acetabulum.

HA 13 horizontal section, 7.5 cm below the antero-superior iliac spines, inferior-superior view. The section passes through the sciatic spine, the apex of the great trochanter, the coccyx, and the upper side of the pubic symphysis.

HA 1

A 1 pericardiophrenic a.
A 2 internal thoracic a.
A 3 inferior vena cava
A 4 fibrosus pericardium
A 5 inferior valve of vena cava
A 6 right atrium
A 7 right coronary a.
A 8 small cardiac v.
A 9 right atrioventricular valve (posterior cusp)
A 10 right atrioventricular valve
A 11 right coronary a. (posterior interventricular branch)
A 12 coronary sinus
A 13 musculophrenic a.
A 14 interventricular septum
A 15 left ventricle
A 16 left coronary a. (anterior interventricular branch)
A 17 10th posterior intercostal a.
A 18 11th posterior intercostal v.
A 19 azygos v.
A 20 intervertebral v.
A 21 thoracic duct
A 22 anterior internal vertebral venous plexus
A 23 thoracic aorta
A 24 11th posterior intercostal a.
A 25 hemiazygos v.
A 26 11th posterior intercostal a.
A 27 short gastric v.

M 1 serratus anterior m.
M 2 intercostal mm.
M 3 external oblique m.
M 4 transverse thoracic m.
M 5 diaphragm (costal portion)
M 6 intercostal mm. (intercostal space 5)
M 7 serratus anterior m.
M 8 latissimus dorsi m.
M 9 serratus posterior inferior m.
M 10 diaphragm (crus)
M 11 iliocostalis m.
M 12 longissimus m.
M 13 transversospinalis m.
M 14 longissimus m.
M 15 diaphragm (crus)
M 16 iliocostalis m.
M 17 latissimus dorsi m.

N 1 anterior vagal trunk
N 2 sympathetic trunk
N 3 greater splanchnic n.
N 4 11th thoracic n.
N 5 spinal medulla
N 6 posterior vagal trunk
N 7 greater splanchnic n.
N 8 sympathetic trunk

O 1 xiphoid process
O 2 vertebral lamina T 11

S 1 right lung (middle lobe, lateral segment)
S 2 liver (right lobe)
S 3 peritoneal cavity
S 4 pleural cavity
S 5 right lung (middle lobe, medial segment)
S 6 esophagus
S 7 pulmonary ligament
S 8 gastrophrenic ligament
S 9 oblique fissure
S 10 left lung (superior lobe)
S 11 right lung (inferior lobe)
S 12 pleural cavity
S 13 pleural cavity
S 14 left lung (inferior lobe)
S 15 perirenal fat
S 16 spleen

HA 2

A 1 right hepatic v.
A 2 musculophrenic a.
A 3 middle hepatic v.
A 4 superior epigastric a.
A 5 inferior vena cava
A 6 inferior phrenic a.
A 7 left hepatic v.
A 8 inferior phrenic a.
A 9 left gastric a. (esophageal branch)
A 10 10th posterior intercostal a.
A 11 11th posterior intercostal a.
A 12 portal v. (right branch)
A 13 azygos v.
A 14 thoracic duct
A 15 internal vertebral venous plexus (anterior)
A 16 subcostal a.
A 17 thoracic aorta
A 18 hemiazygos v.
A 19 12th posterior intercostal a.
A 20 subcostal a.
A 21 11th posterior intercostal v. (dorsal branch)
A 22 left suprarenal v. (branch)
A 23 splenic v.
A 24 short gastric a.
A 25 splenic a. (splenic branch)

M 1 diaphragm (costal portion)
M 2 rectus abdominis m.
M 3 diaphragm (sternal portion)
M 4 diaphragm (esophageal hiatus)
M 5 diaphragm (crus)
M 6 transversus thoracis m.
M 7 rectus abdominis m.
M 8 diaphragm (costal portion)
M 9 external oblique m.
M 10 latissimus dorsi m.
M 11 intercostal mm.
M 12 intercostal mm.
M 13 diaphragm (crus)
M 14 transversospinalis m.
M 15 longissimus m.
M 16 diaphragm (crus)
M 17 thoracolumbar fascia (superficial layer)
M 18 iliocostalis m.

N 1 posterior vagal trunk
N 2 greater splanchnic n.
N 3 12th thoracic n.
N 4 spinal medulla
N 5 spinal n. (ventral ramus)

O 1 xiphoid process

S 1 costodiaphragmatic recess
S 2 right lung (inferior margin, fat body)
S 3 caudate lobe
S 4 lesser omentum
S 5 vestibule of omental bursa
S 6 stomach (cardiac orifice)
S 7 liver (left lobe)
S 8 fibrosus pericardium
S 9 liver (inferior margin)
S 10 left lung (superior lobe)
S 11 oblique fissure
S 12 fundus of stomach
S 13 peritoneal cavity
S 14 left lung (inferior lobe)
S 15 liver (right lobe)
S 16 perirenal fat
S 17 right kidney (superior pole)

S 18 suprarenal gland
S 19 pleural recess
S 20 pleural recess
S 21 suprarenal gland
S 22 left kidney
S 23 perirenal fat
S 24 spleen
S 25 costodiaphragmatic recess

SY 1 joint of head of 12th rib

HA 3

A 1 right hepatic v.
A 2 portal v. (right posterior branch)
A 3 inferior vena cava
A 4 portal v. (left branch)
A 5 proper hepatic a. (left branch)
A 6 left hepatic v. (branch)
A 7 abdominal aorta
A 8 celiac trunk (orifice)
A 9 left gastric a.
A 10 left gastric a.
A 11 inferior phrenic a.
A 12 left suprarenal v.
A 13 6th intercostal a.
A 14 short gastric a.
A 15 11th posterior intercostal v.
A 16 right hepatic v. (branch)
A 17 renal a. (superior anterior segmental a.)
A 18 renal a. (superior segmental a.)
A 19 renal a. (posterior segmental a.)
A 20 superior suprarenal a.
A 21 ascending lumbar v.
A 22 internal vertebral venous plexus
A 23 thoracic duct
A 24 1st lumbar a.
A 25 ascending lumbar v.
A 26 renal a. (superior segmental a.)
A 27 renal a. (posterior segmental a.)
A 28 renal v.
A 29 renal a. (superior anterior segmental a.)
A 30 splenic v.
A 31 splenic a.
A 32 splenic a.

M 1 intercostal mm.
M 2 diaphragm (costal portion)
M 3 rectus abdominis m.
M 4 diaphragm (aortic hiatus)
M 5 diaphragm (crus)
M 6 rectus abdominis m.
M 7 transversus thoracis m.
M 8 external oblique m.
M 9 diaphragm (costal portion)
M 10 latissimus dorsi m.
M 11 iliocostalis m.
M 12 quadratus lumborum m.
M 13 longissimus m.
M 14 diaphragm (lumbar portion)
M 15 diaphragm (crus)
M 16 transversospinalis m.
M 17 quadratus lumborum m.
M 18 thoracolumbar fascia (superficial layer)
M 19 iliocostalis m.
M 20 latissimus dorsi m.
M 21 diaphragm (costal portion)

N 1 celiac ganglion
N 2 cauda equina
N 3 conus medullaris
N 4 sympathetic trunk
N 5 greater splanchnic n.
N 6 celiac ganglion
N 7 12th thoracic n. (ventral ramus)

O 1 costal process L 1
O 2 pedicle L 1

S 1 costodiaphragmatic recess
S 2 liver (right lobe)
S 3 peritoneal cavity
S 4 suprarenal gland
S 5 left hepatic duct
S 6 vestibule of omental bursa
S 7 caudate lobe
S 8 lesser omentum
S 9 liver (left lobe)
S 10 suprarenal gland
S 11 body of stomach
S 12 peritoneal cavity
S 13 inferior omental recess
S 14 gastrosplenic ligament
S 15 costodiaphragmatic recess
S 16 greater curvature of stomach
S 17 perirenal fat
S 18 kidney
S 19 renal pelvis
S 20 renal sinus
S 21 tail of pancreas
S 22 perirenal fat
S 23 spleen
S 24 costodiaphragmatic recess

SY 1 lumbocostal ligament
SY 2 supraspinous ligament
SY 3 ligamentum flavum

HA 4

A 1 7th posterior intercostal a.
A 2 right hepatic v.
A 3 portal v. (right posterior branch)
A 4 proper hepatic a. (right branch, posterior segment)
A 5 proper hepatic a. (right branch, anterior segment)
A 6 proper hepatic a. (right branch)
A 7 proper hepatic a. (right branch)
A 8 proper hepatic a. (left branch)
A 9 portal v. (left branch)
A 10 common hepatic a.
A 11 portal v.
A 12 left renal v.
A 13 left hepatic v.
A 14 superior mesenteric a.
A 15 abdominal aorta
A 16 splenic v.
A 17 left gastric a. (anterior branch)
A 18 left renal a.
A 19 splenic a.
A 20 10th posterior intercostal a.
A 21 renal v. (branch)
A 22 renal a. (branch)
A 23 1st lumbar v. (dorsal branch)
A 24 inferior vena cava
A 25 right renal a.
A 26 internal vertebral venous plexus
A 27 1st lumbar a.
A 28 ascending lumbar v.
A 29 renal v. (branch)
A 30 left testicular v.
A 31 splenic a.
A 32 splenic v.
A 33 renal a. (inferior segmental a.)
A 34 renal v. (branch)
A 35 splenic a.
A 36 short gastric a.
A 37 10th posterior intercostal a.

M 1 diaphragm (costal portion)
M 2 rectus abdominis m.
M 3 linea alba
M 4 sheath of rectus abdominis m. (anterior layer)
M 5 rectus abdominis m.
M 6 sheath of rectus abdominis m. (posterior layer)
M 7 external oblique m.
M 8 diaphragm (costal portion)
M 9 latissimus dorsi m.
M 10 quadratus lumborum m.
M 11 iliocostalis m.
M 12 longissimus m.
M 13 transversospinalis m.
M 14 diaphragm (right crus)
M 15 diaphragm (left crus)
M 16 longissimus m.
M 17 thoracolumbar fascia (superficial layer)
M 18 iliocostalis m.
M 19 quadratus lumborum m.
M 20 thoracolumbar fascia (deep layer)
M 21 latissimus dorsi m.
M 22 diaphragm (costal portion)

N 1 superior mesenteric plexus
N 2 aorticorenal plexus

N 3 cauda equina
N 4 sympathetic trunk

O 1 mamillary process L 2
O 2 inferior articular process L 1
O 3 pedicle L 2

S 1 liver (right lobe)
S 2 liver (quadrate lobe)
S 3 common hepatic duct
S 4 peritoneal cavity
S 5 inferior omental recess
S 6 lesser omentum
S 7 head of pancreas
S 8 falciform ligament
S 9 body of pancreas
S 10 stomach (anterior wall)
S 11 liver (inferior margin)
S 12 body of stomach
S 13 greater curvature of stomach
S 14 gastrosplenic ligament
S 15 peritoneal cavity
S 16 costodiaphragmatic recess
S 17 liver (right lobe)
S 18 perirenal fat
S 19 kidney (pyramid)
S 20 minor calyces
S 21 major calyces
S 22 ureter
S 23 liver (caudate process)
S 24 ureter
S 25 renal sinus
S 26 tail of pancreas
S 27 inferior omental recess
S 28 perirenal fat
S 29 left colic flexure
S 30 spleen

SY 1 interchondral articulation 7–8
SY 2 intervertebral disc L 1–L 2
SY 3 interspinal ligament
SY 4 ligamentum flavum
SY 5 articulating processes L 1–L 2

HA 5

A 1 hepatic v.
A 2 cystic a.
A 3 testicular a.
A 4 superior mesenteric v.
A 5 inferior pancreaticoduodenal a.
A 6 testicular a.
A 7 superior mesenteric a.
A 8 1st jejunal a.
A 9 right gastric a. (anterior branch)
A 10 right gastric a. (posterior branch)
A 11 right gastric a. (branch)
A 12 short gastric a.
A 13 10th posterior intercostal a.
A 14 11th posterior intercostal a.
A 15 right hepatic v.
A 16 portal v. (right posterior branch)
A 17 testicular v.
A 18 inferior vena cava
A 19 2nd lumbar v.
A 20 abdominal aorta
A 21 2nd lumbar a.
A 22 renal v.
A 23 2nd lumbar a.
A 24 inferior mesenteric v.
A 25 left colic v. (branch)
A 26 left colic v. (branch)
A 27 short gastric a.
A 28 10th posterior intercostal a.

M 1 external oblique m.
M 2 intercostal mm.
M 3 diaphragm (costal portion)
M 4 rectus abdominis m.
M 5 linea alba
M 6 sheath of rectus abdominis m. (anterior layer)
M 7 rectus abdominis m.
M 8 sheath of rectus abdominis m. (posterior layer)
M 9 thoracolumbar fascia
M 10 quadratus lumborum m.
M 11 erector spinae m.
M 12 psoas m.
M 13 thoracolumbar fascia (superficial layer)
M 14 transversospinalis m.
M 15 diaphragm (right crus)
M 16 diaphragm (left crus)
M 17 thoracolumbar fascia (deep layer)
M 18 thoracolumbar fascia
M 19 diaphragm (costal portion)

N 1 11th intercostal n.
N 2 2nd lumbar n. (ventral ramus)
N 3 cauda equina
N 4 2nd lumbar n. (ventral ramus)

O 1 inferior articular process L 2
O 2 superior articular process L 3

S 1 liver (right lobe)
S 2 gall bladder
S 3 liver (quadrate lobe)
S 4 pyloric canal
S 5 common bile duct
S 6 pyloric antrum
S 7 head of pancreas
S 8 pancreatic duct
S 9 ligamentum teres
S 10 falciform ligament
S 11 jejunum
S 12 liver (left lobe)
S 13 body of stomach
S 14 greater curvature of stomach
S 15 perirenal fat
S 16 kidney (inferior pole)
S 17 duodenum (superior portion)
S 18 ureter
S 19 ureter
S 20 jejunum
S 21 kidney (inferior pole)
S 22 body of pancreas
S 23 inferior omental recess
S 24 jejunum
S 25 transverse mesocolon
S 26 pancreaticosplenic ligament
S 27 descending colon
S 28 transverse colon
S 29 gastrosplenic ligament

SY 1 ligamentum flavum
SY 2 supraspinous ligament
SY 3 intervertebral disc L 2–L 3 (nucleus pulposus)
SY 4 articulating processes L 2–L 3

HA 6

A 1 right gastroepiploic v.
A 2 gastroduodenal a. (pancreatic branch)
A 3 inferior epigastric a.
A 4 right colic a.
A 5 right colic a. (branch)
A 6 inferior vena cava
A 7 inferior pancreaticoduodenal a.
A 8 portal v. (origin)
A 9 left renal v. (variation)
A 10 superior mesenteric a.
A 11 abdominal aorta
A 12 splenic v.
A 13 jejunal a.
A 14 right gastric a.
A 15 jejunal a.
A 16 right gastric a.
A 17 left colic a.
A 18 left gastroepiploic a. (epiploic branch)
A 19 portal v. (branch)
A 20 testicular v.
A 21 3rd lumbar a.
A 22 internal vertebral venous plexus
A 23 3rd lumbar a.
A 24 left colic a.
A 25 3rd lumbar a.
A 26 left colic a. (branch)
A 27 lumbar a. (branch)
A 28 jejunal a.

M 1 external oblique m.
M 2 internal oblique m.
M 3 transversus abdominis m.
M 4 rectus abdominis m.
M 5 sheath of rectus abdominis m. or rectal sheath

M 6 transversus abdominis m.
M 7 internal oblique m.
M 8 external oblique m.
M 9 quadratus lumborum m.
M 10 erector spinae m.
M 11 psoas major m.
M 12 diaphragm (right crus)
M 13 psoas major m.
M 14 thoracolumbar fascia (superficial layer)
M 15 erector spinae m.
M 16 quadratus lumborum m.
M 17 latissimus dorsi m.

N 1 10th thoracic n. (ventral ramus)
N 2 11th thoracic n. (ventral ramus)
N 3 iliohypogastric n.
N 4 2nd lumbar n. (ventral ramus)
N 5 sympathetic trunk
N 6 3rd lumbar n.
N 7 cauda equina
N 8 3rd lumbar n.
N 9 sympathetic trunk
N 10 2nd lumbar n. (ventral ramus)
N 11 11th thoracic n.

O 1 pedicle L 3
O 2 lamina L 3
O 3 costal process L 3

S 1 liver (right lobe)
S 2 gall bladder
S 3 head of pancreas
S 4 stomach (pyloric portion)
S 5 duodenum (ascending portion)
S 6 ligamentum teres
S 7 lesser curvature of stomach (incisura angularis)
S 8 jejunum
S 9 mesentery
S 10 transverse mesocolon
S 11 inferior omental recess
S 12 body of stomach
S 13 gastrocolic ligament
S 14 perirenal fat
S 15 duodenum (descending portion)
S 16 major duodenal papilla
S 17 ureter
S 18 ureter
S 19 mesentery
S 20 jejunum
S 21 descending colon
S 22 transverse colon

SY 1 posterior longitudinal ligament
SY 2 ligamentum flavum

HA 7

A 1 right colic v.
A 2 right gastroepiploic v.
A 3 right gastroepiploic a.
A 4 right colic a. (branch)
A 5 right colic a. (branch)
A 6 inferior vena cava
A 7 iliocolic a.
A 8 superior mesenteric a. (branch)
A 9 superior mesenteric v.
A 10 superior mesenteric a. (branch)
A 11 abdominal aorta
A 12 jejunal v.
A 13 inferior mesenteric a.
A 14 jejunal a.
A 15 inferior epigastric a.
A 16 right colic a. (branch)
A 17 3rd lumbar a.
A 18 testicular a.
A 19 lumbar v.
A 20 internal vertebral venous plexus
A 21 left colic a.
A 22 inferior mesenteric v.
A 23 jejunal v.
A 24 left colic v.
A 25 lumbar v.
A 26 left gastroepiploic v. (epiploic branch)

M 1 external oblique m.
M 2 internal oblique m.
M 3 transverse abdominis m.
M 4 rectus abdominis m.
M 5 linea alba
M 6 transversus abdominis m.
M 7 internal oblique m.
M 8 external oblique m.
M 9 thoracolumbar fascia
M 10 quadratus lumborum m.
M 11 fat body of lumbar triangle
M 12 thoracolumbar fascia (deep layer)
M 13 psoas major m.
M 14 erector spinae m.
M 15 transversospinal m.
M 16 thoracolumbar fascia (superficial layer)
M 17 psoas major m.
M 18 quadratus lumborum m.

N 1 lumbar splanchnic nn.
N 2 12th thoracic n. (ventral ramus)
N 3 iliohypogastric n.
N 4 2nd lumbar n. (ventral ramus)
N 5 3rd lumbar n. (ventral ramus)
N 6 sympathetic trunk
N 7 spinal dura mater
N 8 spinal nerve roots
N 9 3rd lumbar n. (ventral ramus)
N 10 2nd lumbar n. (ventral ramus)
N 11 genitofemoral n.
N 12 subcostal n.

O 1 costal process L 4
O 2 inferior articular process L 3
O 3 pedicle L 4
O 4 accessory process
O 5 iliac crest

S 1 liver (right lobe)
S 2 transverse colon
S 3 gall bladder
S 4 transverse mesocolon
S 5 ureter
S 6 peritoneal cavity
S 7 duodenum (horizontal portion)
S 8 stomach (pyloric portion)
S 9 ligamentum teres
S 10 greater omentum
S 11 inferior omental recess
S 12 body of stomach
S 13 greater curvature of stomach
S 14 greater omentum
S 15 right colic flexure
S 16 inferior duodenal flexure
S 17 mesentery
S 18 small intestine
S 19 transverse mesocolon
S 20 descending colon
S 21 transverse colon

SY 1 articulating processes L 3–L 4
SY 2 intervertebral disc L 3–L 4
SY 3 anterior longitudinal ligament
SY 4 articulating processes L 3–L 4

HA 8

A 1 inferior epigastric a.
A 2 testicular a.
A 3 ileocolic a.
A 4 ileocolic v.
A 5 intestinal a.
A 6 intestinal v.
A 7 middle sacral a.
A 8 abdominal aorta (bifurcation)
A 9 left colic a.
A 10 inferior mesenteric a.
A 11 inferior mesenteric v.
A 12 testicular a.
A 13 jejunal a.
A 14 right colic a. (branch)
A 15 lumbar v.
A 16 inferior vena cava
A 17 ascending lumbar v.
A 18 lumbar v. (dorsal branch)
A 19 iliolumbar a. (lumbar branch)
A 20 iliolumbar a. (iliac branch)
A 21 deep circumflex iliac a.

M 1 transverse abdominis m.
M 2 external oblique m.
M 3 internal oblique m.
M 4 umbilicus
M 5 rectus abdominis m.

M 6 transverse abdominis m.
M 7 internal oblique m.
M 8 external oblique m.
M 9 gluteus medius m.
M 10 fat body of lumbar triangle
M 11 iliacus m.
M 12 psoas major m.
M 13 erector spinae m.
M 14 transversospinal m.
M 15 thoracolumbar fascia
M 16 transversospinal m.
M 17 psoas major m.
M 18 erector spinae m.
M 19 iliacus m.
M 20 gluteal fascia
M 21 fat body of lumbar triangle
M 22 gluteus medius m.

N 1 superior hypogastric plexus
N 2 iliohypogastric n.
N 3 femoral n.
N 4 obturator n.
N 5 genitofemoral n.
N 6 4th lumbar n.
N 7 cauda equina
N 8 4th lumbar n.
N 9 obturator n.
N 10 femoral n.

O 1 ala of ilium
O 2 costal process L 5
O 3 inferior articulating process L 4
O 4 spinous process L 5
O 5 superior articulating process L 5
O 6 costal process L 5
O 7 iliac crest
O 8 iliac crest

S 1 ascending colon
S 2 transverse colon
S 3 transverse colon
S 4 transverse mesocolon
S 5 greater omentum
S 6 transverse colon
S 7 ureter
S 8 stomach (pyloric portion)
S 9 transverse mesocolon
S 10 root of mesentery
S 11 ureter
S 12 transverse mesocolon
S 13 transverse mesocolon
S 14 transverse colon
S 15 greater omentum
S 16 small intestine
S 17 greater omentum
S 18 ascending colon
S 19 small intestine
S 20 descending colon

SY 1 iliolumbar ligament
SY 2 ligamentum flavum
SY 3 supraspinal ligament
SY 4 intervertebral disc L 4–L 5
SY 5 articulating processes L 4–L 5
SY 6 iliolumbar ligament

HA 9

A 1 ileocolic a. (anterior cecal a.)
A 2 posterior cecal a.
A 3 ileocolic a.
A 4 testicular a.
A 5 inferior epigastric a.
A 6 common iliac a.
A 7 superior mesenteric v.
A 8 superior mesenteric a.
A 9 intestinal a.
A 10 middle sacral a.
A 11 left common iliac v.
A 12 common iliac a.
A 13 jejunal a.
A 14 inferior mesenteric a.
A 15 inferior epigastric a.
A 16 sigmoid a.
A 17 testicular a.
A 18 deep circumflex iliac a.
A 19 ileocolic a. (branch)
A 20 superior gluteal a. (branch)
A 21 iliolumbar a. (spinal branch)
A 22 right common iliac v.
A 23 ascending lumbar v.
A 24 intervertebral v. L 5–S 1
A 25 iliolumbar a. (iliac branch)
A 26 superior gluteal a. (branch)

M 1 external oblique m.
M 2 internal oblique m.
M 3 transverse abdominis m.
M 4 transverse fascia
M 5 linea alba
M 6 rectus abdominis m.
M 7 sheath of rectus abdominis m.
M 8 transverse abdominis m.
M 9 internal oblique m.
M 10 external oblique m.
M 11 fat body of lumbar triangle
M 12 gluteus medius m.
M 13 gluteus maximus m.
M 14 iliacus m.
M 15 psoas major m.
M 16 thoracolumbar fascia
M 17 erector spinae m.
M 18 erector spinae m.
M 19 iliacus m.
M 20 gluteus medius m.
M 21 gluteus maximus m.
M 22 fat body of lumbar triangle

N 1 iliohypogastric n.
N 2 ilioinguinal n.
N 3 lateral femoral cutaneous n.
N 4 genitofemoral n.
N 5 femoral n.
N 6 obturator n.
N 7 4th lumbar n. (ventral ramus)
N 8 5th lumbar n.
N 9 sympathetic trunk
N 10 spinal dura mater
N 11 1st sacral n.
N 12 5th lumbar n.
N 13 4th lumbar n. (ventral ramus)
N 14 obturator n.
N 15 femoral n.
N 16 genitofemoral n.

O 1 iliac tubercle
O 2 iliac crest
O 3 ala of ilium
O 4 middle sacral crest S 1
O 5 lamina S 1
O 6 inferior articulating process L 5
O 7 iliac crest
O 8 ala of sacrum
O 9 ala of ilium
O 10 iliac tubercle

S 1 ascending colon
S 2 ascending colon
S 3 ileum (terminal portion)
S 4 transverse colon
S 5 ureter
S 6 transverse colon
S 7 small intestine
S 8 mesentery
S 9 ureter
S 10 small intestine
S 11 small intestine
S 12 small intestine
S 13 appendices epiploicae
S 14 descending colon
S 15 greater omentum

SY 1 sacroiliac articulation
SY 2 ligamentum flavum L 5–S 1
SY 3 intervertebral disc L 5–S 1
SY 4 posterior longitudinal ligament
SY 5 articulating processes L 5–S 1
SY 6 dorsal sacroiliac ligament

HA 10

A 1 deep circumflex iliac a.
A 2 deep circumflex iliac a. (branch)
A 3 testicular a.
A 4 inferior epigastric a.
A 5 inferior epigastric a.
A 6 testicular a.
A 7 deep circumflex iliac a.
A 8 superior gluteal a. (deep branch)
A 9 superior gluteal v. (branch)
A 10 iliolumbar v.
A 11 common iliac v.

A 12 external iliac a.
A 13 lateral sacral a.
A 14 internal iliac a.
A 15 middle sacral a.
A 16 inferior mesenteric a.
A 17 sigmoid a.
A 18 common iliac v.
A 19 internal iliac a.
A 20 iliolumbar a.
A 21 external iliac a.
A 22 superior gluteal a. (deep branch)

M 1 gluteus minimus m.
M 2 rectus abdominis m.
M 3 psoas m.
M 4 internal oblique m. and transverse abdominis m.
M 5 external oblique m.
M 6 iliacus m.
M 7 gluteus minimus m.
M 8 gluteus medius m.
M 9 gluteus maximus m.
M 10 erector spinae m.
M 11 thoracolumbar fascia
M 12 gluteus maximus m.
M 13 gluteus medius m.

N 1 lateral femoral cutaneous n.
N 2 lateral femoral cutaneous n.
N 3 femoral n.
N 4 genitofemoral n.
N 5 femoral n.
N 6 lateral femoral cutaneous n.
N 7 obturator n.
N 8 lumbosacral trunk
N 9 1st sacral n.
N 10 2nd sacral n.
N 11 3rd sacral n.
N 12 4th sacral n.
N 13 2nd sacral n.
N 14 1st sacral n.
N 15 lumbosacral trunk
N 16 obturator n.

O 1 anterior superior iliac spine
O 2 anterior superior iliac spine
O 3 ala of ilium
O 4 lateral sacral crest
O 5 intervertebral foramen S 1–S 2
O 6 middle sacral crest
O 7 ala of sacrum

S 1 ileocecal valve
S 2 cecum
S 3 vermiform appendix (orifice)
S 4 greater omentum
S 5 vermiform appendix
S 6 ileum (terminal portion)
S 7 ureter
S 8 small intestine
S 9 mesentery
S 10 urachus
S 11 mesentery
S 12 small intestine
S 13 ureter
S 14 small intestine
S 15 small intestine
S 16 descending colon

SY 1 sacroiliac articulation
SY 2 lumbosacral articulation
SY 3 anterior longitudinal ligament
SY 4 sacroiliac articulation

HA 11

A 1 deep circumflex iliac a.
A 2 testicular a.
A 3 external iliac v.
A 4 external iliac a.
A 5 umbilical a.
A 6 inferior epigastric a.
A 7 inferior epigastric a. (branch)
A 8 intestinal v.
A 9 inferior epigastric a.
A 10 testicular a.
A 11 external iliac v.
A 12 external iliac a.
A 13 deep circumflex iliac a.
A 14 superior gluteal a.
A 15 superior gluteal a.
A 16 inferior gluteal a.
A 17 internal pudendal a.
A 18 inferior gluteal v.
A 19 lateral sacral a.
A 20 superior rectal a.
A 21 middle sacral a.
A 22 internal pudendal a.
A 23 superior gluteal a. (anastomosis with inferior gluteal a.)
A 24 superior gluteal a.
A 25 superior gluteal a.
A 26 superior gluteal a. (superficial branch)
A 27 superior gluteal v.
A 28 superior gluteal a. (deep branch)
A 29 superior gluteal a. (deep branch)

M 1 gluteus minimus m.
M 2 iliacus m.
M 3 internal oblique m.
M 4 psoas m.
M 5 rectus abdominis m.
M 6 psoas m.
M 7 inguinal ligament
M 8 iliacus m.
M 9 sartorius m.
M 10 tensor fasciae latae m.
M 11 gluteus medius m.
M 12 gluteus maximus m.
M 13 piriformis m.
M 14 erector spinae m.
M 15 piriformis m.
M 16 gluteus maximus m.
M 17 gluteus minimus m.
M 18 gluteus medius m.

N 1 femoral n.
N 2 genitofemoral n.
N 3 obturator n.
N 4 obturator n.
N 5 genitofemoral n.
N 6 femoral n.
N 7 lumbosacral trunk
N 8 1st sacral n. (ventral ramus)
N 9 2nd sacral n. (ventral ramus)
N 10 3rd sacral n. (ventral ramus)
N 11 2nd sacral n. (ventral ramus)
N 12 1st sacral n. (ventral ramus)
N 13 lumbosacral trunk

O 1 ilium
O 2 arcuate line
O 3 posterior inferior iliac spine
O 4 lateral sacral crest
O 5 sacral canal
O 6 3rd ventral sacral foramen
O 7 greater sciatic foramen

S 1 cecum
S 2 ureter
S 3 ileum
S 4 mesentery
S 5 small intestine
S 6 small intestine
S 7 ureter
S 8 descending colon
S 9 small intestine
S 10 rectum
S 11 sigmoid mesocolon

HA 12

A 1 superficial circumflex iliac a.
A 2 external iliac a.
A 3 external iliac v.
A 4 inferior epigastric a.
A 5 right testicular v.
A 6 obturator a.
A 7 umbilical a.
A 8 inferior epigastric a.
A 9 inferior vesical a.
A 10 external iliac v.
A 11 external iliac a.
A 12 superior gluteal a. (branch)
A 13 superior gluteal a. (branch)
A 14 internal pudendal a.
A 15 inferior gluteal a.
A 16 inferior vesical a. (branch)
A 17 inferior vesical a. (branch)
A 18 superior rectal a.
A 19 superior rectal v. (branch)

A 20 superior rectal a. (branch)
A 21 middle sacral a.
A 22 inferior gluteal a.
A 23 vesical v.
A 24 internal pudendal a.

M 1 tensor fasciae latae m.
M 2 rectus femoris m. (tendon)
M 3 sartorius m.
M 4 iliopsoas m.
M 5 internal oblique m.
M 6 rectus abdominis m.
M 7 external oblique m.
M 8 iliopsoas m.
M 9 rectus femoris m. (tendon)
M 10 sartorius m.
M 11 gluteus minimus m.
M 12 tensor fasciae latae m.
M 13 gluteus minimus m.
M 14 gluteus maximus m.
M 15 piriformis m.
M 16 obturator internus m.
M 17 piriformis m.
M 18 gluteus maximus m.
M 19 gluteus maximus m. (tendon)

N 1 lateral femoral cutaneous n.
N 2 femoral n.
N 3 obturator n.
N 4 femoral n.
N 5 sciatic n.
N 6 pudendal n.
N 7 pudendal n.
N 8 sciatic n.

O 1 anterior inferior iliac spine
O 2 body of ilium
O 3 sacral hiatus
O 4 sacral cornu
O 5 lateral sacral crest
O 6 greater sciatic notch

S 1 ductus deferens
S 2 ductus deferens
S 3 small intestine
S 4 small intestine
S 5 middle umbilical fold (urachus)
S 6 small intestine
S 7 small intestine
S 8 sigmoid colon
S 9 ductus deferens
S 10 small intestine
S 11 mesentery
S 12 rectal ampulla
S 13 sigmoid colon
S 14 sigmoid mesocolon
S 15 sigmoid colon
S 16 sigmoid colon
S 17 ureter

SY 1 sacrotuberous ligament
SY 2 dorsal sacrococcygeal ligament
SY 3 sacrotuberous ligament

HA 13

A 1 lateral femoral circumflex a. (ascending branch)
A 2 femoral a.
A 3 superficial epigastric v.
A 4 femoral v.
A 5 external pudendal v.
A 6 obturator a.
A 7 obturator v.
A 8 pampiniform plexus
A 9 pampiniform plexus
A 10 obturator a.
A 11 obturator v. (acetabular branch)
A 12 superficial epigastric v.
A 13 femoral v.
A 14 femoral a.
A 15 lateral femoral circumflex v. (ascending branch)
A 16 inferior gluteal v.
A 17 inferior gluteal v.
A 18 inferior gluteal a.
A 19 internal pudendal a.
A 20 inferior vesical a.
A 21 superior rectal a. (branch)
A 22 inferior vesical a.
A 23 inferior gluteal a.
A 24 internal pudendal a.

M 1 gluteus minimus m.
M 2 gluteus medius m.
M 3 rectus femoris m.
M 4 sartorius m.
M 5 iliopsoas m.
M 6 cremaster m.
M 7 reflected inguinal ligament
M 8 rectus abdominis m.
M 9 pyramidalis m.
M 10 rectus abdominis m. (tendon)
M 11 obturator internus m.
M 12 pectineus m.
M 13 iliopsoas m.
M 14 sartorius m.
M 15 rectus femoris m.
M 16 tensor fasciae latae m.
M 17 obturator internus m. (tendon)
M 18 gluteus maximus m.
M 19 superior gemellus m.
M 20 coccygeus m.
M 21 obturator internus m.
M 22 obturator internus m. (tendon)
M 23 gluteus maximus m.
M 24 gluteus minimus m.
M 25 gluteus medius m.

N 1 lateral femoral cutaneous n.
N 2 femoral n.
N 3 obturator n.
N 4 femoral n.
N 5 sciatic n.
N 6 sciatic n.
N 7 pudendal n.
N 8 pudendal n.
N 9 sciatic n.

O 1 neck of femur
O 2 pecten pubis
O 3 head of femur
O 4 greater trochanter
O 5 ischial spine
O 6 body of ischium
O 7 neck of femur
O 8 greater trochanter

S 1 ductus deferens
S 2 small intestine
S 3 urinary bladder
S 4 urachus
S 5 supravesical fossa
S 6 small intestine
S 7 ureter
S 8 ductus deferens
S 9 rectovesical space
S 10 rectal ampulla
S 11 sigmoid colon
S 12 ductus deferens
S 13 ureter

SY 1 articular capsule of hip joint
SY 2 acetabular labrum
SY 3 ligamentum teres
SY 4 ligamentum teres
SY 5 acetabular labrum
SY 6 sacrospinous ligament
SY 7 sacrotuberous ligament
SY 8 sacrotuberous ligament
SY 9 acetabular fossa
SY 10 labrum acetabular
SY 11 articular capsule

HP♂
Horizontal sections
Pelvis

40-year-old male

Transverse sections perpendicular to the plane upon which the subject is lying. The plane of reference passes through the antero-superior iliac spines. The sections are spaced 2 cm apart. All of the sections are viewed from the inferior side.

HP♂ 1 horizontal section at the antero-superior iliac spines, inferior-superior view
HP♂ 2 horizontal section, 2 cm below the antero-superior iliac spines, inferior-superior view
HP♂ 3 horizontal section, 4 cm below the antero-superior iliac spines, inferior-superior view
HP♂ 4 horizontal section, 6 cm below the antero-superior iliac spines, inferior-superior view
HP♂ 5 horizontal section, 8 cm below the antero-superior iliac spines, inferior-superior view
HP♂ 6 horizontal section, 10 cm below the antero-superior iliac spines, inferior-superior view
HP♂ 7 horizontal section, 12 cm below the antero-superior iliac spines, inferior-superior view
HP♂ 8 horizontal section, 14 cm below the antero-superior iliac spines, inferior-superior view

HP♂ 1

A 1 deep circumflex iliac a. (branch)
A 2 deep circumflex iliac a.
A 3 testicular a.
A 4 external iliac a.
A 5 common iliac v.
A 6 ileocolic a. (branch)
A 7 ileocolic v. (branch)
A 8 inferior epigastric a.
A 9 superior mesenteric a. (intestinal a.)
A 10 superior mesenteric v. (intestinal v.)
A 11 external iliac a.
A 12 testicular a.
A 13 left colic v. (branch)
A 14 left colic vessels
A 15 deep circumflex iliac v.
A 16 superior gluteal v. (deep branch)
A 17 superior gluteal a. (deep branch)
A 18 iliolumbar a.
A 19 superior gluteal a.
A 20 right internal iliac a.
A 21 right internal iliac v.
A 22 superior mesenteric v. (intestinal v.)
A 23 lateral sacral v. (spinal branch)
A 24 internal vetebral venous plexus
A 25 superior rectal v.
A 26 superior rectal a.
A 27 sigmoid a.
A 28 lateral sacral a. (spinal branch)
A 29 internal iliac v.
A 30 internal iliac a.
A 31 external iliac v.
A 32 iliolumbar a. (branch)
A 33 superior gluteal a.

M 1 gluteus minimus m.
M 2 iliacus m.
M 3 internal oblique m.
M 4 transversus abdominis m.
M 5 psoas major m.
M 6 rectus abdominis m.
M 7 linea alba
M 8 transversus abdominis m.
M 9 internal oblique m.
M 10 external oblique m.
M 11 gluteus minimus m.
M 12 gluteus medius m.
M 13 gluteus maximus m.
M 14 thoracolumbar fascia
M 15 erector spinae m.
M 16 erector spinae m.
M 17 gluteus maximus m.
M 18 gluteus medius m.

N 1 lateral femoral cutaneous n.
N 2 femoral n.
N 3 genitofemoral n.
N 4 femoral n.
N 5 obturator n.
N 6 lumbosacral trunk
N 7 lumbosacral trunk
N 8 1st sacral n. (anterior ramus)
N 9 sympathetic trunk
N 10 2nd sacral n.
N 11 3rd sacral n.
N 12 4th sacral n.
N 13 lumbosacral trunk
N 14 obturator n.

O 1 anterior superior iliac spine
O 2 ala of ilium
O 3 anterior superior iliac spine
O 4 ala of ilium
O 5 middle sacral crest (S 2)
O 6 posterior sacral foramen 2
O 7 ventral sacral foramen 1
O 8 sacrum (lateral portion)

S 1 cecum
S 2 vermiform appendix
S 3 ureter
S 4 small intestine
S 5 ileocecal orifice
S 6 ileum
S 7 mesentery
S 8 small intestine
S 9 small intestine
S 10 ureter
S 11 small intestine
S 12 descending colon

SY 1 dorsal sacroiliac ligament
SY 2 intervertebral disc S 1–S 2
SY 3 anterior longitudinal ligament
SY 4 dorsal sacroiliac ligament
SY 5 sacroiliac articulation

HP♂ 2

A 1 deep circumflex iliac a.
A 2 deep circumflex iliac a. (branch)
A 3 testicular a.
A 4 external iliac a.
A 5 external iliac v.
A 6 posterior cecal a.
A 7 ileal aa.
A 8 inferior epigastric a.
A 9 external iliac v.
A 10 external iliac a.
A 11 left colic a. (branch)
A 12 deep circumflex iliac v.
A 13 superior gluteal v.
A 14 superior gluteal a.
A 15 obturator a.
A 16 obturator v.
A 17 internal pudendal a.
A 18 internal iliac v.
A 19 inferior gluteal a.
A 20 superior rectal v.
A 21 superior rectal a. (branch)
A 22 superior rectal a. (branch)
A 23 sigmoid vessels
A 24 inferior gluteal a.

A 25 internal pudendal a.
A 26 superior gluteal a.
A 27 superior gluteal v.
A 28 superior gluteal vessels (superficial branch)
A 29 superior gluteal v.
A 30 superior gluteal a.

M 1 gluteus minimus m.
M 2 iliacus m.
M 3 external oblique m.
M 4 internal oblique m.
M 5 transversus abdominis m.
M 6 rectus abdominis m.
M 7 linea alba
M 8 rectus abdominis m.
M 9 psoas major m.
M 10 iliacus m.
M 11 sartorius m.
M 12 gluteus minimus m.
M 13 tensor fasciae latae m.
M 14 gluteus maximus m.
M 15 gluteus medius m.
M 16 piriformis m.
M 17 erector spinae m.
M 18 thoracolumbar fascia
M 19 piriformis m.
M 20 gluteus maximus m.
M 21 gluteus medius m.

N 1 femoral n.
N 2 obturator n.
N 3 obturator n.
N 4 femoral n.
N 5 lumbosacral trunk
N 6 1st sacral n. (anterior ramus)
N 7 2nd sacral n. (anterior ramus)
N 8 3rd sacral n.
N 9 4th sacral n.
N 10 2nd sacral n. (anterior ramus)
N 11 1st sacral n. (anterior ramus)
N 12 lumbosacral trunk (root L 5)
N 13 lumbosacral trunk (root L 4)

O 1 ala of ilium
O 2 arcuate line
O 3 arcuate line
O 4 ala of ilium
O 5 posterior superior iliac spine
O 6 lateral sacral crest
O 7 posterior sacral foramen 3
O 8 middle sacral crest
O 9 sacral canal
O 10 lateral sacral crest

S 1 small intestine
S 2 small intestine
S 3 vermiform appendix
S 4 cecum
S 5 ileum
S 6 mesentery
S 7 small intestine
S 8 small intestine
S 9 ureter
S 10 descending colon
S 11 ureter
S 12 sigmoid colon (terminal portion)
S 13 sigmoid mesocolon

SY 1 posterior sacroiliac ligament

HP ♂ 3

A 1 deep circumflex iliac a.
A 2 deep circumflex iliac a. (branch)
A 3 external iliac a.
A 4 external iliac v.
A 5 obturator a.
A 6 inferior epigastric a.
A 7 ileal vessels
A 8 external iliac v.
A 9 external iliac a.
A 10 lateral femoral circumflex a. (ascending branch)
A 11 superior gluteal a. (branch)
A 12 obturator v.
A 13 internal pudendal a.
A 14 inferior vesical v.
A 15 inferior gluteal v.
A 16 inferior gluteal a.
A 17 superior rectal v. (branch)
A 18 superior rectal v.
A 19 lateral sacral a.
A 20 superior rectal v. (branch)
A 21 inferior gluteal a.
A 22 inferior vesical v.
A 23 inferior vesical v.
A 24 internal pudendal a.
A 25 superior gluteal a. (branch)

M 1 tensor fasciae latae m.
M 2 gluteus minimus m.
M 3 sartorius m.
M 4 rectus femoris m. (tendon)
M 5 iliacus m.
M 6 psoas m.
M 7 internal oblique m.
M 8 external oblique m.
M 9 linea alba
M 10 rectus abdominis m.
M 11 psoas major m.
M 12 inguinal ligament
M 13 iliacus m.
M 14 rectus femoris m. (tendon)
M 15 sartorius m.
M 16 gluteus minimus m.
M 17 tensor fasciae latae m.
M 18 gluteus medius m.
M 19 gluteal fascia
M 20 gluteal fascia
M 21 gluteus medius m.
M 22 gluteus maximus m.
M 23 piriformis m.
M 24 obturator internus m.
M 25 obturator internus m.
M 26 piriformis m.
M 27 gluteus maximus m.

N 1 femoral n.
N 2 obturator n.
N 3 obturator n.
N 4 femoral n.
N 5 sciatic n.
N 6 pudendal n.
N 7 pudendal plexus
N 8 sciatic n.

O 1 superior acetabular sulcus
O 2 anterior inferior iliac spine
O 3 body of ilium
O 4 arcuate line
O 5 body of ilium
O 6 anterior inferior iliac spine
O 7 greater sciatic notch
O 8 sacral hiatus
O 9 sacral cornu

S 1 cecum
S 2 small intestine
S 3 urachus
S 4 small intestine
S 5 sigmoid colon
S 6 ureter
S 7 small intestine
S 8 small intestine
S 9 rectum
S 10 mesentery
S 11 small intestine
S 12 ureter

SY 1 sacrotuberous ligament
SY 2 posterior sacrococcygeal ligament

HP ♂ 4

A 1 lateral femoral circumflex a. (ascending branch)
A 2 superficial inguinal lymph node
A 3 external iliac a.
A 4 external iliac v.
A 5 testicular a.
A 6 pampiniform plexus
A 7 obturator v.
A 8 inferior epigastric a. (obturator branch)
A 9 external iliac v.
A 10 external iliac a.
A 11 deep circumflex v.
A 12 deep circumflex iliac a.
A 13 lateral femoral circumflex a. (ascending branch)
A 14 inferior gluteal v.

A 15 inferior vesical v.
A 16 internal pudendal v.
A 17 inferior vesical a.
A 18 superior rectal a. (branch)
A 19 superior vesical a.
A 20 inferior vesical v.
A 21 inferior gluteal v.
A 22 internal pudendal a.
A 23 inferior gluteal v. (branch)
A 24 superior gluteal a. (deep branch)

M 1 tensor fasciae latae m.
M 2 gluteus minimus m.
M 3 rectus femoris m. (tendon)
M 4 sartorius m.
M 5 iliopsoas m.
M 6 pectineus m.
M 7 pyramidalis m.
M 8 rectus abdominis m.
M 9 sheath of rectus abdominis m.
M 10 aponeuroses of external oblique, internal oblique, and transversus abdominis mm.
M 11 iliopsoas m.
M 12 rectus femoris m. (tendon)
M 13 sartorius m.
M 14 gluteus minimus m.
M 15 tensor fasciae latae m.
M 16 gluteus medius m.
M 17 piriformis m.
M 18 gluteus maximus m.
M 19 internal obturator m.
M 20 coccygeus m.
M 21 levator ani m.
M 22 coccygeus m.
M 23 levator ani m.
M 24 internal obturator m.
M 25 piriformis m.
M 26 gluteus maximus m.

N 1 femoral n.
N 2 obturator n.
N 3 femoral n.
N 4 sciatic n.
N 5 pudendal n.
N 6 sciatic n.

O 1 iliopectineal eminence
O 2 arcuate line
O 3 head of femur
O 4 acetabulum
O 5 acetabular fossa
O 6 body of ischium
O 7 greater sciatic foramen
O 8 cornu of coccyx
O 9 acetabular fossa

S 1 ductus deferens
S 2 ductus deferens
S 3 small intestine
S 4 cecum
S 5 umbilical a.
S 6 small intestine
S 7 ductus deferens
S 8 ureter
S 9 ductus deferens
S 10 small intestine
S 11 rectal ampulla
S 12 small intestine
S 13 ureter

SY 1 capsule of hip joint
SY 2 acetabular labrum
SY 3 sacrospinal ligament
SY 4 sacrotuberous ligament
SY 5 sacrococcygeal articulation
SY 6 ligamentum teres
SY 7 acetabular labrum
SY 8 articular capsule

HP ♂ 5

A 1 lateral femoral circumflex a. (ascending branch)
A 2 femoral a.
A 3 femoral v.
A 4 obturator a.
A 5 obturator v.
A 6 inferior vesical v.
A 7 pampiniform plexus
A 8 testicular a.
A 9 obturator v.
A 10 superficial inguinal lymph node
A 11 femoral v.
A 12 superficial epigastric v.
A 13 femoral a.
A 14 medial femoral circumflex v.
A 15 inferior gluteal a. (branch)
A 16 inferior gluteal a. (branch)
A 17 inferior gluteal v. (branch)
A 18 inferior gluteal v. (branch)
A 19 internal pudendal a.
A 20 vesical venous plexus
A 21 inferior vesical a.
A 22 superior rectal a. (branch)
A 23 superior rectal v.
A 24 inferior gluteal v. (branch)
A 25 internal pudendal a.
A 26 inferior gluteal a.
A 27 inferior gluteal a. (branch)

M 1 tensor fasciae latae m.
M 2 rectus femoris m.
M 3 sartorius m.
M 4 iliopsoas m.
M 5 pectineus m.
M 6 reflected inguinal ligament
M 7 rectus abdominis m.
M 8 pyramidalis m.
M 9 pectineus m.
M 10 iliopsoas m.
M 11 sartorius m.
M 12 rectus femoris m.
M 13 tensor fasciae latae m.
M 14 gluteus medius m.
M 15 gluteus minimus m.
M 16 gluteus maximus m.
M 17 gemellus inferior m.
M 18 obturator internus m.
M 19 ischiorectal fossa
M 20 levator ani m.
M 21 ischiorectal fossa
M 22 levator ani m.
M 23 obturator internus m.
M 24 gluteus maximus m.
M 25 gemellus inferior m.
M 26 gluteus medius m.
M 27 gluteus minimus m.

N 1 lateral femoral cutaneous n.
N 2 femoral n.
N 3 obturator n.
N 4 obturator n.
N 5 femoral n.
N 6 sciatic n.
N 7 pudendal n.
N 8 pudendal n.
N 9 sciatic n.

O 1 acetabular notch
O 2 pubis (superior ramus)
O 3 pubis (superior ramus)
O 4 pubic tubercle
O 5 obturator sulcus
O 6 head of femur
O 7 greater trochanter
O 8 neck of femur
O 9 coccyx
O 10 lesser sciatic notch
O 11 articular surface
O 12 greater trochanter

S 1 deferent duct
S 2 urinary bladder (apex)
S 3 retropubic space
S 4 spermatic cord
S 5 seminal vesicle
S 6 ureter
S 7 ampulla of ductus deferens
S 8 rectal ampulla
S 9 ampulla of ductus deferens
S 10 seminal vesicle
S 11 ureter

SY 1 transverse acetabular ligament
SY 2 obturator membrane
SY 3 symphysis pubis
SY 4 ligamentum teres
SY 5 transverse acetabular ligament
SY 6 articular capsule
SY 7 ischiofemoral ligament
SY 8 acetabular labrum
SY 9 sacrotuberous ligament
SY 10 sacrotuberous ligament

SY 11 acetabular labrum
SY 12 ischiofemoral ligament

HP♂ 6

The left femoral artery has been catheterized to allow injection of a preserving fluid.

A 1 lateral femoral circumflex a. (ascending branch)
A 2 femoral a.
A 3 medial femoral circumflex v.
A 4 medial femoral circumflex a.
A 5 femoral v.
A 6 external pudendal v. (branch)
A 7 obturator v.
A 8 external pudendal v.
A 9 obturator a. (branch)
A 10 obturator v. (branch)
A 11 obturator v. (branch)
A 12 testicular a.
A 13 ductus deferens a.
A 14 deep dorsal vein of penis
A 15 superficial dorsal vein of penis (branch)
A 16 prostate venous plexus
A 17 greater saphenous v.
A 18 femoral v.
A 19 medial femoral circumflex a.
A 20 deep femoral a.
A 21 lateral femoral circumflex v. (branch)
A 22 lateral femoral circumflex a. (branch)
A 23 lateral femoral circumflex a. (branch)
A 24 medial femoral circumflex v.
A 25 inferior gluteal a. (branch)
A 26 internal pudendal a. (branch)
A 27 inferior rectal a. (branch)
A 28 internal pudendal a.
A 29 prostate venous plexus
A 30 internal pudendal v.
A 31 internal pudendal a. (branch)
A 32 inferior gluteal a. (branch)
A 33 inferior gluteal a. (branch)
A 34 medial femoral circumflex a.

M 1 tensor fasciae latae m.
M 2 vastus lateralis m.
M 3 rectus femoris m.
M 4 sartorius m.
M 5 iliopsoas m.
M 6 obturator externus m.
M 7 pectineus m.
M 8 adductor brevis m.
M 9 cremaster m.
M 10 adductor longus m.
M 11 obturator externus m.
M 12 adductor brevis m.
M 13 adductor longus m.
M 14 pectineus m.
M 15 iliopsoas m.
M 16 sartorius m.
M 17 rectus femoris m.
M 18 tensor fasciae latae m.
M 19 gluteus medius m.
M 20 gluteus medius m.
M 21 gluteus minimus m. (tendon)
M 22 trochanteric bursa of gluteus maximus
M 23 gluteus maximus m. (tendon)
M 24 quadratus femoris m.
M 25 gluteus maximus m.
M 26 obturator externus m.
M 27 ischiocrural mm. (tendon)
M 28 obturatus internus m.
M 29 ischiorectal fossa
M 30 levator ani m.
M 31 sphincter ani externus m.
M 32 levator ani m.
M 33 ischiorectal fossa
M 34 obturator internus m.
M 35 gluteus maximus m.
M 36 quadratus femoris m.

N 1 lateral femoral cutaneous n.
N 2 femoral n.
N 3 obturator n. (posterior branch)
N 4 obturator n. (anterior branch)
N 5 obturator n. (posterior branch)
N 6 femoral n.
N 7 lateral femoral cutaneous n.
N 8 sciatic n.
N 9 sciatic n.
N 10 pudendal n.
N 11 sciatic n.

O 1 neck of femur
O 2 pubis (inferior ramus)
O 3 neck of femur
O 4 intertrochanteric crest
O 5 tuberosity of ischium
O 6 tuberosity of ischium
O 7 intertrochanteric crest
O 8 greater trochanter

S 1 ductus deferns
S 2 prostate
S 3 urethra
S 4 corpus cavernosum penis
S 5 spermatic cord
S 6 rectal ampulla

SY 1 symphysis pubis
SY 2 arcuate ligament
SY 3 sacrotuberous ligament
SY 4 obturator membrane

HP♂ 7

The left femoral artery has been catheterized to allow injection of a preserving fluid.

A 1 lateral femoral circumflex a. (ascending branch)
A 2 lateral femoral circumflex a. (ascending branch)
A 3 deep femoral a.
A 4 femoral a.
A 5 femoral v.
A 6 greater saphenous v.
A 7 external pudendal v.
A 8 testicular a.
A 9 femoral v.
A 10 deep femoral a.
A 11 lateral femoral circumflex a.
A 12 artery to sciatic n. (sciatic a.)
A 13 medial femoral circumflex a. (deep branch)
A 14 medial femoral circumflex a. (deep branch)
A 15 obturator v. (branch)
A 16 perineal a.
A 17 rectal venous plexus
A 18 inferior gluteal a. (branch)
A 19 medial femoral circumflex a. (deep branch)
A 20 medial femoral circumflex a. (deep branch)

M 1 tensor fasciae latae m.
M 2 vastus intermedius m.
M 3 rectus femoris m.
M 4 iliopsoas m.
M 5 iliopsoas m. (tendon)
M 6 pectineus m.
M 7 adductor magnus m.
M 8 adductor longus m.
M 9 gracilis m.
M 10 bulbospongiosus m.
M 11 gracilis m.
M 12 adductor brevis m.
M 13 pectineus m.
M 14 iliopsoas m.
M 15 sartorius m.
M 16 rectus femoris m.
M 17 vastus intermedius m.
M 18 tensor fasciae latae m.
M 19 greater trochanter (bursa)
M 20 vastus lateralis m.
M 21 quadratus femoris m.
M 22 ischiocrural mm.
M 23 gluteus maximus m.
M 24 obturator externus m.
M 25 ischiocavernosus m.
M 26 superficial transverse perinei m.
M 27 ischiorectal fossa
M 28 bulbospongiosus m.
M 29 sphincter ani externus m.
M 30 ischiorectal fossa

M 31 adductor magnus m.
M 32 bursa of ischial tuberosity
M 33 obturator externus m.
M 34 ischiocrurales mm.
M 35 quadratus femoris m.
M 36 gluteus maximus m.

N 1 lateral femoral cutaneous n. (branch)
N 2 lateral femoral cutaneous n. (branch)
N 3 femoral n. (branch)
N 4 femoral n. (branch)
N 5 obturator n. (anterior branch)
N 6 obturator n. (posterior branch)
N 7 obturator n. (anterior branch)
N 8 obturator n. (posterior branch)
N 9 lateral femoral cutaneous n.
N 10 sciatic n.
N 11 perineal n. (branch)
N 12 sciatic n.

O 1 neck of femur
O 2 neck of femur
O 3 intertrochanteric crest
O 4 ischial tuberosity
O 5 ramus of ischium
O 6 greater trochanter

S 1 tunica vaginalis
S 2 corpus cavernosum penis
S 3 septum penis
S 4 corpus cavernosum penis
S 5 corpus spongiosum penis
S 6 tunica albuginea
S 7 ductus deferens
S 8 tunica vaginalis
S 9 crus penis
S 10 anal canal
S 11 urethra
S 12 bulb
S 13 crus penis

HP♂ 8

The left femoral artery has been catheterized to allow injection of a preserving fluid.

A 1 lateral femoral circumflex v. (branch)
A 2 lateral femoral circumflex a. (descending branch)
A 3 deep femoral a.
A 4 femoral a.
A 5 deep femoral v.
A 6 femoral v.
A 7 great saphenous v.
A 8 femoral v.
A 9 deep femoral v.
A 10 deep femoral a.
A 11 lateral femoral circumflex v.
A 12 lateral femoral circumflex v. (branch)
A 13 lateral femoral circumflex a. (branch)
A 14 v. to sciatic n.
A 15 medial femoral circumflex a. (deep branch)
A 16 v. to sciatic n.
A 17 medial femoral circumflex a. (deep branch)

M 1 tensor fasciae latae m.
M 2 rectus femoris m.
M 3 vastus intermedius m.
M 4 iliopsoas m.
M 5 sartorius m.
M 6 adductor longus m.
M 7 gracilis m.
M 8 gracilis m.
M 9 adductor longus m.
M 10 pectineus m.
M 11 sartorius m.
M 12 rectus femoris m.
M 13 vastus intermedius m.
M 14 vastus lateralis m.
M 15 tensor fasciae latae m.
M 16 vastus lateralis m.
M 17 gluteus maximus m.
M 18 pectineus m.
M 19 biceps femoris m. (long head)
M 20 semimembranosus m.
M 21 adductor brevis m.
M 22 semitendinosus m.
M 23 adductor magnus m.
M 24 ischiorectal fossa
M 25 semitendinosus m.
M 26 semimembranosus m.
M 27 gluteus maximus m.
M 28 vastus lateralis m.

N 1 lateral femoral cutaneous n. (branch)
N 2 lateral femoral cutaneous n. (branch)
N 3 femoral n. (branch)
N 4 femoral n. (branch)
N 5 obturator n. (posterior branch)
N 6 obturator n. (posterior branch)
N 7 lateral femoral cutaneous n.
N 8 sciatic n.
N 9 sciatic n.

O 1 femur
O 2 lesser trochanter
O 3 lesser trochanter

S 1 glans penis
S 2 urethra (pars spongiosa)
S 3 corpus spongiosum penis
S 4 corpus cavernosum penis
S 5 testis
S 6 epididymis
S 7 septum of scrotum
S 8 tunica albuginea
S 9 scrotum

HP♀
Horizontal sections Pelvis

42-year-old female

Transverse sections perpendicular to the plane upon which the subject is lying. The plane of reference passes through the antero-superior iliac spines. The sections are spaced 2.5 cm apart. All of the sections are viewed from the inferior side; some are also viewed from the superior side.

HP♀1 horizontal section, 2.5 cm below the antero-superior iliac spines, inferior-superior view
HP♀2 horizontal section, 5 cm below the antero-superior iliac spines, inferior-superior view
HP♀3 horizontal section, 5 cm below the antero-superior iliac spines superior-inferior view
HP♀4 horizontal section, 7.5 cm below the antero-superior iliac spines, inferior-superior view
HP♀5 horizontal section, 7.5 cm below the antero-superior iliac spines, superior-inferior view
HP♀6 horizontal section, 10 cm below the antero-superior iliac spines, inferior-superior view
HP♀7 horizontal section, 12.5 cm below the antero-superior iliac spines, inferior-superior view
HP♀8 horizontal section, 15 cm below the antero-superior iliac spines, inferior-superior view

HP♀1

A 1 superior gluteal v. (deep branch)
A 2 superficial iliac circumflex v.
A 3 deep circumflex iliac v. (ascending branch)
A 4 deep circumflex iliac v. (deep branch)
A 5 external iliac a.
A 6 external iliac v.
A 7 inferior epigastric v.
A 8 uterine venous plexus
A 9 uterine venous plexus
A 10 intestinal v.
A 11 superficial epigastric v.
A 12 intestinal v.
A 13 superficial epigastric v.
A 14 sigmoid v.
A 15 external iliac v.
A 16 external iliac a.
A 17 superficial iliac circumflex v.
A 18 deep circumflex iliac a.
A 19 deep circumflex iliac v. (branch)
A 20 superior gluteal a. (branch)
A 21 superior gluteal v.
A 22 obturator a.
A 23 obturator v.
A 24 inferior gluteal a.
A 25 lateral sacral v.
A 26 superior rectal v. (branch)
A 27 lateral sacral a.
A 28 superior rectal v.
A 29 sigmoid v.
A 30 inferior gluteal a.
A 31 internal pudendal a.
A 32 superior gluteal a.
A 33 superior gluteal v. (superficial branch)
A 34 superior gluteal v. (deep branch)

M 1 tensor fasciae latae m.
M 2 gluteus minimus m.
M 3 sartorius m.
M 4 iliopsoas m.
M 5 internal oblique m.
M 6 external oblique m.
M 7 rectus abdominis m.
M 8 linea alba
M 9 external oblique m.
M 10 transversus abdominis m.
M 11 iliopsoas m.
M 12 sartorius m.
M 13 tensor fasciae latae m.
M 14 gluteus maximus m.
M 15 gluteus medius m.
M 16 piriformis m.
M 17 erector spinae m.
M 18 piriformis m.
M 19 gluteus maximus m.
M 20 gluteus medius m.
M 21 fat body of lumbar triangle

N 1 femoral n.
N 2 femoral n.
N 3 lateral femoral cutaneous n.
N 4 lumbosacral trunk and 1st sacral n. (ventral ramus)
N 5 obturator n.
N 6 2nd sacral n. (ventral ramus)
N 7 3rd sacral n. (ventral ramus)
N 8 2nd sacral n. (ventral ramus)
N 9 1st sacral n. (ventral ramus)
N 10 lumbosacral trunk
N 11 obturator n.

O 1 body of ilium
O 2 arcuate line
O 3 lateral sacral crest
O 4 sacral canal
O 5 arcuate line
O 6 ala of ilium

S 1 cecum
S 2 small intestine
S 3 mesentery
S 4 small intestine
S 5 descending colon
S 6 sigmoid mesocolon
S 7 ureter (bifid)
S 8 ureter
S 9 broad ligament
S 10 ovary
S 11 uterine tubes
S 12 infundibulum of uterine tube
S 13 peritoneal cavity
S 14 body of uterus
S 15 rectal ampulla
S 16 sigmoid colon

SY 1 dorsal sacroiliac ligament
SY 2 sacroiliac articulation

HP♀2

A 1 superior gluteal a. (deep branch)
A 2 lateral femoral circumflex v. (branch)
A 3 deep circumflex iliac v.
A 4 superficial iliac circumflex v.
A 5 external iliac a.
A 6 inferior epigastric a.
A 7 inferior epigastric a.
A 8 external iliac v.
A 9 iliac lymph node
A 10 inferior vesical a.
A 11 superficial epigastric v.
A 12 inferior epigastric a. (muscular branch)
A 13 superficial epigastric v.
A 14 uterine a.
A 15 inferior epigastric v.
A 16 external iliac v.
A 17 external iliac a.
A 18 superficial iliac circumflex v.
A 19 deep circumflex iliac v.
A 20 internal pudendal a.
A 21 inferior gluteal a.
A 22 uterine v.
A 23 vaginal a.
A 24 uterine a.
A 25 uterine venous plexus
A 26 lateral sacral v.
A 27 middle sacral a.
A 28 uterine venous plexus
A 29 lateral sacral a.
A 30 superior rectal a.
A 31 inferior gluteal v.
A 32 vaginal a.
A 33 internal pudendal a.
A 34 superior gluteal a. (branch)

M 1 gluteus medius m.
M 2 gluteus minimus m.
M 3 tensor fasciae latae m.
M 4 sartorius m.
M 5 rectus femoris m.
M 6 iliopsoas m.
M 7 linea alba
M 8 rectus abdominis m.
M 9 internal oblique m.
M 10 external oblique m.
M 11 iliopsoas m.
M 12 rectus femoris m. (reflected head)
M 13 sartorius m.
M 14 gluteus minimus m.
M 15 gluteus medius m.
M 16 gluteus maximus m.
M 17 piriformis m.
M 18 obturator internus m.
M 19 piriformis m.
M 20 gluteus maximus m.

N 1 femoral n.
N 2 obturator n.
N 3 femoral n.
N 4 sacral plexus
N 5 pudendal n.
N 6 sacral plexus

O 1 anterior inferior iliac spine
O 2 greater sciatic notch
O 3 sacral hiatus
O 4 sacral cornu
O 5 body of ilium

S 1 peritoneal cavity
S 2 small intestine
S 3 small intestine
S 4 descending colon
S 5 ureter
S 6 ovary
S 7 broad ligament
S 8 body of uterus
S 9 rectouterine pouch
S 10 uterine cavity
S 11 rectal ampulla

S 12 appendices epiploicae
S 13 round ligament
S 14 uterine tube
S 15 ovary
S 16 uterine tube
S 17 ureter (bifid)

SY 1 hip joint (capsular and iliofemoral ligaments)
SY 2 lunate surface of acetabulum
SY 3 sacrotuberous ligament

HP♀3

A 1 lateral femoral circumflex v. (branch)
A 2 deep circumflex iliac a.
A 3 external iliac a.
A 4 external iliac v.
A 5 inferior epigastric a.
A 6 umbilical a.
A 7 uterine a.
A 8 superficial epigastric v.
A 9 superficial epigastric v.
A 10 inferior epigastric v.
A 11 external iliac v.
A 12 inferior epigastric a.
A 13 external iliac a.
A 14 superficial circumflex iliac v.
A 15 deep circumflex iliac v.
A 16 superior gluteal a. (deep branch)
A 17 internal pudendal a.
A 18 vaginal a.
A 19 superior rectal a.
A 20 superior rectal a. (branch)
A 21 uterine v.
A 22 lateral sacral a.
A 23 uterine venous plexus
A 24 lateral sacral v.
A 25 uterine venous plexus
A 26 middle rectal a.
A 27 uterine a.
A 28 uterine v.
A 29 inferior gluteal v.
A 30 internal pudendal a.
A 31 obturator a.

M 1 tensor fasciae latae m.
M 2 gluteus medius m.
M 3 sartorius m.
M 4 iliopsoas m.
M 5 internal oblique m.
M 6 rectus abdominis m.
M 7 linea alba
M 8 external oblique m.
M 9 iliopsoas m.
M 10 rectus femoris m.
M 11 sartorius m.
M 12 tensor fasciae latae m.
M 13 gluteus minimus m.
M 14 gluteus medius m.
M 15 fascia lata
M 16 rectus femoris m. (reflected head)
M 17 gluteus minimus m.
M 18 gluteus maximus m.
M 19 piriformis m.
M 20 obturator internus m.
M 21 obturator internus m.
M 22 piriformis m.
M 23 gluteus maximus m.

N 1 femoral n.
N 2 obturator n.
N 3 femoral n.
N 4 sacral plexus
N 5 pudendal n.
N 6 sacral plexus

O 1 anterior inferior iliac spine
O 2 body of ilium
O 3 greater sciatic notch
O 4 sacral cornu
O 5 sacral hiatus
O 6 head of femur

S 1 descending colon
S 2 broad ligament
S 3 round ligament
S 4 small intestine
S 5 urinary bladder
S 6 small intestine
S 7 vesicouterine space
S 8 broad ligament
S 9 small intestine
S 10 ureter (bifid)
S 11 uterine tube
S 12 ovary
S 13 rectouterine space
S 14 uterine cavity
S 15 rectal ampulla
S 16 body of uterus
S 17 ureter

SY 1 lunate surface of acetabulum
SY 2 hip joint (articular capsule and iliofemoral ligament)
SY 3 sacrotuberous ligament
SY 4 dorsal sacrococcygeal ligament

HP♀4

A 1 lateral femoral circumflex v. (branch)
A 2 femoral a. (ascending branch)
A 3 femoral a.
A 4 femoral v.
A 5 superficial circumflex v.
A 6 obturator v.
A 7 superficial epigastric v.
A 8 obturator a.
A 9 femoral v.
A 10 femoral a.
A 11 inferior gluteal v.
A 12 internal pudendal a.
A 13 vaginal a.
A 14 uterine venous plexus
A 15 superior rectal a.
A 16 middle sacral v.
A 17 uterine v.
A 18 superior rectal a.
A 19 vaginal a.
A 20 internal pudendal a.
A 21 inferior gluteal a. (branch)

M 1 gluteus medius m.
M 2 tensor fasciae latae m.
M 3 rectus femoris m.
M 4 sartorius m.
M 5 iliopsoas m.
M 6 pectineus m.
M 7 obturator internus m.
M 8 rectus abdominis m., external oblique m., internal oblique m., transverse abdominis m.
M 9 inguinal canal
M 10 reflected inguinal ligament
M 11 iliopsoas m.
M 12 sartorius m.
M 13 rectus femoris m.
M 14 tensor fasciae latae m.
M 15 gluteus minimus m.
M 16 gluteus medius m.
M 17 fascia lata
M 18 gluteus maximus m.
M 19 inferior gemellus m.
M 20 obturator internus m. (tendon)
M 21 levator ani m.
M 22 coccygeus m.
M 23 obturator internus m. (bursa)
M 24 superior gemellus m.

N 1 femoral n.
N 2 obturator n.
N 3 obturator n.
N 4 femoral n.
N 5 sciatic n.
N 6 pudendal n.
N 7 pudendal n. (part)
N 8 sciatic n.

O 1 neck of femur
O 2 head of femur
O 3 pecten pubis
O 4 obturator sulcus
O 5 greater trochanter
O 6 body of ischium
O 7 spine of ischium

S 1 round ligament
S 2 retropubic space
S 3 urinary bladder
S 4 ureter
S 5 cervix uteri
S 6 rectouterine pouch

S 7 cervical canal
S 8 rectal ampulla
S 9 ureter (bifid)

SY 1 articular capsule, iliofemoral ligament
SY 2 ligamentum teres
SY 3 ligamentum teres
SY 4 acetabular labrum
SY 5 articular capsule, iliofemoral ligament
SY 6 sacrococcygeal ligament
SY 7 acetabular labrum

HP♀5

A 1 deep femoral a.
A 2 femoral a.
A 3 superficial circumflex iliac v.
A 4 femoral v.
A 5 obturator v.
A 6 obturator a.
A 7 superficial epigastric v.
A 8 superficial epigastric v.
A 9 obturator a.
A 10 femoral v.
A 11 femoral a.
A 12 femoral a. (ascending branch)
A 13 lateral femoral circumflex v. (branch)
A 14 inferior gluteal v.
A 15 inferior gluteal a. (branch)
A 16 internal pudendal a.
A 17 vesical venous plexus
A 18 vaginal a.
A 19 superior rectal v.
A 20 uterine venous plexus
A 21 superior rectal v. (branch)
A 22 middle sacral v.
A 23 uterine venous plexus
A 24 perineal a.
A 25 internal pudendal v.
A 26 inferior gluteal v.
A 27 inferior gluteal v.

M 1 fascia lata
M 2 gluteus medius m.
M 3 tensor fasciae latae m.
M 4 gluteus minimus m.
M 5 rectus femoris m.
M 6 sartorius m.
M 7 iliopsoas m.
M 8 obturator internus m.
M 9 pectineus m.
M 10 rectus abdominis m.
M 11 reflected inguinal ligament
M 12 obturator internus m.
M 13 pectineus m.
M 14 iliopsoas m.
M 15 sartorius m.
M 16 rectus femoris m.
M 17 tensor fasciae latae m.
M 18 gluteus medius m.
M 19 gluteus maximus m.
M 20 superior gemellus m.
M 21 bursa of obturator internus m.
M 22 coccygeus m.
M 23 coccygeus m.
M 24 obturator internus m. (tendon)
M 25 inferior gemellus m.

N 1 femoral n.
N 2 obturator n.
N 3 obturator n.
N 4 femoral n.
N 5 sciatic n.
N 6 pudendal n. (portion)
N 7 sciatic n.

O 1 head of femur
O 2 pecten pubis
O 3 body of pubis
O 4 greater trochanter
O 5 spine of ischium
O 6 body of ischium
O 7 neck of femur
O 8 greater trochanter

S 1 internal urethral opening
S 2 retropubic space
S 3 body of bladder
S 4 ureter (bifid)
S 5 rectal ampulla
S 6 ureter
S 7 fornix
S 8 cervical canal
S 9 ureter

SY 1 ligamentum teres
SY 2 acetabular labrum
SY 3 articular capsule and iliofemoral ligament
SY 4 articular capsule and ischiofemoral ligament
SY 5 acetabular ligament
SY 6 sacrotuberous ligament
SY 7 sacrococcygeal joint
SY 8 sacrospinal ligament
SY 9 acetabular fossa

HP♀6

A 1 lateral femoral circumflex v. (branch)
A 2 femoral a. (ascending branch)
A 3 deep femoral a.
A 4 femoral a.
A 5 medial femoral circumflex a.
A 6 femoral v.
A 7 great saphenous v.
A 8 obturator v. (branch)
A 9 superficial epigastric v.
A 10 greater saphenous v.
A 11 femoral v.
A 12 femoral a.
A 13 deep femoral a.
A 14 lateral femoral circumflex v. (ascending branch)
A 15 perineal a.
A 16 internal pudendal v.
A 17 vaginal venous plexus
A 18 vaginal venous plexus
A 19 internal pudendal v.
A 20 inferior gluteal v. (branch)
A 21 inferior gluteal v. (branch)

M 1 fascia lata
M 2 tensor fasciae latae m.
M 3 vastus lateralis m.
M 4 rectus femoris m.
M 5 sartorius m.
M 6 iliopsoas m.
M 7 pectineus m.
M 8 obturator externus m.
M 9 levator ani m.
M 10 adductor longus m.
M 11 obturator externus m.
M 12 pectineus m.
M 13 sartorius m.
M 14 iliopsoas m.
M 15 rectus femoris m.
M 16 tensor fasciae latae m.
M 17 gluteus maximus m.
M 18 quadratus femoris m.
M 19 ischiocrurales mm. (tendon)
M 20 obturator internus m.
M 21 levator ani m.
M 22 obturator internus m.
M 23 quadratus femoris m.

N 1 lateral femoral cutaneous n.
N 2 obturator n.
N 3 obturator n.
N 4 femoral n.
N 5 femoral n.
N 6 sciatic n.
N 7 sciatic n.

O 1 neck of femur
O 2 tuberosity of ischium

S 1 mons pubis
S 2 urethra
S 3 suspensory ligament of clitoris
S 4 retropubic space
S 5 vagina
S 6 urethral sphincter
S 7 rectal ampulla
S 8 ischiorectal fossa

SY 1 iliofemoral ligament
SY 2 hip joint
SY 3 pubic symphysis
SY 4 arcuate pubic ligament

SY 5 obturator membrane
SY 6 sacrotuberous ligament
SY 7 sacrotuberous ligament

HP♀7

A 1 lateral femoral circumflex v. (descending branch)
A 2 deep femoral a.
A 3 femoral a.
A 4 femoral v.
A 5 superficial inguinal lymph node
A 6 great saphenous v.
A 7 vein of vestibular bulb
A 8 anterior labial v.
A 9 femoral v.
A 10 obturator v. (branch)
A 11 deep femoral a.
A 12 lateral femoral circumflex v.
A 13 vein of vestibular bulb
A 14 vaginal venous plexus
A 15 internal pudendal v.
A 16 inferior gluteal v. (branch)

M 1 tensor fasciae latae m.
M 2 vastus lateralis m.
M 3 rectus femoris m.
M 4 sartorius m.
M 5 pectineus m.
M 6 adductor brevis m.
M 7 adductor longus m.
M 8 gracilis m.
M 9 bulbospongiosus m.
M 10 adductor longus m.
M 11 adductor magnus m.
M 12 obturator externus m.
M 13 adductor brevis m.
M 14 pectineus m.
M 15 sartorius m.
M 16 iliopsoas m.
M 17 rectus femoris m.
M 18 vastus intermedius m.
M 19 vastus lateralis m.
M 20 tensor fasciae latae m.
M 21 gluteus maximus m.
M 22 quadratus femoris m.
M 23 semimembranosus m. (tendon)
M 24 biceps femoris m. and semitendinosus m. (tendon)
M 25 sphincter ani internus m.
M 26 sphincter ani externus m.
M 27 ischiocavernosus m.
M 28 biceps femoris m. and semitendinosus m. (tendon)
M 29 semimembranosus m. (tendon)
M 30 gluteus maximus m.
M 31 quadratus femoris m.
M 32 fascia lata

N 1 femoral n. (branch)
N 2 femoral n. (branch)
N 3 obturator n. (posterior branch)
N 4 obturator n. (anterior branch)
N 5 sciatic n.
N 6 perineal n.
N 7 sciatic n.

O 1 ramus of ischium
O 2 ramus of ischium
O 3 tuberosity of ischium
O 4 lesser trochanter

S 1 vestibular bulb
S 2 vagina
S 3 urethra
S 4 clitoris
S 5 vestibular bulb
S 6 labia majora
S 7 ischiorectal fossa
S 8 perineal body
S 9 anal canal
S 10 crus clitoris

HP♀8

The left femoral artery has been catheterized to allow injection of a preserving fluid.

A 1 deep femoral a.
A 2 deep femoral v.
A 3 femoral a.
A 4 femoral v.
A 5 great saphenous v.
A 6 posterior labial v. (branch)
A 7 lateral femoral circumflex v. (descending branch)

M 1 tensor fasciae latae m.
M 2 vastus lateralis m.
M 3 rectus femoris m.
M 4 vastus intermedius m.
M 5 pectineus m.
M 6 sartorius m.
M 7 adductor brevis m.
M 8 adductor longus m.
M 9 gracilis m.
M 10 gracilis m.
M 11 sartorius m.
M 12 vastus lateralis m.
M 13 tensor fasicae latae m.
M 14 fascia lata
M 15 gluteus maximus m.
M 16 gluteus maximus m.
M 17 adductor magnus m.
M 18 tendon of semimembranosus m.
M 19 adductor magnus m.
M 20 semitendinosus m.
M 21 semimembranosus m.
M 22 biceps femoris m. (long head)

N 1 sciatic n.
N 2 sciatic n.

O 1 body of femur
O 2 linea aspera

S 1 pudendal cleft
S 2 labium minora
S 3 labium majora
S 4 genitofemoral sulcus
S 5 anal cleft
S 6 gluteal region (subcutaneous fat)